Bioengineering in Cell and Tissue Research

Gerhard M. Artmann · Shu Chien (Eds.)

Bioengineering in Cell and Tissue Research

With 342 Figures

Prof. Dr. habil. Gerhard M. Artmann
Aachen University of Applied Sciences
Centre of Competence in Bioengineering
Ginsterweg 1
52428 Jülich
Germany

Prof. Dr. Shu Chien
University of California
Department of Bioengineering
and Whitaker Institute
of Biomedical Engineering
San Diego
La Jolla, CA 92093-0412
and
University of California
Department of Medicine
San Diego
La Jolla, CA 92093-0427
USA

The cover figure represents a shape transformed red blood cell (echinocyte) as calculated by Prof. Reinhard Grebe, University of Technology, Compiegne, France.

ISBN-13: 978-3-540-75408-4   e-ISBN-13: 978-3-540-75409-1

DOI 10.1007/978-3-540-75409-1

Library of Congress Control Number: 2007938800

© 2008 Springer-Verlag Berlin Heidelberg

This work is subject to copyright. All rights are reserved, whether the whole or part of the material is concerned, specifically the rights of translation, reprinting, reuse of illustrations, recitation, broadcasting, reproduction on microfilm or in any other way, and storage in data banks. Duplication of this publication or parts thereof is permitted only under the provisions of the German Copyright Law of September 9, 1965, in its current version, and permission for use must always be obtained from Springer. Violations are liable to prosecution under the German Copyright Law.

The use of general descriptive names, registered names, trademarks, etc. in this publication does not imply, even in the absence of a specific statement, that such names are exempt from the relevant protective laws and regulations and therefore free for general use.

Cover design: WMXDesign GmbH, Heidelberg
Typesetting and production: le-tex publishing services oHG, Leipzig, Germany

Printed on acid-free paper

9 8 7 6 5 4 3 2 1

springer.com

Editorial Board

Chiu, Jeng-Jiann, Prof. PhD.[1]
Digel, Ilya, PhD.[2]
Huang, Wei, PhD.[3]
Kassab, Ghassan S., Prof. PhD.[4]
Kaunas, Roland, Prof. PhD.[5]
Kayser, Peter, Dipl.-Ing.[2]
Li, Julie Y.-S., PhD.[3]
Li, Song, Assistant Professor, PhD.[6]
Linder, Peter, Dipl.-Ing. MSc. cand. PhD.[2]
Porst, Dariusz, Dipl.-Ing.[2]
Schmid-Schönbein, Geert, Prof. PhD.[3]
Shyy, John, Prof. PhD.[7]
Sponagel, Stephan, Prof. Dr. rer.nat. Dr.-Ing.[2]
Staat, Manfred, Prof. Dr.-Ing.[2]
Temiz Artmann, Aysegül, Prof, PhD. MD.[2]
Zerlin, Kay F., Dr. rer.nat.[2]

---

[1] National Health Research Institutes, Division of Medical Engineering Research, Taiwan

[2] Aachen University of Applied Sciences, Center of Competence in Bioengineering and Institute of Bioengineering, Germany

[3] University of California, San Diego, Whitaker Institute of Biomedical Engineering, USA

[4] University Indianapolis, Indiana University Purdue, Department of Biomedical Engineering, USA

[5] Texas A&M University Department of Biomedical Engineering and the Cardiovascular Research Institute, USA

[6] University of California, Berkeley, Dept. of Bioengineering, USA

[7] University of California, Riverside, Biomedical Sciences, USA

Introduction

The idea of publishing this book on "*Bioengineering in Cell and Tissue Research*" was originated by Gerhard M. Artmann, with the goal of writing about our dreams and making the reader dream with the authors and be fascinated. The book is meant to have life and spirit, and to become a pioneer in technology and sciences, especially the life science. The chapters in this book are written by excellent scientists on advanced, frontier technology and address scientific questions that need considerable thinking in terms of engineering. The aims are to provide the readers, including students, faculty, and all scientists working in academia and industry, new information on bioengineering in cell and tissue research to enhance their understanding and innovation.

This book is composed of six sections that cover a broad hierarchy from genes to the universe. These sections are Genes, Genome and Information Network; Cell and Tissue Imaging; Regenerative Medicine and Nanoengineering; Mechanics of Soft Tissues, Fluids and Molecules; Bioengineering in Clinical Applications; and Plant and Microbial Bioengineering.

Section I on "*Genes, Genome and Information Network*" contains three chapters:

Chapter 1 on "*Reporter Genes in Cell-based Ultra-High Throughput Screening*" by Stefan Golz presents the processes used to identify target-specific, disease-relevant genes or gene products devoid of side effects with the aid of ultra-high throughput screening (uHTS). It discusses how to set up a drug discovery pipeline starting from target identification to finally delivering molecules for clinical development. This chapter presents many of the large arsenal of technologies available for researchers in industry and academia to generate data in support of a functional link between given genes and a disease state. The author concludes that experimental testing of candidate drug compounds remains the major route for the lead drug discovery process, which has been aided by uHTS using targeted assays together with the design of combinatorial chemistry libraries. Converting the knowledge of the target mechanism and underlying molecular recognition principles into robust and sensitive assays is a prerequisite of successful uHTS screening.

Chapter 2 on "*Gene Arrays for Gene Discovery*" by David Ruau and Martin Zenke provides an overview of the strategies for gene discovery using gene arrays.

A wealth of genetic information has become available due to the determination of the DNA sequences of the entire genomes of human and other biological organisms and the advancement of microarray technology that measures simultaneously the expression of thousands of genes. As a result, bioinformatic tools have been developed to determine transcriptional "signatures" of various cell types, tissues and entire organisms in normal and disease states, to apply data mining strategies for gene discovery, and to synthesize and understand complex gene networks. This chapter reviews the data mining strategies for microarray gene expression data, including data pre-processing, cluster analysis, information retrieval from knowledge-based databases, and their integration into microarray data analysis workflows, as exemplified by studies on antigen-presenting dendritic cells that have been treated with transforming growth factor $\beta 1$.

Chapter 3 on *"Physical Modulation of Cellular Information Networks"* by Sumihiro Koyama and Masuo Aizawa reviews the effects of physical stimulation on the information networks in cells. The modulating factors applied include electrical potential, hydrostatic pressure, electromagnetic field, shear stress, and heat shock application, and the cellular functions investigated include viability, proliferation, differentiation, gene expression, and protein production, as well as morphological changes. The authors conclude that future work in cellular engineering for physical stimuli-induced gene expression will involve the exploration of specific physical stimulus-responsive promoters and investigations on stress-induced expression mechanisms. They also conclude that the effects of physical stimulation on mammalian cells will have a wide range of applications, particularly in cellular engineering, tissue engineering, and medical engineering.

Section II on *"Cell and Tissue Imaging"* contains three chapters:

Chapter 4 on *"Fluorescence Live-Cell Imaging"* by Yingxiao Wang, John Y-J. Shyy, and Shu Chien surveys the principles and technologies used in performing fluorescence live-cell imaging and their applications in mechanobiology. The coverage includes the uses of fluorescent proteins and their derivatives and fluorescent microscopy and its integration with atomic force microscopy, with the purposes of visualizing intraellular localization of organelles, signaling/structural molecules, gene expression, and post-translational modifications. Techniques are presented for the spatiotemporal quantification of these subcellular and molecular events, e.g., by using fluorescence resonance energy transfer (FRET), fluorescence recovery after photobleach (FRAP), and fluorescence lifetime imaging microscopy (FLIM). The impact of these fluorescence technologies on cardiovascular research in relation to mechanobiology is discussed. It is pointed out that studies of cardiovascular specimens from transgenic animals would reveal new information on the target molecules.

Chapter 5 on *"Optical Coherence Tomography (OCT)"* by Gereon Hüttmann and Eva Lankenau addresses this emerging technology for three-dimensional imaging of biological tissues. OCT is a new imaging technology that provides higher imaging depth with a high resolution that goes below 10 micrometers and does not rely on the depth of focus of the imaging lens. The basic theory of OCT and its emerging applications in tissue engineering are presented. Compared to technologies such as ultrasound, X-ray, MRT and electron microscopy, OCT offers a non-contact, non-invasive in-vivo applicability with compact and affordable devices. Coupling with spectroscopy, stains and specially designed probes provides additional means to generate site- and function-specific contrast. Hence, optical imaging and microscopy are well developed technologies now widely used in medicine and biotechnology.

Chapter 6 on *"Ultrasonic Strain Imaging and Reconstructive Elastography for Biological Tissue"* by Walaa Khaled and Helmut Ermert reviews the technology of imaging relevant to biomechanics and summarizes their work in the field of elastography. The authors discuss some basic principles and limitations in the calculation of displacement estimators that are needed to evaluate strain images and the relative elastic moduli of biological tissues, which are illustrated with results from tissues *in-vitro* and *in-vivo*. The quantitative information obtained on the relative shear modulus seems promising for differential diagnosis of lesions in biological tissues. This chapter closes with a discussion on the prospects and applications of ultrasound elastography. This real-time ultrasound strain imaging system has been used in a clinical study for the early detection of prostate cancer during conventional transrectal ultrasound examinations. The authors feel that elastography holds promise in the in vivo diagnosis of prostate cancer and intravascular diseases.

Section III on *"Regenerative Medicine and Nanoengineering"* contains five chapters:

Chapter 7 on *"Embryonic Stem Cell Derived Somatic Cell Therapy"* by Kurt Pfannkuche, Agapios Sachinidis and Jürgen Hescheler discuss the results from transplantation studies of embryonic stem cell derived somatic cells in animal models. The results are promising for ES cell-based cell therapy of degenerative diseases such as cardiac and neurological diseases. However, clinical application of ES cell-based therapies requires resolving the current barriers regarding safety aspects, purity and quantity of the cells, immunological rejection and ethical issues. Tissue engineering in combination with the ES cells might contribute to the development of new therapeutical concepts for treatment of severe degenerative diseases. Indeed, many important studies and concepts exist to develop strategies for generation of tissue engineered heart valves. The authors discuss cardiac tissue engineering with a special focus on embryonic stem cell derived cardiomyocytes.

Chapter 8 on *"Collagen Fabrication for Cell-based Implants in Regenerative Medicine"* by Hwal (Matthew) Suh discusses the use of fabricated collagen for cell-based regenerative medicine as applied to skin regeneration, bone reconstruction, esophagus replacement, and liver regeneration, as well as anti-adhesive matrix that promotes wound healing. This chapter provides an overview for the characterization and fabrication of collagen, and the requirements of materials for cell-based implants and biomaterials in cell hybridization. The author points out that a systemized approach with single cell encapsulation by collagen containing signal transduction agents and ligands to attract specific cell adhesive receptors and with stem cell differentiation and proliferation toward the designated target tissue may contribute to the future progress of stem cell-based implant.

Chapter 9 on *"Tissue Engineering"* by Bernd Denecke, Michael Wöltje, Sabine Neuss and Willi Jahnen-Dechent discusses the approach to tissue engineering by combining cells and biomaterials into functional tissues. The authors recapitulate the basic considerations of cell-material interactions in the development of biological substitutes. The tissue-engineered substitutes are comprised of cells cultured in a natural or nature-like environment that sustain growth and differentiation. The biomaterial scaffolds in the substitute need to provide cell attachment sites and a basic three-dimensional organization resembling the extracellular matrix. This chapter covers the recent advances in material scaffolds and cell-material interactions, and the use of stem cells for tissue engineering. The authors point out that the combination of cells and material scaffolds into tissue-engineered replacements of tissues and organs poses a formidable task because of the hurdles of regulatory approval and commercial viability.

Chapter 10 on *"Micro and Nano Patterning for Cell and Tissue Engineering"* by Shyam Patel, Hayley Lam, and Song Li focuses on micro and nano technologies that have been used to investigate the regulation of cell functions by micro and nano features in matrix distribution, surface topography and three-dimensional microenvironment. The authors discuss the technologies that can be applied to the fabrication of functional tissue constructs and the recent advances in microfabrication and nanotechnology industry that have enabled many new ways of examining, manipulating and engineering biological processes. This chapter provides information on the uses of this emergent micro and nano patterning technology for the study of cell behavior and the creation of new therapies and engineering of tissues that are functionally similar to native tissues.

Chapter 11 on *"Integrative Nanobioengineering"* by Andrea A. Robitzki and Andrée Rothermel presents novel bioelectronic tools for real-time pharmaceutical high-content screening in living cells and tissues. The authors discuss new online and real-time diagnostic systems. Miniaturized biomonitoring systems that utilize nano- and micro systems technology, automatic methods and software programs will allow remote measurements of disease-related biomarkers, thus making it possible to perform risk assessments on the patient before actual symptoms occur. This will allow the implementation of personalized prevention program and early individual therapies for people with an increased risk for a certain disease. The authors propose that future screening platforms should provide several technologies, including predictive tests of incompatibility, proteomics and post-genomics, cell and tissue based microsystems, combination with *in silico* models, and bioelectronical cell-based immunological measurements.

Section IV on *"Mechanics of Soft Tissues, Fluids and Molecules"* contains six chapters:

Chapter 12 on *"Soft Materials in Technology and Biology"* by Manfred Staat, Gamal Baroud, Murat Topcu, Stephan Sponagel gives an introduction to the mechanics of soft materials. The authors pointed out that growing interest in flexible structures has brought biomechanics into the focus of engineering. Elastomers and soft tissues consist of similar networks of macromolecules. The chapter provides a brief introduction to the concepts of continuum mechanics, and presented typical isotropic models of soft materials in technology and biology. The authors discussed the similarities and differences of the thermo-mechanical behavior. For rubber-like materials, a modification of the Kilian network is suggested which greatly simplifies the identification of material parameters. Finally, the dynamical loading of biopolymers and volume changes with phase transitions are considered.

Chapter 13 on *"Modeling Cellular Adaptation to Mechanical Stress"* by Roland Kaunas aims at elucidating the mechanism underlying the adaptive changes of cells in response to mechanical loading. In arteries, atherosclerotic plaques are preferentially located at sites of low and oscillating wall shear stress; bone and skeletal muscle adapt to compressive and tensile forces, respectively. In all these cases, mechanical loading modulates the form and function of tissues. This chapter describes a new approach for developing an adaptive constitutive model of adherent cells subjected to mechanical stretch. Such a model would provide a framework by which the accelerating accumulation of mechanotransduction data can be interpreted, thus providing a clearer understanding of how cells respond to mechanical loading. To illustrate the concepts, the model is applied toward understanding how endothelial cells adapt to cyclic stretch.

Chapter 14 on *"How Strong is the Beating of Cardiac Myocytes?"* by Jürgen Trzewik, Peter Linder and Kay F. Zerlin describes the CellDrum technique, which involves the culture of cell monolayers or thin tissue composites under biaxial load conditions for mechanical evaluation. A CellDrum consists of a plastic cylinder sealed on one end with a thin and biocompatible silicon membrane. The membrane allows cell attachment and proliferation under in vitro cell culture conditions. This device can be used to monitor the stress-strain relationship of the cell-membrane composites and has been applied to studying biomechanical properties of cells cultured on CellDrum membranes, including endothelial cells, fibroblasts and cardiomyocytes (self-contracting in monolayer cultures or embedded in collagen I matrices). The relative displacement of silicon membranes attached to cylindrical wells (diameter 16 mm) is measured with non-contact displacement sensors at a resolution in the μm-range. This system can be adapted for medical applications to determine the mechanical properties of cells/tissues in disease states.

Chapter 15 on *"Mechanical Homeostasis of Cardiovascular Tissue"* by Ghassan S. Kassab addresses the state of mechanical homeostasis in the cardiovascular system, with special attention to the variations of stresses and strains. This chapter critically evaluates the various hypotheses on cardiovascular homeostasis of stresses and strains, viz., uniform wall shear stress, uniform mean circumferential stress and strain, uniform transmural stress and strain, and biaxial stress and strain. The evidence for these hypotheses is considered in the normal, flow-overload, and pressure-overload cardiovascular system, as well as during its development and in atherogenesis. This chapter considers the implications of these hypotheses on mechanotransduction and on vascular growth and remodeling. The author concludes that there is significant evidence that the internal mechanical factors are narrowly bounded and carefully regulated such that a perturbation of mechanical loading causes adaptive responses to restore mechanical homeostasis.

Chapter 16 on *"The Role of Macromolecules in Stabilization and Destabilization of Biofluids"* by Björn Neu and Herbert J. Meiselman focuses on the interactions between RBCs as influenced by the presence of macromolecules in the suspending phase. RBCs are chosen to study cell-cell interaction because they are the most numerous cells in blood, have an important function of oxygen transport, and can raise low-shear blood viscosity and adversely affect microvascular blood flow when undergoing excessive aggregation. The authors deal in detail the depletion of macromolecules near the RBC surface because they believe this phenomenon is the most likely mechanism for RBC aggregation. Stabilization of RBC systems via small polymers or covalent attachment of polymers to the membrane surface is also considered. The authors point out that increased attention to the biophysical aspects of cell-cell interactions will yield important new information that may lead to improved disease diagnosis, patient care and advances in cellular- and bio-engineering.

Chapter 17 on *"Hemoglobin Senses Body Temperature"* by Gerhard M. Artmann, Kay F. Zerlin and Ilya Digel report on a glass-transition like temperature transition occurring under certain conditions in RBCs around body temperature. The chapter was based on an accidental discovery of a temperature transition of RBC passage through small micropipettes many years ago at the University of California, San Diego. Artmann's group followed the biophysical mechanism of this effect since then. The authors' claim that the body temperature of a species was imprinted into the structure of hemoglobin and other proteins has been supported by several recent publications. The writing reflects Dr. Artmann's philosophy and view of life.

Section V on *"Bioengineering in Clinical Applications"* contains five chapters:

Chapter 18 on *"Nitric Oxide in the Vascular System: Meet a Challenge"* by Stefanie Keymel, Malte Kelm, and Petra Kleinbongard reviews several aspects of research on nitric oxide (NO), especially its role in vascular biology and other fields in biology and medicine. Theoretical and experimental studies of NO metabolism and the *in vivo* and *ex vivo* detections of NO by bioassay and biochemical methods are outlined. The authors provide evidence that plasma nitrite is a sensitive marker for eNOS activity and propose that it may be used to monitor the efficacy of therapeutic interventions influencing endothelial function and NO metabolism in future clinical trials. The role of the NOS activity of RBCs as an intravascular source of NO is discussed. The authors point out the importance of using high-resolution intravital microscopy and real-time image acquisition and analysis to visualize the microcirculation, in order to study NO dynamics in relation to RBC rheology in vivo.

Chapter 19 on *"Vascular Endothelial Responses to Disturbed Flow"* by Jeng-Jiann Chiu, Shunichi Usami, and Shu Chien focuses on the role of non-uniform and irregular distribution of wall shear stress at branches and bends of the arterial tree in the preferential distribution of atherosclerosis. While laminar blood flow and sustained high shear stress in the straight part of the arterial tree up-regulates the expression of genes and proteins in endothelial cells (ECs) to protect them against atherosclerosis, disturbed flow and the associated oscillatory and low shear stress up-regulate pro-atherosclerotic genes and proteins. This chapter summarizes the know-ledge on the effects of disturbed flow on ECs in terms of signal transduction, gene expression, cell structure and function, as well as pathologic implications.

Such investigations serve to elucidate the mechanisms underlying the effects of disturbed flow on ECs, thus facilitating the understanding of the etiology of lesion development in the disturbed flow regions.

Chapter 20 on *"Why is Sepsis an Ongoing Clinical Challenge?"* by Aysegül Temiz Artmann and Peter Kayser reports on RBC responses during sepsis. Most *in-vivo* studies conducted by introducing microorganisms such as *Escherichia Coli* into experimental animals showed deleterious effects of the infection on RBCs. The authors found in their own *in vitro* studies with whole blood, partly in contrary to earlier studies, that after adding *E. coli*-derived lipopolysaccharides, RBCs lost their capability of aggregation, and this was accompanied by RBC swelling. The authors formulated a new biophysical hypothesis based on colloid osmotic pressure considerations to explain the phenomenon.

Chapter 21 on *"Bioengineering of Inflammation and Cell Activation"* by Alexander H. Penn, Erik B. Kistler, and Geert W. Schmid-Schönbein discusses the role of inflammation and cell activation in circulatory shock and multi-organ failure. The authors have traced the inflammatory mediators in several forms of shock to the action of digestive enzymes, which are synthesized in the pancreas and act on the intestine. During intestinal ischemia, the mucosal barrier becomes permeable to pancreatic enzymes, allowing their entry into the intestine wall to cause auto-digestion of matrix proteins and tissue cells and production of inflammatory mediators, which are released into the central circulation, lymphatics and the peritoneal cavity to cause multi-organ failure. Inhibition of pancreatic enzymes in the lumen of the intestine serves to attenuate the inflammation in several forms of shock, and this may have major significance in the treatment of a variety of clinical conditions.

Chapter 22 on *"Percutaneous Vertebroplasty"* by Christianne Vant, Manfred Staat, and Gamal Baroud reviews the two intraoperative complications of this interventional radiology procedure used to treat vertebral compression fractures. While this procedure shows promising results, it can cause complications due to excessive injection pressure, which is an extravertebral problem, or extraosseus cement leakage, which is an intravertebral problem. Current solutions for these complications involve the modification of cement delivery devices and procedure and the modulation of the rheological properties of the cement. Testing in a synthetic model demonstrates the existence of conflicting demands on the cement viscosity, i. e., low-viscosity cement is needed to solve the extravertebral problem, but high-viscosity cement is needed to solve the intravertebral problem. The challenge is to develop biomaterials, techniques and/or devices that can optimize the conflicting demands on cement viscosity.

Section VI on *"Plant and Microbial Bioengineering"* contains five chapters:

Chapter 23 on *"Molecular Crowding"* by Ira. G. Tremmel addresses the effects of a crowded physico-chemical environment on the structure, function and evolution of cellular systems in photosynthetic membranes. The impact of crowding on photosynthetic electron transport has been simulated and analysed, taking into account realistic concentrations and shapes of photosynthetic proteins. The effects of macromolecular crowding may depend on many factors (e. g., background molecules) and there may be opposing effects (e. g., decrease of diffusion vs. increase of thermodynamic activities). Diffusion coefficients for both large and small molecules are reduced 3- to10-fold from those in water, but the mobility of some enzymes is large in mitochondria, suggesting the presence of channels to facilitate protein movement. The findings on movement of molecules through crowded environment raises interesting questions in molecular evolution as to how biological macromolecules have evolved to optimize their function in environments that have evolved to become crowded.

Chapter 24 on *"Higher Plants as Bioreactors"* by Fritz Kreuzaler, Christoph Peterhänsel, and Heinz-Josef Hirsch discusses the use of gene technology with C3-type plants to optimize CO_2 fixation for the production of biomass and bio-energy. As a result of photosynthesis utilizing CO_2, ATP is generated to provide the energy to do work in biological organisms. In higher plants, the important biochemical pathways involved in CO_2 reduction and assimilation are C3 and C4. Because the loss of fixed CO_2 by photorespiration is large in C3 plants, but virtually none in C4 plants, it is desirable to modify the plants from C3- to C4-type. The authors discuss the strategies of integrating new biosynthetic pathways into C3 plants by modifying the photorespiratory pathway or the CO_2-fixating enzyme Rubisco in plant chloroplasts. Among the various possibilities to produce more biomass to deal with the progressive increase in demand, the best way is to use gene technology to improve the efficiency of photosynthesis and CO_2-fixation, thus producing more biomass with better quality.

Chapter 25 on *"Controlling Microbial Adhesion"* by Ilya Digel discusses the use of a surface engineering approach to manipulate microbial adhesion to surfaces. Microbes have a tendency to colonize surfaces, and this complex adhesion process is affected by many factors, including the characteristics of the bacteria, the target material surface, and environmental factors such as the presence of macromolecules, ions or bactericidal substances. A better understanding of these factors would enable the control of the adhesion process. The controllable stimulation or inhibition of microbial adhesion can be achieved either by physical adsorption or by chemical grafting of functional groups onto a suitable matrix. The feasibility and applicability of these methods have been proved in ethanol production using laboratory-scale bioreactors. Such surface engineering approach exploiting the propensity of microbes to adhere to surfaces can serve many biotechnological purposes.

Chapter 26 on *"Air Purification by Means of Cluster Ions Generated by Plasma Discharge"* by Kazuo Nishikawa and Matthew Cook discusses ways to purify air pollution by using cluster ions. The authors review the growing need for the removal of harmful molecules in the air due to increasing pollution. The removal of airborne particles allow for an improvement in indoor air quality and a reduction in illnesses caused by airborne viruses, bacteria, and fungi and due to allergic bronchial asthma. The authors discuss the research on air purification through applying a plasma discharge into the atmosphere and creating ozone and radicals of strong chemical reactivity, and they present a novel plasma discharge technology, which can produce positive and negative "cluster" ions at a normal atmospheric pressure and characterize the resultant cluster ions. A series of experiments have been performed to prove the air purification effects of such cluster ions, with close attention paid to airborne harmful microbes and cedar pollen allergens.

Chapter 27 on "*Astrobiology*" by Gerda Horneck discusses this relatively new area that attempts to reveal the origin, evolution and distribution of life on Earth and throughout the Universe in the context of cosmic evolution. A multidisciplinary approach is required involving astronomy, planetary research, geology, paleontology, chemistry, biology and others. Astrobiology extends the boundaries of biological investigations beyond the Earth, to other planets, comets, meteorites, and space at large. Focal points are the different steps of the evolutionary pathways through cosmic history that may be related to the origin, evolution and distribution of life. Increasing data on the existence of planetary systems in our Galaxy support the assumption that habitable zones are not restricted to our own Solar System. From the extraordinary capabilities of life to adapt to environmental extremes, the boundary conditions for the habitability of other bodies within our Solar System and beyond can be assessed. Astrobiology has the potential to give new impulses to biology.

In summary, the twenty-seven chapters in this book on "*Bioengineering in Cell and Tissue Research*" provide state-of-the-art knowledge on Genes, Genome and Information Network; Biological Imaging; Regenerative Medicine and Nanoengineering; Mechanics of Soft Tissues, Fluids and Molecules; Clinical Applications; and Plant and Microbial Bioengineering. The book is intended to stimulate the reader to think about these problems and create innovative solutions. The authors and the editors would be most gratified if these aims are achieved.

Chu Chien

Contents

Part I Genes, Genome and Information Network

1 Reporter Genes in Cell Based ultra High Throughput Screening 3
   1.1 Introduction 3
   1.2 From Gene to Target 4
   1.3 Screening Assay Classes 4
   1.4 Reporter Gene Classes 5
   1.5 Flash-Light Reporter Genes 6
   1.6 Glow-Light Reporter Genes 8
   1.7 Coelenterazine Dependent Luciferases 8
   1.8 Luciferin Dependent Luciferases 10
   1.9 Non-Luciferase Glow-Light Reporter Genes 11
   1.10 Fluorescent Proteins 12
   1.11 Cell Based Assay Formats in ultra High Throughput Screening (uHTS) 14
   1.12 Reporter Genes in uHTS 16
   1.13 Photoprotein Readouts and Cell-Based Assay Development 17
   1.14 Multiplexing Reporter Gene Readouts 18
   1.15 Ultra High Throughput Screening 19

References 21

2 Gene Arrays for Gene Discovery 23
   2.1 Introduction 23
   2.2 Data Processing 25
      2.2.1 Challenges in Data Acquisition and Processing 25
      2.2.2 Data Preprocessing: An Overview 25
   2.3 Gene Discovery by Gene Clustering 27
   2.4 TGF-β1 Signaling in Dendritic Cells Assessed by Gene Expression Profiling 29

	2.4.1	Linking Gene Expression Data to Knowledge-based Databases	32
2.5		Conclusions	32

References 35

3 Physical Modulation of Cellular Information Networks 37
   3.1 Introduction 37
   3.2 Cellular Responses to Physical Stimulation 38
       3.2.1 Electrical Potential 39
       3.2.2 Hydrostatic Pressure 42
       3.2.3 Shear Stress 43
       3.2.4 Heat Shock 44
   3.3 Electrically Controlled Proliferation Under Constant Potential Application 45
       3.3.1 Electrical Potential-Controlled Cell Culture System 45
       3.3.2 Cell Viability Under Constant Potential Application 45
       3.3.3 Electrical Modulation of Cellular Proliferation Rate 47
   3.4 Modulated Proliferation Under Extreme Hydrostatic Pressure 47
   3.5 Electrically Modulated Gene Expression Under Alternative Potential Application 50
       3.5.1 Electrically Stimulated Nerve Growth Factor Production . 50
       3.5.2 Electrically Induced Differentiation of PC12 Cells 52
   3.6 Cellular Engineering to Enhance Responses to Physical Stimulation 54
   3.7 Concluding Remarks 57

References 59

Part II Cell and Tissue Imaging

4 Fluorescence Live-Cell Imaging: Principles and Applications in Mechanobiology 65
   4.1 Introduction 65
   4.2 Fluorescence Proteins 66
       4.2.1 Green Fluorescence Protein (GFP) 66
       4.2.2 Derivatives of GFP with Different Colors 67
       4.2.3 Fluorescence Proteins Derived from DsRed 68
       4.2.4 Small Molecule Fluorescence Dyes 69
   4.3 Fluorescence Microscopy 69
       4.3.1 Epi-fluorescence Microscopy 69
       4.3.2 Confocal Fluorescence Microscopy 69
       4.3.3 Total Internal Reflection Fluorescence Microscopy (TIRF) 70
       4.3.4 Integration with Atomic Force Microscopy (AFM) 70

	4.4	Applications in Mechanobiology	71
		4.4.1 Visualization of Cellular Localization of Signaling Molecules and Expression of Genes	71
		4.4.2 Spatiotemporal Quantification of Post-translational Modifications	74
	4.5	Perspective in Cardiovascular Physiology and Diseases	78

References 81

5 Optical Coherence Tomography (OCT) – An Emerging Technology for Three-Dimensional Imaging of Biological Tissues 85

    5.1 Introduction 85
    5.2 Optical Coherence Tomography (OCT) 87
        5.2.1 The Basic Principle 87
        5.2.2 Image Formation 90
        5.2.3 Figure of Merits for the Performance of OCT 91
        5.2.4 Functional Imaging 94
    5.3 Applications in Tissue Engineering 95
        5.3.1 Visualization of Cells in Scaffolds 96
        5.3.2 Visualization of the Morphology of Artificially Grown Tissues 96
        5.3.3 Measurement of Tissue Function 97
    5.4 Conclusion 98

References 99

6 Ultrasonic Strain Imaging and Reconstructive Elastography for Biological Tissue 103

    6.1 Introduction 103
    6.2 Ultrasound Elastography 105
        6.2.1 Theory for Static Deformations 107
        6.2.2 Imaging Tissue Strain 109
        6.2.3 Displacement Estimation in Strained Tissues 110
        6.2.4 Methods Using Radio Frequency Data 110
        6.2.5 Block Matching Methods: 113
    6.3 Reconstructive Ultrasound Elastography 114
        6.3.1 Tissue-like Material Phantoms 115
        6.3.2 Solution of the Inverse Problem 116
        6.3.3 Simulation and Experimental Results 119
    6.4 Medical Results of Elastography 123
    6.5 Results of an Intravascular Ultrasound Study 124
    6.6 Summary and Conclusion 127

References 129

Part III Regenerative Medicine and Nanoengineering

7 Aspects of Embryonic Stem Cell Derived Somatic Cell Therapy of Degenerative Diseases 135
- 7.1 Introduction 135
- 7.2 Rationale for the Cardiac Tissue Engineering 136
- 7.3 Embryonic Stem Cells as an Unlimited Source for Cardiomyocytes 137
- 7.4 Therapeutical Cloning of Embryonic Stem Cells 140
- 7.5 Stem Cell Derived Cardiomyocytes 143
- 7.6 Cardiac Tissue Slices 145
- 7.7 Bioartificial Heart Tissue Based on Biomaterials 147
- 7.8 Scaffolds for Cardiac Tissue Engineering 148
- 7.9 The Ideal Cell 151
- 7.10 Preparation of Cells for *In-Vitro* Tissue Engineering: Cell Permeable Cre/loxP System 152
- 7.11 Outlook 154

References 157

8 Collagen Fabrication for the Cell-based Implants in Regenerative Medicine 159
- 8.1 Regenerative Medicine 159
- 8.2 The Cell-Based Implants 160
- 8.3 Requirements of Materials for the Cell-Based Implants 161
- 8.4 Biomaterials in the Cell-Hybridization 162
- 8.5 Characteristics of Collagen 163
- 8.6 Fabrication of Collagen 165
- 8.7 Collagen in the Cell-based Implants 170
  - 8.7.1 Skin Regeneration 170
  - 8.7.2 Bone Reconstruction 175
  - 8.7.3 Esophagus Replacement 177
  - 8.7.4 Wound Healing promoting Anti-Adhesive Matrix 182
  - 8.7.5 Liver Regeneration 184
- 8.8 Discussion 187

References 191

9 Tissue Engineering – Combining Cells and Biomaterials into Functional Tissues 193
- 9.1 Introduction 193
- 9.2 The Cells 194
  - 9.2.1 In Search of an Ideal Cell Source for Organ Replacement 194
  - 9.2.2 Stem Cells 195
  - 9.2.3 Pluripotent Embryonic Stem Cells 196
  - 9.2.4 Adult Stem Cells 198
  - 9.2.5 Perspectives 201

	9.3	The Material	202
		9.3.1 Scaffolds – Support Materials to Grow Cells into Tissues	202
		9.3.2 Vascularisation and Blood Supply in Tissue Engineering	202
		9.3.3 Scaffold Material Influences Cell Behaviour	203
		9.3.4 Commercial Tissue Engineered Products	206

References 209

10 Micro and Nano Patterning for Cell and Tissue Engineering 215
    10.1 Overview 215
    10.2 Regulation of Cell Functions by Matrix Patterning 216
        10.2.1 Matrix Patterning 216
        10.2.2 Cell Proliferation and Survival on Micropatterned Matrix 218
        10.2.3 Cell Migration on Micropatterned Matrix 219
        10.2.4 Cell Differentiation on Micropatterned Matrix 220
    10.3 Topographic Regulation of Cell Functions 221
    10.4 Engineering 3D Environments with Micro Features 222
    10.5 Nano Patterning for Cell and Tissue Engineering 225
    10.6 Perspective 227

References 229

11 Integrative Nanobioengineering: Novel Bioelectronic Tools for Real Time Pharmaceutical High Content Screening in Living Cells and Tissues 231
    11.1 Introduction 231
        11.1.1 Preventive Medicine and High Effective Therapies 231
        11.1.2 Follow-up Monitoring After Therapy – Therapeutical Control 232
    11.2 Real Time Monitoring and High Content Screening 232
    11.3 Outlook and Future Aspects 245

References 247

Part IV Mechanics of Soft Tissues, Fluids and Molecules

12 Soft Materials in Technology and Biology – Characteristics, Properties, and Parameter Identification 253
    12.1 Introduction 253
    12.2 Material Description 255
        12.2.1 Why Material Description 255
        12.2.2 What is Rheology and What is the Task of This Discipline? 256
        12.2.3 Macromolecular Substances 258
        12.2.4 Phenomenological Behavior of Soft Substances 260
        12.2.5 Simple Model Formulation of Soft Materials 262

	12.2.6	Basic Tests 264
	12.2.7	Remarks 271

12.3 Basics of Continuum Mechanics 272
    12.3.1 Kinematics 272
    12.3.2 Stress State 274
    12.3.3 Balance Equations 275
    12.3.4 Material Equation 276
    12.3.5 Constraints 277

12.4 Basics of Material Theory 278
    12.4.1 Mathematical Fundamentals 278
    12.4.2 Necessity of Asymptotic Representations 281
    12.4.3 Principle of Fading Memory 282
    12.4.4 Asymptotic Representation of the Hysteresis Term 283
    12.4.5 Examination of the Static Term 285
    12.4.6 Considering Phenomenological Thermodynamics 286
    12.4.7 Equibiaxial Loading 291

12.5 Material Laws for Technical and Biological Polymers 294
    12.5.1 Technical Polymers (Gauß Networks
        and Kilian Networks) 294
    12.5.2 Biological Polymers (Collagen) 297

12.6 Volume Change in Biopolymers 303

12.7 Summary and Outlook 310

References 313

13 Modeling Cellular Adaptation to Mechanical Stress 317

13.1 Introduction 317

13.2 A Brief Review of Stretch-Induced Cell Remodeling 318
    13.2.1 Cell Mechanics and Dynamics 318
    13.2.2 Actin Cytoskeletal Remodeling in Response
        to Different Modes of Stretch 319

13.3 Measurements, Modeling, and Mechanotransduction 322
    13.3.1 Mechanical Testing of Adherent Cells 323
    13.3.2 Non-adaptive Constitutive Models of Adherent Cells 325
    13.3.3 Adaptive Constitutive Models of Adherent Cells 326
    13.3.4 Mechanotransduction 331

13.4 A New Approach for the Study of the Mechanobiology
    of Cell Stretching 332
    13.4.1 An Adaptive, Microstructure-based Constitutive Model
        for Stretched Cells 333
    13.4.2 Biaxial Loading Traction Microscopy 335
    13.4.3 Mechanotransduction Experiments 338

13.5 Illustrative Examples 341
    13.5.1 Step Stretch: Effect of Ramp Rate on the Cell Stress
        and Fiber Organization 341

Contents

13.5.2 Cyclic Stretch: Effect of Frequency on the Cell Stress, Fiber Organization, and Mechanotransduction 343
13.6 Closure 344

References 347

14 How Strong is the Beating of Cardiac Myocytes? – The CellDrum Solution 351
14.1 Introduction 351
14.2 The CellDrum Technique 354
    14.2.1 The Measurement Principle 355
    14.2.2 The Steady State Measurement, a Useful Modification ... 356
14.3 Preparation of Samples 356
    14.3.1 Methods of Cell-Membrane Connection 356
    14.3.2 Visualization of Location and Orientation of the Cells Inside the Gel 359
    14.3.3 Example Cardiac Myocytes 361
    14.3.4 Beating of Cardiac Myocytes 362
    14.3.5 Medical Application in Future 365

References 367

15 Mechanical Homeostasis of Cardiovascular Tissue 371
15.1 Introduction 371
15.2 Shear Stress and Scaling Laws of Vascular System 372
    15.2.1 Variation of Shear Stress 372
    15.2.2 Design of Vascular System: Minimum Energy Hypothesis 372
    15.2.3 Mechanism for Murray's Minimum Energy Hypothesis .. 373
    15.2.4 Uniform Shear Hypothesis 374
15.3 Stress and Strain 374
    15.3.1 Tension 374
    15.3.2 Definitions of Stress and Strain 375
    15.3.3 Spatial Variation of Circumferential Stress and Strain in the CV System 375
    15.3.4 Dynamic Variations of Stress and Strain 376
15.4 Intramural Stress and Strain 376
    15.4.1 Residual Circumferential Strain 376
    15.4.2 Axial Pre-stretch 377
    15.4.3 Radial Tissue Constraint 377
15.5 Perturbation of Mechanical Homeostasis 378
    15.5.1 Flow Reversal 378
    15.5.2 Flow-Reduction and Flow-Overload 379
    15.5.3 Pressure-Overload 380
    15.5.4 Simultaneous Hypertension and Flow-Overload ... 381
    15.5.5 Changes in Axial Stretch 382
    15.5.6 Postnatal Growth and Development 382
    15.5.7 Implications for Atherosclerosis 383

15.6 Limitations, Implications and Future Directions 384
    15.6.1 What is the Stimulus for Mechanotransduction?......... 384
    15.6.2 Strain Homeostasis..................................... 385
    15.6.3 Existence of Gauge Length? 385
    15.6.4 Other Possibilities 386
    15.6.5 Summary and Future Directions 386

References 387

16 The Role of Macromolecules in Stabilization and De-Stabilization of Biofluids 393
16.1 Introduction 393
    16.1.1 Macromolecules as a Determinant of Blood Cell Interaction 394
    16.1.2 The Impact of Red Blood Cell Aggregation on Blood Flow 395
16.2 The Effects of Macromolecules on the Stability of Colloids 396
16.3 Macromolecular Depletion at Biological Interfaces 397
16.4 Cell–Cell Interactions Mediated by Macromolecular Depletion ... 401
    16.4.1 Depletion Interaction Energy 401
    16.4.2 Depletion at Soft Surfaces 403
    16.4.3 Electrostatic Forces 403
    16.4.4 Cell Surface Properties as a Determinant of Depletion Interaction 406
16.5 Stabilization of Bio-Fluids via Macromolecules 406
16.6 Destabilization of Bio-Fluids via Macromolecular Binding 408
16.7 Conclusion & Outlook 409

References 411

17 Hemoglobin Senses Body Temperature 415
17.1 Instead of an Introduction 415
17.2 Physiological Aspects of Thermoregulation in the Body 418
17.3 Red Blood Cells... 420
17.4 Temperature Transition in RBC Passage Through Micropipettes .. 420
17.5 The Molecular Mechanism of the Micropipette Passage Transition 421
17.6 Hemoglobin Viscosity Transition 423
17.7 Circular Dichroism Transition in Diluted Hb Solutions 424
17.8 A RBC Volume Transition Revealed with Micropipette Studies ... 426
17.9 Micropipette Passage Transition in D_2O Buffer................. 428
17.10 NMR T1 Relaxation Time Transition of RBCs in Autologous Plasma 429
17.11 Colloid Osmotic Pressure Transition of RBC Suspended in Plasma 432
17.12 The Temperature Transition Effect so Far..................... 433
17.13 Strange coevals – *Ornithorhynchus anatinus* and *Tachyglossus aculeatus* 434

	17.14	Hb Temperature Transition of Species with Body Temperatures Different from 37 °C 435
	17.15	Molecular Structural Mechanism of the Temperature Transitions .. 438
	17.16	Physics Meets Physiology 440

References 443

Part V Bioengineering in Clinical Applications

18 Nitric Oxide in the Vascular System: Meet a Challenge 451
    18.1 Nitric Oxide: NO 451
    18.2 NO in Vascular Biology 451
    18.3 Key Questions 453
    18.4 Assessment of NO Mediated Vasoactivity 453
    18.5 From the *In-Vivo* and *Ex-Vivo* Detection of NO Effects
          to Biochemical Assessment of NO 455
    18.6 On the Road to a Potential Sensitive Marker
          for NO Formation: Is Nitrite a Candidate? 456
    18.7 More Information About NO Interactions in the Blood 458
    18.8 Intravascular Sources of NO 459
    18.9 The Potential Relevance of RBC NOS Activity 459
    18.10 Outlook... 462

References 465

**19 Vascular Endothelial Responses to Disturbed Flow:
Pathologic Implications for Atherosclerosis** 469
    19.1 Introduction 469
    19.2 Endothelial Dysfunction is a Marker of Atherosclerotic Risk 470
    19.3 Correlation Between Lesion Locations
          and Disturbed Flow Regions of the Arterial Tree 471
    19.4 *In Vitro* Studies on the Effects of Disturbed Flow on ECs 473
          19.4.1 *In Vitro* Models for Studying the Effects
                of Disturbed Flow on ECs 473
          19.4.2 Effects of Disturbed Flow on EC Morphology,
                Cytoskeletal Organization, and Junctional Proteins 476
          19.4.3 Effects of Disturbed Flow on EC Proliferation
                and Migration 478
          19.4.4 Effects of Disturbed Flow on EC Permeability 479
          19.4.5 Effects of Disturbed Flow on EC Signaling
                and Gene Expression 481
    19.5 *In Vivo* Studies on the Effects of Disturbed Flow on ECs 483
    19.6 Summary and Conclusions 485

References 489

20 Why is Sepsis an Ongoing Clinical Challenge? Lipopolysaccharide Effects on Red Blood Cell Volume 497
- 20.1 Introduction 498
- 20.2 Physiopathological Events During Sepsis 499
- 20.3 Markers in Clinical Diagnosis of Sepsis 499
- 20.4 Microcirculation and Sepsis 500
- 20.5 Therapy 500
- 20.6 Activated Protein C 501
- 20.7 Red Blood Cell Behaviour During Sepsis 501
- 20.8 New Perspective 502

References 507

21 Bioengineering of Inflammation and Cell Activation: Autodigestion in Shock 509
- 21.1 Introduction 510
- 21.2 Inflammation in Shock and Multi-Organ Failure 511
- 21.3 The Pancreas as a Source of Cellular Activating Factors and the Role of Serine Proteases 512
- 21.4 Blockade on Pancreatic Digestive Enzymes in the Lumen of the Intestine 513
- 21.5 What Mechanisms Prevent Auto-digestion? 515
- 21.6 Triggers of Shock Increase Intestinal Wall Permeability 515
- 21.7 Intestine as Source of Inflammatory Mediators in Shock 516
- 21.8 Characterization of Protease-Derived Shock Factors 516
- 21.9 Cytotoxic Factors Derived from the Intestine 518
- 21.10 Removal or Blockade of Intestinal Cytotoxic Mediators 519
- 21.11 Conclusions 520

References 523

22 Percutaneous Vertebroplasty: A Review of Two Intraoperative Complications 527
- 22.1 Introduction 527
- 22.2 Vertebroplasty: Minimally Invasive and Cost-Effective Solution 528
- 22.3 Extravertebral Biomechanics: Excessive Delivery Pressure 529
- 22.4 Intravertebral Biomechanics: Risk of Extravasation 531
- 22.5 Injectable Biomaterials 533
- 22.6 Discussion 534

References 537

Part VI Plant and Microbial Bioengineering

23 Molecular Crowding: A Way to Deal with Crowding in Photosynthetic Membranes 543
- 23.1 In the Crowd 543
- 23.2 Macromolecular Crowding 545
  - 23.2.1 Excluded Volume 545
  - 23.2.2 Reaction Equilibria and Aggregation 547
  - 23.2.3 Reaction Rates 549
  - 23.2.4 Evidence for Consequences of Crowding in Living Systems 550
- 23.3 Photosynthesis in a Crowded Environment 552
  - 23.3.1 Basic Principles of Photosynthesis in Higher Plants 552
  - 23.3.2 The Problem of Fast Electron Transport in Thylakoids ... 556
- 23.4 Crowding Effects in Photosynthetic Membranes 558
  - 23.4.1 Restricted PQ Diffusion in Crowded Thylakoids 558
  - 23.4.2 Occupation of Binding Sites 564
  - 23.4.3 Reaction Mechanisms in Crowded Thylakoids 565
- 23.5 Summary and Outlook 570
  - 23.5.1 Summary 570
  - 23.5.2 Treatments of Molecular Crowding *In Vivo*, *In Vitro* and *In Silico* 572

References 575

24 Higher Plants as Bioreactors. Gene Technology with C3-Type Plants to Optimize CO_2 Fixation for Production of Biomass and Bio-Energy 581
- 24.1 Introduction 581
- 24.2 The C3 and C4 CO_2 Fixation Mechanisms 583
  - 24.2.1 The C3 CO_2 Fixation Pathway, the Calvin Cycle and Photorespiration 583
  - 24.2.2 The C4 Photosynthetic Pathway 585
- 24.3 The Metabolism of Glycolate in *Escherichia coli* and in Some Green Algae 587
  - 24.3.1 The Metabolism of Glycolate in *E. coli* 587
  - 24.3.2 The Metabolism of Glycolate in Green Algae 588
- 24.4 Increased Biomass Production in Transgenic *Arabidopsis* Plants Containing the *E. coli* Glycolate Pathway in the Chloroplasts 589
  - 24.4.1 The Strategy to Improve CO_2 Fixation and Hence Photosynthesis 589
  - 24.4.2 Establishment of the *E. coli* Glycolate Pathway in *Arabidopsis* Chloroplasts 590
- 24.5 Analysis of the GT-DEF Transgenic Plants. DNA, RNA, Proteins, Physiology, Growth and Production of Biomass 594

24.5.1 Enhanced Biomass Production in DEF and GT-DEF Plants 596
24.6 Summary 597

References 599

25 Controlling Microbial Adhesion: A Surface Engineering Approach .. 601
25.1 The Lost World of Sessile Microorganisms 601
25.2 Biotechnological Potential of Adhered Microorganisms and Its Limitations 604
25.3 Physicochemical Aspects of Microbial Adhesion 607
25.4 Biological Aspects of Microbial Adhesion 609
25.5 Surface Conditioning as a Tool Facilitating Microbial Adhesion ... 612
25.5.1 Adsorption Facilitation by Transition Metal Ions 614
25.5.2 Surface Preconditioning with Water-soluble Charged Polymers 617

References 621

26 Air Purification Technology by Means of Cluster Ions Generated by Plasma Discharge at Atmospheric Pressure 625
26.1 Ion Generating Device 625
26.2 Characteristics of Positive and Negative Ions 626
26.3 Effect of Removing Airborne Bacteria 627
26.4 Effect of Removing Floating Fungi (Mould) 630
26.5 Effect of Deactivating Floating Viruses 631
26.6 Virus Deactivation Model Using Cluster Ions 633
26.7 Allergen Deactivation Effect 634
26.7.1 Allergen Evaluation Reaction 634
26.7.2 ELISA Method 635
26.7.3 ELISA Inhibition Method 636
26.7.4 Intradermal Reaction and Conjunctival Reaction Tests ... 636
26.8 Conclusion 637

References 639

27 Astrobiology 641
27.1 Introduction 641
27.2 Origin and History of Life on Earth 642
27.3 Impact Scenario and Interplanetary Transport of Life 646
27.4 Strategies of Life to Adapt to Extreme Environments 647
27.5 Signatures of Life 649
27.6 Criteria for Habitability 651
27.7 Planets and Moons in Our Solar System That are of Interest to Astrobiology 653
27.8 Planetary Protection 658

27.9 Search for Life Beyond the Solar System 661
27.10 Outlook 662

References 665

28 Authors 667

Index 681

Part I
Genes, Genome and Information Network

Chapter 1
Reporter Genes in Cell Based ultra High Throughput Screening

Stefan Golz

Bayer Healthcare AG, Institute for Target Research, 42096 Wuppertal, Germany, stefan.golz@bayerhealthcare.com

Abstract Pharma research in most organizations is organized in discrete phases together building a "value chain" along which discovery programs process to finally drug candidates for clinical testing. The process envisioned to identify target-specific modulators lacking several side effects. Following a technical assessment of the targets "drugability", the probability to identify small molecule modulators, and technical feasibility target-specific assays are developed to probe the corporate compound collection for meanful leads. "High-Throughput-Screening" (HTS) started roughly one decade ago with the introduction of laboratory automation to handle the different assay steps typically performed in microtiter plates. Today a large arsenal of screening technologies is available for researchers in industry and academia to set up uHTS or HTS assays. Here the use of reporter genes offer an alternative for following signal transduction pathways from receptors at the cell surface to nuclear gene transcription in living cells.

1.1 Introduction

The modern drug research process has reversed the classical pharmacological strategy. Today, research programs are initiated based on biological evidence suggesting a particular gene or gene product to be a meaningful target for small molecule drugs useful for therapies. The process envisioned to identify target-specific modulators lacking several side effects. Also, it allows setting up a linear drug discovery process starting from target identification to finally delivering molecules for clinical development. One central element is lead discovery through high-throughput screening of comprehensive corporate compound collections. Pharma research in most organizations is organized in discrete phases together building a "value chain" along which discovery programs process to finally drug candidated for clinical testing (Hüser et al. 2006).

This pipeline is fueled by targets suggested from external or in-house generated data suggesting a gene or gene product to be disease relevant. Today a large arsenal of technologies is available for researchers in industry and academia to generate data in support of a functional link between a given gene and a disease state.

1.2 From Gene to Target

Following a technical assessment of the targets "drugability" (Hopkins and Groom, 2002), the probability to identify small molecule modulators, and technical feasibility target-specific assays are developed to probe the corporate compound collection for meanful leads. Lead discovery in the pharmaceutical industry today still depends largely on experimental screening of compound collections. To this end, the industry has invested heavily in expanding their compound files and established appropiate screening capabilities to handle large numbers of compounds within a reasonable period of time. "High-Troughput-Screening" (HTS) started roughly one decade ago with the introduction of laboratory automation to handle the different assay steps typically performed in microtiter plates. HTS technologies during the last decade have witnessed remarkable developments. Assay technologies have advanced to provide a large variety of various cell-based and biochemical test formats for a large spectrum of disease relevant target classes (Walters and Namchuk, 2003). In parallel, further miniaturization of assays volumes and parallelization of processing have further increased the test throughput. The ultra-high-throughput is required to fully exploit big compound files of >1 million compounds and is performed entirely in 1536-well plates with assay volumes between $5-10\,\mu l$. This assay carrier together with fully-automated robotic systems allow for testing in excess of 200,000 compounds per day. The comprehensive substance collection, together with sophisticated screening technologies, have resulted in a clear advantages in lead discovery especially for poorly druggable targets with a poor track record in the past.

1.3 Screening Assay Classes

Today a large arsenal of screening technologies is available for researchers in industry and academia to set up uHTS or HTS assays with high reproducibilty and accuracy. The assay technologies can be divided into three differrent classes:

1. Cellular growth and proliferation assays have been employed to search for therapeutics in anti-infectives and cancer. The ease use, the possibility for adaptation to high-throughput formats and the direct relevance to disease pathology of these growth assays were the reasons for the wide use in drug discovery processes. The difficulty to discriminate pharmacologically from cytotoxic effects is the main problem for an assay principle with unclearly defined targets. As a consequence cellular growth and proliferation assays have a higher risk to fail even in toxicological testing in late stages.

2. Biochemical bioassays are an experimental approach for the testing of isolated enzymes or receptors in which the activty of a purified protein is monitored directly.
3. Cell based bioassays are a different experimental approach for pharmacological assays and are widely used in drug discovery processes. Recently, molecular biology has revolutionized cell based bioassays by providing recombinant cell lines containing readout technologies amenable to ultra high-throughput formats. The functional readout of this assay format allows, to monitor all possible drug-receptor interactions, including allosteric modulation and allow the screening of different pharmacological target classes as G-protein coupled receptors, Ion-channel or transporter. The timescale of the assay type ranges from seconds, e. g. hormone-stimulated Ca^{2+} signals, to few hours for reporter gene readouts. This assay type is target biased and allows the discrimination and differentiation of target-specific signals from general phenotypic effects. Optical assays rely on absorbance, fluorescence or luminescence as readouts. The used readout technologies are comprising fluorescent Ca^{2+} indicators and different reporter genes. The reporter genes used in ultra high-throughput screening can be divided into two different classes: photoproteins & luciferases and fluorescent proteins.

The choice of cellular screening approach has an enormous impact both on the development and implementation of the HTS or uHTS for a target. The availability and behavior of the cells, together with the amplitude and reproducibility of the signal attainable against that cellular background, can all determine whether primary cells or cell lines, both native and engineered, are selected. Primary cells of human origin are arguably the most physiologically relevant model system and several selected primary cell types, human and other species, are commercially available and amenable to HTS or uHTS respectively. In general, primary cells cannot be obtained at the scale necessary for HTS or uHTS, and thus primary cell screens are positioned in the screening paradigm as low-throughput secondary assays. Transformed cell lines of mammalian origin (e. g. CHO, HEK 293) are the most commonly used cell-based uHTS assay formats. The advent of molecular and cell biology techniques to clone and express human proteins has provided access to cell lines with high expression levels of the target of interest. Cell lines can be engineered to express or over-express a target of interest. Expression can be transient or stable and several expression systems can be employed depending on the nature of the cell line and the target. Stable cell lines are most commonly generated by plasmid transfection infection. Stable expression of the target is the approach of choice for HTS and uHTS in the most drug discovery processes.

1.4 Reporter Gene Classes

Bioluminescence is the light produced in certain organism as a result of luciferase- or photoprotein mediated reactions. Numerous marine and terrestrial organisms are bioluminescent, while the biochemistry and molecular biology of the underlying processes has been evaluated recently. The reaction involves the oxidation of a substrate (called luciferin or coelenterazine respectively) by an enzyme (the luciferase) or photoprotein (Wilson and Hastings 1998). Oxygen is usually the oxidant. Bioluminescent organisms are found in a variety of organisms. Common examples are insects, fish, squid, sea cacti, sea pansies, clam, shrimp, and jellyfish. The bioluminescent systems in these organisms are not all evolutionarily conserved, and the genes coding for the proteins involved in bioluminescence are not homologous. The emitted light commonly has one of three functions: defense, offense, and communication.

The bioluminescent systems could be devided into different classes and subclasses: 1. Flash-light reporter gene, 2. Glow-light reporter genes with subclasses coelenterazine-dependent, non-coelenterazine dependent luciferases and non-luciferase glow-light systems, and 4. fluorescent proteins.

In the 1980s, molecular and cell biologists used reporter genes to investigate the 5' untranslated regions of cDNAs to determine which sequences were involved in modulation of gene transcription. In the ensuing decades, the technology and instrumentation have evolved such that many target classes are now amenable to cell-based HTS in reporter formats. Reporter enzymes provide a highly amplified signal, thereby providing sensitivity, and luciferase appears to be the most commonly used reporter enzyme for HTS. Transcriptional regulation assays are configured by linking the natural promoter, or elements of the promoter, of the gene of interest to the coding region of the reporter gene. To take advantage of common signal transduction pathways, synthetic repeats of a particular response element can be inserted upstream of the reporter gene to regulate its expression in response to signaling molecules generated by activation of that pathway (Johnston and Johnston 2002).

1.5 Flash-Light Reporter Genes

Many coelenterates and ctenophores, such as the jelly-fishes Aequorea victoria (Fig. 1.1B) and Clytia gregaria (Fig. 1.1A), the hydroid Obelia longissima (Fig. 1.1D), and the ctenophore Mnemiopsis are bioluminescent (Shimomura 1985). The emission is triggered by calcium, even though the use the same luciferin – called coelenterazine (Jones et al. 1999). Coelenterazine is an imidazolopyrazine (Fig. 1.2), which occurs widely in luminous and non-luminous marine organisms. The first isolated flash-light photoprotein was the the hydrozoan Aequorea system, aequorin.

In the presence of a calcium chelator, one can isolate and purify aequorin, which requires only coelenterazine and calcium for light production. It is now clear that the photoprotein is simply a stable luciferase reaction intermediate to which an oxygenated form of the coelenterazine is already bound, probably as a hydroperoxide.

1.5 Flash-Light Reporter Genes

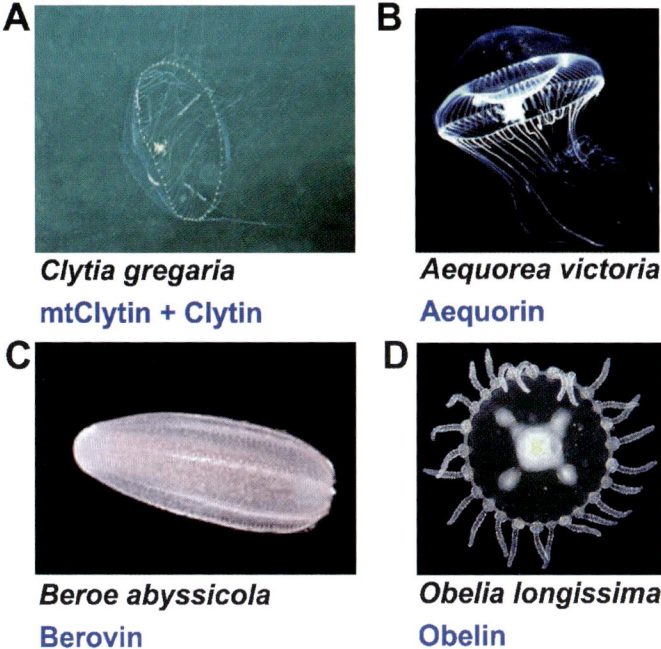

Fig. 1.1 Origins of Flash light photoproteins used in uHTS. Picture of **A** Clytia gregaria (identified photoproteins Clytin and mtClytin), **B** Aequorea Victoria (identified photoprotein: aequorin), **C** Beroe abyssicola (identified photoprotein: Berovin) and **D** Obelia longisima (identified photoprotein: Obelin)

Calcium, for which the protein has three binding sites, triggers the flash by allowing the reaction to go to completion via the dioxetanone intermediate. Thus instead of triggering at the stage of luciferin availability, calcium acts here on a reaction intermediate. The emission is blue (486 nm) when the Aequorea reaction is carried out in vitro, whereas the bioluminescence from the living organism is green (508 nm) because of the presence of the soluble green fluorescent protein, Aequorea GFP. The involvement of energy transfer was first described in the case of Obelia, where, in extracts, photoprotein and GFP are found together in granules (Morin and Hastings 1971). If the cells are mechanically ruptured in seawater containing $MgCl_2$ (a Ca^{2+} antagonist), then centrifuged lightly to remove large cell debris, the supernatant contains both intact granules and the soluble "photoprotein". If calcium is added to the supernatant, it causes a flash of blue light by reacting with the photoprotein. If water is now added, the granules osmotically rupture, and green light is produced as the photoprotein and its associated GFP come in contact with calcium. However, if the order of additions is reversed, first water, then calcium, only blue light is emitted, because the photoprotein-GFP complex has dissociated in dilute solution, thus preventing energy transfer (Predergast 1999).

Recently new flash-light photoproteins could be identified from other organisms: Berovin from Beroe abyssicola (Fig. 1.1C), two different Bolinopsin photopro-

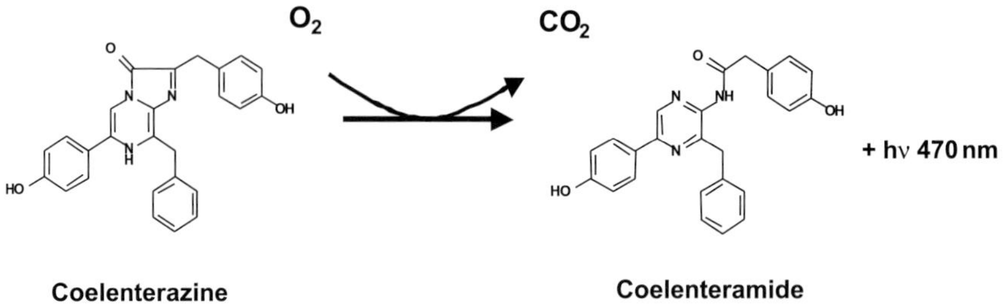

Fig. 1.2 Chemical turnover of Coelenterazine by flash-light reporter genes as Aequorin, Obelin, mtClytin or Berovin ($hv470$ = light, aequorin)

teins from Bolinopsis infundibilum and a new photoprotein from Clytia gregaria (Fig. 1.1A) with a mitochondrial leader sequence.

1.6 Glow-Light Reporter Genes

Many different organisms, ranging from bacteria and fungi to fireflies and fish, are endowed with the ability to emit light, but the bioluminescent systems are not evolutionarily conserved: genes coding for the luciferase proteins are not homologous, and the luciferins are also different, falling into many unrelated chemical classes. Biochemically, all known luciferases are oxygenases that utilize molecular oxygen to oxidize a substrate (luciferin), with formation of a product molecule in an electronically excited state (Hastings 1998). One of the most striking characteristics of luciferases is the very high diversity of mechanisms, structures and functions that bioluminescent organisms have achieved.

1.7 Coelenterazine Dependent Luciferases

Luciferases encompass a wide range of enzymes that catalyze light-producing chemical reactions in living organisms. To date, all known luciferases use molecular oxygen to oxidize their substrates while emitting photons. Different coelenterazine dependent luciferases were identified: Renilla luciferase, Gaussia luciferase and Metridia luciferase.

Gaussia luciferase (Gluc) is the one of the smallest luciferase known and is naturally secreted and was isolated from Gaussia princeps (Fig. 1.3B). This luciferase emits light at a peak of 480 nm with a broad emission spectrum extending to 600 nm. Gaussia luciferase has been cloned, overexpressed in bacteria, and used as a sensitive analytical reporter for hybridization assays and monitoring of cellular expression in culture and *in vivo* (Tannous et al. 2005). The Gaussia lu-

1.7 Coelenterazine Dependent Luciferases

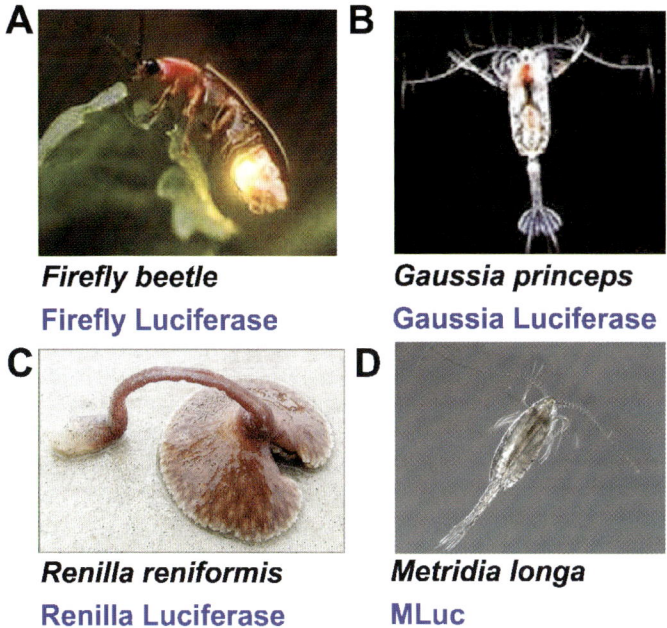

Fig. 1.3 Origins of Glow-light reporter genes used in uHTS. Picture of **A** Firefly Beetle (identified luciferase: Firefly luciferase), **B** Gaussia princeps (identified luciferase: Gaussia luciferase; GLuc), **C** Renilla reniformis (identified luciferase: Renilla luciferase) and **D** Metridia longa (identified luciferase: Metridia luciferase; MLuc)

ciferase is a coelenterazine dependent luciferase which does not require ATP for activity. Gaussia luciferase cDNA was isolated from a cDNA library using expression cloning and the protein was expressed and purified in quantity from bacterial ferments. This enzyme is 188 amino acids in length and encoded in a cDNA of 555 bp. Gaussia luciferase has a broad pH optimum with peak activity at 7.7. Analysis of the gene sequence indicates a secretory signal which is functional in eukaryotes.

The bioluminescent sea pansy Renilla reniformis contains a luciferase-luciferin system (Fig. 1.3C). Renilla luciferase is a monomeric protein that catalyzes the oxidation of coelenterazine to produce light emission at 482 nm. In vivo an energy transfer to a green-fluorscent protein (GFP) occurs and light is emitted at 509 nm. In recent years several secreted proteins have been used as markers of gene expression. The main advantage using a secreted reporter protein, in comparison with an intracellular reporter protein, is the ability to measure gene expression without destroying the cells or tissues. Renilla luciferase gene was engineered to encode a protein product secreted by mammalian cells by fusing the end of a short DNA sequence encoding the signal peptide from human interleukin-2 (IL-2) protein to the renilla luciferase cDNA (Liu et al. 1997). This construct was cloned under transcriptional control of the cytomegalovirus (CMV) promoter or responsive elements in a mammalian expression vector, respectively. Both secreted and non-secreted renilla luciferases are widely used in screening and compound characterization approaches in single and multiplexing readouts.

There are bioluminescent organisms with naturally occurring secreted luciferases, and the cDNAs encoding some of them have been cloned. These include the marine ostracod Vargula hilgendorfii (Thompson et al. 1989), the deep sea shrimp Oplophorus gracilorostris (Yamaguchi 1975), and the marine copepod Gaussia princeps (as described before). The attractiveness of secreted luciferases as reporters is a strong stimulus for the investigation and exploitation of new bioluminescent systems. Metridia longa is a small luminous marine copepod (Fig. 1.3D). The bioluminescence originates as a secretion from epidermal glands located in the head part and abdomen in response to mechanical, electrical, or chemical stimuli. Bioluminescence in Metridia longa may well serve as a defense mechanism against predators; the release of a luminous bolus from the animal is accompanied by rapid swimming that displaces the copepod away from its "glowing phantom". Metridia luciferase (Mluc) was cloned and coelenterazine was identified as substrate (Markova et al. 2004). MLuc is expressable in bacteria and mammalian cells respectively. It is used as a reporter gene in uHTS in single and multiplexing readout. Its use is described in detail in Sect. 1.14.

1.8 Luciferin Dependent Luciferases

Most bioluminescent insects are beetles (Coleoptera), in the families of Elateridae (such as click beetles), Phengodidae (the railroad worm with its red and green lanterns is a spectacular example) (Viviani et al. 2006), and Lampyridae, the fireflies (Fig. 1.3A, origin of Photinus pyralis luciferase). The reaction chemistry is presumably the same or similar for all beetles because their luciferases all react and give light with firefly luciferin. Firefly luciferin is a benzothiazoyl-thiazole, an different substrate from coelenterate luciferin, but again a dioxetanone is the critical energy-rich intermediate in the reaction (reaction is shown in Fig. 1.4). Luciferase first catalyzes the condensation of luciferin with ATP in the presence of Mg^{2+}, followed by the reaction of the adenylate with oxygen and cyclization of the peroxide; ATP provides the good leaving group AMP. The breakdown of the dioxetanone releases the energy. Even though the luciferin is the same in all beetles, their emissions span a wide-wavelength range, from green to orange.

Firefly luciferase is a 62 kDa monomeric protein with no prosthetic group. The firefly luciferase enzyme has had a long history of use in biology, especially for the detection of ATP. The cloning of the firefly luciferase gene and its expression in cells from different organisms has generated a great deal of interest in possible applications of the gene as a tool in biological studies and drug screening (Wood et al. 1989). Its first use was as a reporter for monitoring promoter activity. Though available for only a short time, the gene has already been widely applied in this role, due to the great sensitivity, ease of use, and cost efficiency of the luciferase assay. All beetle luciferases catalyse the conversion of chemical energy into light by a two-step process (Fig. 1.4). This process utilizes ATP, O_2, and beetle luciferin, a unique heterocyclic acid found only in bioluminescent beetles. In the first step, the carboxylate group of luciferin is activated by acylation with the alpha-phosphate of ATP. The luciferyl adenylate is then oxidized with molecular oxygen, in the second step, to yield AMP, carbon dioxide, and oxyluciferin. The oxyluciferin is generated in an electronically excited state which, upon transition to the ground state, emits the photon characteristic of bioluminescence. Under optimal conditions the firefly luciferase emits light whose peak intensity is at 561 nm (yellow-green).

The bioluminescence of the marine ostracod crustacean Vargula (Cypridina) hilgendorfii (sea firefly) is due to a luciferase-luciferin reaction. Both substances are secreted by the organism into the sea water where they combine and produce light at 460 nm. The coding sequence of Vargula luciferase has been cloned and its use as a reporter gene is described. One of the advantages is that the gene product (555 amino acids) is secreted by the cell. Secreted Vargula luciferase is quantitated by the reation with Vargula luciferin and is inhibited by the addition of EDTA or EGTA suggesting a role for divalent ions such as calcium. A disadvantage of Vargula luciferase is that the luciferin required is not overall available (Bronstein et al. 1994).

Bacterial luciferase is a heterdmeric enzyme for which the subunits α and β are coded by the LuxA and LuxB genes, respectively. The active site is on the α unit, but the β unit is essential for the activity. The key difference between the bacterial and mammalian luciferases is their temperature stability. It differs between 7 and 40 °C. Bacterial luciferases are used in biochemical assays in HTS and uHTS as readout for different metabolic pathways coupled to changes in NADH concentration.

1.9 Non-Luciferase Glow-Light Reporter Genes

Non-luciferase reporter genes are well known and widely used in industrial and academic research. Two prominent and representative examples are alkaline phosphatase and β-Galactosidase.

Alkaline phosphatase is a membrane bound protein, but could be secreted from the cell by alterations of the coding region of the membrane localization domain.

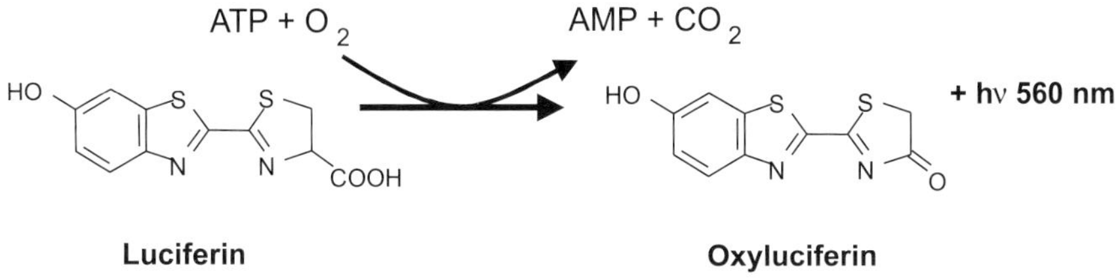

Fig. 1.4 Chemical turnover of Luciferin by glow-light reporter gene Firefly luciferase. ($h\nu 560$ = light)

This truncated form of the engineered gene could be used as a secreted reporter gene for cell based assays. The detection of the secreted gene product is performed while the cell population remains intact for further investigations. Secreted alkaline phosphatase (SEAP) can be measured in a two-step assays using a bioluminescent substrate and is extremly heat stable.

In contrast to alkaline phosphatase the β-Galactosidase assays is based on a b-D-galactopyranoside substrate. β-Galactosidase is one of the most widely used reporter genes in molecular biology. The β-Galactosidase readouts are based on colorimetric or chemiluminescent formats. The chemiluminescent β-Galactosidase assay is highly sensitive, but its usefulness is limited in certain cell lines and tissues due to endogenous enzyme activity.

Both β-Galactosidase and Alkaline Phosphatase (Hiramatsu et al. 2005) assays formats are belonging to the glow bioluminecent reporter genes and could be used as readouts for promotor- or responsive element readout approaches.

1.10 Fluorescent Proteins

Green fluorescent protein (GFP) is responsible for the green bioluminescence from the jellyfish Aequorea. The intense fluorescence of GFP is due to the nature of a chromophore composed of modified amino acids within the polypeptide. Formation of the fluorescent chromophore is species independent and apparently does not require any additional factors. Hence, because the gene product is easily detectable by its intense fluorescence, the GFP cDNA has become a unique reporter system. The advantages of GFP are being exploited in a variety of experimental systems (e. g. Dlctyostellum, plants, Drosophila and mammalian cells). In the last 10 years green fluorescent protein (GFP) has changed from a nearly unknown protein to a commonly used tool in molecular biology, medicine, and cell biology. GFP is used as a biological marker and reporter gene. Green fluorescent proteins are found in numerous organisms such as jellyfish and sea pansies. In 1955 it was first reported that Aequorea (Aequorea victoria) fluoresced green when irradiated with ultraviolet light. Two proteins in Aequorea are involved in its bioluminescence, aequorin (a flash-light type coelenterazine-dependent photoprotein) and the green fluorescent protein (GFP). GFP is a spontaneously fluorescent protein converts the blue chemiluminescence of the primary bioluminescent proteins (as luciferases or photoproteins) into green fluorescent light, presumably to reduce scattering and hence improve penetration of the light over longer distances. The molecular cloning of GFP cDNA from the Pacific jellyfish Aequorea victorea and the demonstration that this GFP can be functionally expressed in bacteria have opened exciting new avenues of investigation in cell, developmental and molecular biology and biotechnological engineering.

Knowledge of the 3D structures of GFP is helping to understand the basic photochemistry and structure of the protein. Crystalline GFP exhibits a nearly identical fluorescence spectrum and lifetime to that in aqueous solution, and fluorescence is not an inherent property of the isolated fluorophore, the elucidation of its 3D structure helps provide an explanation for the generation of fluorescence in the mature protein, as well as the mechanism of autocatalytic fluorophore formation. Furthermore, the development of fluorescent proteins with varied emission and excitation or other characteristics would dramatically expand their biological applications as intracellular reporters (Zimmer 2002).

The structure of the wild-type protein was solved by Ormo et al. (1996). The monomer structure consists of a 11-stranded β-barrel forming cylinder structure. Inside the cylinder resides the fluorescent center of the molecule – a modified tyrosine sidechain and cyclized protein backbone. This motif, with a single α-helix inside a very uniform cylinder of 13-sheet structure, represents a new class of protein fold. The regularity of the 13-can of GFP is quite remarkable. The current dataset of sequenced and spectroscopically characterized GFPs and GFP-like proteins could be classified in four color types: green, yellow, orange-red fluorescent proteins and purple-blue nonfluorescent chromoproteins. These color types apperently posses specific chromophore structures.

The GFP cDNA has proved to be a better reporter gene than expected originally. It was demonstrated that a fluorescent gene product can be easily produced in a heterologues system, because there is no absolute requirement for another factor to form the GFP chromophore. GFP can be detected after fixation, and both N- and C-terminal protein fusions are possible (Wang and Hazelrigg 1994). Furthermore the system has already yielded derivatives that have different spectral properties, which will probably permit simultaneous use of multiple reporter genes. Today GFPs widely used for 1. Identification of transformed cells (e. g. by fluorescence activated cell sorting or fluorescence microscopy), 2. Determination of gene expression *in vivo* and *in vitro*, 3. Localization of fusion proteins, 4. Characterization of intracellular protein traffic, 5. Labeling of unicellular or multicellular organisms, and 6. Identification of organisms released or not to be released to the environment (Prasher 1995). In high throughput screening approaches GFPs today are widely used as co-transferction markers to identify and develop cell lineages expressing recombinant target genes or pathway readouts.

The process of exciting fluorescent probes is not entirely selective and causes unwanted fluorescence, excitation snd light scattering signals from endogcnous cellular and equipment sources, and comprises the background autofluoresescence present in all fluorescent assays. A concern for screening assays is that certain compounds may themselfe be fluorescent and cause false readings in the assay. Despite this concerns the use of GFPs in screening could be divided into three classes (Gonzalez and Negulescu 1998). 1. Intensity-based assays. This type of fluorescent assay utilizes indicators, loaded into membranes or compartments that alter their intensities in response to environmental change. To be practical for cell-based assays, these dyes should be membranc permeable, non-toxic, and well retained in appropriate locations inside the cell. 2. Fluorescence resonance energy transfer. FRET is a well-described phenomenon that results when two fluorophores (one donor and one acceptor) with appropriate spectral overlap are in close proximity. Excitation energy is transferred from the donor (e. g. a luciferase or photoprotein) to the acceptor and results in a decrcase in donor intensity and a increase in acceptor intensity.

The practical advantages of a ratiometric FRET read-out for screening are improved reproducibility and signal-to-noise ratio and less signal perturbation during solution additions to the sample. One limitation of FRET probes is that spectral overlap between the emissions of donor and acceptor can reduce the dynamic range of the assay. The use of GFPs as primary readout in ultra high-throughput screening is limited by the long half-life of GFP and the lower sensitivity compared to enzymatic bioluminescent readouts for target classes as GPCRs, ionchannels, transporters or nuclear receptors.

1.11 Cell Based Assay Formats in ultra High Throughput Screening (uHTS)

Reporter genes offer an alternative assays for following signal transduction pathways from receptors at the cell surface to nuclear gene transcription in living cells. Specific reporter genes are now available for the study of ligand activity at Gαi/o, Gαs and Gαq G-protein-coupled receptors (GPCR).

GPCRs interact with heterotrimeric G proteins to regulate a range of second messenger pathways to enable communication from the cell surface to the nucleus. Upon activation of the receptor, the G protein dissociates into α and $\beta\gamma$ subunits both of which stimulate intracellular signalling pathway. GPCRs coupled to the Gαq family of G proteins stimulate the enzyme phospholipase C (PLC). PLC cleaves membrane phospholipids to produce inositol-1,4,5-trisphosphate (IP3), which mobilises calcium from intracellular storage sites, and diacylglycerol (DAG), which activates protein kinase C (PKC). Gαs-coupled receptors stimulate adenylyl cyclase, which synthesises cAMP from ATP. In contrast, Gαi-coupled receptors inhibit adenylyl cyclase and so reduce cAMP formation. The $\beta\gamma$ subunits from Gαi and other G proteins are able to activate the mitogen-activated protein kinase (MAP kinase) pathway and PLC. These second messenger pathways then activate a range of effector systems to change cell behaviour. In many cases this includes the regulation of gene transcription. Depending on the G-protein coupling of the target GPCR different reporter gene readouts are used for uHTS. The choice of reporter gene and the second messenger pathways of GPCRs are illustrated in Fig. 1.5 (Eglen 2005; Jacoby 2006).

Glow-light reporter genes are often used to detect Gαi/Gαs receptor or nuclear receptor activity in promotor assay approaches. The choice of promoter depends on the nature of the signalling pathway under study. To generate reporter assays specific synthetic promotors, containing a single type of transcription factor binding site only, for single signaling cascades have been developed. Commenly used promotor constucts are the cAMP response element (CRE), serum response element (SRE), the transcription factor NFAT (nuclear factor of activated T cells) and the Gal4 upstream activating sequence.

The CRE system is used for Gαi- and Gαs-coupled GPCRs readouts, whereas an agonist causes a rise in intracellular cAMP (Williams 2004). The decreased cAMP level activates protein kinase A (PKA) and the catalytic unit, of PKA can then move

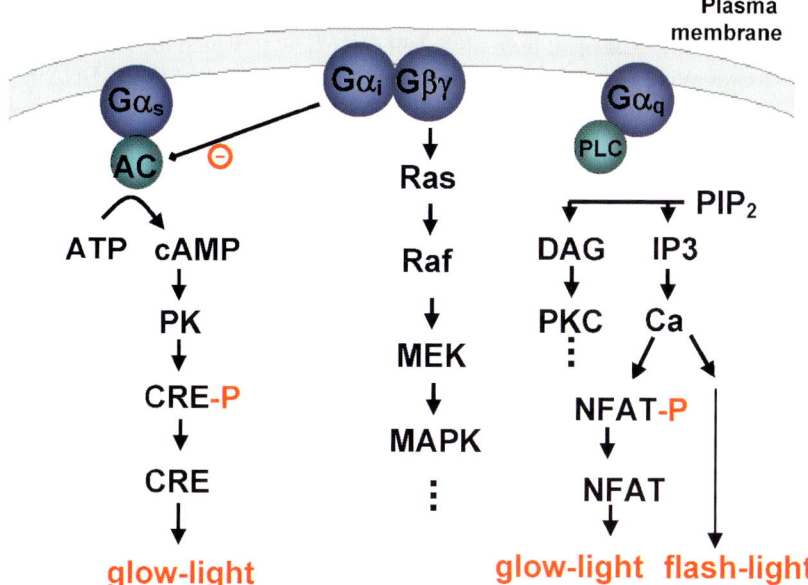

Fig. 1.5 Schematic diagram of luminescent readouts for G-protein coupled receptor signaling pathways. AC: adenylate cyclase, ATP: Adenosine triphosphate, cAMP: cyclic Adenosine monophosphate, PK: Proteinkinase, CRE: cAMP responsive element, CRE-P: phosphorylated CRE binding protein, MEK: MAPK/ERK kinase, MAPK: mitogen activated protein kinase, DAG: diacylglycerol, PKC: protein kinase C, PIP2: inositol-bisphosphate, IP3: inositol-1,4,5-trisphosphate, Ca: Calcium, NFAT: nuclear factor of activated T cells, NFAT-P: phosphorylated NFAT, PLC: phospholipase C, Gαs: G-protein alpha s, GαI: G-protein alpha I, G$\beta\gamma$: G-protein beta gamma

into the nucleus and phosphorylate a CRE binding protein (CREB). The phosphorylated CREB binds to the CRE in the promoter of a target gene, thus increasing transcription.

The NFAT system could be used for Gαq receptor and target which are signaling through a slow kinetic calcium release. Antigenic stimulation of the T-cell receptor stimulates PLC, which then activates the transcription factor NFAT (nuclear factor of activated T cells). Stimulation of PLC increases levels of IP3, and hence intracellular calcium concentration, and diacylglycerol, which activates PKC. Increased intracellular calcium concentration stimulates calcineurin, which dephosphorylates cytoplasmic NFAT allowing it to move into the nucleus. PKC, meanwhile, stimulates the production of the AP-1 immediate–early gene partners Fos and Jun. Once in the nucleus, NFAT binds with Fos and Jun to form a transcriptional complex capable of synergistically activating both the NFAT and AP-1 response elements to stimulate gene expression.

The Gal4 System is widely used to detect nuclear receptor activities. The yeast transcription factor Gal4 binds to the Gal4 upstream activating sequence (UAS) causing transcription of two genes involved in galactose metabolism. Gal4 consists of a transcriptional regulatory domain and a DNA binding and transcription activation domain. Chimeric proteins have been made, in which the transcriptional regulatory domain of mammalian transcription factors are fused to the DNA binding domain of Gal4. If expressed in mammalian cells containing a reporter linked to GAL4-UAS, this chimeric transcription factor can be phosphorylated by second messenger kinases and so increase reporter transcription. This system is widely used in cell-based assays for HTS and uHTS of nuclear receptors respectively.

Flash-light reporter genes generally encoded ion indicators rather than reporter gene and are widely used to detect and monitor calcium release or calcium flux in cell-based assay approaches for G-protein receptors and ion channels. Photoproteins as Aequorin, Obelin, Clytin or Berovin are used as cytosolic or mitochondrial tagged reporters in generic reporter cell lines. Flash-light reporter genes are applicable for agonist and antagonist screening approaches for G-protein coupled receptors and some ion channels. Kinetic as well as integral-based data analysis, allow the differentiation between agonistic or antagonistic test compounds in a single assay approach (see Sect. 1.13).

1.12 Reporter Genes in uHTS

Cell-based reporter gene assays can be designed to detect transcriptional regulation of a particular gene, or they can be configured to detect the activity of a signaling pahway. In reporter assays, the expression of a protein whose enzymatic activity is easily detected is linked to the biological activity of the target of interest. Transcriptional regulation reporter assays are configured by linking the promotor, or elements of the promotor, from the gene of interest to the coding region of the reporter gene. To detect the activity of a signaling pathway, repeats of a particular responsive element are located upstream of the reporter gene and regulate its expression in response to the activation of the pathway.

The advantages of the use of new reporter gene systems in cell-based uHTS are multiple instrumentation platforms, diversity of reagent kits, acceptable cost, greater signal amplitude, assay stability and most importantly, reporter gene assays are highly amenable to miniaturization. Reporter gene assays have their limitations too.

For example, reporter gene assays designed to detect Gs coupled receptor activity are often configured to amplify the signal via multiple copies of the cAMP responsive elements. Reporter gene assays designed to detect the receptor activity via calcium mobilization typically utilize the NFAT responsive element or flash-light reporter gene systems. NFAT mediated gene expression is dependent on both the amplitude and the duration of the calcium flux and therfore NFAT-driven reporter gene assays can not be detect rapid calcium-mediated responses.

1.13 Photoprotein Readouts and Cell-Based Assay Development

Obviously, cell-based assays screens designed to detect calcium mobilization via flash-light reporter gene systems were limited to those receptors that signal through the Gαq pathway. Co-expression of a promiscous (Gα15, Gα16), or chimeric G-protein can switch the signal of the GPCRs to the IP3 pathway and calcium release (see Fig. 1.5). These G-protein switching strategies have made calcium mobilization screening approaches applicable to many more receptors. The calcium mobilization could be detected through a flash-light reporter gene system or by a fluorescent imaging system that collects information about the calcium flux in each well. The intracellular calcium concentrations could be measured by preloading the cells with a calcium-sensitve fluorescent dye, such as Fluo3 (Minta et al. 1989).

Calcium mobilization screens are often used as a single readout suited for the identification of agonist activities. Antagonist screens are typically configured as binding assays using isolated receptors or membrane preparations with a radiolabeled ligand readouts.

Flash-light reporter genes as mitochondially tagged aequorin (mtAequorin) or clytin (mtClytin) are widely used for the identification of agonist and antagonist activities in two step bioluminescent readouts. Generic CHO-based reporter cell lines constitutively expressing the mtClytin or mtAequorin respectively are used to generate recombinant G-protein coupled receptor cell lines. The activation of recombinant GPCR by its ligand (Fig. 1.6, Signal 1) leads to the initiation of a cascade of reactions releasing Ca^{2+}, that triggers the emission of bioluminescence by the mtAequorin or mtClytin present in the system. The amount of light emission correlates directly to the used ligand concentration.

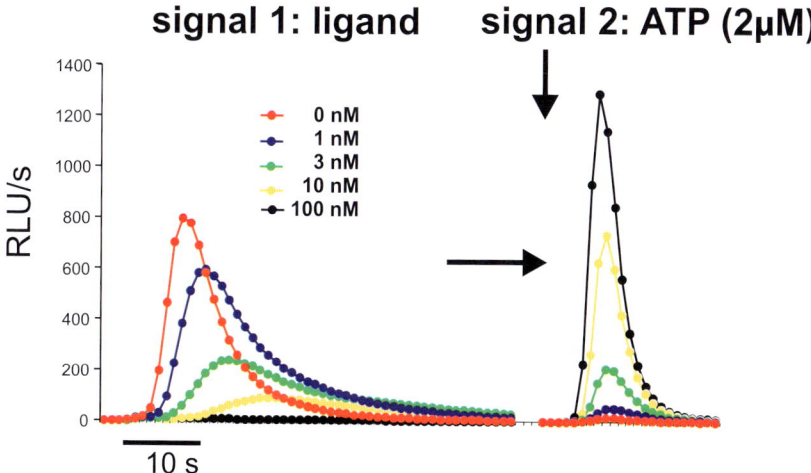

Fig. 1.6 Kinetic readout of flash-light type reporter gene aequorin in 1536-well format. Y: RLU/s:relative light units, X: time [s]

To differentiate between agonists and antagonists a second readout is necessary. Non saturating ligand concentrations of the recombinant GPCR lead to a partial activation and consumption of the coelenterazine-loaded photoprotein amount. The second line activation of an endogenous Gαq GPCR (here P2Y2 receptor, Fig. 1.6, Signal 2) by its ligand triggers the emmsion of bioluminescence of the coelenterazine-loaded photoprotein which was not triggered by the activation of the recombinant GPCR. The amount of light emission of the second signal correlates directly to the ligand concentration of the recombinant GPCR. This principle could be used to differentiate between agonists and antagonist in a two step bioluminescence readout.

1.14 Multiplexing Reporter Gene Readouts

An early drug discovery approach focusing on gene families can benefit from strategies that exploit common signaling mechanisms to more effectively identify and characterize novel chemical lead structures. Multiplexing, defined as the screening of multiple targets or the use of different reporter genes within the same experiment, are examples of this strategy (Miret et al. 2005).

To meet the increasing demands of compound testing created by combinatorial chemistry, technology investments in screening are focused on 3 major areas: miniaturization, automation, and multiplexing. Miniaturization attempts to increase the density of data points per plate of a particular assay format, whereas automation provides equipment to perform these assays with fewer human resources and higher reproducibility. Both of these approaches are applicable to any target class. Multiplexing allows for the parallel processing of multiple targets and promises to increase efficiencies, reduce costs, and improve the quality and content of screening

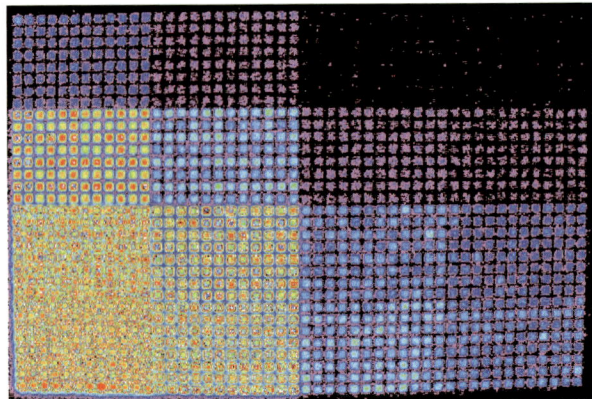

Fig. 1.7 Use of Mluc in cell-based 1536 assay format. CHO cell line with recombinant Metridia luciferase under the control of a CRE regulated promoter. X-axis: increasing concentration of forskoline (*left* to *right*), Y-axis: increasing concentration of coelenterazine (*top* to *bottom*)

data. In addition, it can also simplify the construction and execution of target selectivity panels. The use of multiplexing in hit discovery has been previously reported for kinases, nuclear receptors, and GPCRs. Common strategies of cell-based multiplexing assay approaches are combinations of glow- and flash-light reporter genes or reporter genes with different substrate specificity. Combinations of Firefly luciferase (Glow) with photoproteins as mtAequorin or mtClytin or Firefly luciferase (Glow) with Metridia luciferase (Glow, Fig. 1.7) are mixtures of both approaches.

1.15 Ultra High Throughput Screening

The availability of genomic information significantly increases the number of potential targets available for drug discovery. In a chemical re-search approach, ultra-high throughput screening (uHTS) (a screening robot-system for cellbased assays is shown in Fig. 1.8) of validated targets takes place early in the drug discovery process. Effective implementation of a chemical strategy requires assays that can perform uHTS for a large numbers of targets. Most assay principles are based on receptor competition between a substrate or ligand and the candidate compound measured by either binding or function of the receptor. Assays are tuned for the detecting competitive inhibitors. The rules of receptor theory and enzymology together with the detailed knowledge of the kinetic behaviour of the system dictate the most straightforward conditions for identifying inhibitors or activators.

Recent advances in organic synthesis, and combinatorial chemistry in particular, made available large libraries of small synthetic compounds. The established chemical access to these small molecules offer the possibility for the fast expansion of the chemical space around active compounds by structural derivatization. As a consequence, today most big pharmaceutical companies predominantly rely on libraries

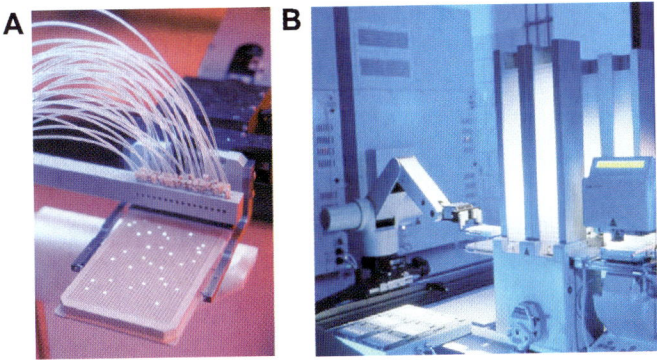

Fig. 1.8 A Bioluminescent readout in cell-based HTS. **B** Picture of a uHTS Robot-System for compound screening (Bayer Healthcare AG, Wuppertal, Germany)

made up of synthetic compounds for lead discovery. The identified lead structures provide entry points into medicinal chemistry programs.

A lead structure displays a desired biological activity but does not yet combine all pharmacodynamic and pharmacokinetic properties required for therapeutic use. The comprehensive list of attributes required from a lead structure explains why the discovery of novel molecules is expected to be a rare event. In most cases, uHTS derived lead candidates display liabilities regarding one or more properties and significant medicinal chemistry efforts are needed to completely assess the potential of a novel compound class.

Experimental testing of candidate drug compounds remains the major route for lead discovery in the drug discovery process. uHTS using targeted assays together with the design of combinatorial chemistry libraries have matured over the last decade to provide a technology platform for lead discovery. Converting the knowledge of the target mechanism and underlying molecular recognition principles into robust and sensitive assays is a prerequisite of successful uHTS screening.

References

1. Bronstein I, Fortin J, Stanley PE, Stewart GS, Kricka LJ (1994) Chemiluminescent and bioluminescent reporter gene assays. Anal Biochem 219(2):169–181
2. Eglen (2005) Functional G-protein coupled receptor assays for primary and secondary screening. Combinatorial Chemsitry $ High Throughput Screening 8:311–318
3. Gonzalez, Negulescu (1998) Intracellular detection assays for high-throughput screening. Current Opinion in Biotechnology 9:624–631
4. Hill JS, Baker JG, Rees S (2001) Reporter-gene systems for the study of G-protein-coupled receptors. Current Opinion in Pharmacology 1:526–532
5. Hiramatsu N, Kasai A, Meng Y, Hayakawa K, Yao J, Kitamura M (2005) Alkaline phosphatase vs luciferase as secreted reporter molecules in vivo. Analytical Biochemistry 339:249–256
6. Hopkins, Groom (2002) The druggable genome. Nat Rev Drug Discov 1(9):727–730
7. Hüser J, Lohrmann E, Kalthof B, Burkhardt N, Brüggemeier U, Bechem M (2006) High-throughput Screening for Targeted Lead Discovery. In: Hüser J (ed) High-Throughput Screening in Drug Discovery. Wiley-VCH Verlag GmbH & Co. KGaA, Weinheim
8. Jacoby E, Bouhelal R, Gerspacher M, Seuwen K (2006) The 7TM G-Protein-Coupled Receptor Target Family. Chem Med Chem 1:760–782
9. Johnston P, Johnston P (2002) Cellular platforms for HTS: three case studies. DDT 7:6
10. Jones K, Hibbert F, Keenan M (1999) Glowing jellyfish, luminescence and a molecule called coelenterazine. TIBTECH 17
11. Liu J, O'Kane DJ, Escher A (1997) Secretion of functional Renilla reniformis luciferase by mammalian cells. Gene 203:141–148
12. Markova SV, Golz S, Frank LA, Kalthof B, Vysotski ES (2004) Cloning and Expression of cDNA for a Luciferase from the Marine Copepod Metridia longa. J Biol Chem 279(5):3212–3217
13. Miret J, Zhang J, Min H, Lewis K, Roth M, Charlton M, Bauer P (2005) Multiplexed G-Protein–Coupled Receptor Ca^{2+} Flux Assays for High-Throughput Screening. Journal of Biomolecular Screening 10(8)
14. Minta A, Kao JP, Tsien RY (1989) Fluorescent indicators for cytosolic calcium based on rhodamine and fluorescein chromophores. J Biol Chem 264:8171–8178
15. Ormo M, Cubitt AB, Kallio K, Gross LA, Tsien RY, Remington SJ (1996) Crystal structure of the Aequorea victoria green fluorescent protein. Science 6;273(5280):1392–1395
16. Prasher (1995) Using GFP to see the light. TIG 11:8
17. Predergast (1999) Biophysics of the green fluorescent protein. Methods cell Biol 58:1–18
18. Shimomura O (1985) Bioluminescence in the sea: photoprotein systems. Symp Soc Exp Biol 39:351–372
19. Tannous BA, Kim DE, Fernandez JL, Weissleder R, Breakefield XO (2005) Codon-optimized Gaussia luciferase cDNA for mammalian gene expression in culture and in vivo. Mol Ther 11(3):435–443

20. Thompson EM, Nagata S, Tsuji FI (1989) Cloning and expression of cDNA for the luciferase from the marine ostracod Vargula hilgendorfii. Proc Natl Acad Sci USA 86(17):6567–6571
21. Viviani VR, Arnoldi FG, Venkatesh B, Neto AJ, Ogawa FG, Oehlmeyer AT, Ohmiya Y (2006) Active-site properties of phrixotrix railroad worm green and red bioluminescence-eliciting luciferases. J Biochem (Tokyo) 140(4):467–474
22. Walters WP, Namchuk M (2003) Designing screens: how to make your hits a hit. Nature Reviews in Drug Discovery 2:259–266
23. Wang S, Hazelrigg T (1994) Implications for bcd mRNA localization from spatial distribution of exu protein in Drosophila oogenesis. Nature 369:400–403
24. Williams (2004) cAMP Detection Methods in HTS: Selecting the Best from the Rest. Nature Review Drug Discovery 3:125–135
25. Wilson T, Hastings JW (1998) Bioluminescence. Annu Rev Cell Dev Biol 14:197–230
26. Wood KV, Lam YA, McElroy WD (1989) Introduction to Beetle Luciferases and their Applications. Journal of Bioluminescence and Chemiluminescence 4:289–301
27. Yamaguchi I (1975) Oplophorus oxyluciferin and a model luciferin compound biologically active with Oplophorus luciferase. Biochem J 151(1):9–15
28. Zimmer (2002) Green Fluorescent Protein (GFP): Applications, Structure, and Related Photophysical Behavior. Chem Rev 102:759–781

Chapter 2
Gene Arrays for Gene Discovery

David Ruau[1,2], Martin Zenke[1,2]

[1] Institute for Biomedical Engineering, Department of Cell Biology, RWTH Aachen University Medical School, Pauwelsstrasse 30, 52074 Aachen, Germany
[2] Helmholtz Institute for Biomedical Engineering, RWTH Aachen University, Pauwelsstrasse 20, 52074 Aachen, Germany
David.Ruau@rwth-aachen.de, martin.zenke@rwth-aachen.de

Abstract The wealth of genetic information available in life sciences now allows the engineering of DNA and RNA molecules, proteins, cells, tissues and even entire organisms with wanted properties. This is due to large scale sequencing projects that determined the DNA sequence of the entire human genome and of the genomes of several animal species, microorganisms and pathogens. At the same time microarray technology advanced to a stage that permits measuring the expression levels of thousand of genes simultaneously in one single experiment. In addition, specific bioinformatic tools have been developed to determine transcriptional "signatures" of various cell types, tissues and entire organisms, both in the normal and pathological state. Furthermore, various data mining strategies are being used for gene discovery and for a systematic and genome wide understanding of complex gene networks.

Here we provide an overview of data mining strategies for microarray gene expression data, including (i) data preprocessing, (ii) methods of cluster analysis and (iii) the retrieval of information from knowledge-based databases and their integration into microarray data analysis workflows. Our analysis shows examples of mining strategies for antigen presenting dendritic cells (DC) that were treated with transforming growth factor β type 1 (TGF-β1).

2.1 Introduction

The development of DNA microarrays about 15 years ago represents one of the most revolutionary inventions in cell biology. DNA microarrays allow measuring the expression levels of thousand of genes simultaneously thereby generating profiles of mRNA expression for the entire transcriptome of an organism. DNA microarrays are made by robotic disposition of complementary DNA (cDNA) in an ordered array on glass or nylon membrane supports. Each spot corresponds to a sequence that is highly specific for a gene of interest and one single DNA microarray can now cover all genes of a given organism. The technologies employed for producing microarrays are diverse: manufacturers like Affymetrix use photolithography to synthesize oligonucleotides directly on the support thereby achieving densities of 1,300,000 oligonucleotides on a 1.3 cm^2 surface for measuring 38,500 genes.

DNA microarrays are now being used in scientific research and medical diagnosis to address a wide range of very diverse questions. In addition, novel applications of DNA microarrays emerged with genome sequencing projects that provided the complete sequence information of a given organism. This genome-wide sequence information now allows measurements that are more than just comparing gene expression levels. For example, DNA microarrays are being designed for determining single nucleotide polymorphisms (SNPs) in order to map and relate individual genuine variations with a particular disease or phenotype. Another newly developed form of DNA microarrays is the tiling array. Such tiling arrays cover the organisms' entire genomic sequence to systematically interrogate genomic regions. For example the seven Affymetrix human tiling arrays contain 45 million oligonucleotide probes tiling the entire human genome at an average resolution of 35 base pairs. Such tiling arrays are most suitable to precisely map novel RNA transcripts and therefore extend the information that is obtained by conventional gene expression arrays. Additionally, tiling arrays are now being used in chromatin immunoprecipitation (ChIP) studies to map the positions of transcription factors and other DNA binding molecules on a global scale. Epigenetics represents an intensive field of research and tiling arrays allow the identification of DNA methylation patterns [26].

Importantly, DNA microarrays emerging as a novel genomic platform technology raised multiple challenges: (i) the genome-wide analysis produces huge amounts of data that need to be processed, stored and managed; (ii) advanced data mining techniques have to be employed to assess and understand the underlying biological events. Frequently, data mining or knowledge discovery involves multiple iterative steps:

1. Data preprocessing;
2. Data mining;
3. Data presentation;
4. Interpretation of the results.

Genome wide microarray studies are systematic and unbiased approaches for a given question and produce results that are often outside of the expertise of the investigator. Therefore, the extracted information needs to be enriched with knowledge from additional sources in order to put data into context. However, integrating additional information adds further dimensions to the data and increases the complexity of finding relevant clusters by traditional clustering techniques, such as k-means or agglomerative hierarchical clustering. Techniques, like principal component analysis (PCA) [35] or singular value decomposition, are often used in gene expression mining. They are known as feature transformation methods that allow to summarize data and to reveal hidden structures. Subspace clustering is also used in microarray analysis for mining of high dimensional data. Subspace clustering selects subsets from all dimensions (e. g. samples or patients) to identify groups of genes that exhibit similar expression profiles in the selected subset of dimensions.

Other data mining strategies perform searches no longer on gene expression values but directly on meta-data such as gene function. Here we will show an example for Gene Ontology (GO) [2] over-representation analysis that became a standard in DNA microarray analysis workflows.

2.2 Data Processing

2.2.1 Challenges in Data Acquisition and Processing

One of the major challenges in microarray technology is to demonstrate the consistency of the data points measured. This issue has been addressed very recently by the MicroArray Quality Control (MAQC) project [29]. This consortium showed that for many manufacturers the most recent generation of DNA microarrays allows robust and precise monitoring of gene expression of a given biological sample. Thus, microarray quality is not a major concern anymore and variability in data points is mainly inherent to the biological samples and their preparation, and to some extent also influenced by microarray scanning and data preprocessing. Factors playing a major role in data consistency can be categorized as follows:

- Precision (defines the reliability of the measurement);
- Accuracy (shows how close signal intensity translates into expression levels for a given gene);
- Specificity (indicates whether the DNA sequence present on the array measures the expression of the gene it has been design for);
- Sensitivity (describes the limitation of detecting low abundant transcripts while maintaining an acceptable accuracy and specificity).

All these criteria rely on the array design, preparation of the biological material and data processing. In the following we will review the currently available methods for preprocessing of raw data and how these data are further processed for gene mining.

2.2.2 Data Preprocessing: An Overview

Here we will focus on the Affymetrix platform that received the most attention from the bioinformatic community and is also one of the most widely used microarray technologies.

The Affymetrix platform uses one-color microarrays as opposed to two-color microarrays. For one-color microarrays one RNA sample is hybridized to a single array while for two-color microarrays two RNA sample are hybridized competitively to the same array. These different technologies imply a difference in signal interpretation. The signal obtained from two-color microarrays is the ratio of abundance of the test sample with respect to the control sample whereas with one-color technology measures are absolute expression values. Signal processing is also different, but some steps are identical, such as image acquisition and primary data analysis that produce the raw expression values.

The first step in data preprocessing is the background noise correction which is particularly difficult. This is because the background noise correction has two contradictory effects: it simultaneously increases accuracy and lowers the precision by increasing the variance between replicates. For more information on background noise corrections we refer to the Affymetrix manual [1] and [5].

In the next step raw expression levels need to be normalized in order to be comparable across samples. There are numerous normalization methods that often rely on platform specific features. For reasons of space we will not cover this topic here in detail but refer to Chap. 2 of Gentleman et al. [10] and also to some excellent reviews [6, 16, 27]. Yet, it is important to point out that a direct and quantitative comparison between the expression levels of different genes on the same microarray is not possible since each probe/gene on the microarray hybridizes with the target sequence with a different specificity. Consequently, the relation between signal and transcript concentration is not linear for low and high transcript concentrations and thus unique for every probe. Microarray technology accurately measures variations in the abundance of a particular transcript in different biological samples and thus up- or down-regulation of genes. The information obtained by comparing the signal intensities of different genes of the same biological sample hybridized to the same array is however rather limited.

Following normalization one would like to identify genes or samples that exhibit a similar expression profile e. g. by performing gene clustering. At this stage it is important to know that the choice for the distance or similarity measure is crucial before any clustering is performed. Thus, gene clustering can lead to different interpretations depending on the similarity measures applied. There are different types of distances that exhibit different properties. Frequently, distance is categorized into three types: (i) Minkowski metrics, (ii) parametric correlation and (iii) non-parametric correlation. Minkowski metrics (1) yields Manhattan and Euclidean distances with $p = 1$ and $p = 2$, respectively.

Minkowski Metrics (2.1)

$$D(x,y) = \sqrt[p]{\sum_{i=1}^{n} |x_i - y_i|^p} . \qquad (2.1)$$

For the Minkowski metrics the distance between samples is the same for relative and absolute expression measures (two-color microarrays). This does not hold for the distance between genes. On the other hand, distances based on the Pearson correlation (2.2) yield the same distance between genes for both relative and absolute measures but not for the distance between samples. Considering these facts and that most people are interested in classifying similar multidimensional gene expression profiles, the most widely used metrics are the correlations, such as Pearson correlation.

The Pearson correlation (2.2) and its related correlations, like the cosine distance used by Eisen et al. [9], give good results when clustering aims at the identification of genes with similar expression profiles independently of scale.

Pearson Correlation Coefficient (2.2)

$$D(x, y) = \frac{\sum_{i=1}^{n}(x_i - \bar{x}_i)(y_i - \bar{y}_i)}{\sqrt{\sum_{i=1}^{n}(x_i - \bar{x}_i)^2 \sum_{i=1}^{n}(y_i - \bar{y}_i)^2}} . \tag{2.2}$$

The Pearson correlation and their derivatives are parametric and able to measure linear relations between gene vectors of expression values but are sensitive to outliers. Pearson correlation can also be used to classify samples but with the assumption that the populations are normally distributed. A normal distribution of expression values is in fact observed in the large majority of microarray experiments performed today [11]. Parametric correlation measure can – due to its sensitivity to outliers – give non-homogeneous cluster solutions. In this case non-parametric correlations, such as Spearman rank correlation or Kendall's τ rank correlation, are preferred.

Further measures of similarity have been specifically designed for gene expression clustering. Often microarray experiments have an underlying structure determined by covariate (supplementary) information such as treatment, cell or tissue type, or time. Those features can be incorporated into the similarity measure as weight to make clusters more relevant [4, 14]. Furthermore, to create even more meaningful clusters, genes can be linked to knowledge-based databases such as literature co-citation or biological pathway information. Text mining on the available online medical literature to extract gene function annotation is a particularly dynamic field of bioinformatics research [24, 33]. The general idea behind text mining for gene co-citation is that genes that are cited in the same paper might be functionally related [17]. Another approach outlined in Maere et al. [23] produces clusters based on GO. GO describes genes according to their cellular component, biological process, molecular function etc. Numerous tools make use of the GO classification and are listed on the GO web page (http://www.geneontology.org/GO.tools.microarray.shtml). The majority of these software tools perform statistical tests on the genes of interest by searching for enriched GO classes. Further methods propose to incorporate knowledge from GO to assess or improve the cluster interpretation [15].

2.3 Gene Discovery by Gene Clustering

Gene clustering aims at identifying groups of genes exhibiting common features in complex data sets. There are many clustering algorithm that essentially can be classified into hierarchical and non-hierarchical clustering (Fig. 2.1). Classical approaches for clustering are the partitioning and hierarchical algorithms, but there is no universal clustering algorithm that fits all applications. So far non-hierarchical clustering has received the most attention (Fig. 2.1). Agglomerative hierarchical

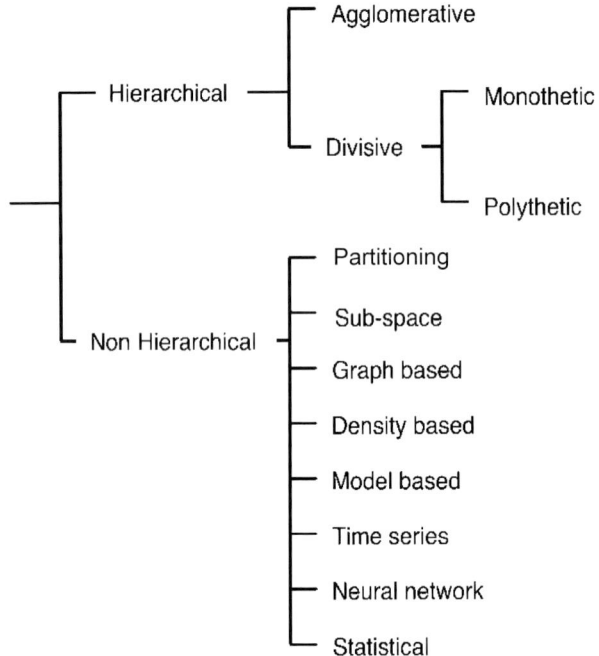

Fig. 2.1 Schematic representation of clustering algorithms

clustering with heatmap representation was one of the first clustering method applied to gene expression profiling published by Eisen et al. [9] in 1998.

There are basically two ways for generating clusters by hierarchical clustering: either agglomerative or divisive, depending whether the tree is built from bottom to top or vice versa. Agglomerative hierarchical clustering is mainly used and gives better results than the divisive methods [19]. Agglomerative hierarchical methods use different linkage methods to cluster the groups when building the tree and most frequently the average linkage method (UPGMA) is used. A list of available linkage methods is given in Table 2.1.

Table 2.1 The different linkage methods commonly used for hierarchical clustering

Methods	Description
Single	Shortest distance or nearest neighbor [25]
Average	Unweighted average distance (UPGMA) also called Group average distance
Weighted	Weighted average distance (WPGMA)
Complete	Furthest distance/neighbor [32]
Mcquitty	Similar to single method [20]
Ward	Inner squared distance; for interval-scaled measurements [13]
Median	Weighted pair-group method using centroid (WPGMC)
Centroid	Unweighted pair-group method using centroid (UPGMC)
Flexible	Generalization of the others [13]

There are three types of non-hierarchical clustering algorithms: (i) partitioning methods, (ii) hybrid methods combining partitioning and hierarchical methods and (iii) further methods such as density based clustering or subspace clustering [13].

Hierarchical clustering methods perform well on data sets with a few dimensions but when the dimensionality increases other clustering techniques should be considered. In this instance principal component analysis (PCA) and subspace clustering, like CLIQUE or PROCLUS, will perform better and find more meaningful clusters than the classical hierarchical clustering. An extensive literature exists on high-dimensional clustering approaches [13]. An interesting method applied in microarray mining is the frequent pattern mining technique, such as biclustering [8]. Biclustering works on the genes and the conditions to produce subclusters [34]. This clustering method finds genes regulated in the same way between just a subset of the samples. For a survey on biclustering techniques see Madeira et al. [22].

In summary, the results obtained by clustering, either hierarchical or non-hierarchical clustering, are highly dependent on the metric (distance or similarity/dissimilarity) used to quantify the difference between the features. It is important to note that all clustering techniques described here are unsupervised methods. Supervised methods, such as those employed in machine learning techniques, are not often used in gene expression mining. Machine learning is however often employed in medical applications where one wants to classify a new sample or patient within a given disease category.

2.4 TGF-β1 Signaling in Dendritic Cells Assessed by Gene Expression Profiling

Conventional strategies in biology analyze the impact of a single gene or its product on a biological phenomenon mainly by employing hypothesis driven approaches. This has now changed by analyzing genes and/or proteins in systematic and unbiased approaches within complex regulatory networks on a global scale, for example by DNA microarrays. The wealth of information obtained requires novel ways of data acquisition, management, mining and the representation of the knowledge. In the following we will demonstrate the power of such genome-wide approaches by analyzing the impact of the multifunctional cytokine TGF-β1 on gene expression in dendritic cells (DC), a key cell of our immune system. This study led to the identification of major players in DC development [12, 18].

DC are professional antigen presenting cells that induce antigen specific T cell responses and control the balance between immunity and immunological tolerance [3, 31]. DCs are derived from hematopoietic stem cells in bone marrow and different cytokines have been implicated in DC development and function. TGF-β1 is a multifunctional cytokine involved in a variety of biological processes including DC development. TGF-β1$-/-$ mice lack for example Langerhans cells (LC), the cutaneous contingent of DC located in epithelial tissues, such as skin [7].

In our previous work we used in vitro culture systems and gene expression profiling with DNA microarrays to search for genes with a decisive function in DC development. These studies identified the helix-loop-helix (HLH) transcription factor Id2 (inhibitor of differentiation/DNA binding 2) as one of the critical factors for LC development [12]. Id2 is up-regulated during DC differentiation and important for DC subset specification [12, 30]. Id2−/− mice lack LC and we demonstrated that TGF-β1 up regulates Id2 expression.

Id2 is a member of the inhibitory HLH transcription factor family that antagonizes the action of the activating HLH factors, such as E2A, and was found to repress B cell genes in DC [12]. This led us to propose a model where the relative abundance of Id2 and for example E2A determines lineage choice by affecting the property of a common progenitor to develop into B lymphoid cells or DC [12]. Thus, high Id2 expression provides an environment where B cell/DC precursors are permissive to the action of factors with an instructive function in DC development. We then proceeded to search for such factor by performing gene expression profiling of TGF-β1 treated cells [18].

DC were obtained by a 2-step culture system where CD34+ stem cells from human cord blood are amplified with a stem cell factor/cytokine cocktail (in the following referred to as hematopoietic progenitor cells, HPC) and then induced to differentiate into DC with GM-CSF and IL-4. HPC and DC were treated with TGF-β1 for 4, 16 and 36 hours, RNA was isolated and subjected to DNA microarray analysis using Affymetrix human HG-U95Av2 GeneChip arrays [18, 28].

The first point to be addressed is "what is the global effect of TGF-β1 treatment"? To answer this question we performed principle component analysis (PCA) on all HPC and DC data sets by considering all genes present on the array irrespective whether they were expressed or not. PCA separated HPC and DC data sets (Fig. 2.2)

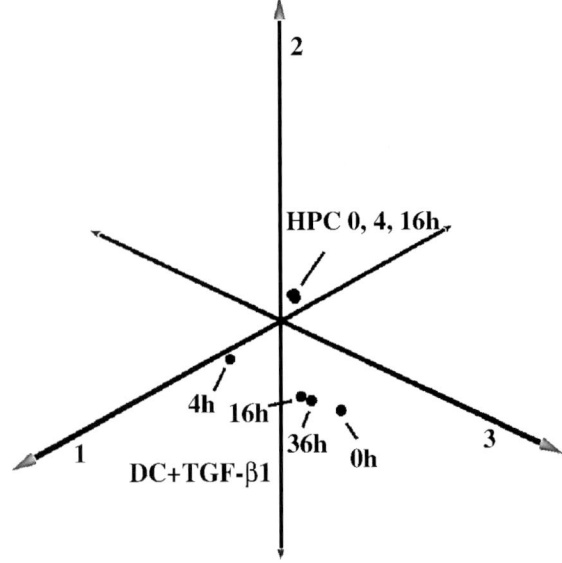

Fig. 2.2 Principal component analysis (PCA) of hematopoietic progenitor cells (HPC) and dendritic cells (DC) treated with TGF-β1 for 4, 16 and 36 hours (h), or left untreated (0 hours) [28]. The three first components, capturing together 99.3% of the variability in the data sets, are shown

2.4 TGF-β1 Signaling in Dendritic Cells Assessed by Gene Expression Profiling

and showed that TGF-β1 effectively influenced gene expression in DC. There was only a minor effect of TGF-β1 on gene expression of HPC.

PCA can also be used to classify genes but when all genes analyzed in a microarray experiment are considered (frequently several thousands) the resulting images are cluttered. Correspondence analysis, a variant of PCA, represents a way to overcome this limitation [21].

Phenotyping of immune cells, including DC, frequently involves determining the expression pattern of cluster of differentiation (CD) molecules. We thus proceeded to investigate changes in the expression of CD molecule in response to TGF-β1 by hierarchical clustering and display in heatmap format (Fig. 2.3). Gene expression values and conditions were compared using Pearson correlation coefficient. The hierarchical clustering was performed by agglomerative bottom-top UPGMA linkage method (Table 2.1) and this approach gave the most interpretable output.

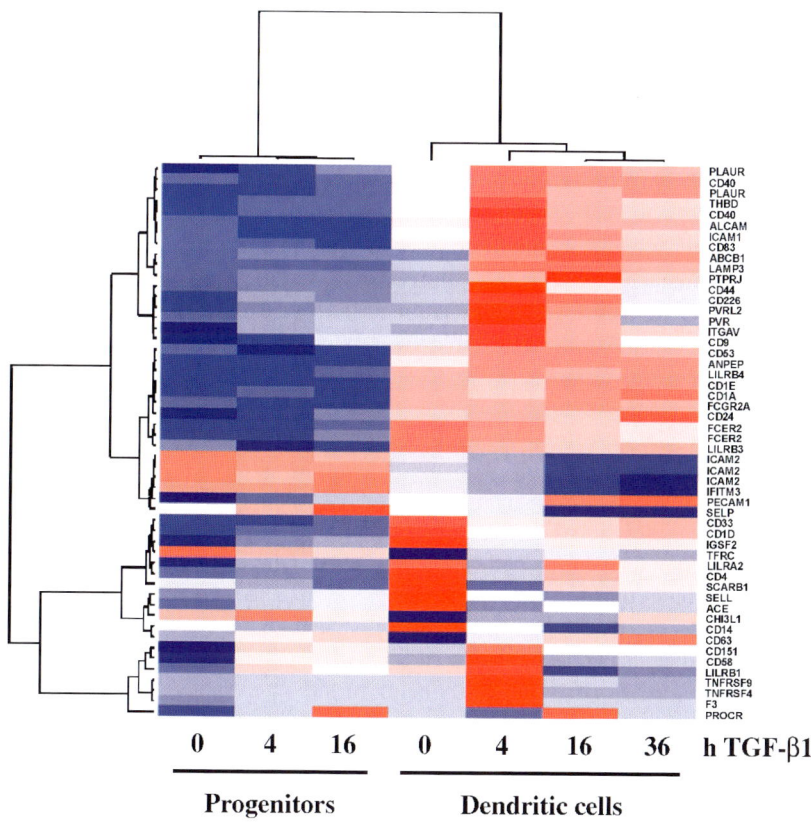

Fig. 2.3 Heatmap view of hierarchical clustering of CD molecules in HPC (progenitors) and dendritic cells (DC) treated with TGF-β1 [28]. Each row represents one gene and color indicates expression variation: *red*, higher expression and *blue*, lower expression in relation to the mean expression value of the gene

There were only minor effects of TGF-β1 on CD gene expression in HPC and the most dramatic changes were observed in DC treated with TGF-β1 for 4 hours (Fig. 2.3). This result is very much in line with the PCA result (Fig. 2.2).

Heatmap representation of hierarchical clustering provides a graphical view on the different gene expression patterns that exist in a given data set. The main advantage of hierarchical clustering over other clustering methods, including neural networks such as self organizing maps (SOMs), is that it is an unsupervised approach that does not requiring further input from the user (e.g. defining the numbers of clusters to be generated).

2.4.1 Linking Gene Expression Data to Knowledge-based Databases

Gene function is described in a controlled vocabulary referred to as Gene Ontology (GO) (www.geneontology.org) that describes genes and their products according to their activity in biological processes, molecular function and cellular component. Thus, further information on the changes in gene expression induced by TGF-β1 was obtained by GO over-representation analysis. In this approach genes are investigated for their over-representation in GO biological process categories thereby identifying the impact of a given treatment on, for example, a particular cellular pathway, process or compartment.

To this end 2204 differentially regulated genes were subjected to GO over-representation analysis and most of the TGF-β1 target genes were found in GO categories related to immune and defense responses, responses to stress, wounding, pathogens etc. (Fig. 2.4). These results provide further information on the gene categories affected by TGF-β1 in DC.

Kyoto Encyclopedia of Genes and Genomes (KEGG) represents a knowledge-based database that contains information on gene networks including a large number of signaling and metabolic pathways. Thus, microarray data can be linked to KEGG database for exploring gene expression data for specific pathways (Fig. 2.5). Therefore, DC expression data were investigated for their contribution to the canonical TGF-β1/Smad signaling pathway (Fig. 2.5). Changes in gene expression are displayed in color code.

As expected the expression of a number of known TGF-β1 target genes was found to be induced such as PAI-1. In addition, the analysis identified several novel TGF-β1 target genes that are currently being further investigated.

2.5 Conclusions

Genome-wide gene expression profiling and advanced bioinformatics provide unique opportunities to decipher complex biological processes and are expected to pave the way for computational prediction of complicated cellular processes. To

2.5 Conclusions

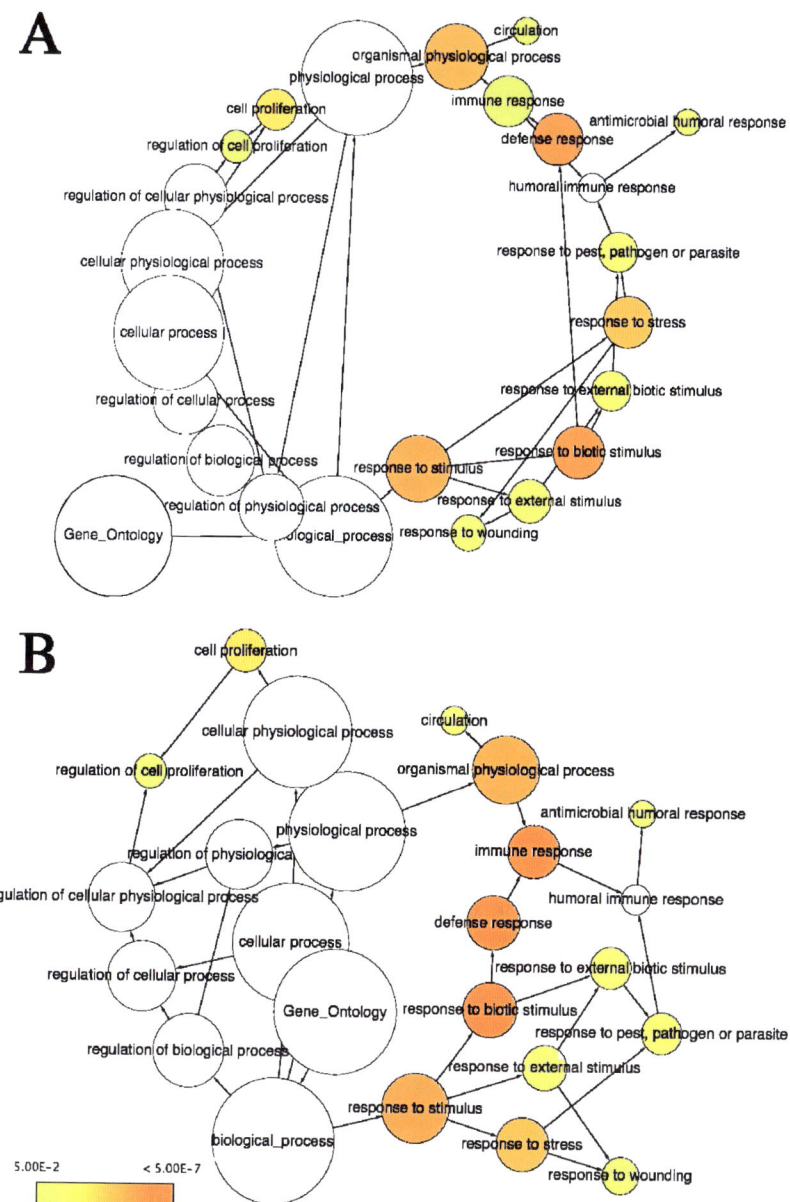

Fig. 2.4 Gene ontology (GO) over-representation analysis of 2204 genes that are differentially regulated upon TGF-β1 treatment of DC [28]. *Color scale* indicates confidence (corrected *p*-value) in the over-representation of the respective GO class; white nodes are not over-represented. Node size is proportional to the number of gene present in the individual category. A circular representation (*panel A*) and an organic network layout (*panel B*) are shown

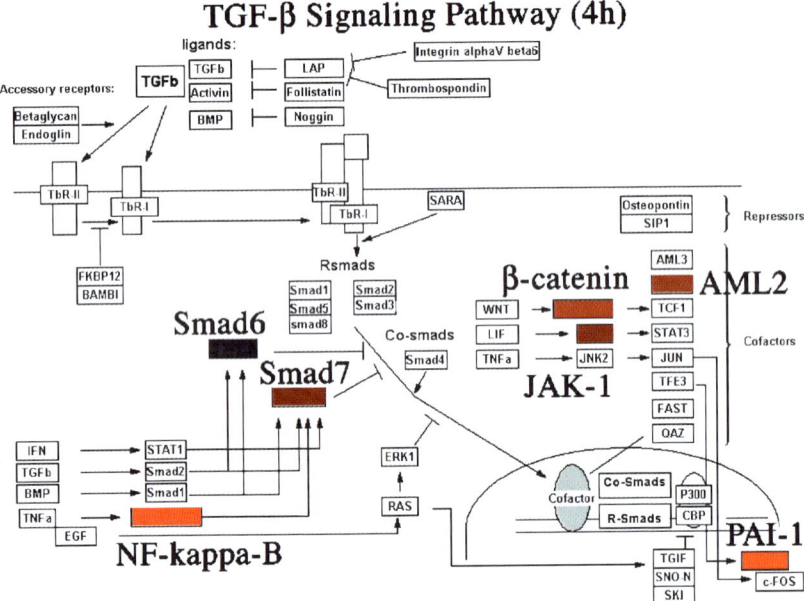

Fig. 2.5 KEGG pathway for TGF-β1 signaling with gene expression variation map. Schematic representation of the TGF-β1 signaling pathway from KEGG database. Changes in gene expression upon TGF-β1 treatment (4 hours) of DC are displayed in *color code* (e. g. *red*, increase in gene expression) and several TGF-β1 target genes are shown

improve the reliability of microarray technology and data management still remains a major challenge. Initiatives, such as the Minimum Information About Microarray Experiment (MIAME) society and more recently the MicroArray Quality Control (MAQC) and the External RNA Controls Consortium (ERCC), represent steps in this direction. Here we provide an overview of classical gene mining strategies by focusing on microarray gene expression data. Many other approaches are currently being developed that will contribute to the expanding field of system biology. The ultimate aim is to construct in silico models that faithfully describe genomic and molecular interactions, complicated cellular processes and finally the behavior of cells. The new discipline of system biology uses concepts and methods from physics, chemistry, statistics, bioinformatics and knowledge from biology to enhance our understanding of complex biological processes, thereby providing opportunities to find solutions for diseases that are out of reach in current biomedicine.

Acknowledgements We would like to thank T. Seidl, M. Hofmann-Apitius, K. Seré and P. Jäntti for critical reading of the manuscript and A. Offergeld for expert secretarial assistance. Part of this work was supported by grants from BMBF, DFG/Ze432/2-3, DFG/SFB542 and the Interdisciplinary Centre for Clinical Research BIOMAT to M.Z.

References

1. Affymetrix manual, www.Affymetrix.Com/support/technical/whitepapers/sadd_whitepaper.Pdf
2. Ashburner M, Ball CA, Blake JA, Botstein D, Butler H, Cherry JM, Davis AP, Dolinski K, Dwight SS, Eppig JT, Harris MA, Hill DP, Issel-Tarver L, Kasarskis A, Lewis S, Matese JC, Richardson JE, Ringwald M, Rubin GM, Sherlock G (2000) Gene ontology: Tool for the unification of biology. The gene ontology consortium. Nat Genet 25(1):25–29
3. Banchereau J, Briere F, Caux C, Davoust J, Lebecque S, Liu YJ, Pulendran B, Palucka K (2000) Immunobiology of dendritic cells. Annu Rev Immunol 18:767–811
4. Bolshakova N, Azuaje F, Cunningham P (2005) A knowledge-driven approach to cluster validity assessment. Bioinformatics 21(10):2546–2547
5. Bolstad B, Irizarry R, Gautier L, Wu Z (2005) Preprocessing high-density oligonucleotide arrays. Bioinformatics and computational biology solutions using r and bioconductor. Gentleman R, Carey V, Huber W, Irizarry R and Dudoit S (ed) Springer, Berlin, Heidelberg, New York
6. Bolstad BM, Irizarry RA, Astrand M, Speed TP (2003) A comparison of normalization methods for high density oligonucleotide array data based on variance and bias. Bioinformatics 19(2):185–193
7. Borkowski TA, Letterio JJ, Farr AG, Udey MC (1996) A role for endogenous transforming growth factor beta 1 in langerhans cell biology: The skin of transforming growth factor beta 1 null mice is devoid of epidermal langerhans cells. J Exp Med 184(6):2417–2422
8. Cheng Y, Church GM (2000) Biclustering of expression data. Proc Int Conf Intell Syst Mol Biol 8:93–103
9. Eisen MB, Spellman PT, Brown PO, Botstein D (1998) Cluster analysis and display of genome-wide expression patterns. Proc Natl Acad Sci USA 95(25):14863–14868
10. Gentleman R, Carey V, Huber W, Irizarry R, Duboit S (2005) Bioinformatics and computational biology solution using r and bioconductor. Springer, Berlin, Heidelberg, New York
11. Giles PJ, Kipling D (2003) Normality of oligonucleotide microarray data and implications for parametric statistical analyses. Bioinformatics 19(17):2254–2262
12. Hacker C, Kirsch RD, Ju XS, Hieronymus T, Gust TC, Kuhl C, Jorgas T, Kurz SM, Rose-John S, Yokota Y, Zenke M (2003) Transcriptional profiling identifies id2 function in dendritic cell development. Nat Immunol 4(4):380–386
13. Han J, Kamber M (2006) Data mining: Concept and techniques, 2nd ed., Morgan Kaufmann
14. Homayouni R, Heinrich K, Wei L, Berry MW (2005) Gene clustering by latent semantic indexing of medline abstracts. Bioinformatics 21(1):104–115
15. Huang D, Pan W (2006) Incorporating biological knowledge into distance-based clustering analysis of microarray gene expression data. Bioinformatics 22(10):1259–1268
16. Irizarry RA, Wu Z, Jaffee HA (2006) Comparison of affymetrix genechip expression measures. Bioinformatics 22(7):789–794

17. Jenssen TK, Laegreid A, Komorowski J, Hovig E (2001) A literature network of human genes for high-throughput analysis of gene expression. Nat Genet 28(1):21–28
18. Ju XS, Ruau D, P J, Sere K, Becker C, Wiercinska E, Bartz C, Erdmann B, Dooley S, Zenke M (2007) Transforming growth factor b1 (TGF-b1) upregulates interferon regulatory factor 8 (IRF-8) during Langerhans cell development. Eur J Immunol 35(7):1174–1183
19. Kaufman LR, P J (1990) Finding groups in data, Wiley
20. Lance GN, and Williams WT (1966) A generalized sorting strategy for computer classifications. Nature 212:218
21. Legendre P, Legendre L (1998) Numerical ecology, Elsevier, Copenhagen
22. Madeira SC, Oliveira AL (2004) Biclustering algorithms for biological data analysis: A survey. IEEE/ACM Trans. Comput Biol Bioinformatics 1(1):24–45
23. Maere S, Heymans K, Kuiper M (2005) Bingo: A cytoscape plugin to assess overrepresentation of gene ontology categories in biological networks. Bioinformatics 21(16):3448–3449
24. Masys DR, Welsh JB, Lynn Fink J, Gribskov M, Klacansky I, Corbeil J (2001) Use of keyword hierarchies to interpret gene expression patterns. Bioinformatics 17(4):319–326
25. McQuitty LL (1966) Similarity analysis by reciprocal pairs for discrete and continuous data. Edu and Psychological Measurement 26:825–831
26. Mockler TC, Chan S, Sundaresan A, Chen H, Jacobsen SE, Ecker JR (2005) Applications of DNA tiling arrays for whole-genome analysis. Genomics 85(1):1–15
27. Quackenbush J (2002) Microarray data normalization and transformation. Nat Genet 32(Suppl):496–501
28. Ruau D, Ju XS, Zenke M (2006) Genomics of TGF-β1 signalling in stem cell commitment and dendritic cell development. Cell. Immunol. 244(2):116–120
29. Shi L, Reid LH, Jones WD, Shippy R, Warrington JA, Baker SC, Collins PJ, de Longueville F, Kawasaki ES, Lee KY, Luo Y, Sun YA, Willey JC, Setterquist RA, Fischer GM, Tong W, Dragan YP, Dix DJ, Frueh FW, Goodsaid FM, Herman D, Jensen RV, Johnson CD, Lobenhofer EK, Puri RK, Scherf U, Thierry-Mieg J, Wang C, Wilson M, Wolber PK, Zhang L, Amur S, Bao W, Barbacioru CC, Lucas AB, Bertholet V, Boysen C, Bromley B, Brown D, Brunner A, Canales R, Cao XM, Cebula TA, Chen JJ, Cheng J, Chu TM, Chudin E, Corson J, Corton JC, Croner LJ, Davies C, Davison TS, Delenstarr G, Deng X, Dorris D, Eklund AC, Fan XH, Fang H, Fulmer-Smentek S, Fuscoe JC, Gallagher K, Ge W, Guo L, Guo X, Hager J, Haje PK, Han J, Han T, Harbottle HC, Harris SC, Hatchwell E, Hauser CA, Hester S, Hong H, Hurban P, Jackson SA, Ji H, Knight CR, Kuo WP, Leclerc JE, Levy S, Li QZ, Liu C, Liu Y, Lombardi MJ, Ma Y, Magnuson SR, Maqsodi B, McDaniel T, Mei N, Myklebost O, Ning B, Novoradovskaya N, Orr MS, Osborn TW, Papallo A, Patterson TA, Perkins RG, Peters EH, Peterson R, Philips KL, Pine PS, Pusztai L, Qian F, Ren H, Rosen M, Rosenzweig BA, Samaha RR, Schena M, Schroth GP, Shchegrova S, Smith DD, Staedtler F, Su Z, Sun H, Szallasi Z, Tezak Z, Thierry-Mieg D, Thompson KL, Tikhonova I, Turpaz Y, Vallanat B, Van C, Walker SJ, Wang SJ, Wang Y, Wolfinger R, Wong A, Wu J, Xiao C, Xie Q, Xu J, Yang W, Zhong S, Zong Y, Slikker W Jr (2006) The microarray quality control (maqc) project shows inter- and intraplatform reproducibility of gene expression measurements. 24(9):1151–1161
30. Spits H, Couwenberg F, Bakker AQ, Weijer K, Uittenbogaart CH (2000) Id2 and id3 inhibit development of CD34(+) stem cells into predendritic cell (pre-DC)2 but not into pre-DC1. Evidence for a lymphoid origin of pre-dc2. J Exp Med 192(12):1775–1784
31. Steinman RM, Hawiger D, Nussenzweig MC (2003) Tolerogenic dendritic cells. Annu Rev Immunol 21:685–711
32. Ward JH, Jr. (1963) Hierarchical grouping to optimize an objective function. Journal of the American Statistical Association 58:236–244
33. Yandell MD, Majoros WH (2002) Genomics and natural language processing. Nat Rev Genet 3(8):601–610
34. Yang J, Wang W, Wang H, Yu P (2002) D-clusters: Capturing subspace correlation in a large data set. ICDE'02 (18th International Conference on Data Engineering (ICDE'02)), p 0517
35. Yeung KY, Ruzzo WL (2001) Principal component analysis for clustering gene expression data. Bioinformatics 17(9):763–774

Chapter 3
Physical Modulation of Cellular Information Networks

Sumihiro Koyama[1], Masuo Aizawa[2]

[1] Extremobiosphere Research Center, Japan Agency for Marine-Earth Science and Technology, 2-15 Natsushima-cho, Yokosuka 237-0061, Japan, skoyama@jamstec.go.jp
[2] Tokyo Institute of Technology, O-okayama, Meguro-ku, Tokyo 152-8550, Japan

Abstract In this chapter, we reviewed the effects of physical stimulation including electrical potential, hydrostatic pressure, electromagnetic field, shear stress, and heat shock application on cellular functions such as proliferation, differentiation, gene expression, and protein production. Prospects for future investigation are also discussed.

3.1 Introduction

It has been known for many years that the alteration of the extracellular environment including chemical and physical factors can trigger the activation of intracellular signaling cascades and gene expressions in a cell. Extensive researches have gradually revealed that chemical and physical stimulation can activate a sort of receptor molecules which induce the gene expressions. For example, the exposure of cells to raised or reduced temperatures or to a wide variety of physical and chemical injuries activates the expression of heat-shock proteins by causing the intracellular accumulation of abnormal or degraded proteins (Lindquist 1986; Lindquist and Craig 1988; Lindquist 1992; Yanagida et al. 2000; Hochleitner et al. 2000; Koyama et al. 2002a; Calini et al. 2003). As another example, the plasma membrane disruption of sea urchin eggs, fibroblasts, and endothelial cells by a mechanical injury induces a rapid burst of localized exocytosis (McNeil 2002). The exocytotic reaction is rapidly evoked in a Ca^{2+}-dependent fashion and is quantitatively related to the disruption magnitude (McNeil 2002). It has also been reported that a wide range of physicochemical stimulation activated growth factors and the immunomodulatory cytokines production in the animal cells (Koyama et al. 1996, 1997, 2002b; Koyama and Aizawa 2002; Lee et al. 1995; Mohamadzadeh et al. 1995; Enk et al. 1996; Wlaschek et al. 1997; Finkel 2001; Fujimori et al. 2005; Park et al. 2006). A fundamental question which remains unanswered is how the physicochemical stimuli trigger the synthesis of the growth factors and the immunomodulatory cytokines. The chemical stimulation induced growth factors and cytokines productions have been extensively studied and clarified for many years. Meanwhile, it should be emphasized at this point that, compared with the chemically induced growth factors and cytokines production, physical stimulation induced mechanisms of growth factors and cytokines productions have not yet been elucidated.

Aizawa and coworkers have looked at the physical stimulation-induced cellular protective action for inspiration and developed a novel method for electrically modulated cellular functions of animal cells (reviewed in Aizawa et al. 1999). When animal cells are cultured directly on an electrode surface and then exposed to a weak electrical potential, no electrochemical reaction is induced. The cell/electrode-interactive effects have been reported to involve cell proliferation control, gene expression, protein production, cell differentiation, etc. Animal cells can remain alive if the electrode potential is controlled properly (Yaoita et al. 1989, 1990; Kojima et al. 1991). On the other hand, Koyama and colleagues (Koyama et al. 2002b; Koyama and Aizawa 2002; Koyama et al. 2005) discovered novel modulation methods for the regulation of animal cell functions using extremely high hydrostatic pressure exposure for a short period, during which the cells remained alive. Hydrostatic pressure exposure of the animal cells triggers the activation of protein kinase C (PKC)-dependent intracellular signaling cascades (Koyama and Aizawa 2002), gene expression, and secretion of immunomodulatory cytokines (Koyama et al. 2002b; Koyama and Aizawa 2002). Recently, the modulation of the proliferation rate of animal cells due to hydrostatic pressure has also been reported (Koyama et al. 2005).

In this chapter, we review the effects of physical stimulation including electrical potential, hydrostatic pressure, electromagnetic field, shear stress, and heat shock application on cellular functions such as proliferation, differentiation, gene expression, and protein production. Most of the findings were discovered by our research group. Prospects for future investigation are also discussed.

The dotted arrows indicated indirect or direct activation of the targets. G = G protein; PLC = phospholipase C; PIP_2 = phosphatidylinositol 4,5-bisphosphate; IP3 = inositol triphosphate; DAG = diacylglycerol; NFκB = nuclear factor κB; IκB = inhibitor of NFκB; MAPK = mitogen activated protein kinase; MAPKK = MAPK kinase; MAPKKK = MAPKK kinase; ERK = extracellular regulated kinase; JNK = c-Jun NH$_2$-terminal kinase; Ash = abundant Src homology; Grb2 = growth factor receptor bound protein2; SOS = son of sevenless; PI3 kinase = phosphatidylinositol 3-kinase; ECM = extracellular matrix.

3.2 Cellular Responses to Physical Stimulation

To activate useful cellular functions with the help of physical stimulation, it is necessary to understand the stimulation-induced intracellular signaling cascades and the gene expression subsequently triggered. This section introduces various types of physical stimuli-activated intracellular signaling cascades and cytokine production. Figure 3.1, Table 3.1 and 3.2 summarized intracellular signaling pathways, gene

3.2 Cellular Responses to Physical Stimulation

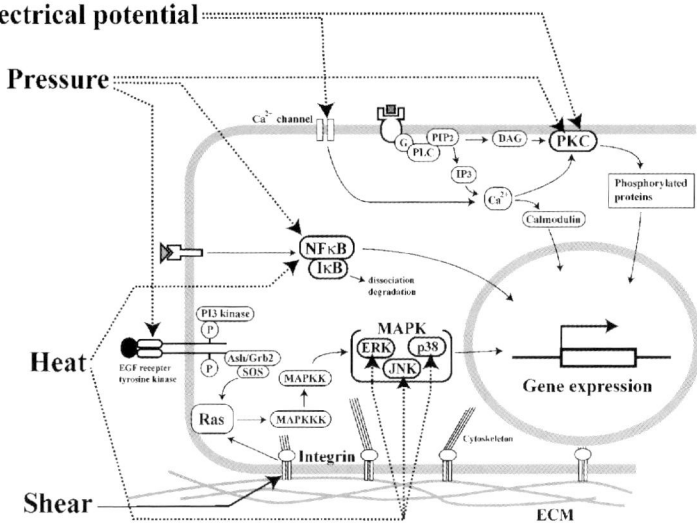

Fig. 3.1 Schematic diagram showing the intracellular signaling pathways in response to the physical stimuli

expression, and protein production activated by physical stimulation. A detailed explanation is described in the below subsections from 2.1 to 2.5. Because heat shock stress influences both membranous and cytoplasmic proteins, it is generally thought that physical stimulation can affect not only membrane functions but also halfway along the intracellular signaling cascades (Fig. 3.1).

3.2.1 Electrical Potential

We have investigated the effects of low voltage of electrical potential stimulation on gene expression and protein production in mammalian cells which cultured on an optically transparent electrode surface (Fig. 3.2). Koyama and coworkers (1996, 1997) have discovered that a 10 Hz sine wave potential application induces the nerve growth factor (NGF) production in mouse astroglial cells. Astroglial cells had been cultured for 7 days in a serum-free Dulbecco's modified eagle medium (DMEM) to synchronize a resting stage of astrocytes, which cells scarcely produce NGF. The cells were subjected to another 1 hr culture along with the application of the sine wave potential at varying amplitudes. NGF was maximally secreted at an upper potential of +0.3 V and was increased by six-fold over the control.

The intracellular levels of NGF remained unchanged by the electrical stimulation. Furthermore, we confirmed that the electrical potential induced NGF production involves gene expression mechanisms activated by protein kinase C (PKC) – activator protein-1 (AP-1) intracellular signaling pathway (Fig. 3.1 and Table 3.2; Details are described in Sect. 3.5.1).

Table 3.1 Physical modulation of gene expression and protein production

Physical modulation	Stimulation type	Cell type	Result	Reference
Hydrostatic pressure	Statical stimulation	Human skin fibroblasts	↑ IL-6, ↑ IL-8, ↑ PKC	Koyama and Aizawa (2002)
Hydrostatic pressure	Statical stimulation	Human skin fibroblasts	↑ IL-6, ↑ IL-8, ↑ IL-6 mRNA, ↑ IL-8 mRNA, ↑ IL-1α mRNA, ↑ IL-1β mRNA, ↑ IL-12 mRNA, ↑ MCP-1 mRNA	Koyama et al. (2002b)
Hydrostatic pressure	Statical stimulation	Human skin fibroblasts	↑ IL-6, ↑ IL-8, ↑ PKC	Koyama and Aizawa (2002)
Hydrostatic pressure	Statical stimulation	Human optic nerve astrocytes	↑ NOS-2, ↑ NOS-2 mRNA	Liu et al. (2001)
Hydrostatic pressure	Fluctuated stimulation	Osteoarthritis	↑ NOS-2 mRNA, ↑ NO	Fermor et al. (2001)
Hydrostatic pressure	Statical stimulation	Human optic nerve astrocytes	↑ NOS-2, ↑ NFκB, ↑ EGFR tyrosine kinase	Neufeld and Liu (2003)
Heat shock	Statical stimulation	Human skin fibroblasts	↑ IL-6, ↑ ERK, ↑ JNK	Park et al. (2006)
Heat shock	Statical stimulation	Human monocytes	↑ TLR2 and 4, ↑ ERK, ↑ p38 ↑ NFκB	Zhou et al. (2005)
Shear stress	Statical stimulation	Endothelial cells	↑ MCP-1, ↑ MCP-1 mRNA, ↑ Integrin - Ras - MAPK signaling pathway	Reviewed by Chien (2003)

3.2 Cellular Responses to Physical Stimulation

Table 3.2 Electrical modulation of cell functions

Electrical modulation	Stimulation type	Cell type	Result	Reference
+0.7 V (vs. Ag/AgCl) potential	Statical stimulation	HeLa cells	↑ Necrosis	Yaoita et al. (1988)
+0.6 ~ 0.65 V (vs. Ag/AgCl) potential	Statical stimulation	HeLa cells	Decrease of proliferation rate	Yaoita et al. (1988 and 1990)
+0.6 V (vs. Ag/AgCl) potential	Statical stimulation	HeLa cells	Decrease of membrane fluidity	Yaoita et al. (1988)
+0.4 V (vs. Ag/AgCl) potential	Statical stimulation	MKN45 cells	Decrease of membrane fluidity	Kojima et al. (1991)
+0.4 V (vs. Ag/AgCl) potential	Statical stimulation	MKN45 cells	Stop of proliferation	Kojima et al. (1991 and 1992)
Sine wave potential	Fluctuated stimulation	Mouse astroglial cells	↑ NGF, ↑ NGF mRNA, ↑ c-fos mRNA, ↑ c-jun mRNA	Koyama et al. (1996)
Sine wave potential	Fluctuated stimulation	Mouse astroglial cells	↑ NGF, ↑ PKC	Koyama et al. (1997)
Rectangular potential	Fluctuated stimulation	PC12 cells	↑ Cell differentiation, ↑ Calcium influx, ↑ Tyrosine kinase	Kimura et al. (1998a)
Rectangular potential	Fluctuated stimulation	PC12 cells	↑ Cell differentiation, ↑ SA calcium channel, ↑ c-fos mRNA, ↑ PKC	Kimura et al. (1998b)
Sine wave potential	Fluctuated stimulation	3T3-HSP cells NIE-115 cells	↑ hsp70 promoter	Yanagida et al. (2000) Mie et al. (2003)

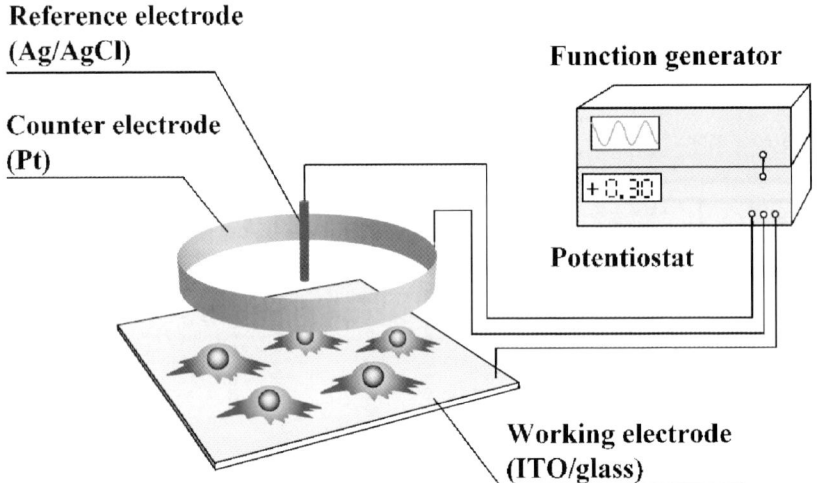

Fig. 3.2 Schematic illustration of electrical potential-controlled cell culture system. Mammalian cells are cultured on the surface of an ITO/glass electrode. The electrode potential is controlled with an Ag/AgCl reference electrode through a potentiostat and a function generator

As another example for an electrically stimulated activation of gene expression, Kimura and colleagues (1998a, b) have discovered a rectangular wave potential application induces cell differentiation of PC12 cells. The electrically induced differentiation was completely inhibited by both of these inhibitors including lanthanum ion, a calcium channel blocker, genistein, a tyrosine kinase inhibitor, and chelerythrine, a specific PKC inhibitor, respectively (Fig. 3.1 and Table 3.2). The authors mentioned that the electrically induced PC12 differentiation might involve a stretch-activated (SA) calcium ion channel, because the calcium ion induction due to the electrical stimulation was completely inhibited by a nifedipine or gadolinium ion treatment (Details are described in Sect. 3.5.2).

Electrical potential stimulation is useful to modulate intracellular signaling pathways and cytokine production at the single-cell level. Therefore, electrical potential stimulation is utilized to control protein production in a lab-on chip device.

3.2.2 Hydrostatic Pressure

In our previous study, extremely high hydrostatic pressure exposure triggered interleukin (IL)-6 and -8 expression and secretion in normal human dermal fibroblasts (Koyama and Aizawa 2002; Koyama et al. 2002b). More than 90% confluence of the cells was pressurized at up to 70 MPa (0.1 MPa = 1 bar) for 20 min and they remained alive after the pressurization (Fig. 3.3). The mechanisms of pressure-induced IL-6 production involve PKC activation (Fig. 3.1; Koyama and Aizawa 2002).

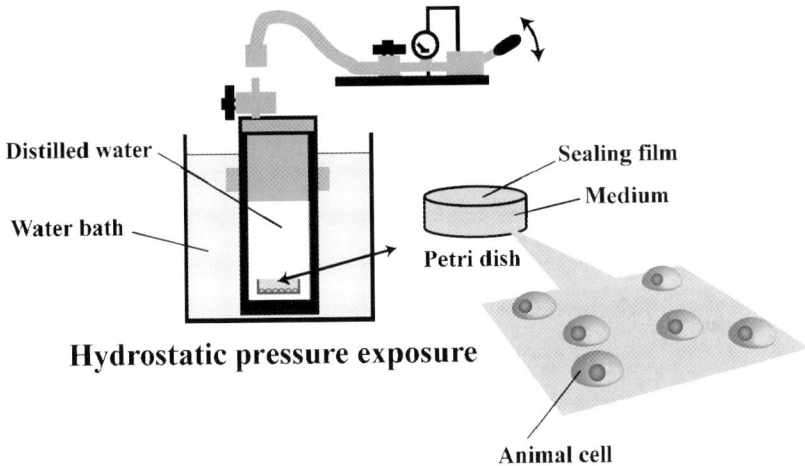

Fig. 3.3 Schematic illustration of hydraulic apparatus system

Several researchers have reported that weak hydrostatic pressure can induce gene expression of nitric oxide synthase (NOS)-2 in chondrocytes and human optic nerve astrocytes (Fermor et al. 2001; Liu and Neufeld 2001). Studying intracellular signaling cascades using nerve astrocytes, 1 kPa of hydrostatic pressure for 12 h or 48 h induced the activation of epidermal growth factor receptor (EGFR) tyrosine kinases and nuclear factor (NF)κB, in which NOS-2 is involved (Fig. 3.1; Neufeld and Liu 2003). Because the pressure exposure activated EGFR tyrosine kinases in the astrocytes, the application of pressure might also induce the type of cytokine production that activates EGFR tyrosine kinases. Hydrostatic pressure is useful to stimulate numerous cells rather than the single-cell level because the pressure is uniformly transmitted throughout the solvent. Therefore, pressure-activated gene expression is useful to improve the protein production in large volumes.

3.2.3 Shear Stress

The effects of shear stress on the MCP-1 expression have been studied in cultured human umbilical vein endothelial cells (Fig. 3.4; reviewed in Chien 2003). The application of a shear stress of 12 dynes/cm^2 causes an increase in the monocyte chemotactic protein-1 (MCP-1) gene expression to approximately 2.5 fold in about 1.5 h. Studies on the promoter region of MCP-1 showed that the TPA responsive element (TRE) is a critical cis-element in the shear stress activation of MCP-1. The transcription factor for TRE is the activator protein-1 (AP-1), which is a dimmer composed of c-Jun – c-Fos or c-Jun – c-Jun. The activation of c-Jun and c-Fos is mediated by the mitogen activated protein kinase (MAPK) signaling pathways (Fig. 3.1; reviewed in Chien 2003).

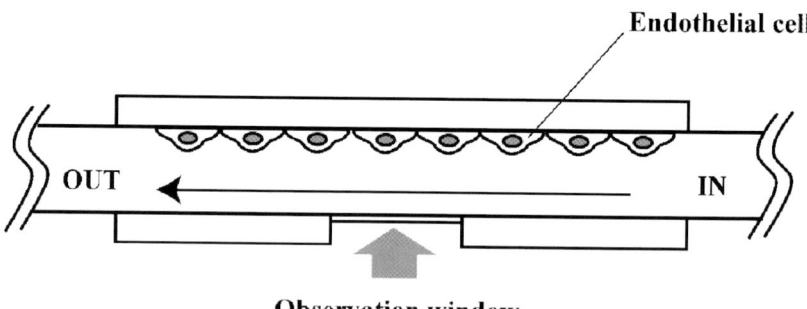

Fig. 3.4 Schematic illustration of shear stress flow chamber system

The application of shear stress (12 dynes/cm^2) to cultured bovine aortic endothelial cells activates the Ras pathway including ERK and JNK (Fig. 3.1; reviewed in Chien 2003). Activation of Ras, which become GTP-bound style, occurs within seconds after the beginning of shearing, and the activation of the downstream signaling molecules and gene expression follows sequentially (Fig. 3.1).

Integrins are transmembrane receptors that link the intracellular cytoskeletal proteins with the proteins in the ECM to provide two-way communication between the cell and its extracellular matrix (ECM). The integrins play a significant role in the initiation of signaling in response to shear stress (Fig. 3.1). There are more than 20 types of integrins, each of them is composed of two subunits, α and β. Shear stress causes the association of $\alpha_v\beta_3$ with Shc and the subsequent activation of the Ras pathway (Fig. 3.1) only when the endothelial cells are cultured on ECM composed of vitronectin or fibronectin (reviewed in Chien 2003). Another endothelial cell derived integrin is $\alpha_6\beta_1$ which has laminin as its cognate ECM molecule. Shear stress can activate $\alpha_6\beta_1$ in endothelial cells only when they are cultured on laminin.

The shear stress is useful to modulate the cytokine production at the single-cell level and it can be utilize to modulate the protein production in microfluidic devices.

3.2.4 Heat Shock

The heat shock proteins (hsps) are strongly induced by high and low temperatures, starvation, and a wide variety of other stresses, conditions which repress the synthesis of most cellular proteins. The hsps are induced by a myriad of stresses, including ethanol, anoxia, inhibitors of electron transport, amino-acid analogs, virus infections, arsenite and cadmium (Lindquist and Craig 1988; Lindquist 1986; Lindquist 1992). Several hsps, or their close relatives, are expressed at normal temperatures and play vital roles in the cell growth as well as in the stress tolerance. Their induction is often accompanied by tolerance of these stresses.

Besides the heat shock induced hsps, several researchers have reported that a heat shock at 43 °C for 24 h increases the synthesis and release of IL-6 into the culture media in human skin fibroblasts (Park et al. 2006). Moreover, the heat shock-induced expression of matrix metalloproteinase (MMP)-1 and MMP-3 is mediated via an IL-6-dependent autocrine mechanism, in which ERK and JNK play an important role (Fig. 3.1). As another example, Zhou and coworkers (2005) have reported that heat shock activated p38 kinase, ERK and NFκB signal pathways in monocytes (Fig. 3.1). The p38 pathway takes part in a heat shock induced up-regulated Toll-like receptor (TLR) 2 and TLR4 in monocytes (Table 3.1). The induction of TLRs was prior to that of Hsp70. This suggested that the up regulation of TLR2 and TLR4 might be independent of the induction of Hsp70.

In the same way as hydrostatic pressure, the long-term exposure of cells to heat shock stress can be used to improve the large-volume protein production.

3.3 Electrically Controlled Proliferation Under Constant Potential Application

3.3.1 Electrical Potential-Controlled Cell Culture System

Aizawa and coworkers have developed a novel method for electrically modulating cellular functions of animal cells (reviewed in Aizawa et al. 1999). Animal cells are directly cultured on an electrode surface and then exposed to a weak electrical potential which does not induce any electrochemical reaction. Figure 3.2 shows a schematic illustration of an electrical potential-controlled cell culture system (reviewed in Aizawa et al. 1999). To avoid electrochemical reactions, the application of the electrical potential to the animal cells was carried out using a three-electrode system that included working, counter-, and 0-V reference electrodes. An indium tin oxide (ITO) optically transparent working electrode is placed on the bottom of a culture dish with the counter- (Pt) and reference (Ag/AgCl) electrodes. Animal cells are directly plated onto the ITO electrode surface. This is followed by filling the culture dish with a culture medium. A potentiostat and a function generator control the ITO working electrode potential. Living animal cells remain alive on the ITO electrode surface if the electrical potential is properly controlled (Yaoita et al. 1988, 1989; Kojima et al. 1991).

3.3.2 Cell Viability Under Constant Potential Application

Using the potential-controlled cell culture system (Fig. 3.2), the viability of animal cells was examined under a constant potential application between -0.4 V to $+1.2$ V, a potential range that did not induce water electrolysis and electrochemical reactions (Yaoita et al. 1988, 1989). HeLa cells were cultured on the optically transparent working electrode for a few days, followed by a potential application (Yaoita et al. 1988). After the potential application, the HeLa cells were examined using the trypan blue dye-exclusion viability test (Fig. 3.5). Almost all of the HeLa cells remained alive until the electrical potential reached +0.65 V vs. Ag/AgCl for 5 days of cultivation. At constant potential application of +0.7 V (vs. Ag/AgCl), the necrosis of the HeLa cells was gradually induced, and almost all of the cells died after a 12-h exposure (Table 3.2). At +1.0 V (vs. Ag/AgCl) and greater constant potential application, the necrosis of the HeLa cells occurred within 2 h (Fig. 3.5).

The relationship between constant electrical potential and the morphology of animal cells is schematically illustrated in Fig. 3.6 (reviewed in Aizawa et al. 1999). At a potential of less than −0.2 V (vs. Ag/AgCl), the animal cells on the working electrode became globular and then gradually swelled outward, projecting the intracellular contents (Fig. 3.6). At +0.4 V (vs. Ag/AgCl) and greater positive potential application, the animal cells on the electrode became rounded (Fig. 3.6). Despite the

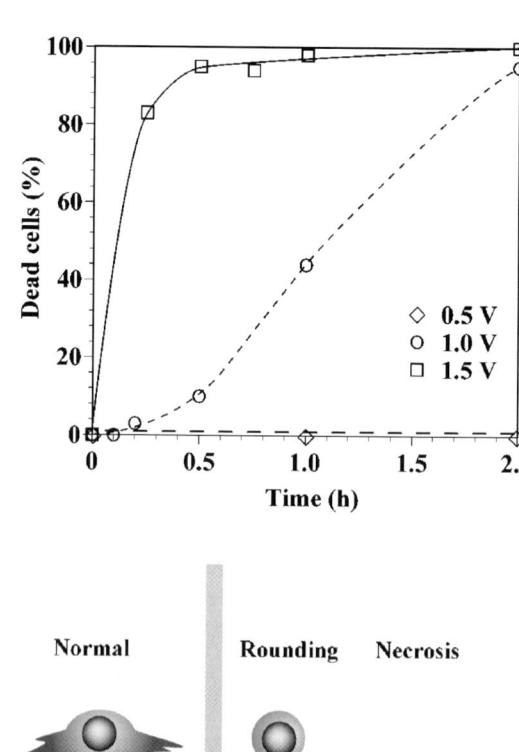

Fig. 3.5 Time course of damaged HeLa cells on the electrode surface at constant potential. The percentage of dead HeLa cells was determined by staining with trypan blue dye

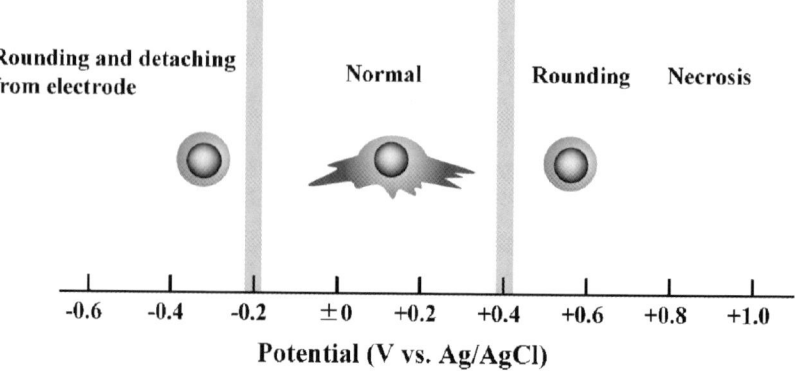

Fig. 3.6 Morphology of cells cultured at constant potential

fact that the electrical potential-induced death of cells depends on the type of animal cells, the electric effect on cell morphology tends to differ only slightly among various animal cells (Aizawa et al. 1999).

3.3.3 Electrical Modulation of Cellular Proliferation Rate

Since mammalian cells are one of the major devices for producing functional proteins, controlled cell proliferation techniques are very attractive for the application in cellular engineering. When applying a constant potential to the electrode surface on which animal cells are cultured, these cells proliferate at various rates depending on the electrode potential (Yaoita et al. 1988, 1990; Kojima et al. 1991, 1992). The proliferation of HeLa cells cultured on the electrode surface is markedly inhibited and the cells undergo morphological change (Yaoita et al. 1988, 1990). At potentials of +0.6 and +0.65 V vs. Ag/AgCl, the proliferation rates of HeLa cells decreased to 70% and 20%, respectively, compared with cells to which no potential was applied (Table 3.2). With a constant potential application of +0.6 V, the HeLa cells became globular in shape without disruption of actin filament structures. Almost all of the HeLa cells remained alive in case of a constant potential application of +0.65 V vs. Ag/AgCl for 5 days.

As another example, human carcinoma MKN45 cells stopped proliferating and showed morphological changes at +0.4 V (vs. Ag/AgCl) potential and grew normally again when the electrode potential shifted to +0.1 V vs. Ag/AgCl (Table 3.2; Fig. 3.4; Kojima et al. 1991, 1992). More than 90% of the MKN45 cells remained alive until the constant potential of +0.4 V vs. Ag/AgCl was applied for 100 h of cultivation. Simultaneously with the proliferation rate experiment, Kojima and coworkers (1991) examined the membrane fluidity of MKN45 cells by measuring the fluorescent anisotropy. After a potential application of +0.4 V vs. Ag/AgCl for 60 h, the fluorescent anisotropy increased by 10% compared with the MKN45 cells to which no potential was applied (Table 3.2; Kojima et al. 1991). HeLa cells on the potential-controlled electrode showed the same reactions, and thus the fluorescent anisotropy increased by 10% compared with the control cells after a potential of +0.6 V vs. Ag/AgCl was applied for 1 h (Table 3.2; Yaoita et al. 1988). An increase in the fluorescent anisotropy means a decrease in the membrane fluidity, which is one of the keys to solving the mechanisms of the inhibition of cell proliferation.

3.4 Modulated Proliferation Under Extreme Hydrostatic Pressure

Since high hydrostatic pressure exposure decreases the membrane fluidity in animal cells (Gibbs 1997), the application of pressure can also modulate cell proliferation rates in the same manner as constant an electrode potential application (Fig. 3.7;

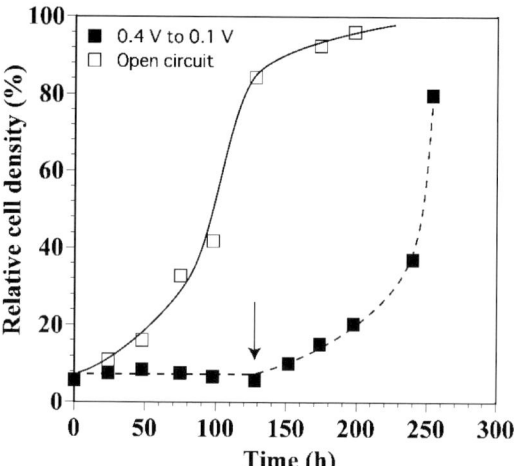

Fig. 3.7 Controlled growth rate of MKN45 cells on the electrode surface. MKN45 cells were cultured at 0.4 V vs. Ag/AgCl for 100 h, and then the applied potential was shifted to 0.1 V vs. Ag/AgCl (*arrow*)

Yaoita et al. 1988; Kojima et al. 1991). Koyama and colleagues (2005) confirmed this hypothesis and reported that the application of high hydrostatic pressure modulates the cell proliferation rate changes in mouse 3T3-L1, conger eel (*Conger myriaster*), and deep-sea eel (*Simenchelys parasiticus*; habitat depth, 366–2630 m) cells. They investigated the survival rates of mouse 3T3-L1, conger eel, and deep-sea eel cells after the exposure to high hydrostatic pressure for 20 min (Fig. 3.8; Koyama et al. 2005). The cell viability was investigated using double staining with both calcein-AM and EthD-1. The majority of the mouse 3T3-L1 cells at 37 °C and all of the conger eel cells at 25 °C remained alive until pressure reached 60 MPa (0.1 MPa = 1 bar, Fig. 3.8). At 15 °C, conger eel cells began to sustain damage or die at pressure of only 5 MPa (Fig. 3.8). All of the deep-sea eel cells remained alive at 150 MPa, although no mouse and conger eel cells remained alive at pressures of 130 MPa and greater (Fig. 3.8).

Figure 3.9 shows the cell proliferation rates of each cell type under pressure conditions (Koyama et al. 2005). From the results in Fig. 3.8, Koyama and colleagues (2005) set the temperature conditions for mouse, conger eel, and deep-sea eel cells at the growth optima of 37 °C, 25 °C, and 15 °C, respectively. When no statistically significant differences in cell density between 0 and 100 h was detected, the cell growth rate was considered to be zero. The conger eel cells were sensitive to high hydrostatic pressure and did not grow at 10 MPa. On the other hand, the mouse 3T3-L1 cells grew more rapidly at pressure of 5 MPa than at atmospheric pressure and stopped growing at 18 MPa. Even though surface-dwelling organism-derived cells such as mouse and conger eel cells did not grow at a pressure greater than 18 MPa, deep-sea eel cells were capable of growth at high hydrostatic pressure of up to 25 MPa. Mouse 3T3-L1, conger eel, and deep-sea eel cells became rounded after 2–3 days of cultivation under the pressures of 18, 10, and 30 MPa, respectively.

The cell proliferation rate of mouse 3T3-L1 cells increased by 27% at the pressure of 5 MPa compared with the rate under atmospheric pressure (Fig. 3.9). Several researchers reported that the application of weak pressure, such as gas pressure of

3.4 Modulated Proliferation Under Extreme Hydrostatic Pressure

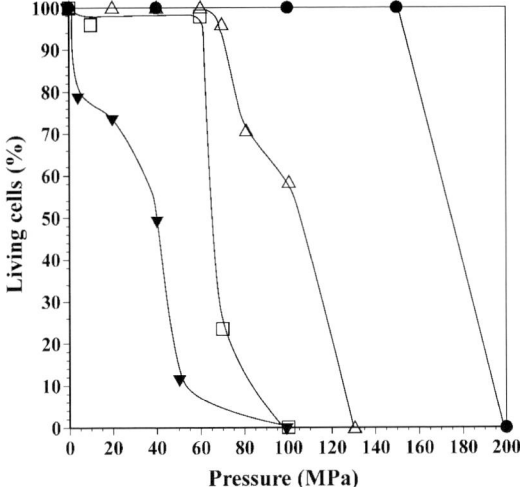

Fig. 3.8 Survival rates of mouse 3T3-L1, conger eel, and deep-sea eel cells under elevated pressure conditions. E, Deep-sea eel cells at 15 °C; H, conger eel cells at 25 °C; G, mouse 3T3-L1 cells at 37 °C; S, conger eel cells at 15 °C. Each cell type was grown to $1-5 \times 10^3$ cells/cm^2 and subjected to hydrostatic pressure for 20 min. After pressurization, the viability of more than 100 cells was investigated using the calcein-AM and EthD-1 double-staining method

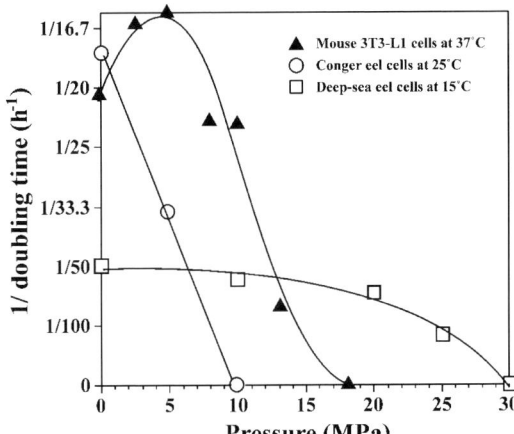

Fig. 3.9 Properties of cell growth under elevated hydrostatic pressure conditions. When no statistically significant differences in cell density between 0 and 100 h was detected, the cell growth rate was considered to be zero

4–12 kPa and H$_2$O pressure of 0.4 kPa, promoted the cell proliferation rates of rat mesangial cells and human umbilical vein endothelial cells, respectively (Kawata et al. 1998; Schwartz et al. 1999). Hydrostatic pressure-induced cytokine gene expression and secretion involve the PKC activation in normal human dermal fibroblasts (Koyama et al. 2002b; Koyama and Aizawa 2002). Because some cell types can be stimulated to proliferate in culture when PKC is activated (Alberts et al. 2002), the mechanisms of pressure-stimulated cell proliferation in mouse 3T3-L1 cells might involve the activation of PKC-signaling cascades.

3.5 Electrically Modulated Gene Expression Under Alternative Potential Application

3.5.1 Electrically Stimulated Nerve Growth Factor Production

Since controlled gene expression techniques are of great interest for the application in cellular engineering, our efforts were concentrated on determining the electric effects of the alternative potential application in activating gene expression processes. First, we investigated the glial cell-driven nerve growth factor (NGF) production (Table 3.2; Koyama et al. 1996, 1997), which restores neuronal function in the brain (Levi-Montalcini 1987; Whittemore and Seiger 1987). Mouse astroglial cells that had been cultured on an electrode (Fig. 3.2) for 7 days in a serum-free Dulbecco's modified Eagle's medium (DMEM) were subjected to the application of 10-Hz sine-wave electrical potential at varying amplitudes for another 1 h of culture. After the electrical stimulation, the cells were placed in fresh DMEM containing 1% (w/v) bovine serum albumin and cultured for another 24 h. After the incubation, we determined the levels of secreted and stored NGF proteins. The electrical stimulation induced NGF secretion, the although intracellular NGF levels remained unchanged after the stimulation (Fig. 3.10). The NGF was maximally secreted by an upper potential of +0.3 V vs. Ag/AgCl. The NGF secretion was increased by six- and three-fold compared with the control at +0.3 V ($p < 0.001$) and +0.4 V ($p < 0.005$), respectively. However, the astrocytes began to be partially detached from the growth surface of the ITO working electrode at a stimulation of +0.4 V. This was the first direct demonstration that the NGF protein production is enormously enhanced by stimulating resting-stage astrocytes with a low-frequency sine-wave potential (Koyama et al. 1996, 1997).

The expression of the NGF gene correlates with several protooncogenes encoding proteins of the Fos and Jun families. These proteins can form homodimers or heterodimers, referred to as AP-1, which behave as transcriptional factors (Halazonetis et al. 1988; Rauscher et al. 1988; Smeal et al. 1989; Zerial et al. 1989). In fibroblasts, the NGF mRNA expression is mediated via the interaction of the *c-fos* protooncogene with the AP-1 binding site in the first intron of the NGF gene (Hengerer et al. 1990; D'Mello and Heinrich 1991). Furthermore, the levels of NGF transcripts changed corresponding to the levels of *c-jun* transcripts (Jehan et al. 1995). The results in those reports suggested that the NGF secretion might also be induced by a sine-wave potential application in association with AP-1 complexes. To confirm this hypothesis, we extensively assayed electrically stimulated astroglial cells for NGF, *c-fos*, and *c-jun* expression to elucidate the molecular mechanisms.

Figure 3.11 shows the time course of the NGF, *c-fos*, and *c-jun* mRNA expression. In the *c-fos* mRNA expression, the mRNA level in electrically stimulated astroglial cells reached a maximum at 30 min, followed by a decrease to the control level at 2 h. On the other hand, the time course of the *c-jun* expression gradually increased, as shown in Fig. 3.11, and was similar to that of the NGF expression. These results indicate that the electrical sine-wave potential application induced the *c-fos*

3.5 Electrically Modulated Gene Expression Under Alternative Potential Application

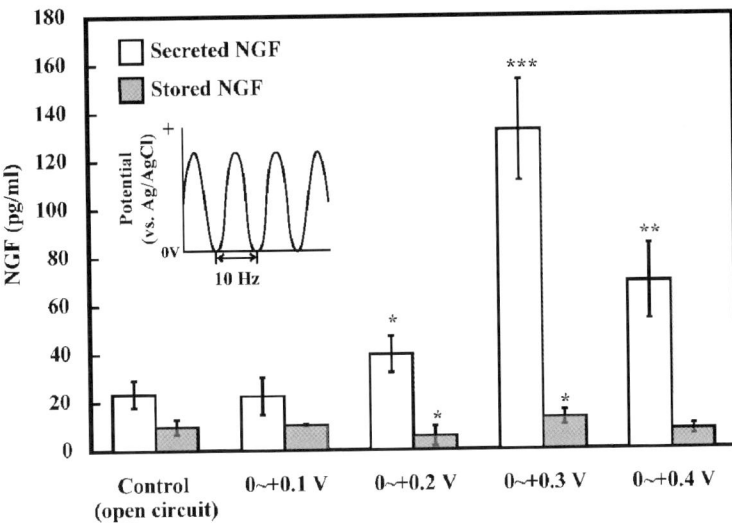

Fig. 3.10 NGF secretion at different upper potentials of stimulation. Media and homogenized cell solutions were collected after 24 h of incubation and assayed for NGF using ELISA (*$p < 0.05$, **$p < 0.005$, ***$p < 0.001$; compared with secreted or cellular NGF of control). Values are mean ± S.D. of three independent determinations

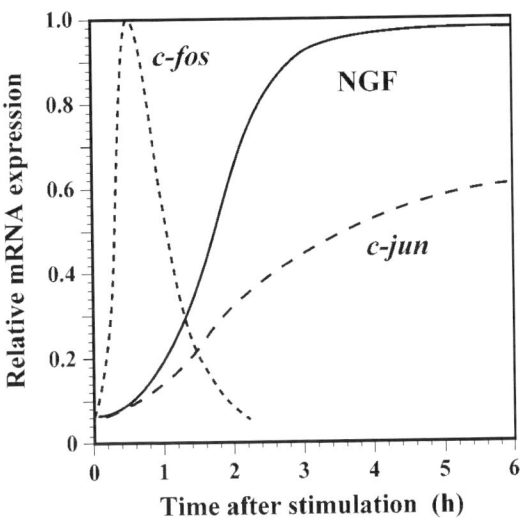

Fig. 3.11 Time courses of NGF, *c-jun*, and *c-fos* mRNA expression after 1-h sine-wave potential application in which the upper and lower potentials were set at +0.3 V and 0 V, respectively, vs. Ag/AgCl

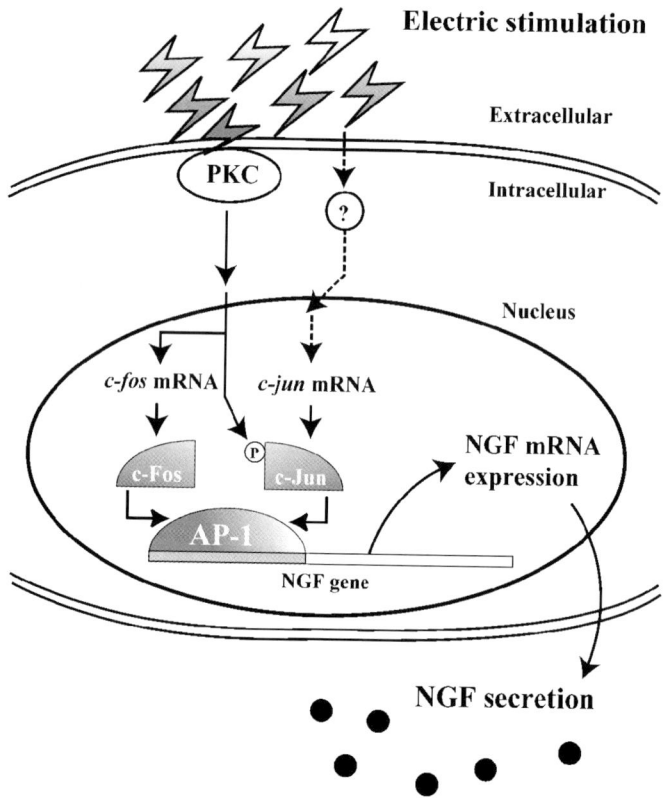

Fig. 3.12 Assumed scheme of the electrically stimulated activation of NGF production

and the *c-jun* mRNA expression, and consequently the NGF mRNA was expressed to release NGF proteins. As the electrically stimulated NGF secretion is inhibited by either the PKC inhibitor staurosporine or the PKC down-regulator phorbol 12-myristate 13-acetate (Koyama et al. 1996), the electrically stimulated NGF expression and secretion are modulated by the PKC–AP-1 signaling pathway in astroglial cells (Fig. 3.12).

3.5.2 Electrically Induced Differentiation of PC12 Cells

As another example of an electrically stimulated activation of gene expression, Kimura and coworkers (1998a, b) succeeded in the differentiation of PC12 cells via a rectangular-wave potential application-induced *c-fos* mRNA expression (Table 3.2). PC12 is a cell line originating from pheochromocytoma in the rat adrenal medulla and it differentiates into sympathetic nerve-like cells with NGF treatment (Greene and Tischler 1976), extending long neurites that have been used as good markers to investigate the differentiation.

3.5 Electrically Modulated Gene Expression Under Alternative Potential Application

PC12 cells were cultured on an ITO electrode, as shown in Fig. 3.2, and subjected to a rectangular peak-to-peak potential of 100 mV with a frequency of 100 Hz for 5 – 60 min every 24 h, repeated 3 times, followed by an incubation for 2 days at 37 °C. The differentiation assay was performed in randomized areas by counting the percentage of cells that extended neurites of more than 10 μm in length. Figure 3.13 shows the differentiation ratio of PC12 cells. With an electrical application for 5 min repeated 3 times, PC12 cells differentiated slightly but proliferated as well as normally cultured cells. After being subjected to the electrical potential for 30 min and this being repeated 3 times, the cells differentiated at a rate of $16.2 \pm 0.26\%$ ($n = 3$). When PC12 cells were treated for 60 min every 24 h, repeated 3 times, and then incubated for 2 days, the differentiation ratio was $9.9 \pm 0.71\%$ ($n = 3$). However, a rectangular potential stimulation for more than 60 min may be harmful because some PC12 cells sustain damage or die after the application. These results demonstrate that low-frequency potential stimulation induces the differentiation of PC12 cells without any growth factor.

Calcium-ion influx (Morgan and Curran 1986; Manivannan and Terakawa 1993) and the activation of the *c-fos* mRNA expression (Kruijer et al. 1985) have been known to induce the differentiation of PC12 cells. Kimura and colleagues (1998b) showed that the rectangular potential stimulation-induced differentiation of PC12 cells was involved in the activation. The treatment with lanthanum ion, a calcium ion-channel blocker, completely inhibited the potential-induced differentiation of PC12 cells, although the cells treated with both lanthanum ion and NGF extended long neurites in a manner similar to normal NGF-treated cells, as shown in Fig. 3.14. Kimura et al. also examined the effects of chelerythrine, a specific PKC inhibitor, on the potential application-induced differentiation (Fig. 3.14). In the presence of both NGF and chelerythrine, the neurite outgrowth was slightly reduced (Fig. 3.14). On the other hand, the electrically stimulated differentiation was completely inhibited by the chelerythrine treatment (Fig. 3.14). The electrically induced differentiation

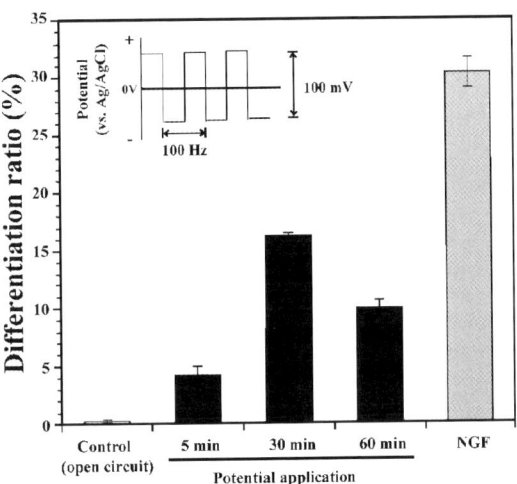

Fig. 3.13 Differentiation rate of PC12 cells subjected to electrical potential for 5, 30, or 60 min every 24 h, repeated 3 times, and then incubated for another 2 days. NGF, the cells were treated with 50 ng/ml of 2.5S-NGF for 96 h. All experiments were performed more than three times

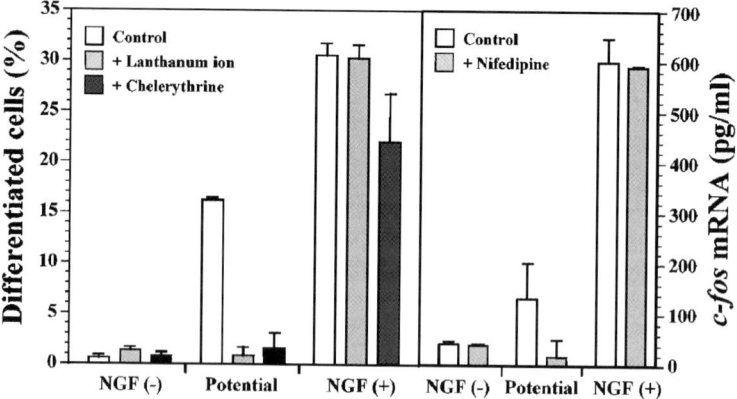

Fig. 3.14 Differentiation and *c-fos* mRNA expression in PC12 cells treated with electrical stimulation or NGF in the presence or absence of chemical compounds. The cells were treated with 20 μM of nifedipine, a specific L-type calcium channel blocker, 1 μM of chelerythrine, a PKC inhibitor, or 100 μM of LaCl$_3$, a nonspecific calcium-ion channel blocker. The cells were subjected to the electrical potential for 30 min every 24 h, repeated 3 times, and then incubated for another 2 days (Potential). PC12 cells were treated with 50 ng/ml of 2.5S-NGF for either 30 min in the *c-fos* experiment or 90 min in the differentiation assay (NGF(+)). Control cells were treated with neither the electrical potential nor NGF (NGF(−))

might involve a calcium-dependent PKC-signaling cascade. Moreover, the potential application-induced calcium-ion influx measured with indo-1 was completely inhibited by the treatment with either the L-type calcium ion-channel blocker nifedipine or the stretch-activated channel blocker gadolinium ion. Nifedipine also completely inhibited the electrically induced *c-fos* mRNA expression, although it did not inhibit the NGF-induced one (Fig. 3.14). These results indicate that the electrical stimulation-induced calcium-ion influx occurred via an L-type calcium channel as a stretch-activated channel. As mentioned above, the constant potential application decreases the membrane fluidity of MKN45 cells (Kojima et al. 1991) and HeLa cells (Yaoita et al. 1988). Because high hydrostatic pressure exposure also decreases membrane fluidity in animal cells (Gibbs 1997), the repetitive potential shift might cause oscillation of the cellular membrane due to the electrical interaction between the working ITO electrode (Fig. 3.2) and the charged membrane molecules.

3.6 Cellular Engineering to Enhance Responses to Physical Stimulation

Because controlled gene expression techniques are of great interest in cellular engineering applications, several researchers have developed methods for instantly converting the physical stimulation into an useful gene expression using hsp70 promoter activity (Yanagida et al. 2000; Mie et al. 2003). The Hsp70 gene expression is induced by various types of physical stimuli including elevated temperature (Lindquist

3.6 Cellular Engineering to Enhance Responses to Physical Stimulation

1986; Lindquist and Craig 1988), hydrostatic pressure (Osaki 1998; Koyama et al. 2002a), magnetism (Goodman and Blank 2002), UV-C radiation (Niu et al. 2006), ionizing radiation (Calini et al. 2003), and electrical stimulation (Yanagida et al. 2000). Therefore, a wide range of physical stimuli can induce an useful protein-encoded gene expression in mammalian cells using an hsp70 promoter in a constructed plasmid.

Yanagida and coworkers (2000) constructed a plasmid in which an hsp70 promoter was connected with a firefly luciferase and emitted the luminescence from transfected mouse 3T3-L1 cells (3T3-HSP cells), which were cultured on a working ITO electrode (Fig. 3.2). Figure 3.15 shows the time course of the hsp70 promoter-induced luciferase activity in electrically and heat-stimulated 3T3-HSP cells.

The cells were exposed to a 10-Hz electrical sine-wave potential, in which the upper and lower peak potentials were set at $+0.3$ V and 0 V, respectively, vs. Ag/AgCl, at $37\,^\circ$C for 1 h. The electrically stimulated 3T3-HSP cells induced a greater luciferase activity compared with that induced by the heat treatment of 3T3-HSP cells at $42\,^\circ$C for 2 h (Fig. 3.15). After the heat treatment at $42\,^\circ$C for 2 h, the luciferase activity increased and reached a maximum at 8 h of incubation after the termination of the stimulation. On the other hand, the sine wave potential-induced luciferase activity appeared to occur more slowly than with heat stimulation and reached the maximum activity at 24 h of incubation. This result indicates that alternative potential stimulation can modulate the hsp70 promoter activity. Using the electrically modulated hsp70 promoter activity, Mie and coworkers (2003) demonstrated that the electrical stimulation induced neural differentiation of mouse N1E-115 neuroblastoma cells in areas required to construct neural networks. They constructed a plasmid containing a mouse NeuroD2 gene under the hsp70 promoter and transfected it into the mouse neuroblastoma cell line N1E-115. The stably transfected cells were cultured on a working ITO electrode and subjected to a 100-Hz sine-wave potential in which the upper and lower potentials were kept at $+0.4$ V and -0.4 V vs. the resting potential of the electrode for 90 min. Under these conditions, the cell morphology and the proliferation rate did not change. After a 72-h incubation, the electrically stimulated cells differentiated as neural cells. This raises the

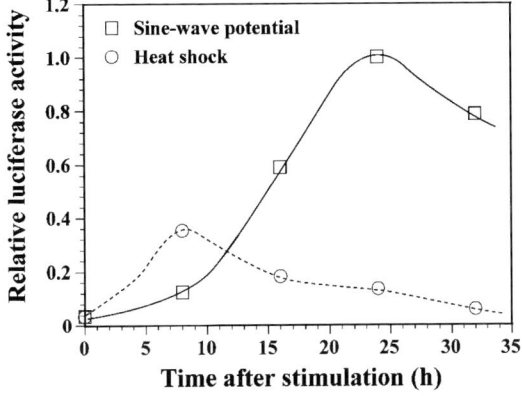

Fig. 3.15 Time course of luciferase activity in electrically and heat-stimulated 3T3-HSP cells. (G) Cells were electrically stimulated with a 10-Hz sine-wave potential of which the upper and lower peak potentials were maintained constant at $+0.3$ V and 0 V, respectively, vs. Ag/AgCl at $37\,^\circ$C for 1 h. (E) Cells were heated at $42\,^\circ$C for 2 h

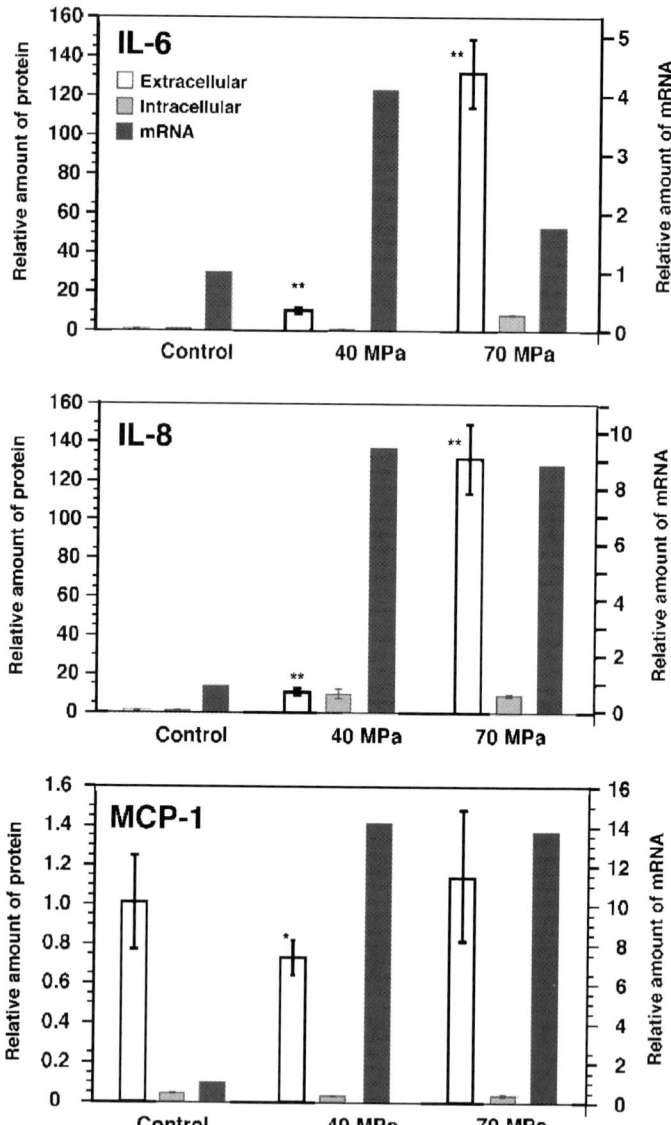

Fig. 3.16 Relative rate of gene expression and protein production of IL-6, IL-8, and MCP-1. The cytokine mRNA expression after 4 h of incubation were assayed, and values were normalized to that of L32. Media and homogenized cell solutions were collected after 24 h of incubation and assayed using ELISA. Values are mean ± S.D. of four independent experiments ($^*p < 0.01$, $^{**}p < 0.001$, compared with control)

hope that it will be feasible to apply electrical stimulation to cells in order to form neural networks in microelectronic devices.

As another example of controlled gene expression, Koyama and colleagues (2002b) found that extremely high hydrostatic pressure stress induced the expression of a variety of immunomodulatory cytokine mRNA and protein production in normal human dermal fibroblasts. Normal human dermal fibroblasts were found to survive and be active in producing IL-6 and -8, and monocyte chemoattractant protein-1 (MCP-1) under extremely high hydrostatic pressure of up to 70 MPa for 20 min. Hydrostatic pressure application at 70 MPa for 20 min markedly enhanced IL-6 and IL-8 secretion by about 130-fold compared with the control (Fig. 3.16). It should be noted that the secretion levels of both IL-6 and -8 at 70 MPa increased by about 13-fold compared with those induced by pressure of 40 MPa without transcriptional enhancement (Fig. 3.16). Conversely, MCP-1 secretion decreased at hydrostatic pressure of 40 MPa, although the mRNA level increased (Fig. 3.16). With the application of 70 MPa, MCP-1 secretion did not change compared with the basal level (Fig. 3.16). Although the induction of IL-1α, IL-1β, and IL-12 mRNA occurred under high hydrostatic pressure conditions, no translation of IL-1α, IL-1β, and IL-12 proteins was found. The results in Fig. 3.16 indicate that hydrostatic pressure stress triggers posttranscriptional regulation mechanisms that modulate the immunomodulatory cytokine production.

Future work in cellular engineering for the physical stimuli-induced gene expression will involve exploring specific physical stimulus-responsive promoters and investigating stress-induced expression mechanisms.

3.7 Concluding Remarks

A variety of effects of electric stimulation, including the halting of the cell proliferation (Table 3.2; Fig. 3.7; Yaoita et al. 1988, 1990; Kojima et al. 1991, 1992), the decrease in plasma membrane fluidity (Table 3.2; Yaoita et al. 1988; Kojima et al. 1991), the changes in cell morphology and the cytoskeletal distribution (Fig. 3.6; Yaoita et al. 1989, 1990), the gene expression (Table 3.2; Figs. 3.11 and 3.14; Koyama et al. 1996; Kimura et al. 1998b), the activation of the hsp70 promoter (Table 3.2; Fig. 3.13; Yanagida et al. 2000; Mie et al. 2003), the protein secretion (Table 3.2; Fig. 3.10; Koyama et al. 1996, 1997), and the cell differentiation (Table 3.2; Fig. 3.14; Kimura et al. 1998a,b), have been reported in mammalian cells cultured on an ITO working electrode (Fig. 3.2). Among these effects, the constant potential-induced inhibition of the cell proliferation and the decrease in plasma membrane fluidity are especially similar to those of hydrostatic pressure. High hydrostatic pressure exposure also decreases the membrane fluidity (Gibbs 1997) and stops the proliferation in animal cells (Fig. 3.9; Koyama et al. 2005). Similarly, extremely high hydrostatic pressure also triggers intracellular PKC-signaling cascades (Table 3.1; Fig. 3.1; Koyama and Aizawa 2002) and consequently induces the immunomodulatory cytokine gene expression and the protein secretion (Fig. 3.16) as does alternative potential stimulation (Figs. 3.10, 3.11, and 3.14). The rectangular-wave potential application induced calcium-ion influx occurs via an L-type calcium channel as a stretch-activated channel (Table 3.2; Figs. 3.1 and 3.14; Kimura et al. 1998a, b). Therefore, a repetitive potential shift might cause oscillation of the cellular membrane due to electrical interaction and result in a variety of the phenomena seen in mammalian cells.

The effects of physical stimulation on mammalian cells are certain to find a wide range of applications, particularly in cellular engineering, tissue engineering, and medical engineering.

References

1. Aizawa M, Koyama S, Kimura K, Haruyama T, Yanagida Y, Kobatake E (1999) Electrically stimulated modulation of cellular function in proliferation, differentiation, and gene expression. Electrochem 67:118–125
2. Alberts B, Johnson A, Lewis J, Raff M, Roberts K, Walter P (2002) Molecular biology of the cell, 4th edn. Garland Science, NewYork
3. Calini V, Urani C, Camatini M (2003) Overexpression of HSP70 is induced by ionizing radiation in C3H 10T1/2 cells and protects from DNA damage. Toxicol In Vitro 17:561–566
4. Chien S (2003) Molecular and mechanical bases of focal lipid accumulation in arterial wall. Prog. Biophys. Mol. Biol. 83:131–151
5. D'Mello SR, Heinrich G (1991) Structural and functional identification of regulatory regions and cis elements surrounding the nerve growth factor gene promoter. Brain Res Mol Brain Res 11:255–264
6. Enk CD, Mahanty S, Blauvelt A, Katz SI (1996) UVB induces IL-12 transcription in human keratinocytes in vivo and in vitro. Photochem Photobiol 63:854–859
7. Fermor B, Weinberg JB, Pisetsky DS, Misukonis MA, Banes AJ, Guilak F (2001) The effects of static and intermittent compression on nitric oxide production in articular cartilage explants. J Orthop Res 19:729–737
8. Finkel T (2001) Reactive oxygen species and signal transduction. IUBMB Life 52:3–6
9. Fujimori A, Okayasu R, Ishihara H, Yoshida S, Eguchi-Kasai K, Nojima K, Ebisawa S, Takahashi S (2005) Extremely low dose ionizing radiation up-regulates CXC chemokines in normal human fibroblasts. Cancer Res 65:10159–10163
10. Gibbs AG (1997) Biochemistry at depth. In: Randall DJ, Farrell AP (eds) Deep-Sea Fishes, Academic Press, San Diego, pp 239–277
11. Goodman R, Blank M (2002) Insights into electromagnetic interaction mechanisms. J. Cell. Phys. 192:16–22
12. Greene LA, Tischler AS (1976) Establishment of a noradrenergic clonal line of rat adrenal pheochromocytoma cells which respond to nerve growth factor. Proc Natl Acad Sci USA 73:2424–2428
13. Halazonetis TD, Georgopoulos K, Greenberg ME, Leder P (1988) c-Jun dimerizes with itself and with c-Fos forming complexes of different DNA binding affinities. Cell 55:917–924
14. Hengerer B, Lindholm D, Heumann R, Rüther U, Wagner EF, Thoenen H (1990) Lesion-induced increase in nerve growth factor mRNA is mediated by c-fos. Proc Natl Acad Sci USA, 87:3899–3903
15. Hochleitner BW, Hochleitner EO, Obrist P, Eberl T, Amberger A, Xu Q, Margreiter R, Wick G (2000) Fluid shear stress induces heat shock protein 60 expression in endothelial cells in vitro and in vivo. Arterioscler Thromb Vasc Biol 20:617–623
16. Jehan F, Neveu I, Naveilhan P, Wion D, Brachet P (1995) Interactions between second messenger pathways influence NGF synthesis in mouse primary astrocytes. Brain Res 672:128–136

17. Kimura K, Yanagida Y, Haruyama T, Kobatake E, Aizawa M (1998a) Electrically induced neurite outgrowth of PC12 cells on the electrode surface. Med Biol Eng Comput 36:493–498
18. Kimura K, Yanagida Y, Haruyama T, Kobatake E, Aizawa M (1998b) Gene expression in the electrically stimulated differentiation of PC12 cells. J Biotech 63:55–65
19. Kojima J, Shinohara H, Ikariyama Y, Aizawa M (1992) Electrically promoted protein production by mammalian cells cultured on the electrode surface. Biotech Bioeng 39:27–32
20. Kojima J, Shinohara H, Ikariyama Y, Aizawa M, Nagaike K, Morioka S (1991) Electrically controlled proliferation of human carcinoma cells cultured on the surface of an electrode. J Biotech 18:129–139
21. Koyama S, Horii M, Yanagida Y, Kobatake E, Miwa T, Aizawa M (2002a) Pressure sensitivity of HSP70 promoter in transformed 3T3-L1 cells. JAMSTEC J Deep Sea Res 20:53–58
22. Koyama S, Fujii S, Aizawa M (2002b) Post-transcriptional regulation of immunomodulatory cytokines production in human skin fibroblasts by intense mechanical stresses. J Biosci Bioeng 93:234–239
23. Koyama S, Aizawa M (2002) PKC-dependent IL-6 production and inhibition of IL-8 production by PKC activation in normal human skin fibroblasts under extremely high hydrostatic pressure. Extremophiles 6:413–418
24. Koyama S, Yanagida Y, Haruyama T, Kobatake E, Aizawa M (1996) Molecular mechanisms of electrically stimulated NGF expression and secretion by astrocytes cultured on the potential controlled electrode surface. Cell Eng 1:189–194
25. Koyama S, Haruyama T, Kobatake E, Aizawa M (1997) Electrically induced NGF production by astroglial cells. Nat Biotech 15:164–166
26. Koyama S, Kobayashi H, Inoue A, Miwa T, Aizawa M (2005) Effects of the piezo-tolerance of cultured deep-sea eel cells on survival rates, cell proliferation, and cytoskeletal structures. Extremophiles 9:449–460
27. Kruijer W, Schubert D, Verma IM (1985) Induction of the proto-oncogene fos by nerve growth factor. Proc Natl Acad Sci USA 82:7330–7334
28. Lee YJ, Galoforo SS, Berns CM, Erdos G, Gupta AK, Ways DK, Corry PM (1995) Effect of ionizing radiation on AP-1 binding activity and basic fibroblast growth factor gene expression in drug-sensitive human breast carcinoma MCF-7 and multidrug-resistant MCF-7/ADR cells. J Biol Chem 270:28790–28796
29. Levi-Montalcini R (1987) The nerve growth factor 35 years later. Science 237:1154–1162
30. Lindquist S (1986) The heat shock response. Annu Rev Biochem 55:1151–1191
31. Lindquist S, Craig EA (1988) The heat shock proteins. Annu Rev Genet 22:631–677
32. Lindquist S (1992) Heat-shock proteins and stress tolerance in microorganisms. Curr. Opin. Genet. Dev. 2:748–755
33. Liu B, Neufeld AH (2001) Nitric oxide synthase-2 in human optic nerve head astrocytes induced by elevated pressure in vitro. Arch Ophthalmol 119:240–245
34. Manivannan S, Terakawa S (1993) Rapid filopodial sprouting induced by electrical stimulation in nerve terminals. Jpn J Physiol 43:217–220
35. McNeil PL (2002) Repairing a torn cell surface: make way, lysosomes to the rescue. J Cell Sci 115:873–879
36. Mie M, Endoh T, Yanagida Y, Kobatake E, Aizawa M (2003) Induction of neural differentiation by electrically stimulated gene expression of NeuroD2. J. Biotech. 100:231–238
37. Morgan JI, Curran T (1986) Role of ion flux in the control of *c-fos* expression. Nature 322:552–555
38. Mohamadzadeh M, Takashima A, Dougherty I, Knop J, Bergstresser PR, Cruz PD Jr (1995) Ultraviolet B radiation up-regulates the expression of IL-15 in human skin. J Immunol 155:4492–4496
39. Neufeld AH, Liu B (2003) Comparison of the signal transduction pathways for the induction of gene expression of nitric oxide synthase-2 in response to two different stimuli. Nitric Oxide 8:95–102
40. Niu P, Liu L, Gong Z, Tan H, Wang F, Yuan J, Feng Y, Wei Q, Tanguay RM, Wu T (2006) Overexpressed heat shock protein 70 protects cells against DNA damage caused by ultraviolet C in a dose-dependent manner. Cell Stress Chaperones 11:162–169

41. Osaki J, Haneda T, Kashiwagi Y, Oi S, Fukuzawa J, Sakai H, Kikuchi K (1998) Pressure-induced expression of heat shock protein 70 mRNA in adult rat heart is coupled both to protein kinase A-dependent and protein kinase C-dependent systems. J Hypertens. 16:1193–1200
42. Park HJ, Kim HJ, Kwon HJ, Lee JY, Cho BK, Lee WJ, Yang Y, Cho DH (2006) UVB-induced interleukin-18 production is downregulated by tannic acids in human HaCaT keratinocytes. Exp Dermatol 15:589–595
43. Rauscher FJ 3rd, Voulalas PJ, Franza RB Jr, Curran T (1988) Fos and Jun bind cooperatively to the AP-1 site: reconstitution in vitro. Genes Dev 2:1687–1699
44. Schwartz EA, Bizios R, Medow MS, Gerritsen ME (1999) Exposure of human vascular endothelial cells to sustained hydrostatic pressure stimulates proliferation: involvement of the αv integrins. Circ Res 84:315–322
45. Smeal T, Angel P, Meek J, Karin M (1989) Different requirements for formation of Jun:Jun and Jun:Fos complexes. Genes Dev 3:2091–2100
46. Wlaschek M, Wenk J, Brenneisen P, Briviba K, Schwarz A, Sies H, Scharffetter-Kochanek K (1997) Singlet oxygen is an early intermediate in cytokine-dependent ultraviolet-A induction of interstitial collagenase in human dermal fibroblasts in vitro. FEBS Lett 413:239–242
47. Whittemore SR, Seiger A (1987) The expression, localization, and functional significance of β-nerve growth factor in the central nervous system. Brain Res 434:439–464
48. Yanagida Y, Mizuno A, Motegi T, Kobatake E, Aizawa M (2000) Electrically stimulated induction of hsp70 gene expression in mouse astroglia and fibroblast cells. J Biotech 79:53–61
49. Yaoita M, Shinohara H, Aizawa M, Hayakawa Y, Yamashita T, Ikariyama Y (1988) Potential-controlled morphological change and lysis of HeLa cells cultured on an electrode surface. Bioelectrochem Bioenerg 20:169–177
50. Yaoita M, Aizawa M, Ikariyama Y (1989) Electrically regulated cellular morphological and cytoskeletal changes on an optically transparent electrode. Expl Cell Biol 57:43–51
51. Yaoita M, Ikariyama Y, Aizawa M (1990) Electrical effects on the proliferation of living HeLa cells cultured on optically transparent electrode surface. J Biotech 14:321–332
52. Zerial M, Toschi L, Ryseck RP, Schuermann M, Müller R, Bravo R (1989) The product of a novel growth factor activated gene, fos B, interacts with jun proteins, enhancing their DNA binding activity. EMBO J 8:805–813
53. Zhou J, An H, Xu H, Liu S, Cao X (2005) Heat shock up-regulates of Toll-like receptor-2 and Toll-like receptor-4 in human monocytes via p38 kinase signal pathway. Immunology 114:522–530

Part II
Cell and Tissue Imaging

Chapter 4
Fluorescence Live-Cell Imaging: Principles and Applications in Mechanobiology

Yingxiao Wang[1], John Y-J. Shyy[2], Shu Chien[3]

[1] Department of Bioengineering and Beckman Institute for Advanced Science and Technology, University of Illinois, Urbana-Champaign, Urbana, IL 61801, yingxiao@uiuc.edu
[2] Division of Biomedical Sciences, University of California, Riverside, CA 92521, shyy@ucr.edu
[3] Department of Bioengineering and Whitaker Institute of Biomedical Engineering, University of California, San Diego, La Jolla, CA 92093, shuchien@ucsd.edu

Abstract It has become clear that mechanical force plays critical roles in regulating a variety of cellular functions. However, the molecular mechanism by which cells perceive mechanical cues remains elusive. With the development of novel fluorescence probes and optical microscopy, there has been significant advancement in the spatiotemporal characterization of signaling transduction. In this chapter, we describe the recent progress in the development of fluorescence proteins and biosensors capable of visualizing signaling cascades in live cells. Special emphasis is placed on how these biosensors are applied to image the subcellular localization of organelles and signaling/structural molecules, and the transcriptional regulation of target genes. Several technologies, including fluorescence resonance energy transfer (FRET), fluorescence recovery after photobleaching (FRAP), and fluorescence lifetime imaging microscopy (FLIM), are highlighted to demonstrate their utility and efficacy in live cell imaging of post-transcriptional modifications in response to mechanical stimulation. The impact of these fluorescence technologies on cardiovascular research in relation to mechanobiology is also discussed. In summary, we overview the research progress in fluorescence technologies and their applications in mechanobiology.

4.1 Introduction

The use of fluorescence probes tagged specifically to molecules of interest and their detection with fluorescence microscopy provide powerful technologies for live-cell imaging of molecular events at cellular and molecular levels. The application of such fluorescence microscopic technology to live cells can further elucidate molecular events with spatial and temporal resolutions.

Cells in the body are exposed to physiological and pathophysiological stimuli that encompass both chemical and mechanical factors, and it is important to understand how these factors interact to modulate functions at cellular and organ levels. Compared to the large amount of information on cellular or organ responses to chemical factors, there is a paucity of knowledge on the effects of mechanical factors and their interaction with chemical factors. During the past decade, tremendous research advances have been achieved on mechanobiology, i.e. how cells perceive mechanical cues, and transmit them into intracellular chemical signals for the regulation of cellular functions. Much of the knowledge on mechanobiology was obtained by studying the responses of cells in the cardiovascular system to mechanical forces due to hemodynamic factors. For example, vascular endothelial cells (ECs) are exposed to shear stress that play important roles in maintaining physiological functions of the vascular wall and also in contributing to patho-physiological changes in disease. The contraction of cardiac myocytes is essential for cardiac outputs, and abnormalities in contractility during several disease states can cause cardiac hypertrophy or heart failure.

At the cellular and subcellular levels, mechanical forces have been shown to activate a variety of membrane receptors and ion channels. These mechano-sensors initiate many downstream molecular events via cytoskeleton-dependent or -independent pathways. Such signal transduction processes involve post-translational regulation, including phosphorylation, methylation, sumolation, etc. The mechanotransduction not only changes the structure and function of these signaling molecules, but also their intracellular locations. Ultimately, the mechanically initiated cues reach the nucleus to control the transcription program with subsequent regulations of gene expression and cellular functions.

Recent advances of fluorescence microscopy make it a very useful tool for elucidating the mechanotransduction processes; the state-of-the-art technologies for live-cell imaging of signaling is particularly valuable for investigating the spatial and temporal aspects of molecular mechanisms in mechanobiology.

4.2 Fluorescence Proteins

4.2.1 Green Fluorescence Protein (GFP)

GFP was first discovered by Shimomura [1]. This protein is accompanied by aequorin in Aequorea jellyfish. The excitation of aequorin leads to a blue emission peaked at 470 nm, which trans-activates GFP to emit a bright green fluorescence light. Different kinds of GFPs exist in Aequorea, Obelia, Phialidium, and Renilla. In this article, GFP only refers to that from Aequorea unless otherwise specified. The gene sequence encoding GFP was first obtained by Prasher [2]. The fusion of the GFP coding sequence with genes encoding various signaling molecules allows an dynamic visualization of these target molecules [3, 4] (Fig. 4.1).

The chromophore of the wild-type GFP is a p-hydroxybenzylidene-imidazolinone encompassing Ser-Tyr-Gly (65–67), which is protected by a shell consisting

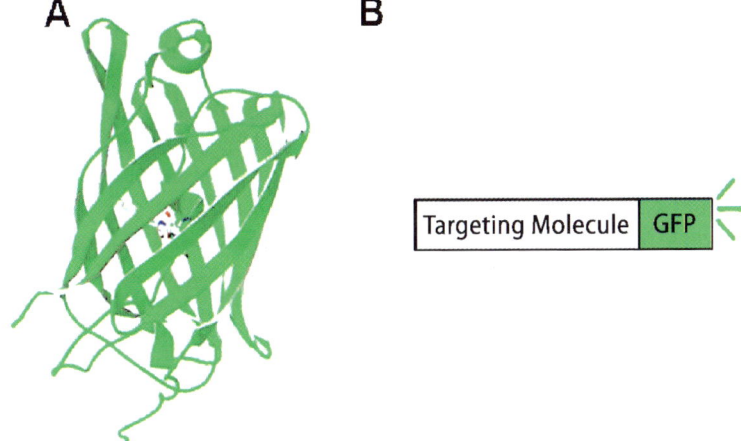

Fig. 4.1 The structure of GFP and its labeling of targeting molecules by fusion. **A** The domain structure of GFP; **B** a scheme of a recombinant protein with a targeting molecule fused with GFP to trace the positions of the targeting molecule

of 11 strands of β-barrels and an α-helix [5,6] (Fig. 4.1A). Because of the protection of this shell, GFP is relatively stable in conditions with different pHs [7]. Several amino acids located in the N- and C-terminals of GFP are flexible and hence can be truncated without perturbing its chromophore and the emission fluorescence [8]. The wild-type GFP consists of two chemically different populations that can be excited at 395 nm and 475 nm (a minor peak). The excitation of 395 nm induces an emission at 508 nm, whereas the excitation of 475 nm causes an emission at 503 nm [9]. This wild-type GFP does not fold efficiently at 37 °C and hence has limited biological applications. There have been tremendous efforts to create GFP mutants to improve its properties. An enhanced version of GFP has been developed by introducing two mutations at F64L and S65T [10, 11]. This improved version of GFP from Tsien's group can fold fairly well at 37 °C and hence is widely used in modern cell biology [11].

4.2.2 Derivatives of GFP with Different Colors

Mutations have been introduced into GFP to obtain various forms of fluorescence proteins generating distinct colors. The introduction of mutations F64L, Y66H, and Y145F results in a blue fluorescence protein (BFP) with excitation peak at 383 nm and emission peak at 447 nm. This BFP is spectrally distinctive from GFP and can be used in combination with GFP to double-label the same cells [12]. The mutation of Tyr 66 to Trp (Y66W) shifts the excitation and emission maxima to 436 and 476 nm, respectively. Additional mutations F64L, S65T, N146I, M153T, and V163A restore the brightness of the Y66W mutation and create the cyan fluorescence protein (CFP) [13]. A new version of CFP, CyPet [14], has been developed through directed evolutionary screening method which allows a higher fluorescence resonance energy transfer (FRET) between a coupled YFP. Although folding well at room temperature, CyPet protein is not stable at 37 °C and hence not suitable for studies in cells [15]. Yellow fluorescence protein (YFP) was created by stacking an aromatic ring next to the phenolate anion of the chromophore. The mutation at 203 (T203Y) introduces an aromatic ring and causes a 20 nm shift of excitation and emission peaks. Together with mutations at other sites (i. e., S65G, V68L, Q69K, S72A), this new YFP can be distinguished from GFP [12]. The recently developed versions of YFP, Venus and YPet have further improved fluorescence properties, including brightness and photostability [15].

4.2.3 Fluorescence Proteins Derived from DsRed

With its red color, the fluorescence protein DsRed has been popular to cell biologists. However, DsRed has a very severe disadvantage in that it forms tetramers, i. e. four copies of expressed DsRed aggregate in cells. The formation of DsRed tetramer often causes the artificial oligomerization of the targeting molecules fused to DsRed and alters their native structure and functions [16]. Hence, the potential of DsRed as a fluorescence labeling tag is very much limited. Many mutations have been introduced to DsRed through a directed evolution strategy to yield mRFP1, which is a monomer and retains the red fluorescence properties of DsRed [16]. However, mRFP1 has poor extinction coefficient, fluorescence quantum yield, and photostability [17]. Recently, Tsien's group at UC San Diego created a wide range of fluorescence proteins based on mRFP1 by using directed evolution methods and iterative somatic hyper-mutation [17, 18]. These newly developed fluorescence proteins have relatively long wavelengths in excitation and emission and have been named with different fruit names according to their emission colors. These fluorescence proteins can also be easily differentiated from those derived from GFP and thus allow labeling different targeting molecules with different colors within the same cells. Among these colorful fluorescence proteins, mOrange, mCherry, mPlum, and Tdtomato are of great interest to cell biologists [15]: mOrange (excitation: 548 nm; emission: 562 nm) is the brightest monomer and has excellent extinction coefficient and quantum yield; mCherry with long wavelengths (excitation: 587 nm; emission: 610 nm) is attractive because it has excellent photostability, fast maturation rate, and high resistance against varying pHs; mPlum (excitation: 590 nm; emission: 649 nm) is the only far-red fluorescence protein with reasonable brightness and photostability; tdTomato (excitation: 554 nm; emission: 581 nm) is a dimer, but has the highest brightness. Hence, tdTomato can be a good choice for molecular labeling of red colors if the size of the fusion tag is not a major concern for the targeting molecule.

4.2.4 Small Molecule Fluorescence Dyes

Because the fluorescence proteins are relatively large in size (about 30 KD), it is of interest to develop alternative dyes with smaller molecular weights that can be tagged to the target proteins. One example is the biarsenical-tetracysteine system, which consists of a cell-membrane-permeable biarsenical dye and a tetracysteine motif fused with a gene encoding the targeting protein. The biarsenical dye is non-fluorescent until binding to its coupling partner, the tetracysteine motif. Two biarsenical dyes, the green FlAsH and red ReAsH, have been widely used to stain different molecules in live cells [19–21]. A potential problem for FlAsH and ReAsH systems is the non-specific high level of background staining. Recently, improved tetracysteine sequences, HRWCCPGCCKTF and FLNCCPGCCMEP, have been identified through random mutagenesis and fluorescence activated cell sorting (FACS) [22]. These sequences have higher fluorescence quantum yield and affinity for biarsenical dyes, hence enhancing the signal/background staining contrast.

4.3 Fluorescence Microscopy

4.3.1 Epi-fluorescence Microscopy

Epi-fluorescence Microscopy is widely used for live-cell imaging. In general, a dichroic mirror reflects the excitation light with a shorter wavelength toward the specimen through an objective. The emission light with a longer wavelength from the specimen passes the same objective and dichroic mirror to the camera or eye pieces. A complete set of filters/mirrors for an epi-fluorescence microscopy consists of excitation filters, dichroic mirrors, and emission filters. For multicolor time-lapse imaging systems, multiple excitation and emission filters and dichroic mirrors are automatically controlled by filter controllers and dichroic exchangers, which are synchronized by a computer and imaging acquisition software. This allows the automated time-lapse imaging acquisition of multiple colors in live cells.

4.3.2 Confocal Fluorescence Microscopy

In a conventional fluorescence microscope, the entire specimen is excited by fluorescence light from a light source. The emission lights from different focal planes of the specimen are collected by the detecting device, either a camera or eyepieces. Such an imaging approach decreases the signal/noise ratios for signals from the focal planes of interest. In a confocal fluorescence microscope, a pinhole is introduced in front of the imaging device, which allows only lights from focal points of the objective lens to pass and be collected by the imaging device. Hence, confocal fluorescence microscopy can efficiently exclude lights from planes out of focus, thus providing clearer images than epi-fluorescence microscopy. The stacking of images from different layers of focal planes can provide a clear 3D view of distributed molecules within the cell. Because most confocal microscopes utilize lasers as the light sources to provide focused and bright excitation light, the strong excitation lights may cause severe photobleaching of the fluorescent proteins/dyes and thus damage the molecular features in the image. This is especially important when using confocal microscopy for live-cell imaging.

4.3.3 Total Internal Reflection Fluorescence Microscopy (TIRF)

Many molecular events, e. g. membrane protein activities and cell-ECM adhesion, occur only at a very thin layer within the cell or at the cellular interface. The fluorescence signals of interest from these thin layers are often overwhelmed by the background fluorescence if observation is made with conventional epi-fluorescence, or even confocal, microscopy. Total Internal Reflection Fluorescence Microscopy (TIRF) can overcome this problem and detect fluorescence signals emitted from a very thin layer ($\sim$200 nm). When an excitation light is directed into a cover glass underneath a specimen with a well-calculated angle, the excitation light can be completely reflected and travels only in the glass. Some of the light energy will propagate a short distance ($\sim$200 nm) into the specimen and generate an evanescent wave. All fluorescent molecules of the specimen layer within this evanescent wave distance, but not those from other layers, will be excited and emit the desirable fluorescence signals. Hence, TIRF microscopy allows the imaging of only the molecular events occurring within this short distance and eliminates the interfering signals from layers outside of this range.

4.3.4 Integration with Atomic Force Microscopy (AFM)

Atomic Force Microscopy allows the imaging of structural features of molecules/organelles within cells at sub-nanometer resolution. Furthermore, AFM can also measure the mechanical properties of inter- and intra-molecular bonds. Because AFM differentiates different types of molecules according to their size or shape, it is relatively less powerful in detecting untagged molecules with similar size/shape. Since fluorescent microscopy can easily distinguish molecules by tagging them with fluorescence proteins/dyes with different colors, the combination of these two imaging modalities can provide images of cellular/molecular structures with high-resolution, as well as excellent molecular specificity [23].

4.4 Applications in Mechanobiology

4.4.1 Visualization of Cellular Localization of Signaling Molecules and Expression of Genes

Cells in the cardiovascular system are constantly exposed to biomechanical forces. Flow channels and stretch chambers, which allow the precise control of chemical and mechanical factors, have been used as in vitro models to investigate the mechanisms by which ECs, VSMCs and cardiac myocytes convert mechanical cues to biochemical signals, which in turn modulate the expression of genes regulating cardiovascular functions. By using such systems, we and others showed that signaling molecules located in the membrane, lipid raft, and focal adhesions are activated by various mechanical forces. For example, shear stress activates focal adhesion kinase (FAK) in ECs, which was revealed by the increased tyrosine phosphorylation and clustering with vitronectin receptor [24]. IκB kinase (IKK) was also shown to be activated by shear stress, which resulted in the nuclear translocation of transcription factor NF-κB [25].

The cDNAs encoding GFP and other fluorescence proteins (FPs) can be conveniently fused to targeting genes to generate a recombinant gene sequence, which encodes the targeting molecules covalently coupled together with a fluorescence protein tag (Fig. 4.1B). FPs are in general inertial and should not perturb the native functions of the fused targeting molecules if the expressed fusion protein is correctly assembled. The positions of this targeting molecule of interest can be visualized with high spatiotemporal resolutions in live cells when the recombinant gene is expressed in host cells. A wide range of the known signaling molecules have been labeled with FPs [7]. Many of these fluorescence-labeled signaling molecules have been used in mechanobiology study to monitor the signaling relay, deformation and locomotion of organelles, and cellular structures responding to mechanical forces.

4.4.1.1 Subcellular Changes of Organelle-Positions

Because some of the signaling molecules can specifically localize to certain subcellular organelles, these signaling molecules have been labeled with FPs to highlight their subcellular positions and monitor the changes of organelles. Cytochrome-c oxidase, which stays at the inner mitochondrial membrane, has been fused with yellow fluorescence protein (YFP) to visualize the deformation of mitochondria. When a mechanical torque is applied through a RGD-coated magnetic bead adhered on a human airway smooth muscle cell surface, large displacement of mitochondria can be observed proximal and distal from the bead-attaching site [26, 27]. GFP has also been used to label nucleus and study its volume changes upon micropipette aspiration. Results from such experiments indicate that cells lacking emerin, an inner nuclear membrane protein, are less deformable than normal cells when mechanical force was generated through micropipette aspiration [28]. Recently, cell-permeable small molecule compounds have been developed, e. g. Mitotrackers (Invitrogen), to stain mitochondria and observe the cell deformation upon mechanical loading [27, 29, 30]. Although these cell-membrane-permeable dyes are easier to handle during staining experiments, FPs can be more conveniently fused to signal peptide targeting toward different organelle-localization [31]. These FP-fused proteins can be used to report the deformation of different organelles responding to mechanical stimulation.

4.4.1.2 Localization of Signaling/Structural Molecules

The live cell imaging with GFP-fused proteins is a very useful method in studying mechanobiology. The deformation and hence the mechanical properties of cellular structures in response to mechanical stimulation can be monitored in a dynamic fashion. GFP-fused microtubules or actins have helped to reveal that cells can be separated into three mechanically different regions: an elastic nucleus, a viscoelastic-fluid-like cytoplasm, and an elastic cortical layer upon mechanical force application through micropipettes [32]. Shear stress has been shown to cause a heterogeneous deformation of intermediate filaments visualized by GFP-fused vimentin [33]. GFP-fused actin has also been used to visualize the shape deformations and migration patterns of wounded endothelial monolayer under shear stress [34]. With GFP-fused microtubules, cells were visualized to behave like a globally-interconnected network in response to mechanical stimulation, suggesting that the local mechanical stimulation can be rapidly transmitted to cause large deformation in distal regions in the cell [26].

GFP-fused proteins have also been widely utilized to study molecular dynamics in mechanobiology. Utilizing the GFP-fused proteins involved in focal adhesion, i. e., GFP-paxillin and GFP-tensin, Zamir et al. revealed that focal contacts and fibrillar adhesions, two types of cell-ECM contacts, have different dynamic characteristics and may provide a molecular switch mechanism to perceive different mechanical micro-environments [35]. The application of GFP-zyxin further revealed that small and nascent focal contacts at the leading edge exert stronger forces on the substrate than those in the large and mature focal adhesion sites. In another study, it was shown that the associated traction forces diminish as the nascent focal contacts start to mature and grow into focal adhesion sites [36]. GFP-fused zyxin and vinculin were also used to assess the dynamics of focal adhesion assembly/disassembly under mechanical stress [37]. Actin fused with GFP was used to visualize podosomes on gel substrate with different stiffness. The results from such experiments suggest that the mechanical stiffness of the substrate controls the spatio-temporal patterns of the podosome structures [38]. In other studies, paxillin [39] and FAK [40] fused to GFP were applied to visualize the motility of focal adhesion complex upon mechanical force application.

Another major application of GFP-fused proteins in mechanobiology is to monitor the translocation of specific targeting molecules among different subcellular organelles and locations. Shear stress has been shown to induce the nuclear translocation of endothelial glucocorticoid receptor (GR), visualized by an expressed GFP-

GR [41]. Mechanical stretch induced a plasma-membrane-translocation of GFP-fused RhoA [42]. Mechanical stretch on the plasma membrane induced by osmosis also caused a translocation of GFP-fused Pleckstrin-homology domain of PLCδ from the membrane to cytosol, reflecting a mechanically induced PIP2 hydrolysis [43]. GFP-fused IκBα has been used to visualize the decreased level of IκBα in the cytosol upon flow application [44]. GFP has also been fused to TRPM7, an ion channel and a protein kinase. Shear stress in this study was shown to induce a plasma membrane translocation of GFP-fused TRPM7, which resulted in the concurrent increase of TRPM7 current in VSMCs [45].

The functionality and biophysical properties of specific proteins can be examined by GFP-fused proteins. Mechanical perturbation of HeLa cells by glass pipette was shown to induce a Ca^{+2} wave propagating between neighboring cells. GFP-fused connexins were used to show that the expression level of connexin determines the distance range of these wave propagations, suggesting the involvement and importance of tight junctions for this mechanotransduction process [46]. With multiple copies of GFP fused together with fibronectin domains, the folding/unfolding biophysical properties of fibronectin were also studied utilizing single molecule force spectroscopy (SMFS) [47].

4.4.1.3 Transcriptional Activation of Specific Genes

Mechanical stimulation plays a crucial role in regulating expression of many genes in various tissues. Results from many laboratories, including ours, show that shear stress regulates many genes and their products, including vasodilators (e. g., NO), vasoconstrictors (e. g., endothelin-1), growth factors (e. g., platelet-derived growth factors), adhesion molecules (e. g., intercellular adhesion molecule-1), chemoattractants (e. g., monocyte chemotactic protein-1), and molecules involved in survival and apoptosis (e. g., growth arrest and DNA damage inducible protein 45 (GADD45), p53, and p21cip1) [48, 49]. GFP is a very useful marker to monitor the level of these expressed proteins. Usually, GFP is fused with the promoter region of the gene to be studied. Cells harboring this reporter system can then be subjected to various types of mechanical stimuli, and the induction and/or reduction of the gene can be monitored by the level of expressed GFP.

GFP fused with a promoter derived from smooth muscle actin gene revealed that mechanical stretch induced an increase of smooth muscle actin gene expression, which is dependent on the intact actin and microtubules [50]. GFP has also been fused to cis-regulatory regions of dentin matrix protein 1 (DMP1) to report their transcription activities. This system was shown to be up-regulated by mechanical stretch and correlated with local mechanical strain where cells are seeded [51]. Mechanical stretch also induced an NF–κB-dependent increase of iex-1 expression visualized by the GFP-fused iex-1 promoter [52]. A promoter of osteopontin (OPN) fused to GFP revealed that mechanical stretch can induce the expression of the OPN gene and bone remodeling [53]. Similarly, shear stress-induced NF–κB gene expression was monitored by a NF–κB promoter sequence fused together with GFP [44].

The use of GFP-fused eNOS revealed that the shear stress magnitude is in proportional to the eNOS protein expression level in transgenic mice *in vivo* [54].

4.4.2 Spatiotemporal Quantification of Post-translational Modifications

Post-translational modifications of molecules play crucial roles in regulating cellular signaling processes. For example, phosphorylation has been shown to be the key mechanism for the rapid and orchestrated transmission of signals inside the cells [55]. Acetylation/deacetylation is involved in regulating chromatin structure and function [56]. The enzymatic activation of matrix metalloproteinases (MMPs) is also crucial for cell migration and cancer development [57]. These active molecular events are important for mechanotransduction and mechanobiology [58]. There is an emerging need to develop imaging reagents and methods to visualize the post-translational modifications of signaling molecules in live cells upon mechanical stimulation.

Fluorescence resonance energy transfer (FRET), fluorescence recovery after photobleaching (FRAP), and fluorescence lifetime imaging (FLIM) provide powerful means in this respect. The implementation of these technologies involves the integration of several facets including the developments of molecular sensors, optical microscopic systems, and imaging analysis methods. The focus of this chapter is on the development of molecular sensors and their applications in mechanobiology.

4.4.2.1 Fluorescence Resonance Energy Transfer (FRET)

FRET is a phenomenon of quantum mechanics. When two fluorophores, with the emission spectrum of one fluorophore (the donor) overlapping the excitation spectrum of the other (the acceptor), are close to each other with favorable relative orientations, the excitation of the donor can elicit a sufficient energy transfer to the acceptor and produce emission from the acceptor. The modulation of the distance or relative orientations between the fluorophores can affect the FRET efficiency. At present, CFP and YFP are the favorite FRET pair comparing to BFP and GFP, since CFP has better extinction coefficient, quantum yield and photo-stability than BFP [59]. FRET imaging of molecular activities has high spatio-temporal resolutions and is therefore suitable for molecular imaging in live cells. Upon activation, a typical FRET biosensor would undergo a conformational change leading to an alteration of the relative distance/orientation of the donor and acceptor. Hence, the change of FRET signal of these biosensors can represent the activation status of the targeting molecules. Based on this principle, many biosensors have been developed to visualize molecular activities in live cells, e.g. protease activities, calcium dynamics, Ras, Rho small GTPases activation, and tyrosine/serine/threonine kinases, phosphor-lipid dynamics, etc. [60].

4.4 Applications in Mechanobiology

Many of these FRET-based techniques have been applied to visualize signal transduction in response to mechanical stimulation. For example, GFP-fused Rac and Alexa568-p21-binding domain of PAK1 (PBD) were used to monitor the Rac activation in live cells by measuring FRET between GFP to Alexa568 [61]. With this FRET-based biosensor, shear stress was shown to induce a directional activation of Rac concentrated at the leading edge of the cell along flow direction [62]. Shear stress has also been shown to induce a polarized Cdc42 activation along flow direction visualized by the FRET between a GFP-Cdc42 and an Alexa568-PBD [63]. A separated pair of ECFP-fused relA and EYFP-fused IκBα was used to monitor the interaction of relA and IκBα. The FRET efficiency between ECFP-relA and EYFP-IκBα decreased upon shear stress application, indicating a mechanical-force-induced dissociation of relA and IκBα [44]. In our laboratory at UC San Diego, a Src biosensor has been recently generated with ECFP and EYFP covalently con-

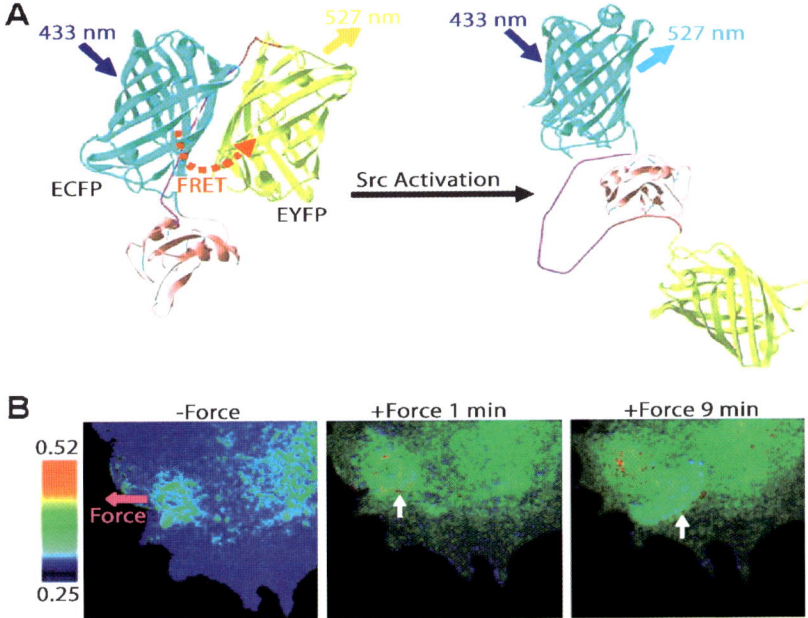

Fig. 4.2 A FRET-based Src biosensor and its application in detecting the mechanical-stimulated Src activation. **A** A cartoon scheme depicting the activation mechanism of the Src biosensor. When Src kinase is inactive, ECFP and EYFP are positioned proximal to each other and have strong FRET. The excitation of the biosensor at 433 nm results in the emission from EYFP at 527 nm. When Src kinase is activated to phosphorylate the substrate peptide in the biosensor, the biosensor will undergo a conformational change and separate EYFP from ECFP, which results in the decrease of FRET. The excitation of the biosensor at 433 nm then results in the emission from ECFP at 476 nm. Hence, the emission spectra of the biosensor represent the activation status of Src kinase. **B** FRET response of a cell with clear directional wave propagation away from the site of mechanical stimulation introduced by laser-tweezers. The color bar on the left indicates ECFP/EYFP emission ratios, with cold color representing low ratios and hot color representing high ratios. The *pink arrow* represents the site of force application and the force direction. The *white arrows* point to the front edge of activated Src wave. This figure is adapted from Wang et al. [64]

catenated with a phosphor-amino-acid binding domain and a substrate peptide specific for Src kinase. The activation of Src can phosphorylate this biosensor to cause changes in conformation and FRET between ECFP and EYFP. Hence, FRET levels of the biosensor represent different activation status of the Src kinase (Fig. 4.2A). When a local mechanical force is introduced to a bead adhered on a live endothelial cell by a laser-tweezer, a rapid distal Src activation and a slower wave propagation of Src activation can be observed [64] (Fig. 4.2B). The results demonstrate how biochemical signals are initiated and transmitted in live cells upon mechanical stimulation. CFP and YFP have also been fused to human B2 bradykinin receptor, a G protein-coupled receptor (GPCR), to detect the activation of GPCR. Shear stress was shown to activate B2 bradykinin GPCR within 2 min, which can be inhibited by B2-selective antagonist [65]. These results suggest that B2 bradykinin GPCR may serve as a mechano-sensing molecule in response to shear stress.

4.4.2.2 Fluorescence Recovery After Photobleach (FRAP)

Fluorescence recovery after photobleaching (FRAP) is a procedure to selectively photobleach the fluorescence signals within a region of interest and monitor the recovery of fluorescence in this region over a period of time (Fig. 4.3). This simple technique can be used to determine the kinetic characteristics of targeting fluorescence molecules, e.g. their diffusion coefficient or transport rate [66]. Naturally, FRAP has been used in live cell analysis of dynamics of mechanobiology. It is conceivable that cellular membrane is the primary site where mechanical cues are converted to biochemical signals because of its proximity to extracellular environment. Although other subcellular organelles such as cytoskeleton and mitochondria have been shown to be mechanical sensitive, many plasma membrane-associated signaling proteins and receptors are still considered to be crucial for mechanobiology. A line of evidence supporting such a hypothesis is the modulation of cell membrane fluidity by mechanical forces [67, 68], which may greatly affect the activity of membrane-bound proteins and receptor. By using confocal laser scanning microscope analyzing FRAP, Butler et al. studied the effects of shear stress on the spatial distribution and persistence of the shear-induced increases in fluidity of plasma membranes [69]. The results suggest that the lipid layer of plasma membrane can serve as a mechano-sensing element and show that the upstream and downstream portions of plasma membrane respond differently under certain flow applications. Other intracellular signaling/structural molecules have also been studied using FRAP. Osborn et al. [70] recently showed that shear stress, while inhibiting cell motility, caused an increased F-actin turnover in ECs at early stages (1 – 2 hr) upon flow application. FRAP has also been applied to assess the dissociation rates of different focal adhesion molecules zyxin and vinculin, before and after the mechanical stress dissipation by modulating actin cytoskeleton with cytochalasin D or laser-scissors. The results indicate that mechanical forces regulate the focal adhesion assembly by modulating the kinetic characteristics of zyxin [37]. Finally, FRAP has been used to assess the speed of intercellular Ca^{2+} wave propagation upon the mechanical stretch [71, 72].

4.4 Applications in Mechanobiology

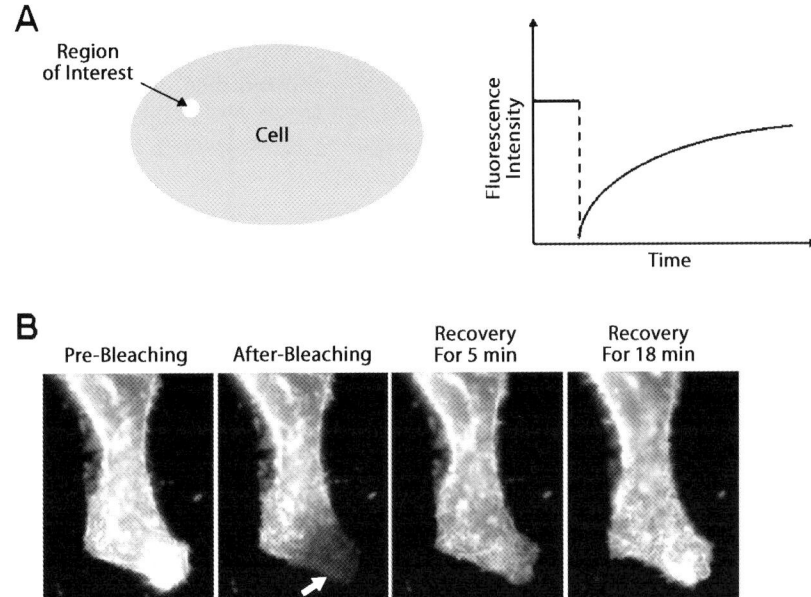

Fig. 4.3 FRAP principles and examples. **A** The principles of FRAP. The fluorescence of a region of interest in a cell can be photo-bleached and monitored for its recovery. The diagram shown on the right represents a typical time course of fluorescence intensity of a selected region on a FRAP experiment. **B** A sample FRAP experiment. A HeLa cell was transfected with a membrane-targeted Src biosensor [64]. EYFP of the biosensor was photo-bleached and the fluorescence recovery was monitored on a region highlighted by the *white arrow*

4.4.2.3 Fluorescence Lifetime Imaging Microscopy (FLIM)

Fluorescence lifetime imaging microscopy (FLIM) is a technique to visualize the life time of excitation state of spatially distributed fluorescence molecules. Methods to measure fluorescence lifetime can be separated into two categories: frequency domain and time domain. For the frequency domain method, the excitation light is in a sinusoidal format and the emission fluorescence is also a sinusoidal wave with the same frequency as the excitation light. The delay or phase-shift of emission fluorescence comparing to the excitation light is used to calculate the lifetime of the fluorescence probe (Fig. 4.4A). For the time domain method, the specimen is excited with a pulse light with its duration much shorter than the emission lifetime of the fluorescence probe. The time course of the emission fluorescence intensity captured is used to calculate the lifetime characteristics (Fig. 4.4B).

Because FLIM is independent of the local concentration of fluorescence molecules and the excitation intensity, this method provides relatively more reliable signals comparing with other technologies based on fluorescence intensity. When a fluorescence protein/probe serving as a donor interacts with its acceptor fluorescence molecule during FRET, the lifetime of interacting donor changes. Hence, FLIM can separate the population of interacting fluorescence molecules from those

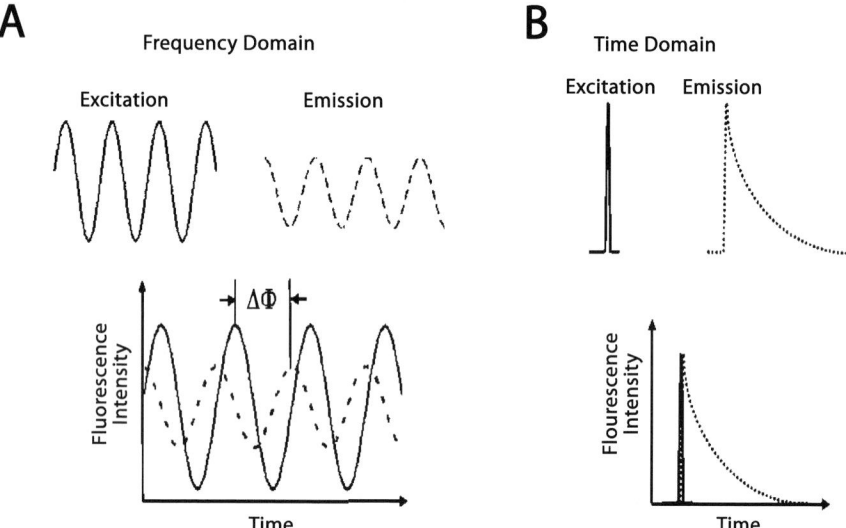

Fig. 4.4 Schemes of two methods to measure fluorescence lifetime. **A** Frequency domain. **B** Time domain

of non-interacting ones based on lifetime distribution. Therefore, when integrated with FRET, FLIM can help to separate signals of donor-molecules having FRET from those non-FRET donors, thus enhancing the signal/noise ratio of FRET measurement [73]. Although FLIM is very useful and has been applied to study cellular imaging in general, e. g. FLIM was used to visualize the lateral signal propagation from ErbB1 receptor in live cells [74], it is a largely unexplored field in mechanobiology.

4.5 Perspective in Cardiovascular Physiology and Diseases

Mechanobiology is crucial for the regulation of function and structure of cells in the cardiovascular system. Located at the interface between the blood and vessel wall, ECs are constantly exposed to shear stress. The circumferential stresses resulting from blood pressure are mainly born by the VSMCs in the artery. In the heart, cardiac myocytes undergo regular contractions to drive the circulation of blood. Since various biomechanical forces are crucial for cardiovascular functions, impaired cellular responses to the mechanical forces or dysfunctional mechanical outputs are pathophysiological processes that can lead to various cardiovascular diseases. Thus, the study of mecahnobiology at the levels of molecules, cells, and organs have been important topics for interdisciplinary research, including bioengineering, bioimaging, cardiology, physiology, and other fields.

Imaging techniques such as MRI, X-ray, and ultrasound have been widely used to monitor cardiovascular structures and functions *in vivo*. To date, these techniques

4.5 Perspective in Cardiovascular Physiology and Diseases

are mainly used for diagnosis and research at the organ level. The current fluorescence live-cell imaging technology has the capability to study mechanobiology at the cellular and subcellular levels in single cells, but has its limitations in investigating cardiovascular physiology and pathophysiology in live experimental animals and human patients. Heart and blood vessels are buried in the body and it is difficult to introduce fluorescence probe into cells in the cardiovascular system *in vivo*. Nevertheless, these advanced imaging technologies can be potentially important in several areas related to the mechanobiology in the cardiovascular system. First, the dynamic rolling and adhesion of white blood cells in the microcirculation can be studied by intravital fluorescence microscopy. The white blood cells can be labeled with GFP to increase their visibility. The homing of bone marrow-derived endothelial progenitor cells (EPCs) in repairing injured vessel wall and myocardium has been a popular research topic recently. Labeling of EPCs with GFP or using EPCs harboring GFP isolated from transgenic mice in conjunction with fluorescence live-cell imaging would be a fruitful research direction. As GFP can be introduced into various molecules through genetic approach, observations of cardiovascular specimens from transgenic animals would also reveal new information on the target molecules.

References

1. Shimomura O, Johnson FH, Saiga Y (1962) Extraction, purification and properties of aequorin, a bioluminescent protein from the luminous hydromedusan, Aequorea. J Cell Comp Physiol 59:223–239
2. Prasher DC, Eckenrode VK, Ward WW, Prendergast FG, Cormier MJ (1992) Primary structure of the Aequorea victoria green-fluorescent protein. Gene 111:229–233
3. Chalfie M, Tu Y, Euskirchen G, Ward WW, Prasher DC (1994) Green fluorescent protein as a marker for gene expression. Science 263:802–805
4. Inouye S, Tsuji FI (1994) Aequorea green fluorescent protein. Expression of the gene and fluorescence characteristics of the recombinant protein. FEBS Lett 341:277–280
5. Ormo M, Cubitt AB, Kallio K, Gross LA, Tsien RY, Remington SJ (1996) Crystal structure of the Aequorea victoria green fluorescent protein. Science 273:1392–1395
6. Yang F, Moss LG, Phillips GN Jr (1996) The molecular structure of green fluorescent protein. Nat Biotechnol 14:1246–1251
7. Tsien RY (1998) The green fluorescent protein. Annu Rev Biochem 67:509–544
8. Dopf J, Horiagon TM (1996) Deletion mapping of the Aequorea victoria green fluorescent protein. Gene 173:39–44
9. Heim R, Prasher DC, Tsien RY (1994) Wavelength mutations and posttranslational autoxidation of green fluorescent protein. Proc Natl Acad Sci USA 91:12501–12504
10. Cormack BP, Valdivia RH, Falkow S (1996) FACS-optimized mutants of the green fluorescent protein (GFP). Gene 173:33–38
11. Heim R, Cubitt AB, Tsien RY (1995) Improved green fluorescence. Nature 373:663–664
12. Cubitt AB, Woollenweber LA, Heim R (1999) Understanding structure-function relationships in the Aequorea victoria green fluorescent protein. Methods Cell Biol 58:19–30
13. Heim R, Tsien RY (1996) Engineering green fluorescent protein for improved brightness, longer wavelengths and fluorescence resonance energy transfer. Curr Biol 6:178–182
14. Nguyen AW, Daugherty PS (2005) Evolutionary optimization of fluorescent proteins for intracellular FRET. Nat Biotechnol 23:355–360
15. Shaner NC, Steinbach PA, Tsien RY (2005) A guide to choosing fluorescent proteins. Nat Methods 2:905–909
16. Campbell RE, Tour O, Palmer AE, Steinbach PA, Baird GS, Zacharias DA, Tsien RY (2002) A monomeric red fluorescent protein. Proc Natl Acad Sci USA 99:7877–7882
17. Shaner NC, Campbell RE, Steinbach PA, Giepmans BN, Palmer AE, Tsien RY (2004) Improved monomeric red, orange and yellow fluorescent proteins derived from Discosoma sp. red fluorescent protein. Nat Biotechnol 22:1567–1572
18. Wang L, Jackson WC, Steinbach PA, Tsien RY (2004) Evolution of new nonantibody proteins via iterative somatic hypermutation. Proc Natl Acad Sci USA 101:16745–16749
19. Adams SR, Campbell RE, Gross LA, Martin BR, Walkup GK, Yao Y, Llopis J, Tsien RY (2002) New biarsenical ligands and tetracysteine motifs for protein labeling in vitro and in vivo: synthesis and biological applications. J Am Chem Soc 124:6063–6076

20. Griffin BA, Adams SR, Tsien RY (1998) Specific covalent labeling of recombinant protein molecules inside live cells. Science 281:269–272
21. Nakanishi J, Nakajima T, Sato M, Ozawa T, Tohda K, Umezawa Y (2001) Imaging of conformational changes of proteins with a new environment-sensitive fluorescent probe designed for site-specific labeling of recombinant proteins in live cells. Anal Chem 73:2920–2928
22. Martin BR, Giepmans BN, Adams SR, Tsien RY (2005) Mammalian cell-based optimization of the biarsenical-binding tetracysteine motif for improved fluorescence and affinity. Nat Biotechnol 23:1308–1314
23. Vickery SA, Dunn RC (2001) Combining AFM and FRET for high resolution fluorescence microscopy. J Microsc 202:408–412
24. Li S, Kim M, Hu YL, Jalali S, Schlaepfer DD, Hunter T, Chien S, Shyy JY (1997) Fluid shear stress activation of focal adhesion kinase. Linking to mitogen-activated protein kinases. J Biol Chem 272:30455–30462
25. Bhullar IS, Li YS, Miao H, Zandi E, Kim M, Shyy JY, Chien S (1998) Fluid shear stress activation of IkappaB kinase is integrin-dependent. J Biol Chem 273:30544–30549
26. Wang N, Naruse K, Stamenovic D, Fredberg JJ, Mijailovich SM, Tolic-Norrelykke IM, Polte T, Mannix R, Ingber DE (2001) Mechanical behavior in living cells consistent with the tensegrity model. Proc Natl Acad Sci USA 98:7765–7770
27. Hu S, Chen J, Fabry B, Numaguchi Y, Gouldstone A, Ingber DE, Fredberg JJ, Butler JP, Wang N (2003) Intracellular stress tomography reveals stress focusing and structural anisotropy in cytoskeleton of living cells. J Am Physiol Cell Physiol 285:C1082–1090
28. Rowat AC, Lammerding J, Ipsen JH (2006) Mechanical properties of the cell nucleus and the effect of emerin deficiency. Biophys J 91:4649–4664
29. Knight MM, Bomzon Z, Kimmel E, Sharma AM, Lee DA, Bader DL (2006) Chondrocyte deformation induces mitochondrial distortion and heterogeneous intracellular strain fields. Biomech Model Mechanobiol 5:180–191
30. Bomzon Z, Knight MM, Bader DL, Kimmel E (2006) Mitochondrial dynamics in chondrocytes and their connection to the mechanical properties of the cytoplasm. J Biomech Eng 128:674–679
31. Baker A, Kaplan CP, Pool MR (1996) Protein targeting and translocation; a comparative survey. Biol Rev Camb Philos Soc 71:637–702
32. Heidemann SR, Kaech S, Buxbaum RE, Matus A (1999) Direct observations of the mechanical behaviors of the cytoskeleton in living fibroblasts. J Cell Biol 145:109–122
33. Helmke BP, Goldman RD, Davies PF (2000) Rapid displacement of vimentin intermediate filaments in living endothelial cells exposed to flow. Circ Res 86:745–752
34. Albuquerque ML, Flozak AS (2001) Patterns of living beta-actin movement in wounded human coronary artery endothelial cells exposed to shear stress. Exp Cell Res 270:223–234
35. Zamir E, Katz M, Posen Y, Erez N, Yamada KM, Katz BZ, Lin S, Lin DC, Bershadsky A, Kam Z, Geiger B (2000) Dynamics and segregation of cell-matrix adhesions in cultured fibroblasts. Nat Cell Biol 2:191–196
36. Beningo KA, Dembo M, Kaverina I, Small JV, Wang YL (2001) Nascent focal adhesions are responsible for the generation of strong propulsive forces in migrating fibroblasts. J Cell Biol 153:881–888
37. Lele TP, Pendse J, Kumar S, Salanga M, Karavitis J, Ingber DE (2006) Mechanical forces alter zyxin unbinding kinetics within focal adhesions of living cells. J Cell Physiol 207:187–194
38. Collin O, Tracqui P, Stephanou A, Usson Y, Clement-Lacroix J, Planus E (2006) Spatiotemporal dynamics of actin-rich adhesion microdomains: influence of substrate flexibility. J Cell Sci 119:1914–1925
39. Mack PJ, Kaazempur-Mofrad MR, Karcher H, Lee RT, Kamm RD (2004) Force-induced focal adhesion translocation: effects of force amplitude and frequency. J Am Physiol Cell Physiol 287:C954–962
40. Li S, Butler P, Wang Y, Hu Y, Han DC, Usami S, Guan JL, Chien S (2002) The role of the dynamics of focal adhesion kinase in the mechanotaxis of endothelial cells. Proc Natl Acad Sci USA 99:3546–3551

References

41. Ji JY, Jing H, Diamond SL (2003) Shear stress causes nuclear localization of endothelial glucocorticoid receptor and expression from the GRE promoter. Circ Res 92:279–285
42. Smith PG, Roy C, Zhang YN, Chauduri S (2003) Mechanical stress increases RhoA activation in airway smooth muscle cells. J Am Respir Cell Mol Biol 28:436–442
43. Zhang L, Lee JK, John SA, Uozumi N, Kodama I (2004) Mechanosensitivity of GIRK channels is mediated by protein kinase C-dependent channel-phosphatidylinositol 4,5-bisphosphate interaction. J Biol Chem 279:7037–7047
44. Ganguli A, Persson L, Palmer IR, Evans I, Yang L, Smallwood R, Black R, Qwarnstrom EE (2005) Distinct NF-kappaB regulation by shear stress through Ras-dependent IkappaBalpha oscillations: real-time analysis of flow-mediated activation in live cells. Circ Res 96:626–634
45. Oancea E, Wolfe JT, Clapham DE (2006) Functional TRPM7 channels accumulate at the plasma membrane in response to fluid flow. Circ Res 98:245–253
46. Paemeleire K, Martin PE, Coleman SL, Fogarty KE, Carrington WA, Leybaert L, Tuft RA, Evans WH, Sanderson MJ (2000) Intercellular calcium waves in HeLa cells expressing GFP-labeled connexin 43, 32, or 26. Mol Biol Cell 11:1815–1827
47. Abu-Lail NI, Ohashi T, Clark RL, Erickson HP, Zauscher S (2006) Understanding the elasticity of fibronectin fibrils: unfolding strengths of FN-III and GFP domains measured by single molecule force spectroscopy. Matrix Biol 25:175–184
48. Chien S, Shyy JY (1998) Effects of hemodynamic forces on gene expression and signal transduction in endothelial cells. Biol Bull 194:390–391; discussion 392–393
49. Gimbrone MA Jr, Topper JN, Nagel T, Anderson KR, Garcia-Cardena G (2000) Endothelial dysfunction, hemodynamic forces, and atherogenesis. Ann NY Acad Sci 902:230–239; discussion 239–240
50. Wang J, Su M, Fan J, Seth A, McCulloch CA (2002) Transcriptional regulation of a contractile gene by mechanical forces applied through integrins in osteoblasts. J Biol Chem 277:22889–22895
51. Yang W, Lu Y, Kalajzic I, Guo D, Harris MA, Gluhak-Heinrich J, Kotha S, Bonewald LF, Feng JQ, Rowe DW, Turner CH, Robling AG, Harris SE (2005) Dentin matrix protein 1 gene cis-regulation: use in osteocytes to characterize local responses to mechanical loading in vitro and in vivo. J Biol Chem 280:20680–20690
52. Schulze PC, de Keulenaer GW, Kassik KA, Takahashi T, Chen Z, Simon DI, Lee RT (2003) Biomechanically induced gene iex-1 inhibits vascular smooth muscle cell proliferation and neointima formation. Circ Res 93:1210–1217
53. Fujihara S, Yokozeki M, Oba Y, Higashibata Y, Nomura S, Moriyama K (2006) Function and regulation of osteopontin in response to mechanical stress. J Bone Miner Res 21:956–964
54. Cheng C, van Haperen R, de Waard M, van Damme LC, Tempel D, Hanemaaijer L, van Cappellen GW, Bos J, Slager CJ, Duncker DJ, van der Steen AF, de Crom R, Krams R (2005) Shear stress affects the intracellular distribution of eNOS: direct demonstration by a novel in vivo technique. Blood 106:3691–3698
55. Karin M, Hunter T (1995) Transcriptional control by protein phosphorylation: signal transmission from the cell surface to the nucleus. Curr Biol 5:747–757
56. Narlikar GJ, Fan HY, Kingston RE (2002) Cooperation between complexes that regulate chromatin structure and transcription. Cell 108:475–487
57. Genis L, Galvez BG, Gonzalo P, Arroyo AG (2006) MT1-MMP: universal or particular player in angiogenesis? Cancer Metastasis Rev 25:77–86
58. Chien S, Li S, Shiu YT, Li YS (2005) Molecular basis of mechanical modulation of endothelial cell migration. Front Biosci 10:1985–2000
59. Miyawaki A, Llopis J, Heim R, McCaffery JM, Adams JA, Ikura M, Tsien RY (1997) Fluorescent indicators for Ca^{2+} based on green fluorescent proteins and calmodulin. Nature 388:882–887
60. Giepmans BN, Adams SR, Ellisman MH, Tsien RY (2006) The fluorescent toolbox for assessing protein location and function. Science 312:217–224
61. Kraynov VS, Chamberlain C, Bokoch GM, Schwartz MA, Slabaugh S, Hahn KM (2000) Localized Rac activation dynamics visualized in living cells. Science 290:333–337

62. Tzima E, Del Pozo MA, Kiosses WB, Mohamed SA, Li S, Chien S, Schwartz MA (2002) Activation of Rac1 by shear stress in endothelial cells mediates both cytoskeletal reorganization and effects on gene expression. Embo J 21:6791–6800
63. Tzima E, Kiosses WB, del Pozo MA, Schwartz MA (2003) Localized cdc42 activation, detected using a novel assay, mediates microtubule organizing center positioning in endothelial cells in response to fluid shear stress. J Biol Chem 278:31020–31023
64. Wang Y, Botvinick EL, Zhao Y, Berns MW, Usami S, Tsien RY, Chien S (2005) Visualizing the mechanical activation of Src. Nature 434:1040–1045
65. Chachisvilis M, Zhang YL, Frangos JAG (2006) protein-coupled receptors sense fluid shear stress in endothelial cells. Proc Natl Acad Sci USA 103:15463–15468
66. Reits EA, Neefjes JJ (2001) From fixed to FRAP: measuring protein mobility and activity in living cells. Nat Cell Biol 3:E145–147
67. Sato M, Ohshima N, Nerem RM (1996) Viscoelastic properties of cultured porcine aortic endothelial cells exposed to shear stress. J Biomech 29:461–467
68. Haidekker MA, L'Heureux N, Frangos JA (2000) Fluid shear stress increases membrane fluidity in endothelial cells: a study with DCVJ fluorescence. J Am Physiol Heart Circ Physiol 278:H1401–H1406
69. Butler PJ, Tsou TC, Li JY, Usami S, Chien S (2002) Rate sensitivity of shear-induced changes in the lateral diffusion of endothelial cell membrane lipids: a role for membrane perturbation in shear-induced MAPK activation. Faseb J 16:216–218
70. Osborn EA, Rabodzey A, Dewey CF Jr, Hartwig JH (2006) Endothelial actin cytoskeleton remodeling during mechanostimulation with fluid shear stress. J Am Physiol Cell Physiol 290:C444–452
71. Himpens B, Stalmans P, Gomez P, Malfait M, Vereecke J (1999) Intra- and intercellular Ca^{2+} signaling in retinal pigment epithelial cells during mechanical stimulation. Faseb J 13(Suppl):S63–S68
72. Gomes P, Malfait M, Himpens B, Vereecke J (2003) Intercellular Ca(2+)-transient propagation in normal and high glucose solutions in rat retinal epithelial (RPE-J) cells during mechanical stimulation. Cell Calcium 34:185–192
73. Wallrabe H, Periasamy A (2005) Imaging protein molecules using FRET and FLIM microscopy. Curr Opin Biotechnol 16:19–27
74. Verveer PJ, Wouters FS, Reynolds AR, Bastiaens PI (2000) Quantitative imaging of lateral ErbB1 receptor signal propagation in the plasma membrane. Science 290:1567–1570

Chapter 5
Optical Coherence Tomography (OCT) – An Emerging Technology for Three-Dimensional Imaging of Biological Tissues

Gereon Hüttmann[1,2], Eva Lankenau[1,2]

[1] Institute of Biomedical Optics, University Lübeck, Peter-Monnik-Weg 4, 23562 Lübeck, Germany, huettmann@bmo.uni-luebeck.de
[2] Medical Laser Center Lübeck GmbH, Peter-Monnik-Weg 4, 23562 Lübeck, Germany

Abstract Optical imaging technologies offer high resolution non-contact visualization of tissues with a large variety of contrast modalities. However tissues are optically inhomogeneous due to microscopic variations of the refractive index. The resulting scattering limits resolution and imaging depth. Optical coherence tomography (OCT) is a relatively new approach for obtaining 3-D high resolution imaging of tissue which is successfully used in the diagnosis of the eye diseases. However, OCT is not limited to medical applications. This review will cover theory and technology of OCT and provide an overview over current use and future prospects in tissue engineering.

5.1 Introduction

Biological tissues are composed of 3-dimensional structures whose sizes cover about nine orders of magnitude, ranging from nanometers to meters. Consequently tools were developed for the visualization and quantification of tissue morphology on these scales. Today, from X-ray diffraction for the analysis of biomolecules to whole body MRT, biological structures can be imaged. Optical imaging is not only very attractive because of it's resolution down to a subcellular level. Compared to competing technologies like ultrasound, X-ray, MRT, and electron microscopy it offers a non-contact, non-invasive in-vivo applicability with compact and affordable devices. Spectroscopy, stains and specially designed probes provide additional means to generate site- and function-specific contrast. Consequently optical imaging and microscopy are well developed technologies which are widely used in medicine and biotechnology.

One of the main disadvantage of visible and near infrared light for the investigation of complex tissue structures is the strong scattering, which is caused by a random variation of the refractive index. Consequently in the past imaging was limited to 2-dimensional surfaces or thin tissue sections. However, biology is inherently 3-dimensional and therefore several methods were developed to provide a depth resolved imaging of tissue layers. The challenge here is to provide a good depth resolution and additionally an efficient rejection against scattered photons, which degrade the image contrast. The two principal ways to go are the use of a non-linear interaction between light and tissue, which limits the optical response (e. g. fluorescence, scattering) to the high irradiance at the focal point or the rejection of out of focus light by optical means. Two-photon excited fluorescence microscopy, second harmonic, and CARS imaging are examples for the first, confocal microscopy (CM) for the second approach. CM utilizes a point illumination and a confocal detector, which limits the depth of focus and rejects the out-of-focus light. Invented by Minsky in 1961 [44], CM is now a standard technique in microscopy and is also used for in-vivo imaging in medical diagnosis [6, 8, 26]. A good depth sectioning capability requires objectives with a large numerical aperture (NA) and a good optical quality of the sample, which can not be assured for all applications. Additionally the imaging depth is limited to 100 – 200 μm.

A new imaging technology which provides higher imaging depth combined with a resolution below 10 micrometer is optical coherence tomography (OCT). Instead of relying on the depth of focus of the imaging lenses, OCT uses the same working principle as ultrasound or radar. The propagation time of light from the objective to the object and back is measured. Due to the high speed of light and the small distances involved, the pulse-echo principle is not applicable for OCT and interferometry is used instead. Thereby OCT effectively decouples the depth resolution from the imaging NA and achieves a longitudinal resolution of only a few micrometers even with low NA objectives. Figure 5.1 shows the resolution and maximal imaging depth of OCT in comparison with ultrasound, MRT, and microscopy.

OCT was first successfully applied to high resolution retina imaging, which is especially hampered by the low NA of the human lens system [29]. Starting from

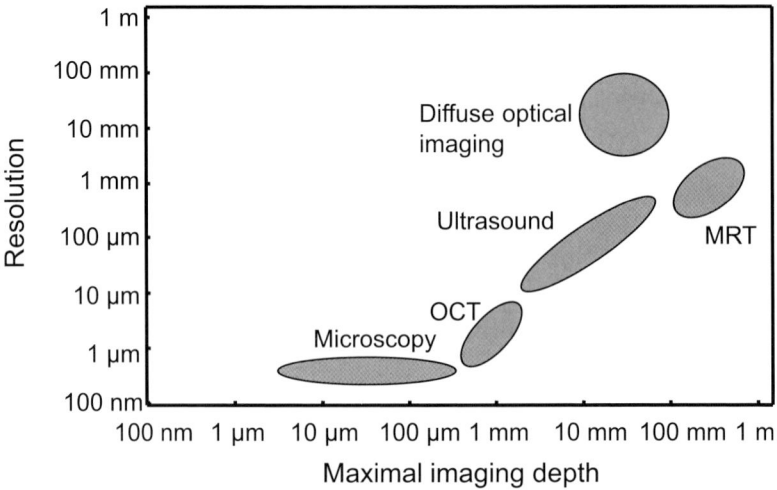

Fig. 5.1 Resolution and maximal imaging depth of different imaging modalities (modified after [5])

here it is now spreading to other medical specialties [4, 11, 19, 20, 24, 25, 31, 40, 48, 57, 67, 74] in search of diagnostic applications and to non-medical fields, which include displacement measurements [50], investigation of scattering polymer composites [17], as well as paints and coatings [71]. Within medicine, OCT has established itself as a valuable imaging modality with a combination of resolution and imaging depth, which is unmatched by other methods. These features will also be of value for biotechnological applications when OCT is used for 3-dimensional imaging of cultured cells and tissues in research as well as in process control.

This chapter will address the theory of OCT and emerging applications in tissue engineering. Excellent reviews on OCT technology and medical applications have been published [5, 21, 53, 64]. Here we will concentrate on the basic theory of OCT in order to give the reader a feeling what he can expect from the OCT technology. In the second part, the use of OCT in the field of tissue engineering will be reviewed.

5.2 Optical Coherence Tomography (OCT)

5.2.1 The Basic Principle

OCT was developed in an endeavor to increase the depth resolution in 3-dimensional medical imaging over the depth of focus of the imaging device, which is limited by the maximal focusing angle or more precisely the numerical aperture (NA). For classical imaging transverse resolution Δx and longitudinal resolution Δz scale differently with the NA [33]

$$\Delta x \propto \frac{\lambda}{\text{NA}}$$
$$\Delta z \propto \frac{\lambda}{\text{NA}^2} \quad . \tag{5.1}$$

Δx decreases linearly, Δz goes with the square. Only for a large NA in the range of 1.0 to 1.3, the focal volume has comparable lateral and longitudinal dimensions [65]. Reducing the NA for example to 0.05 dramatically decrease the depth resolution at 800 nm wavelength to 320 μm with moderate effects on the lateral resolution ($\Delta x = 16$ μm). 3-dimensional imaging with low NA will always result in a poor depth discrimination and a considerably better lateral resolution.

Interferometry measures distances without being limited by the focusing angle. The phase $2\pi/\lambda\, z_S$ of the field $E_S = \hat{E}_S \cos(2\pi/\lambda\, z_S)$ of the reflected light contains information on the propagation distance z_S. Unfortunately the phase can not be measured directly, but it can be revealed by an interferometer which superimposes E_S with a reference wave E_R (Fig. 5.2). Constructive and destructive interference at the output of the interferometer will result in an intensity $I(z_S, \lambda)$ which depends on z_S and on the wavelength λ

$$I(z_S, \lambda) = I_S + I_R + 2\sqrt{I_S I_R} \cos(2\pi/\lambda(z_S - z_R)) \quad . \tag{5.2}$$

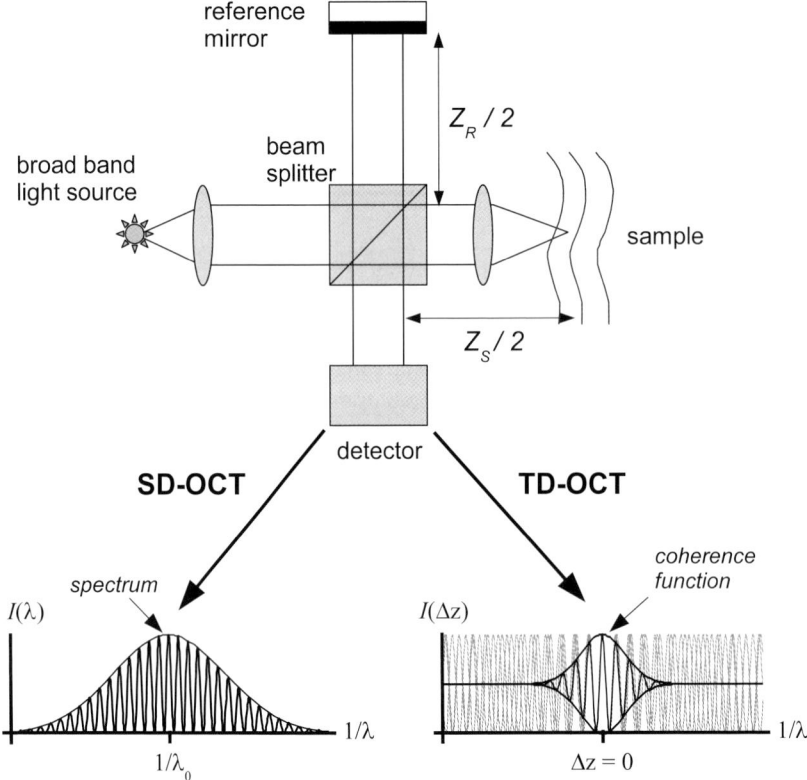

Fig. 5.2 The principle of optical coherence tomography (OCT). Interferometry with spectrally broad light source measures distances of scattering structures in tissues in order to construct images with high depth resolution. Either the spectrum, which contains the depth information as intensity modulations (SD-OCT), or the total intensity when changing z_R (TD-OCT) is measured at the output of the interferometer

With a single wavelength, the intensity $I(z_S, \lambda)$ changes periodically with z_S and a unique depth information is only obtained within the range of one wavelength. However the dependence of the phase $2\pi/\lambda(z_S - z_R)$ and the interference pattern $I(z_S, \lambda)$ on the reciprocal wavelength $1/\lambda$ gives a unique depth information. The difference of the interferometer arm lengths $z_{RS} = z_S - z_R$ is the slope of the linear relation of the phase versus $2\pi/\lambda$. There are two ways to recover the depth information from the interferogram. The first, which is called spectral-domain OCT (SD-OCT), uses a spectral resolved detection of the output of the interferometer, the second, time-domain OCT (TD-OCT) measures the wavelength integrated output as a function of the reference arm length z_R. Both use spectrally broad and therefore low coherent light sources.

In SD-OCT, the spectrum $I(z_S, \lambda)$ is measured versus $1/\lambda$. According to Eq. (5.2), the spectrum is modulated with a frequency which is proportional to the path length difference z_{RS}. Therefore the frequency content of spectrum contains the distances of all scattering structures and an A-scan can is easily reconstructed

5.2 Optical Coherence Tomography (OCT)

by a Fourier transform. SD-OCT uses either a broad band light source together with a spectrometer or a rapidly tunable laser source to record the spectrogram of the superimposed sample and reference radiation at the output of the interferometer.

SD-OCT was proposed by Fercher for ophthalmologic measurements [22] and later demonstrated with a broad band light source and a spectrometer under the name "spectral radar" by the group of Häusler [3]. Tunable laser sources were already used for distance measurements of scattering sites in fibers [18] and OCT based on rapidly wavelength-swept lasers has recently drawn under the name swept-source OCT [9] or optical frequency-domain imaging (OFDI) [12] the interest of several groups, because extremely high A-scan rates are possible.

SD-OCT can measure only path length differences, i.e. it cannot discriminate between positive and negative z_{RS}. Additionally self interference from strong reflections inside the tissue causes spurious signals which may obscure the image information.

These problems can be circumvented by the TD-OCT principle, which was the traditional OCT approach early proposed for high resolution imaging of retinal structures [23, 60]. If the length of the reference arm is changed, each wavelength causes a periodically intensity modulation at the output of the interferometer due to constructive and destructive interference. The intensity modulations have different frequencies for different optical wavelengths λ. They are only in phase and can provide a measurable modulation at nearly equal arm lengths ($z_{RS} \approx 0$). For larger z_{RS} the modulation is lost due to averaging over the different phase-shifted wavelength components (see Fig. 5.2). The z_{RS} range, over which a modulation is observed, is called the coherence length of the light source [28]. The wavelength integrated interference pattern is described mathematically by introducing the coherence function γ, which gives the contrast of the interference pattern for a certain path length difference, and the central wavelength λ_0 of the light source:

$$I(\Delta z) = I_S + I_R + 2\sqrt{I_S I_R}\gamma(z_{RS})\cos(2\pi/\lambda\ z_{RS}) \ . \tag{5.3}$$

Since γ and the power spectrum of the radiation are Fourier pairs, a broad emission spectrum of the light source corresponds to a short temporal coherence length, i.e. an narrow coherence function. A-scans are measured by moving the "coherence gate" formed by γ trough the sample when the reference arm is scanned over the desired depth range. Back reflections from the sample are identified by measuring the modulation of the output signal. From the modulation amplitude the scattering intensity, from the reference arm length at maximal modulation the depth are inferred.

Since the interference pattern is usually measured in a time-dependent fashion as z_R is continuously changed, this OCT approach was named time domain OCT. TD-OCT does not suffer from the signal ambiguities or the self interference of sample signals which are known from SD-OCT. However the sensitivity is significantly reduced because only a fraction of the photons returning from the sample contribute to the modulation and therefore to the OCT signal. A further technological problem of fast TD-OCT is the delay line in the reference arm. For each A-scan the optical path length has to be changed with constant velocity. Large affords were made to build fast delay lines and A-scan rates of more than 4000 per second were achieved by rotating cubes [5] or gratings [51]. As a way to overcome moving parts in TD-OCT it was suggested to measure the whole interference pattern in parallel with a line detector [27, 35, 36].

At the exit of the interferometer the backscattered photons give rise to a signal which – according to Eq. (5.2) – consists of the background $I_S + I_R$ and an information carrying modulation with the amplitude $A = 2\sqrt{I_S I_R}$. The interference with the reference wave provides effectively an amplification of the signal from the probe, which is able to lift the modulation A out of the detector noise. Under optimized conditions nearly shot noise limited performance is reached. For a quantification of the sample intensity, the modulation amplitude is squared and displayed on a logarithmic scale.

5.2.2 Image Formation

Image formation in OCT is similar to ultrasound. A measurement of a depth scan at certain position of the tissue gives the A-scan. By scanning the beam or translating the sample in one direction B-scans (x–z image), by scanning in two directions C-scans (3-dimensional x–y–z image stacks) are constructed. For TD-OCT there is also the possibility to acquire en-face OCT images by fast lateral scanning at fixed reference arm positions. Images at different z_R are afterwards combined to a virtual depth scan.

Tissue samples scramble phase of reflected and transmitted light due to a random arrangement of scatters. On the detector waves with statistical phases differences will interfere and produce random intensity fluctuations which are usually called speckles [10]. When superimposed with the reference beam a predictable phase relation exists only inside of each speckle. If the detector is larger than the average speckle size, which corresponds to the diffraction limited resolution of the imaging system, averaging over a number of speckles will decrease the modulation and destroy the depth information. Therefore the detector diameter has to be reduced to less than one speckle, which effectively results in a confocal detection. In general, interferometry with rough or scattering objects requires detection, and, if good light efficiency is anticipated, also illumination with high spatial coherence, i.e. small emitting and detecting areas are needed. Mono mode fibers which are commonly used in the OCT interferometer automatically fulfill this condition.

When the tissue is sampled the speckle pattern moves over the detector and modulates the measured signal. Therefore the OCT image are degraded by a salt and paper noise with a contrast of approximately 0.5 [21]. The average sizes of the speckle grains corresponds in lateral and longitudinal directions to the diffraction limited resolution [56]. Speckle noise is multiplicative and drastically reduces the resolution, especially for low contrast objects. It is one of the main disadvantages of OCT compared to confocal or other incoherent imaging modalities. Several approaches, mostly based on averaging a number of images with uncorrelated speckle patterns, were proposed to improve the image quality [55].

In summery, OCT combines low NA confocal microscopy with coherence gating, which provides a drastically increased depth resolution and rejection of scattered photons. Additionally a fast z-scan is possible since no movements of the sample or the optics are involved, and the interferometric detection enables nearly shot noise limited sensitivity. However on the flip-side of the coin we have a degradation of the images by speckle noise, which is not present in incoherent imaging.

5.2.3 Figure of Merits for the Performance of OCT

An OCT device is composed of four main components: light source, interferometer, detector, and application system. The overall performance depends on all four components. The impressive progress of OCT in the last years was only possible by technological developments in all four fields. For comparison of different OCT devices the following figures of merit are usually used.

5.2.3.1 Depth Resolution

In SD-OCT the depth resolution Δz_{OCT} depends on the accuracy by which the spectral modulation frequency can be determined. The broader the spectrum, the more fringes can be used and the smaller is the error. In TD-OCT Δz_{OCT} is the width of the coherence function γ, which is directly connected to the spectral width $\Delta \lambda$ of the light source. Hence for both SD- and TD-OCT depth resolution is inverse proportional to $\Delta \lambda$ and is described by the same equation:

$$\Delta z_{OCT} \approx \frac{\lambda_0^2}{\Delta \lambda}. \qquad (5.4)$$

Interestingly the depth resolution increases with the square of the center wavelength λ_0 which favors shorter wavelengths for higher resolution. Unfortunately shorter wavelengths have to fight with stronger tissue scattering and reduced tissue penetration.

5.2.3.2 Measurable Depth Range

In SD-OCT a large depth of the imaged tissue structure results in high frequency modulations of the spectrum (see Eq. (5.2)). Therefore the spectral resolution $\delta \lambda$ of the spectrometer or the instantaneous line width of the swept laser source determine the maximal depth range z_{max} which can be covered

$$z_{max} \approx \frac{1}{4} \frac{\lambda_0^2}{\delta \lambda}. \qquad (5.5)$$

In TD-OCT the pathlength changes, which can be realized in the reference arm limit z_{max}. Therefore available technology, e. g. stability of the reference arm and, for a given A-scan rate, the velocities by which z_R can be changed, limit z_{max}. If the fringe pattern is recorded by a line detector essentially the number of detector elements defines the depth range [35]. In practice the measuring depth is also limited by the depth of focus of the imaging optics. However, this limitation can be overcome by dynamic focus tracking [54], image fusion techniques [15], or special imaging optics [39].

5.2.3.3 Signal to Noise Ratio

The OCT signal is the squared modulation of the measured signal, which is proportional to both sample and reference intensity. Noise consists of three main parts: the shot noise of the sample intensity, detector noise, and relative intensity noise (RIN) which increases with the reference intensity [58]. The signal to noise ratio (SNR) is optimal, when the reference intensity I_R is that large that due to the multiplication in the modulation term shot noise from the sample intensity I_S dominates the two other noise components [52]. In this case the SNR equals the number of detected sample photons. Under optimal conditions OCT is able to detect nearly every photon from the sample. However, if the reference intensity is too high, RIN will degrade the performance because it increases stronger with I_R than the OCT signal [49].

5.2.3.4 Sensitivity

The sensitivity S of an OCT device is defined as the minimal reflectivity in the sample which can be measured at a SNR of one. At the shot noise limit, the sensitivity is given by the reciprocal of the number of photons falling on the sample during a single pixel integration time τ multiplied by the detection efficiency of the detector. For SD-OCT τ is the inverse of the A-scan rate. All photons from all reflecting structures give rise to a modulation of the spectrum which is detected during the full A-scan acquisition time. In time domain OCT photons reflected from a certain structure cause a modulation and hence an OCT signal only within the coherence gate. Photons outside the coherence gate are lost for OCT. Therefore the effective integration time is reduced be the ratio between z-range and z-resolution ($z_{max}/\Delta z_{OCT}$). Usually there is an advantage of 100 to 1000 times in SNR and sensitivity for SD-OCT compared to TD-OCT, which can be used for higher speed, lower optical power on the sample, or higher effective imaging depth.

5.2.3.5 A-Scan Rate

The A-scan rate determines not only the imaging speed, but also the sensitivity to sample movements. Due to the interferometric nature of OCT even small sample displacements cause fringe washout in the interference pattern. During the integra-

5.2 Optical Coherence Tomography (OCT)

tion time sample movements have to stay below one wavelength. The parallel detection make spectral radar and TD-OCT with a line detector especially susceptible to fringe washout, because the integration time is the full A-scan acquisition time [77]. High A-scan rates are preferable for rapid measurements of large imaged fields or volumes, but in general are limited by the maximal attainable light intensity on the sample, the data acquisition time, and the sweep rate of the reference arm (TD-OCT) or of the laser source (swept source OCT). With respect to imaging speed SD-OCT is today far superior to TD-OCT due to a higher SNR and the lack of moving parts in the interferometer. Line detectors suitable for spectral radar can support more than 30,000 A-scans per second. Swept laser sources with more 300,000 A-scans/second were successfully used for OCT [30]. With that high measurement rates volumes of $1000 \times 1000 \times 1000$ pixels are acquired within 3 seconds.

5.2.3.6 Effective Imaging Depth

The effective imaging depth, i.e. the maximal depth in which structures can be visualized with a certain SNR or image quality, depends not only on the OCT device but also on the tissue investigated. Main limiting factors are the sensitivity S and attenuation coefficient μ_{eff}. As a rule of the thumb the amount of back reflected light scales according to Lambert–Beer's law and the imaging depth scales with

$$z_{\text{eff}} \propto \frac{\ln(S)}{2\mu_{\text{eff}}}. \tag{5.6}$$

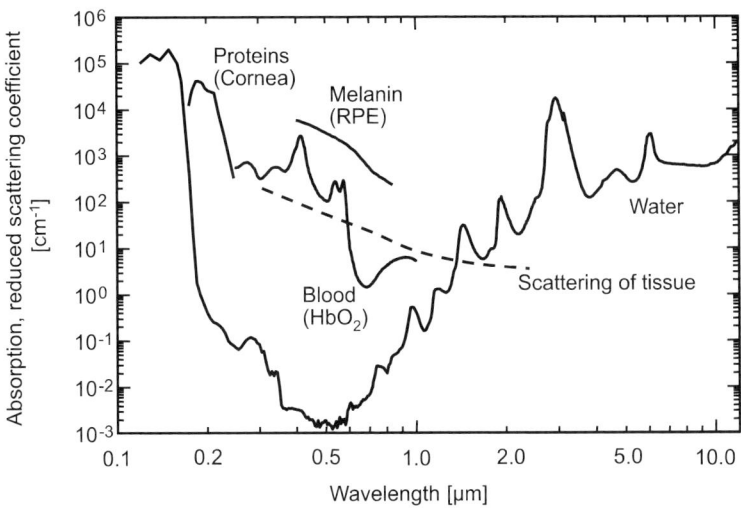

Fig. 5.3 The wavelength dependence of absorption (*solid line*) and scattering (*dashed line*) of biological tissue (modified after [66])

Light attenuation by the tissue depends on absorption as well as on scattering. Below a wavelength of 700 nm the absorption of proteins, blood, and melanin is the limiting factor (Fig. 5.3). Between 700 nm and 1500 nm, a range which is called the optical window of tissue, scattering is dominant until above 1500 nm strong water absorption kicks in again. In general, z_{eff} increases for scattering biological tissue from 400 nm to 1400 nm with increasing wavelength.

Not only a lack of returning photons limits the imaging depth. Depending on the scattering properties of the tissue, there is a certain chance for multiple scattered photons to reach the detector [47]. This gives rise to a background signal which reduces the image contrast in deeper layers.

5.2.4 Functional Imaging

OCT only detects scattered or reflected light which primarily carries information on tissue morphology. However, there is a strong desire in imaging tissue functions. Therefore any OCT imaging which relies on additional parameters of the detected light is dubbed functional OCT imaging.

5.2.4.1 Spectroscopic OCT

Light reemitted from the tissue may not only be changed in its phase (according to the propagation time), but also in its spectral shape by wavelengths dependent reflection, scattering, or absorption [69]. The reemitted light therefore carries spectroscopic information about the tissue. For recovering this information either the spectrum of the OCT light source is divided (physically or mathematically) in different subbands which are than used for separate reconstructions of OCT images in the different spectral ranges or an OCT interferometer with several light sources of different wavelength is used [32,45,70]. Spectroscopic OCT can be used to visualize the main tissue chromophores water, blood, haemoglobin, and melanin.

5.2.4.2 Polarization Sensitive OCT

When tissue is illuminated with polarized light the birefringence of certain tissues or depolarizing scattering will alter the polarization state of the reemitted light [13]. An OCT with two orthogonal polarization channels can detect the residual degree of depolarization and the rotation of the polarization axis in a depth resolved fashion. Polarization contrast was successfully demonstrated for skin, cornea (collagen), and the retinal nerve fiber layer [14].

5.2.4.3 Doppler OCT

Flowing of blood or other movements of tissue structures introduce small frequency shifts in the reemitted radiation (Doppler shifts), which are proportional to the velocity [43]. When interfering on the detector these Doppler shifts lead to a time-dependent change of the interference pattern. Velocities can be calculated when the phases of the interference patterns at two time points, which are separated by a certain delay, are compared. The minimal measurable velocity depends on the SNR of the OCT signal, the delay time, and phase noise in the interferometer. Spectral radar and TD-OCT with a linear detector are especially suited for Doppler OCT because they possess a very high phase stability due to the lack of moving parts, which are prone to phase jitter. The maximal measurable velocity is limited by an ambiguity as the phase difference exceeds 2π. This limit can be extended by special algorithms used in ultrasound Doppler imaging [75]. By choosing an adequate time delay between the phase measurements, OCT can be adapted to certain velocity ranges. For example, at 10,000 A-scans per second a range from $10\,\mu m/s$ to $2\,mm/s$ was demonstrated for a spectral radar system [38].

5.2.4.4 Contrast Agents for OCT

In microscopy the use of fluorescent contrast agents has dramatically improved the information obtained from tissues and cells. OCT visible markers which could be functionalized with antibodies or other molecules for selective targeting would enlarge the applications of OCT enormously. Either a strong absorption or an efficient scattering is needed to make visible changes in the OCT images. Due to their high cross-sections micro- and nanoparticles (e. g. microbubbles, microcrystals, and gold nanoparticles) are especially attractive [1, 37]. Magnetically moved particles were also demonstrated as a contrast agent. They use the high phase sensitivity of OCT to generate a Doppler contrast [46]. This ongoing research field may in future lead to a molecular contrast OCT [73].

5.3 Applications in Tissue Engineering

Cell-based biotechnology, i. e. stem-cell research, industrial cell culture, and tissue engineering still uses technologies developed in biological research labs. These are characterized by labor-intensive handling of cells und tissues, a lack of standardization, and a limited control of the involved processes. In contrast to established industrial production, the manufacturing of differentiated cell lines or specialized tissues is more complex and usually not fully understood. High quality standardized products can not simply be guaranteed by a strict control of all process parameter, but often the literally "right touch" of the lab technician is necessary for success. This calls for a tight supervision of cell or tissue growth during production. Optical imaging is especially attractive, because it is well established in biology and medicine, and a not-too-expensive, non-destructive technology, which works without contact to the sample [41, 42]. In principle an on-line monitoring of cell and tissue growth in a high throughput fashion is possible. For cell culture and thin samples below 200 µm thickness, confocal or two-photon microscopy give high resolution 3-dimensional images. However, for thicker tissues the penetration depth of the classical microcopy is too low. Conceptually OCT can fill the gap between microscopy on one side and ultrasound or MRT on the other side with an imaging depth of a 1–2 millimeters. In this thickness range are a wide selection of important tissues like epithelial tissues (skin, mucous membranes), cartilage, vessels, heart valves, bladder or lung tissue.

Astonishingly, rather few application of OCT in tissue engineering, all in very early states, were published [2, 16, 34, 41, 61, 62, 72, 76]. They can be assigned into three mayor groups:

5.3.1 Visualization of Cells in Scaffolds

Tan et al. used an 800 nm OCT and CM to investigate GFP transfected fibroblasts in chitosan scaffolds [62]. Cell-free and cell-filled pores of the scaffold and changes in cell density were observed over nine days. These changes included an inhomogeneous distribution of cells and a blockage of superficial pores, which may have led to a deprivation of deeper lying cells. The combination with CM gave additional information on the GFP stained cytoskeleton. In a similar work the OCT was used to calculate the occupation of chitosan microchannels by the implanted cells [2]. Here CM imaging was limited to a depth of 150 µm. With a 1300 nm OCT bone cell-line in a poly(l-lactic acid) (PLLA) matrix were imaged [76]. The structure of the empty scaffold and the cells inside the scaffold were observed down to a depth of 1 mm. A direct discrimination between cells and scaffold was not possible even though microparticles with a high refractive index were used as a scattering contrast agent. Therefore OCT served for larger volume, larger depth observations, whereas microscopy gave high resolution images of individual cells, which were easily distinguishable with microscopy from the scaffold down to a depth of a few hundred micrometer.

A recently published work demonstrates that high resolution OCT can also visualize cell migration, cell proliferation, cell adhesion, and cell matrix interaction in 3-dimensional tissue models [61].

5.3.2 Visualization of the Morphology of Artificially Grown Tissues

OCT is not only useful for investigating and improving the cell growth and cells interaction in scaffolds. An important application of OCT may also be the supervision of the growth of artificial tissue at a production stage. In-vitro grown cartilage samples were successfully monitored by OCT. Inhomogeneities and the formation of

5.3 Applications in Tissue Engineering

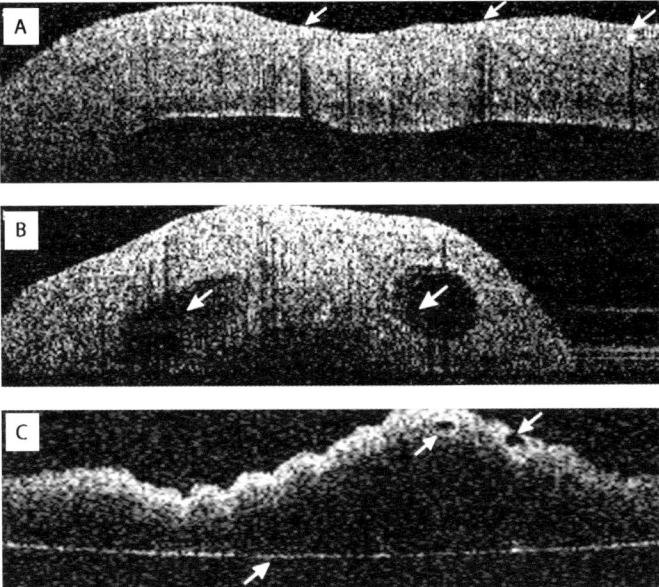

Fig. 5.4 OCT image of in-vitro grown cartilage tissue, which is used for transplantation (image size 6 mm × 1.6 mm). **A** Cartilage sample of medium quality having some inhomogeneities at the surface (*arrows*). **B** Cartilage, which was previously judged of having a medium quality by superficial inspection. Large hidden cavities were seen with OCT (*arrows*). **C** Cartilage in an early growth stage with beginning cavity formation (*arrows*)

cavities were detected in early stages of cultivation (Fig. 5.4). In proof-of-principle experiment OCT was used to image a model for an artificial coronary artery in a dedicated bioreactor [41]. Tissue thickness and, via Doppler-OCT, flow profiles in the lumen were quantified. The use of OCT for monitoring the uniformity of the wall thickness, and the epithelization of the inner surface at the end of the growing process were proposed. Additionally, Doppler OCT may provide valuable information on the state of flow, i. e. turbulent, transitional, or laminar flow.

Successful visualization of the different layers of engineered skin equivalents including stratum corneum, epidermal and dermal layer as well as the basement membrane zone was demonstrated with a high resolution OCT [59].

5.3.3 Measurement of Tissue Function

The ultimate goal of OCT imaging is to include information about tissue function into the OCT images. Unfortunately spectral information is hidden under tissue scattering which exceeds absorption by far in the optical window. With the exception of water, tissues have no strong absorption differences in the spectral range of 750 nm to 1500 nm, which is used for OCT. Though, changes of the OCT signal were reported after deliberately altering the redox state of cytochrome oxidase in 3-dimensional cultured cells [72]. However, these changes were not consistent with biochemical measurements of the redox state. Probably parallel changes in the scattering obscured the absorption changes. A strong influence of tissue scattering on the OCT signals was also observed, when spectroscopic OCT at 1300 nm and 1500 nm was used to quantify the water content of skin [68].

The high phase sensitivity of OCT provides different ways for functional imaging. Doppler OCT, which was already used successfully in medical diagnosis [7], is a means to probe function of artificial vascular structures [40]. Additionally, small displacements in the tissue can be measured and quantified by OCT even though they lead only to minute changes of the phase or speckle position. Ko et al. used a speckle correlation technique to visualize mechanical properties of engineered tissues by monitoring deformations in response to external forces [34]. Differences in microscopic displacements and shear stress were observed during the growth process which could be related to changing cell–cell and cell–matrix adhesion.

5.4 Conclusion

OCT provides unique imaging possibilities which will by useful for monitoring cell and tissue growth. However, up to now this potential was barely exploited. The reason may lie partly in former limitations of the OCT technology which used to be quite complicated, expensive, and slow for 3-dimensional imaging and partly in lack of understanding of the OCT technology by researchers and companies involved in cell and tissue engineering. Last but not least the limited availability of OCT technology prohibited wider use. We anticipate, that this will change as technology and application of OCT in other fields progress rapidly. Recently OCT devices with several hundred thousands A-scans were demonstrated, prices below 10,000 $ are possible, and general purpose OCT devices are now commercially available [63]. Progress in functional OCT imaging (polarization sensitive and Doppler OCT, contrast agents) will further enlarge the applicability of OCT in tissue engineering.

Acknowledgements We like to acknowledge support by the Government of Schleswig-Holstein (ISH 2004-12-HWT) and the German Ministry of Research (InnoNet 16IN0250) and the help of Peter Koch (Thorlabs-HL AG) and Ralf Engelhardt (Heidelberg Engineering AG).

References

1. Agrawal A, Huang S, Wei Haw Lin A, Lee MH, Barton JK, Drezek RA, Pfefer TJ (2006) Quantitative evaluation of optical coherence tomography signal enhancement with gold nanoshells. J Biomed Opt 11:041121
2. Bagnaninchi PO, Yang Y, Zghoul N, Maffulli N, Wang RK, Haj AJ (2007) Chitosan microchannel scaffolds for tendon tissue engineering characterized using optical coherence tomography. Tissue Eng 13:323–331
3. Bail MA, Haeusler G, Herrmann JM, Lindner MW, Ringler R (1996) Optical coherence tomography with the "spectral radar": fast optical analysis in volume scatterers by short-coherence interferometry. In: Benaron DA, Chance B, Mueller GJ (eds) Photon Propagation in Tissues II. SPIE pp 298–303
4. Böhringer HJ, Boller D, Leppert J, Knopp U, Lankenau E, Reusche E, Hüttmann G, Giese A (2006) Time-domain and spectral-domain optical coherence tomography in the analysis of brain tumor tissue. Lasers in Surgery and Medicine 38:588–597
5. Bouma BE, Tearney GJ (2002) Handbook of Optical Coherence Tomography. Marcel Dekker, New York Basel
6. Branzan AL, Landthaler M, Szeimies RM (2006) In vivo confocal scanning laser microscopy in dermatology. Lasers Med Sci 10.1007/s10103–006-0416–8:online first
7. Cense B, Chen TC, Nassif N, Pierce MC, Yun SH, Park BH, Bouma BE, Tearney GJ, de Boer JF (2006) Ultra-high speed and ultra-high resolution spectral-domain optical coherence tomography and optical Doppler tomography in ophthalmology. Bull Soc Belge Ophtalmol 302:123–132
8. Chiou AG, Kaufman SC, Kaufman HE, Beuerman RW (2006) Clinical corneal confocal microscopy. Surv Ophthalmol 51:482–500
9. Choma MA, Sarunic MV, Yang CH, Izatt JA (2003) Sensitivity advantage of swept source and Fourier domain optical coherence tomography. Opt express 11:2183–2189
10. Dainty JC (1984) Laser Speckle and Related Phenomena. Springer, New York
11. Daniltchenko D, König F, Lankenau E, Sachs M, Kristiansen G, Hüttmann G, Schnorr D (2006) Anwendung der optischen Kohärenztomographie (OCT) bei der Darstellung von Urotherlerkrankungen der Harnblase. Radiologe DOI 10.1007/s00117–005-1250-x (on-line first)
12. de Boer JF, Cense B, Park BH, Pierce MC, Tearney GJ, Bouma BE (2003) Improved signal-to-noise ratio in spectral-domain compared with time-domain optical coherence tomography. Opt Lett 28:2067–2069
13. de Boer JF, Milner TE (2002) Review of polarization sensitive optical coherence tomography and Stokes vector determination. J Biomed Opt 7:359–371
14. de Boer JF, Srinivas SM, Nelson JS, Milner TE, Ducros MD (2002) Polarization-Sensitive Optical Coherence Tomography. In: Bouma BE, Tearney GJ (eds) Handbook of Optical Coherence Tomography Marcel Dekker, New York, Basel, pp 237–274

15. Drexler W, Morgner U, Kärtner FX, Pitris C, Boppart SA, Li XD, Ippen EP, Fujimoto JG (1999) Invivo ultrahigh-resolution optical coherence tomography. Opt Lett 24:1221–1223
16. Dunkers J, Cicerone M, Washburn N (2003) Collinear optical coherence and confocal fluorescence microscopies for tissue engineering. Opt express 11:3074–3079
17. Dunkers JP, Parnas RS, Zimba CG, Peterson RC, Flynn KM, Fujimoto JG, Bouma BE (1999) Optical coherence tomography of glass reinforced polymer composites. Composites Part A: Applied Science and Manufacturing 30:139–145
18. Eickhoff W, Ulrich R (1981) Optical frequency domain reflectometry in single-mode fiber. Appl Phys Lett 39:693–695
19. Escobar PF, Belinson JL, White A, Shakhova NM, Feldchtein FI, Kareta MV, Gladkova ND (2004) Diagnostic efficacy of optical coherence tomography in the management of preinvasive and invasive cancer of uterine cervix and vulva. Int J Gynecol Cancer 14:470–474
20. Familiari L, Strangio G, Consolo P, Luigiano C, Bonica M, Barresi G, Barresi V, Familiari P, D'Arrigo G, Alibrandi A, Zirilli A, Fries W, Scaffidi M (2006) Optical coherence tomography evaluation of ulcerative colitis: the patterns and the comparison with histology. Am J Gastroenterol 101:2833–2840
21. Fercher AF, Drexler W, Hitzenberger CK, Lasser T (2003) Optical coherence tomography – principles and applications. Rep Prog Phys 66:239–303
22. Fercher AF, Hitzenberger C, Juchem M (1991) Measurement of intraocular optical distances using partially coherent laser light. J Mod Opt 38:1327–1333
23. Fercher AF, Hitzenberger CK, Drexler W, Kamp G, Sattmann H (1993) In vivo optical coherence tomography. Am J Ophthalmol 116:113–114
24. Geerling G, Muller M, Winter C, Hoerauf H, Oelckers S, Laqua H, Birngruber R (2005) Intraoperative 2-dimensional optical coherence tomography as a new tool for anterior segment surgery. Arch Ophthalmol 123:253–257
25. Giese A, Böhringer HJ, Leppert J, Kantelhardt SR, Lankenau E, Koch P, Birngruber R, Hüttmann G (2006) Non-invasive intraoperative optical coherence tomography of the resection cavity during surgery of intrinsic brain tumors. In: Kollias N, Zeng H, Choi B, Malek RS, Wong BJ, Ilgner JF, Trowers EA, Riese WTd, Gregory KW, Tearney GJ, Marcu L, Hirschberg H, Madsen SJ, Lucroy MD, Tate LP (eds) Photonic Therapeutics and Diagnostics II. SPIE, San Jose, California USA, pp 495–502
26. Goetz M, Hoffman A, Galle PR, Neurath MF, Kiesslich R (2006) Confocal laser endoscopy: new approach to the early diagnosis of tumors of the esophagus and stomach. Future Oncol 2:469–476
27. Hauger C, Worz M, Hellmuth T (2003) Interferometer for optical coherence tomography. Appl Opt 42:3896–3902
28. Hecht E (2001) Optics. Addison Wesley
29. Huang D, Swanson E, Lin C, Schuman JS, Stinson WG, Chang W, Hee MR, Flotte T, Gregory K, Puliafito CA, Fujimoto JG (1991) Optical Coherence Tomography. Science 254:1178
30. Huber R, Adler DC, Fujimoto JG (2006) Buffered Fourier domain mode locking: unidirectional swept laser sources for optical coherence tomography imaging at 370,000 lines/s. Opt Lett 31:2975–2977
31. Jackle S, Gladkova N, Feldchtein F, Terentieva A, Brand B, Gelikonov G, Gelikonov V, Sergeev A, Fritscher-Ravens A, Freund J, Seitz U, Soehendra S, Schrodern N (2000) In vivo endoscopic optical coherence tomography of the human gastrointestinal tract – toward optical biopsy. Endoscopy 32:743–749
32. Jeon SW, Shure MA, Rollins AM, Huang D (2004) Corneal hydration imaging using dual-wavelength optical coherence tomography In: Tuchin VV, Izatt JA, Fujimoto JG (eds) Coherence Domain Optical Methods and Optical Coherence Tomography in Biomedicine VIII. SPIE, Vol 5316, pp 113–118
33. Jonkman JEN, Stelzer EHK (2002) Resolution and contrast in confocal and two-photon microscopy. In: Diaspro A (ed) Confocal and Two-photon Microscopy: Foundations, Applications, and Advances. Wiley-Liss, Inc., New York, pp 101–125
34. Ko HJ, Tan W, Stack R, Boppart SA (2006) Optical coherence elastography of engineered and developing tissue. Tissue Eng 12:63–73

References

35. Koch P, Hellemanns V, Hüttmann G (2006) Linear OCT System with extended measurement range. Opt Lett 31:2882–2884
36. Koch P, Hüttmann G, Schleiermacher H, Eichholz J, Koch E (2004) Linear optical coherence tomography system with a downconverted fringe pattern. Opt Lett 29:1644–1646
37. Lee TM, Oldenburg AL, Sitafalwalla S, Marks DL, Luo W, Toublan FJ-J, Suslick KS, Boppart SA (2003) Engineered microsphere contrast agents for optical coherence tomography. Opt Lett 28:1546–1548
38. Leitgeb R, Schmetterer L, Wojtkowski M, Hitzenberger CK, Sticker M, Fercher FA (2002) Flow Velocity Measurements by Frequency Domain Short Coherence Interferometry. In: Tuchin VV, Izatt JA, Fujimoto JG (eds) Coherence Domain Optical Methods in Biomedical Science and Clinical Applications VI. SPIE, Vol 4619, pp 16–21
39. Leitgeb RA, Villiger M, Bachmann AH, Steinmann L, Lasser T (2006) Extended focus depth for Fourier domain optical coherence microscopy. Optics Letters 31:2450–2452
40. Löning M, Lankenau E, Strunck C, Krokowski M, Hillbricht S, Diedrich K, Hüttmann G (2003) Optische Kohärenztomographie – ein neues hochauflösendes Schnittbildverfahren als Ergänzung zur Kolposkopie. Geburtshilfe Frauenheilkd 63:1158–1161
41. Mason C, Markusen JF, Town MA, Dunnill P, Wang RK (2004) Doppler optical coherence tomography for measuring flow in engineered tissue. Biosens Bioelectron 20:414–423
42. Mason C, Markusen JF, Town MA, Dunnill P, Wang RK (2004) The potential of optical coherence tomography in the engineering of living tissue. Phys Med Biol 49:1097–1115
43. Milner TE, Izatt JA, Yazdanfar S, Rollins AM, Lindmo T, Chen Z, Nelson JS, Wang X (2002) Doppler Optical Coherence Tomography. In: Bouma BE, Tearney GJ (eds) Handbook of Optical Coherence Tomography Marcel Dekker, New York, Basel, pp 203–236
44. Minsky M (1961) Microscopy apparatus. Patent US 3,013,467
45. Morgner U, Drexler W, Kärtner FX, Li XD, Pitris C, Ippen EP, Fujimoto JG (2000) Spectroscopic optical coherence tomography. Opt Lett 25:111–113
46. Oldenburg AL, Toublan FJ-J, Suslick KS, Wei A, Boppart SA (2005) Magnetomotive contrast for in vivo optical coherence tomography. Opt express 13:6597–6614
47. Pan Y, Birngruber R, Engelhardt R (1997) Contrast limits of coherence-gated imaging in scattering media. Appl Opt 36:2979–2983
48. Pinto TL, Waksman R (2006) Clinical applications of optical coherence tomography. J Intervent Cardiol 19:566–573
49. Podoleanu AG, Jackson DA (1999) Noise Analysis of a Combined Optical Coherence Tomograph and a Confocal Scanning Ophthalmoscope. Appl Opt 38:2116–2127
50. Rao YJ, Ning Y, Jackson DA (1993) Synthesized source for white-light sensing systems. Opt Lett 18:462
51. Rollins A, Yazdanfar S, Kulkarni M, Ung-Arunyawee R, Izatt J (1998) In vivo video rate optical coherence tomography. Opt express 3:219–229
52. Rollins AM, Izatt JA (1999) Optimal interferometer designs for optical coherence tomography. Opt Lett 24:1484–1486
53. Schmitt JM (1999) Optical coherence tomography (OCT): a review. IEEE Journal of Selected Topics in Quantum Electronics 5:1205–1215
54. Schmitt JM, Knüttel A (1997) Model of optical coherence tomography of heterogeneous tissue. J Opt Soc Am A 14:1231–1242
55. Schmitt JM, Xiang SH, Yung KM (1999) Speckle in Optical Coherence Tomography. J Biomed Opt 4:95–105
56. Schmitt JM, Xiang SH, Yung KM (2002) Speckle Reduction Techniques. In: Bouma BE, Tearney GJ (eds) Handbook of Optical Coherence Tomography Marcel Dekker, New York, Basel, pp 175–201
57. Sommer K, Lankenau E, Thorns C, Fedder C, Hüttmann G, Wollenberg B (2004) Optical coherence tomography – a new non-invasive high resolution imaging for cancer and precancerous lesions of the upper aerodigestive tract. In: Papaspyrou (ed) 5th European Congress of Oto-Rhino-Laryngology Head and Neck Surgery. Medimond, Bologna, pp 401–404
58. Sorin WV, Baney DM (1992) A simple intensity noise reduction technique for optical low-coherence reflectometry. IEEE Photonics Technology Letters 4:1404–1406

59. Spoler F, Forst M, Marquardt Y, Hoeller D, Kurz H, Merk H, Abuzahra F (2006) High-resolution optical coherence tomography as a non-destructive monitoring tool for the engineering of skin equivalents. Skin Res Technol 12:261–267
60. Swanson EA, Izatt JA, Hee MR, Huang D, Lin CP, Schuman JS, Puliafito CA, Fujimoto JG (1993) In-vivo retinal imaging by optical coherence tomography. Opt Lett 18:1864–1866
61. Tan W, Oldenburg AL, Norman JJ, Desai TA, Boppart SA (2006) Optical coherence tomography of cell dynamics in three-dimensional tissue models. Opt express 14:7159–7171
62. Tan W, Sendemir-Urkmez A, Fahrner LJ, Jamison R, Leckband D, Boppart SA (2004) Structural and functional optical imaging of three-dimensional engineered tissue development. Tissue Eng 10:1747–1756
63. Thorlabs (2006) Spectral Radar OCT Imaging System. In: http://www.thorlabs.com
64. Tomlins PH, Wang RK (2005) Theory, developments and applications of optical coherence tomography. Journal of Applied Physics D: Applied Physics 38:2519–2535
65. Vogel A, Noack J, Hüttmann G, Paltauf G (2005) Mechanisms of femtosecond laser nanosurgery of cells and tissues. Appl Phys B 81:1015–1047
66. Vogel A, Venugopalan V (2003) Mechanisms of Pulsed Laser Ablation of Biological Tissues. Chem Rev 103:577–644
67. Welzel J (2001) Optical coherence tomography in dermatology: a review. Skin Res Technol 7:1–9
68. Welzel J (2006) Personal communication
69. Xu C, Carney PS, Boppart SA (2005) Wavelength-dependent scattering in spectroscopic optical coherence tomography. Opt express 13:5450–5462
70. Xu C, Kamalabadi F, Boppart SA (2005) Comparative performance analysis of time–frequency distributions for spectroscopic optical coherence tomography. Appl Opt 44:1813–1822
71. Xu F, Pudavar HE, Prasad PN, Dickensheets D (1999) Confocal enhanced optical coherence tomography for nondestructive evaluation of paints and coatings. Opt Lett 24:1808–1810
72. Xu X, Wang RK, El Haj A (2003) Investigation of changes in optical attenuation of bone and neuronal cells in organ culture or three-dimensional constructs in vitro with optical coherence tomography: relevance to cytochrome oxidase monitoring. Eur Biophys J 32:355–362
73. Yang C (2005) Molecular Contrast Optical Coherence Tomography: A Review. Photochem Photobiol 81:215–237
74. Yang VX, Tang SJ, Gordon ML, Qi B, Gardiner G, Cirocco M, Kortan P, Haber GB, Kandel G, Vitkin IA, Wilson BC, Marcon NE (2005) Endoscopic Doppler optical coherence tomography in the human GI tract: initial experience. Gastrointest Endosc 61:879–890
75. Yang VXD, Gordon ML, Qi B, Pekar J, Lo S, Seng-Yue E, Mok A, Wilson BC, Vitkin IA (2003) High speed, wide velocity dynamic range Doppler optical coherence tomography (Part I): System design, signal processing, and performance. Opt express 11:794–809
76. Yang Y, Dubois A, Qin XP, Li J, El Haj A, Wang RK (2006) Investigation of optical coherence tomography as an imaging modality in tissue engineering. Phys Med Biol 51:1649–1659
77. Yun SH, Tearney G, Boer Jd, Bouma B (2004) Motion artifacts in optical coherence tomography with frequency-domain ranging. Opt express 12:2977–2998

Chapter 6
Ultrasonic Strain Imaging and Reconstructive Elastography for Biological Tissue

Walaa Khaled[1,2] and Helmut Ermert[1,2]

[1] Institute of High Frequency Engineering, Ruhr-University Bochum, IC 6, D-44780 Bochum, Germany, walaa.khaled@rub.de
[2] Ruhr Center of Excellence for Medical Engineering, KMR eV Bochum, Germany

Abstract In the field of medical diagnosis, there is a strong need to determine mechanical properties of biological tissue, which are of histological and pathological relevance. In order to obtain non-invasively mechanical properties of tissue, we developed a real-time strain imaging system for clinical applications. The output data of this system allows an inverse approach leading to the spatial distribution of the relative elastic modulus of tissue. The internal displacement field of biological tissue is determined by applying quasi-static compression to the considered tissue. Axial displacements are calculated by comparing echo signal sets obtained prior to and immediately following a small compression, using a cross-correlation technique. Strain images representing mechanical tissue properties in a non quantitative manner are displayed in real time mode. For additional quantitative imaging, the stiffness distribution is calculated from the displacement field assuming the investigated material to be elastic, isotropic, and nearly incompressible. Different inverse problem approaches for calculating the shear modulus distribution using the internal displacement field have been implemented and compared using tissue-like phantoms. One of the important applications for diagnosis using ultrasound elastography is the coronary atherosclerosis, which is a common disease in industrialized countries. In this work some clinical in-vivo results are presented using intravascular ultrasound elastography. In the field of tumor diagnosis, the results of an ongoing clinical study with more than 200 patients show, that our real time strain imaging system is able to differentiate malignant and benign tissue areas in the prostate with a high degree of accuracy (Sensitivity $= 76\%$ and Specificity $= 89\%$). The reconstruction approaches applied to the strain image data deliver quantitative tissue information and seem promising for an additional differential diagnosis of lesions in biological tissue.

6.1 Introduction

Palpation is one of the oldest clinical examination methods and was practiced by the ancient Egyptians in 3000 BC. The first treatise in the book of the heart at the Edwin Smith and Ebers Papyrus is entitled "Beginning of the secret of the physician". Breasted 1930 described how the ancient physician followed the same steps in the process of examination as in our modern medical practice: interrogation of the patient as a first step then followed by the classical steps, inspection, palpation of the body and diseased organs. According to Sakorafas 2001, this papyrus included the first written evidence suggestive of breast cancer, which was detected by palpation. Palpation is still one of the standard screening procedures for the detection of breast, thyroid, prostate, liver abnormalities et cetera. Palpation is used to measure swelling, detect bone fracture, find and measure the pulse, or to locate changes in the pathological state of tissue and organs. However, palpation is not very accurate, because of its poor sensitivity with respect to small and deeply located lesions as well as to its limited accuracy in the morphological localization of lesions. The standard medical practice of soft tissue palpation is based on qualitative assessment of the low-frequency stiffness of tissue to define mechanical property changes of biological tissue, which represent important diagnostic information and are of histological and pathological relevance (Ophir et al. 2002). The elastic properties of soft tissues depend on their molecular building blocks and on the microscopic and macroscopic structural organization of these blocks (Fung 1993). Pathological changes are generally correlated with changes in tissue stiffness as well.

Although many researchers have proposed imaging the stiffness distribution in tissue to enhance diagnosis of cancer disease, as shown in Anderson 1977 and Wellman 1999, current medical practice routinely uses sophisticated diagnostic tests through magnetic resonance imaging (MRI), computed tomography (CT) and ultrasound (US) imaging, which cannot provide direct measure of tissue elasticity. However, early detection is one of the primary requirements of successful cancer treatment especially in breast and prostate cancer. Thus, early detection through screening methods, such as mammography (female breast), is considered central to cancer surveillance programs throughout the world. In spite of the unquestionable successes, there remains an urgent need to improve both sensitivity and specificity of cancer imaging modalities. In the USA and other developed countries, cancer is responsible for about 25% of all deaths. On a yearly basis, 0.5% of the population is diagnosed with cancer, especially breast and prostate cancer, which present about 33% of all common cancer cases for females and males respectively (Jemal et al. 2005). For example, Adenocarcinoma of the prostate is the most prevalent malignant cancer and the second cause of cancer-specific death in men. It is estimated, according to the American Red Cross Prostate Cancer Statistics, that in 2007 in the USA 218,890 men will be diagnosed (new case) and 27,050 men will die of cancer of the prostate (for details see, http://www.seer.cancer.gov/ National Cancer Institute). The probability of developing prostate cancer from birth to death is 1 to 6. Its annual incidence is approximately 185,000 in Europe, 52,200 in the UK and 56,160 in Germany (Ferlay et al. 2002). Its therapy is more effective when cancer is diagnosed at an early stage, but this carcinoma is usually asymptomatic and therefore reliable diagnostic modalities are required. Many cancers, such as cancers of the breast or in the prostate appear as extremely stiff nodules (Anderson 1977). Accurate assessment of the local extent of the disease is fundamentally important in the selection of appropriate local treatment modalities.

In order to obtain non-invasively the needed mechanical properties of tissue, we developed at the Institute of High Frequency Engineering at the Ruhr-University in Bochum a novel real-time strain imaging system for clinical applications (Pesavento et al. 2000; Lorenz et al. 2002). Initial experiences with this system in biopsy guidance and prostate cancer detection show that it is possible to detect prostate cancer with a high degree of sensitivity using real-time elastography in conjunction with conventional diagnostic methods (König et al. 2004, 2005).

Another important application for diagnosis using elastography could be the coronary atherosclerosis (de Korte 2002), which is a common disease in industrialized countries. Acute coronary syndromes are associated with a high mortality rate. They are usually caused by a sudden occlusion of the coronary lumen due to rupture of unstable plaques in the vessel wall, often with less than 50% stenosis. The use of computer tomography, X-ray or ultrasound to determine plaque morphology does not give sufficient information for determining the risk of an acute syndrome. However, the mechanical properties of vulnerable coronary plaques were shown to be different from other plaque types. Therefore, IVUS strain imaging can be an important imaging tool for risk assessment of plaques (Doyley 2001).

In this chapter, we review some relevant background material from the field of biomechanics and summarize our work in the field of elastography. We then discuss some basic principles and limitations that are involved in the calculation of displacement estimators needed to evaluate strain images or the relative elastic modulus of biological tissues. Results from biological tissues *in vitro* and *in vivo* are shown to demonstrate this point. Most of the findings were discovered by our research group and the Ruhr-University hospitals, Department of Urology, Marienhospital, Herne and the Department of Cardiology, Bergmannsheil, Bochum. Prospects and results for different applications of ultrasound elastography are shown and discussed at the end of this chapter.

6.2 Ultrasound Elastography

Elastography is an imaging technique which was developed over the past two decades for imaging soft tissue elastic modulus (Ophir et al. 1991) and may offer new information about tissue diagnosis. This technique was developed as a more quantitative alternative to manual palpation, which is commonly used to detect cancerous tissue. The goal of elastography is to use the ultrasound machine to determine mechanical properties of tissue from which a pathological reference can be proven. Elastography is based on the static deformation of a linear isotropic elastic material and generates several new kinds of images, called elastograms. We differentiate here between strain images in which only strain distributions are shown and modulus images in which the relative shear or elastic modulus of biological tissue is presented. As such, the properties of elastograms are different from the familiar properties of sonograms. Sonograms are related to the local acoustic backscattered energy from tissue components, elastograms relate to local strains.

Since the echogenicity and stiffness of tissues are not directly related (Ophir 2002), it could be expected that imaging tissue stiffness, or a related parameter such as local tissue strain, will provide new information compared to sonograms which relate to tissue morphology and architecture. Figure 6.1 shows a tissue mimicking phantom with a hard rigid body inside, which is not visible in the conventional ultrasound image, because of equal backscattering properties, but clearly seen in the strain image. These phantom results have now been confirmed in different medical applications as in the prostate diagnosis (König et al. 2005) and in the breast diagnosis (Garra et al. 1997; Hiltawsky et al. 2001; Insana et al. 2004) and in animal studies (Bilgen et al. 2003).

In the first step of elastography the internal displacement field of biological tissue is determined using an ultrasound elastography system by applying quasi-static compression to the considered tissue (Fig. 6.1). Axial displacements are calculated by comparing echo signal sets obtained prior to and immediately following less than 1% compression, using a fast phase-sensitive technique. Strain images representing mechanical tissue properties in a non-quantitative manner are displayed in real time.

Tissue strain imaging elastography methods based on ultrasound fall currently into three main groups:

1. Methods where a quasi-static compression is applied to the tissue and the resulting components of the strain tensor are estimated (Ophir et al. 1991; O'Donnell et al. 1994).
2. Methods where a low frequency vibration (<1 kHz) is applied to the tissue, and the resulting tissue behavior is inspected by ultrasonic or audible acoustic means (Lerner and Parker 1987; Lerner et al. 1990; Fatemi et al. 1999; Taylor et al. 2000), and

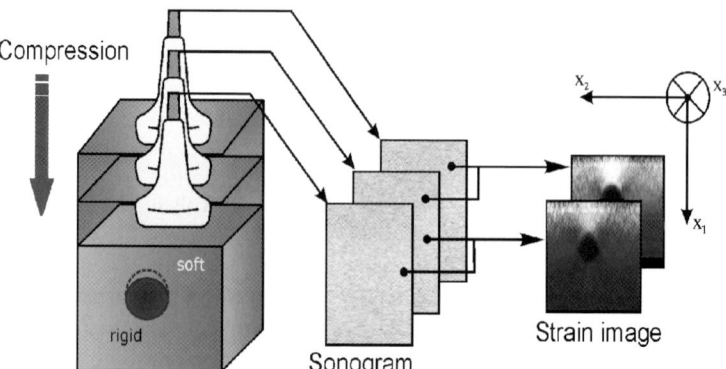

Fig. 6.1 A soft tissue mimicking phantom with a hard inclusion is slightly compressed using the US transducer. The isoechoic rigid body with spherical shape in the phantom is not visible in the conventional B-mode image *on the left* because of equal backscattering properties. The same body can be clearly seen in the strain image *to the right*, where artifacts are seen above the inclusion

3. Methods involving the measurement of displacements induced by the LF propagation (50 to 200 Hz) of pulsed shear waves using a time-resolved 2-D transient elastography system (Sandrin et al. 1999; Catheline et al. 1999).

To understand the mechanical properties of tissue, it is necessary to understand the basic elasticity theory for Static Compressions. The next sections will introduce various types of elastography methods: strain imaging and reconstructive ultrasound elastography. The next subsection summarizes quasi-static deformations and the basic theory of static compression. In the last section we present some strain imaging phantom and *in vivo* results.

6.2.1 Theory for Static Deformations

For a large number of solids, the measured strain is proportional to the load over a wide range of loads. This linear relationship is known as Hooke's law, which states that each of the components of the state of stress at a point is a linear function of the components of the state of strain at the point (Saada 1989). Mathematically, this is expressed as a constitutive equation, which may be written in tensor notation as:

$$\sigma_{ij} = C_{ijmn}\varepsilon_{mn} \quad \{i,j,m,n = 1,2,3\}, \tag{6.1}$$

where C_{ijmn} are elastic constants comprising the elements of a stiffness matrix and σ_{ij} is the stress component in a plane perpendicular to the axis x_i and parallel to the axis x_j.

The components of the matrix C_{ijmn} are elastic constants that are intrinsic properties of the material. Furthermore, using the 2D-scan conditions from a conventional ultrasound 2D-machine and deformations seen in Fig. 6.1 all structures and boundary conditions can be reduced to a two dimensional case. The constants C_{ijmn} characterize a general anisotropic, linearly elastic material. In order to calculate adequate material properties, some assumptions with respect to the material and the geometry are needed. In general, soft tissue can be well modeled as a visco-elastic, anisotropic, and incompressible material. However, in the context of elastography the material behavior is assumed to be linear elastic for small deformations, isotropic and nearly incompressible. While the first simplification can be achieved using a suitable experimental set-up, the second assumption excludes the investigation of anisotropic tissue such as muscle. Since the elastic properties of an isotropic material are independent of the orientation of the axes, the equations may be expressed in terms of only two independent parameters known as the Lame's constants λ and μ. Linear elastic materials obey the generalized Hooke's law which is expressed above and use the following 3 elastic constants:

$$C_{1122} = \lambda \tag{6.2}$$

$$C_{1111} = \lambda + 2\mu \tag{6.3}$$

$$C_{1212} = \frac{1}{2}(C_{1111} - C_{1122}) = \mu. \tag{6.4}$$

The constant μ, is referred to as the shear modulus. The volume change per unit volume due to spherical stress is dependent on the bulk compressional modulus K, which is related to the Lame's constants by:

$$K = \frac{3\lambda + 2\mu}{3}. \qquad (6.5)$$

There are also two other engineering parameters commonly used to characterize the mechanical properties of solid materials: Young's modulus, E, and Poisson's ratio, v. These are related to K and μ by the following expressions:

$$K = \frac{E}{3(1-2v)} \qquad (6.6)$$

and

$$\mu = \frac{E}{2(1+v)}. \qquad (6.7)$$

Soft tissues contain both solid and fluid components and therefore may have mechanical properties that fall somewhere between both (Sarvazyan et al. 1995). The ratio between the shear modulus and the compressional modulus μ/K is close to a few tenths for solid materials, while it equals zero for liquids. Many soft tissues are nearly incompressible with Poisson's ratios ranging from 0.4900 to 0.4999 (Rychagov et al. 2003), which make them mechanically similar to liquids. Note that from equation (Eqs. (6.6)–(6.7)) for incompressible tissues ($v \cong 0.5$), the relationship $E \cong 3\mu$ holds. This means that for incompressible materials there exists a simple proportionality between the shear and the Young's moduli. The bulk compressional modulus K may be estimated from the propagation speed of compressional waves c and the material density ($K \cong \rho c^2$).

Since it is well known from the literature that the changes of the speed of sound as well as the density in different soft tissue types are small (Christensen 1988), the difference of the bulk modulus of tissues is small as well. Hence in traditional medical ultrasound imaging, the speed of sound in tissue is assumed to be a constant 1540 m/s. Thus, the ability to create properly scaled sonograms of soft tissues is in part due to this small variability in the speed of sound and the attenuation in different tissue layers. It has been shown that the shear modulus of normal and abnormal soft tissues may span much more than an order of magnitude (Krouskop et al. 1998) which can not be explained based on variations of the speed of sound or the bulk modulus only, because of the small change of this modulus for different soft tissue types and the concomitant low contrast to noise ratios of imaging results.

Indeed, the elasticity analysis using the linear elastic Hookean behavior is possible only for very small static deformations. Within each subset the elasticity (Young's modulus) can be well approximated as a constant permitting reconstruction based on a linear elastic model. However, since each subset is associated with a different overall internal strain, in other words, different preload deformations, nonlinear elastic properties of tissue are evaluated (Reichling et al. 2005). Note that the real material behavior of tissue is non-linear and the visco-elastic properties play an important role.

6.2.2 Imaging Tissue Strain

If a constant uniaxial load deforms an elastic medium, all points in the medium experience a level of longitudinal strain. The main component of this strain is along the axis of compression. The longitudinal axial strain is estimated in one dimension from the analysis of ultrasonic backscattered signals obtained from standard diagnostic ultrasound equipment. This means that ultrasonic speckle may be used for this purpose, and no discrete resolvable targets must be present. A stiffer tissue element will generally experience less strain than a softer one. In the experimental setup (Fig. 6.2) the ultrasonic RF-data were acquired from scatterers contained in the tissue region of interest, and then digitized.

The data set is acquired while compressing the tissue with the ultrasonic transducer (or with an extra compressor combination seen in Fig. 6.2). Along the ultrasonic radiation axis a small surface deformation is applied, generally about 1% or less of the total tissue depth. Echo lines are acquired before and after compression from the same region of interest.

The main goal of elastography is to use the difference in elastic modulus of tissue types to distinguish them. The basic steps to create elastograms are: The first step is to subject the specimen, for example as seen in Fig. 6.2 a uniform target containing a simple circular pattern of higher elastic modulus, to a deformation. The second step includes the calculation of speckle motion using the ultrasound and a displacement estimator to assume the uniaxial displacement of this target. The third step is to calculate the strain as the gradient of the measured displacement.

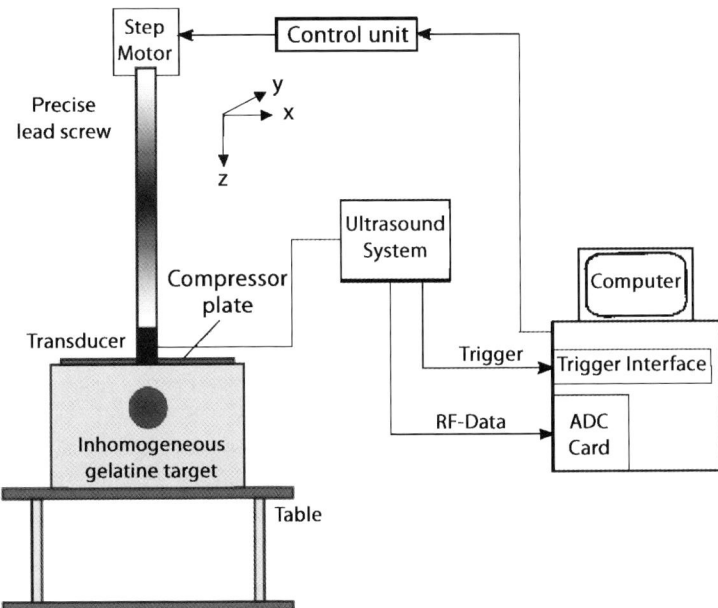

Fig. 6.2 Schematic diagram of the strain imaging experiment

Measuring and predicting displacement and strain fields from the known target modulus is also defined as the forward elasticity problem. In general, the displacement field may be determined using ultrasound or nuclear magnetic resonance. In the case of ultrasound the speckle images (echo lines) and the displacements are obtained under static conditions, and effort is made to keep the deformations small. As a result the deformations are governed by the equations of equilibrium of an incompressible, linear elastic solid undergoing small quasi-static deformation. In the case of magnetic resonance imaging the displacements are calculated from the phase of the measured magnetic field and are required to be time harmonic (Sinkus et al. 1999). Thus the governing equations are time harmonic and sensitive to visco-elastic properties of tissue, which are negligibly small in the quasi-static case.

The strain distribution calculated is not an intrinsic tissue property. It is dependent on both internal and external boundary conditions, as well as on the distribution of shear modulus in the tissue. The external boundary conditions depend on the 3D-shape and the relative size of the compressor compared to the tissue, as well as on the degree of friction between the internal and external contact surfaces. The shape and type of the tissue components determine the internal boundary conditions. Therefore, a map of the strain distribution in tissue reveals not only information about tissue shear modulus distributions, but also about tissue connectivity (interfaces between tissue components) and other geometrical considerations. An incorrect interpretation of the stiffness distribution as a direct map of strain images may result in some known image artifacts due to stress concentrations that may be misinterpreted as areas of low modulus or stress decay inside the tissue layers (Kallel et al. 1996).

6.2.3 Displacement Estimation in Strained Tissues

Several algorithms have been developed to estimate displacement and strain in one and two dimensions either from ultrasound radio frequency-data or demodulated data. These approaches can be summarized in estimates with ultrasound speckle tracking or block matching techniques (O'Donnell 1994; Kaluzynski 2001), correlation based algorithms (Ophir 1991, 1996) and optical flow algorithms (Lorenz 1999; Khalil 2005).

6.2.4 Methods Using Radio Frequency Data

The methods using RF-data cross-correlation are the most effective and common methods to determine the tissue motion. Using ultrasound RF-data the digitized congruent echo lines before and after compression are segmented into small temporal windows that are compared pair-wise using one of a variety of possible time-delay estimation techniques such as cross-correlation, from which the change in arrival time of the echoes before and after compression can be estimated. Due to the small

6.2 Ultrasound Elastography

magnitude of the applied compression, there are only small distortions of the echo lines, and the changes in arrival times are also small. The windows are usually translated in small overlapping steps along the temporal axis of the echo line, and the calculation is repeated for all depths (Fig. 6.3).

The fundamental assumption is that speckle motion adequately represents the underlying tissue motion for small uniaxial compressions. This assumption appears to be reasonable as long as the distributions of the scatterers before and after compression remain highly correlated. Therefore is the time delay estimation (TDE) a very important aspect for the quality of elastography. The local tissue displacements are estimated from the time delays of gated pre- and post-compression echo signals (Eq. (6.4)).

The auto- and cross-correlation functions of two segments of the pre- and post-compression echo signals are defined as:

$$r_{jk}(\tau) = \frac{1}{T_c} \int_{t_i-T_c/2}^{t_i+T_c/2} e_j(t+\tau)e_k(t)\,dt \qquad (6.8)$$

$$\text{for} \quad |\tau| < \frac{\pi}{\omega_0} \ll T_c \qquad (6.9)$$

where T_c is the window length of the segments being searched e_1 and e_2 (seen in Figs. 6.3 and 6.4) and $i \neq k$ (or $i = k$) for the cross- and the auto-correlation function respectively. Therefore the cross-correlation coefficient will be defined as

$$\rho(\tau) = \frac{r_{12}(\tau)}{\sqrt{r_{11}(0)}\sqrt{r_{22}(0)}} \ . \qquad (6.10)$$

Time delays are estimated from the lag of the peak of the cross-correlation function $\tau(t)$ between the pre- and post-compression gated echo signals, as seen in Eq. (6.4). Using the constant speed of sound we can then calculate the displacement Δz as follows:

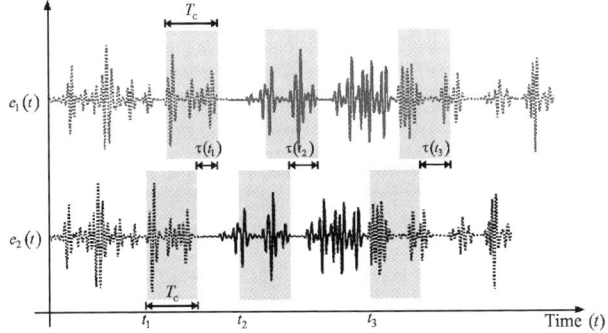

Fig. 6.3 Definition of segments and steps using RF-data from a compressed and an uncompressed Echo with determined time lag. (T_c: Window length)

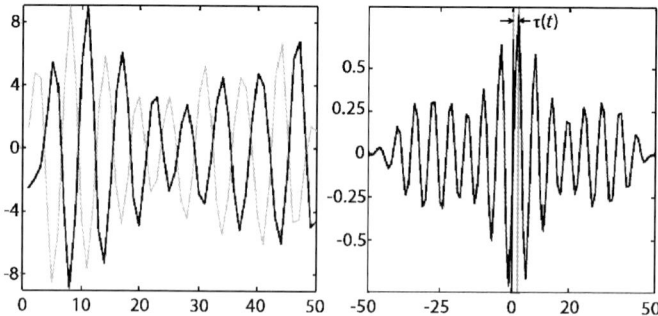

Fig. 6.4 Echo signals from pre- and post-compression (*left*) and the calculated correlation coefficient (*right*), where the time delay $\tau(t)$ is calculated using the maximum of the cross correlation function

$$\Delta z(t) = \frac{1}{2} c \cdot \tau(t) . \qquad (6.11)$$

However this method is time consuming. It is also possible to use other estimators of time delay, such as the frequency shift of the cross-spectrum (Konofagou et al. 1999), or the point of zero phase of the cross-correlation function (PRF-Algorithm) (Pesavento et al. 1999; Lorenz et al. 2001). The quality of elastograms is highly dependent on the optimality of the time delay estimation procedure, which is mainly degraded by two factors:

1. Random noise, which introduces errors in the TDE and is especially disruptive for small time delays.
2. The decorrelation, which results from tissue compression and occurs when the post-compression signal is distorted such that it is no longer an exact delayed version of the pre-compression signal. This error increases with increasing strain and is independent of the signal to noise ratio of the system but mainly arises from new non linear effects, for more details see Erkamp et al. 2004.

Any phenomenon (such as lateral and elevational motion) that degrades the precision of the time-delay estimates will also degrade the strain estimates, thus introducing additional noise into the elastogram. Echo signal decorrelation is one of the major limiting factors in strain estimation and imaging. For small strains, it has been shown that temporal stretching of the post-compression signal (or temporal compression of the pre-compression signal) by the appropriate factor, entirely compensate for signal decorrelation, for more details see the work of Lindop et al. 2006. The quality measure of elastograms is denoted SNR_e, which has previously been defined by (Cespedes et al. 1993) and can be measured experimentally in images where the underlying strain field is known to be homogeneous.

$$SNR_e = \frac{\widehat{\mu}_s}{\widehat{\sigma}_s} \qquad (6.12)$$

$\widehat{\mu}_s$ is the mean strain estimate and $\widehat{\sigma}_s$ is the standard deviation. When the post-compression echo signal is stretched, it realigns all the scatterers within the correla-

6.2 Ultrasound Elastography

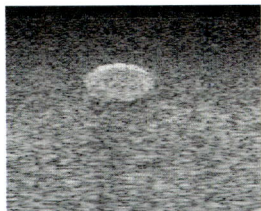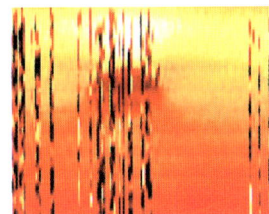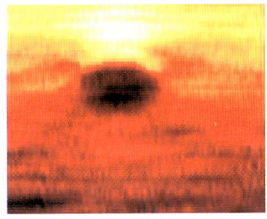

Fig. 6.5 Tissue like phantom with a stiff cylindrical inclusion: B-scan (*left*), strain image calculated using the RF-data with 1.5% applied deformation (*middle*) and using 2D-Block Matching algorithm (*right*) with 3% applied deformation

tion window. Global uniform stretching was found to significantly improve the elastographic signal-to-noise ratio (SNR_e) and the strain dynamic range in elastograms (Alam and Ophir 1997). Thus, a global uniform stretching of the post-compression A-line prior to the displacement estimation is highly advisable, unless the applied compression is very small (<1%) and thus stretching may not be necessary. If high strains (large deformations) are applied, significant decorrelation occurs, which cannot effectively be compensated by stretching. Axial stretching is mandatory in the presence of intermediate strains; otherwise the elastograms become so noisy (Khaled et al. 2006) that they are practically useless as seen in Fig. 6.5.

6.2.5 Block Matching Methods:

The RF-based time delay estimation was shown to be accurate but heavily depending on the compression magnitude and echo decorrelation. In case of deformations larger than 1%, fast cross-correlation methods fail to estimate the exact displacement of the deformed tissue, as seen in Fig. 6.6. A number of other techniques for the estimation of displacement based on *Block Matching* (e. g. Sum Absolute Differences (SAD)) have been explored. As radio-frequency data is rarely available in conventional ultrasound machines we proposed strain imaging using base-band envelope data (Lesniak et al. 2005). The envelope signals can be digitized at a lower sampling rate and they require less storage and processing time. However, the feasibility and accuracy of envelope data based strain imaging has not been extensively explored yet. The use of *Block Matching* techniques for displacement estimation is in case of large deformations crucial. The reference image and the deformed image are divided in overlapping equal blocks, as seen in Fig. 6.6, from which the displacement is calculated using for example the *Sum Absolute Differences* (SAD), which is presented to perform as precisely as the normalized cross correlation, which reduces computation complexity.

$$\text{SAD}_{\min}(i,j) = \min_{v} \left[\sum_{n=1}^{N} |x_{1D}(i,j) - x_{2D}(i, j+v)| \right]. \quad (6.13)$$

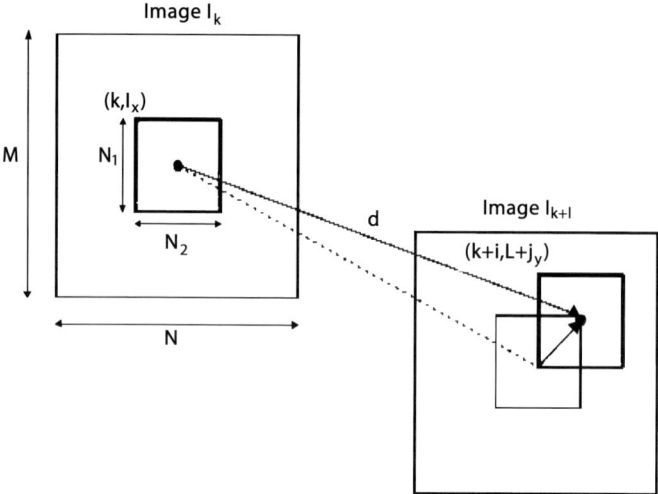

Fig. 6.6 Scheme of SAD block matching

Therefore, to estimate displacements of local tissue regions motion compensation methods based on block matching, mainly global block matching and based on SAD were used.

From Eq. (6.13) x_{1D} and x_{2D} denote 2D-blocks from two consecutive interpolated windows with the size (N_1 by N_2) the values (i, j) denotes the block index, v is varied in a certain interval. For each window position the axial and lateral displacements of the best matching pair is recorded, resulting in (M by N) displacement values per image block. Finally, a global axial displacement matrix is determined for each image pairs (Khaled et al. 2006b).

The disadvantage of this method is that it is very time consuming, therefore other methods using variable size block matching, like diamond search, 3-step search, et cetera, which improve results of video codec applications, are tested to determine better and faster displacement estimation results.

6.3 Reconstructive Ultrasound Elastography

As described in the previous section measuring and predicting strain fields from a known target E-modulus is also defined as the forward elasticity problem. In this problem we were given the material properties and the boundary data and were asked to calculate the displacement field. In the inverse problem the situation is reversed. We are now given the displacement field (or the related measurement) and the boundary data and are asked to calculate the material properties. While this situation is typical for inverse problems in other fields such as ultrasound tomography, the inverse elasticity problem is distinct in that the measured data is known only on a significant subset of the domain and sometimes the boundary data too.

6.3.1 Tissue-like Material Phantoms

In the frame of reconstructive elastography it was important to solve the forward elasticity problem, so it is possible to understand more about material properties and boundary data, which will be applied to solve the inverse problem. Thus in freehand elastography, the compression is manually induced by the conducting physician and is therefore operator dependent, leading to unstable image sequences. On the one hand it was important to present a new designed phantom to help physicians get used to real time elastography systems and on the other hand it was important to make phantoms with known target E-modulus and boundary data so we can solve the forward and inverse problems.

Tissue-mimicking phantoms used for elastographic experiments are mostly made from a mixture of agar and gelatin. The solution of agar and gelatin has effectively revealed interesting properties, since gelatin contributes to the elastic character and agar insures stiffness and cohesion (Hall et al. 1997). The acoustical scattering of such phantoms can be adjusted by simply varying the scatterer concentration. The independent control of acoustical and mechanical properties of such phantoms made agar gelatin phantoms attractive for elastographic experiments. However, the disadvantage is the long-term change in geometric and physical properties, as well as that agar-gelatin phantoms tend to rupture easily under increasing radial stress. For solving the inverse problem and testing the prostate diagnostic using ultrasound elastography and a rectal probe these phantoms could not simulate the ruggedness and elasticity needed.

For those reasons, a new material has been tested for this specific application purpose, which has a high breaking strength and is able to emulate tissue structures accurately. This material is Poly Vinyl Alcohol (PVA). It acquires its properties by freeze-thaw cycle processes in its cryogel form (PVA-c). This gel transition makes a thermoreversible gel, which is a matrix of a physically crosslinked polymer containing uncrosslinked ones and water. Normally cryogel prosperities depend on several factors such as the molecular weight of the uncrosslinked polymers, concentration of aqueous solution, temperature and time of freezing, and number of freezing cycles (Peppas et al. 1991). Most of the gel types are stable in room temperature; they could be extended up to six times of their initial size, which show their rubbery and elastic properties and also their high mechanical strength.

Chu et al. 1997 have studied this material and found that the acoustical and elastic characteristics of the PVA-c are within the range of those of soft tissues. Therefore for the production of phantoms with variable realistic elastic, ultrasound and even also magnetic resonance properties, we need a set of materials to represent fat, glandular and the perineal membrane. These materials consist of ethyl alcohol in PVA-c with TLC Silica gel 60 H dispersions with different concentrations and different thaw and freeze cycles (Khaled et al. 2005). Production of molds for making phantoms with geometries suitable for current ultrasound elastography with internal structures was achieved including simulated tumors. Some niceties in mechanical properties could be simulated also in the phantoms for the appropriate experiment. The variation of the elastic modulus according to thaw and freeze cycles was quantified using a compression experiment at the Institute of Mechanics, Ruhr University Bochum. Preliminary results established a range of 26.3 ± 2.6 kPa–86.9 ± 11.8 kPa for 2–5 freeze-thaw cycles and 10% PVA-solution concentration, with a 19 hour freeze and 4 hours thaw cycle duration (Arnold 2006; Reichling 2007). Different studies of the Young's moduli the ultrasound speeds and ultrasound attenuation coefficient were calculated. The results show that the stress-strain relationship of PVA-phantoms is close to that of tissue, and that the ultrasound propagation speed is in the range of [1498 m/s–1560 m/s] at 22 °C room temperature.

6.3.2 Solution of the Inverse Problem

Unfortunately, strain images alone do not represent a quantitative measure of elasticity, because they do not take account of how the stresses are distributed in the medium and therefore provide only approximate measures of tissue elasticity, which could be misinterpreted as explained before. Better images can be obtained by exploring the feasibility of reconstructing tissue elasticity within the framework of solving the inverse problem. However, inverting measured responses, such as strain distribution, prove to be difficult due to stability problems and it requires the complete understanding of the equivalent forward problem using a finite element (FE) simulation technique.

Almost any tissue material has a more complex behavior than the simplifications given above. A few soft tissues obey Hooke's law in a very limited range of temperature, stress and strain, but usually soft tissue can be well modeled as a visco-elastic, anisotropic, and incompressible material. There are three additional properties found in many tissue types, which also affect the deformation character of tissue, as explained by (Fung 1993): The first property is called stress relaxation, which means decreasing the corresponding induced stress in a body if this body is suddenly strained and the strain is maintained constant, the second is called creep, which happens when the body is suddenly stressed and this stress is maintained constant, and the body is continuing its deformation. The third feature is called hysteresis; it means there is a difference in the stress-strain relationship between the loading and the unloading process. These three phenomena are called features of visco-elasticity. Mechanical models are often used to discuss the visco-elastic behavior of materials, such as the Maxwell model, the Voigt model, and the Kelvin model (Fung 1993). These models all consist of a combination of a linear spring and dashpots with coefficients of viscosity. The ideal linear spring is deformed proportionally to the load. The dashpot is supposed to produce a velocity proportional to the load at any instant. Therefore the total force produces a displacement in the spring and a velocity in the dashpot, which defines the mechanical behavior of all models.

In spite of these real properties, tissue and tissue-like materials are usually assumed that they behave as linear, elastic, and isotropic materials to simplify and study them (Ophir et al. 2002). These assumptions are likely to be reasonable for small strains, short duration load application, and a spatial scale that is large compared to the relative correlation length of the elastic variability in the tissue sample.

6.3 Reconstructive Ultrasound Elastography

Furthermore, using the assumption of a plane strain deformation, all structures and boundary conditions can be reduced to a two dimensional case. Using these simplifications, the inverse problem of elastography can be defined as an estimation of the shear modulus distribution μ from the measured axial displacement component $u_1(x_1)$. In order to solve this inverse problem, two completely different classes of methods can be distinguished. On the one hand there are so called *direct approaches*, e.g. (Sumi et al. 1995) and on the other hand there are the *iterative approaches*, e.g. (Oberai et al. 2003). Generally in the direct method the strong form of equilibrium equations is used, neglecting the body force (f_i) (Skovoroda et al. 1997).

$$\frac{\partial \sigma_{ij}}{\partial x_j} + f_i = 0 \quad \{i, j = 1, 2\} \tag{6.14}$$

$$f_i \simeq 0,$$

where σ_{ij} is one component of the 2nd ranked stress tensor and f_i is the body force per unit volume acting on the body in the x_i direction (Saada et al. 1989). Since most tissues can be treated as incompressible, a value close to 0.5 can be assumed for the Poisson's ration ν (Cespedes et al. 1997). In addition the pseudo-constitutive law for incompressible materials can be used as in the following equation:

$$\sigma_{ij} = p\delta_{ij} + 2\mu\varepsilon_{ij}, \tag{6.15}$$

where δ_{ij} is the Kronecker delta, p is the mean normal stress, μ is the shear modulus defined above in Eqs. (6.6)–(6.7) and E is Young's modulus and in the case of incompressible soft tissue the relationship $E \cong 3\mu$ holds. The Hooke's law, (Eq. (6.1)), and the Eqs. (6.14)–(6.15) are then rearranged to yield a linear partial differential equation for μ:

$$\frac{\partial \mu}{\partial x_3} = 0 \tag{6.16}$$

and

$$\mathbf{A} \left(\frac{\partial \mu}{\partial x_1} \cdot \frac{1}{\mu}, \frac{\partial \mu}{\partial x_2} \cdot \frac{1}{\mu} \right)^{\mathrm{T}} = -\mathbf{B}. \tag{6.17}$$

$\mathbf{A}$ and $\mathbf{B}$ are matrices depending on the strain and the strain derivatives in x_1 and x_2 direction as follows:

$$\mathbf{A} = \begin{pmatrix} 2\varepsilon_{11} + \varepsilon_{22} & \varepsilon_{12} \\ \varepsilon_{21} & \varepsilon_{11} + 2\varepsilon_{22} \end{pmatrix} \tag{6.18}$$

and

$$\mathbf{B} = \begin{pmatrix} \frac{\partial (2\varepsilon_{11} + \varepsilon_{22})}{\partial x_1} & \frac{\partial \varepsilon_{12}}{\partial x_2} \\ \frac{\partial \varepsilon_{21}}{\partial x_1} & \frac{\partial (\varepsilon_{11} + 2\varepsilon_{22})}{\partial x_2} \end{pmatrix}. \tag{6.19}$$

Using Eq. (6.15) we can obtain the derivative of the relative shear modulus except in the points where the determinant of **A** is zero as follows:

$$\text{div}(\ln \mu) = -\mathbf{A}^{-1}\mathbf{B}. \quad (6.20)$$

The relative shear modulus can then determined using a line integration at any point in the plane (x_1, x_2) closed with the surface c, relatively to the known shear modulus of a starting point (a_1, a_2).

$$\frac{\mu(x_1, x_2)}{\mu(a_1, a_2)} = \exp\left(-\int_c \left(\mathbf{A}^{-1}\mathbf{B}\right)\right). \quad (6.21)$$

The benefit of these direct solution methods is that they are time-saving compared to the iterative methods. A disadvantage is that all components of the strain tensor are assumed to be known. Furthermore, in order to estimate the strain field, noisy data have to be differentiated. Better solutions could be achieved using iterative methods solving the inverse problem of elasticity, e. g. (Oberai et al. 2003; Reichling et al. 2007). These algorithms generally use the weak form of the equilibrium equations. In these methods the inverse problem is regarded as a minimization problem. Figure 6.7 and Eq. (6.22) show the optimization procedure in which a functional (π) is defined, in order to minimize the difference between the measured and the simulated deformation, which is equivalent to minimizing the difference between the corresponding displacements. The second part of the sum is a Tikhonov regularization parameter (Bertero et al. 1998), needed for the stability of inverse problems. The Tikhonov parameter is essentially a trade-off between fitting the data and reducing a norm of the solution. It helps to regulate the numerical solution in terms of noisy measured data and is chosen according to the theory of residues due to Morozov (see Isakov 1998 for example). This is defined as in the following equation:

$$\text{Find } \mu = \mu(x) : \pi(\mu) = \| T(u) - T(u^m) \|_\Omega^2 + \frac{\alpha}{2} \to \min!, \quad (6.22)$$

where u^m is the measured and u is the calculated displacement field from appropriate finite element simulations, while T denotes a projection tensor onto the axial direction in the space Ω and π the resulting error-function to be minimized.

We consider the solution of this inverse problem by using a class of optimization algorithms (Reichling et al. 2007), that require the value of the functional and its derivative (gradient). φ is defined as the solution of the direct problem and $D_\mu \pi$ the derivative of the error function π to the variable μ. Several quasi-Newton algorithms exist, such as the steepest descent, BFGS (Broyden–Fletcher–Goldfarb–Shanno) algorithm, see Zhu et al. 1997 for more details.

Figure 6.7 shows the block diagram of the BFGS algorithm used to solve the inverse problem of elasticity. The main advantages of these methods are the robustness with respect to the noisy measured data u^m and the fact that only a one dimensional component of the displacement is needed. The disadvantage is that these algorithms are comparatively time-consuming.

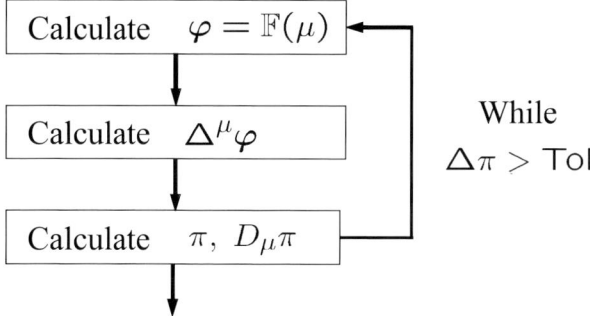

Fig. 6.7 Optimization Algorithm used to solve the inverse problem. φ is defined as the solution of the direct problem and $D_\mu \pi$ the derivative of the error function π to the variable μ

6.3.3 Simulation and Experimental Results

The above mentioned methods were implemented and applied to a number of numerical simulations and experimental data. To reconstruct the elastic properties of tissue a simulation procedure for effective modeling is needed. This procedure contains three composites, and interactive parts: (1) a mechanical tool using finite element simulations giving the simulated axial displacement and stress at each of its nodes after the application of the external compression on the specimen; (2) an acoustical tool measuring the acoustic RF-signals and images inside the phantom before and after compression and (3) and image formation technique giving estimates of the displacement field, strain distribution and reconstructing the elasticity image of the tissue. Numerical simulations were performed using a standard finite element program. To simulate the mechanical behavior of the tissue-mimicking phantoms, a two-dimensional mesh consisting of 2236 eight-node, isoparametric, arbitrary quadrilateral elements, designed for incompressible or nearly incompressible plane strain applications, was used. As only the axial displacement can be measured with an adequate accuracy, only one-dimensional simplification of the direct approach Eq. (6.21) is used. In the simulation, the experiment shown in Fig. 6.8

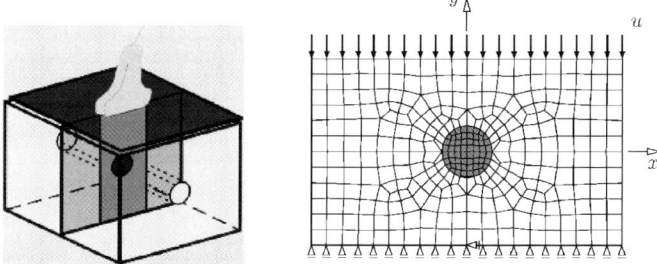

Fig. 6.8 Schematic diagram of the experimental setup and the simulated phantom finite element mesh

(left) is modeled as a plane strain state. The simulated phantom consists of a hard cylindrical inclusion in a softer matrix.

Figure 6.8 (right) shows a sketch of the finite element discretization with applied boundary conditions. In order to reconstruct the shear modulus appropriate boundary data were applied to the model: all nodes on the upper and lower boundary can move freely in x-direction, whereas the movement in y-direction is prescribed, all other nodes can move in x- and y-direction. In all FE-analyses, the Young's modulus of the inclusion was higher than surrounding material. The Poisson's ratio of both parts was set to 0.495 defining nearly incompressible tissue-like materials. Corresponding to a finite element solution of the elasticity problem, the equation for determining the nodal displacement u in Fig. 6.8 (right) is given by:

$$\sum_{j=1}^{n} K_{ij}(\mu) u_j = f_i . \qquad (6.23)$$

K is the stiffness matrix depending on μ, which is the vector containing the nodal values of the shear modulus and the vector f, which appears on the right-hand side contains contributions from prescribed external force and displacement boundary conditions. In this study, we have used the finite element formulation for a nearly incompressible linear elastic solid described in Hughes (2000). As seen in Fig. 6.9 (left) on the top of the specimen a compressor plate is shown, where the transducer is integrated in the real experiment, which permits the assumption of the plane strain state. The ratio of the shear modulus between the inclusion and the matrix is set to $\mu_{inc}/\mu_{mat} = 3/1$ and both materials are chosen to be nearly-incompressible. To simulate a measurement situation, the noisy displacement field is obtained by adding white Gaussian noise with a noise level of 1% to the calculated displacement field.

Figure 6.9 shows a comparison between numerical simulation results of calculating the relative shear modulus using noisy data with the direct and the iterative approaches described earlier. Both methods are capable of finding the hard inclusion but with different results. The comparison to the right (Fig. 6.10, right) shows that the iterative method is much better than the direct one. However, an error is still found between the real and the calculated values of the shear modulus, because of the simulated noisy data. The disadvantage of the iterative method is the higher computation time which is about tens times or higher than the time used in the direct

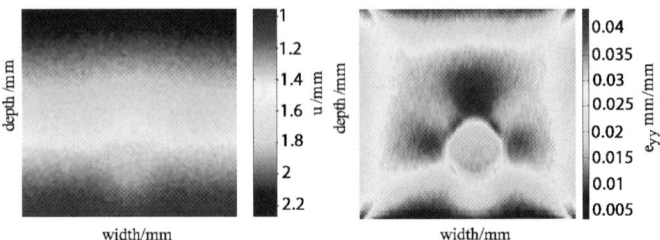

Fig. 6.9 Results of finite element analyses for $\mu_{inc}/\mu_{mat} = 3/1$: displacement in y-direction (*left*) and (c) strain ε_{yy} (*right*)

6.3 Reconstructive Ultrasound Elastography

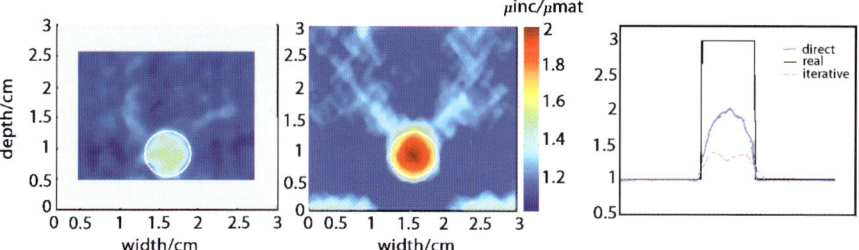

Fig. 6.10 Numerical simulations of an experiment: calculated relative shear modulus distribution using noisy data with the direct (*left*) and the iterative method (*middle*), path plot showing the shear modulus ratio along the path in 1 cm depth for the real simulation and the inverse results (*right*)

method. For all cases, the overall time taken to solve the problem is reasonable (for the 80×60 mesh about 5 min using a 1.7 GHz Pentium processor). In Fig. 6.10 the thin white circle marks the position of the inclusion in the scanned region.

In the experimental setup (Fig. 6.11) the ultrasonic RF-data were acquired using a commercially available ultrasound system (Siemens SONOLINE Omnia) modified to get RF-signals, equipped with a 9 MHz linear array probe, a trigger interface, and sampled by a conventional analog to digital conversion card (12 Bit; sample frequency: 50 MHz) and a desktop PC. To measure the stress-strain relation we used a pressure sensor and sensitive digital position encoders combined with a multi-channel electronic PC measurement unit for parallel, dynamic measurement data acquisition using a second desktop computer.

In the experiment we reconstruct the elastic properties of the ultrasonic tissue mimicking PVA-phantom, seen in Fig. 6.12, containing a harder 1.4 mm diameter cylindrical inclusion with more freeze thaw cycles than the surrounding material.

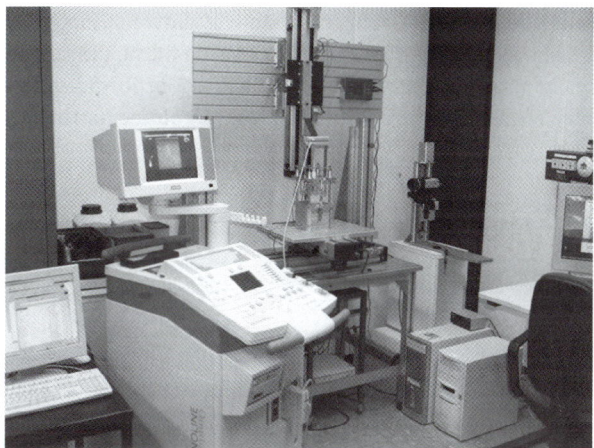

Fig. 6.11 The experimental setup to measure and reconstruct the shear modulus of phantom materials using ultrasound elastography

Fig. 6.12 PVA-c tissue mimicking ultrasound phantom with a cylindrical inclusion and the ROI of the ultrasound image

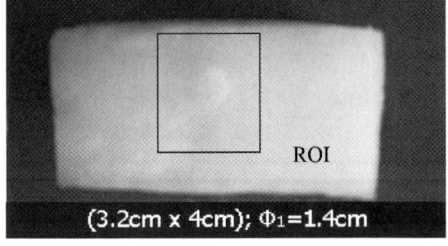

We used PVA-c phantoms (10% concentration) with 2 and 5 freeze and thaw-cycles to emulate the surrounding material (soft tissue) and the hard inclusion (tumor tissue) respectively.

The tissue-like phantom was imaged using the ultrasound elastography system and a stepper motor with a compressor plate to generate a plane strain state and measure stress-strain relation as seen in Fig. 6.13. Acquired experimental results using a tissue mimicking phantom (Fig. 6.13) were similar to simulated results in Fig. 6.13.

In order to simplify the visualization of the inclusion and to compare the numerical results, the chosen test phantom was not isoechoic. In this experiment small displacements are determined between ultrasonic image pairs, which are acquired under varying small axial compression, using a cross-correlation analysis of corresponding echo-lines within RF-data sets as detailed earlier. The derivative of displacement field is equal to the strain in tissue. The strain imaging system uses the fast phase root algorithm (Pesavento et al. 2000) for strain estimation leading to a frame rate of 30 frames per second.

After acquiring a series of 2D elastograms, the volume is created by placing each image at the proper location in the volume. The position data acquired with each 2D elastogram determines the particular location of the image. Using the three-dimensional data to reconstruct an equivalent virtual object could help in calculating the volume of the object imaged. This procedure is shown in Fig. 6.14.

Some physicians assert that the correct estimation of the tumor volume (*in vivo*) for example in the prostate cancer would be helpful in formulating treatment for the disease, since (ex vivo) determination of tumor volume has been shown to cor-

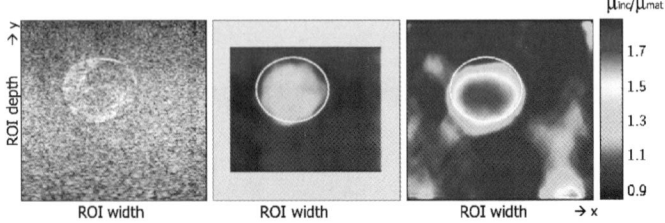

Fig. 6.13 Experimentally obtained results of the phantom: the B-mode image (*left*), reconstructed relative shear modulus using the direct method (*middle*) and the calculated relative shear modulus using the optimization iterative method (*right*)

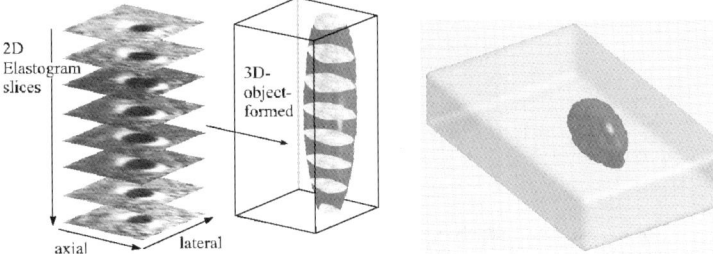

Fig. 6.14 Sequential acquisition of parallel slices using elastography, combined with image segmentation, enables the reconstruction of 3D image to the *right*

relate with the progression of the disease after radical prostatectomy (Taylor et al. 2000). Other applications for elastography could be in the development of haptic devices. These could be used in many application fields, such as in virtual reality, intra-operative navigation, in teaching and in palpation perception representation for pathological tissues or organs in medical examinations (Khaled et al. 2004).

6.4 Medical Results of Elastography

In principle, elastography may be applied to any tissue system that is accessible ultrasonically and which can be subjected to a small static (or dynamic) compression. The compression may be applied externally or internally as shown in case of intravascular applications.

In this section a summary of results from some applications is presented. By comparison, the companion sonograms do not provide a clear visualization of the tumors shown in the elastography images of the prostate. It should be noted that each elastogram was derived from two or more of very similar sonograms, one of which is shown. The first *in vivo* application of elastography was for imaging the breast and skeletal muscle (Cespedes et al. 1993).

More recently, the first real time ultrasound elastography was developed at the Ruhr-University to determine strain images in the prostate side by side with the B-mode scan on a conventional ultrasound machine (Voluson 730, Kretz, GE), using a transrectal transducer with a middle frequency of 7.5 MHz. During the ongoing clinical study on prostate tumor diagnosis at the Urology Hospital of the Ruhr-University Bochum (Marien-Hospital Herne), more than 216 patients have undergone clinical examinations. It has been shown that our system for real time ultrasound elastography is able to detect the prostate carcinoma with a high degree of accuracy, reaching a sensitivity of 76% and a specificity of 84%, compared with only 34% using ultrasound B-Mode images alone (König et al. 2004, Kühne et al. 2003 and Scheipers et al. 2003). Figure 6.15 shows an example, where prostate slices with histological diagnosis following radical prostatectomies act as reference. Cancerous areas have been stained and marked on the prostate slices.

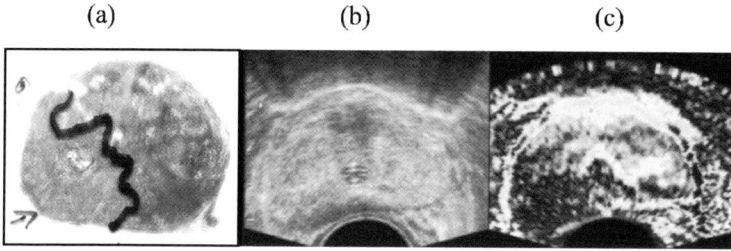

Fig. 6.15 *In vivo* results of a human prostate: **a** Histology, tumors have been stained, malign and benign tissue areas have been marked by pathologist. The tumor is in the *lower left side*. **b** B-mode image: tumor nearly not visible. **c** Strain image: The tumor is clearly visible as a *dark area* on the *left side*. (In cooperation with the Urology Hospital, Ruhr Univ. Bochum, Marien-Hospital Herne and LP-IT Innovative Technologies GmbH)

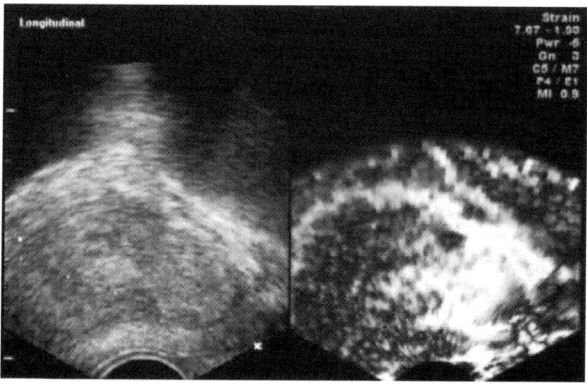

Fig. 6.16 *In vivo* results of a human prostate during a biopsy guided with real time strain imaging in the longitudinal mode. (In cooperation with the Urology Hospital, Ruhr Univ. Bochum, Marien-Hospital Herne and LP-IT Innovative Technologies GmbH)

In a second study we examined 56 patients (preliminary results) suspected of suffering cancer according to either the digital rectal examination (DRE) and/or the PSA-test (blood-test) values and/or transrectal ultrasound (TRUS). The patient underwent sextant needle biopsy assisted with real time strain imaging, where the "gold standard" is the needle biopsy result, 85% of the cancer patients were correctly recognized and 61% of needle specimen were correctly classified (König et al. 2004, 2005) as seen in Fig. 6.16.

6.5 Results of an Intravascular Ultrasound Study

Recently, the method of ultrasound elastography became of major interest for plaque discrimination (de Korte 2002). The stiffness of the wall tissue is evaluated to characterize vulnerable plaques, usually soft plaques regions exhibiting an increased

6.5 Results of an Intravascular Ultrasound Study

strain, which are characterized by a thin fibrous cap covering an eccentric soft lipid pool. Any physiological phenomena, such as pulsating arteries or respiration, could be used as a source of tissue compression.

In an experiment an intravascular ultrasound (IVUS) scanner (*GalaxyII*, Boston Scientific) is used for *in vivo* experiments in the catheter lab at the Ruhr-University Hospital (Bergmannsheil, Cardiology Department). Single element rotating transducers with a center frequency of 40 MHz are used for the acquisitions. Custom made hardware is used for triggering and signal amplification. *In vivo* imaging is performed and full frame RF data are sampled ($f_s = 400$ MHz), digitized and recorded with informed patient's consent during IVUS examinations in the catheter lab. Simultaneously, intracoronary pressure signal is recorded along with an electrocardiogram. Pressure guide wires were used for pressure measurements and evaluated.

Pressure measurements were used to align the RF data of consecutive diastolic frames. The analysis of the acquired image series reveals that the image correlation reaches a maximum during the diastolic phase of the heart cycle (Fig. 6.17).

For strain image calculations only frames in the late diastole are considered, since in this part of the heart cycle vessel motion and thus decorrelation effects are supposed to be minimal. However, motion is still present even in the diastole and often degrades strain estimation results. During a heart cycle only a small number of image frames are suitable for strain analysis. Another problem is the rotating single element system, which reduces signal correlation according to non uniform rotational distortion (NURD) and has a low frame rate of about 30 frames per second. Therefore real-time strain calculations are not feasible. Contour mapping procedures and block matching methods reduces the error of decorrelation and can improve strain images.

The strain of the surface of the plaque is lower than the surrounding lumen (red) under this pressure. The plaque layers are clearly recognized at several places with different strain values. Several layers are also seen inside the plaque, which are having a higher or a lower strain locally. The Figs. 6.19–6.20 show results from different patients. B-mode images with 50 dB dynamic range are displayed along with the strain images calculated from two consecutive frames of data in the diastolic phase

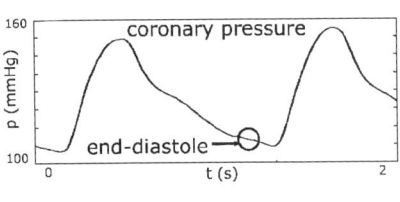
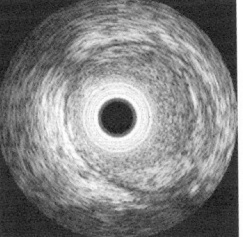

Fig. 6.17 *In vivo* results of a 40 MHz IVUS scan of the left coronary artery (*right*). Full frame RF data is recorded for three seconds. Simultaneously, an intracoronary pressure signal is recorded (*left*). For strain calculations only frames in late diastole are considered. The arrow in the pressure plot indicates the acquisition time of the two consecutive data sets

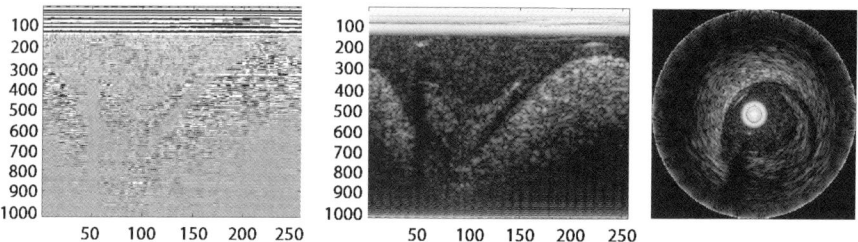

Fig. 6.18 *In vivo* data of the left coronary artery of a 71 years old patient. RF-data collected (*left*) and B-mode image (1024 × 256) samples ($f_s = 200$ MHz) for a rotating single Element in polar coordinates. To the *right* is the B-mode image in Cartesian coordinates

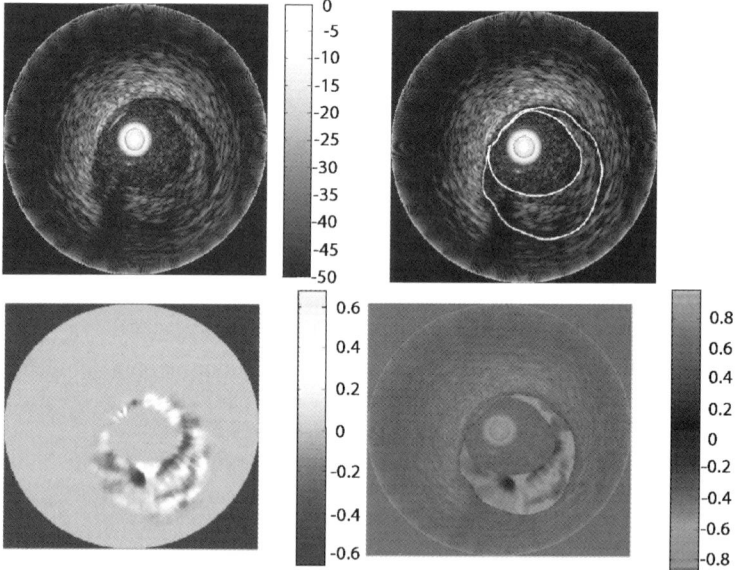

Fig. 6.19 B-mode image (*upper left*), results of contour detection of the plaque surface (*upper right*), strain images determined inside this contour (*lower left*), the virtual addition of both the strain and the B-mode image (*lower right*) [soft = light; hard = dark]

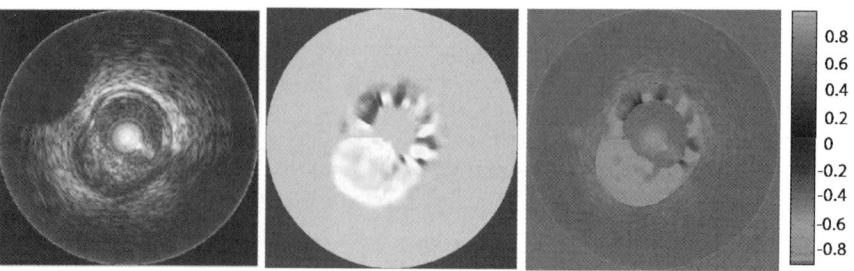

Fig. 6.20 Results of B-scan and contour detection of the plaque surface (*left*), strain images determined inside this contour (*center*), the virtual addition of both the normal strain and the normal B-scan (*right*) [soft = light; hard = dark]

of the heart cycle. The strain is only displayed in the contours around the luminal border. With increased depth the signal correlation is reduced and strain calculation is not feasible, this area is blackened out. The B-mode image in Figs. 6.18–6.19 and Fig. 6.20 are showing a thickened intima with a plaque formation. The strain images show that parts of this plaque are hard (dark) on the cap of the plaque and other parts are soft inside the plaque (yellow or green).

6.6 Summary and Conclusion

Ultrasound elastography is a newly developed imaging modality based on the fact that different tumors and cancerous tissue are significantly harder than normal tissue. The goal of this modality is to determine non-invasively elastic properties of examined tissue using the displacement and strain calculated according to a small deformation.

We presented in this chapter a novel real-time ultrasound strain imaging system, which allows the use of elastography in a clinical study for the early detection of prostate cancer during conventional transrectal ultrasound examinations. We have shown that elastography holds promise in the in vivo diagnosis of prostate cancer and intravascular disease. The fact that free-hand real-time strain imaging for the deformation is used, leads to decorrelations in RF-signals, which increases images noise.

We presented a method to determine tissue deformation using block-matching, which improves strain images calculated in the large deformation case. Thus, the displacement estimation can be improved if we take into account the non-linear effects. While strain imaging artifacts are fairly well understood, their possibly ambiguous role for the detection of lesion and diagnosis remain unknown. Therefore we have developed and implemented an efficient formulation to solve the inverse problem of elasticity using the reconstructive elastography. It delivers additional quantitative information about the relative shear modulus and seems promising for differential diagnosis of lesions in biological tissue. Although, only one-dimensional displacements were calculated for the quasi-static problem, it has been shown that for a given range of SNR_e a relative modulus reconstruction procedure is possible using a fast direct method and an accurate iterative method. These methods can be further improved by calculating the lateral displacements and the non-linear axial distortions in finite deformations. The strain imaging and the reconstructive elastography methods are able to localize inclusions, where only the iterative reconstructive approach can be used for an acceptable non-invasive quantitative identification of the shear modulus distribution of tissue.

Finally, in spite of the progression happened in the past several years in elastography, much progress has yet to be made in order to develop elastography towards a viable clinical tool. The results of strain imaging of tissue are certain to find a wide range of applications, particularly in cancer diagnosis, intravascular diagnosis, tissue engineering, and other medical engineering applications.

Acknowledgements A project of the Ruhr Center of Excellence for Medical Engineering (KMR), Bochum, Germany, supported by the German Federal Ministry of Education and Research (BMBF), No. 13N8079. The authors would like also to acknowledge funding by the German Federal Ministry of Education and Research (BMBF) through the research project "Haptisches Sensor-Aktor-System (HASASEM)" grant 01IRA14B and the German Research Foundation (DFG) through the projects ER 94/30-1 and BR 580/32-1. This support is gratefully acknowledged.

References

1. Alam SK, Ophir J (1997) Reduction of signal decorrelation from mechanical compression of tissues by temporal stretching: Applications to elastography. Ultrasound in Medicine and Biology 23:95–105
2. Anderson WAD (1977) Pathology, 7 edition, Saint Louis, Mosby Co
3. Arnold A (2006) Identifikation von Materialparametern an homogenen Zylinderproben aus Polyvinylalkohol. MS.E. Thesis. Institute of mechanics, Ruhr-University, Bochum
4. Bertero M, Boccacci P (1998) Introduction to inverse problems in imaging. Institute of Physics Publishing, Bristol, UK
5. Bilgen M, Srinivasan S, Lachman LB, Ophir J (2003) Ultrasonic elastography imaging of small animal oncology models: a feasibility study. Ultrasound Med Biol 29(9):1291–1296
6. Breasted JH (1930) The Edwin Smith surgical papyrus. Chicago: University of Chicago Press, pp 403–406
7. Catheline S, Thomas JL, Wu F, Fink M (1999) Diffraction field of a low-frequency vibrator in soft tissues using transient elastography, IEEE Transactions on Ultrasonics, Ferroelectrics, and Frequency Control 46(4):1013–1020
8. Cespedes I, Ophir J (1993) Reduction of image noise in elastography. Ultrasonic Imaging, 15:89–102; Varghese T, Ophir J (1997) A theoretical framework for performance characterization of elastography: the strain filter. IEEE Transactions on Ultrasonics, Ferroelectrics, and Frequency Control 1:164–172
9. Cespedes I, Ophir J, Ponnekanti H et al. (1993) Elastography: elasticity imaging using ultrasound with application to muscle and breast in vivo. Ultrason Imaging 15:73–88
10. Christensen DA (1988) Ultrasound Bioinstrumentation, John Wiley & Sons, Inc., NY, Chapter 4
11. Chu KC, Rutt BK (1997) Polyvinyl Alcohol Cryogel: An Ideal Phantom Material for MR studies of Arterial Flow and Elasticity, Magnetic Resonance in Medicine 37(2):314–319
12. de Korte CL, Carlier SG, Mastik F, Doyley MM, van der Steen AFW, Bom N (2002) Morphologic and mechanic information of coronary arteries obtained with intravascular elastography: a feasibility study *in vivo*. European Heart Journal 23(5):405–413
13. Erkamp RQ, Emelianov SY, Skovoroda A, O'Donnell M (2004) Nonlinear Elasticity Imaging: Theory and Phantom Study IEEE Transactions on Ultrasonics, Ferroelectrics, and Frequency Control 51(5):532–539
14. Fatemi M, Greenleaf JF (1999) Vibro-acoustography: An imaging modality based on ultrasound-stimulated acoustic emission. Proc National Academic Science USA 96(12):6603–6608
15. Ferlay J et al., GLOBOCAN (2002) Cancer Incidence, Mortality and Prevalence Worldwide IARC CancerBase No.5, Version 2.0. IARCPress, Lyon
16. ftp://svr-ftp.eng.cam.ac.uk/pub/reports/lindop_tr550.pdf

17. Fung YC (1993) Biomechanics: Mechanical Properties of Living Tissues 2nd Edition. Springer Verlag, NY
18. Gann P et al. (1997) Interpreting Recent Trends in Prostate Cancer. Incidence and Mortality Epidemiology 8:117–119
19. Gao L, Parker KJ, Leruer RM et al. (1996) Imaging of the elastic properties of tissue – A review. Ultrasound in Medicine and Biology 22:959–977
20. Garra BS, Cespedes EI, Ophir J et al. (1997) Elastography of breast lesions: Initial clinical results. Radiology 202:79–86
21. Hall TJ, Bilgen M, Insana MF, Krouskop TA (1997) Phantom materials for elastography. IEEE transactions UFFC 44(6):1355–1364
22. Hiltawsky KM, Krüger M, Starke C, Heuser L, Ermert H, Jensen A (2001) Freehand Ultrasound Elastography of Breast Lesions: Clinical Results. Ultrasound in Medicine and Biology 27(11):1451–1460
23. Hughes TJR (2000) The Finite Element Method Linear Static and Dynamic Finite Element Analysis, Mineola NY, Dover
24. Insana MF, Barakat CP, Sridhar M, Lindfors KK (2004) Viscoelastic imaging of breast tumor microenvironment with ultrasound. Journal of Mammary Gland Biology and Neoplasia 9(4):393–404
25. Isakov V (1998) Inverse problems for partial differential Equations, 1st edition. Springer Verlag, New York
26. Jemal A, Murray T, Ward E, Samuels A, Tiwari R, Ghafoor A, Feuer E, Thun M (2005) Cancer statistics. CA Cancer Journal for Clinicans 55(1):10–30
27. Kallel F, Bertrand M (1996) Tissue Elasticity Reconstruction Using Linear Perturbation Method. In: IEEE Transactions on Medical Imaging 15(3):299–313
28. Kaluzynski K, Chen X, Emelianov SY, Skovoroda AR, O'Donnell M (2001) Strain rate imaging using two-dimensional speckle tracking, IEEE Transactions on Ultrasonics, Ferroelectrics, and Frequency Control 48(4):1111–1123
29. Khaled W, Bojara W, Lindstaed M, Ermert H (2006b) Strain Imaging with Intravascular Ultrasound: An In Vivo Study IEEE Ultrasonics Symposium Proceedings, Vancouver, Canada, pp 1313–1316
30. Khaled W, Reichling S, Bruhns OT, Ermert H et al. (2004) Palpation Imaging using a Haptic System for Virtual Reality Applications in Medicine. Proceedings of the 12th Annual Medicine Meets Virtual Reality Conference: Building a Better You: The Next Tools for Medical Education, Diagnosis, and Care. Newport Beach, California, USA, pp 147–153
31. Khaled W, Neumann T, Stapf J, Ermert H (2005) PVA Prostate Phantom for Ultrasound and MR Elastography, Proceedings of the 4th International Conference on Ultrasonic Measurement and Imaging of Tissue Elasticity, Lake Travis, Austin, Texas USA, p 105
32. Khaled W, Reichling S, Bruhns OT, Ermert H (2006) Displacement Estimators in Ultrasonic Elastography for Finite Deformations. Proceedings of the 5th International Conference on Ultrasonic Measurement and Imaging of Tissue Elasticity, Snowbird, Utah USA, p 72
33. Khalil AS, Chan RC, Chau AH, Bouma BE, Kaazempur-Mofrad MR (2005) Tissue elasticity estimation with optical coherence elastography: Toward mechanical characterization of in vivo soft tissue. Annals Biomedical Engineering 33(11):1631–1639
34. König K, Pannek J, Scheipers U, Ermert H, Phillippou S, Noldus J, Senge T (2004) Ultrasonic multifeature tissue characterization for prostate cancer diagnostics. Journal of Urology 171(4):476–476
35. König K, Scheipers U, Pesavento A, Lorenz A, Ermert H, Senge T. Initial experiences with real-time elastography guided biopsies of the prostate. Journal of Urology 174(1):115–117
36. Krouskop TA, Wheeler TM, Kallel F, Garra BS, Hall T (1998) Elastic moduli of breast and prostate tissues under compression. Ultrasonic Imaging 20(4):260–274
37. Kuehne K, Sommerfeld HJ, Schuermann et al. (2003) Real-time elastography: First experience of prostate needle biopsy findings using the ultrasound multicompression strain imaging. Journal of Urology 169(4):433–433
38. Lerner RM, Huang SR, Parker KJ (1990) Sono-elasticity images derived from ultrasound signals in mechanically vibrated tissues. Ultrasound Medicine and Biology 16:231–239

39. Lesniak B, Khaled W, Perrey C, Ermert H (2005) Comparison of two displacement estimators in ultrasonic elastography. Biomedizinische Technik 50(1):635–636
40. Lindop JE, Treece GM, Gee AH, Prager RW (2006) Estimation of displacement location for enhanced strain imaging. University of Cambridge Online publications, pp 1–28
41. Lorenz A, Schäpers G, Sommerfeld H-J, Garcia-Schürmann M, Philippou S, Senge T, Ermert H (1999) On the use of a modified optical flow algorithm for the correction of axial strain estimates in ultrasonic elastography for medical diagnosis. Proceedings of the FORUM ACUSTICUM Berlin, published on CD-ROM, pp 4aBB1–4aBB4
42. Lorenz A, Ermert H, Siebers S, Senge T, Phillipou S (2002) Ultrasonic Real Time Strain Imaging: A new diagnostic Modality for the early Detection of Prostate Cancer. Fortschritte der Akustik – DAGA, Bochum 28:685–686
43. Doyley MM, Mastik F, de Korte CL, Carlier SG, Céspedes EI, Serruys PW, Bom N. van der Steen AFW (2001) An automated approach for clinical intravascular ultrasound palpation. Ultrasound in Medicine and Biology 27(11):471–480
44. Oberai AA, Gokhale NH, Feijoo GR (2003) Solution of inverse problems in elasticity imaging using the adjoint method. Inverse Problems 19(2):297–313
45. O'Donnell M, Skovoroda AR, Shapo BM et al. (1994) Internal Displacement and Strain Imaging Using Ultrasonic Speckle Tracking IEEE Trans UFFC 41:314–325
46. Office for National Statistics (2005) Cancer Statistics registrations: Registrations of cancer diagnosed in 2003. England Series MB1 No. 34. London: National Statistics
47. Ophir J, Alam SK, Garra BS, Kallel F, Konofagou EE, Krouskop T, Merritt CRB, Righetti R, Souchon R, Srinivasan S, Varghese T (2002) Elastography: Imaging the Elastic Properties of Soft Tissues with Ultrasound, in: Journal of Medical Ultrasonics 29(4):155–171
48. Ophir J, C'espedes I, Ponnekanti H, Yazdi Y, Li X (1991) Elastography: A quantitive method for imaging the elasticity of biological tissues. In: Ultrasonic Imaging 13:111–134
49. Peppas NA, Stauffer SR (1991) Reinforced uncrosslinked poly (vinyl alcohol) get produced by cyclic freezing-thawing processes: a short review. Journal of Controlled Release 16:305–310
50. Perrey C (2005) Signal processing concepts fort he assessment of coronary plaques with intravascular ultrasound. PhD Thesis. Ruhr-University Bochum, logos Verlag, Berlin
51. Pesavento A, Perrey C, Kruger M, Ermert H (1999) A time-efficient and accurate strain estimation concept for ultrasonic elastography using iterative phase zero estimation. IEEE Trans UFFC pp 1057–1067
52. Pesavento A, Lorenz A, Ermert H (1999) System for real-time elastography, Electronics Letters 35(11):941–942
53. Pesavento A, Lorenz A, Siebers S, Ermert H (2000) New real-time strain imaging concepts using diagnostic ultrasound: Physics in Medicine and Biology 45:1423–1435
54. Reichling S (2007) Das inverse Problem der quantitativen Ultraschallelastografie unter Berücksichtigung großer Deformationen. PhD Thesis. Institute of mechanics, Ruhr-University, Bochum in press
55. Reichling S, Khaled W, Bruhns OT, Ermert H (2005) Shear modulus identification in soft tissue-like materials. PAMM 5(1):511–512
56. Ries LAG et al. SEER Cancer Statistics Review, 1975–2002, 2005, NCI: Bethesda MD
57. Rychagov MN, Khaled W, Reichling S, Bruhns O, Ermert H (2003) Numerical modeling and experimental investigation of biomedical elastographic problem by using plain strain state model // Fortschritte der Akustik. DAGA 2003. Proceedings. Aachen (Germany), pp 586–589
58. Saada AS (1989) Elasticity, theory and applications. New York: Pergamon Press: 3nd Edition, pp 377–399
59. Sakorafas GH (2001) Breast Cancer Surgery – Historical Evolution, Current Status and Future Perspectives, Acta Oncologica 40(1):5–18
60. Sandrin L, Catheline S, Tanter M, Hennequin X, Fink M (1999) Time-resolved pulsed elastography with ultrafast ultrasonic imaging, Ultrason. Imaging 21:259–272
61. Sarvazyan AP, Skovoroda AR, Emelianov SY et al. (1995) Biophysical bases of elasticity imaging. Acoustical Imaging 21:223–240

62. Scheipers U, Lorenz A, Pesavento A, Ermert H, Sommerfeld H-J, Garcia-Schürmann M, Kühne K, Senge T, Philippou S (2001) Ultrasonic multifeature tissue characterization for the early detection of prostate cancer. IEEE Ultrasonics Symposium, pp 1265–1268
63. SEER Cancer Statistics Review National Cancer Institute http://www.seer.cancer.gov/
64. Sinkus R, Lorenzen J, Schrader D, Lorenzen M, Dargatz M, Holz D (1999) MR-Elastography applied to in-vivo MR-mammography. Proceedings of the International Society for Magnetic Resonance in Medicine, 7th Scientific Meeting and Exhibition, Philadelphia, Pennsylvania USA, pp 22–28
65. Skovoroda AR, Emelianov SY, O'Donnell M, Gokhale NH, Feijoo GR (1995) Tissue Elasticity Reconstruction Based on Ultrasonic Displacement and Strain Images. IEEE Trans UFFC 42(4):1057–1067
66. Sumi C, Suzuki A, Nakayama K (1995) Estimation of shear modulus distribution of soft tissue from strain distribution. IEEE Transactions on Biomedical Engineering 2:193–202
67. Svensson WE, Amiras D (2006) Ultrasound elasticity imaging. Cambridge University Press, Breast Cancer Online, pp 1–7
68. Taylor LS, Porter BC, Rubens DJ, Parker KJ (2000) Three-dimensional sonoelastography: principles and practices. Phys Med Biol 45:1477–1494
69. Wellman PS (1999) Tactile Imaging. PhD Thesis. Harvard University
70. Wellman PS, Howe R, Dalton E, Kern KA (1999) Breast tissue stiffness in compression is correlated to histological diagnosis. Tech Rep, Harvard BioRobotics Laboratory, Harvard University, Cambridge, Mass, USA
71. Zhu C, Byrd RH, Nocedal J (1997) L-BFGS-B: Algorithm 778: LBFGS-B, Fortran routines for large scale bound constrained optimization ACM. Trans Math Software 23:550–60

Part III
Regenerative Medicine and Nanoengineering

Chapter 7
Aspects of Embryonic Stem Cell Derived Somatic Cell Therapy of Degenerative Diseases

Kurt Pfannkuche, Agapios Sachinidis, Jürgen Hescheler

University of Cologne, Center of Physiology and Pathophysiology, Institute of Neurophysiology, Robert Koch Str. 39, 50931 Cologne, Germany,
a.sachinidis@uni-koeln.de

Abstract Results from transplantation studies of embryonic stem cell derived somatic cells in animal models are promising for ES cell-based cell therapy of degenerative diseases such as heart and neurological diseases. However, clinical application of ES cell-based therapies will become possible after resolving the current barriers regarding safety aspects, purity and quantity of the cells, immunological rejection and ethical issues. Tissue engineering is an emerging research field that in combination with the ES cells might contribute to the development of new therapeutical concepts for treatment of severe degenerative diseases. Indeed, many important studies and concepts exist to develop strategies for generation of tissue engineered heart valves. Also, which heart cell type is optimal for cell therapy of cardiac insufficiency is controversially discussed. In this article we discuss interesting aspects of cardiac tissue engineering with special focus of embryonic stem cell derived cardiomyocytes. In addition, barriers of the tissue engineering approach as well as of the ES cells in general are discussed that should be conquered before ES-cell-derived therapies can be considered for therapeutical application.

7.1 Introduction

The institute of Neurophysiology of Cologne is mainly focused on investigating embryonic stem (ES) cell derived cardiomyocytes from murine (mES) and human stem (hES) cells. The functional characterisation of these cells is still ungoing to unravel details of their specific physiology. In addition, the complete transcriptome of the mES derived pure cardiomyocytes has been identified (Doss MX et al. 2007). Gene ontology analysis from the transcriptome shows a specific pattern of gene expression in the ES cells-derived cardiomyocytes that reflect the biological, physiological and functional processes occurring in mature cardiomyocytes. More over, cardiac specific signal transduction pathways were identified (Doss MX et al. 2007).

 In the present manuscript we are highlighting aspects of cardiac tissue engineering with special focus on ES cell derived cardiomyocytes. The generation of cardiomyocytes from ES cells is discussed as well as special topics such as cardiac progenitor cells, reprogramming of somatic cells the use of cell-permeable proteins, in vitro culture of viable heart tissue slices, recent advantages and technical hitches in cardiac tissue engineering.

7.2 Rationale for the Cardiac Tissue Engineering

During embryonic development the heart is the first organ of mesodermal origin. Initialized from a simple tube like structure it undergoes major morphological changes, recapitulating a sophisticated pattern, to form the four-chambered muscle pump, driving blood flow by continuous contractions through all vessels and capillaries of the human body. The heart wall is build up mainly from blood vessels, cardiomyocytes and fibroblasts. While cardiomyocytes are responsible for the contraction fibroblasts promote the electrical coupling of the cells and build up the extracellular matrix.

During fetal development cardiac myocytes rapidly proliferate but in the postnatal life proliferation ceases. The terminally differentiated cardiomyocyte has evaded cell cycle and has therefore lost its capacitance to duplicate. Even though there are reports that adult cardiomyocytes can increase DNA content and might also duplicate under some conditions this has to be considered as a very rare event. The mechanisms by which cardiomyocytes are terminating cell cycle are not completely understood but down-regulation of cyclins and cyclin-dependent kinases is found in adult cardiomyocytes. There is also evidence that the high expression of retinoblastoma susceptibility gene (rb) in differentiated cardiomyocytes is playing a critical role for the cell cycle exit. In addition the minimal incidence of cardiac tumors indicates for a possible tight and multilayered regulation of cell cycle in cardiomyocytes of higher vertebrates (for a detailed review see Ahuja et al. 2007). In contrast, adult cardiomyocytes of amphibians can contribute to tissue regeneration by mechanisms that are not fully investigated. In newt, a dedifferentiation of a subpopulation of cardiomyocytes was found after tissue injury. This dedifferentiation is accompanied by proliferation of the cells and tissue restore. The dedifferentiated cells are not restricted to the cardiac fate any longer and when transferred to the limbs they transdifferentiate into chondrocytes and skeletal muscle (Laube et al. 2006).

Although putative cardiac stem cells and side populations of cells are discussed to retain a potential to expand and to adopt a cardiac phenotype, it is a fact that the contractile tissue damage occurring during cardiac infarction cannot be repaired neither by regenerative functions of the organ itself nor by circulating stem or progenitor cells.

Cardiac infarction is the outcome of a dramatically reduced oxygen supply, mainly as a result of an occlusion of atherosclerotic vessels and ends up with a rapid loss of contractile cells. During the acute phase an inflammatory process is initiated by macrophages, monocytes and neutrophils migrating into the infarcted tissue. The extracellular matrix is damaged by matrix metalloproteases, resulting in a thinning of the ventricle wall. During the recovery process, a remodelling of the injured area occurs. A scar tissue is formed, mostly consisting of fibroblasts that are forming large quantities of extracellular matrix proteins, namely collagen, building a fibrous scar tissue with implemented mechanical properties and reduced elasticity.

Cardiac tissue engineering may be useful for repairing of damaged myocardium to improve contractile function. In principle, two approaches are promising to engineer cardiac tissue: First, it may be possible to build up an artificial tissue *in vitro* and transplant it as a functional patch of beating tissue or second, to perform *in vivo* tissue engineering by introduction of cells as pure preparations or in combination with matrix compounds that can repair the damaged organ directly.

7.3 Embryonic Stem Cells as an Unlimited Source for Cardiomyocytes

During the past decade there is an emerging field of embryonic stem cell biology. These cells have the fascinating properties to be able to proliferate infinitely ("self-renewal") and to differentiate into any cell type of the body (pluripotency). Murine ES (Fig. 7.1) cells can be easily modified *in vitro* to express transgenes. Genetically engineered vectors for labelling and selection purposes are of interest in the view of tissue engineering.

ES cells differentiation into cardiomyocytes is spontaneous in the absence of leukaemia inhibitory factor (LIF). For murine ES cell lines large scale bioreactor systems (Schroeder et al. 2005) have been developed to achieve high numbers of cardiomyocytes. In contrast, human ES cells have a lower capability to generate cardiomyocytes by spontaneous differentiation and further development is neces-

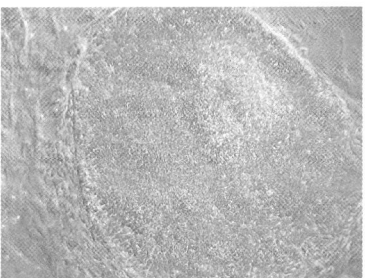
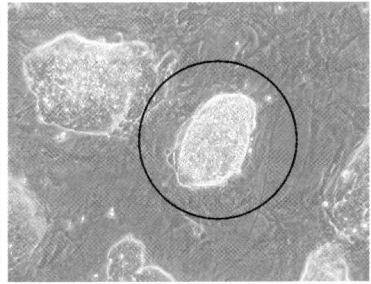

Fig. 7.1 Phase contrast images of embryonic stem cell colonies. The colonies are grown on feeder cells. Feeder cells are mitotically inactivated fibroblasts, generated typically from murine embryos (there are also human foreskin fibroblasts in use for human ES). The *left panel* shows a colony of human ES cells (H1 line), the *right panel* shows murine ES cells (D3 line). A colony with typical morphology is marked by the *black circle*. It appears in *elliptical shape*. Colonies with irregular, flattened borders and cellular outgrowth are of minor quality with high degree of differentiated cells. Some murine ES cell lines are adapted for feeder free cultivation, but today there is no sufficient method to keep human ES cells feeder free for long culture periods

sary to generate human stem cell derived cardiomyocytes in numbers that could be used for transplantation experiments.

To initiate the differentiation of the ES cells, three-dimensional aggregates, the "embryoid bodies", are formed by cultivation of the ES cell suspension on a non-adherent dish under continuous agitation or by cultivation in hanging drops. Applying the hanging drop technique the cell suspension is dispensed in drops of 20 μl, containing around 400 mES cells/drop, onto the lid of a non-adherent bacteriological plastic dish and placed upside down to collect the cells in the tip of the drop. During a 2-days incubation the mES cells form ball shaped aggregates, the embryoid bodies (EBs) and differentiation is initiated. After two days, the hanging drop EBs are collected and further incubated in cell culture medium, containing fetal calf serum, at a concentration of 200 to 1000 EBs per 10 cm dish. Within an EB all three germlayers are formed during this differentiation process and many different types of cells are developing including skeletal and cardiac myocytes, smooth muscle cells, neurons, hepatocytes, endothelial cells, chondrocytes, hematopoietic precursors, keratinocytes, osteoblasts and fibroblasts. After eight to nine days of differentiation, mEBs start contracting rhythmically indicating the appearance of cardiomyocytes. These cardiomyocytes are not suitable for tissue engineering or transplantation purposes since they still contain some undifferentiated stem cells that possess a high tumorigenic potential *in vivo*. Since there is no specific surface epitope known, that is specifically expressed on cardiomyocytes, antibody-based enrichment strategies like magnetic cell sorting (MACS) cannot be applied. As an alternative lineage selection strategies have been applied. Zandstra and colleagues described a transgene that consists of the α-cardiac myosin heavy chain promoter driving the expression of the aminoglycoside phosphotransferase (neomycin resistance) gene (Zandstra et al. 2003). The transgenic ES cells can be differentiated and cardiomyocytes are highly enriched by application of the antibiotic agent geneticin. Another method to label and purify mES-derived cardiomyocytes was developed by Eugen Kolossov (Kolossov et al. 2006). He decided to use the cardiomyocyte-specific alpha-cardiac myosin heavy chain promoter to control the expression of enhanced green fluorescent protein (eGFP) and puromycin acetyl transferase (PAC). The selection transgene contains a bicistronic expression cassette: The eGFP and PAC genes are coupled via an internal ribosome entry site (IRES) that allows the expression of both transgenes from a single messenger RNA. The eGFP enables monitoring of the upcoming cardiomyocytes in the embryoid body due to its bright green fluorescence and it indicates the time point when selection can start. The PAC gene contributes a resistance against the strong antibiotic puromycin. After one week of selection only cardiomyocytes have survived the puromycin treatment (Fig. 7.2). At this time point the cardiomyocytes form vigorously beating clusters of bright green fluorescent cells reaching beating frequencies of several hundred contractions per minute like they would do in a native murine heart. Even though a purity of 99% could be shown after antibiotic selection (Zandstra et al. 2003; Kolossov et al. 2006), and Oct-4 as a marker for undifferentiated cells is not detected after puromycin-selection for 10 days there is still a risk of contaminating undifferentiated cells. Since ES cells are not limited in there proliferation capacity

7.3 Embryonic Stem Cells as an Unlimited Source for Cardiomyocytes

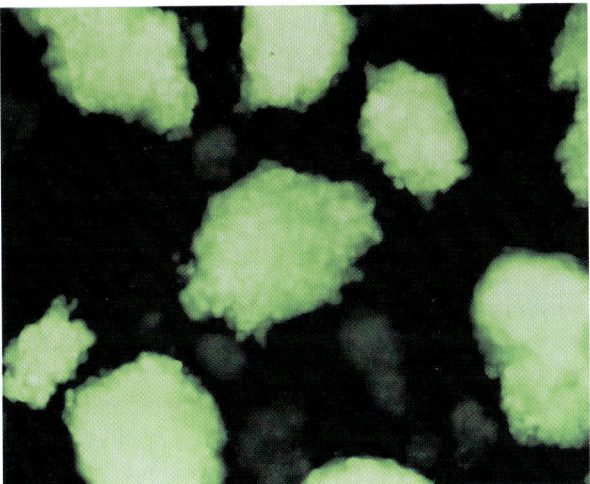

Fig. 7.2 Puromycin selected cardiomyocytes from murine embryonic bodies. The cardiomyocytes express transgenes for enhanced green fluorescent protein and puromycin-acetyltransferase. After selection all non-cardiomyocytes have been killed and disappeared from the spheroid. A compact cluster of beating and green fluorescent cells remains

a single surviving cell can expand rapidly if the selection pressure is taken away, by use of the purified cells for transplantation or *in vitro* tissue engineering. The consequence for a patient treated with ES cell derived cells could be the development of terratomas originating from impurities in the cell preparation. Therefore safety is a major aspect when cells derived from embryonic stem cells are considered for tissue engineering. A second barrier for a therapeutical use of ES derived somatic cells might arise from immunological intolerance. Early differentiation stages of ES cell derived cells and ES cells themselves are not eliminated by the immune system when transplanted in an allogeneic animal. The most likely explanation namely the absence of major histocompatibility complex molecules (MHC) on the cell surface turned out to be not true. Although the expression of MHC class I on mES cells is below the detection threshold for flow cytometry the transcripts could be identified by RT-PCR and ES cells that were transfected with the lymphocytic choriomeningitis virus (LCMV) are recognized by virus-specific $CD8^+$ cytotoxic lymphocytes (CTLs) (Abdullah et al. 2006). Even though recognized by the lymphocytes the LCMV infected ES cells are not lysed. The reason for this immune privilege was found in the expression of serin-protease-inhibitor 6 (SPI-6) in mES cells and their derivatives, making these cells tolerant against granzyme B attack by cytotoxic lymphocytes. It could be estimated that the mES cells and cells from embryoid bodies are expressing SPI-6 in levels comparable to CTLs, which have to protect themselves against the toxic effects of the granzyme B they secret. Knockdown of the SPI-6 transcript by RNA interference abolished protection of ES cells against CTL-mediated lysis. How tolerance against immune attack is altered in mature ES cell derived cells is unknown and the use of cells that are not recognized as foreign tissue would be preferable. In the light of safety and tumorigenic potential cells that can evade the surveillance by the immune system are truly problematical. Permanent knock-down of genes like the SPI-6 and safety mechanisms may tackle this task.

7.4 Therapeutical Cloning of Embryonic Stem Cells

Generation of ES cells that are identical with the patient to be treated is still only a theoretical possibility. Since cloning of ES cells by nuclear transfer from a somatic cell to an oocyte is of very low efficiency and the generation of an embryo to derive ES cells is for ethical reasons no option for human applications alternative ways to generate cloned ES cells have to be found. For applications apart from usage to generate cloned human ES cells cloning has become reality. In 2003 Cesare Galli (Galli et al. 2003) was the first to succeed in generation of a cloned horse. Meanwhile the French company Cryozootech offers cloning services for valuable horses. The first horse to be cloned by Cryozootech was the stallion Calvaro, successful in jumping events. From 2000 oocytes used for nuclear transfer only 22 embryos could be generated for the transfer in foster-mares and finally a healthy clone was born. Others will follow to retain valuable individuals for the breed.

To produce ES cells that are genetically identical with the donor cell it is not likely that cloning will ever become the method of choice for the following reasons. First, it will not be ethically accepted to generate human embryos for the extraction of ES cells. Second, the process is very inefficient and a sufficient number of oocytes can't be obtained for large-scale application. Third, the costs of this technique are extremely high, due to its low efficiency. As a further point it might be argued that there are known problems concerning preterm aging of cloned animals, but it was reported that mES cells derived from cloned embryos do not show any abnormalities pointing to a selection mechanism during the ES cell derivation process (Brambrink et al. 2006).

It is known for many years that the fusion of an embryonic stem cell with a differentiated (somatic) cell can result in the reprogramming of the somatic nucleus (Tada et al. 1997, 2001). In this context, reprogramming describes a process that reverts the epigenetic program of the somatic nucleus to a stem cell like state. The memory of the somatic nucleus, that is mainly fixed as a pattern of DNA methylation, regulating gene activities, is erased during the reprogramming process. Common interest in these experiments was initiated by the publication of Cowan and colleagues (Cowan et al. 2005), demonstrating the reprogramming of human fibroblasts by polyethylengycol-mediated fusion with human embryonic stem cells. The fusion cells are selected by application of a double selection system, based on two different antibiotic-resistances expressed in the hES cells and the somatic cells. By this simple double selection approach tetraploid cells can be enriched, carrying the chromosomes of the ES cell and the somatic cell that is going to be reprogrammed by factors originating from the ES cell nucleus. The reprogrammed cells recapture the ability of ES cells with respect to proliferation and differentiation qualities. The

7.4 Therapeutical Cloning of Embryonic Stem Cells

disadvantage of this technique is the presence of the nucleus from the ES cell that was fused to the somatic cell to initiate reprogramming. As a consequence the cell fusion approach is first of all a proof of principle, demonstrating that a somatic cell can be transformed into an ES cell under appropriate conditions. For therapeutical applications the reprogrammed fusion cells offers not an alternative since the nuclei of the fusion partners are fused to one nucleus that is initially containing a tetraploid chromosome set. Current focus of research is addressing several lines to solve the problem. One approach is focused on the fusion between ES cells and somatic cells. It was found that the tetraploid fusion cells can eliminate chromosomes while kept in culture but this appears to happen randomly. Consequently scientists are investigating novel techniques to perform directed elimination of chromosomes from a cells nucleus by application of the Cre/loxP system (Matsumura et al. 2006).

The Cre recombinase is widely used as a DNA modifying enzyme. It can detect 34 basepare consensus sequences the "loxP sites". Between two loxP sites the Cre recombinase performs recombination that can result in deletion of a DNA segment if the loxP sites are placed in line to flank a specific region. If loxP sites are placed in opposite directions the recombination can result in inversion of the flanked target, but no deletion of the loxP-flanked region occurs (Fig. 7.3). To perform directed chromosome elimination Matsumura and colleagues designed a chromosomal elimination cassette (CEC) that contains an expression vector for green fluorescent protein and a puromycin-acetyltransferase driven by the ubiquitous active CAG Promoter. The expression cassette is flanked by loxP sites that are located in opposite directions. When stable transfection is performed the CEC integrates randomly in the genome. Matsumura chose two mES clones where the CECs were located on chromosome 11 respectively chromosome 12. The transfected cells can be selected with puromycin and express GFP. During the cell cycle each chromosome is replicated and both copies of the DNA strand are attached at the centromer. Before segregation of the sister chromatids occurs during anaphase of the cell cycle the CEC is present on both arms of the chromosome. If Cre recombinase appears in the cells after transient transfection with a Cre expressing plasmid the recombinase can recombine the CECs on the sister chromatids like it is demonstrated in (Fig. 7.4). The recombination results in a circular dicentric chromosome and a acentric fragment. Both products are lost during cell cycle, resulting in the loss of the complete chromosome. The modified cells lose the GFP fluorescence and can be enriched by fluorescence activated cell sorting as the GFP negative population. Cells that do not contain duplicated CEC are not altered by the Cre recombinase since the loxP sites are placed in opposite directions and recombination can only result in inversion, but not in deletion of the CEC. By this novel technique the deletion of single chromosomes and also chromosome pairs could be demonstrated and it might be possible to adapt it for the deletion of several chromosomes, but still it is not possible to remove a complete chromosome set.

Coming back to the previous point additional approaches aim to reprogram somatic cells by transfection with expression vectors for factors that are believed to realize reprogramming. Takahashi transfected embryonic and adult fibroblast with expression vectors for Oct-3/4, Sox2, c-myc and Kfl4. As a result the differentiation

ATAACTTCGTATA GCATACAT TATACGAAGTTAT

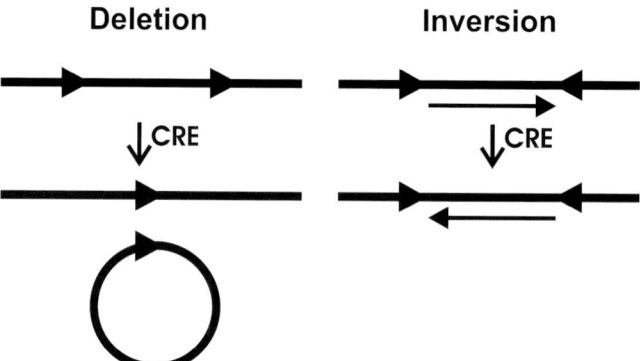

Fig. 7.3 The Cre recombinase detects the loxP sites, consisting of 34 basepares (*upper line*). Depending on the orientation of two loxP site (*triangles*) in the genome a recombination occurs: LoxP sites that are placed in the same direction lead to deletion of the flanked target whereas the flanked region is only inverted if the loxP site are in opposite direction (since the inversion can take place for several rounds the result of the recombination is unpredictable in this case)

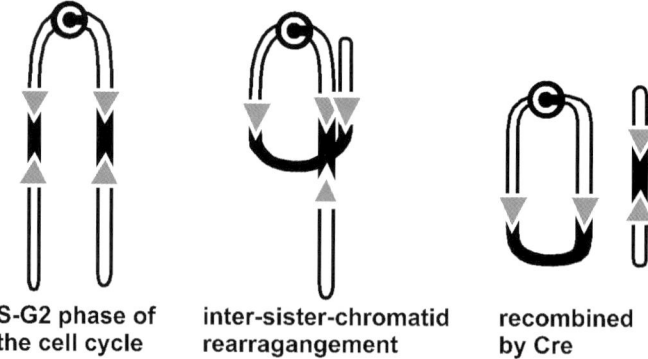

Fig. 7.4 Chromosomal deletion by Cre recombinase. The *left panel* shows a chromosome in S to G2 phase of the cell cycle. The cell has already replicated the DNA and the sister chromatids of each chromosome are associated at the centromer. The chromosome elimination cassette (CEC) is represented by a *black bar with triangles* for the loxP sites. If Cre recombinase is now introduced in the cell a recombination between the two chromatids can occur. This results in a circular fragment and an accentric, linear one. Both products are lost during the cell cycle

pattern of the fibroblasts was abrogated and the resulting cells exhibit ES cell like properties (Takahashi 2006). To circumvent the disadvantages of cell transfections, like the possibility of stable vector integration, in the genome and the low transfection efficiencies faced with many primary cells, cell permeable proteins could be the solution. The TAT domain from a HIV virus protein was shown to mediate protein passage through cell membranes. We will elucidate some details about the use of cell permeable proteins later in this article when cell permeable Cre recombinase is introduced. Briefly the cell permeable proteins are expressed as fusions of the target protein and the TAT domain in bacterial expression systems. The use of cell permeable pluripotency factors could simplify the application of these factors for means of reprogramming. The workgroup of Frank Edenhofer (University of Bonn, Germany) is addressing this option by evaluating cell permeable pluripotency factors like TAT-Nanog (Edenhofer F, submitted). It is reasonable that future reprogramming can be performed by isolation and treatment of somatic cells with a cocktail of cell permeable pluripotency factors resulting in the formation of pluripotent stem cells.

7.5 Stem Cell Derived Cardiomyocytes

Stem cell derived cardiomyocytes from murine systems can be generated and purified relatively easily in numbers of several million cells to perform tissue engineering studies and transplantation experiments in murine model systems. The differentiation of hES cells appears more challenging today, but it can be expected in the near future that technologies will emerge enabling a large scale production of hES cell derived cardiomyocytes as well. ES cell derived cardiomyocytes contract rhythmically without electrical pacing. Furthermore, applying electrophysiological studies action potential shapes of pacemaker – as well as atrial – or ventricular-like cells could be identified. RT-PCR and quantitative real time (qRT)-PCR analysis have shown that cardiomyogenesis in EBs resembles that of the normal myocardial development. The earliest cardiac transcription factor NKX2.5 is followed by the expression of atrial natriuretic factor (ANF), myosin light chain and then α- and β-myosin heavy chain (α-MHC and β-MHC). The early ES cell derived cardiomyocytes are structurally less organised than cardiomyocytes in the neonatal heart. Sarcomeric α-cardiac actinin is found as a striated pattern when beating cells appear but striated areas are often not covering the whole cell and have varying spatial orientations in different regions of the cell. In Fig. 7.5 a stack of optical sections through a day 9 EB is shown. The EB was fixated and stained with anti cardiac α-actinin. The striated staining pattern demonstrates the organisation of structural proteins in these very early ES cell derived cardiomyocytes. Figure 7.6 shows a plated, puromycin purified cardiomyocyte, stained for cardiac α-actinin.

To analyse the capability of these cells to engraft *in vivo*, transplantations into healthy tissue and more important into cryo-damaged hearts have been performed. For these studies a liquid nitrogen cooled copper stamp is used to induce a cryo-injury to the ventricular wall, resulting in large areas, depleted of cardiomyocytes and forming a scar tissue. These cryo-injuries can serve as a model for scar formation in cardiac infarction allowing to study the following interesting aspects:

- Do the transplanted cells engraft and survive?
- Will they be rejected by the immune system?
- Do they undergo electrical coupling to the surrounding tissue or do they remain electrical separated?

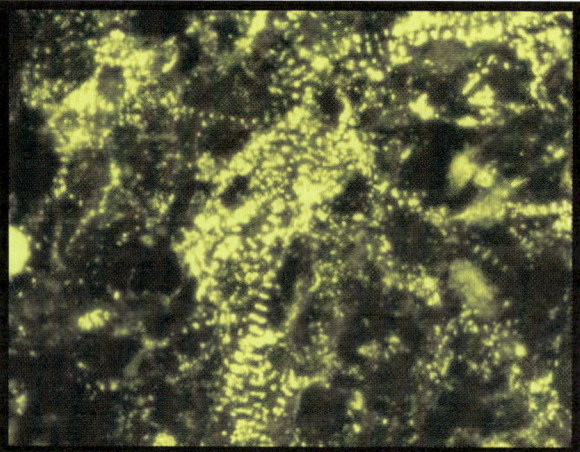

Fig. 7.5 Superimposition of optical sections through a day 9 murine embryoid bodies stained for α-cardiac actinin. The striated patterns indicate a premature orientation of the early cardiomyocytes

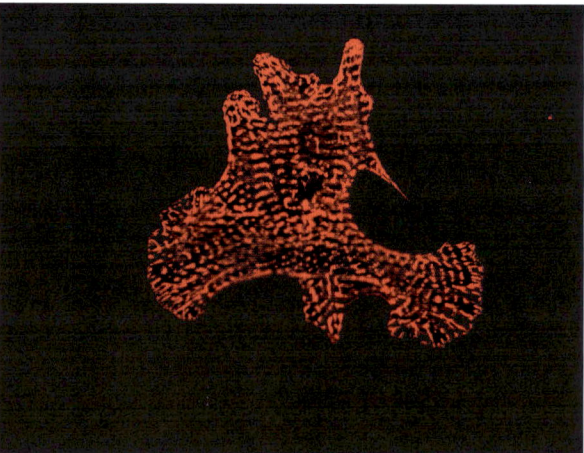

Fig. 7.6 Puromycin selected ES cell derived cardiomyocyte. The cell was plated on a glascoverslip and stained for α-cardiac actinin. The typical striation pattern can be found

- Will the cardiac function improved after the transplantation?
- Will transplanted cells induce arrhythmias?

Till now many studies have addressed different aspects of ES cell derived cardiomyocytes. Most of them with murine cells, but also some data for hES cell derived cardiomyocytes have been collected. It has turned out that the stem cell derived cardiomyocytes indeed are able to engraft in the cryo-damaged tissue but the number of surviving cells is rather low. A beneficial effect for cardiac function has been shown, demonstrated by an improved left ventricular ejection fraction and reduced enddiastolic volume (Kolossov et al. 2006) but it is not clear which components give rise to the improvement. Beside active mechanical contribution of transplanted cardiomyocytes passive effects like enhanced elasticity and reduced wall thinning of the cryo-injured ventricle as well as paracrine effects have to be considered (Gnecchi et al. 2005, 2006).

One very important aspect of the transplantation of contractile cells is their electrical coupling to the surrounding cells. ES derived cardiomyocytes represent a relatively early developmental stage, they are self contracting and they differ in their electrophysiological properties from adult cells. Especially if the cells are not dense and equal distributed after transplantation into the infarct zone they may have limited contact to the surrounding tissue and might be unable to form connexin 43 junctions permitting exaltation spreading. The lack of sufficient contact via gap junctions will probably lead to exaltation-contraction uncoupling in the transplanted heart because the transplanted cells will execute spontaneous self-contractions if coupling to the surrounding myocard is not adequate. An insufficient coupling might be undetected at lower heart frequencies but leads to uncoupling during phases of increased heart frequency. As a consequence electrophysiological parameters like impuls propagation, excitation-contraction coupling, action potential maturation and also the response to drugs and hormones have to be analysed in detail. The risk of transplanting cells that do not couple to the surrounding myocard has become obvious in early studies with skeletal myoblasts that were performed as clinical trials. Even though some improvements in function could be reported the transplantation of myoblasts from satellite cells resulted in ventricular tachycardia or cardiac dead in 10 out of 22 patients included in the first studies (Menasche et al. 2003; Smith et al. 2003).

7.6 Cardiac Tissue Slices

Several of these parameters can be addressed by a very sophisticated *in vitro* transplantation model that has been recently developed using the preparation of viable heart slice cultures (Pillekamp et al. 2005; Halbach et al. 2006). Some time after transplantation the heart is isolated and embedded directly into low melt agarose to encapsulate the whole organ in agarose without warming. Using a vibratome the agarose embedded tissue can be sliced into slices of $150-300\,\mu m$ and be cultured *in vitro*. Due to the low thickness oxygen and nutrients can diffuse (penetrate) through enabling the cardiac slice preparations to stay viable and contractile for several days (depending on the age of the donor heart) for further downstream applications. Multielectrode arrays (MEAs) with 8×8 electrodes, and an electrode distance of $200\,\mu m$ (Multi Channel Systems, Reutlingen, Germany), Fig. 7.7 are used to asses field potentials of the slices and to generate maps characterizing the excitation spread based on the temporal delay between field potentials ablated from electrodes at the origin of an exaltation and more distal electrodes of the MEA.

Very recently, to assess the properties of transplanted cells the slicing technique to achieve viable tissue slices for electrophysiological measurements on single cells

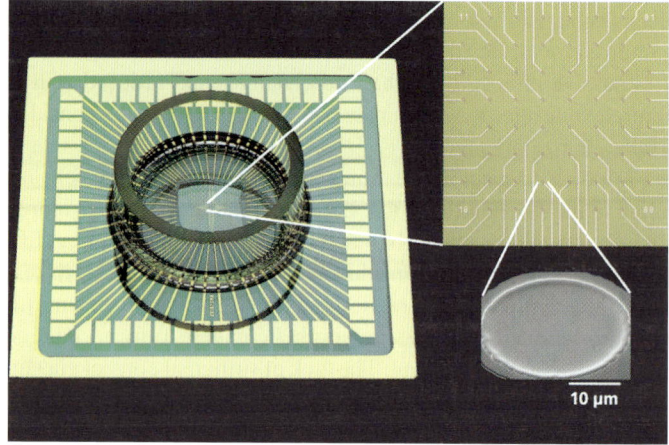

Fig. 7.7 Multi-electrode array (MEA) for the detection of field potentials. These arrays are used for example to address exaltation spread in heart slice preparations and plated clusters of cells. (With permission of Multi Channel Systems, Reutlingen, Germany)

inside a functional tissue has been applied (Halbach et al., submitted). Since the slices are relatively thin, it is possible to locate green fluorescent, transplanted cells inside the slice on an inverted fluorescence microscope (Fig 7.8). With a sharp electrode setup a thin glas electrode is introduced into an individual cell and action potential traces are recorded. By this technique it is possible to measure electrophysiological capabilities in one single transplanted cell (Halbach et al., submitted). Further additional fascinating possibilities are: The slices can be paced by a defined stimulus to beat at a chosen frequency – and it is possible to analyse how the transplanted cells are responding by the sharp electrodes. By this approach questions can be answered whether the transplanted cells can take over the rhythm of the surrounding tissue or whether they are able to held pace when contraction frequency is rising. This approach enables a detailed insight into the electrical coupling after cell

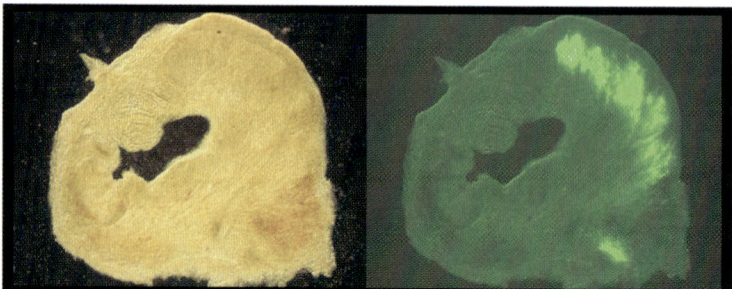

Fig. 7.8 Cardiac slice preparation. The *left panel* shows a transmission light image of a murine adult heart slice. Green fluorescent cells that had been transplanted to the heart earlier can be detected under fluorescent light (*right panel*). Foto: Marcel Halbach

transplantation. Coupling of the beating cells is one of the most important hallmarks of successful restoration of cardiac tissue. Apart from the analysis of transplanted hearts, with respect to electrical coupling and maturation of the transplanted cells, the slicing technique enables scientist to derive functional heart tissue slices and also cardiac matrix for *in vitro* studies. The latter are prepared by a controlled oxygen withdrawal applied to the slice preparation. This results in an elimination of contractile cells from the matrix. These matrices are ideal to analyse mechanisms of cell integration *in vitro* since they can serve as a perfect model for a native cardiac matrix since it is prepared from the heart itself (Pillekamp et al. 2007).

Indeed, the most prominent feature of a transplanted stem cell derived cardiomyocyte is their ability to contract and to build up force. Based on de-cellularized, non-contractile cardiac slices it is possible to directly measure the force that can be generated by stem cell derived cardiomyocytes when these cells are cultivated on the slice matrix. Since the heart is having two ventricles every ventricular heart slice preparation is bearing two holes. Frank Pillekamp has build up a setup where the slice can be placed on two thin steel needles each going through one whole of the preparation. One needle is fixed, the second one is attached to a sensitive force transducer. Keeping the slice in a bath of tempered medium the contraction force of the slice or of stem cell derived cardiomyocytes, seeded on a de-cellularized slice, can be directly measured. The system is equipped with an electrical stimulation device to pace the preparation with a defined frequency to assay the response of the contractile cells on different beating frequencies.

7.7 Bioartificial Heart Tissue Based on Biomaterials

Transplantation of cells is one way of addressing tissue repair. But is it possible to build up tissues *in vitro*? Indeed, today's biotechnology is far away from the possibility to create a whole organ by tissue engineering techniques. But thinking of future technology one might speculate that generation of small bioartifical components becomes reality. Today, scientists are generating neurons growing on semiconductor chips to evaluate possible interactions, creating interfaces between living cells and electronic components. But will it be possible to build up a pump just from cells and proteins? Maybe an *in vitro* construction of the heart is far away from our reach, but a simple pump consisting of a contractile tube with valves to direct liquid flow might be reality soon (Kubo et al. 2007). Living organs have some great properties: Some can regenerate themselves, by cell proliferation and replacement of damaged cells. Remarkable, some organs are able to serve their specific function for more than 150 years without interruption – just thinking of a turtles heart. Even though it is only speculation one of the most fascinating aspects of cardiac tissue engineering, beyond clinical applications, is the fascination of creating contractile modules just from custom made proteins or scaffolds and embryonic stem cell derived cells that might give rise to bioartificial components like pumps and force producing units one day.

7.8 Scaffolds for Cardiac Tissue Engineering

When *in vitro* cardiac tissue engineering is discussed, the optimal cell composition and the appropriate matrix have to be considered. With respect to the matrix, there are two different strategies: The cells can be cultivated on protein or polymer matrices or they can be entrapped in gels.

Gels have one great advantage above other matrices: The seeding of the cells is relatively easy since the cells can be equally distributed in the gel by resuspending them before the gel is cast. Collagen I polymerises rapidly in non-acidic solutions at 37 °C and forms gels and by the use of molding forms it is possible to determine the gels shape. Major effort on the field of collagen gel based tissue engineering was performed by the workgroups of Wolfram-Hubertus Zimmermann and Thomas Eschenhagen. They developed a so-called engineered heart tissue (EHT) from neonatal rat heart cell preparations. The cells are dissociated and a gel composed of collagen I and matrigel (extracellular matrix from Engelbreth–Holm–Swarm tumor) is formed. Ring shaped collagen gels are cast and cultured *in vitro*. The cells within the gel keep contracting and organization and maturing of the embedded cells can be achieved by stretching the collagen rings rhythmically. The EHT can survive in culture for several weeks without loosing its contractile abilities and it could be found to survive likewise *in vivo*. The ring shaped gel can be used to measure contraction forces and it could be shown with this model, that controlled mechanical stimulation, by rhythmical stretching of the gel, results in an ultrastructural maturation of the cells, pointing to the assumption that mechanical stimulation is one parameter involved in cardiomyocyte maturation (Fink et al. 2000).

Transplantation experiments with EHTs showed strong vascularisation and signs of terminal differentiated grafts (Zimmermann et al. 2002). Most interesting the EHTs can be stacked to larger structures and they are able to fuse and to synchronize their beating frequencies within seven days of culture. To generate a contractile patch to be used for an *in vivo* experiment five ring shaped gels were stacked crosswise on a custom made holding devices and cultured for fusion to generate a single EHT. The multiloop EHTs were transplanted to rats with large myocardial infarcts, generated by left descending coronary artery ligation 14 days prior to the EHT transfer. Non-myocyte and formaldehyde fixed grafts were used as controls. An analysis of the transplanted animals was performed 4 week following grafting. At this time point the grafts could be clearly distinguished from the native tissue due to their pale colour, indicating a high content of connective tissue and a relative low density of contractile cells compared to native tissue. Structural analysis revealed the formation of compact and well-differentiated cardiac muscle fibres covering the infarcted myocardium and the formation of blood vessels in the EHT. In addition, evidence for an electrical coupling between the EHT and the recipient heart could be found. Most interestingly functional measurements based on echocardiography, catheterization and magnetic resonance imaging demonstrate an improvement to the diastolic and systolic function in EHT treated hearts (Zimmermann et al. 2006).

As an alternative to protein-gel based approached solid matrices can be used for tissue engineering. These matrices are consisting of proteins that are involved in the formation of the extracellular matrix or of customized biopolymers like Poly-L-lactic Acid. To serve as an appropriate skeleton for the use in cardiac engineering, the matrix needs to be elastic but not stiff like materials that are used for example for bone engineering. Collagen type I is a frequently used matrix component in tissue engineering approaches. It can be engineered to form porous sponges by freeze drying procedures. During this process a solution of collagen is frozen with a defined temperature gradient and vacuum dried. The resulting pores are around 50 μm in diameter depending on the cooling gradient. Tremendous efforts of companies such as matricel (Aachen, Germany (www.matricel.de)) resulted in the optimization of this process and in the production of the collagen sponges with high quality and reproducibility. The collagen sponges for basic research are usually fabricated from collagen of animal sources but the technologies to utilize pure human proteins have already evolved: Companies such as FibroGen (San Francisco, USA (www.fibrogen.com)) have established the large scale production of recombinant human collagens in yeast expression systems. These human collagens are largely used for a broad spectrum of applications like tissue engineering, dermal augmentation in plastic surgery, wound healing and as haemostats to control wound bleeding.

Recently, we used freezed dryed collagen type I sponges from Matricel inc. and seed mES cell derived cardiomyocytes onto this 3D-scaffold. While trying different seeding techniques it turned out to be very challenging to achieve equal distribution and dense cell attachments. Purified cardiomyocytes from mES cells are difficult in attachment and the adhesion on a cell culture surface can take several hours. To seed the cells a dense cell suspension can be used and agitated with the matrix free floating in the suspension. By this technique a homogenous distribution of the cells on the surface can be expected, but high numbers of cells are needed. Another option is to build a funnel that concentrates the cells or cell clusters by sedimentation on the matrix. To seed ES derived cardiomyocytes on the collagen sponge the cell suspension was concentrated to a very high cell number of around 200,000 cells per 20 μl total volume. The suspension is directly placed within a single drop of medium onto the sponge and incubated for cell attachment. Anyway due to the slow adhesion kinetics many cells are washed out when more culture medium is added to the preparation. In Fig. 7.9 a collagen sponge is shown. The photograph was taken one day after seeding with green fluorescent cardiomyocytes that were used as clusters of cells for this experiment. For the evaluation of the seeding procedure the use of these GFP labelled cardiomyocytes turned out to be essential for monitoring cell adhesion. Without any staining it is possible to observe the morphology of seeded cells in the upper layers of the construct. Due to the strong GFP fluorescence, the shape of the cells is visible on an inverted fluorescence microscope. Round shaped morphologies point to non-satisfying attachment whereas elongated cells can be taken as properly attached and the contraction of the cells as well as the contraction of the surrounding matrix material is visible. At day four of seeding we could observe contractions of the whole collagen sponge. When fixed and assayed for sarcomeric α-cardiac actinin we found a very organized striation pattern of the seeded cells indicating a mature phenotype and more interestingly it could be observed that large numbers of ES derived cardiomyocytes cluster together to form fibre like contractile structures with parallel orientated cells (Fig. 7.10). On the other hand the cells have no tendency to migrate into deeper regions of the sponge but are concentrated on the surface.

In vivo most tissues are infiltrated by a dense network of capillaries. Those are maintaining a sufficient oxygen pressure and nutrient supply to the cells. On the other hand the blood flow takes up metabolic residues like CO_2 and maintains physiological pH values. Tissues with high metabolic turnover like the heart have dense capillary networks. In rats the distance between single capillaries in the epicardium is below 20 μm, resulting in a density of 2800–3200 capillaries per square mm tis-

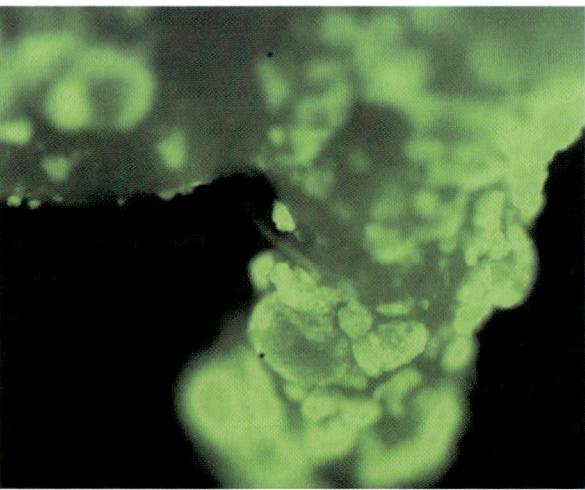

Fig. 7.9 Puromycin selected ES cell derived cardiomyocytes were seeded as clusters on a artificial collagen matrix. The image was taken one day after seeding, indicating that cells show loose attachment

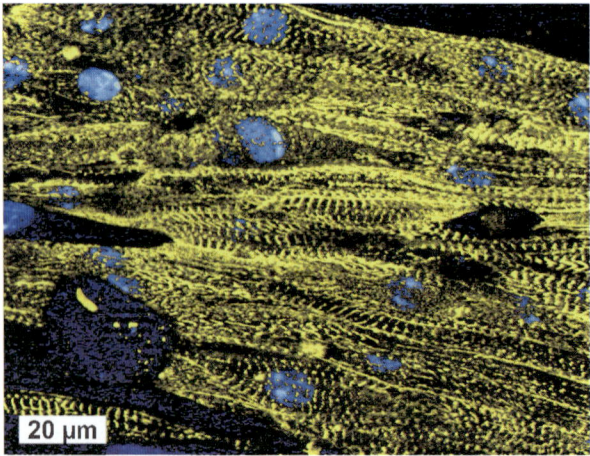

Fig. 7.10 ES cell derived cardiomyocytes seeded on an artificial collagen scaffold. One week after seeding the cells have formed fibre like structures with parallel orientation

sue (Korecky et al. 1982). In an *in vitro* engineered tissue it is until now not possible to mimic the circulation system that maintains supply of tissue *in vivo*. Therefore medium exchange can take place only through pores of the matrix material strongly limiting the size of the engineered tissue. This phenomenon is less problematic when tissues such as cartilage are considered that have low oxygen consumptions but it limits the construction of tissue with high metabolic rates such as the heart muscle. Practically, the impaired tissue supply results in cell densities per volume of engineered heart tissue that are quite low compared to native heart. For cells that have a high capacitance of proliferation this might be overcome by *in vivo* vascularisation. In several studies, a vascularisation of transplanted engineered tissues could be found. By *in vivo* vascularisation seeded cells might be expanded to achieve higher cell densities. Since expansion of the cardiomyocytes is not expected this option is less helpful in cardiac tissue engineering. Trabecularization, seen in the engineered heart tissue (EHT) described by Zimmermann and Eschenhagen, can resolve this problem partially. In the EHT the myocyte strands form a loose network of fibres that are usually 30–50 μm (sometimes up to 100 μm) in diameter permitting liquid transition between the fibres but limiting the density of the contractile cells and therefore the force developed.

7.9 The Ideal Cell

The idea of tissue engineering is to build up a tissue from its components including in particular the appropriate cells. Because the heart wall is based on cardiomyocytes and fibroblasts it seems to be obvious to use those cells for regeneration processes. When ES cells are differentiated in embryoid bodies they recapitulate processes of early embryonic development. Before cardiomyocytes appear around day 8 in murine embryoid bodies progenitor cells are developed. In contrast to the undifferentiated embryonic stem cells that have the capability to develop any somatic cell type, progenitor cell populations are restricted to differentiate in to some tissue specific cell lines. More recently many research projects are focused on such progenitor cells. In this regard, several progenitor cell lines have been identified. For example the brachyury-progenitor cells that are characterized by the expression of the t-box transcription factor brachyury. Expression of brachyury starts at day 2 of differentiation of mES and peaks at day four or five depending on the ES cell strain and culture conditions. Within 2 days a strong expression of brachyury can be observed declining rapidly to very low levels, indicating the short lifetime of the progenitor cells that undergo further differentiation processes. Since the first mesodermal cells are characterised by brachyury these progenitors display a rather large plasticity and are potentially able to differentiate into all mesodermal lineages and give rise also to endodermal lineages (Kubo et al. 2004).

A more restricted progenitor cell population is characterized by the expression of the LIM homeodomain transcription factor Isl1 (islet 1). In the murine embryo, Isl1 marks a subset of undifferentiated cardiac progenitor cells, originating from the secondary heart field, substantially contributing to the developing heart. It could be demonstrated that the Isl1 progenitor cells hold the potential to differentiate into cardiomyocytes as well as into endothelial cells and smooth muscle. With respect to cardiac tissue engineering a cell type that is able to proliferate and to differentiate into cardiomyocytes as well as into cells that build vessels and capillaries might be an ideal candidate for future experiments. Beside the advantages of embryonic stem cell based technologies Isl1 progenitor cells can also be found in neonatal hearts of rodents and potentially also of human hearts. By lineage tracing experiments in rats it was found that 30–40% of the myocardium is build from Isl1 progenitor cells, but the number of progenitors decreases dramatically during the embryonic development and shortly after birth the heart of a rat contains around 500–600 remaining Isl1 positive cells (Laugwitz et al. 2005). These cells are potentially remnants of the fetal Isl1 population that have the capability to re-enter cell cycle when isolated and to differentiate into myocytes when dissociated and cultured in the presence of differentiated myocytes. Anyway, before Isl1 cells from human donors can be considered as a source for tissue repair, it has to be evaluated if there are still Isl1 cells present in adult heart tissue and if these cells can be isolated and expanded largely *in vitro*. However, Isl1 cells from embryoid bodies represent interesting candidates for heart tissue engineering.

7.10 Preparation of Cells for *In-Vitro* Tissue Engineering: Cell Permeable Cre/loxP System

Cells that are used for tissue engineering are often modified genetically for different reason. Embryonic stem cell derived cells might carry vector constructs for lineage selection to serve the requirements in cells purity. Cells that are derived from adult stem cells or from primary cell cultures can be expanded by the expression of telomerase or viral proteins that enable them to overcome limitations in cell cycling capacities. These transgenes are usually undesired to be present in cells for therapeutical application since they might cause immunological intolerance or might bear safety risks. To overcome these limitations strategies must be developed to allow efficient removal of the transgenes prior to transplantation. To address this task transgenes can be flanked by loxP sites. This strategy can facilitate the excision of the transgene in terminally selected cells that are ready for cell transplantation or *in vitro* tissue engineering, by application of the Cre recombinase. The recombined cells can be enriched subsequently if the loxP flanked transgenes that have to be removed, are in alliance with a negative selection marker. The classic negative selection marker is thymidine kinase (TK) – if the loxP-flanked region contains a TK expression cassette, cells that are not recombined are eliminated upon addition of ganciclovir, which is converted to a toxic ametabolite by the TK. If several transgenes are located in the genome of the cell the use of modified lox sites is possible. Modified lox sites like loxA or loxB sites have altered nucleotide compositions of the lox site but they are still identified and recombined by the Cre recombinase.

The advantage is that a lox site is only recombined with its counterpart: loxA site only with loxA site and loxP site only with loxP site. The use of different lox sites for different transgenes has the advantage that no undesired recombination between different loci can occur and only the target is eliminated.

For more sophisticated tasks like the deletion of two transgenes at different time points the use of a second (or third) recombinase system might be necessary. A common system except Cre/loxP is the FLP/FRT system. The Flip recombinase can perform recombination similar to Cre but between different recognition sites, the FRT-sites.

How can Cre recombinase be entered into the cell? Some types of cells are easily transfected with a Cre-expressing vector resulting in the deletion of the target. However, several cell types such as cardiomyocytes can be transfected only with a poor efficiency. This is not acceptable for the valuable cell preparations, since most untransfected cells will not eliminate the target and will be killed by the negative selection step. For a growing number of primary cells the bottleneck of very low transfection efficiencies is optimised by an electroporation method developed by the amaxa company. Amaxa is developing buffer compositions and electroporation protocols that are specially tailored for individual cell types, yielding high transfection efficiencies. However, transfection reagents and buffers can be used only for a limited number of cells and therefore are LESS cost effective. Also toxicity effects of the reagents cannot be excluded.

As an alternative to the transfection with Cre-expressing plasmids the Cre recombinase might be used as a cell permeable protein in a less invasive way. When designed as a fusion between the TAT domain, that was identified from a transcription transactivator of the HIV virus, and the Cre recombinase a protein is formed that can enter cells from the culture medium. A nuclear localisation sequence (NLS) is further added to enhance the power of the recombinase and a hexamer of histidine residues (His_6) serves as a tag for the purification of the recombinant protein.

By an easy and standardised procedure the His_6-TAT-NLS-Cre (HTNC) protein is expressed in *E. coli* bacteria with a yield of up to 40 mg recombinant protein per litre of culture volume in a simple flask culture. The bacteria are lysed by sonification and enzymatic (lysozyme) digestion and a cleared lysate is prepared by centrifugation. The lysate is incubated with agarose beads that are coated with complex bound nickel or cobald ions. These beads are commercially available from several companies. The histidine tagged protein bind to the nickel coated beads. Binding is antagonized by addition of imidazole that is structurally similar to histidine. By the addition of imidazole unspecific binding can be prevented and rising concentrations of imidazole in the buffers are used to wash the beads after collection in a column and to elute the purified protein. By this process the HTNC protein can be obtained in quantities of up to 100 mg with standard laboratory equipment. After dialysing against cell culture medium or appropriate protein storage buffers the HTNC can be frozen in aliquots for long time storage (Peitz et al. 2002).

To apply the cell permeable Cre the cells are incubated with serum free media in monolayer or dissociated and plated with HTNC to achieve optimal recombination efficiencies. For fibroblasts and mES cells a recombination efficiency of up to

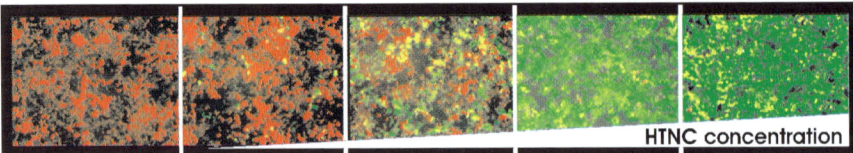

Fig. 7.11 Design of the Cre reporter vector. The ubiquitous CAG promoter drives the expression of red fluorescent protein (RFP). The RFP coding sequence is flanked by loxP sites depicted by triangles. When Cre recombination occurs the RFP sequence is eliminated and the expression of enhanced green fluorescent protein (eGFP) is enabled

Fig. 7.12 HEK293 cells were stable transfected with the Cre reporter vector (Fig. 7.11). Without Cre activity the cells express exclusively RFP (*left panel*). By the application of rising concentrations of cell permeable Cre (HTNC – increasing concentrations from *left* to *right*) in an overnight incubation experiment the GFP expression is turned on in different numbers of cells

96% could be shown (Peitz et al. 2002). To assay the quality of the HTNC protein we have designed a double fluorescent HEK293 reporter cell line. The reporter cells contain an expression cassette for red fluorescent protein (RFP) where the RFP coding sequence is flanked by loxP sites and followed by GFP (Fig. 7.11). In the unrecombined state the cells are brightly red fluorescent but do not show green fluorescence. Upon recombination, as a result of addition of cell-permeable HTNC protein, the RFP coding sequence is excised allowing the expression of GFP. For the assay shown in Fig. 7.12 the Cre reporter cells were treated with different concentrations of HTNC overnight. As a result an increasing fraction of the cells has underwent recombination of the reporter vector with an efficiency near to 100% at higher concentrations of HTNC. Taken together the use of cell permeable Cre has several advantages compared to transfection:

- Easy to produce in large amounts and therefore cost effective;
- Purification of the HTNC protein can be performed with standard laboratory equipment;
- No unknown components;
- No introduction of plasmids that might integrate into the genome with undesired effects;
- Perfect transient action of the Cre recombinase for the lifetime of the protein;
- Very high recombination efficiencies in different cell types.

7.11 Outlook

ES cell derived cardiomyocytes are promising candidates for cell therapy of severe heart diseases. Scientists are now focussed on strategies implicating conversion of a somatic cell to an ES cell like cell with multipotent or pluripotent potential.

7.11 Outlook

Cell permeable pluripotency factors or sophisticated fusion procedures, transfections with genes coding for stem cell factors, or novel tools may resolve the limitations currently existing. Based on these technologies it might become possible to create custom made cells with perfectly matching immunological patterns for each individual patient. Future applications of cardiac tissue engineering might be based on the transplantation of different cell types, including progentior cells, to the site of tissue damage rather than *in vitro* generation of contractile tissue that might be transplanted. Transplanted cells are able to engraft into the given tissue structures requiring less invasive operation procedure than tissue exchange. However, the lack of vascularisation remains a challenge for *in vitro* tissue engineering.

References

1. Abdullah Z, Saric T, Kashkar H, Baschuk N, Yazdanpanah B, Fleischmann BK, Hescheler J, Kronke M, Utermohlen O (2007) Serpin-6 expression protects embryonic stem cells from lysis by antigen-specific CTL. J Immunol 178(6):3390–3399
2. Ahuja P, Sdek P, MacLellan WR (2007) Cardiac myocyte cell cycle control in development, disease, and regeneration. Physiol Rev 87(2):521–544
3. Brambrink T, Hochedlinger K, Bell G, Jaenisch R (2006) ES cells derived from cloned and fertilized blastocysts are transcriptionally and functionally indistinguishable. Proc Natl Acad Sci USA 103(4):933–938
4. Cowan CA, Atienza J, Melton DA, Eggan K (2005) Nuclear reprogramming of somatic cells after fusion with human embryonic stem cells. Science 309(5739):1369–1373
5. Doss MX, Winkler J, Chen S, Hippler-Altenburg R, Sotiriadou I, Halbach M, Pfannkuche K, Liang H, Schulz H, Hummel O, Hubner N, Rottscheidt R, Hescheler J, Sachinidis A (2007) Global transcriptome analysis of murine embryonic stem cell-derived cardiomyocytes. Genome Biol 8(4):R56
6. Fink C, Ergun S, Kralisch D, Remmers U, Weil J, Eschenhagen T (2000) Chronic stretch of engineered heart tissue induces hypertrophy and functional improvement. FASEB J 14(5):669–679
7. Galli C, Lagutina I, Crotti G, Colleoni S, Turini P, Ponderato N, Duchi R, Lazzari G (2003) Pregnancy: a cloned horse born to its dam twin. Nature 424(6949):635. Erratum in: Nature 2003 425(6959):680
8. Gnecchi M, He H, Liang OD, Melo LG, Morello F, Mu H, Noiseux N, Zhang L, Pratt RE, Ingwall JS, Dzau VJ (2005) Paracrine action accounts for marked protection of ischemic heart by Akt-modified mesenchymal stem cells. Nat Med 11(4):367–368
9. Gnecchi M, He H, Noiseux N, Liang OD, Zhang L, Morello F, Mu H, Melo LG, Pratt RE, Ingwall JS, Dzau VJ (2006) Evidence supporting paracrine hypothesis for Akt-modified mesenchymal stem cell-mediated cardiac protection and functional improvement. FASEB J 20(6):661–669
10. Halbach M, Pillekamp F, Brockmeier K, Hescheler J, Muller-Ehmsen J, Reppel M (2006) Ventricular slices of adult mouse hearts – a new multicellular *in vitro* model for electrophysiological studies. Cell Physiol Biochem 18(1–3):1–8
11. Kolossov E, Bostani T, Roell W, Breitbach M, Pillekamp F, Nygren JM, Sasse P, Rubenchik O, Fries JW, Wenzel D, Geisen C, Xia Y, Lu Z, Duan Y, Kettenhofen R, Jovinge S, Bloch W, Bohlen H, Welz A, Hescheler J, Jacobsen SE, Fleischmann BK (2006) Engraftment of engineered ES cell-derived cardiomyocytes but not BM cells restores contractile function to the infarcted myocardium. J Exp Med 203(10):2315–27
12. Korecky B, Hai CM, Rakusan K (1982) Functional capillary density in normal and transplanted rat hearts. Can J Physiol Pharmacol 60(1):23–32

13. Kubo A, Shinozaki K, Shannon JM, Kouskoff V, Kennedy M, Woo S, Fehling HJ, Keller G (2004) Development of definitive endoderm from embryonic stem cells in culture. Development 131(7):1651–1662
14. Kubo H, Shimizu T, Yamato M, Fujimoto T, Okano T (2007) Creation of myocardial tubes using cardiomyocyte sheets and an *in vitro* cell sheet-wrapping device. Biomaterials 8(24):3508–3516
15. Laube F, Heister M, Scholz C, Borchardt T, Braun T (2006) Re-programming of newt cardiomyocytes is induced by tissue regeneration. J Cell Sci 119(Pt 22):4719–4729
16. Laugwitz KL, Moretti A, Lam J, Gruber P, Chen Y, Woodard S, Lin LZ, Cai CL, Lu MM, Reth M, Platoshyn O, Yuan JX, Evans S, Chien KR (2005) Postnatal isl1+ cardioblasts enter fully differentiated cardiomyocyte lineages. Nature 433(7026):647–53. Erratum in: Nature 2007 446(7138):934
17. Matsumura H, Tada M, Otsuji T, Yasuchika K, Nakatsuji N, Surani A, Tada T (2007) Targeted chromosome elimination from ES-somatic hybrid cells. Nat Methods 4(1):23–5
18. Menasche P, Hagege AA, Vilquin JT, Desnos M, Abergel E, Pouzet B, Bel A, Sarateanu S, Scorsin M, Schwartz K, Bruneval P, Benbunan M, Marolleau JP, Duboc D (2003) Autologous skeletal myoblast transplantation for severe postinfarction left ventricular dysfunction. J Am Coll Cardiol 41:1078–1083
19. Peitz M, Pfannkuche K, Rajewsky K, Edenhofer F (2002) Ability of the hydrophobic FGF and basic TAT peptides to promote cellular uptake of recombinant Cre recombinase: a tool for efficient genetic engineering of mammalian genomes. Proc Natl Acad Sci USA 99(7):4489–4494
20. Pillekamp F, Reppel M, Dinkelacker V, Duan Y, Jazmati N, Bloch W, Brockmeier K, Hescheler J, Fleischmann BK, Koehling R (2005) Establishment and characterization of a mouse embryonic heart slice preparation. Cell Physiol Biochem 16(1–3):127–132
21. Pillekamp F, Reppel M, Rubenchyk O, Pfannkuche K, Matzkies M, Bloch W, Sreeram N, Brockmeier K, Hescheler J (2007) Force measurements of human embryonic stem cell-derived cardiomyocytes in an *in vitro* transplantation model. Stem Cells 25(1):174–180
22. Schroeder M, Niebruegge S, Werner A, Willbold E, Burg M, Ruediger M, Field LJ, Lehmann J, Zweigerdt R (2005) Differentiation and lineage selection of mouse embryonic stem cells in a stirred bench scale bioreactor with automated process control. Biotechnol Bioeng 92(7):920–933
23. Smits PC, van Geuns RJ, Poldermans D, Bountioukos M, Onderwater EE, Lee CH, Maat AP, Serruys PW (2003) Catheter-based intramyocardial injection of autologous skeletal myoblasts as a primary treatment of ischemic heart failure: clinical experience with six-month follow-up. J Am Coll Cardiol 42:2063–2069
24. Tada M, Tada T, Lefebvre L, Barton SC, Surani MA (1997) Embryonic germ cells induce epigenetic reprogramming of somatic nucleus in hybrid cells. EMBO J 16(21):6510–6520
25. Tada M, Takahama Y, Abe K, Nakatsuji N, Tada T (2001) Nuclear reprogramming of somatic cells by *in vitro* hybridization with ES cells. Curr Biol 11(19):1553–1558
26. Takahashi K, Yamanaka S (2006) Induction of pluripotent stem cells from mouse embryonic and adult fibroblast cultures by defined factors. Cell 126(4):663–676
27. Zandstra PW, Bauwens C, Yin T, Liu Q, Schiller H, Zweigerdt R, Pasumarthi KB, Field LJ (2006) Scalable production of embryonic stem cell-derived cardiomyocytes. Tissue Eng 9(4):767–778
28. Zimmermann WH, Didie M, Wasmeier GH, Nixdorff U, Hess A, Melnychenko I, Boy O, Neuhuber WL, Weyand M, Eschenhagen T (2002) Cardiac grafting of engineered heart tissue in syngenic rats. Circulation. 24:106(12 Suppl 1):I151–I157
29. Zimmermann WH, Melnychenko I, Wasmeier G, Didie M, Naito H, Nixdorff U, Hess A, Budinsky L, Brune K, Michaelis B, Dhein S, Schwoerer A, Ehmke H, Eschenhagen T (2006) Engineered heart tissue grafts improve systolic and diastolic function in infarcted rat hearts. Nat Med 12(4):452–458

Chapter 8
Collagen Fabrication for the Cell-based Implants in Regenerative Medicine

Hwal (Matthew) SUH

Yonsei University, Seoul 120-752, Republic of Korea, hwal@yumc.yonsei.ac.kr

Abstract Though transplantation of cells, tissue or organ has been regarded as an ideal approach, scarcity of donor is a practical barrier in clinics. Current progresses in cell engineering has opened a new era, providing tools for host-regeneration by implanting manipulated cells in forms of cell therapy, which includes delivery of single cells or multicellular structural support of hybridized cells, as a representative individualized treatment method. This chapter mainly concerns on the cell-based implant made of cells and collagen, the main structural protein in extracellular matrix in mammalian tissue, as it has been regarded as a promising method for manufacturing a biologically mimicked artificial tissues.

8.1 Regenerative Medicine

Regenerative medicine is to repair, replace and/or modify the disordered or damaged human body, either from a disease and/or an injury, via functional regeneration of the host cells, tissues and organs by placing appropriate cells of optimal quantity into the damaged body and maintaining the cellular functions that provide expected efficacy. Outcome of the regenerative medical treatment completely depends on the viability of delivered cells.

Embedding and planting, referred to as implantation, of devices made of biocompatible materials has been widely applied. Metals, ceramics, synthetic and natural polymers are the fundamental biomaterials that are convenient for processing and fabricating into various forms of tissue supporting structures. However, lack of their biological function is too far from the ideal goal for treatment.

From this perspective, transplantation which involves transferring procured cells, tissue or organ from a donor and planting into a recipient to replace the damaged lesion has been regarded as an ideal approach that provides biological restoration with recovery of physiological functions.

Nevertheless, the opportunity for selection of the completely perfect donor for a patient is critically limited in practice. The gene-homogeneity between the donor and recipient, mainly in human leukocyte antigen (HLA) types, is the first factor to avoid any post-operative complication oriented from the immune responses. Although autologous tissue is the most appropriate if damage or defect is not related to immune disease, applicable size and volume of donor site are extremely limited. Allogeneic tissue has been the second choice for transplantation since several immune-suppressive agents have been introduced, but these agents also lead to other immune-depressed diseases by breaking the natural immune homeostasis. Xenogeneic tissue and organ transplantation have been suggested by a number of researchers but longevity difference between human and donor animal is a great barrier that cannot be overcome at present.

An approach to treat gene-defect oriented congenital disease by delivering correct genes directly to a patient, known as gene therapy, has been introduced as an individualized treatment tool. Selected genes hybridized with a carrier vector are injected into a patient, and the vector may infect patient's cells and leave the selected genes within patient's cells. Then, the delivered genes may express its genetic characteristics necessary to cure the disease. However, the gene therapy bears several problems: (1) there is no perfect method to deliver genes into appropriate cells, (2) efficiency of gene expression varies, and (3) the gene delivering viral vector may induce unpredictable and unknown complications in body.

Another individualize treatment tool called cell therapy is planting therapeutic cells. Instead of transplanting tissue or organ, therapeutic cells are delivered in forms of either transplantation of natural cells or implantation of artificially manipulated cells. In addition, resource of applicable tissues from which therapeutic cells could be prepared is expanded. Conventional transplantation mainly consists of matured tissue or organ, but it is now possible to obtain therapeutic cells from any tissue at any growth stage, including embryonic blastocyst. Through *in vitro* gene modification of cells, it is possible to select only the cells which possess appropriate genes for treatment purposes from outside of the human body.

Theoretically, in case of autologous cell therapy, patient's cells are harvested through autopsy or procurement, and the treatment gene for correction is transfected into the cells. After culture, only the completely gene-transfected, healthy cells are collected, and the selected cells are further cultured to obtain abundant number of cells to deliver in forms of cell suspension for injection. As the whole procedure is performed *in vitro* environment, only perfectly modified therapeutic cells are selected. In case of allogeneic cell therapy, donor's HLA could be also replaced by the recipient's to escape from the post-operative immune rejection. Furthermore, risk of inducing uncontrollable oncogenesis oriented from the mutated cells is avoidable during the cell selecting procedure *in vitro* [1].

8.2 The Cell-Based Implants

To deliver the selected cells into the selected site in recipient's body, it is necessary to fabricate cell delivery vehicle with biomaterials that supply agents to maintain cell-viability and act as probes for piloting cells toward appropriate site. The

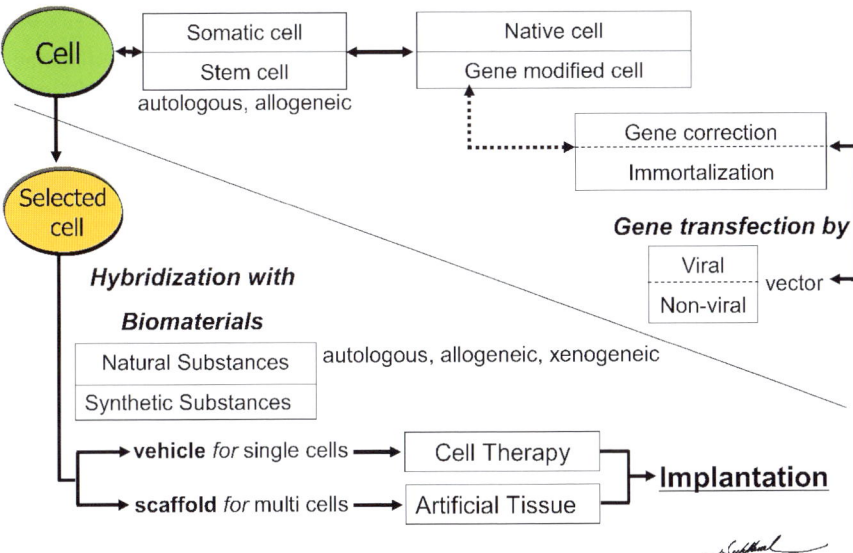

Fig. 8.1 Cell Preparation for Implantation

cell-based implant is defined as any implant in combination of cells and biomaterials which are intended to repair, modify and/or regenerate human body, through expected regulation and control of cellular functions and/or behaviors after implantation. The cell-based implant consists of artificially manipulated therapeutic cells hybridized with biomaterials *in vitro*. For cell therapy, each cell is individually hybridized with biomaterials, but in forms of tissue engineered implants, cells are hybridized with scaffold-biomaterials that act as structural support in the same way as the extracellular matrix in tissue does [2] (Fig. 8.1).

8.3 Requirements of Materials for the Cell-Based Implants

Substances applied to control biological events of cells may have biological risks. In manipulating cells, various kinds of substances functioning chemically at molecular level are used in cell-modifying procedures. Biological agents, such as enzymes in the nucleic acid recombination procedure, cytokines in cell differentiation and proliferation control, and peptides in cell culture media, are regarded as biologic substances, and they must not produce any biological hazard during the manipulation procedures.

In gene modification process, to avoid adverse reactions oriented from viral vectors, such as retrovirus, adenovirus and adeno-associate virus, non-viral vectors are designed to avoid biological risks and to manipulate easily. Capability of non-viral vectors is mainly dependent on endocytosis mechanism, but the extent of development in an artificial vector that provides efficient endosomal escape to deliver the gene finally into nucleus is still questionable at present.

Biocompatibility is the primary requirement for any material used in manufacturing implant. In the cell-based implants, the material hybridized with cells shall either biologically or mechanically be compatible with the neighboring cells at the delivered site. For cell therapy, each cell would be hybridized with nano-sized particles that have specific affinity to the treatment target site and self-driving ability. Particles shall be dissociated from the delivered cell after landing at the site and be completely excreted without accumulation in the recipient's body and any side effect. Size of a tissue cell is about 10×10^{-5} m ($= 10\,\mu$m), a virus is 10×10^{-7} m ($= 100$ nm), diameter of a DNA is 10×10^{-8} m ($= 10$ nm), diameter of a hydrogen atom is 10×10^{-9} m ($= 1$ nm), and hybridizing substances are smaller than a μm. However, there is no accurate tool to detect these extremely small, nano-scale particles that exist in living body for safety evaluation of a substance at present.

The cell-fabricating operator might be exposed to the substances during the whole manipulating procedure, and further investigation is necessary to avoid hazards. Otherwise, currently introduced biomaterials for cell encapsulation are designed to protect therapeutic cells from immune rejection while providing the excretion of biologics from the cell, and mostly to manufacture macromolecules with less difficulty to characterize [3].

8.4 Biomaterials in the Cell-Hybridization

In producing multicellular structural cell-based implants, so called tissue engineered implants, various biomaterials of either synthetic or natural substance are applied. Basically, conventionally available biomaterials can also be adopted as long as the permanent biocompatibility is approved.

Synthetic polymeric biomaterials are the representatives, and can be non-biodegradable or biodegradable after implantation. Non-biodegradable materials are usually intended for use in cell therapy where cell-encapsulation is required to provide and maintain optimal cellular function (e. g., alginate, liposome, etc.) by protecting the cells from host immune reaction, and/or for a tissue engineered implant, which requires the physiological load-bearing compliance (e. g., polyurethane scaffolds for blood vessels, tendons, ligaments, etc.) after implantation. They permanently remain at the planted place in recipient's body. Meanwhile, biodegradable materials are usually intended for use in implants which restore the histological structure and replace the cellular function of recipients. They are gradually degraded in recipient's body through hydrolysis or enzymatic function after implantation (e. g., poly L-lactic acid, poly glycolic acid, etc.).

In cases of using synthetic polymeric biomaterials, behavior of hybrid cells is mainly dependent on surface characteristics of material. Especially, the cellular events such as adhesion, differentiation, proliferation and migration on the non-biological synthetic biomaterial are important for maintaining viability of cells implanted in bio-inert materials. Furthermore, an increased regional acidity induced by dissolved acidic component through hydrolysis from the biodegradable materials demonstrates the limitation in mimicking natural biological environment with synthetic biomaterials.

Natural biomaterials are mainly composed of extracellular matrix (ECM) components (including structural components and biomolecules) originated from autologous, allogeneic and xenogeneic tissues of mammalians. Collagen, elastin, chondroitin-6-sulphate and hyaluronic acid are structural components.

Biomolecules, such as peptides, fibronectin, laminin, vitronectin and fibrin, and cytokines as growth factors and apoptosis signal promoters, are biologically active substances produced by nature. Non-mammalian substances such as silk fibroin, crab chitin and chitosan and agar are also included in this criteria. At this scope, it can be recognized that, although any xenogeneic cells are still not permitted, xenogeneic ECM components are permitted for the "trans-" or "im-"plantable biomaterials. For cell-based implants that utilize xenogeneic tissues or their derivatives as biomaterials, secure risk controls shall be applied on sourcing, collecting and handling xenogeneic ECM, on the validation of elimination and/or inactivation of adventitious agents, such as Transmissible Spongiform Encephalopathy (TSE) agents in case of using bovine tissue, in products.

Non-comparable biological superiority of natural ECM components are applied for improving biocompatibility of synthetic polymeric biomaterials as grafting materials onto the surface of material. Also, bioactive agents (including biologics, antibiotics, and antimicrobials) and/or synthetic drugs can be medicinal components in biomaterials, and they shall be assessed, in the context of their integration with the cell-based implant, according to pharmaceutical principles. This assessment shall consider the effects of medicinal components on product and vice versa. Furthermore, the medicinal components could be an additive to treat the recipient's disease [4].

8.5 Characteristics of Collagen

In mammalians, collagen comprises about 30% of total proteins and exists as a main structural component in ECM and supports anatomical morphology of every tissue and organ. Although more than 20 types of them are informed, type I collagen, which has a specific molecule of the super-coiled triple helical peptide chains, is the most abundant in body. In brief, each peptide chain is a specific left-handed helix of 100,000 molecular weight and consists of repeating "-Glycine-X-Y-" amino acid sequence. As the arginine–glycine–aspartic acid (RGD) sequence is a typical cell adhesive ligand, collagen demonstrates strong cell adhesion property.

3 left-handed helical peptide chains are integrated by intramolecular bindings of hydrogen bonds at glycines to glycines and hydroxyl bonds at hydroxyprolines to hydroxyprolines in each chain to form a right-handed triple helical collagen

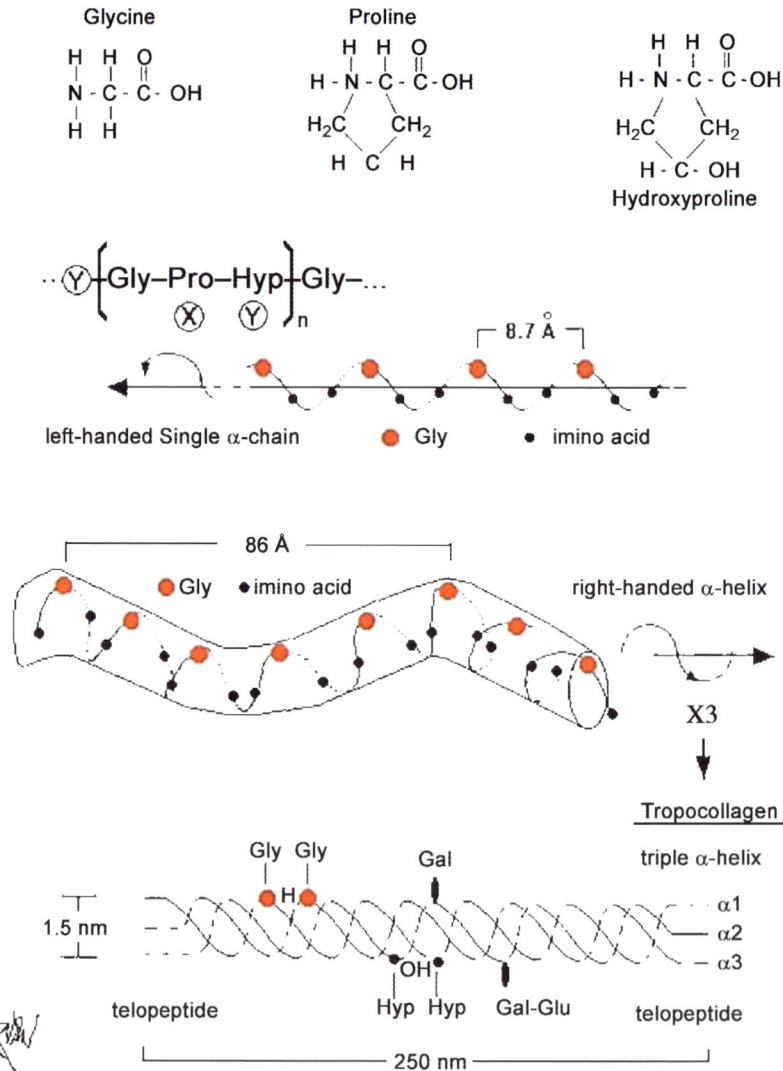

Fig. 8.2 Structure of Type I collagen molecule

molecule. In a chain, 1012 amino acids are involved in the formation of a molecule, and each free-end of a chain that is not integrated with the others, is made of 12–17 amino acids and named as telopeptides. Each of these at the C- and N- terminals of molecule binds to another and forms a long linear chain. Collagen is easily denatured at temperature over 37 °C through disintegration of the intermolecular bonds, and becomes gelatinized to form a randomly coiled chain with less viscosity than that of collagen. Once denatured, gelatin is amorphous and does recover the intermolecular bonds in nature and is more easily digested by metaloprotease than collagen.

A type I collagen fibril consists of 5 collagen linear chains with strong intermolecular hydrogen bonds at lysine to lysine in each molecule.

In nature, ε-NH_2 of lysines and hydroxylysines at the both extremities of telopeptide is converted to aldehydes by lysineoxidase, and the Schiff base formation by binding aldehyde with residual ε-NH_2 or by aldol condensation between aldehyde to aldehyde occur. This strong molecular binding is called as the crosslinking and introduces strong mechanical characteristics to collagen to act as the structural matrix in tissue and provide organ morphology.

As hydrogen concentration of body fluid is pH 7.4, collagen is regarded as a weak base substance, and is generally extracted from mammalian tissue.

Collagen is the hydrophobic protein and is generally insoluble in a neutral solution, but non-crosslinked collagen is soluble in the neutral base solution such as NaCl or Na_2HPO_4. On the other hand, weak acid such as HCl, citric acid and acetate breaks the intermolecular Schiff base bonds and the acid soluble collagen molecules are extracted. After extracting acid soluble collagen, non-acid soluble remnants remain, and these are strongly crosslinked collagen fibrils. When additional process using pepsin that digests the linear intermolecular bonds at each telopeptide is performed, telopeptide-free collagen molecules, known as the atelocollagen, can be obtained from remaining collagen fibrils. Because telopeptide demonstrates individual gene-characteristics, these atelocollagen molecules are recognized as immune-free substance and are applicable to medical and pharmaceutical purposes.

High concentration of bases such as NaCl, Na_2SO_4, $Na_2 HPO_4$ aggregate the acid-soluble collagen molecules in solution, and this phenomenon is applied to produce collagen membranes, threads and hollow fibers. However, randomly reconstructed intramolecular and/or intermolecular bonds do not provide sufficient mechanical properties compared to natural collagen in tissue. Hence, in order to fabricate collagen with higher mechanical strength, re-crosslinking methods are introduced [5] (Fig. 8.2).

8.6 Fabrication of Collagen

Collagen has been regarded as the first candidate for hybridization with cells as it exists in every part of body as the main extracellular matrix component. Although implantation of xenogeneic cells is not practically permitted, immunefree atelocollagen is generally extracted from mammalians for medical use. Bovine is the most popular resource provided if cows are grown at the officially recognized region that is free from TSE.

As atelocollagen has no telopeptides that integrate with other atelocollagen molecules, basic technology to produce intramolecular and intermolecular bonds for construction of the extracellular matrix with adequate mechanical properties required by tissue where the artificial cell-based implant is delivered is the re-crosslinking method using chemical reagents or physical dynamics.

Most common chemical reagent adopted for this process is glutaraldehyde that introduces stable covalent NH_2 to NH_2 bindings between the molecules. As amine to amine (NH-NH) bonding exists in every amino acid in molecules, the chemically crosslinked collagen demonstrates high strength, but irregular patterns of the crosslinked fibers and protracted resorption are general disadvantages. Furthermore, in case of using glutaraldehyde, complications due to residual aldehydes often conduct calcification that directly leads to the loss of mechanical strength. To overcome these disadvantages, 1-ethyl-3(3-dimethylaminopropyl) carbodiimide (EDC), which introduces NH_2 to COOH covalent bonding and has no toxicity, has been applied to the procedure.

To avoid any disadvantage arising from the use chemical reagents, cross-linking by physical method is employed. Irradiating with gamma or ultraviolet rays produces radicals in the form of unpaired electrons in the nuclei of aromatic residues

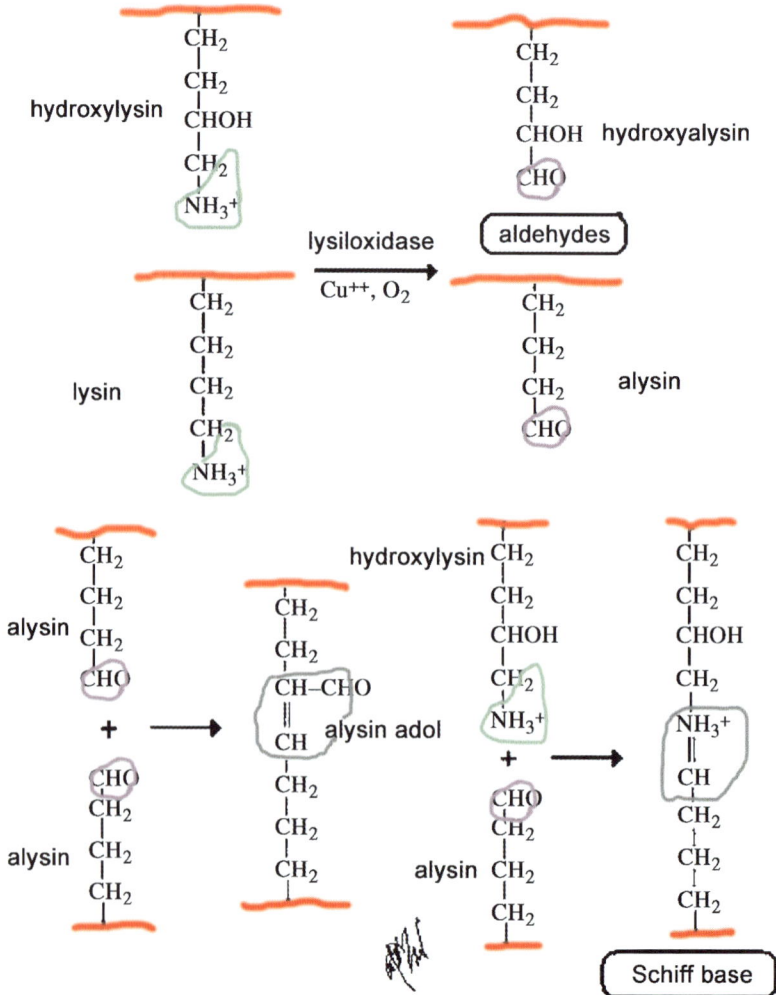

Fig. 8.3 Crosslinking of collagen molecule occurs by Schiff base formation in nature

8.6 Fabrication of Collagen

such as those in tyrosine and phenylalanine, and binding of these radicals results in the collagen cross-linking through random NH_2 to COOH bonding. Even though the physical method is safe as it does not involve the use of chemicals, stability of the crosslink is less than that of chemically induced crosslink.

Collagen can be fabricated into various forms of gel, fiber, membrane with or without pores, and it can even be grafted onto the non-viable metal, ceramic and synthetic biomaterials to introduce biological layer on surface.

Atelocollagen gel is easily prepared by dispersing in weak acidic solutions. In general, collagen is dispersed in a weak acid solution of pH 6–6.5 and finally adjusted to required pH by adding base for medical purpose. Additive substance to

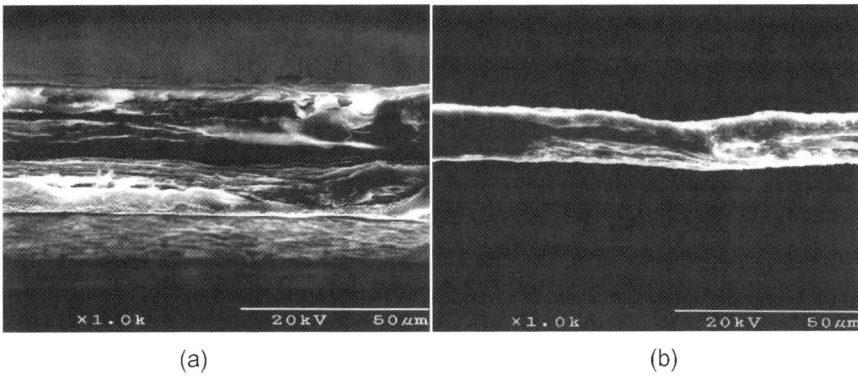

Fig. 8.4 Two hours of UV ray (wave length 245 nm) irradiation cross-linked collagen matrix. **a** before, and **b** after irradiation

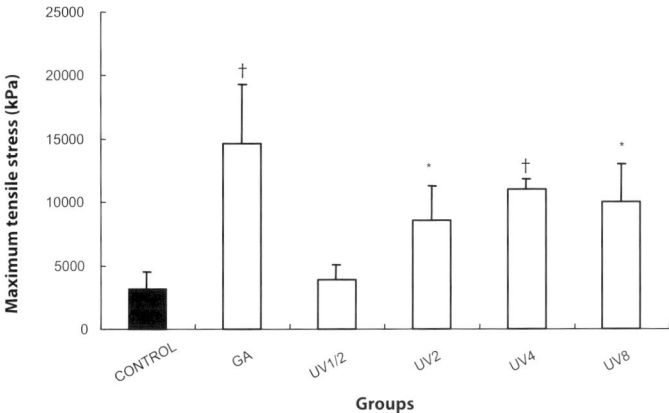

Fig. 8.5 Ultimate tensile stress of dense membrane in relation to UV irradiation time. Each value represents the mean ± SD in five samples, * and †: significantly different compared to the control non-treated group for each type of membrane, *: $p < 0.05$ and †: $p < 0.01$. CONTROL : non-treated group, UV1/2: UV-irradiated group for 30 mins, UV2: UV-irradiated group for 2 hours, UV4: UV-irradiated group for 4 hours, UV8: UV-irradiated group for 8 hours, GA: 0.625% glutaraldehyde pretreated group for 24 hours (Y-axis: Maximum tensile stress (kPa))

provide and/or control the biological function of the collagen can be introduced at this stage utilizing the high viscosity property.

Biochemical property of collagen molecule can be modified by varying the surface electric charge by altering the molecular side chain which leads to the adjustment of hydrophilicity-hydrophobicity balance. Succinylization of amines (NH–CO(CH$_2$)$_2$-COOH) by using anhydrate succinyl acid provides abundant (−) charges, and the succinylated collagen becomes soluble in neutral solution, which finally results in translucent viscous collagen gel with hydrophilicity. Meanwhile esterization (–COOCH$_3$) of carboxyl group by methanol produces abundant (+) charges on the collagen and provides a favorable hydrophobic niche for protein adhesion.

Atelocollagen gel is fabricated into fiber form by using an electro spinning. For example, a collagen solution dissolved in 1,1,1,3,3,3-hexafluoro-2-propanol (HFP) having concentration over 5% is useful to produce a diameter controlled nanofiber by passing it through a diameter adjustable nozzle under the condition of a high voltage at 25 kV, flow velocity at 2.5m/h, metal collector of 2 cm width with rotating speed at 300 rpm, and distance between the metal collector and spinner of 15 cm, and finally removing organic solute for 48 hours by drying in a vacuum chamber. The produced nanofiber is crosslinked by either EDC or UV irradiation later in order to reinforce the fiber strength, and demonstrates the typical triple helical structure of collagen molecule.

Freeze-drying the gel in vacuum condition is a simple procedure for fabrication of a porous membrane, as lower concentration and higher freezing temperature leads to smaller and bigger pores [6–10], (Figs. 8.3–8.8).

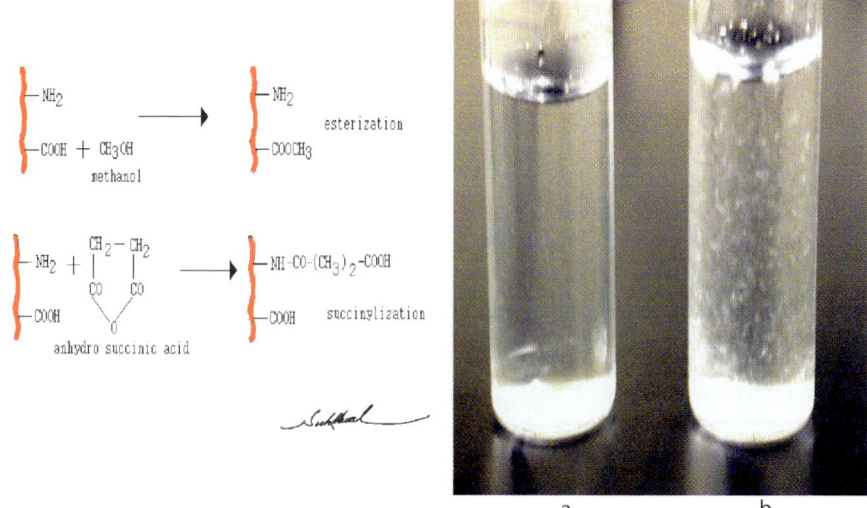

Fig. 8.6 Modification of collagen molecule to increase hydrophobicity by esterization or hydrophilicity by succinylation. Succinylated collagen (**a**) demonstrates higher solubility than normal collagen gel (**b**) in distilled water with hydrophilicity

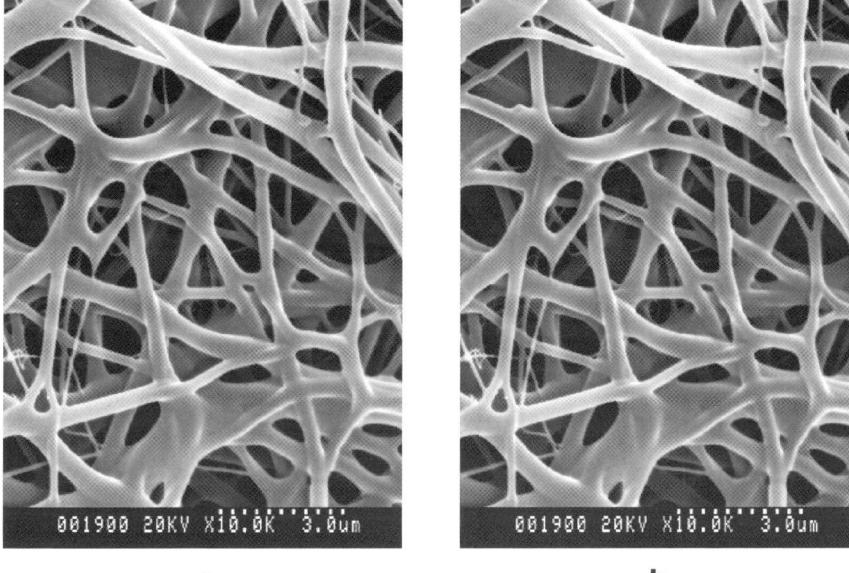

Fig. 8.7 A scanning electron microccopic surface view of an electrospun collagen nanofibrous membrane crosslinked by 1-ethyl-(3-3-dimethylaminopropyl) acrbodiimide hydrochloride (EDC). Crosslinking increased diameter of the nanofibers. **a** before crosslinking, d: 448 nm, **b** after crosslinking, d: 618 nm

Fig. 8.8 Morphological observation of typeI atelocollagen porous membrane: **a** freezing dry at $-20\,°C$, **b** freezing dry at $-70\,°C$, **c** freezing dry at $-196\,°C$. The lower temperature produces higher porosity

8.7 Collagen in the Cell-based Implants

Collagen contains much of the cell adhesive RGD sequences in molecule, thus has cell conductive characteristics and is applied either in the independent form, in form of hybridization with biofunctional agents and cells, or, occasionally, in form of copolymer by introduction of synthetic polymers which is mainly aimed to enforce mechanical property.

8.7.1 Skin Regeneration

Collagen gel is usually applied to cell culture. A cell culture plate whose bottom surface is coated by collagen gel and dried makes it possible for suspended cells to produce a monolayered cell cluster by attaching them to the coated collagen layer. This technique is directly applied to produce skin wound dressing membranes.

Exposure of dermis to the open air, the grade 3 skin defect, in which the basal epithelial layer that prevents direct contact of inner body to the outer bodily environment and supports epithelium as the bed for keratinocytes is destroyed from burn, and sore spots induced by diabetes mellitus or accident are critical emergency in clinics. They evolve the direct contamination of dermis which usually progresses toward fatal septicemia or skin necrosis. The conventional treatment procedure mainly consists of complete irrigation and debridement of the wound, topical administration of broad spectrum antibacterial and antifungal agents, covering the wound by oily ointment to protect it from air contact, painful daily dressing change, and observation of the auto-regeneration of wound by proliferation of both dermal fibroblasts and epithelial keratinocytes and basement membrane reconstruction. Promoting proliferation of these skin cells with protection from air contact until complete healing is the key technology, and collagen membrane has been applied for this. For example, freeze dried atelocollagen in vacuum forms a membrane with random pores, and the pore size into which the cells may proliferate is controlled by the gel concentration and temperature at which it is frozen. Freeze dried collagen gel of 1, 2, and 5% in concentration at -20, -40, and $-80\,°C$ for 1/2, 1, 2, and 4 hours demonstrated that higher concentration produced less and smaller sized pores. Also, the lower temperature and prolonged freezing time decreased viscosity. To produce a collagen membrane with optimal pore size of over $120\,\mu m$ which permits penetration and proliferation of dermal fibroblast proliferation, the 2% collagen gel freeze dried at $-40\,°C$ for 2 hours and crosslinked by EDC or UV irradiation for 2 hours was found to be recommendable to use as a dermal cell conductive membrane that provides the appropriate viscosity for initial anchorage, and any inflammatory exudate escapes through pores.

A collagen bi-layered membrane is applied to skin in forms of either wound dressing or extracellular structural supporting matrix for artificial skin. Laminin, which is a cell adhesive protein and a main component of the basement membrane

8.7 Collagen in the Cell-based Implants

between dermis and epithelium, is mixed with 2% collagen gel. A porous collagen-laminin membrane is fabricated and freeze-dried as previously described and additional 4% collagen gel is coated on the porous membrane and crosslinking is promoted by EDC treatment or UV irradiation. Through this procedure, a bi-layered collagen matrix that consists of dense collagen layer overlaid on the prefabricated porous collagen-laminin membrane was produced. In use as a wound dressing, the upper dense layer prohibits direct wound contact with the open air and plays a role as a bed for keratinocyte proliferation, while the porous layer conducts dermal fibroblast proliferation. Thus, the painful daily dressing change becomes unnecessary. Antibacterial agents encapsulated by hyaluronic acid can be incorporated into the collagen gel and the drug may be released from the collagen membrane to avoid periodic topical drug application.

Fabrication of the bi-layered collagen membrane is the basic technology applicable to manufacture of artificial skin. In the 1980s, Bell introduced a method in

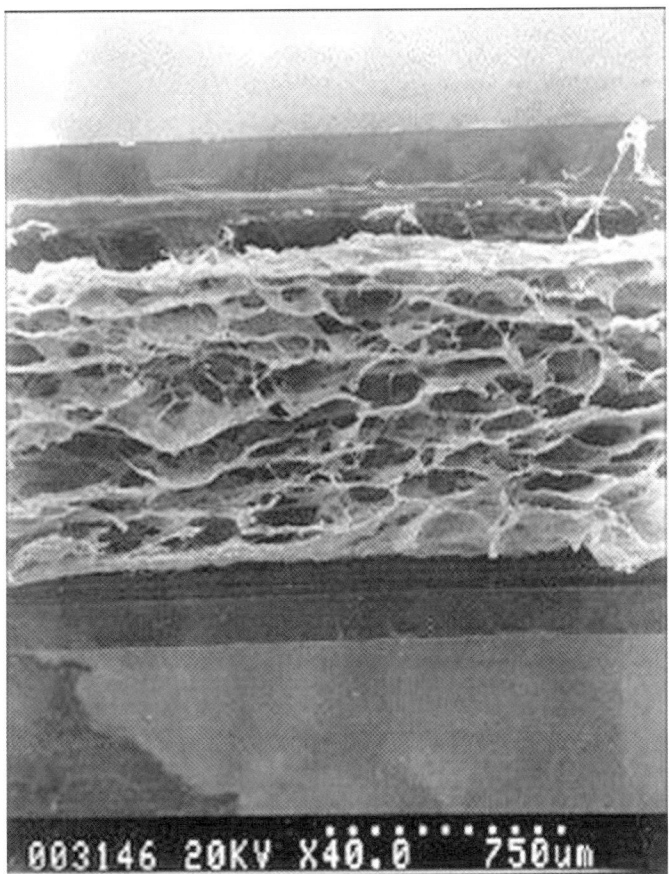

Fig. 8.9 Collagen-Laminin hybridized porous membrane for dermis

which dermal fibroblasts and collagen gel mixture were cultured at 37 °C, and when the cells were confluent in the gradually shrinked gel, the collagen-fibroblast composite is produced. Keratinocytes were overlaid on the composite to mimic anatomy of the dermal cellular structure, but it required about 20 days for the manufacture to be completed and it was therefore not applicable to patients with grade 3 skin wound. On the other hand, Green introduced an advanced method in which dermal fibroblasts were cultured to form monolayer on a collagen coated culture plate and then keratinocyte was directly co-cultured on the fibroblast monolayer in a media

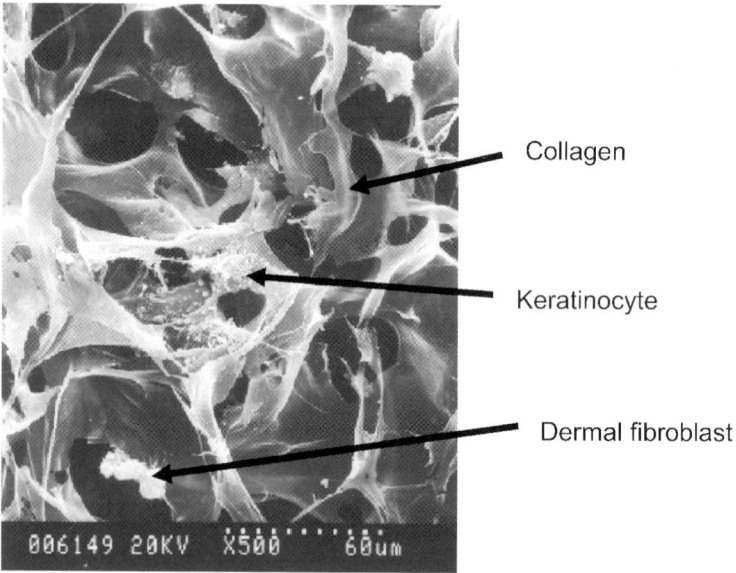

Fig. 8.10 Cells attached onto the porous collagen matrix

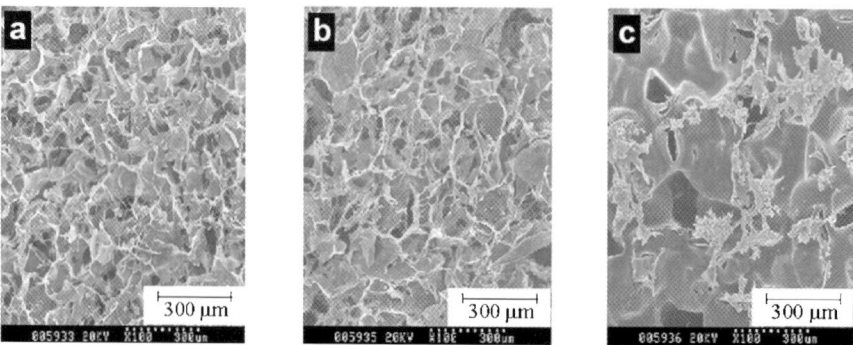

Fig. 8.11 Scanning electron micrographs of cells attached on a collagen matrix (**a**), a collagen matrix containing 9.6% HA (**b**) and a PU matrix (**c**). (magnification × 100)

containing epithelial growth factor. Nonetheless, this method still required at least a week for the complete fabrication, and employment of growth factors holds a risk of oncogenesis induction.

To produce an artificial skin, autologous dermal fibroblasts of rat were seeded into the EDC crosslinked porous collagen-laminin membrane and cultured in minimum essential medium (MEM) for 3 days to provide the cell-niche adaptation period. 3-day culture is not enough for complete proliferation of cells into pores, but cells were firmly attached and anchored onto the superficial layer of the porous membrane due to specific cell adhesive characteristics of –RGD-sequences in col-

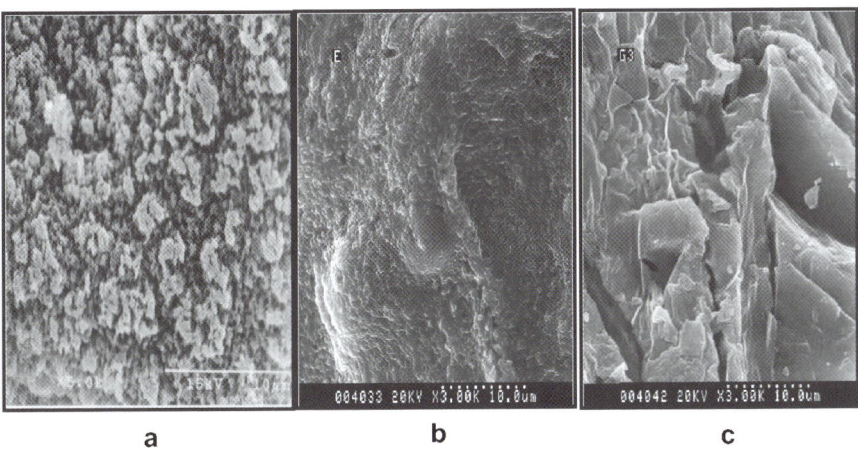

Fig. 8.12 Surface characteristics of HA microparticles obtained by encapsulation method (**a**: with only HA, X5K, **b**: with addition of collagen and antibiotics, X3K) and granulated method (**c**, with addition of collagen and antibiotics, X3K)

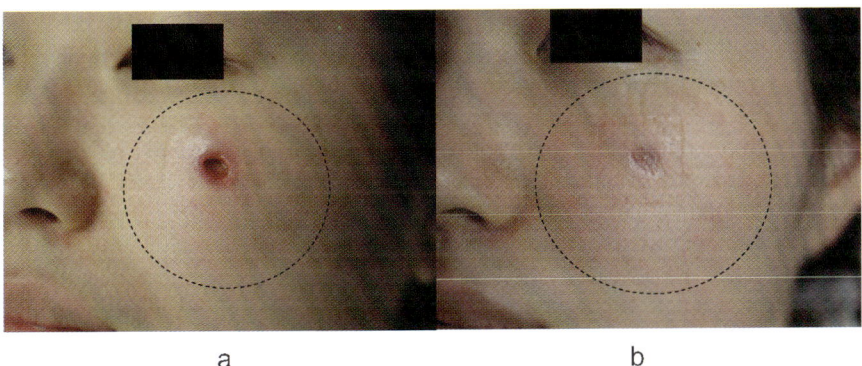

Fig. 8.13 a A facial subcutaneous lesion (*dot circle*) was augmented by a porous antibiotics encapsulated hyaluronic acid hybridized type I atelocollagen implant and dense-porous double layered collagen membrane was applied as a wound dressing. **b** After 3 weeks, subcutaneous dermal wound was completely recovered, and epithelial regeneration was conducted

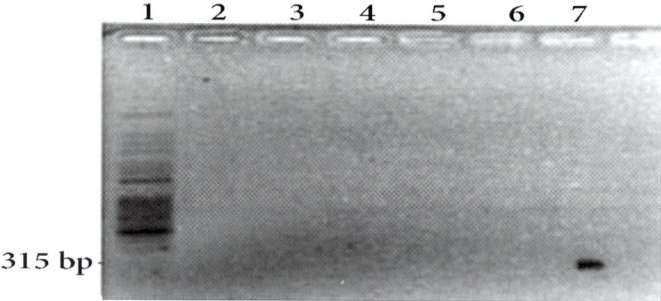

Fig. 8.14 Electrophoretic analysis of mycoplasma polymerase chain reaction (PCR). Supernatants from cell cultures on PLGA or collagen nanofibrous membrane were subjected to the sample preparation. Lane 1: size marker (100 bp DNA ladder), lane 2: collagen nanofiber with cells, lane 3: PLGA nanofiber with cells, lane 4: medium with cells, lane 5: medium without cells, lane 6: negative control (water) and lane 7: positive control

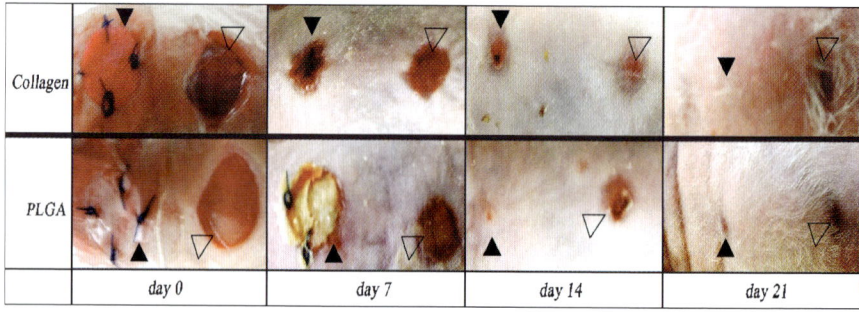

Fig. 8.15 Macroscopic observation of wounds treated with fibroblast seeded collagen nanofibrous or PLGA nanofibrous membrane at day 0, day 7, day 14 and day 21 after implantation. (*Black arrow head*: membrane placed area. *Blank arrow head*: non-treated control area)

lagen and laminin. After 4 weeks of coverage onto a grade 3 skin wound, dermis was completely replaced, and basement membrane lined beneath the dense layer as the seeded dermal fibroblasts were cultured *in situ*. Epithelial regeneration upon the dense layer surface was poor and partially occupied by the keratinocytes, but full coverage by epithelium was observed after 6 weeks. This phenomenon can be understood as the porous collagen based membrane can play as a vehicle for dermal autologous cell delivery, and the reconstructed basement membrane, which is a keratinocyte supplying bed in nature, has driven epithelial conduction. Later, a collagen-elastin nanofiber was fabricated as previously described, and the extruded fibers were brought onto collagen gel and crosslinked by EDC. Using nanofiber made it possible to produce a lattice controlled matrix, and uniformly aligned lattices conducted cells to proliferate into the designed lattice pattern [11–17] (Figs. 8.9–8.16).

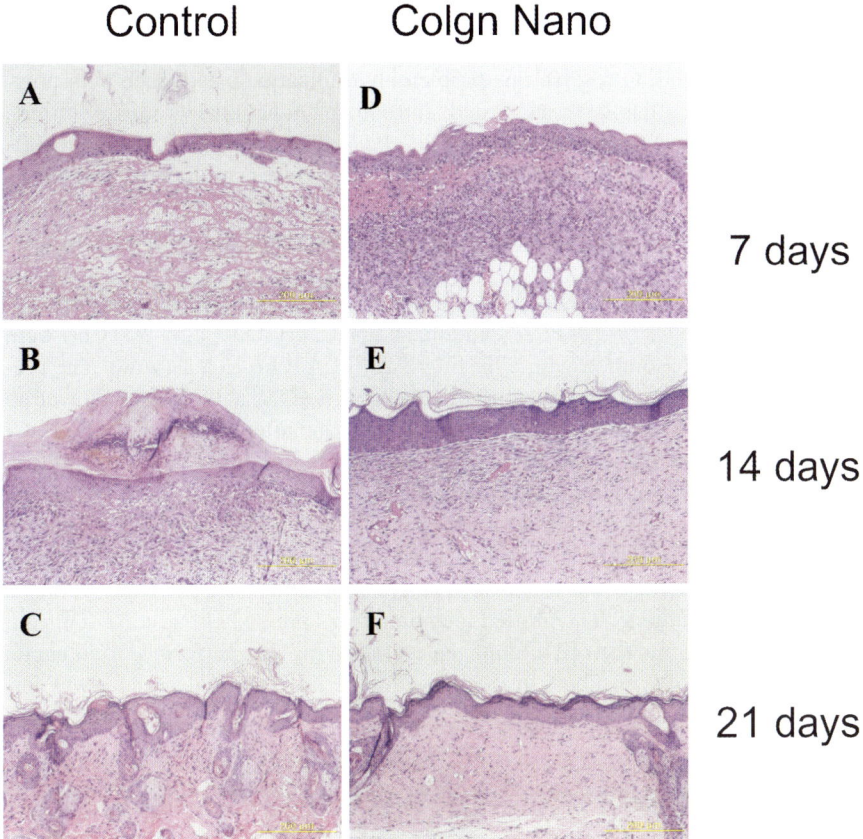

Fig. 8.16 Photomicrographs of biopsy specimens from a wound treated with fibroblast seeded collagen nanofibrous membrane on the postoperative 7th day (**A, D**), 14th day (**B, E**), and 21th day (**C, F**). Nanofibrous membrane implantion (**D, E, F**) and sham-operation controls (**A, B, C**) were compared (Magnification 40X)

8.7.2 Bone Reconstruction

Collagen gel hybridized with calcium phosphate is applicable as a bone conductive substitute. Apatite ($Ca_{10}(PO_4)_6^{++}$), a typical inorganic component of skeleton, was synthesized and heated at 980 °C after which apatite crystals were not sintered but degradable by hydrolysis. 1% collagen gel was mixed with the apatite in 12:88 w/v, and the mixture was crosslinked. Then the produced apatite-collagen pellet was implanted into rabbit's resected tibiae. 4 weeks after implantation, the resected defect was completely regenerated by host bone with cortical bone continuity and cancellous bone stroma, and after that, typical bone remodeling process followed to produce natural bone. However, mechanical strength of the regenerated area was less than general bone.

Although collagen based gel is favorable to fabricate cell conductive substitute, weak mechanical property is a barrier for application in the physiological stress bearing tissues, and, to resolve this problem, hybridization of collagen with polymeric biomaterials has been suggested. Introducing functional groups on esters, such as polyurethane (PU) and poly lactic-glycolic acid (PL-GA), through treatment with ozone induces surface oxidization that produces carboxyl groups on the surface which react with amines in collagen.

As the apatite-collagen provided insufficient strength despite promotion of a favorable osteogenesis, collagen was grafted onto the biodegradable PLA membrane to reinforce the strength. PLA of M.W. 50,000 was resolved by 99% chloroform solution to produce 9% (W/V) PLA solution, and NaCl crystals (425 – 500 μm) were mixed to the PLA solution with adjusting NaCl:PLA ratio to 9:1 (w/w). The prepared solution was cast into an appropriate shape, and the solvent was allowed to evaporate over 24 hrs to produce a porous membrane after leaching out the NaCl particles with distilled water. To induce molecular bindings between the collagen within apatite-collagen composite and the PLA membrane surface, PLA membrane was oxidized by ozone to produce reactive carboxyl and hydroperoxide groups on the PLA surface. After the ozone treatment at 60 V for 60 min, apatite-collagen was delivered onto the ozone treated PLA membrane to produce an apatite-collagen grafted PLA membrane. The grafted membrane was pressed (1 bar at 37 °C) from vertical direction, and then EDC collagen cross-linking was performed to integrate the mixed atelocollagen fibers in the grafted material. The final product was im-

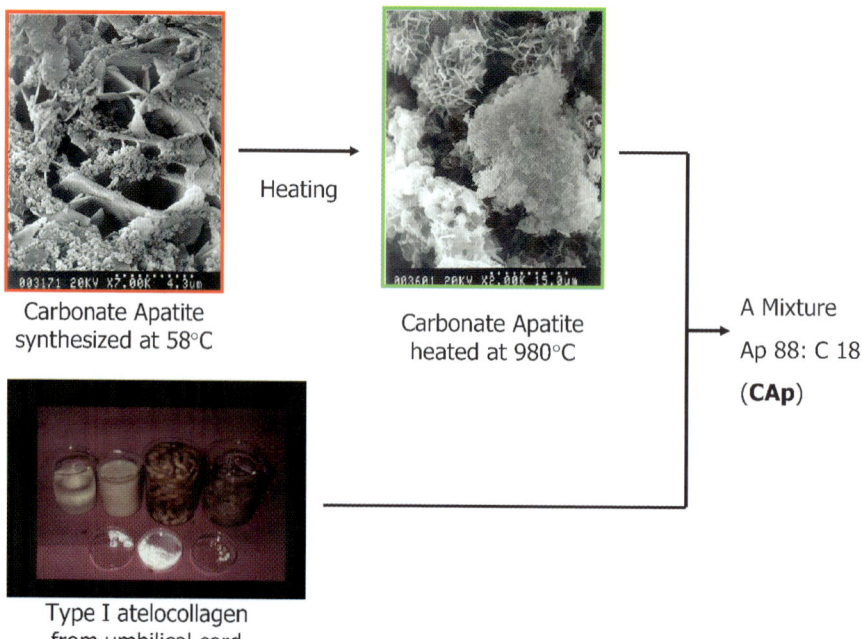

Fig. 8.17 Scheme of producing carbonate apatite-collagen substitute mimicking bone

8.7 Collagen in the Cell-based Implants

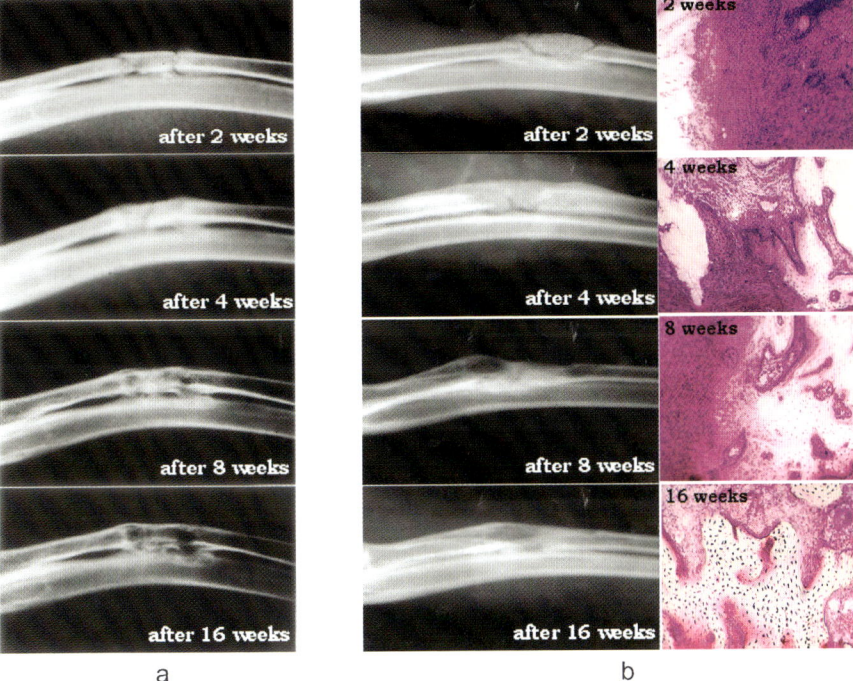

Fig. 8.18 Radiological view of the carbonate apatite-collagen implanted in the resected tibiae of rabbits for 16wks. **a** autograft, **b** apatite-collagen implant and histological view

planted into oval defects induced on rabbit's calvarial bone. The bone was regenerated in a similar manner to that of apatite-collagen implanted model, and the supportive PLA membrane was gradually degraded as the bone was replaced [18–25] (Figs. 8.17–8.23).

8.7.3 Esophagus Replacement

Collagen grafted synthetic polymer is applicable to soft tissue replacement as well. Except the air passages such as trachea and bronchus which require almost no repeating dilatation-shrinkage action of compliance, most of the tubular-shaped tissue requires various level of compliance against the mass passing through lumen, as arteries to pulsating blood flow and esophagus to swallowing dietary mass. In general, tubular walls of those tissues consist of multiple-layered smooth muscle cells to resist against pressure at luminal wall exerted by transporting mass, and appropriate flexibility is required.

To replace esophagus, polyurethane, a bioinert biomaterial with mechanical advantage of high durability against continuous bending stresses, was employed.

An approach is presented for graft copolymerization of type I atelocollagen onto surface of polyurethane (PU) treated with ozone. Through surface oxidization by ozone to modify the PU surface, peroxide groups are easily generated. Those peroxides are broken down by redox-polymerization and provide active species which initiate graft polymerization by reacting with amines in the collagen molecules. Ozone oxidation time and voltage could readily control the amount of peroxide production. Maximum concentration of peroxide was about $10.20 \times 10^{-8}\,\mathrm{mol/cm^2}$ when ozone oxidation was performed at 60 V for 30 min. After the reaction of PU by ozone oxidation, type I atelocollagen gel was graft copolymerized onto the PU. All the physical measurements on the collagen grafted surface indicated that the PU surface was effectively covered with type I atelocollagen. Interaction of the collagen grafted PU surface with fibroblasts could be greatly enhanced by the surface graft polymerization with type I atelocollagen. Attachment and proliferation of fibroblasts on the grafted type I atelocollagen were significantly enhanced, and it is assumed that the atelocollagen matrix supported the initial attachment and growth of cells. In the early stage of proliferation, collagen synthesis in fibroblasts was not activated and remained at a relatively low level due to the grafted type I atelocollagen, increasing only fibroblast differentiation. The mechanical property of tubular tissue was oriented from the alignment of cells that consists of repeated overlayers of longitudinal and circumferential layers and forms wall thickness.

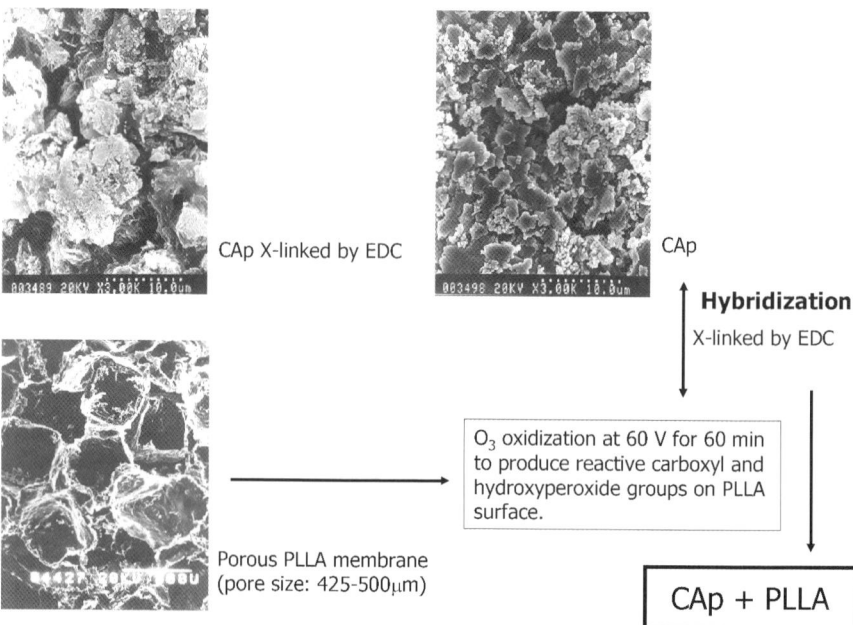

Fig. 8.19 Production of a carbonate apatite-collagen composite hybridized with a porous poly-(L-lactic acid) membrane as a membranous bone implant

8.7 Collagen in the Cell-based Implants

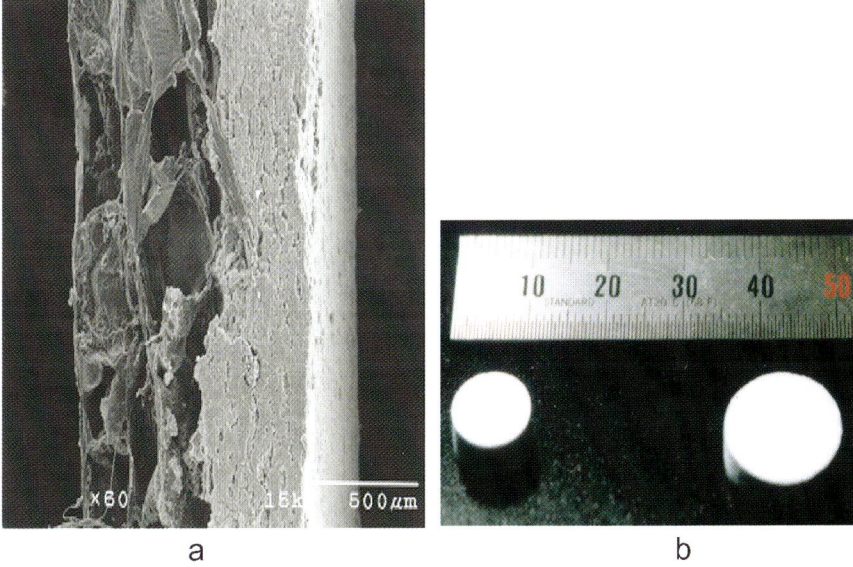

Fig. 8.20 A collagen-apatite-PLLA substitute (SEM) (**a**), and produced pellet (**b**)

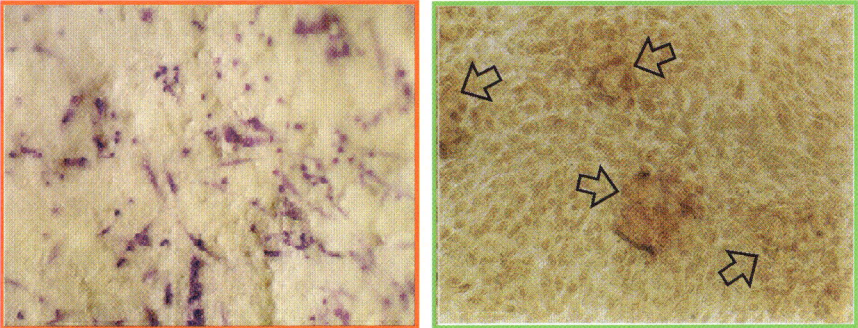

Fig. 8.21 Osteoblasts attached onto CAp+PLLA membrane. Arrows indicate calcium deposit after 3 days of culture

Regarding this result, a double layered PU tubular scaffold grafted with collagen was fabricated to produce an artificial esophagus. PU purification was performed by diluting PU particles in DMAC(dimethyl-acrylamide) solution then precipitating in non-solvent methanol, and drying the precipitate under vacuum for 2 days to remove residual methanol. After purification, PU was dissolved in DMAC solution at the final concentration of 13%w/w. Glass rod, with diameter of 3 mm and length of 15 cm, was worn with 13% PU solution and then air-dried for 6 hours. This process was repeated twice. After wearing, an end part of the glass rod was cut, and then the rod was worn again in reverse direction by the same process to be uniformly worn. It was dried under vacuum for 12 hours afterward. Completely dried sam-

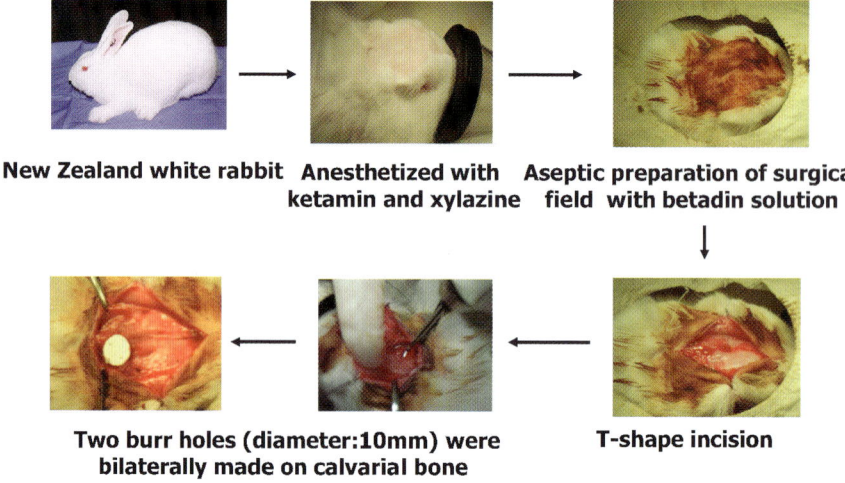

Fig. 8.22 In vivo experiment of the CAp - PLLA on to the rabbit calvarial bone defect bone model

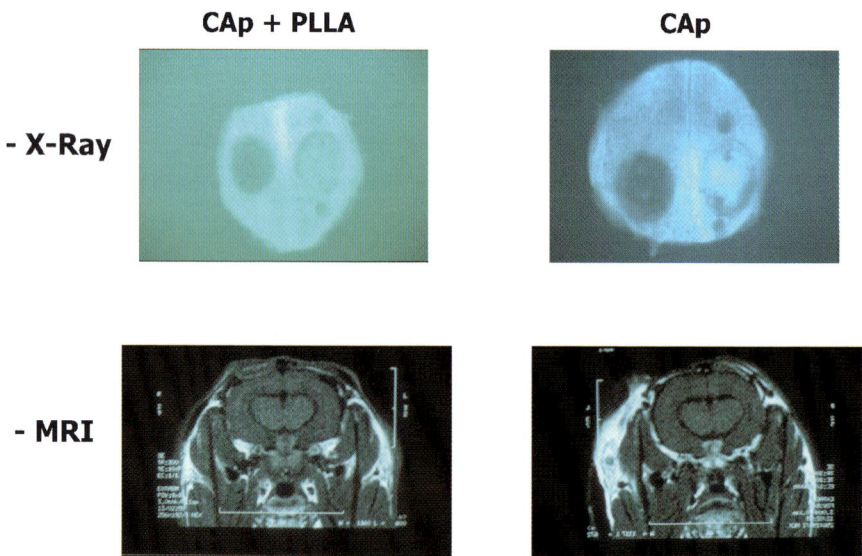

Fig. 8.23 Both implanted lesion were completely replaced by the host bone, and on the bone-remodeling process. CAp+PLLA membrane implant demonstrated a higher structural integration than CAp

ple was swelled in benzene solution for 1 hour and washed with distilled water for 2 hours and ethyl alcohol for 1 hour. After swelling and wearing process, a porous PU tube was separated from glass rod and dried under vacuum. To graft collagen, ozone treatment was performed as described above. The tube was immediately immersed in 1% type I atelocollagen gel, and Mohr's salt [$FeSO_4(NH_4)2SO_46H_2O$]

8.7 Collagen in the Cell-based Implants

was added to the gel to decompose peroxides. Then polymerization was allowed to proceed at 35 °C and pH 8.0. Type I atelocollagen grafted porous PU tube was dried in vacuum at 25 °C. The tube was dipped in a 2% collagen gel and crosslinked by EDC later. By this procedure, a porous collagen-grafted tube with inner diameter of 3 mm, thickness of 0.7 mm covered by dense collagen layer was produced. Rabbit's esophageal smooth muscle cells were seeded and cultured in a mechanically stressful environment of 10% strain magnitude and 1 Hz frequency. Culturing for 18 hours of mechanically stretching condition and 6 hours of stationary condition on every 24-hour period, the cultured cells aligned in the perpendicular direction to the strain direction after 2 days. The collagen grafted PU tube with cell alignment control was subcutaneously implanted in nude mice for 4 weeks, and a histological finding demonstrated that a well-aligned smooth muscle cells reconstructed the tube which is applicable for replacement of esophagus [26–30] (Figs. 8.24–8.26).

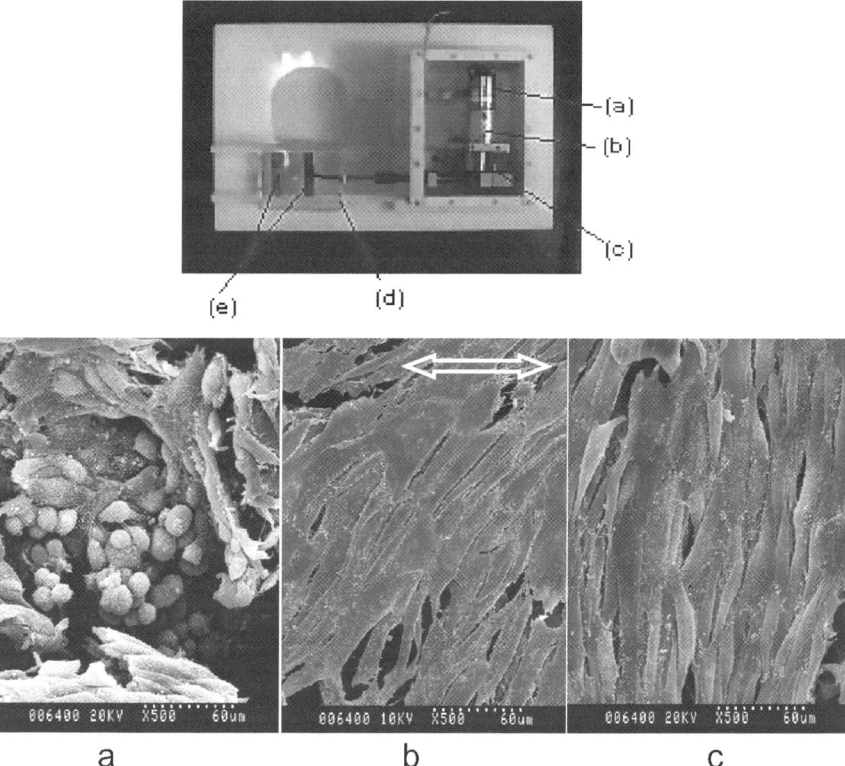

Fig. 8.24 Cyclic 10% strain magnitude and 1 Hz frequency was applied by a self-designed stretching chamber. *a* DC motor, *b* Gear box, *c* Cam system, *d* Quartz chamber, *e* Forcepses. Cell cultured under *white arrow* on **b** indicates the strain direction. **a** rabbit esophageal smooth muscle cell seeded into a collagen grafted porous polyurethane tube, **b** cultured for 12 hrs, **c** cells aligned in perpendicular to the strain axis after 24 hours of culture

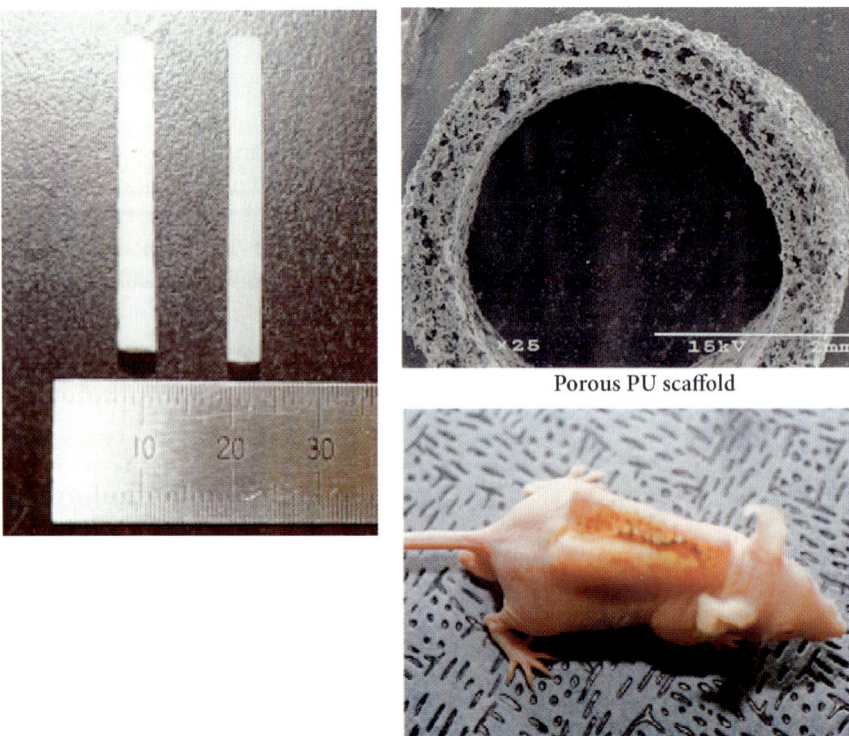

Fig. 8.25 An artificial esophagus consists of rabbit smooth muscle cells dynamically cultured in a type I atelocollagen grafted porous polyurethane tube was subcutaneously implanted in a nude mouse

8.7.4 Wound Healing promoting Anti-Adhesive Matrix

The idea of collagen grafting was also applied to produce a healing-promoting anti-adhesive membrane that is particularly necessary in peritoneal surgery to prevent postoperative adhesion. At first, glycolide and D,L-lactide were recrystallized from ethyl acetate and dried under vacuum before use. Lactide and glycolide, at a molar ratio of 75:25, were put into a glass ampoule, and mPEG was added for preparing the mPEG-PLGA block copolymer through ring-opening polymerization. 0.05%(w/w) of stannous octoate added to the solution as a catalyst. The ampoule was evacuated by a vacuum pump, sealed with a torch, and was heated in an oil bath at 130 °C for 12 hrs. After the reaction was complete, resulting polymers were purified by dissolving in methylene chloride and then precipitated in excess methanol.

Obtained polymers were dried in vacuum. For the preparation of polymer film, 8% solution of mPEG-PLGA in chloroform was cast on a glass plate, and the solvent was evaporated in a vacuum oven for 2 days. On the other side, a porous collagen-hyaluronic acid (HA-Col) membrane was fabricated and crosslinked. 1% of hyaluronic acid (HA) (sodium salt, Mw = 120,000 – 150,000) as an aqueous so-

8.7 Collagen in the Cell-based Implants

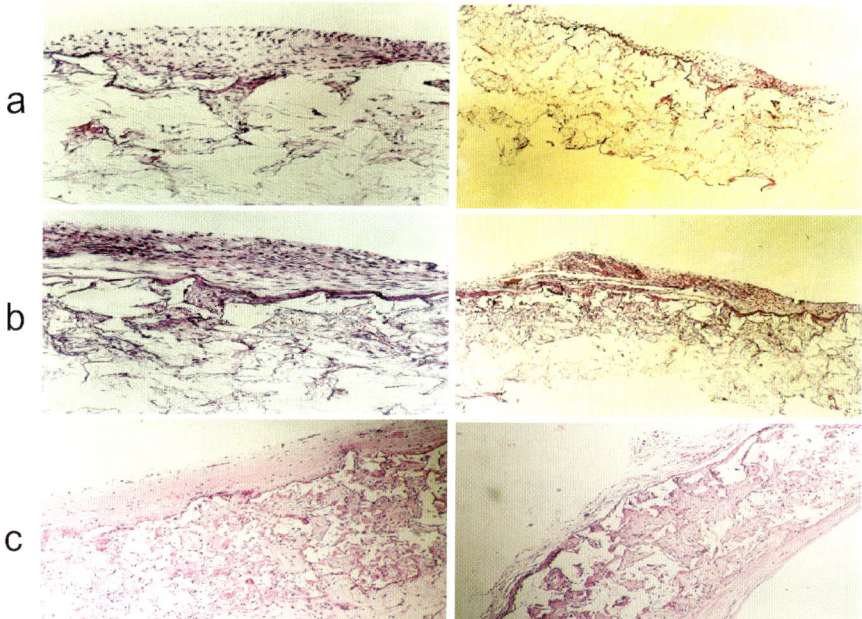

Fig. 8.26 Histological observation of the implanted artificial esophagus demonstrated a well aligned smooth muscle cells reconstructing the tube

lution was added to 1% Type I atelocollagen gel, forming 20%(w/w) HA/collagen solution and homogenized at 8000 rpm for 3 minutes at 4 °C. The resulting slurry was poured into a well plate, frozen at −70 °C, and then lyophilized at −50 °C. The fabricated porous HA-Col membranes obtained were immersed in 50 mM of EDC solution (H_2O-ethanol = 5:95) for 24 hrs. Obtained membranes were washed in distilled water by a sonicator, and then re-lyophilized at −50 °C. To promote cell adhesion, the HA-Col membranes coated with various concentrations of fibronectin(FN) were prepared by applying 1% FN solution over the HA-Col membranes at doses of 50 μg/cm^3 and incubation at 37 °C under humid condition for 5 hours.

To prepare HA-Col and mPEG-PLGA bi-layered composite membrane, chloroform was sprayed on the surface of mPEG-PLGA film, and then cross-linked HA-FN-Col membrane was loaded on the slightly dissolved surface of mPEG-PLGA film. *In vitro* adhesion test revealed that fibroblasts attached better on HA-Col membrane compared to those on mPEG-PLGA film, PLGA film or oxidized cellulose film. mPEG-PLGA film had the lowest cell adhesive property. In confocal microscopic observation, the actin filaments were significantly more polymerized when 50 or 100 μg/cm^3 fibronectin was incorporated on the HA-Col membranes. After 7-day culture, fibroblasts penetrated throughout the HAFN-Col network and the cell density increased whereas very few cells were found attached on surface of the mPEG-PLGA film. *In vivo* evaluation by implantation test for 7 days in rabbit's peritoneal wound showed that the composite membrane could protect tissue

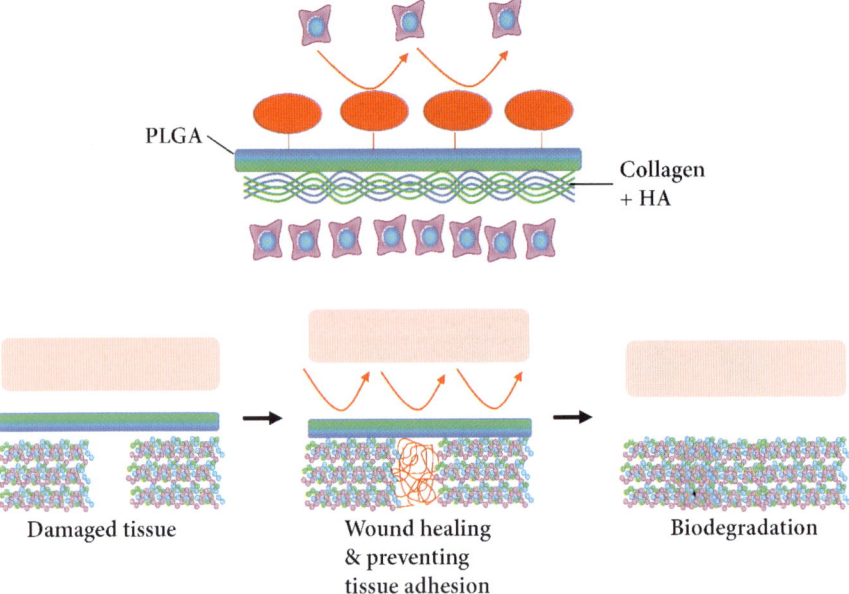

Fig. 8.27 While the hydrophilic polyethylene glycol molecules on the biodegradable polylactic-glycolic acid (PLGA) membrane inhibit the adhesion of neighboring cells, the collagen and hyaluronic acid containing fibronectin porous layer grafted on the other side of PLGA membrane promotes wound healing. The PLGA membrane may gradually be degraded

adhesion during the critical period of peritoneal healing and did not provoke any inflammation or adverse tissue reaction [31, 32] (Figs. 8.27–8.32)

8.7.5 Liver Regeneration

Collagen based biomaterial was also applied as a mesenchymal stem cell (MSC) delivery vehicle. Liver hepatocyte is a representative stable cell that has tremendous proliferative capacity and ability to differentiate and proliferate when injury or damage occurs in liver. In case of no-proliferation of hepatocytes, oval cells are stimulated to divide and eventually differentiate into mature hepatocytes; therefore, an oval cell is regarded as a compensatory cell in liver injury, and has been concerned to be equivalent to liver stem/progenitor cells. Oval cell is oriented from bone marrow mesenchymal cell, and is a facultative bipotential precursor cell that differentiates into hepatocytes or bile duct cells. Human MSCs harvested from tibial bone marrow and co-cultured with hepatocytes for 3 days in medium with additive hepatocyte growth factor (HGF) produced a confluent mixture of undifferentiated MSCs and oval cells. To purify the oval cells from nonparenchymal cells, cell cloning method was applied. Characterization of oval cells was performed by a double immuno-

8.7 Collagen in the Cell-based Implants

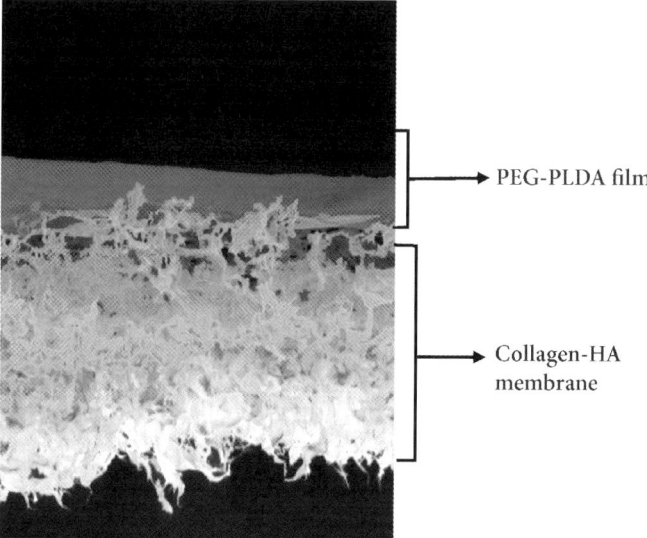

Fig. 8.28 The SEM observation of the composite matrix, which consists of a PEG-PLGA film and porous collagen-HA matrix containing fibronectin

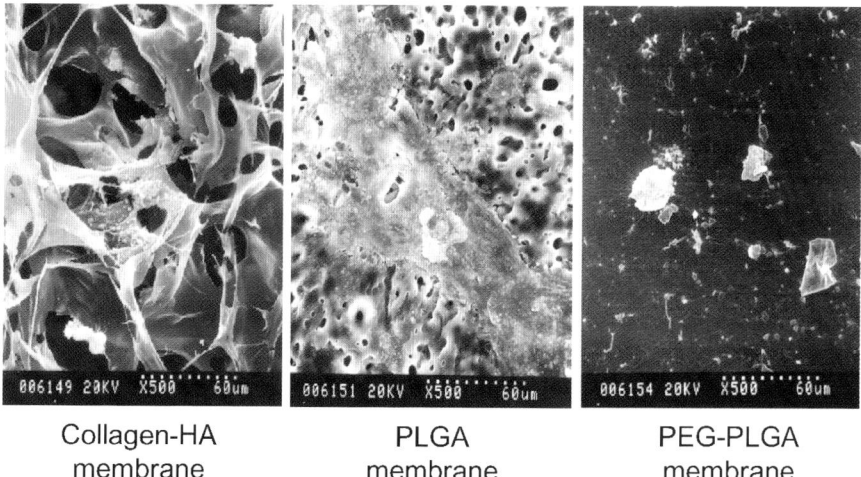

Fig. 8.29 Observed fiblasts on the PEG-PLGA surface were significantly fewer than on the PLGA membrane surface. Collagen-HA membrane demonstrated high affinity to cell attachment

fluorescence method using alpha-fetoprotein (AFP) and cytokeratin-19 monoclonal antibody expecting co-expression. Isolated human MSCs oriented oval cells were seeded on an EDC crosslinked porous type I atelocollagen matrix, and cultured in medium containing insulin, dexamethasone, and hydrocortisone as the hormone stimulators and additive cytokines of HGF and epidermal growth factor (EGF). For

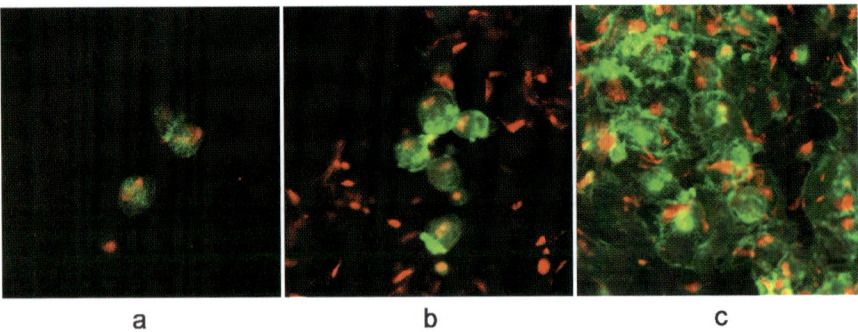

Fig. 8.30 Confocal microscopic images of FITC-phalloidin stained F-actin (*green*) and PI stained nucleus (*red*) in fibroblasts cultured on the PEG-PLGA film (**a**), collagen-hyaluronic acid membrane (**b**), 100 μg/cm^3 fibronectin coated collagen-HA membranes (**c**)

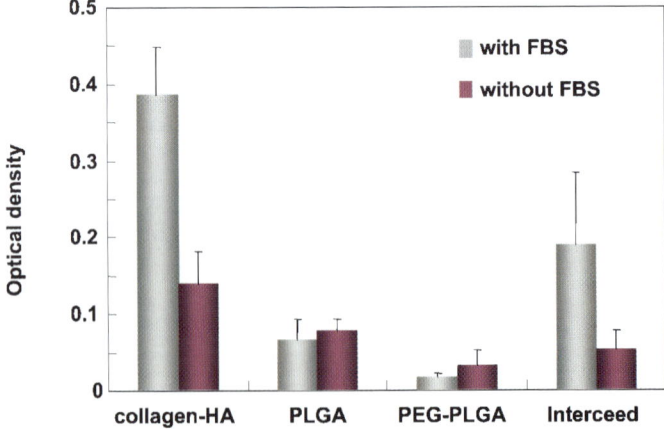

Fig. 8.31 Attachment of fetal human dermal fibroblasts on the various types of membranes 4 hours after seeding: Effect of fetal bovine serum in the culture medium. * Interseed (Johnson & Johnson Co.) is a commercially available anti-adhesive membrane made by oxidized cellulose

3 weeks, albumin secretion and urea detoxification rate continuously increased either in the hormone additive or the cytokine additive groups, and this was directive knowledge that avoidance of using growth factors in committing MSC-oriented oval cell toward hepatocyte can escape from risky induction to liver cancer. The collagen scaffold was able to foster long-term viability and protection of the cells, and this 3-dimensional culture of oval cells is considerable for designing a cell-delivering tool for hepatic disease [33] (Figs. 8.33–8.38).

8.8 Discussion

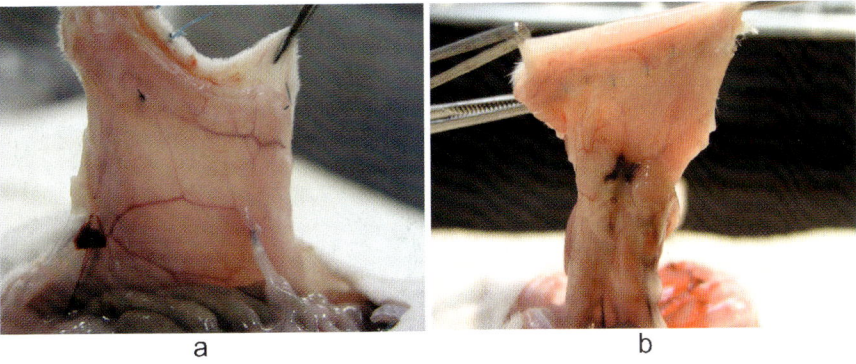

Fig. 8.32 *In vivo* Experiement: rat abdominal sidewall injury model Photographs of **a** Col-FN grafted PEG-PLGA membrane treated group and **b** control group (*right*) after 7 days. After an abdominal midline incision, a+2 cm × 1 cm defect on the anterior abdominal wall of 30 female Sprague-Dawley rats was created by a template and scalpel. The Col-FN grafted PEG-PLGA membrane (3 cm × 2 cm) was placed as the Col-FN grafted side could face the injured peritoneal wall. Asymmetric collagen-fibronectin grafted PEG-PLGA copolymer membrane was appeared to successfully reduce the incidence of postoperative adhesion formation

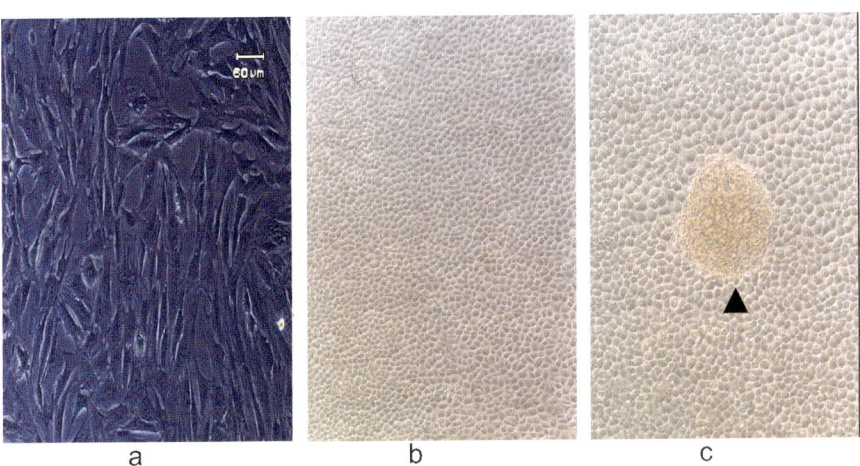

Fig. 8.33 Microscopically observed **a** mesenchymal stem cells co-cultured with hepatocytes on day 0, **b** oval shaped cells on day 3, **c** colony piled oval cells on day 24 in DMEM/F-10 culture media (×100)

8.8 Discussion

Although collagen is a favorable biomaterial to be employed in a wide range of fabrication procedures with cells, there is no established tool to isolate and produce absolute atelocollagen, the immune-free substitute, at present. Even a single remaining collagen dimer or trimer may cause immune reactions in future. The resource of collagen, especially for clinical applications is also important, because unknown

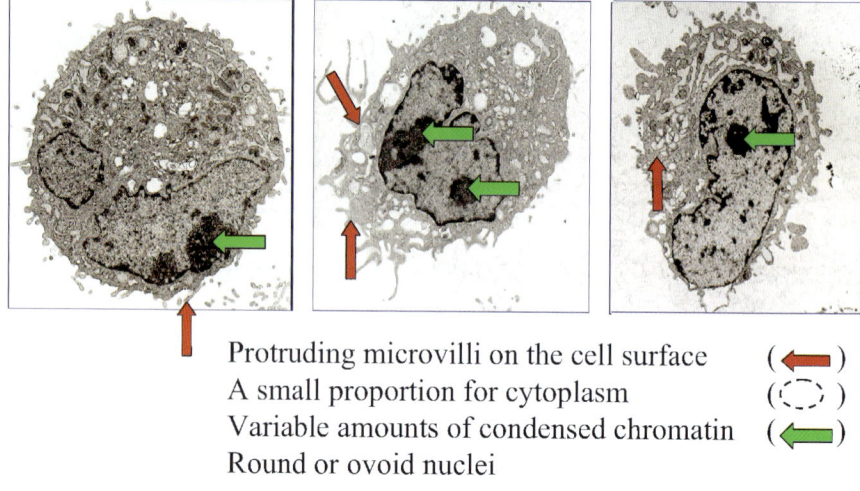

Protruding microvilli on the cell surface (⬅)
A small proportion for cytoplasm (⬭)
Variable amounts of condensed chromatin (⬅)
Round or ovoid nuclei

Fig. 8.34 The oval cells obtained from mesenchymal stem cells through co-culture with hepatocytes demonstrated typical cytomorphorlogcal characteristics

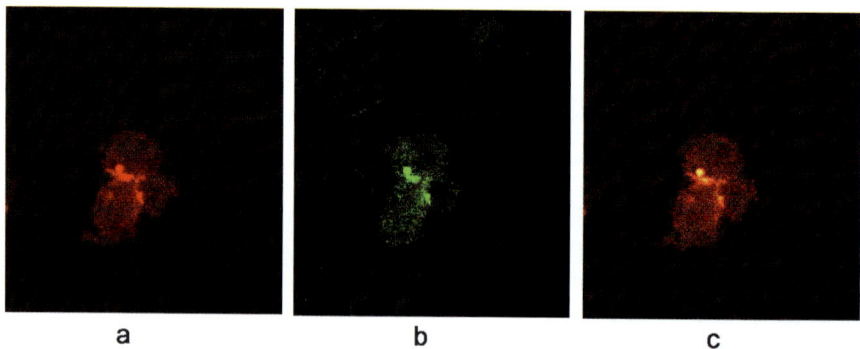

Fig. 8.35 Characterization of the Oval cell mediated from mesenchymal stem cell through co-culture with hepatocyte by double staining using alpha-fetoprotein (AFP) and cytokeratin (CK) 19. **a** AFP – *red* **b** CK 19 – *green* FP and **c** CK 1

viruses that produce future unpredictable diseases from xenogeneic extracellular matrix may exist. If single-cell encapsulation by collagen containing signal transduction agents and ligands attracting specific cell adhesive receptors is systemized, commitment of stem cell differentiation and proliferation toward the designated target tissue will be achieved, and this may contribute to the future progress of stem cell-based implant.

8.8 Discussion

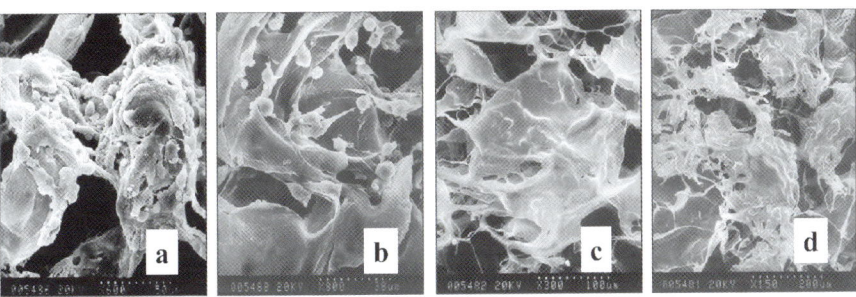

Fig. 8.36 SEM observation of oval cells seeded in porous collagen scaffold. **a** Oval cells in scaffold supplemented with cytokines at day 4, **b** Oval cells in scaffold supplemented with hormones at day 4, **c** Oval cells in scaffold supplemented with cytokines at day 7, **d** Oval cells in scaffold supplemented with hormones at day 7

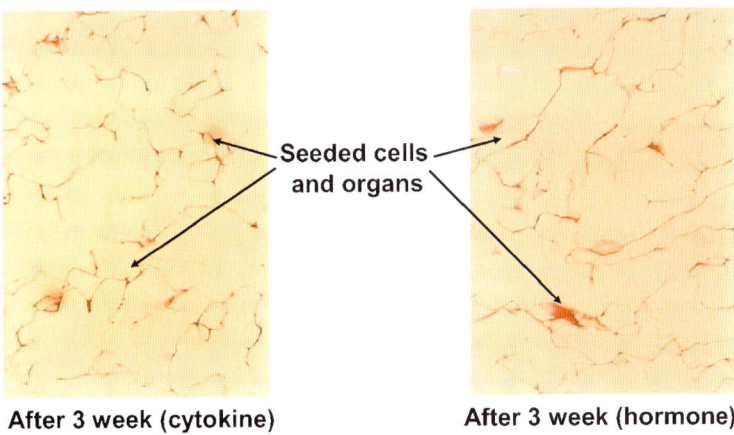

Fig. 8.37 Morphology of oval cells in scaffold. Masson & Trichrome Staining

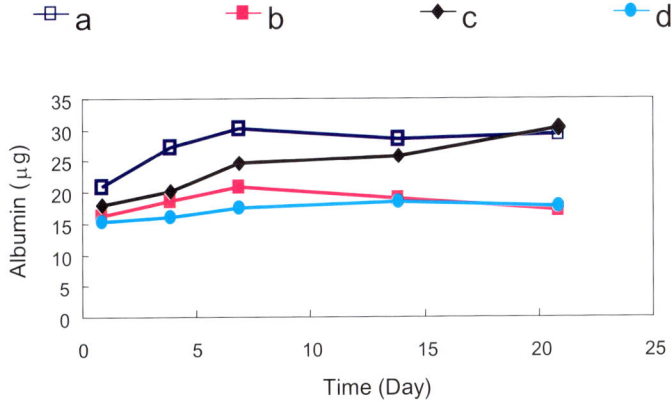

Fig. 8.38 Albumin secretion in scaffold. **a** Cultured with cytokines in scaffold, **b** Cultured with cytokines on tissue culture plate (TCP), **c** Cultured with hormones in scaffold, **d** Cultured with hormones on TCP

References

1. Suh H (2000) Tissue Restoration, Tissue Engineering and Regenerative Medicine. Yonsei Medical Journal 41(6):681–684
2. Suh H (1998) Fundamental Concepts for Tissue Engineering. Biomaterials Research 2(1):1–7
3. Braybrook JH (1997) Biocompatibility assessment of medical devices and materials. John Wiley and Sons Ltd., West Sussex UK
4. Suh H (1998) Recent Advances in Biomaterials. Yonsei Medical Journal 39(2):87–96
5. Ruoslahti E, Engvall E (1994) Extracellular Matrix Components. Academic Press Inc., San Diego, CA.
6. Suh H, Suh S, Min B (1994) Anti-infection Treatment of a Transcutaneous Device by a Collagen-Rifampicine Composite. ASAIO Journal 40(3):406–411
7. Suh H (1999) Treatment of Collagen for Tissue Regenerative Scaffold. Biomedical Egineering Application and Basis Communication 11(4):167–173
8. Kim S-H, Lee J-H, Yun S-Y, Yoo J-S, Jun C-H, Chung K-Y, Suh H (2000) Reaction Monitoring of Succinylation of Collagen in Matrix-Assisted Desorption/Ionization Mass Spectrometry. Rapid Communication of Mass Spectrometry 14(12):2125–2128
9. Lee J-E, Park J-C, Hwang Y-S, Kim JK, Kim J-G, Suh H (2001) Characterization of UV-irradiated Dense/Porous Collagen Membranes: Morphology, Degradation, and Mechanical Properties. Yonsei Medical Journal 42(2):172–179
10. Park S-N, Park J-C, Hae Ok Kim, Min Jung Song, Hwal Suh (2002) Characterization of Porous Collagen/Hyaluronic Acid caffold Modified by 1-ethyl-3(3-dimethylaminopropyl)carbodiimide Cross-linking. Biomaterials 23(4):1205–1212
11. Lee J-E, Park J-C, Kim J-G, Suh H (2001) Preparation of Collagen Modified Hyaluronan Microparticles as Antibiotics Carrier. Yonsei Medical Journal 42(3):291–298
12. Suh H, Lee J-E (2002) Behavior of fibroblasts on a porous hyaluronic acid incorporated collagen matrix. Yonsei Medical Journal 43(2):193–202
13. Lee J-E, Park J-C, Lee KH, Oh SH, Suh H (2002) Laminin modified infection-preventing collagen membrane containing silver sulfadiazine-hyaluronan microparticles. Artificial Oragns 26(6):521–528
14. Lee J-E, Park J-C, Lee KH, Oh SH, Kim J-G, Suh H (2002) An infection-preventing bilayered collagen membrane containing antibiotic-loaded hyaluronan microparticles: physical and biological properties. Artificial Oragns 26(7):636–646
15. Park S-N, Lee HJ, Lee KH, Suh H (2003) Biological characterization of EDC-crosslinked collagen-hyaluronic acid matrix in dermal tissue restoration. Biomaterials 24(9):1631–1641
16. Park S-N, Kim JK, Suh H (2004) Evaluation of antibiotic-loaded collagen-hyaluronic acid matrix as a skin substitute. Biomaterials 25(17):3689–3698
17. Park S-N, Kim JH, Kim IH, Sul AR, Suh H (2006) Electrospun nanofibrous membrane for the engineering of cultured skin substitutes. Biomaterials Research 10(2):82–88

18. Suh H, Lee C (1995) Biodegradable Ceramic-Collagen Composite Implanted in Rabbit Tibiae. ASAIO Journal 41(3):652–656
19. Suh H, Hwang Y-S, Song MJ, Lee WS, Han CD, Park J-C (2000) Behaviors of Osteoblasts-like Cell (MC3T3-E1) on Collagen Grafted Poly L-lactic Acid (PLLA) Membranes with Various Pore Sizes. Biomaterials Research 4(2):37–44
20. Park J-C, Han D-W, Suh H (2000) A Bone Replaceable Artificial Bone Substitute: Morphological and Physicochemical Characterization. Yonsei Medical Journal 41(4):468–476
21. Suh H, Hwang Y-S, Lee J-E, Han CD, Park J-C (2001) Behavior of Osteoblasts on type I Atelocollagen Grafted Ozone Oxidized Poly L-lactic Acid Membrane. Biomaterials 22(2):219–230
22. Suh H, Park J-C, Han D-W, Lee DH, Han CD (2001) A Bone Replacable Artificial Bone Substittue: Cytotoxicity, Cell Adhesion, Proliferation, and Alkaline Phosphatase Activity. Artificial Organs 25(1):14–21
23. Suh H, Han D-W, Park J-C, Lee DH, Lee WS, Han CD (2001) A Bone Replaceable Artificial Bone Substitute: Osteoinduction by combining with Bone Inducing Agent. Artificial Organs 25(6):459–466
24. Kim H, Lee J-H, Suh H (2003) Interaction of Mesenchymal Stem Cells and Osteoblasts for in vitro Osteogenesis. Yonsei Medical Journal 44(2):187–197
25. Suh H, Song MJ, Ohata M, Kang Y-B, Tsutsumi S (2003) Ex Vivo Mechanical Evaluation of Carbonate Apatite-Collagen Grafted Porous PLLA Membrane in Rabbit Calvarial Bone. Tissue Engineering 9(4):635–643
26. Suh H, Hwang Y-S, Lee JE, Kim KT, Park JC, Park KD, Kim YH (1998) Type I Atelocollagen Grafting on Polyurethane Tube and its Mechanical Property. Biomaterials Research 2(4):158–162
27. Suh H, Hwang Y-S, Kang Y-B, Nakai R, Tsutsumi S Park J-C (2000) Compliance of surface Modified Polyurethane Tubular Scaffold for Artificial Esophagus. Biomaterials Research 4(1):8–12
28. Park J-C, Hwang Y-S, Lee J-E, Park KD, Matsumura K, Hyon S-H, Suh H (2000) Type I Atelocollagen Grafting onto Ozone treated Polyurethane Films: Cell Attachment, Proliferation, and Collagen Synthesis. Journal of Biomedical Materials Research 52(4):669–677
29. Cha JM, Park S-N, Noh SH, Suh H (2006) Time-dependent modulation of alignment and differentiation of smooth muscle cells seeded on a porous substrate undergoing cyclic mechanical strain. Artificial Organs 30(4):250–258
30. Cha JM, Park S-N, Park G-O, Kim JG, Suh H (2006) Construction of functional soft tissues from premodulated smooth muscle cells using a bioreactor system. Artificial Organs 30(9):704–707
31. Suh H, Park S-N, Kim JH (2003) Evaluation of tissue adhesion preventive surface modified natural and synthetic polymeric materials. Materials Science Forum 426–432(2003):3255–3260
32. Park SN, Jang HJ, Choi YS, Cha JM, Son SY, Han SH, Kim JH, Lee WJ, Suh H (2007) Preparation and characterization of biodegradable anti-adhesive membrane for peritoneal wound healing. Journal of Materials Science: Materials in Medicine 18:475–482
33. Suh H, Song MJ, Park YN (2003) Behaviors of isolated rat oval cells in porous collagen scaffold. Tissue Engineering 9(3):411–420

Chapter 9
Tissue Engineering – Combining Cells and Biomaterials into Functional Tissues

Bernd Denecke[1], Michael Wöltje[1], Sabine Neuss[1,2], Willi Jahnen-Dechent[1,3]

[1] Interdisciplinary Center for Clinical Research on Biomaterials and Material-Tissue Interaction of Implants "IZKF BIOMAT."
[2] Institute of Pathology
[3] Institute of Biomedical Engineering, Biointerface Laboratory RWTH Aachen University, Pauwelsstraße 30, 52074 Aachen, Germany, Willi.jahnen@rwth-aachen.de

Abstract Tissue engineering is an interdisciplinary field applying the principles of engineering to life science to generate living tissue and organ replacements. Typically, cells are seeded onto a biomaterial scaffold to be integrated into a specific tissue. Both biological and synthetic biomaterials are currently employed including metals, ceramics and polymers. The cell/biomaterial interactions are complex and hardly predictable. Inevitably this requires the assessment of cytotoxicity and biocompatibility in each novel combination of cells and biomaterials. A cell/biomaterial biohybrid should mimic the natural equivalent as close as possible – "biomimetic". Researchers match biomaterial rigidity, stability, porosity, degradability etc. to the extracellular matrix of a given organ to control cell adhesion, proliferation, and differentiation in a predictable fashion. Examples of this interface biology and methods to assess the cell-material interactions are discussed in this chapter.

Stem cells as a source of cells for tissue engineering continue to play an important role in tissue engineering. Pluripotent embryonic and multipotent adult stem cells alike are extensively used in experimental tissue engineering applications and cell-based therapies. Recent developments in stem cell biology are discussed.

The engineering of cell/biomaterial hybrids for tissue replacement in three dimensions is a critical goal of tissue engineering posing a formidable task in itself. Adequate tissue vascularization, functional tissue integration *in vivo*, regulatory approval and economical viability however, must take center stage in any tissue engineering approach seriously interested in commercialization.

9.1 Introduction

In 1993 Robert Langer and Joseph Vacanti defined tissue engineering as "an interdisciplinary field that applies the principles of engineering and the life sciences toward the development of biological substitutes that restore, maintain, or improve tissue function" (Langer and Vacanti 1993). These biological substitutes combine cells cultured in a natural or nature-like environment sustaining growth and differentiation. The environment comprises biomaterial scaffolds providing cell attachment sites and a basic three-dimensional organisation resembling extracellular matrix (ECM). A tissue engineered substitute is meant to mimic the differentiated structure and function of its native counterpart in as many ways as possible.

Many reviews have dealt with tissue engineering of specialized tissues including skin, bone, cartilage, ligament etc. Several monographs have been published on this topic. Here we attempt to recapitulate basic considerations of cell-material interactions. In particular we will discuss the use of stem cells for tissue engineering, recent advances in material scaffolds, and cell-material interactions.

9.2 The Cells

9.2.1 *In Search of an Ideal Cell Source for Organ Replacement*

One primary consideration in tissue engineered organ replacement is the choice of cells and the cell source. It is important to stress that any tissue engineered product should contain a cell mass far exceeding the cell mass of the biopsy originally taken. This is not always the case in experimental therapies. In these unfortunate cases tissue engineering actually results in net loss of tissue. Therefore cells for tissue engineering must possess a high intrinsic proliferation capacity. Despite a high proliferation rate, the cells must be able to eventually stop proliferation and enter terminal differentiation into a desired cell type. Most tissue engineering approaches include *ex vivo* steps to expand and differentiate cells in culture. In principle however, the cell selection and proliferation step might as well be accomplished in the body using biomaterials with exquisite control of mobilization, proliferation and differentiation of cells *in situ*.

Organ function in the body is primarily determined by the differentiated cell type(s) constituting a given organ. To date skin epidermis is one of the few tissues where this process is understood in any appreciable detail (Clayton et al. 2007). Roughly 200 distinguishable cell types are associated with specialized tissues like connective tissue (fibroblasts), muscle (myocytes), liver (hepatocytes), cartilage (chondrocytes) and bone (osteoblasts). It should be recalled that almost every organ is made up of more than one cell type and that any piece of tissue larger than a cubic millimetre (1 microliter volume) requires vascularisation to ensure nutrient and waste exchange. Thus cell plasticity, the ability of a precursor cell to form most or all cells of a given organ and angiogenesis, the sprouting and growth of new blood vessels continue to be important issues in tissue engineering. For obvious reasons cell plasticity is highest in embryonic tissues. Hence a lot of our knowledge about cell plasticity was learned from embryo-derived stem cells, ES cells.

9.2.2 Stem Cells

Stem cells have two major distinguishing properties: i) They are undifferentiated cells that renew themselves for the entire life span of an organism through cell division and ii) they have a remarkable capacity to develop from a common precursor into multiple cell types with specialized functions such as the insulin producing cells of the pancreas, blood cells, nerve cells or beating cardiomyocytes (Fig. 9.1).

Protocols to obtain stem cells from early mouse embryos were developed more than 20 years ago (Evans and Kaufman 1981; Martin 1981). Since 1998 stem cells from human embryos can also be isolated and grown in the laboratory (Thomson et al. 1998). These so-called ES cells were isolated from fertilized oocytes of in vitro fertilisation patients. In the mouse ES cells are derived from blastocysts 3–5 days after conception. They give rise to specialized cell types of all three germ layers and are thus called "pluripotent". Adult tissues also contain stem cells, e. g. the bone marrow. These "adult stem cells" may replace cells that are lost through normal turnover, injury, or disease. In order to use these cells in tissue engineering or in cell based therapies researchers intensively studied the fundamental properties of stem cells. These studies are designed to i) determine precisely how stem cells remain unspecialized and self-renewing for many years, ii) identifying the signals that cause stem cells to become specialized cells, iii) gain knowledge about how an organism develops from a single cell, and iv) how healthy cells replace damaged cells in adult organisms.

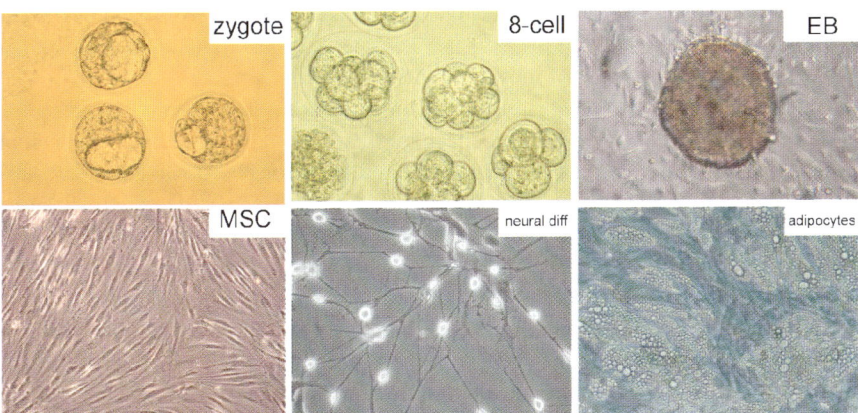

Fig. 9.1 Mouse stem cells, their origin and derivatives. All cells originate from fertilized oocytes, zygotes. Multiple cell divisions into 2-, 4-, 8-cell stages etc. eventually form complex tissue and whole animals. Bulk tissue formation is mimicked in embryoid bodies (EB), which spontaneously differentiate into cells of all three germ layers. Adult bone marrow stromal cells (MSC) can be differentiated e. g. into neuon-like cells following neural differentiation or into adipocytes following adipogenic differentiation

9.2.3 Pluripotent Embryonic Stem Cells

The history of pluripotent ES cells started in the seventies and is based on observations in teratocarcinomas. Solter et al. (Solter et al. 1970) as well as Stevens (Stevens 1970) observed that teratocarcinomas developed after transplantation of early mouse embryos into adult mice. A teratocarcinoma is a malign tumor which is composed of differentiated cells of all three germ layers as well as a population of undifferentiated cells causing the malignity. These undifferentiated cells, embryonic carcinoma, EC cells, can be expanded in cell culture (Andrews 2002). During establishment of an EC cell line the cells undergo oncogenic transformation. As a result these cells have highly unstable karyotypes resulting in chromosomal abnormalities. Moreover, EC cells can contribute to all three germ layers in chimeric mice. In general, EC cells have only a limited differentiation potential and due to the chromosomal abnormalities they cannot undergo meiosis. The origin of EC cells is the embryoblast. It could be shown that in transplantation experiments only embryoblast-derived cells were able to generate teratocarcinomas. Grafts originated from other regions of the early embryo lacked teratogenicity (Diwan and Stevens 1976).

Furthering the research on EC cells, embryonic stem cells, ES cells were isolated from the inner cell mass of blastocyst stage mouse embryos in 1981 (Evans and Kaufman 1981; Martin and Evans 1975). From these cells undifferentiated cell lines could be established. Given the right culture conditions ES cells showed a remarkable genetic stability even at high passage numbers. ES cells give rise to all tissues of the adult animal and unlike EC cells can form also germ cells (Bradley et al. 1984). The ability to contribute to the germline formed the basis of gene targeting technology in mice (Doetschman et al. 1987; Thomas and Capecchi 1987).

The remarkable features of murine ES cells sparked interest in similar cell types of human origin. This interest is based mainly on four reasons (Smith 2001):

i) Use of these cells to examine aspects of the embryonic development which is otherwise not accessible. ii) Use of the cells as a basis for functional genomic analyses in diploid human cells to examine and/or manipulate specific gene functions. iii) Use of these cells as a source of large numbers of phenotypic identical human cells for pharmacological tests (e. g. toxicity tests). iv) Use of the cells for tissue engineering and regenerative medicine.

Human pluripotent cells were first described in 1998 (Thomson et al. 1998). The definition of ES cells maintains that these cells are able to generate chimaeric offspring and especially that they contribute to the germline. In human beings this definition can not be tested for obvious ethical reasons. Despite such limitations in human ES cell research important differences in the biology of human ES cells and murine ES cells have been established. One major difference between mouse and human ES cells concerns the regulation of pluripotency. Mouse ES cells are usually maintained on fibroblast feeder cells. The feeder cells produce leukemia inhibitory factor, LIF as the major factor preventing differentiation while maintaining high proliferation. Mouse ES cells grow without feeder cells when sufficient LIF is added to the culture medium (Smith et al. 1988; Williams et al. 1988). *Vice*

9.2 The Cells

versa it was shown that feeder cells lacking LIF are unable to prevent the differentiation of ES cells (Stewart et al. 1992). LIF belongs to the interleukin family of cytokines. Following binding to its cognate receptor LIF triggers an intracellular signal cascade resulting in activation of the transcription factor STAT3 (Burdon et al. 1999a; Burdon et al. 1999b). By using a constitutively active variant of STAT3 it could be shown that this activation was sufficient to maintain self renewal of ES cells (Matsuda et al. 1999). In the absence of a STAT3 signaling, LIF activates the mitogen activated protein kinase, MAPK pathway. This pathway in turn activates extracellular signal regulated kinase, ERK. ERK activation alone promotes the differentiation of ES cells (Burdon et al. 1999b). Thus continuous LIF activity and prolonged inhibition of the MAPK cascade enable the establishment of ES cell lines (Buehr et al. 2003). This example shows that the maintenance of self renewal and pluripotency depend on a fine balance of inhibitors and activators. Small amounts of bone morphogenetic protein 4, BMP-4, are also required for ES cell renewal. This quantity of BMP-4 is included in the serum added to the culture medium (Ying et al. 2003). BMPs were originally detected in association with bone growth, hence their name. For the self renewing of ES cells the cooperation of LIF and BMP is necessary. BMP inhibits neuroectodermal differentiation In the presence of LIF. Without LIF and STAT3 activity, BMP induces the differentiation into endodermal and mesodermal cell types. Thus the LIF cascade acts as a molecular master switch changing the action of BMP from an inductor to an inhibitor of differentiation.

In human ES cells unlike in murine ES cells, LIF is dispensable for maintaining pluripotency (Daheron et al. 2004; Reubinoff et al. 2000; Thomson et al. 1998). Furthermore, adding of BMP to human ES cells does not maintain pluripotency, but induces the differentiation into trophectoderm (Xu et al. 2002). In human ES cells, LIF is replaced by basic fibroblast growth factor, bFGF/FGF2. Further essential factors necessary to maintain pluripotency are the transcription factors Oct4 (Niwa et al. 2000; Pesce et al. 1998; Ying et al. 2003), and Nanog (Chambers et al. 2003; Mitsui et al. 2003).

Generally ES cells remain pluripotent because they employ several mechanisms preventing differentiation. Intriguingly ES cells have an unusual cell cycle in that they lack signal pathways crucial for controlling the cell cycle in most other cells (Burdon et al. 2002). Interestingly STAT3 activity can influence the cell cycle. One of the first reactions of mouse ES cells following LIF withdrawal is a change in cell cycle and an immediate start of differentiation. It is increasingly evident that epigenetic mechanisms can profoundly regulate the expression profile of a cell, e.g. by affecting chromatin structure and genome wide methylation patterns (Chen et al. 2003; Rasmussen 2003; Rugg-Gunn et al. 2005). How do the key genetic factors and the epigenetic mechanisms act together? On a practical note this question is very important, because the mechanisms rendering ES cells pluripotent may be identical with the mechanisms returning or "reprogramming" a differentiated cell to a pluripotent cell (Takahashi and Yamanaka 2006). Practical knowledge in this area will advance therapeutic cloning and cell replacement therapy alike.

9.2.4 Adult Stem Cells

By definition an adult stem cell is an undifferentiated cell found among differentiated cells in a tissue or organ which can renew itself or differentiate to yield the major specialized cell types of that tissue or organ. Research on adult stem cells started about 40 years ago. In the 1960s, researchers discovered that the bone marrow contains at least two kinds of stem cells. One population, called hematopoietic stem cells, HSCs, forms all the types of blood cells in the body. This knowledge is applied daily throughout the world when bone marrow or HSC transplants are administered. HSCs from bone marrow have been used in transplants for more than 30 years (Gatti et al. 1968).

A second population, called bone marrow stromal cells, MSCs, was discovered a few years later (Friedenstein 1968). MSCs are a mixed cell population generating bone, cartilage, fat, and fibrous connective tissue. In the 1960s, scientists studying rats discovered two regions of the brain containing mitotic cells generating neurons. Despite these early reports, most scientists continued to believe that neurons never regenerated in the adult brain, but were strictly post-mitotic. Only in the 1990s scientists realized that the adult brain indeed harbors stem cells able to generate astrocytes and oligodendrocytes as well as neurons.

The primary role of adult stem cells is tissue repair. Unlike ES cells, which have a clearly defined origin tissue (the inner cell mass of the blastocyst), the exact location of adult stem cells in mature tissues is unknown. Adult stem cells have been isolated from many organs and tissues including brain, bone marrow, peripheral blood, blood vessels, skeletal muscle, skin and liver. It is generally believed that stem cells reside in a specific niche tissue within bulk (Blanpain et al. 2004; Dao et al. 2007; Lin 2002). This stem cell niche ensures that stem cells remain quiescent for many years until they are activated by disease or tissue injury. We have determined that hepatocyte growth factor can potentially mobilize adult stem cells from bone marrow (Neuss et al. 2004).

Adult stem cells show remarkable plasticity in that they differentiate into a number of different cell types, given the right conditions (Blau et al. 2001). Bone marrow stromal cells (MSCs) give rise to a variety of cell types like bone cells (osteocytes), cartilage cells (chondrocytes), fat cells (adipocytes), and other kinds of connective tissue cells such as those in tendons (Pittenger et al. 1999) depending on the kind of differentiation protocol employed (Figs. 9.2 and 9.3).

Neural stem cells in the brain give rise to nerve cells (neurons) and two categories of non-neuronal cells – astrocytes and oligodendrocytes (Rietze et al. 2001). Epithelial stem cells in the lining of the digestive tract occur in deep crypts and give rise to absorptive cells, goblet cells, Paneth cells, and enteroendocrine cells. Skin stem cells occur in the basal layer of the epidermis and at the base of hair follicles (Toma et al. 2001). The epidermal stem cells give rise to keratinocytes, which migrate to the surface of the skin and form a protective layer. The follicular stem cells can give rise to both the hair follicle and to the epidermis.

A substantial number of publications have thus suggested that adult stem cells are multipotent. This ability has been questioned (Wagers and Weissman 2004),

9.2 The Cells

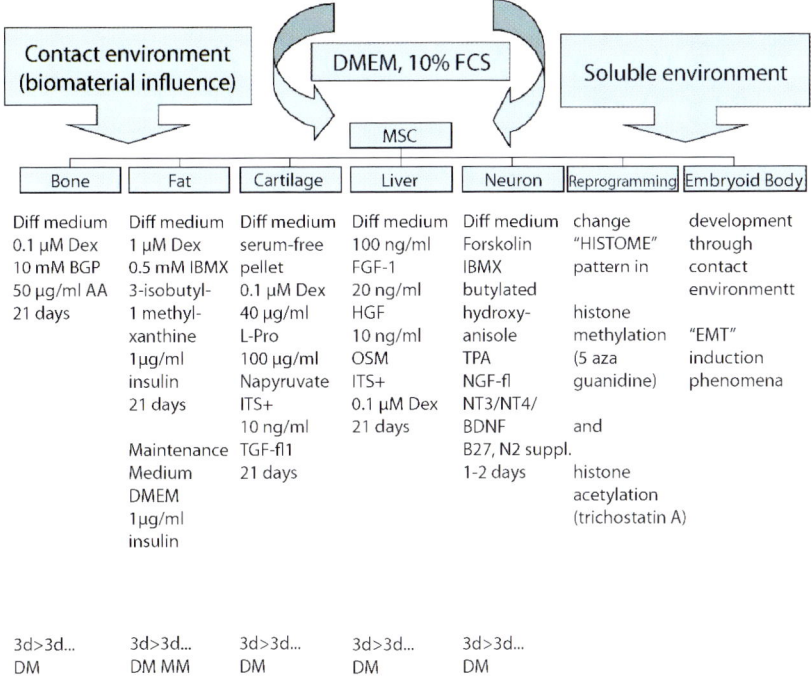

Fig. 9.2 Common differentiation protocols for adult mesenchymal stromal cells. Basal medium is Dulbeccos modified Eagles medium (DMEM) including 10% fetal calf serum. Media supplements and treatments are as listed. ITS is 5 mg/ml insulin, 5 mg/ml transferrin and 5 μg/ml selenic acid. DM, differentiation medium, MM, maintenance medium

because several alternative explanations for the observed "transdifferentiated" cell phenotypes also exist.

Thus the plasticity of adult stem cells is still a hotly debated issue (Aranguren et al. 2007; Dao et al. 2007; Jahagirdar and Verfaillie 2005; Pelacho et al. 2007; Ross et al. 2006; Ross and Verfaillie 2007; Serafini and Verfaillie 2006; Snykers et al. 2006). It is unknown how many kinds of adult stem cells exist in the body and where. Some data suggest that the "plasticity" of adult stem cells is largely an artifact of extended culture periods. Furthermore cell fusion (Ying et al. 2002) of genetically or chemically labeled implanted cells with recipient cells have fooled several researchers into believing that the implanted cells had attained characteristics of the recipient tissue (Fig. 9.4). These issues need to be resolved, before a final statement about plasticity of adult stem cells can be made.

9.2.4.1 What Distinguishes Embryonic from Adult Stem Cells?

In summary, embryonic and adult stem cells each have advantages and disadvantages regarding their potential use for cell based regenerative therapies. Firstly MSC and ES cells differ in the number and type of differentiated cell types they can be-

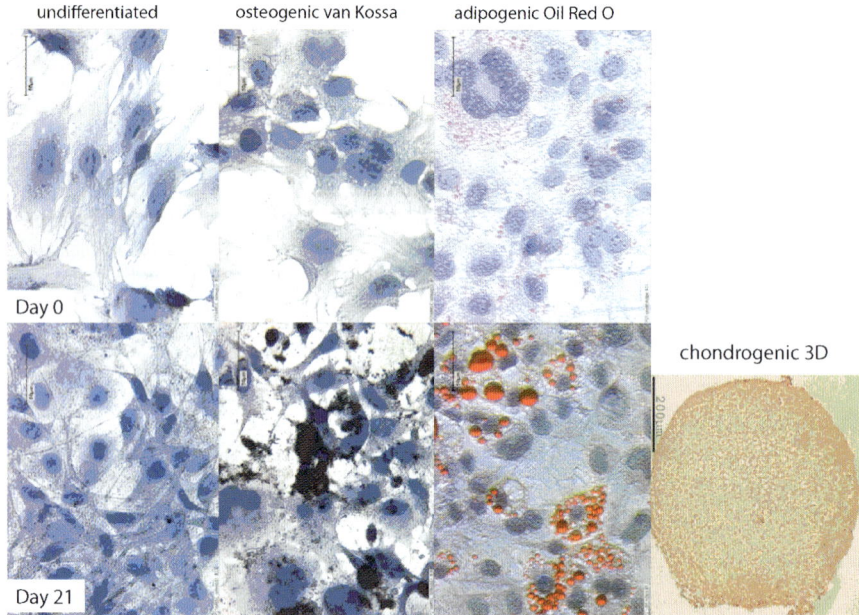

Fig. 9.3 Differentiation of MSC into osteoblasts, adipocytes and chondrocytes. The distinguishing phenotype of each cell type is stained by von Kossa stain (mineral deposits), Oil Red O (fat vacuoles). Chondrocytes develop only in 3D cultures formed by compacting a cell pellet by centrifugation

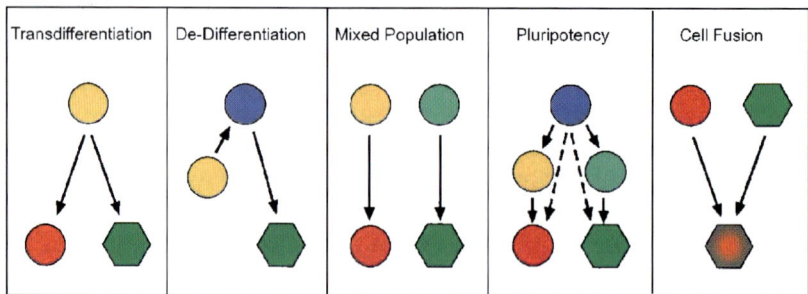

Fig. 9.4 Alternative explanations of "plasticity" and "transdifferentiation" of cells. In transdifferentiation, a cell becomes several cell types without de-differentiating. Pluripotent cells have even higher differentiation potential. "New" cell phenotypes can however, also arise from impure, mixed cell populations or from cell fusion

come. Pluripotent ES cells can potentially become all cell types of the body. Adult stem cells are generally limited to differentiating into different cell types of their tissue of origin.

ES cells can be grown to large numbers relatively easy, while MSCs are rare and methods for substantially expanding their numbers in cell culture have not yet been worked out. This is an important distinction, as large numbers of cells are required for regenerative medicine and tissue engineering procedures.

One important advantage of MSCs over ES cells is that the patient's own cells could be expanded in culture and subsequently administered to this patient. Using a patient's own MSCs should avert immunological rejection, because of the autologous source. In contrast, any form of allograft or xenograft transplantation would require the use of life-long immunosuppression. The major perceived disadvantage of ES cells is teratogenicity. This is the natural down-side of their exceedingly high proliferation and plasticity. Any ES cell based tissue engineering would hence require strict culling of undifferentiated precursors.

9.2.5 Perspectives

Dedifferentiation and reprogramming of cells are very powerful in rejuvenating adult cells into an ES cell-like state. One of the earliest successful strategies along these lines was cloning by nuclear transfer (Paterson et al. 2003; Wilmut et al. 1997). Transferring the cell nucleus of a fully differentiated body cell into an enucleated oocyte creates a zygote that is capable of developing into a clone of the nuclear donor animal like in the case of the sheep "Dolly". This "reproductive cloning" procedure can be terminated after embryo implantation and novel ES cells can be isolated from the blastocysts that developed from the oocytes. ES cells and tissues derived that way are immunologically identical to the nuclear donor and should thus be an ideal source for therapeutic purposes in that donor. Hence the term "therapeutic cloning" for this procedure to generate donor-matched ES-cells and tissues. Reprogramming can also be achieved by cellular fusion (Tada et al. 2003), admixing of cell extracts, and by extended *in vitro* cell culture (Collas and Taranger 2006; Collas et al. 2006; Hakelien et al. 2005; Taranger et al. 2005). The crucial factors required for reprogramming of mouse cells have recently been identified in an elegant study. Two Japanese researchers successfully rejuvenated cells of adult mice to a cell type similar to ES cells (Takahashi and Yamanaka 2006) by introducing just four genes necessary and sufficient to convert adult cells into pluripotent cells: Oct3/4, Sox2, c-Myc, and Klf4.

Several adult stem cell populations, called multipotent adult progenitor cells, MAPCs (Breyer et al. 2006; Verfaillie 2005) or unrestricted somatic stem cells, USSC (Kogler et al. 2004) are published. Last not least CD117 positive stem cells were isolated from amniotic fluid, AFS cells. These cells could produce cell types representing the three primary embryonic lineages mesoderm, ectoderm and definitive endoderm (De Coppi et al. 2007). AFS cells represent about 1% of the cells found in the amniotic fluid. The cells could be expanded for over 250 doublings without loss of telomere length. Moreover AFS cells expressed markers of pluripotency, Oct4 and the stage specific embryonic antigen 4, SSEA-4. Other markers for ES cells were not detected. These cells also expressed markers characteristic for mesenchymal and neuronal stem cells, but no markers of hematopoietic stem cells, HSCs. Importantly, unlike ES cells, AFS cells never produced teratomas when transplanted to animals. In summary AFS cells differ from pluripotent ES cells and from multipotent MSCs therefore may represent a new class of stem cells. They are easily grown without feeder cells and are readily available from amniocentesis samples. The obvious advantages of AFS cells may render these cells useful for regenerative medicine, tissue engineering, gene therapy and drug screening.

9.3 The Material

9.3.1 Scaffolds – Support Materials to Grow Cells into Tissues

Scaffold materials suitable for tissue engineering vary in porosity, composition, and biodegradability to best mimic the requirements of the organ to be replaced. Furthermore the physiological tissue morphology and function can be fine tuned by generating increasingly complex and structurally organised implants, which are aligned to the target organ structure. Bone tissue engineering is a prime example where this has been thoughtfully considered. Bone is a complex tissue with highly porous morphology in the spongy bone and with solid compact structure in the cortical bone. The tissue engineering of bone requires strategies to design 3D scaffolds that closely mimic the anatomical organization of bone and its tissue matrix. Changes in scaffold geometry may impact the flow of medium across the scaffold and thus affect the supply with gases and nutrients and the removal of metabolites (Detamore and Athanasiou 2003). For any given porosity, different geometries will lead to different effective stiffness (Hollister 2005). An increase in pore size in various scaffolds was associated with increased vascularisation and osteointegration *in vivo*, but the increased pore size reduced mechanical stability of the scaffold (Karageorgiou and Kaplan 2005). In order to create a mechanically and biologically functional implant, a compromise was found making the best use possible of the available materials (Muschler et al. 2004).

9.3.2 Vascularisation and Blood Supply in Tissue Engineering

Proper blood supply is a major problem that must be solved to achieve a successful organ replacement by a tissue engineered construct. The importance of vascularisation is illustrated by examples from several pathological conditions including ischaemic heart disease (Fukuda et al. 2004) and diabetic ulcers (Bennett et al. 2003). Impaired wound healing in the case of diabetic ulcers occurs due to the lack perfusion resulting in oxygen and nutrient supply as well as inadequate removal of waste products (Patel and Mikos 2004). Stimulation of angiogenesis much improved healing of diseased tissues (Hughes et al. 2004). The process of vascularisation requires angiogenesis, the formation of new vessels from endothelial precursors. One strong angiogenic stimulus is hypoxia (Bicknell and Harris 2004). The best-known angiogenic factor is vascular endothelial growth factor, VEGF, which stimulates cells to produce matrix metalloproteinases, MMPs, degrading the surrounding extracellular matrix. This in turn creates a highly pro-angiogenic environment prompting endothelial cell migration and proliferation. Subsequently, pericytes proliferate and migrate towards newly formed vessel sprouts and induce maturation by forming a single cell layer around the vessel sprout (Bruick and McKnight 2001; Hoeben et al. 2004). A second growth factor with high angiogenic potential is basic fibroblast growth factor, bFGF/FGF2. This factor prompts endothelial cells to produce both MMPs and VEGF and increases VEGF receptor expression thus enforcing the VEGF signal and locking endothelial precursor cells into a mature phenotype. Like VEGF, FGF2 stimulates endothelial cell migration, pericyte attraction and matrix deposition (Presta et al. 2005). Both growth factors act synergistically and may be applied together to achieve fully developed mature blood vessels (Asahara et al. 1995; Giavazzi et al. 2003; Laschke et al. 2006).

Biomimetic principles are applied to optimize cell performance on the scaffold material. For instance, to increase the vascularisation of tissue engineered constructs the pore size of a scaffold has been varied to match an optimum diameter for cellular adhesion and migration ($\sim$100 µm) (O'Brien et al. 2005). Furthermore, endothelial cells and fibroblasts were combined in gelatine coated polystyrene scaffolds in order to kick-start robust angiogenesis *in vitro*, prior to transplantation (Rickert et al. 2003). The addition of glycosaminoglycans and growth factors likewise increased angiogenesis *in vivo* (Pieper et al. 2002). Finally, the addition of VEGF and FGF2 bound to its natural ligand, heparin (Nillesen et al. 2007) or to hyaluronan hydrogels (Pike et al. 2006) greatly enhanced the vascularization of implanted scaffolds in experimental animals.

9.3.3 Scaffold Material Influences Cell Behaviour

Materials themselves can greatly influence cell behaviour through chemical composition or surface topology. Taking 2D cell culture to 3D-culture of cells can also greatly influence cell behaviour due to compartmentalisation. Our knowledge of material control of cell behaviour and the influence of surface chemistry, initial protein and cell adhesion/morphology and ultimately differentiation is still limited. The physicochemical properties of the bulk material, including topography, chemistry, and surface energy modulate protein adsorption in terms of adsorbed species density and biological activity (Garcia and Keselowsky 2002). Substrate dependent differentiation translates into altered cellular functions, including adhesion, spreading, migration, and differentiation (Chen et al. 1998; Garcia et al. 1999; Gorbet and Sefton 2001; Grinnell and Feld 1982; Shen and Horbett 2001). Cell adhesion to adsorbed proteins is primarily mediated by integrin receptors (Hynes 2002). Integrins represent a widely expressed family of heterodimeric transmembrane receptors that bind to adhesive motifs present in various extracellular matrix proteins, including fibronectin, vitronectin, laminin, and collagen (Hynes 2002; Ruoslahti and Pierschbacher 1986). Following ligand binding, integrins cluster and associate with cytoskeletal elements to form so-called "focal adhesions", supramolecular assemblies of structural and signalling proteins that provide anchorage forces and activate signalling cascades regulating cell cycle progression and differentiation (Geiger et al. 2001). Many biomaterials are designed to prevent non-specific protein adsorption and to favour specific interactions e. g. by presenting short biomimetic motifs, such as the tripeptide sequence of the amino acids RGD (arginine, glycine, aspartic acid), to promote cell adhesion (Hubbell 2003; Langer and Tirrell 2004). The use of self-assembled monolayers, SAMs to create a well-defined surface chemistry with selective binding of integrins increases the ligand density and specificity. SAMs are ordered molecular assemblies formed by the adsorption of an active surfactant onto a solid surface (Ulman 1996). These thin crystalline films are an ingeniously simple, yet powerful approach to modifying the surface properties of a material. A primary advantage of SAMs is that fine-tuning the terminal chemistry can greatly alter the macroscopic properties of the surface. Such surface properties include wettability, cytotoxicity and protein/cellular adhesion. Using alkanethiol SAMs with well-defined chemistries (OH, CH_3, NH_2 and COOH) and fibronectin-coating sustained the growth of myoblasts and osteoblasts, which are hard to grow otherwise. The modifications were found to modulate integrin binding to the absorbed fibronectin. Thus material properties direct protein adsorption and thus binding of specialized cell types (Keselowsky et al. 2005; Keselowsky and Garcia 2005; Lan et al. 2005).

Development of novel biomaterials is incremental and involves rational design and repeated testing to improve performance. Contemporary approaches focus on the development of parallel, combinatorial strategies, and the development of large libraries of polymeric biomaterials (Anderson et al. 2003; Brocchini 2001; Brocchini et al. 1998). Using an array spotter and simple acrylate chemistry Anderson et al. developed a platform enabling nanoliter scale synthesis and cell-based screening of 1728 individual polymeric spots of microarrayed biomaterials in contact with human ES cells. Using this approach the authors identified polymers that control ES cell attachment and spreading, cell type specific growth, and growth factor specific proliferation. Surprisingly, a variety of materials mediated differentiation of ES cells into epithelial-like cells (Anderson et al. 2004). This method enabled the rapid screening of diverse biomaterials within one chemical class. The method can potentially be further adapted to allow for high throughput screening of polymers of increasing complexity. As a proof of principle an array of blends of well characterised biodegradable polymers assembled in 3456 spots was tested with regards to human MSC compatibility, bovine articular chondrocyte and murine neural stem cell growth (Anderson et al. 2005).

Further studies combined high throughput screening methods with automatic high content image analysis to rapidly assay for material effects on cell behavior (Abraham et al. 2004; Ghosh et al. 2007). Genome wide expression profiles will undoubtedly further boost our knowledge of biomaterials design to produce tailor made scaffolds for tissue engineering. Genome wide alterations of gene expression will detect early changes in cell adhesion, metabolism, proliferation, and differentiation. One simple example how a material can influence the genome-wide expression pattern of cells is shown in Fig. 9.5.

9.3 The Material

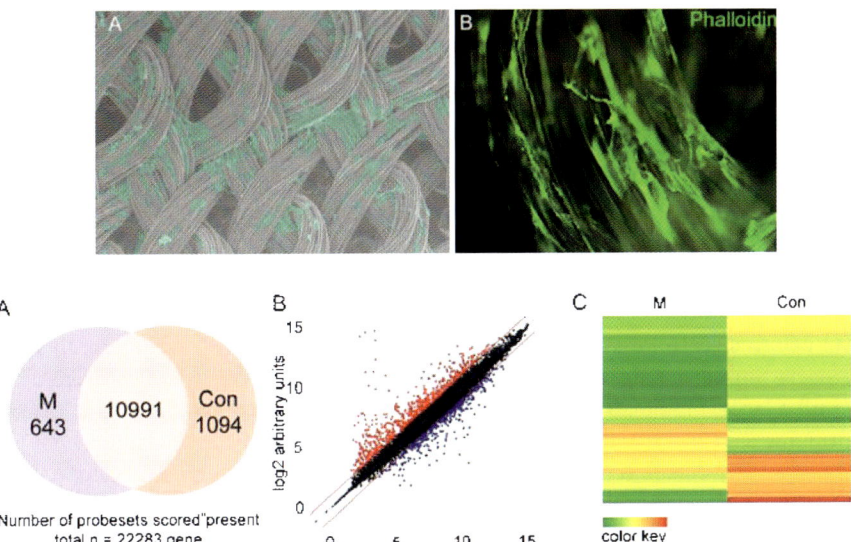

Fig. 9.5 Human mesenchymal stromal cells (MSC) were grown on PVDF mesh. *Top*: **A** Scanning micrograph of cells on scaffold. **B** fluorescence micrograph of cells stained with FITC-phalloidin depicting the actin cytoskeleton. Actin fibres show the cell orientation in relation to the scaffold. *Bottom*: The cells were analyzed by genome-wide RNA expression profiling. Cells were grown on PVDF mesh (M) or on tissue culture plastics (Con). Differentially expressed genes are illustrated in a Venn diagram (**A**), a scatter plot (**B**) and a hierarchical cluster (**C**) illustrating the entirety of genes analyzed (**A**), the most highly regulated genes (**B**) and their kins (**C**). The expression patterns varied with the material

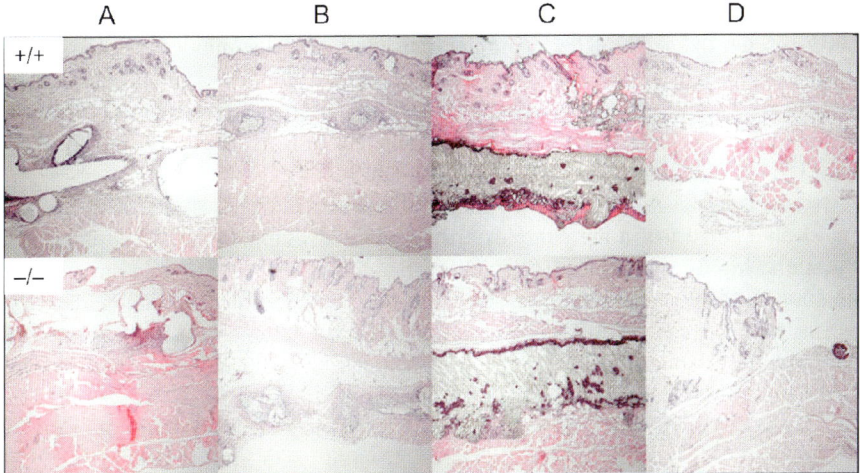

Fig. 9.6 Biomaterial sheets and meshes were implanted subcutaneously in wildtype (+/+) or fetuin-A knockout mice (−/−). Explanted materials were stained with hematoxilin/eosin. Dark red staining indicates calcified lesions. Note that material C was highly prone to dystrophic calcification, while material B did not calcify at all in this stringent animal model

The implantation model shown in Fig. 9.6 was chosen to test the calcification of materials in a very stringent way that can otherwise only be achieved by long term implantation. Animal testing should also include disease models in which the implanted tissue engineered substitute will compensate for lost organ function. For example tissue engineered bone or cartilage constructs should be tested in a critical size defect precluding regeneration by endogenous repair mechanisms. Examples for successful integration of tissue engineered bone may be found e.g. in the work of Bruder 1998 and Livingston 2002 (Bruder et al. 1998; Livingston et al. 2002). Both groups implanted cell seeded 3D scaffolds into critical size defects. The tissue engineered constructs were able to bridge the bone defects in canine and rat femora, but cell free scaffolds were not, proving the critical role of cell seeding in this approach. Similar repair was achieved when tissue engineered constructs were used to fill a large, high load bearing osteochondral defect in rabbits (Shao et al. 2006). Bridging a critical size defect is also important in tissue engineering approaches for the repair of spinal cord injuries. In the field of neurological repair functional recovery is best tested in behavioral studies, which yield information about sensory and motor reflexes. Several attemps were made to fill spinal cord defects with cell seeded 3D scaffolds to build a conduit directing growth of axons (Prang et al. 2006; Rochkind et al. 2006; Sykova et al. 2006).

Morphological regeneration was routinely reported but unfortunately, not all of those studies included behavioural tests precluding a final assessment of outcome.

Animal models can be made highly discriminating by using transgenic animals. For example, mice deficient in fumarylacetoacetate hydrolase (Fah), a metabolic enzyme catalyzing the final step of tyrosine catabolism, die within 12 hr after birth from hypoglycemia and liver dysfunction (Grompe et al. 1993). Implanting functional pancreatic cells (Wang et al. 2001), or bone marrow (Vassilopoulos et al. 2003) rescues this lethal defect. Another transgenic model employed the mouse major urinary protein-urokinase-type plasminogen activator fusion transgene, which induces diffuse hepatocellular damage at 3 weeks of age (Heckel et al. 1990). In this model it was shown that implanted murine ES cells augmented liver regeneration (Heo et al. 2006).

In conclusion, valid *in vivo* test systems must be used to show function of a tissue engineered construct in the complex environment of living animals. These models should be highly standardized to keep results consistent and comparable. Testing function and morphology as the readout of proper integration of tissue engineering organs far exceeds the limitations of cell based testing, which is highly dependent on cell culture conditions, cell types and sources, different donor animal strains, different scaffolds, choice of biomaterials etc.

9.3.4 Commercial Tissue Engineered Products

Few commercial tissue engineered products have entered the market. Biological wound dressings for the treatment of burns, chronic ulcers, and surgical wounds still are the only products with US Food and Drug Administration, FDA approval.

9.3 The Material

Two well known products in this area are Apligraf® and Dermagraft®. Apligraf® is bi-layered mimicking the structure of skin. This skin equivalent comprises a dermal equivalent derived from neonatal foreskin fibroblasts in contracted type I collagen matrix and an epidermis generated by neonatal keratinocytes seeded onto the dermal equivalent (Sabolinski et al. 1996). Dermagraft® is composed of neonatal foreskin fibroblasts, cultured on Polyglactin 910® a bioresorbable scaffold made of a co-polymer of glycolic acid and lactic acid, PGA/PLA. The product does not have an epidermis equivalent (Marston et al. 2003). Another innovative skin substitute, OrCel® is not currently commercially available. Like Apligraf®, OrCel® is a bilayer dressing resembling normal skin. The product is composed of type I collagen in which epidermal keratinocytes and dermal fibroblast, derived from neonatal foreskin tissue, are cultured in two layers. Fibroblasts are placed within the porous sponge while keratinocytes are seeded on the nonporous side of the matrix (Ruszczak 2003). Apligraf® and Dermagraft® took considerable time from development to approval. The time between initiating clinical trials in leg ulcer repair and product launch was 68 months for Apligraf®. Earning approval for one additional application in diabetic foot ulcers took another 51 months. This illustrates the challenges ahead in getting FDA approval for tissue engineered products.

Another example for successful tissue engineered implants is an autologous bladder construct for patients with end-stage bladder disease reported by Atala and colleagues in 2006. Bladder biopsies ($1-2\,\text{cm}^2$) were taken to obtain urothelial and muscle cells. After six weeks in culture about 7×10^8 cells of each cell type were obtained to construct one tissue engineered bladder. The exterior surface of a polymeric scaffold was seeded with smooth muscle cells. 48 hours after the smooth muscle cells seeding urothelial cells were added to the inside of the scaffold. Two different types of scaffolds were used. One was made of homologous cell-free bladder submucosa and the second was a biodegradable composite scaffold made of collagen and poly glycolic acid, PGA. The scaffold was shaped into a three-dimensional bladder using a computer-generated cast created specifically for each of the patients. The total thickness of the scaffold was about 2 mm. All scaffolds were sterilized with ultraviolet light followed by ethylene oxide. The entire tissue engineering process took seven to eight weeks. The engineered bladders improved the function of the residual bladder for a period of at least five years. Additional studies are underway to test the long term function and safety of tissue engineered bladders (Atala et al. 2006).

Regenerating teeth is another area where tissue engineering showed promising results in animal models. Murine embryonic teeth were used to form dental structures. Tooth bud cells were seeded onto PGA and polyglycolide-co-lactide, PLGA scaffolds. The constructs were implanted into adult rats and formed tooth tissue including primary and reparative dentin and enamel (Duailibi et al. 2004; Young et al. 2005). The next major advance was reconstruction of teeth and its associated periodontal root. To this end swine postnatal tooth stem cells including cells from root apical papilla and periodontal ligament stem cells were seeded onto a root shaped block of hydroxyapatite/tricalciumphospate, HA/TCP with an inner channel space that would allow subsequent mounting of a porcelain crown. This construct was implanted into the cavity of an extracted tooth. The HA/TCP block was coated with protective foam. Three months after implantation mineralized root-like tissue had formed and the periodontal ligaments had regenerated. Using this primitive "tooth stump" as a basis a porcelain crown was affixed. Four weeks onward the root/periodontal structure had regenerated. This newly grown root significantly improved compressive strength of the entire artificial tooth when compared to simple HA/TCP blocks (Sonoyama et al. 2006). These promising results in a pig model raise hopes for tissue engineered human tooth and root restoration because of close similarities between swine and human dental tissue.

In summary the combination of cells and material scaffolds into tissue engineered tissue replacements or even organs poses a formidable tasks. Apart from biological and technical problems, which – despite all ingenuity – do not easily yield, two really big hurdles remain at the end of any development: regulatory approval (for safety reasons) and commercial viability (for lack of health insurance compensation). These hurdles were certainly out of scope in this contribution, they may be out of reach for most basic scientists, but one should never loose them out of sight.

References

1. Abraham VC, Taylor DL, Haskins JR (2004) High content screening applied to large-scale cell biology. Trends Biotechnol 22:15–22
2. Anderson DG, Levenberg S, Langer R (2004) Nanoliter-scale synthesis of arrayed biomaterials and application to human embryonic stem cells. Nat Biotechnol 22:863–866
3. Anderson DG, Lynn DM, Langer R (2003) Semi-automated synthesis and screening of a large library of degradable cationic polymers for gene delivery. Angew Chem Int Ed Engl 42:3153–3158
4. Anderson DG, Putnam D, Lavik EB, Mahmood TA, Langer R (2005) Biomaterial microarrays: rapid, microscale screening of polymer-cell interaction. Biomaterials 26:4892–4897
5. Andrews PW (2002) From teratocarcinomas to embryonic stem cells. Philos Trans R Soc Lond B Biol Sci 357:405–417
6. Aranguren XL et al. (2007) In vitro and in vivo arterial differentiation of human multipotent adult progenitor cells. Blood 109:2634–2642
7. Asahara T, Bauters C, Zheng LP, Takeshita S, Bunting S, Ferrara N, Symes JF, Isner JM (1995) Synergistic effect of vascular endothelial growth factor and basic fibroblast growth factor on angiogenesis in vivo. Circulation 92:II365–II371
8. Atala A, Bauer SB, Soker S, Yoo JJ, Retik AB (2006) Tissue-engineered autologous bladders for patients needing cystoplasty. Lancet 367:1241–1246
9. Bennett SP, Griffiths GD, Schor AM, Leese GP, Schor SL (2003) Growth factors in the treatment of diabetic foot ulcers. Br J Surg 90:133–146
10. Bicknell R, Harris AL (2004) Novel angiogenic signaling pathways and vascular targets. Annu Rev Pharmacol Toxicol 44:219–238
11. Blanpain C, Lowry WE, Geoghegan A, Polak L, Fuchs E (2004) Self-renewal, multipotency, and the existence of two cell populations within an epithelial stem cell niche. Cell 118:635–648
12. Blau HM, Brazelton TR, Weimann JM (2001) The evolving concept of a stem cell: entity or function? Cell 105:829–841
13. Bradley A, Evans M, Kaufman MH, Robertson E (1984) Formation of germ-line chimaeras from embryo-derived teratocarcinoma cell lines. Nature 309:255–256
14. Breyer A, Estharabadi N, Oki M, Ulloa F, Nelson-Holte M, Lien L, Jiang Y (2006) Multipotent adult progenitor cell isolation and culture procedures. Exp Hematol 34:1596–1601
15. Brocchini S (2001) Combinatorial chemistry and biomedical polymer development. Adv Drug Deliv Rev 53:123–130
16. Brocchini S, James K, Tangpasuthadol V, Kohn J (1998) Structure-property correlations in a combinatorial library of degradable biomaterials. J Biomed Mater Res 42:66–75
17. Bruder SP, Kraus KH, Goldberg VM, Kadiyala S (1998) The effect of implants loaded with autologous mesenchymal stem cells on the healing of canine segmental bone defects. J Bone Joint Surg Am 80:985–996

18. Bruick RK, McKnight SL (2001) Building better vasculature. Genes Dev 15:2497–2502
19. Buehr M, Nichols J, Stenhouse F, Mountford P, Greenhalgh CJ, Kantachuvesiri S, Brooker G, Mullins J, Smith AG (2003) Rapid loss of Oct-4 and pluripotency in cultured rodent blastocysts and derivative cell lines. Biol Reprod 68:222–229
20. Burdon T, Chambers I, Stracey C, Niwa H, Smith A (1999a) Signaling mechanisms regulating self-renewal and differentiation of pluripotent embryonic stem cells. Cells Tissues Organs 165:131–143
21. Burdon T, Smith A, Savatier P (2002) Signalling, cell cycle and pluripotency in embryonic stem cells. Trends Cell Biol 12:432–438
22. Burdon T, Stracey C, Chambers I, Nichols J, Smith A (1999b) Suppression of SHP-2 and ERK signalling promotes self-renewal of mouse embryonic stem cells. Dev Biol 210:30–43
23. Chambers I, Colby D, Robertson M, Nichols J, Lee S, Tweedie S, Smith A (2003) Functional expression cloning of Nanog, a pluripotency sustaining factor in embryonic stem cells. Cell 113:643–655
24. Chen CS, Mrksich M, Huang S, Whitesides GM, Ingber DE (1998) Micropatterned surfaces for control of cell shape, position, and function. Biotechnol Prog 14:356–363
25. Chen T, Ueda Y, Dodge JE, Wang Z, Li E (2003) Establishment and maintenance of genomic methylation patterns in mouse embryonic stem cells by Dnmt3a and Dnmt3b. Mol Cell Biol 23:5594–5605
26. Clayton E, Doupe DP, Klein AM, Winton DJ, Simons BD, Jones PH (2007) A single type of progenitor cell maintains normal epidermis. Nature 446:185–189
27. Collas P, Taranger CK (2006) Toward reprogramming cells to pluripotency. Ernst Schering Res Found Workshop, pp 47–67
28. Collas P, Taranger CK, Boquest AC, Noer A, Dahl JA (2006) On the way to reprogramming cells to pluripotency using cell-free extracts. Reprod Biomed Online 12:762–770
29. Daheron L, Opitz SL, Zaehres H, Lensch WM, Andrews PW, Itskovitz-Eldor J, Daley GQ (2004) LIF/STAT3 signaling fails to maintain self-renewal of human embryonic stem cells. Stem Cells 22:770–778
30. Dao MA, Creer MH, Nolta JA, Verfaillie CM (2007) Biology of umbilical cord blood progenitors in bone marrow niches. Blood 110:74–81
31. De Coppi P et al. (2007) Isolation of amniotic stem cell lines with potential for therapy. Nat Biotechnol 25:100–106
32. Detamore MS, Athanasiou KA (2003) Structure and function of the temporomandibular joint disc: implications for tissue engineering. J Oral Maxillofac Surg 61:494–506
33. Diwan SB, Stevens LC (1976) Development of teratomas from the ectoderm of mouse egg cylinders. J Natl Cancer Inst 57:937–942
34. Doetschman T, Gregg RG, Maeda N, Hooper ML, Melton DW, Thompson S, Smithies O (1987) Targetted correction of a mutant HPRT gene in mouse embryonic stem cells. Nature 330:576–578
35. Duailibi MT, Duailibi SE, Young CS, Bartlett JD, Vacanti JP, Yelick PC (2004) Bioengineered teeth from cultured rat tooth bud cells. J Dent Res 83:523–528
36. Evans MJ, Kaufman MH (1981) Establishment in culture of pluripotential cells from mouse embryos. Nature 292:154–156
37. Friedenstein AY (1968) Induction of bone tissue by transitional epithelium. Clin Orthop Relat Res 59:21–37
38. Fukuda S, Yoshii S, Kaga S, Matsumoto M, Kugiyama K, Maulik N (2004) Angiogenic strategy for human ischemic heart disease: brief overview. Mol Cell Biochem 264:143–149
39. Garcia AJ, Keselowsky BG (2002) Biomimetic surfaces for control of cell adhesion to facilitate bone formation. Crit Rev Eukaryot Gene Expr 12:151–162
40. Garcia AJ, Vega MD, Boettiger D (1999) Modulation of cell proliferation and differentiation through substrate-dependent changes in fibronectin conformation. Mol Biol Cell 10:785–798
41. Gatti RA, Meuwissen HJ, Allen HD, Hong R, Good RA (1968) Immunological reconstitution of sex-linked lymphopenic immunological deficiency. Lancet 2:1366–1369
42. Geiger B, Bershadsky A, Pankov R, Yamada KM (2001) Transmembrane crosstalk between the extracellular matrix–cytoskeleton crosstalk. Nat Rev Mol Cell Biol 2:793–805

43. Ghosh RN, Lapets O, Haskins JR (2007) Characteristics and value of directed algorithms in high content screening. Methods Mol Biol 356:63–81
44. Giavazzi R, Sennino B, Coltrini D, Garofalo A, Dossi R, Ronca R, Tosatti MP, Presta M (2003) Distinct role of fibroblast growth factor-2 and vascular endothelial growth factor on tumor growth and angiogenesis. Am J Pathol 162:1913–1926
45. Gorbet MB, Sefton MV (2001) Leukocyte activation and leukocyte procoagulant activities after blood contact with polystyrene and polyethylene glycol-immobilized polystyrene beads. J Lab Clin Med 137:345–355
46. Grinnell F, Feld MK (1982) Fibronectin adsorption on hydrophilic and hydrophobic surfaces detected by antibody binding and analyzed during cell adhesion in serum-containing medium. J Biol Chem 257:4888–4893
47. Grompe M, al-Dhalimy M, Finegold M, Ou CN, Burlingame T, Kennaway NG, Soriano P (1993) Loss of fumarylacetoacetate hydrolase is responsible for the neonatal hepatic dysfunction phenotype of lethal albino mice. Genes Dev 7:2298–2307
48. Hakelien AM, Gaustad KG, Taranger CK, Skalhegg BS, Kuntziger T, Collas P (2005) Long-term in vitro, cell-type-specific genome-wide reprogramming of gene expression. Exp Cell Res 309:32–47
49. Heckel JL, Sandgren EP, Degen JL, Palmiter RD, Brinster RL (1990) Neonatal bleeding in transgenic mice expressing urokinase-type plasminogen activator. Cell 62:447–456
50. Heo J, Factor VM, Uren T, Takahama Y, Lee JS, Major M, Feinstone SM, Thorgeirsson SS (2006) Hepatic precursors derived from murine embryonic stem cells contribute to regeneration of injured liver. Hepatology 44:1478–1486
51. Hoeben A, Landuyt B, Highley MS, Wildiers H, Van Oosterom AT, De Bruijn EA (2004) Vascular endothelial growth factor and angiogenesis. Pharmacol Rev 56:549–580
52. Hollister SJ (2005) Porous scaffold design for tissue engineering. Nat Mater 4:518–524
53. Hubbell JA (2003) Materials as morphogenetic guides in tissue engineering. Curr Opin Biotechnol 14:551–558
54. Hughes GC, Biswas SS, Yin B, Coleman RE, DeGrado TR, Landolfo CK, Lowe JE, Annex BH, Landolfo KP (2004) Therapeutic angiogenesis in chronically ischemic porcine myocardium: comparative effects of bFGF and VEGF. Ann Thorac Surg 77:812–818
55. Hynes RO (2002) Integrins: bidirectional, allosteric signaling machines. Cell 110:673–687
56. Jahagirdar BN, Verfaillie CM (2005) Multipotent adult progenitor cell and stem cell plasticity. Stem Cell Rev 1:53–59
57. Karageorgiou V, Kaplan D (2005) Porosity of 3D biomaterial scaffolds and osteogenesis. Biomaterials 26:5474–5491
58. Keselowsky BG, Collard DM, Garcia AJ (2005) Integrin binding specificity regulates biomaterial surface chemistry effects on cell differentiation. Proc Natl Acad Sci USA 102:5953–5957
59. Keselowsky BG, Garcia AJ (2005) Quantitative methods for analysis of integrin binding and focal adhesion formation on biomaterial surfaces. Biomaterials 26:413–418
60. Kogler G et al. (2004) A new human somatic stem cell from placental cord blood with intrinsic pluripotent differentiation potential. J Exp Med 200:123–135
61. Lan MA, Gersbach CA, Michael KE, Keselowsky BG, Garcia AJ (2005) Myoblast proliferation and differentiation on fibronectin-coated self assembled monolayers presenting different surface chemistries. Biomaterials 26:4523–4531
62. Langer R, Tirrell DA (2004) Designing materials for biology and medicine. Nature 428:487–492
63. Langer R, Vacanti JP (1993) Tissue engineering. Science 260:920–926
64. Laschke MW, Elitzsch A, Vollmar B, Vajkoczy P, Menger MD (2006) Combined inhibition of vascular endothelial growth factor (VEGF), fibroblast growth factor and platelet-derived growth factor, but not inhibition of VEGF alone, effectively suppresses angiogenesis and vessel maturation in endometriotic lesions. Hum Reprod 21:262–268
65. Lin H (2002) The stem-cell niche theory: lessons from flies. Nat Rev Genet 3:931–940
66. Livingston T, Ducheyne P, Garino J (2002) In vivo evaluation of a bioactive scaffold for bone tissue engineering. J Biomed Mater Res 62:1–13

67. Marston WA, Hanft J, Norwood P, Pollak R (2003) The efficacy and safety of Dermagraft in improving the healing of chronic diabetic foot ulcers: results of a prospective randomized trial. Diabetes Care 26:1701–1705
68. Martin GR (1981) Isolation of a pluripotent cell line from early mouse embryos cultured in medium conditioned by teratocarcinoma stem cells. Proc Natl Acad Sci USA 78:7634–7638
69. Martin GR, Evans MJ (1975) Differentiation of clonal lines of teratocarcinoma cells: formation of embryoid bodies in vitro. Proc Natl Acad Sci USA 72:1441–1445
70. Matsuda T, Nakamura T, Nakao K, Arai T, Katsuki M, Heike T, Yokota T (1999) STAT3 activation is sufficient to maintain an undifferentiated state of mouse embryonic stem cells. Embo J 18:4261–4269
71. Mitsui K, Tokuzawa Y, Itoh H, Segawa K, Murakami M, Takahashi K, Maruyama M, Maeda M, Yamanaka S (2003) The homeoprotein Nanog is required for maintenance of pluripotency in mouse epiblast and ES cells. Cell 113:631–642
72. Muschler GF, Nakamoto C, Griffith LG (2004) Engineering principles of clinical cell-based tissue engineering. J Bone Joint Surg Am 86-A:1541–1558
73. Neuss S, Becher E, Woltje M, Tietze L, Jahnen-Dechent W (2004) Functional expression of HGF and HGF receptor/c-met in adult human mesenchymal stem cells suggests a role in cell mobilization, tissue repair, and wound healing. Stem Cells 22:405–414
74. Nillesen ST, Geutjes PJ, Wismans R, Schalkwijk J, Daamen WF, van Kuppevelt TH (2007) Increased angiogenesis and blood vessel maturation in acellular collagen-heparin scaffolds containing both FGF2 and VEGF. Biomaterials 28:1123–1131
75. Niwa H, Miyazaki J, Smith AG (2000) Quantitative expression of Oct-3/4 defines differentiation, dedifferentiation or self-renewal of ES cells. Nat Genet 24:372–376
76. O'Brien FJ, Harley BA, Yannas IV, Gibson LJ (2005) The effect of pore size on cell adhesion in collagen-GAG scaffolds. Biomaterials 26:433–441
77. Patel ZS, Mikos AG (2004) Angiogenesis with biomaterial-based drug- and cell-delivery systems. J Biomater Sci Polym Ed 15:701–726
78. Paterson L, DeSousa P, Ritchie W, King T, Wilmut I (2003) Application of reproductive biotechnology in animals: implications and potentials. Applications of reproductive cloning. Anim Reprod Sci 79:137–143
79. Pelacho B et al. (2007) Plasticity and cardiovascular applications of multipotent adult progenitor cells. Nat Clin Pract Cardiovasc Med 4(1):S15–S20
80. Pesce M, Wang X, Wolgemuth DJ, Scholer H (1998) Differential expression of the Oct-4 transcription factor during mouse germ cell differentiation. Mech Dev 71:89–98
81. Pieper JS, Hafmans T, van Wachem PB, van Luyn MJ, Brouwer LA, Veerkamp JH, van Kuppevelt TH (2002) Loading of collagen-heparan sulfate matrices with bFGF promotes angiogenesis and tissue generation in rats. J Biomed Mater Res 62:185–194
82. Pike DB, Cai S, Pomraning KR, Firpo MA, Fisher RJ, Shu XZ, Prestwich GD, Peattie RA (2006) Heparin-regulated release of growth factors in vitro and angiogenic response in vivo to implanted hyaluronan hydrogels containing VEGF and bFGF. Biomaterials 27:5242–5251
83. Pittenger MF et al. (1999) Multilineage potential of adult human mesenchymal stem cells. Science 284:143–147
84. Prang P et al. (2006) The promotion of oriented axonal regrowth in the injured spinal cord by alginate-based anisotropic capillary hydrogels. Biomaterials 27:3560–3569
85. Presta M, Dell'Era P, Mitola S, Moroni E, Ronca R, Rusnati M (2005) Fibroblast growth factor/fibroblast growth factor receptor system in angiogenesis. Cytokine Growth Factor Rev 16:159–178
86. Rasmussen TP (2003) Embryonic stem cell differentiation: a chromatin perspective. Reprod Biol Endocrinol 1:100
87. Reubinoff BE, Pera MF, Fong CY, Trounson A, Bongso A (2000) Embryonic stem cell lines from human blastocysts: somatic differentiation in vitro. Nat Biotechnol 18:399–404
88. Rickert D, Moses MA, Lendlein A, Kelch S, Franke RP (2003) The importance of angiogenesis in the interaction between polymeric biomaterials and surrounding tissue. Clin Hemorheol Microcirc 28:175–181

89. Rietze RL, Valcanis H, Brooker GF, Thomas T, Voss AK, Bartlett PF (2001) Purification of a pluripotent neural stem cell from the adult mouse brain. Nature 412:736–739
90. Rochkind S et al. (2006) Development of a tissue-engineered composite implant for treating traumatic paraplegia in rats. Eur Spine J 15:234–245
91. Ross JJ et al. (2006) Cytokine-induced differentiation of multipotent adult progenitor cells into functional smooth muscle cells. J Clin Invest 116:3139–3149
92. Ross JJ, Verfaillie CM (2007) Evaluation of neural plasticity in adult stem cells. Philos Trans R Soc Lond B Biol Sci
93. Rugg-Gunn PJ, Ferguson-Smith AC, Pedersen RA (2005) Epigenetic status of human embryonic stem cells. Nat Genet 37:585–587
94. Ruoslahti E, Pierschbacher MD (1986) Arg-Gly-Asp: a versatile cell recognition signal. Cell 44:517–518
95. Ruszczak Z (2003) Effect of collagen matrices on dermal wound healing. Adv Drug Deliv Rev 55:1595–1611
96. Sabolinski ML, Alvarez O, Auletta M, Mulder G, Parenteau NL (1996) Cultured skin as a 'smart material' for healing wounds: experience in venous ulcers. Biomaterials 17:311–320
97. Schäfer C, Heiss A, Schwarz A, Westenfeld R, Ketteler M, Floege J, Müller-Esterl W, Schinke T, Jahnen-Dechent W (2003) The serum protein alpha 2-Heremans-Schmid glycoprotein/fetuin-A is a systemically acting inhibitor of ectopic calcification. Journal of Clinical Investigation 112:357–366
98. Serafini M, Verfaillie CM (2006) Pluripotency in adult stem cells: state of the art. Semin Reprod Med 24:379–388
99. Shao XX, Hutmacher DW, Ho ST, Goh JC, Lee EH (2006) Evaluation of a hybrid scaffold/cell construct in repair of high-load-bearing osteochondral defects in rabbits. Biomaterials 27:1071–1080
100. Shen M, Horbett TA (2001) The effects of surface chemistry and adsorbed proteins on monocyte/macrophage adhesion to chemically modified polystyrene surfaces. J Biomed Mater Res 57:336–345
101. Smith AG (2001) Embryo-derived stem cells: of mice and men. Annu Rev Cell Dev Biol 17:435–462
102. Smith AG, Heath JK, Donaldson DD, Wong GG, Moreau J, Stahl M, Rogers D (1988) Inhibition of pluripotential embryonic stem cell differentiation by purified polypeptides. Nature 336:688–690
103. Snykers S, Vanhaecke T, Papeleu P, Luttun A, Jiang Y, Vander Heyden Y, Verfaillie C, Rogiers V (2006) Sequential exposure to cytokines reflecting embryogenesis: the key for in vitro differentiation of adult bone marrow stem cells into functional hepatocyte-like cells. Toxicol Sci 94:330–341; discussion 235–239
104. Solter D, Skreb N, Damjanov I (1970) Extrauterine growth of mouse egg-cylinders results in malignant teratoma. Nature 227:503–504
105. Sonoyama W et al. (2006) Mesenchymal stem cell-mediated functional tooth regeneration in Swine. PLoS ONE 1:e79
106. Stevens LC (1970) The development of transplantable teratocarcinomas from intratesticular grafts of pre- and postimplantation mouse embryos. Dev Biol 21:364–382
107. Stewart CL, Kaspar P, Brunet LJ, Bhatt H, Gadi I, Kontgen F, Abbondanzo SJ (1992) Blastocyst implantation depends on maternal expression of leukaemia inhibitory factor. Nature 359:76–79
108. Sykova E, Jendelova P, Urdzikova L, Lesny P, Hejcl A (2006) Bone marrow stem cells and polymer hydrogels-two strategies for spinal cord injury repair. Cell Mol Neurobiol 26:1111–1127
109. Tada M, Morizane A, Kimura H, Kawasaki H, Ainscough JF, Sasai Y, Nakatsuji N, Tada T (2003) Pluripotency of reprogrammed somatic genomes in embryonic stem hybrid cells. Dev Dyn 227:504–510
110. Takahashi K, Yamanaka S (2006) Induction of pluripotent stem cells from mouse embryonic and adult fibroblast cultures by defined factors. Cell 126:663–676

111. Taranger CK, Noer A, Sorensen AL, Hakelien AM, Boquest AC, Collas P (2005) Induction of dedifferentiation, genomewide transcriptional programming, and epigenetic reprogramming by extracts of carcinoma and embryonic stem cells. Mol Biol Cell 16:5719–5735
112. Thomas KR, Capecchi MR (1987) Site-directed mutagenesis by gene targeting in mouse embryo-derived stem cells. Cell 51:503–512
113. Thomson JA, Itskovitz-Eldor J, Shapiro SS, Waknitz MA, Swiergiel JJ, Marshall VS, Jones JM (1998) Embryonic stem cell lines derived from human blastocysts. Science 282:1145–1147
114. Toma JG, Akhavan M, Fernandes KJ, Barnabe-Heider F, Sadikot A, Kaplan DR, Miller FD (2001) Isolation of multipotent adult stem cells from the dermis of mammalian skin. Nat Cell Biol 3:778–784
115. Ulman A (1996) Formation and Structure of Self-Assembled Monolayers. Chem Rev 96:1533–1554
116. Vassilopoulos G, Wang PR, Russell DW (2003) Transplanted bone marrow regenerates liver by cell fusion. Nature 422:901–904
117. Verfaillie CM (2005) Multipotent adult progenitor cells: an update. Novartis Found Symp 265:55–61; discussion 61–55, 92–57
118. Wagers AJ, Weissman IL (2004) Plasticity of adult stem cells. Cell 116:639–648
119. Wang X, Al-Dhalimy M, Lagasse E, Finegold M, Grompe M (2001) Liver repopulation and correction of metabolic liver disease by transplanted adult mouse pancreatic cells. Am J Pathol 158:571–579
120. Williams RL et al. (1988) Myeloid leukaemia inhibitory factor maintains the developmental potential of embryonic stem cells. Nature 336:684–687
121. Wilmut I, Schnieke AE, McWhir J, Kind AJ, Campbell KH (1997) Viable offspring derived from fetal and adult mammalian cells. Nature 385:810–813
122. Xu RH, Chen X, Li DS, Li R, Addicks GC, Glennon C, Zwaka TP, Thomson JA (2002) BMP4 initiates human embryonic stem cell differentiation to trophoblast. Nat Biotechnol 20:1261–1264
123. Ying QL, Nichols J, Chambers I, Smith A (2003) BMP induction of Id proteins suppresses differentiation and sustains embryonic stem cell self-renewal in collaboration with STAT3. Cell 115:281–292
124. Ying QL, Nichols J, Evans EP, Smith AG (2002) Changing potency by spontaneous fusion. Nature 416:545–548
125. Young CS, Abukawa H, Asrican R, Ravens M, Troulis MJ, Kaban LB, Vacanti JP, Yelick PC (2005) Tissue-engineered hybrid tooth and bone. Tissue Eng 11:1599–1610

Chapter 10
Micro and Nano Patterning for Cell and Tissue Engineering

Shyam Patel*, Hayley Lam*, Song Li

Department of Bioengineering, University of California, Berkeley,
song_li@berkeley.edu
* Equal contribution to this work

> Science may set limits to knowledge, but should not set limits to imagination.
>
> Bertrand Russell

Abstract One of the goals of bioengineers today is to answer the basic question of how individual cells, which together comprise an entire living being, communicate and interact with one another and the physical and chemical environment in which they reside.

In this chapter, we will focus on the micro and nano technologies that have been used to investigate the regulation of cell functions by micro and nano features in matrix distribution, surface topography and three-dimensional (3D) microenvironment, and explore the technologies that can be applied to the fabrication of functional tissue constructs.

10.1 Overview

One of the goals of bioengineers today is to answer the basic question of how individual cells, which together comprise an entire living being, communicate and interact with one another and the physical and chemical environment in which they reside. Beginning with a single cell, these processes proceed with repeatable and redundant elegance and cells divide, differentiate, migrate, and morph into a complete and functional organism. There has always been a fascination with this process, and the evidence of primitive microscopes dates back over 400 years. However, it is not until recently that we have the technology to not only observe, but also manipulate and change the micro and nano environment around cells, and determine what impact the surrounding environment has on cellular processes.

Beyond that, another lofty goal of many bioengineers today is to develop functional tissues for patients suffering from various diseases. A seminal 1997 paper showed a human size 'ear' on the back of a mouse [1]. Ten years later, tissue-engineered skin has been used in clinics, and we are still striving to engineer more functional tissue and organ replacements for clinical applications. While the problem of immune rejection is a huge issue in the field, another problem has been forming fully developed tissues and organs in the first place. In particular, it has proven very difficult to 'grow' in the laboratory the more complex tissues in the body, such as arteries, cardiac tissue and neural tissues. It has become clear that simply seeding cells in various scaffold materials may not be enough to make a functional tissue, and there are many problems with tissue remodeling over time, matching mechanical properties, and even just simply matching cell organization to native tissue. In order to guide and control cell function for tissue engineering, we need to have more in-depth understanding of cell interactions with physical and chemical factors in the microenvironment and also need to develop new technologies for tissue fabrication.

Recent technological advances in microfabrication and nanotechnology industry over the past decade have made many new ways of examining and engineering biological processes possible. Cells themselves are on micron scale, ranging from about 10 microns to 100 microns in diameter. Even smaller, proteins range from nanometers (e. g., integrins) to the hundreds of nanometers (e. g., collagen fibers). Cellular behavior and processes take place on this micro and nano scale, and now that modern technology has caught up with nature, we are able to probe, examine, manipulate, and wonder at the amazing machinery that maintains life. Not only are we able to use this technology to study cell behavior, but we can also utilize this technology as a way to create new therapies and engineer tissues that are functionally similar to native tissues.

In this chapter, we will focus on the micro and nano technologies that have been used to investigate the regulation of cell functions by micro and nano features in matrix distribution, surface topography and three-dimensional (3D) microenvironment, and explore the technologies that can be applied to the fabrication of functional tissue constructs.

10.2 Regulation of Cell Functions by Matrix Patterning

10.2.1 Matrix Patterning

Extracellular matrix (ECM) molecules such as fibronectin, collagen, laminin and elastin are large multi-domain glycoproteins with a multitude of functions. They not only bind and support cells via integrins but also are actively involved in signaling cells and sequestering as well as modulating the function of other signaling molecules. Micropatterning of matrix molecules on cell culture substrates allows for the specific control of cell size, shape, growth, migration and differentiation. Such studies have elucidated the effects of both chemical signaling and spatial ECM patterning on cell functions.

Matrix micropatterning involves fabricating two-dimensional (2D) cell culture substrates with cell adhesive regions and cell repulsive regions. The simplest method relies on the use of polymer molds that allow either physical or fluidic patterning of

10.2 Regulation of Cell Functions by Matrix Patterning 217

molecules on cell culture substrates [2]. To fabricate the molds for micropatterning, a micropatterned silicon wafer is made using photolithography, a technique commonly used to fabricate microelectronic devices. The photolithography process converts a geometric shape in a mask into topographical features on the surface of a silicon wafer that is pre-coated with photoresist. Then, the micropattern on the silicon wafer is transferred to a elastomeric polymer mold such as a poly(dimethylsiloxane) (PDMS) stamp using soft lithography techniques. The polymer mold can be designed for microcontact printing or microfluidic patterning, the two most commonly used methods of micropatterning.

For microcontact printing, the polymer mold has "posts" in the shapes and sizes of the desired micropattern (Fig. 10.1A). The molds are inked with molecules to be

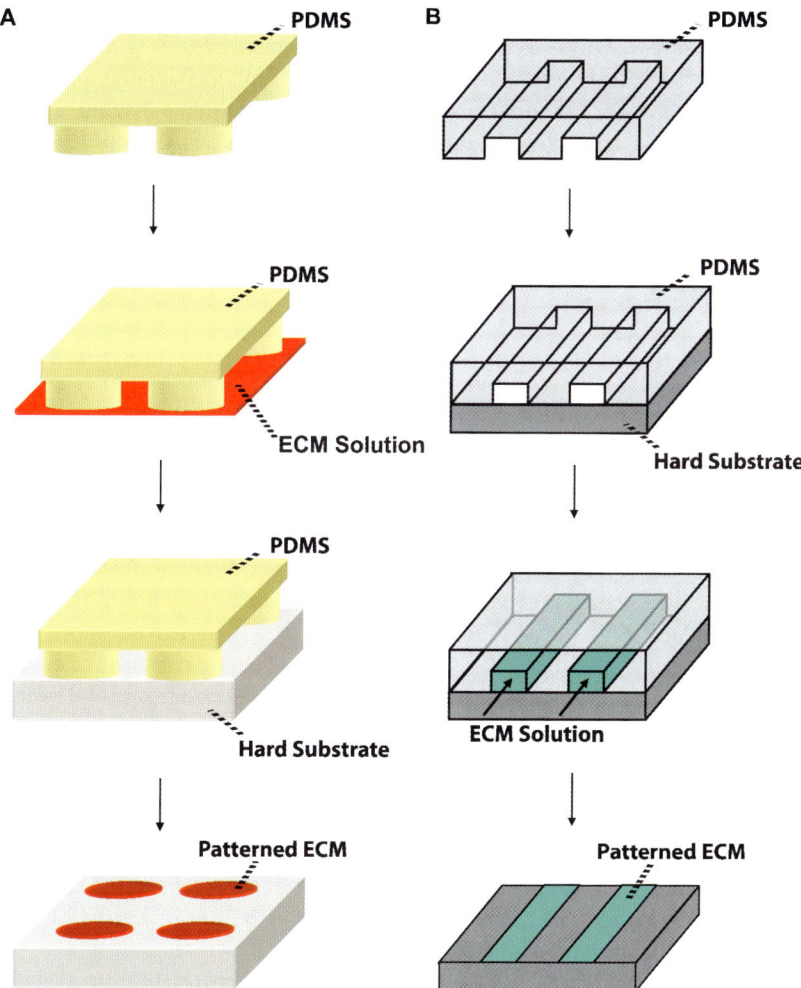

Fig. 10.1 Two techniques for micropatterning ECM. **A** Microcontact printing. **B** Microfluidic patterning

patterned, which can either be the matrix molecules themselves or other molecules that attract matrix molecules. After the polymer mold has been inked, it is brought in conformal contact with the cell culture substrate to transfer the molecules from the mold to the cell culture substrate in the desired patterns. The remaining regions on the cell culture substrate are then blocked with appropriate molecules (e. g., polyethylene glycols) to prevent cell adhesion to those regions. Instead of microcontact printing the ECM molecules themselves, other molecules such as alkanethiols may be printed. By fine-tuning the chemical composition of the alkanethiols, matrix-attractive alkanethiols are printed into specific shapes and the remaining regions are then blocked with matrix-repulsive alkanethiols. The cell culture substrates are then incubated with solutions containing ECM molecules. The ECM molecules will preferentially attach to the micropatterned matrix-attractive alkanethiol regions.

The technique of microfluidic patterning (Fig. 10.1B) relies on the use of polymer molds with microchannels to control the flow or contact of fluids with cell culture substrates. In such techniques, a polymer mold with microchannels is fabricated using soft lithography. The mold is sealed onto the cell culture substrate. Then the solution containing cell adhesive molecules (such as ECM) is flowed through the microchannel formed between polymer mold and the substrate, depositing a specific micropattern of cell adhesive molecules onto the cell culture substrate. Using two or more inlets with a series of mixing steps downstream, it is possible to use fluidic micropatterning to engineer gradients of soluble factors or bound ECM molecules on cell culture substrates [3].

10.2.2 Cell Proliferation and Survival on Micropatterned Matrix

Micropatterning techniques have been used to manipulate cell spreading and morphology, which have profound effects on the ability of the cell to survive and/or proliferate. On 2D culture substrates *in vitro*, adherent cells tend to spread out more than they do *in vivo* and exhibit higher proliferation rates. With the aid of matrix micropatterning, the relationship between cell spreading area and cell proliferation has been clearly defined. In general, adherent cells allowed to spread out more tend to proliferate at higher rates than cells confined to small adhesive regions. For example, hepatocytes cultured on squares of ECM regions ranging in size from 1600 microns^2 to 10,000 microns^2 are less proliferative and secrete more albumin when confined to smaller regions as compared to bigger regions [2]. Endothelial cells (ECs) cultured on very small square islands (100 microns^2) of ECM undergo apoptosis at a rate similar to their apoptotic rate in suspension culture. Increasing the cell spreading area increases the rate of EC proliferation on micropatterned matrices. By micropatterning small islands of ECM and allowing cells to spread across them, the same study proved that it is, in fact, cell spreading area rather than the area of cell-ECM contact that is responsible for modulating cell growth [4].

A potential clinical application for cell size control via matrix micropatterning is to limit the proliferation of smooth muscle cells (SMCs) on vascular implants. Hyperproliferation of SMCs on vascular implants such as stents and vascular grafts leads to restenosis. In vitro, SMCs seeded on micropatterned strips of ECM adopt a more elongated morphology and exhibit a significantly reduced rate of proliferation as compared to SMC cultured on unpatterned substrates [5]. Thus, the concept of modulating cell shape to control proliferation may, in one instance, be translated to vascular implant design in order to prevent hyperproliferation of SMC in clinical applications.

10.2.3 Cell Migration on Micropatterned Matrix

Cell migration is important for tissue morphogenesis, repair and regeneration. In many cell and tissue engineering applications, it is necessary to control the migration of cells to ensure proper cell organization and tissue formation. For example, the patency of vascular grafts is greatly improved when an EC monolayer forms on the luminal surface and separates the implant surface from blood. If an acellular vascular graft is implanted within the body, the long-term patency of the graft relies on the ability of the host ECs to migrate across the luminal surface of the graft and establish an organized monolayer. Cell migration is also important for wound healing where quick and efficient cell infiltration and organization into the wound is critical for wound closure and prevention of infections.

Matrix micropatterning can influence cell migration in a number of ways. It may be used to dictate the direction a cell migrates and the extent of its migration. The shape of the cell plays a major role in influencing its migratory behavior. In fact, a cell seeded on teardrop shaped ECM regions migrates forward from its blunt end rather than its sharp end (Fig. 10.2A). The polarity induced by the teardrop shape translates down to the cytoskeleton as well as the migration machinery, thus committing the cell toward a specific direction [6]. The concept of cell polarity influencing and enhancing cell migration is also further shown by the migratory behavior of cells on thin strips of ECM. For example, ECs assume a more polarized morphology when restricted to very thin ECM strips (e. g., 10–20 microns wide) [7]. This polarized morphology translates to a polar organization of focal adhesions thus gearing the cell toward migration along the length of the strip (Fig. 10.2B). The polar organization of focal adhesions also increases the speed of EC migration as compared to their migration on wider strips of ECM.

Matrix micropatterning may also be used to direct cell migration through haptotaxis – the cell migration towards the regions of higher ECM density. By patterning regions or gradients of ECM, the cell's direction and/or speed of migration may be controlled (Fig. 10.2C). For example, by patterning alternated matrix strips with high and low densities of collagen, step-gradients of collagen are created [8] (Fig. 10.2D). The crosstalk of haptotactic factors and other environmental factors (e. g., fluid shear stress) can be studied in this system. In addition, either step gradients or continuous gradients of substrate bound ECM may be used to attract and organize cells.

A cellular function related to cell migration is neurite extension from neuron cell bodies. Matrix micropatterning may be used to control axon specification and di-

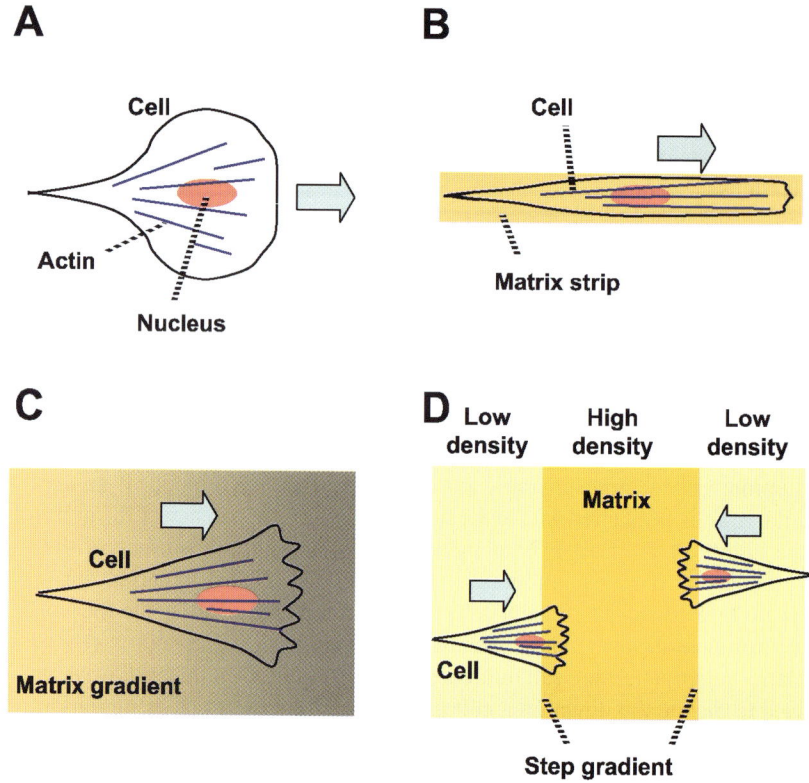

Fig. 10.2 Effects of micropatterned matrix on cell migration. **A** Cell shape affects the direction of cell migration. **B** Cell migration on thin matrix strips. **C** Cell migration on a continuous ECM gradient. **D** Cell migration on step-gradients of ECM proteins. *Arrows* indicate the direction of cell migration

rect neurite extension from neurons. Axon specification of neurons is controlled by micropatterning gradients of ECM such as laminin [9]. In this instance, the neuron sprouts axons from its cell body where it experiences the highest density of laminin molecules. Neurite extension may be specifically guided by culturing neurons on very thin strips of ECM and physically restricting the growth of neurites along the narrow strips [10]. These concepts have important implications for nerve graft design and may help improve the guidance of regenerating nerves across large injury gaps.

10.2.4 Cell Differentiation on Micropatterned Matrix

Controlling cell shape and size via matrix micropatterning can also have profound effects on cell differentiation. Adult bone marrow mesenchymal stem cells have the potential to differentiate toward various mesenchymal lineages such as bone, fat, cartilage and tendon/ligament. By controlling the shape and spreading area of MSC on cell culture substrates, it is possible to direct them toward fat or bone lineages [11]. MSC that are allowed to adhere, spread and flatten out on large micropatterned squares preferentially differentiate toward bone lineages while those allowed to adhere but not spread out on very small micropatterned squares preferentially differentiate toward fat lineages. In these instances, the shape of the cell directs its lineage commitment by influencing the expression and activity of specific intracellular signaling molecules.

Matrix micropatterning has also been used to screen ECM densities and compositions for stem cell differentiation. In such systems, matrix micropatterning allows for high throughput control over cell densities with low consumption of ECM. In an exemplary study, microarray technology was used to study the effect of varying compositions and densities of various ECM molecules on hepatocyte function and mouse embryonic stem cell differentiation [12]. This high throughput matrix micropatterning approach was used to identify various combinations and densities of ECM that specifically enhanced hepatocyte function and encouraged the differentiation of mouse embryonic stem cells toward the hepatocyte lineage. This technology may be used in future applications to screen for factors that enhance or direct differentiation of adult and embryonic stem cells toward specific lineages.

10.3 Topographic Regulation of Cell Functions

The physical structure of the tissue can act as guidance for tissue morphogenesis and remodeling. At the nano to micro level, the topography of ECM plays an important role in regulating cell functions. Collagen, one of the most ubiquitous proteins in the human body, is found mostly in the form of nanometer to micrometer scale fibers. Thus, cells not only interact with ECM proteins and growth factors, but also respond to nano and microscale topographical features in the microenvironment. For example, the porous, stiff structure of bone that osteoblasts encounter is very different from the soft, elastic environment that cells in arteries experience. Even within a given tissue, the local microenvironment changes quickly on a microscale. In the artery, the intima consists of an EC layer aligned longitudinally along the artery. In the media, SMCs are arranged circumferentially. The physical organization of cells within a tissue is important in dictating the function of the cells.

Microscale topographic features have been shown to regulate many aspects of cell functions. Microtopographic patterning uses soft lithography techniques to generate various microtopographical features on PDMS or other polymer membranes. For example, one simple pattern involves channels that can have varying widths, depths and spacing. Similar to the micropatterned matrix strips, the microchannels provide topographical cues that modify cell morphology and, thus, modulate cell functions such as proliferation. For example, SMC cultured on 20-micron wide micropatterned channels have more elongated cell morphology, less spreading, and a significant decrease in cell proliferation compared to SMCs on unpatterned surfaces [5].

Microscale topographic features can also modulate cell and tissue organization. For skeletal muscle, changing the topographical environment of the myoblasts directs differentiation and organization of myotubes. In native tissue, the differentiation of myoblasts into fused, multinuclear myotubes results in organized parallel bundles of muscle fibers that can contract efficiently. *In vitro*, however, myoblasts tend to differentiate and orient randomly, and display no organization. Myoblasts cultured *in vitro* on PDMS membranes patterned with 10 micron wide grooves spaced 10 microns apart differentiate into organized, parallel myotubes with decreased proliferation and increased myotube length compared to the myoblasts on unpatterned surfaces [13]. Actin filaments and muscle fiber striation also align with the microgrooves.

It is well accepted that cell migration can be regulated by topographic cues. Cells can sense the topography of the ECM (e. g., aligned fibrillar matrices, shape, texture, etc.) with dimensions from a few nanometers to hundreds of microns, and the migration is regulated by topographic guidance (or contact guidance) [14–16]. Similar to cell migration on micropatterned matrix strips, the control of migration direction by topography is likely through actin polymerization. The aligned fibril or grooves may promote the actin polymerization and protrusion in the parallel direction, which will result in aligned FAs and traction force in the same direction. Topographic guidance has great potential in tissue engineering applications. Micropattened polymers and aligned nanofibers can be used to control the structure and organization of engineered tissues (see Sect. 10.4).

Furthermore, topographic patterning can be used to measure the deformation of the cell adhesion substrate and thus the forces exerted on the substrate. For example, micropatterning has been used to make an array of the fluorescent dots in a flexible substrate for the measurement of traction forces exerted on ECM by migrating cells. The displacements of fluorescence dots are mapped during cell migration, and the traction forces are computed from the displacements of the dots and the mechanical properties of the elastic substrate [17]. Another approach uses a microfabricated array of vertical elastic posts coated with ECM proteins [18]. Local traction forces at multiple subcellular positions could then be measured based upon the deflection of these posts, which act like elastic beams under pure bending.

Similar to matrix patterning, topographic patterning can be used to control cell functions such as proliferation, organization and migration. While matrix patterning is more appropriate for cell culture studies *in vitro*, topographic patterning is relatively easier to be applied to the fabrication of biomaterials with micro and nano features on the surfaces or in 3D structures.

10.4 Engineering 3D Environments with Micro Features

The *in vivo* environment is a 3D space, and thus 3D *in vitro* examination of cell structures and tissues is essential in identifying important structural signals that dictate tissue function. Micro and nanotechnology make it possible to fabricate biomimetic tissue that can be finely controlled, resulting in 3D tissue models for basic research or better grafts that can be easily integrated into the native tissue.

One approach is to fabricate complex 3D tissue structures using the 3D printing technique, which produces 3D structures layer by layer from a computer-aided design model. Binder materials can be delivered by ink-jet printing to define the shape for each layer (Fig. 10.3A); alternatively, laser beam can be used to cure the photosensitive resin. This technique can be used to create structure of any geometry, and

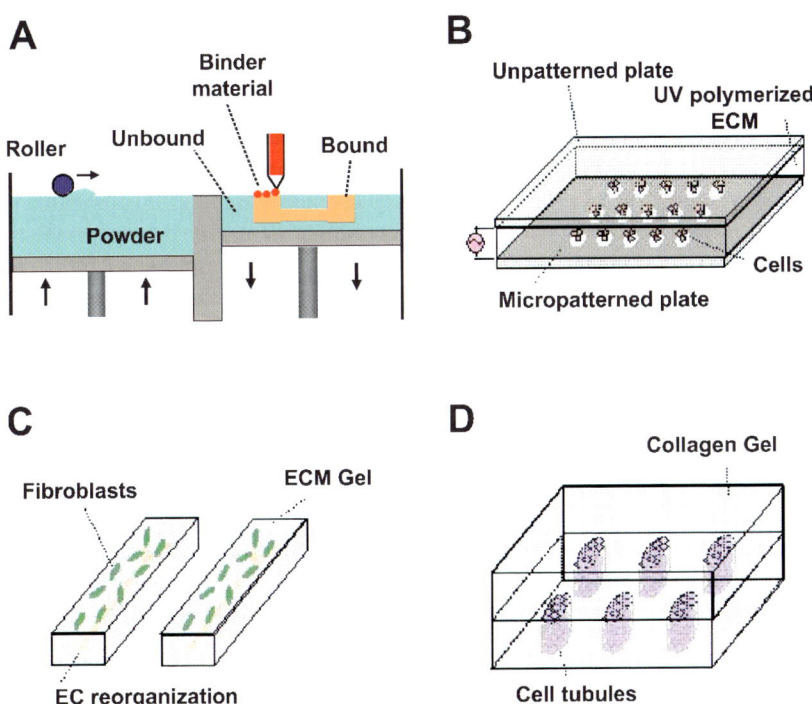

Fig. 10.3 Micropatterning 3D microenvironment. **A** 3D printing. From a computer-aided design model of the desired structure, a slicing algorithm draws detailed information for every layer. Each layer begins with a thin distribution of powder spread over the surface of a powder bed. Using a technology similar to ink-jet printing, a binder material selectively joins particles where the object is to be formed. A piston that supports the powder bed lowers so that the next powder layer can be spread and selectively joined. This layer-by-layer process repeats until the structure is completed. Unbound powder is removed, leaving the fabricated structure. **B** Dielectrophoretic patterning of cells in 3D matrix. Cells in a matrix gel are sandwiched between two conductive plates, one of which is patterned with photoresist (*lower plate*). The setup is subjected to an alternating current, during which time the cells move into the areas where there is no photoresist. The gel is subsequently polymerized with UV to hold the cells in position. **C** Microfluidic patterning of a co-culture system. Matrix gels with randomly seeded endothelial cells are formed using microfluidic patterning, followed by seeding of fibroblasts on top of the culture. After several days, the ECs align themselves along the length of the gels, with early sprouting structures. **D** 3D patterning of cells using microstamping. A 3D polymer microstamp in a matrix gel creates cavities in which cells can be seeded. Another layer of matrix is then polymerized on top, forming tubule shaped clumps of cells

out of any material, including ceramics, metals, polymers and composites. Furthermore, it can exercise local control over the material composition, microstructure, and surface texture. For example, regionally selective adhesion area and microstructure can be printed using biocompatible and biodegradable polymers [19]. However, due to the biocompatibility issues (binder materials, laser beam) during 3D printing process, it is a challenge to print cells into the 3D structure directly.

Another interesting use of microtechnology in 3D combines dielectrophoretic (DEP) forces and photopolymerizable gels to control the arrangement of cells in complex patterns [20]. Manipulation of particles, such as cells, subjected to a non-uniform electric field, has been useful in cell separation and trapping (Fig. 10.3B). This method utilizes this concept by patterning a metallic substrate with SU-8 photoresist to form a conductive plate. A mix of cells in an unpolymerized gel is sandwiched between two conductive plates (one patterned with conductive areas), and an alternating current is applied to the plates, forming areas of high electric field strength where the insulating photoresist layer is not present. The cells move into the areas with high DEP forces within a few minutes, and are then trapped in the arranged configuration through UV polymerization of the gel. Subsequent layers of cells (in the same or different patterns) can then be organized and polymerized sequentially in a similar fashion. This method can be useful in forming a complex, multilayer and multicellular tissue.

The highly organized structure of capillary networks in tissue is very difficult to reproduce *in vitro*. It is possible to fabricate simple 3D channels of ECM (collagen, chitosan, fibronectin) via microfluidic patterning [21]. These 3D gels can be embedded with ECs, and then seeded with fibroblasts (Fig. 10.3C). This 3D directed co-culture results in a uniform reorganization of the ECs into a stem shape with sprouting structures, and cell orientation that is parallel to the channel or sprouts. These ECs are also supported by the fibroblasts in the surrounding ECM. While these structures are not stable in the long term, it offers a proof of concept that directional patterning of 3D structures *in vitro* is possible.

Similarly, a recent study used micropatterned 3D molds to examine the branching morphogenesis of mammary gland epithelial cells [22]. PDMS molds were used to form elongated cavities in collagen gel, in which mouse mammary epithelial cells were seeded (Fig. 10.3D). Upon induction of branching morphogenesis via addition of epidermal growth factor, it was observed that there was differential branching at the ends of the tubules, and not the sides. Further testing concluded that the position of the cells determined the signaling necessary to promote or inhibit branching. Induction factors are mitigated along the stem of the tubules due to the autocrine release of inhibitory morphogen, TGF-β. Overexpression of active TGF-β inhibited branching entirely, and rearrangement of the distance between tubules also affected branching – the tubule ends that were most distant showed branching, but branching was inhibited at tubule ends that were in close proximity to other tubules. This simple model demonstrates that the position of cells dictates the concentration of local growth factor signals that can regulate tissue morphogenesis.

These examples of 3D use of microtechnology show the different approaches that can be taken in this field. One method may be more useful than the other, depending on the particular goals, but all are useful ways of observing cell behavior in 3D. Further studies with 3D tissues will further our understanding of how these tissues develop, grow, and maintain themselves *in vitro* and *in vivo*, and potentially lead to improved tissue replacement options.

10.5 Nano Patterning for Cell and Tissue Engineering

While cells as a whole are in the microscale range, subcellular structures and proteins are much smaller, down to nanometers. Recent advancements in technology have resulted in fabrication of devices and scaffolds with nanoscale features, and with this knowledge a growing field dedicated to nanotechnology has emerged. The development of novel nanotechnologies will allow us to not only investigate cellular processes on the nanoscale, but also develop new therapies for medical applications.

Nanofibers are nanoscale fibers composed of biological and/or synthetic polymers that can be patterned into various orientations and shapes to influence cell and tissue behavior. Nanofibers can be used to mimic the structure, morphology and size of native ECM fibers such as collagen fibrils. A common method for fabricating nanofiber scaffolds for tissue engineering is electrospinning (Fig. 10.4A). Electrospinning is a versatile technique that may be used for both biological and synthetic polymers. Many biocompatible and biodegradable polymers can be made into nanofibrous structure with this technique, which offers tremendous opportunities to engineer various scaffolds for many tissue engineering applications. The basic electrospinning setup involves the controlled delivery of a polymer solution to the tip of a spinneret that is placed with a specific distance from a grounded collector substrate. A power supply is used to charge the polymer solution and create an electrical field between the spinneret and the grounded collector substrate. The technique relies upon the ability of the electrical field to overcome the surface tension of the charged droplet of polymer solution at the tip of the spinneret. The result is a jet of polymer solution that travels from the edge of the droplet toward the collector substrate. As the jet travels through the air the solvent evaporates, resulting in the deposition of thin polymer fibers on the collector substrate. By varying the collector substrate, it is possible to produce nanofiber scaffolds in various shapes (e.g., tubes) and with varying degrees and directions of alignment. For example, anistropic nanofiber scaffolds may be fabricated by using a rotating drum as a collector. Collector substrates with air gaps also induce alignment of fibers that deposit across the air gap [23].

Cells seeded on anisotropic nanofibers align their cytoskeleton parallel to the fiber orientation (Fig. 10.4B). Anisotropic nanofibers may be used to restrict cell size and shape. For example, SMCs cultured on aligned nanofibers adopt an elongated morphology. Such morphology could influence SMC proliferation similarly to the effect of ECM micropatterning or microtopographic patterning. Anisotropic nanofibers also have profound effects on cell migration and neurite extension from neurons. Neurons seeded on these aligned fibers preferentially extend their neurites in directions parallel to the fiber orientations [24].

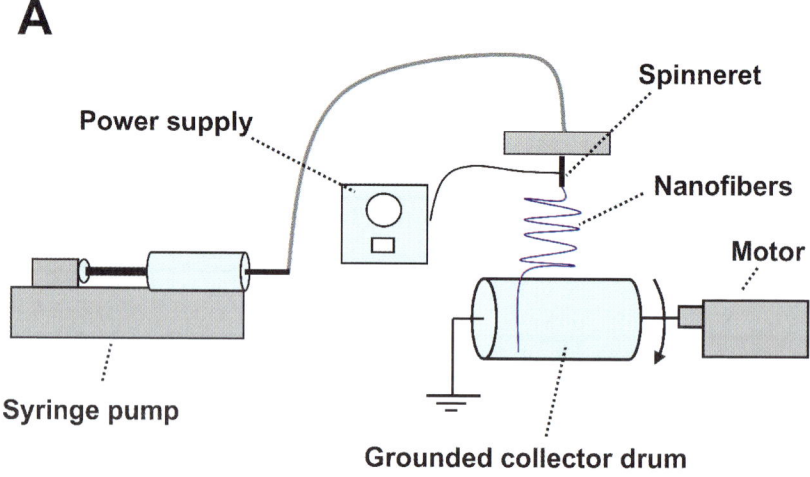

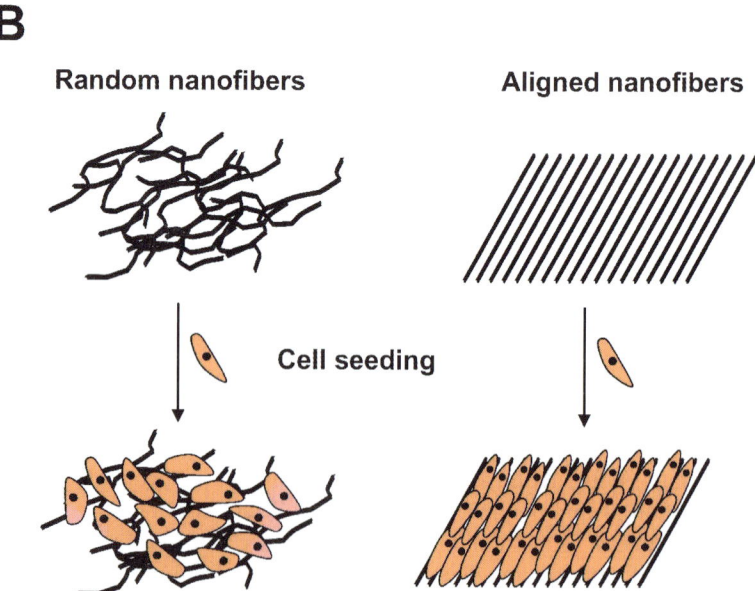

Fig. 10.4 Electrospun nanofibers for tissue engineering. **A** Electrospinning setup. **B** The use of anisotropic nanofibers to organize cells. Cells can respond to nanoscale features in the substrate. Cells on random nanofibers show random orientation and are disorganized, while cells on aligned nanofibers have organized cell-cell interactions and tissue-like structure

Anisotropic nanofiber scaffolds have the greatest potential for regenerating highly organized tissues in the body such as tendons, ligaments, blood vessels, muscle and nerve. In these applications, the aligned nanofibers may influence cell and ECM alignment thus allowing the regenerated tissue to match the morphology of the native tissue. In fact, anisotropic nanofibers induce, enhance and specifically orient the extension of neurites from sensory neurons within dorsal root ganglion nerve tissue. The scaffolds also enhance the fusion of myoblasts into myotubes that organize in parallel orientation to the nanofiber alignment [13].

Anisotropic nanofibers influence cells similarly to matrix micropatterning or micro topographical patterning. However, unlike micropatterned matrices, anisotropic nanofibers are not limited to 2D substrates or by distinct regions of cell adhesive and cell repulsive regions. Three-dimensional biocompatible and biodegradable tissue engineering scaffolds composed entirely of anisotropic nanofibers may influence and organize cells and tissues more effectively than microfabricated scaffolds.

10.6 Perspective

In the past decade, many studies have been performed to examine the regulation of cellular functions by the micro and nano features in the microenvironment. While we expect to see more in-depth investigations of cell responses to micro and nano features in different types of tissues, we are well equipped to explore the use of micro and nano technologies to fabricate biomimetic materials for clinical applications such as tissue regeneration. Beyond the scope of this chapter, there are many other aspects of micro and nano technologies with potential for basic research and medical applications. For example, micro bioreactors can be used for high throughput screening of cell culture conditions and factors, nanoparticles can be used for probing and imaging applications, and nanostructured materials can be used for drug delivery. With the efforts, creativity and imagination of scientists working in this exciting field, micro and nano technologies will have significant impact on science and technology development in the future.

Acknowledgements We would like to thank Rahul Thakar and Kyle Kurpinski for their help in figure preparation. This work was supported in part by grants (HL078534 and HL083900) from National Heart, Lung and Blood Institute.

References

1. Cao Y, Vacanti JP, Paige KT, Upton J, Vacanti CA (1997) Plast Reconstr Surg 100:297–302; discussion 303–304
2. Singhvi R, Kumar A, Lopez GP, Stephanopoulos GN, Wang DI, Whitesides GM, Ingber DE (1994) Science 264:696–698
3. Li Jeon N, Baskaran H, Dertinger SK, Whitesides GM, Van de Water L, Toner M (2002) Nat Biotechnol 20:826–830
4. Chen CS, Mrksich M, Huang S, Whitesides GM, Ingber DE (1997) Science 276:1425–1428
5. Thakar RG, Ho F, Huang NF, Liepmann D, Li S (2003) Biochem Biophys Res Commun 307:883–890
6. Jiang X, Bruzewicz DA, Wong AP, Piel M, Whitesides GM (2005) Proc Natl Acad Sci USA 102:975–978
7. Li S, Bhatia S, Hu YL, Shiu YT, Li YS, Usami S, Chien S (2001) Biorheology 38:101–108
8. Hsu S, Thakar R, Liepmann D, Li S (2005) Biochem Biophys Res Commun 337:401–409
9. Dertinger SK, Jiang X, Li Z, Murthy VN, Whitesides GM (2002) Proc Natl Acad Sci USA 99:12542–12547
10. Clark P, Britland S, Connolly P (1993) J Cell Sci 105(1):203–212
11. McBeath R, Pirone DM, Nelson CM, Bhadriraju K, Chen CS (2004) Dev Cell 6:483–495
12. Flaim CJ, Chien S, Bhatia SN (2005) Nat Methods 2:119–125
13. Huang NF, Patel S, Thakar RG, Wu J, Hsiao BS, Chu B, Lee RJ, Li S (2006) Nano Lett 6:537–542
14. Turner DC, Lawton J, Dollenmeier P, Ehrismann R, Chiquet M (1983) Dev Biol 95:497–504
15. Nakatsuji N, Johnson KE (1984) Nature 307:453–455
16. Curtis A, Wilkinson C (1999) Biochem Soc Symp 65:15–26
17. Balaban NQ, Schwarz US, Riveline D, Goichberg P, Tzur G, Sabanay I, Mahalu D, Safran S, Bershadsky A, Addadi L, Geiger B (2001) in Nat Cell Biol 3:466–472
18. Tan JL, Tien J, Pirone DM, Gray DS, Bhadriraju K, Chen CS (2003) Proc Natl Acad Sci USA 100:1484–1489
19. Park A, Wu B, Griffith LG (1998) J Biomater Sci Polym Ed 9:89–110
20. Albrecht DR, Underhill GH, Wassermann TB, Sah RL, Bhatia SN (2006) Nat Methods 3:369–375
21. Tan W, Desai TA (2003) Tissue Eng 9:255–267
22. Nelson CM, Vanduijn MM, Inman JL, Fletcher DA, Bissell MJ (2006) Science 314:298–300
23. Li D, Ouyang G, McCann JT, Xia Y (2005) Nano Lett 5:913–916
24. Yang F, Murugan R, Wang S, Ramakrishna S (2005) Biomaterials 26:2603–2610

Chapter 11
Integrative Nanobioengineering: Novel Bioelectronic Tools for Real Time Pharmaceutical High Content Screening in Living Cells and Tissues

Andrea A. Robitzki, Andrée Rothermel

Centre for Biotechnology and Biomedicine, University of Leipzig, Germany,
andrea.robitzki@bbz.uni-leipzig.de

Abstract At present, drug development and validation is still a time consuming and cost intensive process. For highly efficient target validation, drug safety as well as drug screening, innovative cell and tissue based sensors provide powerful tools that can be applied in a very early phase of the drug development cycle and that significantly reduce costs by minimizing the failure rate of drug candidates at later stages. The combination of Micro Systems Technology (MST) and Nanotechnology (NT) as well as their integration with living cells and tissues allows the development of highly efficient biological Micro-Electrode-Mechanical-Systems (bioMEMS) or biosensors for functional real-time monitoring. The following chapter gives an overview of various different types of biosensors and describes several recording techniques and the practical application of biosensors for screening of multiple physiological parameters under high-content and high-throughput conditions.

11.1 Introduction

11.1.1 Preventive Medicine and High Effective Therapies

New diagnostic real time test systems making use of nano- and microsystems technology to better measure disease-related biomarkers could and therefore offer individual risk assessments before actual symptoms occur. Based on such an online and real time analysis, patients could be recommended with an increased risk to take up a personalized prevention program. People with an increased risk for a certain disease could benefit from a regular medical check-up schedule to monitor changes in the pattern of their relevant biomarkers or could be individually treated with individual therapies at an early time range.

If the preventive medical investigations had found an indication or already possible symptoms for a certain disease, more specific diagnostic procedures are needed. Miniaturized biomonitoring systems will make it possible to perform bioelectronic-based diagnostics everywhere. Automatic methods and software programs will give easy diagnostic results without an expert at site. Conceptually novel methods, combining biochemical techniques with advanced biomonitoring by e. g. bioimpedance spectroscopy provide insight to the behavior of single diseased cells and their microenvironment for the individual patient. This could lead to personalized treatment and medication tailored to the specific needs of a patient and to more effective drugs.

11.1.2 Follow-up Monitoring After Therapy – Therapeutical Control

Medical reasons may call for an ongoing monitoring of the patient after completing the acute therapy. This might be a regular check for reoccurrence, or, in case of chronic diseases, a frequent assessment of the actual disease status and medication planning. Continuous medication could be made more convenient by microimplants which controlled drug releasing or diluting modes over an extended length of time. In-vitro as well as in-vivo diagnostic techniques play an important role in this part of the care-process for a systematically monitoring to pick up early signs of reoccurrence of a disease or disorder. Neurology, cardiology, and oncology are of the areas where these techniques are already tested today. Some types of tumors can be controlled by continuous medication extending the life expectancy. However, in case the tumor gets resistant to a certain medication, signs of disease progression can be immediately picked up and alternative treatments can be prescribed.

11.2 Real Time Monitoring and High Content Screening

The future of an innovative and efficient drug and therapy development will dependent on technology platforms which will provide a high content and high throughput real time screening using cell and tissue based disease models. In neurological and cardiovascular research during the last years various novel therapies and drug candidates have been developed. Nevertheless the problems are the extensive time frame and intensive and increasing costs for the development and validation of drugs and therapies because of a more complex drug development and a prolonged chain of economic value added. Since 1990 [1] the pharmaceutical industry expended thrice of economic funds whereas since 1996 the amount of applications for approval and accreditations for New Medical Entities (15 NME in 2005) at the FDA (Food and Drug Administration, USA) decreased dramatically [2, 3]. The costs for the development of novel drugs in comprehension to the economic deficit of all other drug candidates is add up to approximately 897 Mio. US $ [1]. Today the average developmental period for drugs until its readiness for marketing and accreditation is 10 to 15 years whereby the failure rate increased dramatically e. g. a drug candidate in phase I clinical trial has a chance of approximately 8% becoming accredited [3, 4]. The celerity for identification of lead structures and targets by genomics and proteomics is not correlated with the subsequent tests for drug activity and mechanisms of drug effects. Preclinical trials are not sufficient for drug safety, effects, and toxicology. Therefore, there is a need for novel cell and tissue based high content screening systems with a modular and fractal technology platform concept for better, ultra-fast and reliable predictions. High Content Screening (HCS) using organotypic (pathologic) cell and tissue models in combination with a real time monitoring would drastically reduce (i) the cost of drug development and (ii) the quote of failures in preclinical trials (approx. 20–30%, e. g. a reduction of 10% corresponds to 100 Mio. US $ [2]). For a high efficient target validation, drug safety as well as drug screening and therapeutical control in innovative medicament development, biosensors combined with cells and tissues as biological targets are in the focus of pharmaceutical assays.

Since microfabrication techniques expand into the field of biology and medicine more and more innovative screening methods became applicable for screening of manifold cellular parameter. The detection of cellular alterations by means of non-invasive bioelectronic screening techniques has fundamental advantages in comparison to classical methods such as molecular, histochemical or proteinchemical analyses of single cells, tissues, organs, and other biological samples. In general bioelectronic screening of physiological parameter can be achieved under labeling-free and non-destructive conditions. Based on the non-invasive nature of these techniques long term measurements can be performed without influencing cellular behavior. Hence, the cellular read out reflects the real time without disturbing effects due to complex and long lasting physical procedures. Although these methods are well suited to study a broad spectrum of biological and medical problems, in many cases the real cellular information dropped away since, e. g. staining artifacts, makes it difficult or completely impossible to interpret the extracted cellular data. In principle, tracing of biological processes in living cell can be performed with modern labeling techniques (generation of fluorescence labeled fusion proteins) but hold the risk to falsify data due to the positioning of foreign substance within the cell itself.

To achieve non-invasive real-time monitoring in living cells and tissues so-called bioMEMS (biological Micro-Electro-Mechanical-Systems) or biosensors are promising tools for a feasible and reliable reading out of multiple physiological parameters. Substantial advances in both Micro Systems Technology (MST) and Nanotechnology (NT) as well as their combination have open new avenues for a broad field of application. Miniaturization of devices and surface modifications at micro- and nanoscale enables a variety of inexhaustible applications in diagnostic and therapeutic medicine, pharmaceutical industry, basic and advanced research (Fig. 11.1). Typical examples for the application of Biosensors and bioMEMS are (i) multimicroelectrode arrays for high content and/or high throughput drug screening and an improved drug safety, (ii) implantable micro devices for controlled and precise drug delivery and analysis of body functions, (iii) lab on a chip for multi-parametric analysis of different cellular parameters of human body fluids, (iv) neuroprosthesis for stimulation and sensing of neuronal signals within the central nervous system (CNS) and peripheral nervous system (PNS), (v) neuronal regeneration (vi) biomaterials and tissue engineering for organ or tissue replacement.

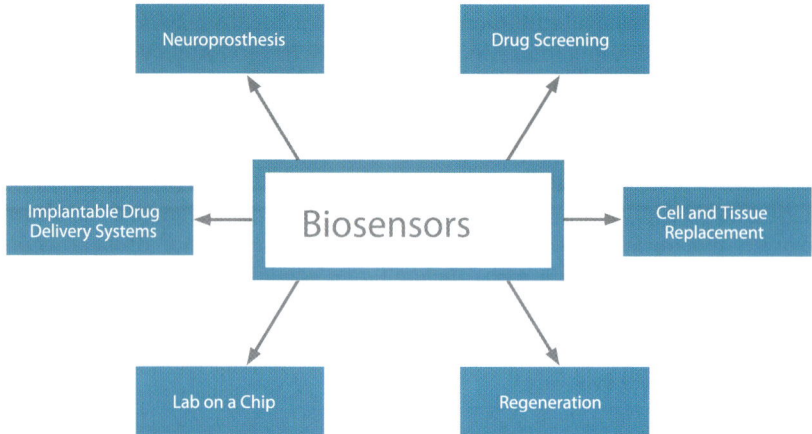

Fig. 11.1 Advances in new fabrication technologies such as Micro-Systems Technology (MST), Nanotechnology (NT) and Micro-machining enables a variety of innovative applications of biosensors and bioMEMS

Fabrication of biosensors or bioMEMS is founded on manufacturing techniques originally developed by the semi-conductor industry. By using micro-fabrication processes like thin film deposition, lithography, etching, substrate bonding and other surface modifying techniques allow the generation of precise micro- and nanometer structures (e. g. microelectrodes, microchannels, microcavities) on glass, silicon and plastic substrates. A textbook example for application of such devices are cell based biosensors also known as microelectrode arrays. These planar silicon or glass based microarrays are well-suited for extracellular recording of electrogenic cells or complex tissues via multiple substrate integrated microelectrodes, mostly consist of noble metals such as gold, platinum, or titanium [5, 6]. In addition to planar microelectrodes, field effect transistors (FETs) embedded on glass or silicon substrates fulfill similar and versatile functions [7–9].

Recording of electrophysiological properties by microelectrode arrays have several advantages, especially when this method is compared with classical patch clamp techniques. For instance, extracellular recording of field potentials by microelectrode arrays represents a non-destructive technique that allows long-term experiments of either single and multiple cells or complex tissues (Fig. 11.2). Another convincing argument for this method is represented by measuring alterations in membrane potentials in parallel at 60 or more microelectrodes at the same time. However, one of the exceptional advantages of microelectrode array based recording is its time-saving and cost-effective experimental requirements. In this respect, multi site recording via planar electrodes point out novel tools for realizing an automated and non-invasive screening, especially in terms of detecting ion channel associated diseases and testing drugs according to their effects and side effects. Although these biosensors are often used for stimulation and recording of neuronal cells an outstanding example for extracellular recording of field potential are cardiomyocytes since this cell type gives adequate signals based on its spontaneous

11.2 Real Time Monitoring and High Content Screening

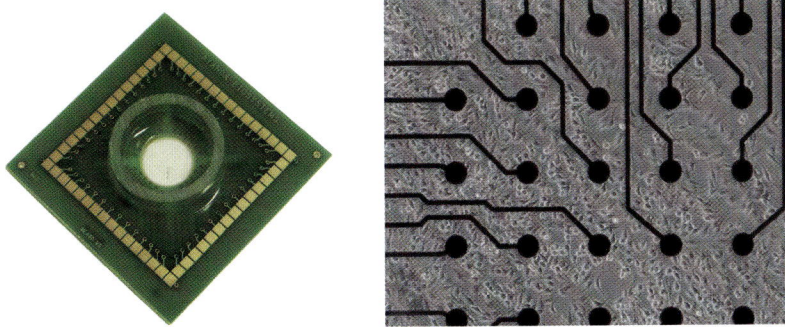

Fig. 11.2 Cells can be cultured on a microelectrode array (*left*) consisting of 60 planar gold electrodes, each with a diameter of 30 μm (*right*). Electrodes are connected via thin conductor paths in order to transfer the active cellular signal to the analyser unit

electric activity. Basically, for passive recording, cardiomyocytes can be simply cultured as monolayer on the top of a microelectrode array for subsequent detection of alterations of extracellular potentials (Fig. 11.2). Here, electric alterations caused by ion inward and outward currents are measured between a small working and large reference electrode. Thereafter, the signal is amplified, processed by an analog-to-digital converter, analyzed and displayed by a computer-based software (Fig. 11.3).

In the broadest sense this type of recording is comparable to the whole cell configuration of the traditional patch clamp technique. Which type of ion channels is involved in the generation of field potentials has been shown by using a number of selective ion channel antagonists (Fig. 11.4; unpublished by Robitzki et al.). The negative field potential (FPmin) mainly results from the voltage dependent Na^+ and L-type Ca^{2+} channels, whereas the repolarisation phase is predominately caused by K^+ and I_{KR}-type channels. Based on the curve progression it is possible to detect influencing positive or negative effects of drugs, biologic active substances such

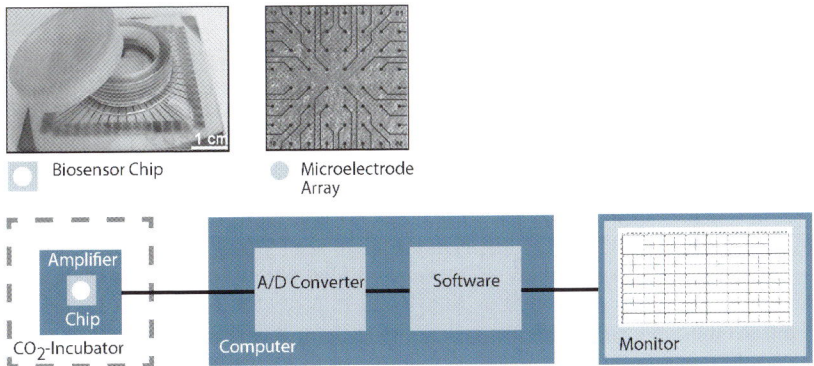

Fig. 11.3 Typical setup for extracellular recording of physiological parameters of cardiomyocytes by means of microelectrode arrays

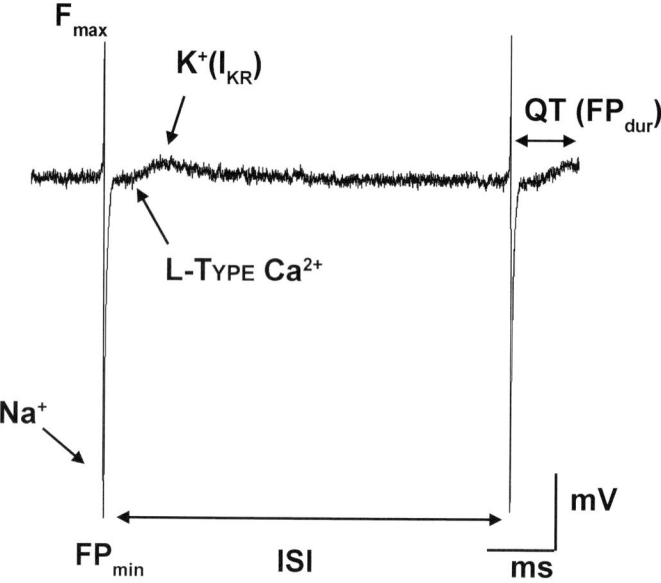

Fig. 11.4 Formation of field potential recorded with microelectrode arrays. The *curve* represents the extracellular recording of membrane potential alterations of cardiomyocytes at a single electrode. For additional information details see text

as recombinant proteins. Modifications in the amplitude of field potentials (F_{max}, F_{min}), duration of the inter spike interval (ISI), as well as the QT interval of the field potential (Fdur) allow an effective screening of potential leads associated with cardiovascular diseases.

In this context, a number of studies using cardiomyocyte based microelectrode arrays could identify chronotropic and QT modulating compounds [10–18]. Rothermel et al. [17] have demonstrated the high sensitivity of microelectrode arrays by detecting angiotensin II (Ang II) associated alterations in contraction frequency of cardiomyocytes. Ang II is a component of the renin-angiotensin system (RAS) and is involved in a number of physiological function. An improved coupling of cardiomyocytes on microelectrode arrays enables the detection of Ang II in a concentration of only 10^{-11} M (Fig. 11.5).

These findings are of significant importance since signaling of Ang II via the angiotensin II type 1 receptor (AT1 receptor) regulates a variety of physiological processes in heart, brain, lung, and kidney as well as dysfunctions in AngII/AT1 signaling system results in the development of a broad spectra of diseases mostly associated with the cardiovascular system [19]. Hence, it becomes clear that this type of screening system is a promising tool in functional screening of active components that can be used in the treatment of AngII/AT1 related cardiovascular diseases.

Besides these pharmaceutical applications in drug development, drug testing, and drug safety recently it has been shown that cardiomyocyte based biosensors are also of great benefit for diagnostic investigations. An outstanding example is the detection of AT1 receptor autoimmune antibodies (AT1AA) in sera of pregnant women

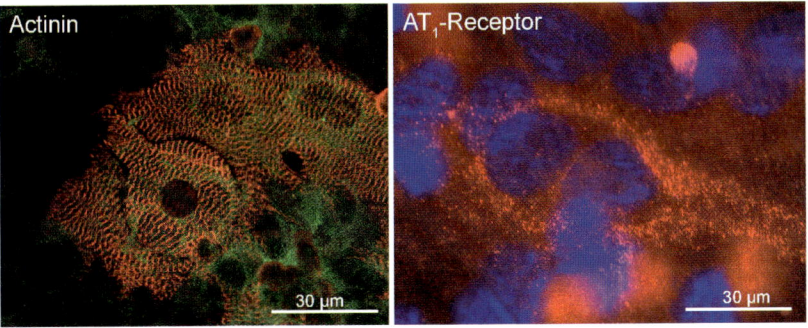

Fig. 11.5 *Left*: Confocal laser scanning microscopy of cardiomyocytes grown on a nano-columnar titanium nitrite microelectrode that were stained with an actinin specific antibody (*red*). *Right*: Expression of angiotensin II type 1 receptor (AT1) in cardiomyocytes (*red*) allows screening of Ang II/AT1 related cardiovascular diseases (DAPI stained cell nuclei, *blue*)

suffering from preeclampsia. Preeclampsia is a disorder that occurs after 20 weeks of pregnancy and affects both the mother and the fetus. It is characterized by high blood pressure and protein in the urine. Although the exact mechanism of AT1AA is still matter of debate it is obvious that this autoimmune antibody is crucial in the clinical manifestation of preeclampsia. Based on the fact that the AT1AA is present at very low concentrations in the sera of pregnant preeclamptic women the conventional ELISA techniques were not successful in detecting the relevant biomedical target. A solution for this problem has been found by using cardiomyocyte based microelectrode arrays. Application of partially purified IgG fractions from sera of non-preeclamptic and preeclamptic women to cardiomyocyte-based microelectrode arrays revealed an increase in contraction frequency in AT1AA positive blood samples, demonstrating the sensitivity of the used cell biosensor [17]. Hence, further development of these microelectrode array based biosensors to 96- and 384-well formats will provide a key technology for functional high throughput screening of all sorts of biomedical relevant substances in electric active cells.

Although multi site-based extracellular recording with microelectrode arrays belongs to the technique of choice for functional high throughput or high content screening of electric active cells, this recording method is inapplicable when non-electric active cells should be used for analysis of cellular parameters. In this case electric impedance spectroscopy (EIS) on biological systems provide an alternative opportunity to solve these problems. Similar to extracellular recording, bioimpedance spectroscopy represents a non-destructive method for measuring various cellular properties under real time conditions. The frequency dependent changes of passive electrical parameters of single cells or complex tissues can be easily analyzed by applying varying small alternate currents to living biological systems. The bioimpedance of single cells or complex tissues is composed of various dielectric parameters of the cell itself and its surrounding environment. Among these cellular properties, the capacitance and resistance of the cell membranes as well as intracellular membranes of organelles, the resistance of the extracellular medium and intrinsic cytoplasm, the extracellular matrix and the contact between cell and electrode contribute to the overall cellular impedance. To analyze alterations of impedance in living cells, an alternate voltage current is applied to a biological sample. Based on the dielectric properties of sub-cellular compartments and molecules the applied current can flow from an active working electrode through the cell or tissue whereby the remaining current is collected by a counter electrode. Depending on the frequency of the applied voltage current alterations of certain cellular compartments can be identified. In this context it is possible to discriminate cellular behavior according to their dispersions that are divided in α-, β-, and γ-dispersion. The α-dispersion ranges from 1 Hz to 1 kHz and results from counter ions, glycocalyx, and from ion channels, whereas the β-dispersion (1 kHz–100 MHz) is due to the cytoplasm membrane, intracellular membranes (organelles), cytosol and proteins. Additionally, the γ-dispersion (100 MHz–100 GHz) is defined by the dielectric properties of free and bound water, relaxation of charged subgroups, and partially by protein-protein interactions [19–22]. In particular, the frequency dependent measurement of manifold cellular alterations of both electric and non-electric active cells under non-destructive real-time conditions point out the infinite possibilities of this innovative technique. There are several promising approaches demonstrating the effectiveness of impedance recording of cellular parameters. First evidences for the feasibility of high throughput impedance recording has been made by Ciambrone et al. [23]. They performed impedance recordings by using 96-well microplates with interdigitated electrodes at the bottom of the device. Application of substances that specifically affects signal transduction of different G-protein coupled receptors (Gs, Gq, Gi) and protein tyrosine kinase receptors showed an impedimetric detection of activated intracellular signaling pathways under non-invasive and labeling free, high throughput conditions.

Although impedance spectroscopy has unique advantages compared to other functional screening methods, up to now, this technique is still not widely used for high throughput application, what is in all probability based on the limited commercial availability of cost-effective, feasible and reliable chip-based microelectrode arrays. To overcome this problem and to make this technique more accessible to a broader scientific, pharmaceutical and medical community the group of A. Robitzki has tested whether commercial available microelectrode arrays that are originally developed for extracellular recording of electric active cells (see above) can be used for impedance recording [24]. For this purpose MCF-7 cells derived from human breast cancer cells were cultured as a monolayer on microelectrodes arrays and treated with PMA (phorbol 12-myristate 13-acetate) to induce intra- and extracellular changes by activation of PKC signaling pathways. To perform feasible, reliable, and cost effective impedimetric measurements a commercial impedance gain-phase analyzer was combined with microelectrode arrays consisting of 60 substrate integrated microelectrodes (Multi Channel System, Reutlingen). Impedance recordings in a range of 10 Hz to 1 MHz revealed that multi arrays with nanocolumnar structured titanium nitrate electrodes are the material of choice since the electrode impedance is extremely low when compared with traditional used gold electrodes (Fig. 11.6; top).

11.2 Real Time Monitoring and High Content Screening

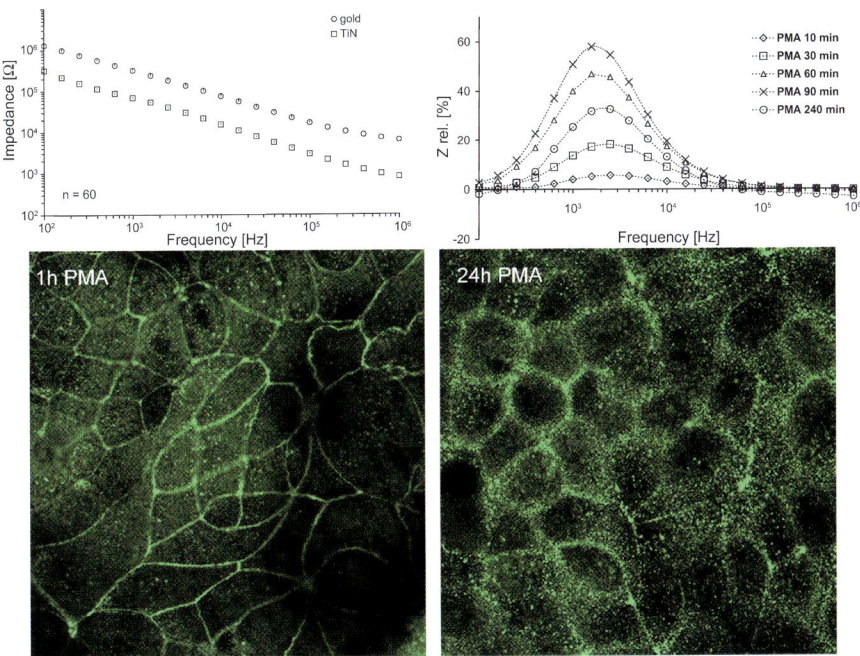

Fig. 11.6 Detection of PMA induced cellular alterations by Microelectrode Array-based Impedance Spectroscopy (MAIS). Nanocolumnar structured titanium nitrite microelectrodes have a remarkable low impedance in comparison to traditional used gold electrodes (*top, left*). PMA (phorbol 12-myristate 13-acetate) induced stimulation of PKC signalling pathways caused a time-dependent increase of the overall cellular impedance in MCF-7 cells (*top, right*) long before the first morphological alterations are detectable via conventional immunhistochemical staining methods as revealed by the detection of the tight junction protein occludin (*below*)

Application of 0.3 µM PMA to MCF-7 cultures resulted in a predominant increase of the overall cellular impedance within the first 90 min. Later on the impedance decreased after 240 min (Fig. 11.6; top, right). However, one of the most striking findings in this study was that impedimetric alterations could be detected already after 10 minutes after PMA stimulation, and were recordable long before the first morphological alterations became detectable by conventional immunhistochemical staining methods (Fig. 11.6; below). Moreover, the standard derivation of all 60 microelectrodes at one array is extremely low and therefore, Microelectrode Array-based Impedance Spectroscopy (MAIS) represents a highly sensitive, feasible, reliable, and reproducible gold standard technique for detecting cellular modifications. A similar microelectrode array based impedance biosensor has been realized by parallel screening of suspension cells on planar microelectrodes. The innovative design of this biosensor allows the parallel micro-fluidic positioning of single cells through micro-holes that are integrated in the centre of planar electrodes [25]. The advantage of this biosensors is its fast and parallel impedance recording since the time consuming pre-cultivation of cells as monolayer cultures on the top of microelectrodes is not needed.

However, the development of biosensors for in vitro testing of cellular functionality under three dimensional (3D) in vivo-like conditions is still a challenge for scientists working in the field of basic or applied sciences. Already existing in vitro tissues such as explants or artificial generated tissues or organs are crucial in the application for impedance based high throughput screening because of their unstable behavior concerning environmental changes and their laborious handling. Moreover, synthetic scaffolds that have been mostly used for the assembly of complex tissues can interfere with impedance recording. For this purpose the improvement of conventional tissue engineering methods or the establishment of new technologies for the generation of 3D in vitro tissues are of great importance. In this context an increasing progress has been made by introducing a scaffold-free rotation-mediated generation of sphere-like histotypic in vitro tissues. This technique is based on self assembly of cells, cell-recognition and cell sorting out processes. Histotypic spheres can be obtained from single cells of various tissues, including tumors [26–28]. A number of tissue-like spheroids have been produced from primary cells of the retina, cerebellum, heart, liver, skin tumors and different cell lines [28–32]. Especially the formation of cell–cell and cell–matrix contacts in all three dimensions are essential for intercellular communication which in turn regulates fundamental cellular processes such as growth, proliferation, differentiation, migration and programmed cell. Another advantage of 3D spheroids is their well-defined size that allows e. g. the monitoring of tissue penetration of drugs and other biological active substances. However, coupling of 3D spheres on microelectrode arrays requires an improvement of the existing sensor design. One of the central problem for this application is the exact positioning of free floating spheroids in close proximity to the measuring and counter electrodes. To solve this topological problem two microelectrode arrays have been design and fabricated for tissue-based impedance screening. The first approach is realized by a parallel hydrodynamic positioning and impedimetric measuring of spheroids in micro-capillary based sensor chips (Fig. 11.7). The functionality of the biohybrid sensor has been tested by the transfection of tumor spheroids derived from a breast cancer cell line with an antisense butyrylcholinesterase (BChE) expression vector construct that is known to inhibit cellular proliferation and to induce apoptosis. Hydrodynamic positioning of treated and non treated spheroids in micro-capillary array with subsequent impedance recording showed a decrease of the extracellular resistance that was obviously induced by the suppression of proliferation and programmed cell death [26, 27].

However, since controlling of micro-fluidic systems for hydrodynamic positioning of spheres is laborious and carrying the risk of interfering with the impedance measuring setup, other strategies for a more simple positioning have been developed. The idea was to integrate micro-cavities on silicon, glass or plastic substrates (Fig. 11.8; unpublished by Robitzki et al.). By using these micro-structured cavities it was possible to fix spheres in cavities simply by gravity forces. Moreover, when the chip consists of cavities with different sizes (e. g. 100, 200, 300, 500 and 1000 μm) spheroids of various diameters can be recorded at once, which is not only of interest for tissue penetration studies of leads and drugs but also to correlate drug efficiency with the overall size of tissues. Implementation of one electrode at the

11.2 Real Time Monitoring and High Content Screening

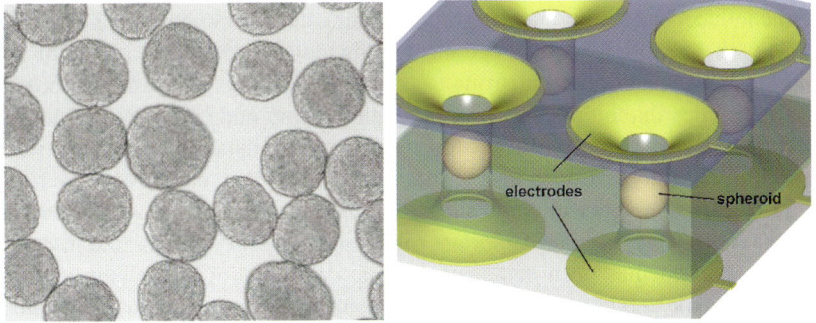

Fig. 11.7 Tumour spheroids can be easily produced by a scaffold free rotation-mediated reaggregation technique (*left*). To determine effects of anti-cancer drugs tumour spheroids are hydrodynamically positioned in micro-capillary and recorded via impedance spectroscopy (*right*; according to [27, 37])

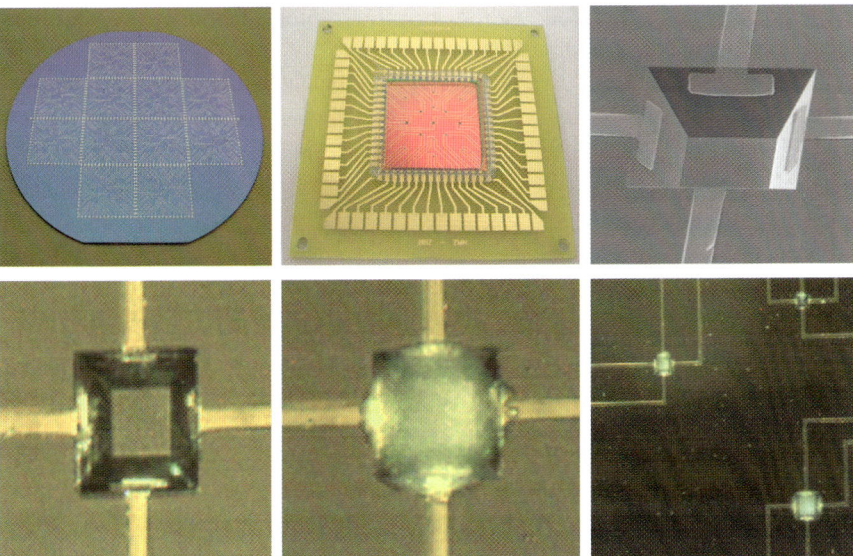

Fig. 11.8 Structure and application of tissue-based MCA® (Microelectrode Cavity Arrays) for impedance measurement. The biosensor consists of multiple cavities with different diameters to facilitate positioning of spheroids of different sizes. Electrodes at the edge of each side of the cavity allow impedimetric four point measurement (according to Robitzki et al., University of Leipzig in collaboration with Fischer et al., TU Ilmenau)

edge of each side of the cavity allows four point measurement that dramatically improve the signal-to-noise relation and thus, reduce the interference liability of impedance spectroscopy.

First tests of this microelectrode cavity array (MCA®) with spontaneous beating 3D heart cell spheroids showed promising results that may open new fields of application especially in terms of analyzing ischemic cardiomyopathy in vitro. Ischemic induced cardiomyopathies are the leading cause of congestive heart failure (CHF) with 15 million affected patients worldwide and a 5-year mortality of 50%. Congestive heart failure results from loss of cardiomyocytes and ventricle remodeling after myocardial infarction. Depending on the amplitude of infarction dramatic structural changes in the ventricle occurred. Ventricle dilatation, scar formation, fibrosis as well as hypertrophy of non-infarcted myocardium can be observed. Over the past years, a number of studies have shown that besides necrosis apoptosis of cardiomyocytes play also an important role during and after myocardial infarction. In this context, it has been demonstrated that cardiomyocytes undergo apoptosis during the acute and chronic stage of myocardial infarction. Here, a significant increase in the number of apoptotic cells at the borders of sub-acute and old infarcts has been observed. On the basis of these observations it has been suggested that the loss of cardiomyocytes via apoptosis may contribute to the progression of ischemia induced heart failure. However, so far, an effective 3D in vitro system for the non-invasive real time monitoring of complex intra- and intercellular processes occurring during ischemia induced cardiomyopathy is not available. To close this gap, tissue based MCA® in combination with the cellular component (cardiomyocyte spheroids) have been used to analyze ischemia induced apoptosis via impedance spectroscopy (Fig. 11.9).

Impedance recording of ischemic versus non-ischemic 3D heart cell spheroids was carried out on MCAs® over a period of 24 hours after induction of ischemia. Treated and non-treated spheroids showed an overall cellular resistance in a range of 1 to 1000 kHz with a maximum peak at 200 kHz (Fig. 11.9, top). Long term recording of an individual spheroid on MCAs® showed a predominant decrease of the relative impedance after application of ischemia inducing culture medium. The corresponding data from vital dye stainings (Fig. 11.9, middle and below; green: living cells; red: apoptotic and necrotic cells) correlates perfectly with the recorded decrease in cellular impedance. At this reason spheroid-base MCAs® are well-suited to analyze apoptotic processes during ischemia. Based on this 3D biohybrid MCAs representing emerging gold standard screening tools for non-invasive functional drug testing under real time conditions. Besides screening of electric non-active cells by impedance spectroscopy, MCAs® can also be utilized for extracellular recording of field potentials of electric active 3D cardiomyocyte spheroids. By using extracellular recording with MCAs® the contraction frequency of spontaneously contracting rat cardiomyocyte spheroids could be analyzed (Fig. 11.10). Taken together, both the MCA® based extracellular recording of contraction frequencies as well as the functional recording of non-electric active cells by impedance spectroscopy contributing to an improved drug safety as it has been claimed recently by the Food and Drug Administration in the United States (FDA) and the European Agency for the Evaluation of Medical Products (EMEA) in the European Union.

A further fascinating technology that might be highly profitable for a large number of different scientific disciplines is the laser-based manipulation of cells and tissues. However, probably based on the missing automation capacity, this powerful technique is not well-accepted in the pharmaceutical industry for high throughput functional screening. In this respect great efforts had been made by our group

11.2 Real Time Monitoring and High Content Screening

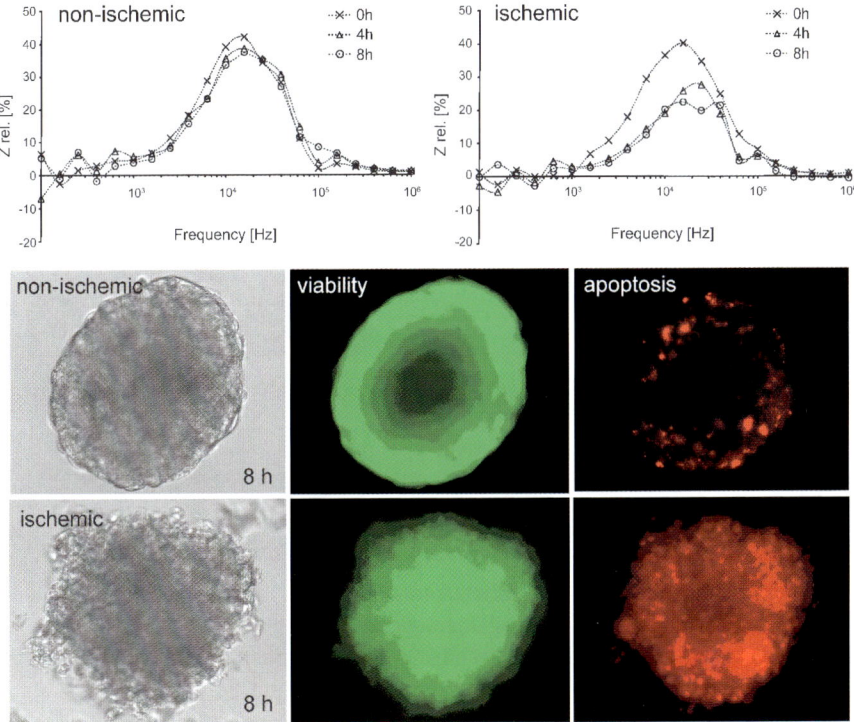

Fig. 11.9 3D tissue based MCAs® are used for detecting ischemia induced alterations in cardiomyocyte spheroids via impedance spectroscopy (*top*) and correlated with conventional analytical methods (cell viability assay, fluorescein diacetate, *green*; propidium iodide, *red*)

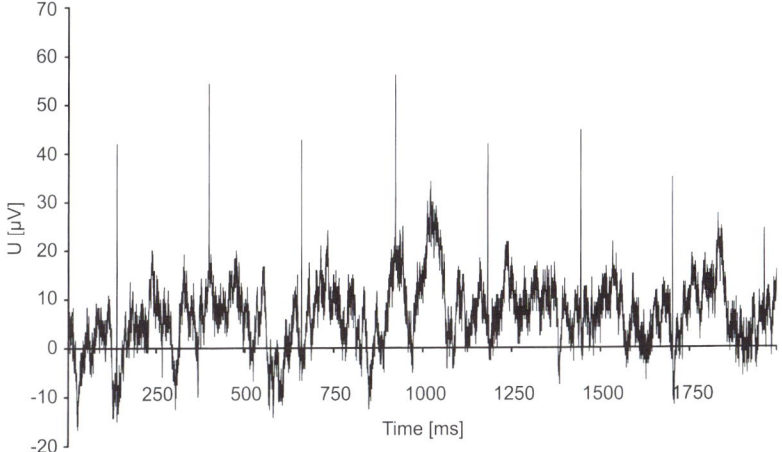

Fig. 11.10 Extracellular recording of 3D in vitro cardiomyocyte spheroids with MCAs®. The positive amplitudes represent the coordinated contraction of all cells within a cardiomyocyte spheroid (according to Robitzki et al.)

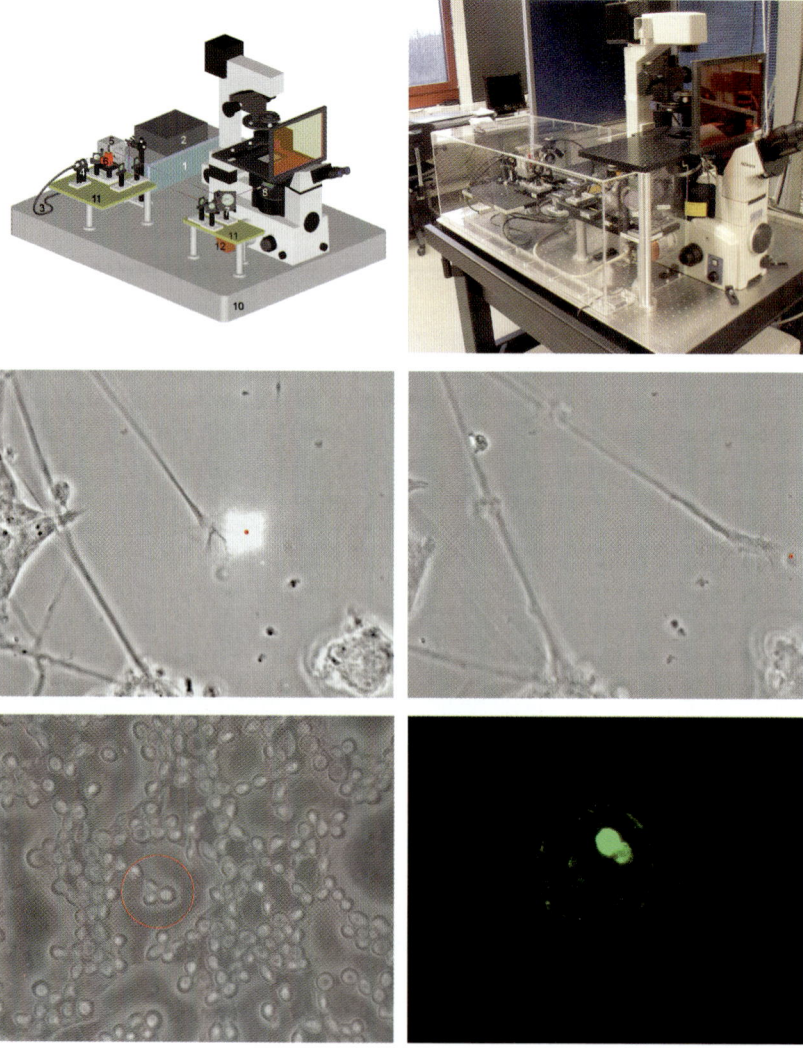

Fig. 11.11 Laser-based micro-dissection and manipulation platform with an implemented microelectrode array monitoring work station for high performance drug screening (*top*, scheme, *left*; *right*; according to Robitzki et al.). (*Top, left*) *1*: unpolarised pulsed ultraviolet nitrogen laser; *2*: linearly polarized continuous wave infrared ytterbium fibre laser; *3*: IR laser resonator; *4*: IR laser mount; *5*: galvanometer mirrows; *6*: acousto-optical deflector; *7*: beam combiner; *8*: coupling mirrow; *9*: safety glass; *10*: table; *11*: sideboards; *12*: CCD camera; *13*: adapter for microelectrode array implementation. (*Middle*) A retinal neuronal fibre can be guided by an IR (infrared) laser (*red spot*). (*Below*) An oligodedrocyte of a rat brain can be selected (*red circle*), dissected and catapulted by an UV (ultraviolet) laser onto a substrate (membrane or micro-array) and still survives this microdissection-catapulting procedure (*below, right*; viability assay using fluorescein diacetate, *green*)

to develop a new laser technology platform that enables the integration of planar microelectrode arrays or e. g. to MCA® and therefore can be used for functional high content and high throughput screening of cellular parameters before and/or after laser-based manipulation [33, 34]. The laser platform consists of two laser each working with a different wavelength (Fig. 11.11). A pulsed UV-A laser (337 nm) is used for dissecting cells and selecting them by laser based pressure catapulting. The second IR laser (1064 nm) is used as optical tweezers to direct and promote growth of neurons and thereby generate artificial neuronal networks on microelectrode arrays. By this means extracellular recording of neuronal activity can be studied in detail by modifying the design of neuronal networks.

Furthermore, the laser platform can be also applied for testing the regeneration capacity of drugs on neurons of the central nervous system and peripheral nervous system. For instance, the targeted cutting of neuronal processes of dorsal root ganglions by laser dissection and the regeneration behavior in the presence of drugs can be monitored optically or by electronic microelectrode recording.

11.3 Outlook and Future Aspects

In this report some aspects for innovative biolectronic cell and tissue based drug screening, diagnosis, and therapeutical control have been introduced and their applicability was demonstrated. Low and high frequency impedance spectroscopy and electrophysiological recording using viable cells and/or tissues for a real time monitoring of toxicological (side) effects, compatibility studies and of metabolic effects are the most innovative measurement methods in high content and high throughput screening of drugs or stem cell quality control [35–39]. Therefore, future technological screening platforms and improvements should be provided for the following applications: (i) technologies for predictive tests of incompatibility (e. g. liver toxicology, the initiation of irregular heart beat (arrhythmia), vascular alterations), (ii) following appendages of proteomics, post-genomics, pharmaco-genomics, pharmacokinetics, and toxico-genomics, (iii) cell and tissue based microsystems, (iv) combination with in silico models for theoretical predictions, and (v) bioelectronical cell-based measurement methods for predictions of autoimmune reactions and foreign antigens.

References

1. Service RF (2004) Surviving the blockbuster syndrome. Science 303:1796–1799
2. Challenge and Opportunity on the critical path to new medical products. FDA 2004
3. EJ – 21–06-2006 BIOPRO Baden Württemberg GmbH, Germany
4. Drug Development Science – Obstacles and opportunities for collaboration academic, industry and government. In: Korn D, Stanski DR (eds) AAMC & FDA, January 13–14, 2005 Washington DC (USA)
5. Thomas CA Jr, Springer PA, Loeb GE, Berwald-Netter Y, Okun LM (1972) A miniature microelectrode array to monitor the bioelectric activity of cultured cells. Exp Cell Res 74(1):61–66
6. Gross GW, Rieske E, Kreutzberg GW, Meyer A (1977) A new fixed-array multi-microelectrode system designed for long-term monitoring of extracellular single unit neuronal activity in vitro. Neurosci Letters 6(2–3):101–105
7. Fromherz P, Kiessling V, Kottig K, Zeck G (1999) Membrane transistor with giant lipid vesicle touching a solicon chip. Appl Phys A 69:571–576
8. Offenhäusser A, Knoll W (2001) Cell-transistor hybrid systems and their potential applications. Trends Biotechnol 19(2):62–66
9. Ingebrandt S, Yeung CK, Staab W, Zetterer T, Offenhäusser A (2003) Backside contacted field effect transistor array for extracellular signal recording. Biosens Bioelectron 18(4):429–435
10. Meyer T, Leisgen C, Gonser B, Gunther E (2004a) QT-screen: high-throughput cardiac safety pharmacology by extracellular electrophysiology on primary cardiac myocytes. Assay Drug Dev Technol 2(5):507–514
11. Halbach M, Egert U, Hescheler J, Banach K (2003) Estimation of action potential changes from field potential recordings in multicellular mouse cardiac myocyte cultures. Cell Physiol Biochem 13(5):271–284
12. Stett A, Egert U, Guenther E, Hofmann F, Meyer T, Nisch W, Haemmerle H (2003) Biological application of microelectrode arrays in drug discovery and basic research. Anal Biomol Chem 377(3):486–495
13. Reppel M, Boettinger C, Hescheler J (2004) Beta-adrenergic and muscarinic modulation of human embryonic stem cell-derived cardiomyocytes. Cell Physiol Biochem 14(4–6):187–196
14. Sproessler C, Denyer M, Britland S, Knoll W, Offenhausser A (1999) Electrical recordings from rat cardiac muscle cells using field-effect transistors. Phys Rev E Stat Phys Plasmas Fluids Relat Interdiscip Topics 60(2 PtB):2171–2176
15. Meyer T, Boven KH, Gunther E, Fejtl M (2004) Micro-electrode arrays in cardiac safety pharmacology: a novel tool to study QT interval prolongation. Drug Saf 27(11):763–772
16. Kurz R, Rothermel A, Rüffer M, Urban C, Jahnke H-G, Weigel W, Robitzki A (2004) A functional cardiomyocyte based biosensor for prediagnostic monitoring: an angiotensin

II angiotensin II study; IFMBE, Medical & Biological Engineering & Computing IFMBE 6, ISBN 88–7780-308–8, ISSN: 1727–1983
17. Rothermel A, Kurz R, Rüffer M, Weigel W, Jahnke HG, Sedello A, Stepan H, Faber R, Schulze-Forster K, Robitzki A (2005) Cells on a Chip – the Use of Electric Properties for Highly Sensitive Monitoring of Blood-Derived Factors Involved in Angiotensin II Type 1 Receptor Signalling. Cell Physiol Biochem 16(1–3):51–58
18. Guo L, Guthrie H (2005) Automated electrophysiology in the preclinical evaluation of drugs for potential QT prolongation. J Pharmacol Toxicol Methods 52(1):123–135
19. Thielecke H, Mack A, Robitzki A (2001) Biohybrid microarrays – impedimetric biosensors with 3D in vitro tissues for toxicological and biomedical screening. Fresenius J Anal Chem 369(1):23–29
20. Gimsa J, Wachner D (1998) A unified resistor-capacitor model for impedance, dielectrophoresis, electrorotation, and induced transmembrane potential. Biophys J 75(2):1107–1116
21. Foster KR, Schwan HP (1989) Dielectric properties of tissues and biological materials: a critical review. Crit Rev Biomed Eng 17(1):25–104
22. Schwan HP (1957) Electrical properties of tissue and cell suspensions. Adv Biol Med Phys 5:147–209
23. Ciambrone GJ, Liu VF, Lin DC, McGuinness RP, Leung GK, Pitchford S (2004) Cellular dielectric spectroscopy: a powerful new approach to label-free cellular analysis. J Biomol Screen 9(6):467–480
24. Rothermel A, Nieber M, Müller J, Wolf P, Schmidt M, Robitzki A (2006) Real time measurement of PMA-induced cellular alterations by microelectrode array-based impedance spectroscopy (MAIS). Biotechniques 41(4):445–450
25. Thielecke H, Stieglitz T, Beutel H, Matthies T, Ruf HH, Meyer JU (1999) Fast and precise positioning of single cells on planar electrode substrates. IEEE Eng Med Biol Mag 18:48–52
26. Reininger-Mack A, Thielecke H, Robitzki AA (2002) 3D-biohybrid systems: applications in drug screening. Trends Biotechnol 20(2):56–61
27. Thielecke H, Mack A, Robitzki A (2001a) A multicellular spheroid-based sensor for anticancer therapeutics. Biosens Bioelectron 16(4–5):261–9
28. Layer PG, Robitzki A, Rothermel A, Willbold E (2002) Of layers and spheres: the reaggregate approach in tissue engineering. Trends Neurosci 25(3):131–134
29. Takezawa T, Inoue M, Aoki S, Sekiguchi M, Wada K, Anazawa H, Hanai N (2000) Concept for organ engineering: a reconstruction method of rat liver for in vitro culture. Tissue Eng 6:641–650
30. Tziampazis E, Sambanis A (1995) Tissue engineering of a bioartificial pancreas: modelling the cell environment and device function. Biotechnol Prog 1:115–126
31. Shea LD, Wang D, Franceschi RT, Mooney DJ (2000) Engineered bone development from a pre-osteoblast cell line on three-dimensional scaffolds. Tissue Eng 6:605–619
32. Steinhoff G, Stock U, Karim N, Mertsching H, Timke A, Meliss RR, Pethig K, Haverich A, Bader A (2000) Tissue engineering of pulmonary heart valves on allogenic acellular matrix conduits: in vivo restoration of valve tissue. Circulation 102(19 Suppl 3):50–55
33. Stuhrmann B, Jahnke HG, Schmidt M, Jähn K, Betz T, Müller K, Rothermel A, Käs J, Robitzki AA (2006) Versatile optical manipulation system for inspection, laser processing, and isolation of individual living cells. Rev Sci Instrum 77:063116
34. Stuhrmann B, Jahnke HG, Schmidt M, Jähn K, Betz T, Müller K, Rothermel A, Käs J, Robitzki AA (2006) Versatile optical manipulation system for inspection, laser processing, and isolation of individual living cells. Virtual Journal of Biological Physics Research, July 1
35. Heuschkel MO, Fejtl M, Raggenbass M, Bertrand D, Renaud P (2002) A three-dimensional multi-electrode array for multi-site stimulation and recording in acute brain slices. J Neurosci Methods 114(2):135–148
36. Hutzler M, Fromherz P (2004) Silicon chip with capacitors and transistors for interfacing organotypic brain slice of rat hippocampus. Eur J Neurosci 19(8):2231–2238

References

37. Bartholomä P, Impidjati, Reininger-Mack A, Zhang Z, Thielecke H, Robitzki AA (2005) More Aggressive Breast Cancer Spheroid Model Coupled to an Electronic Capillary Sensor System for a High-Content Screening of Cytotoxic Agents in Cancer Therapy: 3-Dimensional In Vitro Tumor Spheroids as a Screening Model. J Biomol Screen 10(7):705–14
38. Arndt S, Seebach J, Psathaki K, Galla HJ, Wegener J (2004) Bioelectrical impedance assay to monitor changes in cell shape during apoptosis. Biosens. Bioelectron 19(6):583–594
39. Wegener J, Abrams D, Willenbrink W, Galla HJ, Janshoff A (2004) Automated multi-well device to measure transepithelial electrical resistances under physiological conditions. Biotechniques 37(4):590, 592–594, 596–597

Part IV
Mechanics of Soft Tissues, Fluids and Molecules

Chapter 12
Soft Materials in Technology and Biology – Characteristics, Properties, and Parameter Identification

M. Staat, G. Baroud*, M. Topcu, S. Sponagel

Institute of Bioengineering and Biomechanics Laboratory,
Aachen University of Applied Sciences, Campus Jülich, Ginsterweg 1,
52428 Jülich, Germany,
m.staat@fh-aachen.de
* Laboratoire de biomécanique de l'Université de Sherbrooke, Canada

Abstract The growing interest in flexible structures has also brought biomechanics into the focus of engineers. Elastomers and soft tissues consist of similar networks of macromolecules. After a brief introduction to the concepts of continuum mechanics, typical isotropic models of soft materials in technology and biology are presented. Similarities and differences of the thermo-mechanical behavior are discussed. For rubber-like materials a modification of the Kilian network is suggested which greatly simplifies the identification of material parameters. Finally the dynamical loading of biopolymers and volume changes with phase transitions are considered.

12.1 Introduction

Rigid materials have shaped and characterized the technical world to a strong extent. Bridges, skyscrapers or the Eiffel tower are not allowed to deform noticeably if forces are acting. Therewith a conception of engineering and technical mechanics has arisen that turns the attention to rigid constructions. Engineers think of technical mechanics as soon as they hear the word mechanics. Technical mechanics is the mechanics of the civil engineer as it has been created by August Föppel and the mathematician Felix Klein.

The necessity to look at rigid constructions also led to materials and constructions that conform to these requirements. Further findings were always connected to this in mechanics (statics). The world of engineers is the world of rigid constructions made of metal, concrete or masonry. Only the industrial mobility, i. e. automotive engineering allowed reconsidering these classical views. One recognized that soft materials and constructions are necessary in order to reduce impact forces and to design them bearable. Therewith, bearing and vibration engineering as well as safety constructions have originated. In the case of safety constructions, one thinks of crumple zones of cars, the airbag and the safety belt. Man himself as a transported good also became the focus of attention of engineers because it was necessary to adapt the vibrations of cars to passengers and to get to know the critical limits at which they suffer damages.

In addition to the automotive industry, areas like the military and aerospace industries acted as driving forces for the reconsideration which has just been described. In this dynamic world, areas like biology, medicine and technology get very close and interact with each other. Until recently biomechanics in the static world has not attracted much attention, but nowadays has been awakened by the demands of the technology, the variety of biological materials and structures. These influential and far reaching changes can be followed through the work of Yuan-Cheng Fung, who after a brilliant 20 years career in aeroelasticity and aeronautical engineering [14] changed completely to the field of biomedical engineering around 1965 and became the father of modern biomechanics [17]. He made important original contributions to virtually all questions of biomechanics on the levels of organs, tissues, and cells. His book [16] became a classic, enduring reference.

The similarities between soft biological tissues and rubberlike materials have already been observed as early as 1880 in the context of the mechanic of the arterial wall [2]. Both soft tissues and rubberlike materials consist of macromolecular networks with same bonds. They show entropic elasticity in addition to the energetic elasticity of metals [20]. Therefore modern continuum theories for the thermomechanical behavior of soft materials borough ideas from the physics of macromolecules. The modern material theory, as it has been created by analytical mechanists and mathematicians, has been very helpful for the development of constitutive modeling and it exists nowadays in an improved way. To the engineer the material theory is what the structure of matter is for the physicist [24, 49, 50].

This article intents to deal sketchily with this development. Starting from observed phenomena a material theory is developed which takes into account the dynamic behavior of the materials in a satisfactory way. Furthermore, in the case of statics, the behavior of soft, technical and biological substances will be described and the development of a measuring technique which permits to check certain assumptions about the material behavior will be shown. In the end, chosen examples will be discussed in order to confirm the usability of the developed ideas.

To the non-expert in continuum mechanics the references [33, 36, 44, 45, 49] are recommended. The book [44] concentrates on discussing the solutions of over 100 exercises. Results that cannot be found in other textbooks are particularly emphasized and relevant literature is quoted. Some advanced book on biomechanics also give a good introduction to continuum mechanics [10, 16, 25].

12.2 Material Description

12.2.1 Why Material Description

Mechanics and thermodynamics deliver universally valid balance equations according to which all events in the laboratory, in technology, and in the surrounding nature can be described. The mentioned equations are the

- mass balance,
- momentum balance,
- energy balance, and
- entropy inequality.

The balance equations are almost free of prerequisites about material properties. They are valid for solids that undergo finite deformations under applied stress and for liquids or gases (fluids) which flow unlimitedly under shear stress. Since different materials react differently to the same external field of force, the balance equations being conditional equations for the fields of interest cannot be sufficient. The missing equations are the mentioned material laws (constitutive equations). If confining oneself to a purely mechanical point of view and if neglecting for the moment thermodynamics, these equations establish a connection between the stresses acting in the continuum and the movements of the continuum. These equations establish a connection between the terms *force* and *motion*. One immediately sees that a complete system of equations arises with the momentum balance which also connects the notions motion and force with each other.

Materials according to continuum mechanics are mathematical models that describe approximately the mechanical behavior of substances occurring in nature or used in technology under defined external conditions. The terms "material" and "substance" differ in a way that the term material represents a mathematical idealization whereas the term substance describes the physical reality. Let us, for example, imagine the material steel which behaves under small deformations linearly elastic and thus can be described by Hooke's law. Hooke's law forms the basis for all ranges of engineering science in which the components may not suffer any considerable deformation. The behavior of steel differs if the load or deformation becomes large. In this range steel flows plastically and the appropriate material law is of different nature. The appropriate material law forms the basis of plastic metal working. With this illustration, one can see that material descriptions have to be chosen not only *material dependent*, but also *problem dependent*. The most general material law that may be able to describe a material in all areas of application and in all stress ranges does not exist.

Handling non-classic materials, like synthetic materials, elastomers, and biological substances that receive more and more attention in technology, medicine, and biology underlines this problem even more. Examples for this are process engineering and plastics technology and additionally areas in medicine, biology, and cell biophysics. In all areas the interaction between structural characteristics and material behavior has to be understood, evaluated, and collected quantitatively. However,

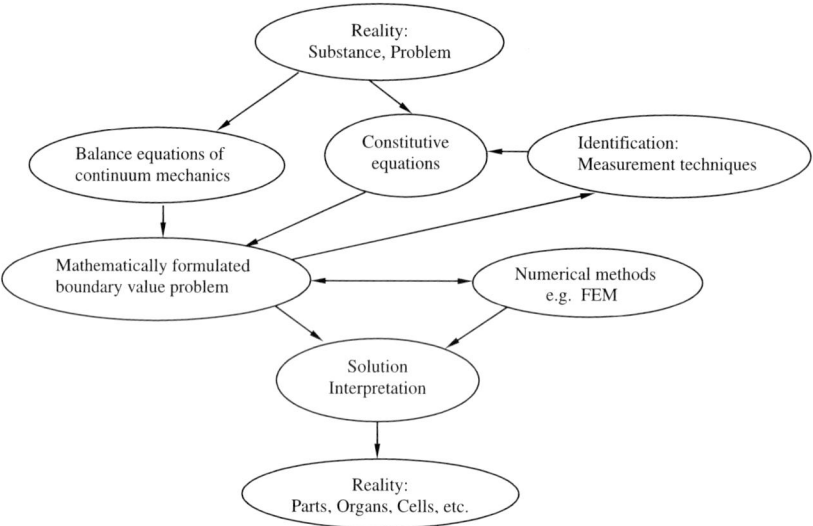

Fig. 12.1 Selection scheme for material laws

especially the quantitative analysis requires a material law. Based on his problem, the involved scientist has to choose the appropriate material law and the adapted measuring technology to an increasing degree. Every problem and every group of problems requires a new consideration of the expected motion, of the material properties important to the type of motion and of the required measuring technology in order to grasp the material properties. The scheme in Fig. 12.1 clearly describes these problems once again.

What remains is the question about the discipline that deals with the provision of material laws. The branch spoken about is rheology which is unfortunately often mixed up with the mere study of the flow of matter.

12.2.2 What is Rheology and What is the Task of This Discipline?

Nowadays rheology can be considered an independent, scientific discipline. Within the last decades it has found a demarcation to other disciplines. Rheology is the science dealing with the description, explanation and measurement of phenomena occurring during the *deformation* and the *flow* of matter.

In the given definition is clearly mentioned that rheology *covers more* than just the flow of matter. Included are phenomena of solids, of liquids, and of hybrid substances that can be assigned to both areas. With the given definition, rheology can be considered a part of physics and at the same time its close linkage with chemistry and engineering sciences is mentioned.

Among the materials examined until this day there are two classes whose rheological treatment is full of specific difficulties. This resulted in the creation of sep-

12.2 Material Description

arate branches. For instance, the deformation-processes in the earth's crust are allotted to "geo-rheology" while the flow-processes in the organisms are objects of "bio-rheology".

The close relation of rheology to mechanics and thermodynamics has to be pointed out. This applies especially to the phenomenological rheology that is often practiced either in the framework of *rational mechanics* as a nonlinear continuum theory or in the framework of *thermodynamics of irreversible processes* as a linear continuum theory or as a theory of linear viscoelastic bodies.

In contrast, structural-rheology may be related closely to statistical mechanics and thermodynamics (theory of transport processes). However, there is an even stronger connection of the micro-rheology of complicated materials (e. g. macro molecular solutions) to physical chemistry and to the investigation of particular classes of substances (e. g. research on colloids, polymers, elastomers, wood, concrete, bricks, glass, metal) done in its sphere of influence.

Rheometry primarily makes use of the measuring methods of mechanics and in addition to that of calorimetric, electrical, and visual methods. According to its objective, it is closely connected to materials testing, to the measurement technology of chemical engineering, and to process engineering. The scheme in Fig. 12.2 summarizes the relation of rheology to other disciplines:

Because of its great economic importance, engineering science deserves special consideration. Applied rheology is here sufficient as a constructive discipline extends over process engineering, polymer sciences, machine design, and structural design. In the area of machine design, this will be clear in the case of the design of components consisting of polymeric or elastomeric materials. Especially elastomeric materials show specific features during processing and during their utilization as machine parts that earn special consideration. Due to this and because of

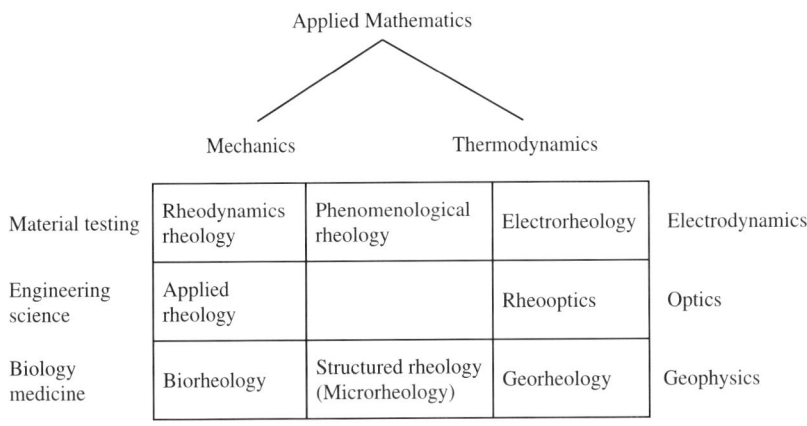

Fig. 12.2 Relation of rheology to other sciences

the economic importance of this material it is appropriated to emphasize elastomer-rheology alongside with geo-rheology and bio-rheology.

This work will deal with substances that are described in technology as elastomers (rubber) and in biology as biopolymers. In order to reduce the linguistic confusion that is obviously present, the very basic differences of these material classes will be described.

12.2.3 Macromolecular Substances

Macro molecular substances have, unlike crystalline ones, no individual molecules but linear polymers as basic structures. The best known technical polymer is rubber and therefore a natural product, too. Only the cross-linking reaction which is carried out over sulfur bridges is artificial. The molecular formula of rubber is C_5H_8. The monomer is isoprene (2-methyl-1,3-butadiene)

$$CH_2 = \underset{\underset{CH_3}{|}}{C} - CH = CH_2$$

and via the loss of energy building blocks are connected to each other forming the macromolecules

$$\left(\underset{\underset{H}{|}}{\overset{\overset{H}{|}}{-C}} - \underset{}{\overset{\overset{CH_3}{|}}{C}} = \underset{\underset{H}{|}}{\overset{}{C}} - \underset{\underset{H}{|}}{\overset{\overset{H}{|}}{C}} - \right)_n$$

The methyl group CH_3 is hydrophobic (nonpolar) and the macromolecule is in a strongly knotted and unordered state. The situation is different for biopolymers. The building blocks are the amino acids

$$H_2N - \underset{\underset{R_k}{|}}{\overset{\overset{HO}{\overset{|}{}}}{C}} - \overset{\overset{O}{\|}}{C} - OH$$

12.2 Material Description

and by losing water they form the macromolecules

$$\begin{pmatrix} \overset{O}{\underset{\|}{}} & \overset{H}{\underset{|}{}} & \overset{H}{\underset{|}{}} & \\ -C & -N & -C & - \\ & & | & \\ & & R_k & \end{pmatrix}_n$$

called proteins. Here R_k ($k \in \{1, \ldots, n\}$) are side chains and depending on the type of these side chains different biopolymers with several different characteristics may form. These side chains can be hydrophilic ("water-loving") or hydrophobic ("water-fearing"). The type of these groups also determines the spatial arrangement of these macromolecules. They can be helical or laminar. In contrast to rubber substantial differences arise in the structure. Macromolecules of cross-linked rubber are strongly knotted and without structure, whereas structures develop in biopolymers. Figure 12.3 schematically shows such structures.

The difference between technical and biological polymers is very distinctive. Technical polymers only have one single and fixed side chain, whereas the side chains of biopolymers differ in order, species and quantity. This leads to a macro molecular cord which resembles a Morse signal – it has information.

The type and arrangement of side chains not only determines the species of the emerging biopolymers, their behavior towards water, and their inner structure but also their mechanical properties. While cross-linked rubber is available as a pure substance and anhydrously, a mixture saturated by water exists for substances consisting of biopolymers. The stress-strain relation for cross-linked rubber is typically degressive while it shows a progressive behavior for biopolymers. In literature this is called S-shaped and J-shaped, respectively (Fig. 12.24).

a) Rubber   b) Biopolymers

Fig. 12.3 Cross-linked rubber and biopolymers

In this contribution, except in the last section, both will be regarded as pure substances, yet their different mechanical behavior will be considered. Both substances have one characteristic in common, viscoelasticity. They show considerable rate dependence during strain tests.

12.2.4 Phenomenological Behavior of Soft Substances

Soft substances clearly show rate dependence during tensile tests. The stress-strain relation is not clear but depends on the test conditions, for instance the strain rate. Figure 12.4 shows this rate dependence.

Curve I shows a real test which takes place with finite strain rate, and curve II shows the limiting case if the test is carried out quasistatically (infinitely slow). If one continues the real test (curve I) up to the point A and then keeps the strain constant, the stress will diminish until point B is reached. This phenomenon is called "stress relaxation". If one keeps the stress constant in point A, the strain will grow further up to point C. This phenomenon is called "strain retardation" or creep. Rubbery materials further show strong internal material damping. In a tensile test where the strain varies periodically between two fixed limits, the stress-strain relation will be given as a closed curve in the (σ, ε) diagram. In Fig. 12.5 such hystereses are outlined.

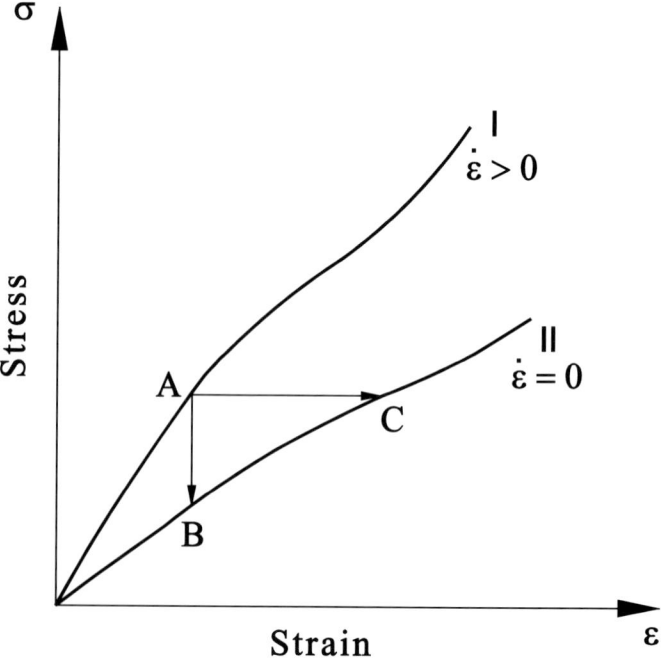

Fig. 12.4 Stress-strain curves depending on strain rate

12.2 Material Description

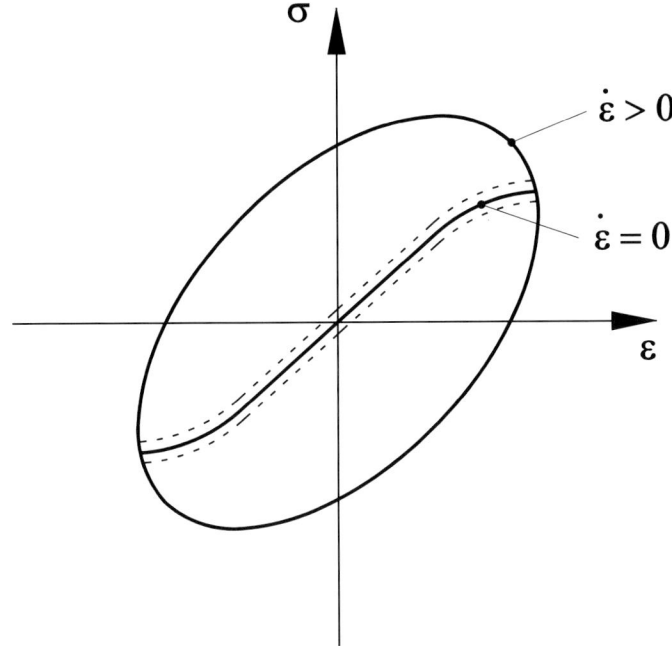

Fig. 12.5 Stress-strain hystereses during periodic strain

The size of the area is a measure for the amount of the damping generally depending strongly on the test speed. Two limit cases are possible in the limit case of a quasistatic test:

1. The substance behaves purely viscoelastic, i. e. the width of the hysteresis disappears with vanishing strain rate.
2. The substance shows irreversible effects partly behaving plastically, i. e. the width of the hysteresis and therewith the inner damping does not disappear despite vanishing test speed.

Since these irreversible (plastic) processes can be partly explained by reorientations in the molecule construction and by strong internal friction processes of the fillers respectively they are not comparable to the plastic yielding well known from metals. They correspond rather to (dry) Coulomb friction events and can be neglected in most cases. In the following, it suffices to restrict oneself to purely viscoelastic material behavior. Plastic properties are excluded from the consideration. Considering Fig. 12.3 once again, it will become clear that these phenomena of stress relaxation, creep, and rate-dependent damping can all be ascribed to entangled macromolecules i. e. the liquid properties of the substance. One can imagine the substance as a *viscoelastic liquid* in which a *weak network* has developed. This image also makes clear why these substances can be deformed considerably with relatively small forces and why they are almost incompressible.

12.2.5 Simple Model Formulation of Soft Materials

The simplest rheological model which can adequately explain viscoelastic properties is the Poynting–Thomson material as it is outlined in Fig. 12.6.

Strain $\varepsilon = \lambda - 1$ and stretch $\lambda = \varepsilon + 1$ are equivalent strain measures. The quantities E_1 and E_2 denote the modulus of elasticity and η is the coefficient of viscosity. Examining this spring-dashpot model, one obtains the following relation between the stress σ and strain ε

$$\dot{\sigma} + \sigma \frac{1}{T} = (E_1 + E_2)\left(\dot{\varepsilon} + \varepsilon \frac{1}{T^*}\right) \tag{12.1}$$

with the abbreviations

$$T = \frac{\eta}{E_1}; \quad T^* = T\left(1 + \frac{E_1}{E_2}\right). \tag{12.2}$$

T is called the relaxation time, and T^* is called retardation time. With the abbreviation $\sigma = (E_1 + E_2)\bar{\sigma}$ one can rewrite Eq. (12.1) in order to obtain a fully symmetrical equation

$$\dot{\bar{\sigma}} + \bar{\sigma}\frac{1}{T} = \dot{\varepsilon} + \varepsilon\frac{1}{T^*}. \tag{12.3}$$

For very slow processes one may neglect the time derivatives in Eq. (12.3) and one obtains Hooke's law with $\bar{\sigma} = T/T^*\varepsilon$ and $\sigma = E_2\varepsilon$ respectively. For slow motions, only the upper spring is effective, the lower spring is constantly relaxed. For very fast processes one may neglect time-independent terms and one obtains Hooke's law again with $\dot{\bar{\sigma}} = \dot{\varepsilon}$ and $\dot{\sigma} = (E_1 + E_2)\dot{\varepsilon}$ respectively. In this limit case the material again behaves purely elastic but it is stiffer since both springs are in action and the dashpot is not moving. If the strain $\varepsilon(t)$ is prescribed, Eqs. (12.1) and (12.3) are

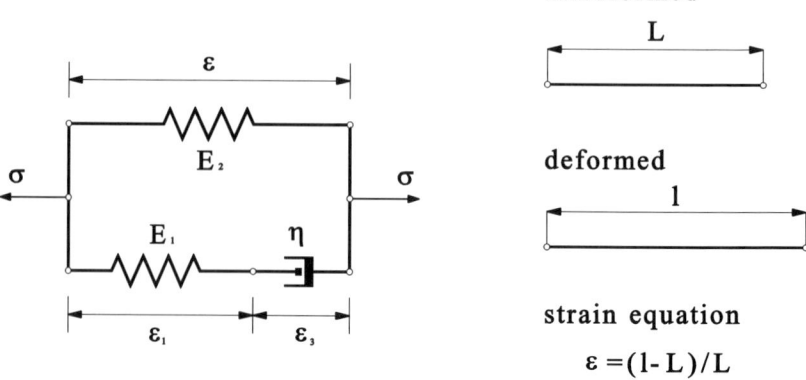

Fig. 12.6 Mechanical model for polymers

12.2 Material Description

linear differential equations for stress. These equations can generally be integrated and an easy modification leads to the following equations

$$\sigma(t) = E_2 \varepsilon(t) - \frac{E_1}{T} \int_0^\infty e^{\frac{-s}{T}} (\varepsilon(t-s) - \varepsilon(t))\, ds, \tag{12.4}$$

$$\bar{\sigma}(t) = \frac{T}{T^*} \varepsilon(t) + \left(\frac{T}{T^*} - 1\right) \frac{1}{T} \int_0^\infty e^{\frac{-s}{T}} (\varepsilon(t-s) - \varepsilon(t))\, ds \tag{12.5}$$

that are increasingly utilized in modern rheology and continuum mechanics literature. In Eqs. (12.4) and (12.5) s describes a time coordinate which leads from the present time (now $\stackrel{\triangle}{=} t$) back into the past. Figure 12.7 represents this time coordinate.

The material laws Eqs. (12.4) and (12.5) respectively have replaced Hooke's law. This shows that the instantaneous stress not only depends on the instantaneous strain values but also on the whole strain history. The instantaneous stress is a linear functional of the strain history $(\varepsilon(t-s) - \varepsilon(t))$, $0 \le s \le \infty$. The dependence of the stress on the instantaneous value of strain is expressed in the integral free term and the dependence on former values of strain in the integral. If one considers the symmetry of Eq. (12.3), one can immediately invert Eq. (12.5). Since Eq. (12.3) transforms into itself due to the symmetry-operation

$$s \to \varepsilon, \tag{12.6}$$

$$T \to T^*, \tag{12.7}$$

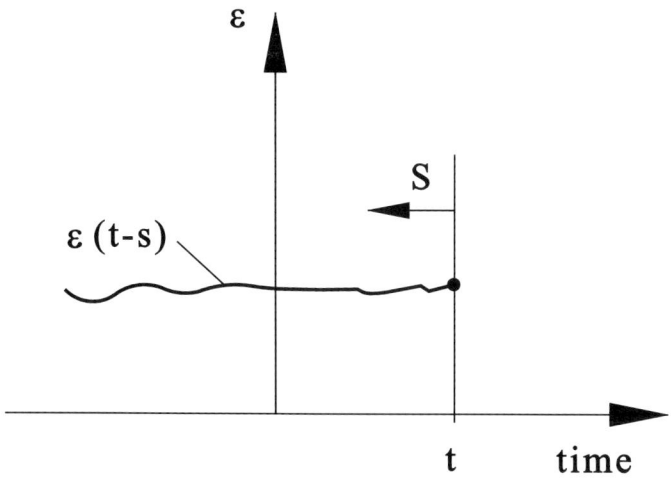

Fig. 12.7 Strain history as a function of time

the inversion follows from Eq. (12.5) under the same symmetry-operation

$$\varepsilon(t) = \frac{T^*}{T}\bar{\sigma}(t) + \left(\frac{T^*}{T} - 1\right)\frac{1}{T^*}\int_0^\infty e^{\frac{-s}{T^*}}(\bar{\sigma}(t-s) - \bar{\sigma}(t))\,ds\,. \tag{12.8}$$

This is a very nice example for making clear the importance of symmetries. If one further considers the meaning of the normalized stress $\bar{\sigma}$, relaxation time T and retardation time T^* respectively, one obtains

$$\varepsilon(t) = \frac{1}{E_2}\sigma(t) + \frac{E_1}{E_2(T^*(E_1 + E_2))}\int_0^\infty e^{\frac{-s}{T^*}}(\sigma(t-s) - \sigma(t))\,ds\,. \tag{12.9}$$

In order to study the behavior of the Poynting–Thomson model basic experiments to which the material is subjected will be discussed. Either the strain-history is given and one asks for the related stress response or vice versa.

12.2.6 Basic Tests

12.2.6.1 Relaxation Test

In this test, the strain history is prescribed in form of a step function, as can be seen in Fig. 12.8.

$$\varepsilon(t-s) = \begin{cases} \varepsilon_0 & 0 \le s \le t \\ 0 & t < s \end{cases}. \tag{12.10}$$

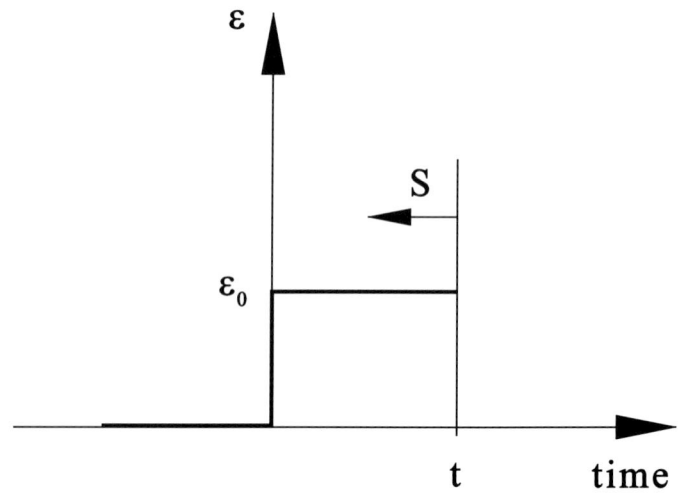

Fig. 12.8 Strain jump, relaxation test

12.2 Material Description

From Eqs. (12.4) or (12.5) one calculates the so-called relaxation curve

$$\sigma(t) = \begin{cases} 0 & t < 0 \\ \left(E_2 + E_1 e^{\frac{-s}{T}}\right)\varepsilon_0 & t \geq 0 \end{cases} \quad (12.11)$$

$$\bar{\sigma}(t) = \begin{cases} 0 & t < 0 \\ \left(\frac{T}{T^*} + \left(1 - \frac{T}{T^*}\right) e^{\frac{-t}{T}}\right)\varepsilon_0 & t \geq 0 \end{cases} \quad (12.12)$$

It is an expression for the linearity of the Poynting–Thomson material that the height of the jump ε_0 is only given as a factor. The quotient of the relaxation curve $\sigma(t)$ and ε_0 is described as *relaxation function*

$$E(t) = E_2 + E_1 e^{\frac{-s}{T}} \quad \text{for} \quad t \geq 0. \quad (12.13)$$

From Eq. (12.11) one can see that the stresses decrease exponentially from the initial value $(E_1 + E_2)\varepsilon_0$ to their asymptotic value $E_2\varepsilon_0$, where the relaxation time T is defined as the decay time parameter (Fig. 12.9). This phenomenon is generally denoted *stress relaxation*.

The examination of the relaxation function $E(t)$ gives reason to generalize the Poynting–Thomson material. If one replaces in Eq. (12.13) t with s, Eq. (12.4) can be written in the following form

$$\sigma(t) = E_\infty \varepsilon(t) + \int_0^\infty E'(s)\left(\varepsilon(t-s) - \varepsilon(t)\right) ds. \quad (12.14)$$

E_∞ is the elasticity limit of the relaxation function for $s \to \infty$, in the case of the Poynting–Thomson material this is E_2, and $E'(s)$ is the time derivative of the relaxation function, in the case of the Poynting–Thomson material this is $-(E_1/T)e^{-s/T}$. $E'(s)$ is denoted the relaxation kernel and E_∞ the equilibrium stiffness. Equation (12.14) is one-dimensional and the most general material law for linear viscoelastic materials, if the relaxation function is left arbitrary. With quite

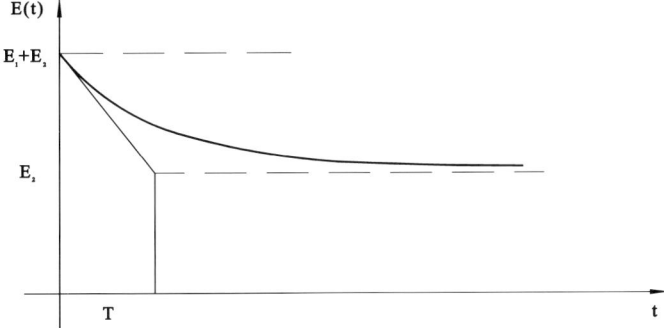

Fig. 12.9 Relaxation function $E(t)$ of the Poynting–Thomson material

plausible assumptions it can be shown that the relaxation function is a Prony series, e. g. it is given by the following equation

$$E(t) = \sum_{k=0}^{N} E_k e^{\frac{-t}{T_k}} \quad \text{for } t \geq 0. \tag{12.15}$$

N may become very large but it is still finite [8, 34, 35].

12.2.6.2 Creep Test

In this test, the stress history is prescribed in form of a step function, as can be seen in Fig. 12.10.

$$\sigma(t-s) = \begin{cases} \sigma_0 & 0 \leq s \leq t \\ 0 & t \leq s \end{cases}. \tag{12.16}$$

Let $\bar{\sigma}_0 = \sigma_0/(E_1+E_2)$. From Eqs. (12.9) or (12.8) one calculates the so-called creep curve

$$\varepsilon(t) = \begin{cases} 0 & t < 0 \\ \left(\frac{1}{E_2} - \frac{E_1}{E_2(E_1+E_2)}\right) e^{\frac{-t}{T^*}} \sigma_0 & t \geq 0 \end{cases} \tag{12.17}$$

$$\varepsilon(t) = \begin{cases} 0 & t < 0 \\ \left(\frac{T^*}{T} + \left(1 - \frac{T^*}{T}\right) e^{\frac{-t}{T^*}}\right) \bar{\sigma}_0 & t \geq 0 \end{cases}. \tag{12.18}$$

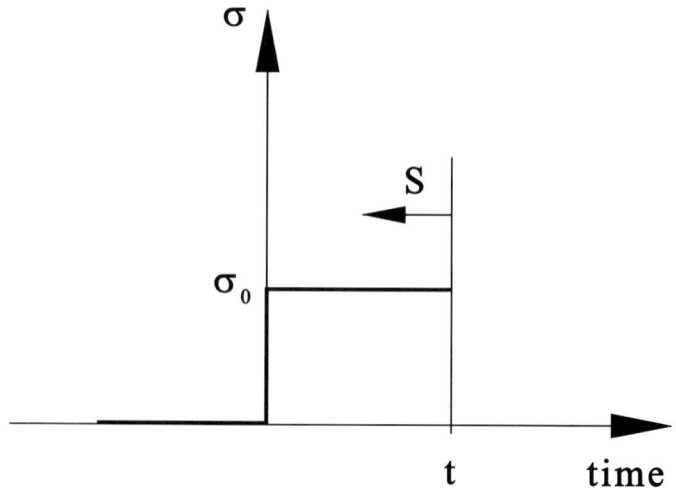

Fig. 12.10 Stress jump, creep test

12.2 Material Description

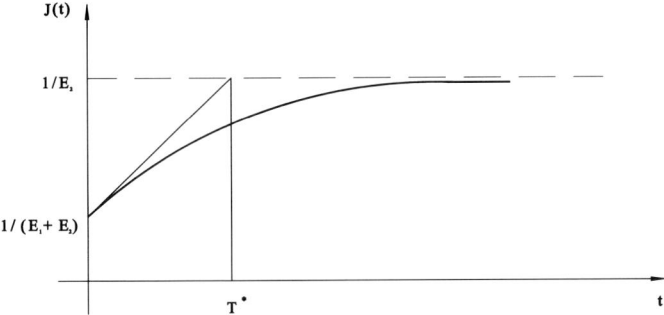

Fig. 12.11 Creep compliance $J(t)$ of the Poynting–Thomson material

If one makes use of the symmetry transformation Eqs. (12.6) and (12.7), one obtains the Eq. (12.18) directly from Eq. (12.12). The ratio of the creep curve $\varepsilon(t)$ to σ_0 is called the *creep compliance*

$$J(t) = \left(\frac{1}{E_2} - \frac{E_1}{E_2(E_1 + E_2)} \right) e^{\frac{-t}{T^*}} \quad \text{for} \quad t \geq 0. \quad (12.19)$$

From Eq. (12.17) one sees that the strain increases exponentially with the retardation time T^* from the initial value $\sigma_0/(E_1 + E_2)$ to the final value σ_0/E_2 (Fig. 12.11). This phenomenon is called *creep* or *stretch retardation*.

Analogously to the last paragraph the analysis of the creep compliance also permits a generalization of the material law. Equation (12.9) can be written with the defined creep compliancy in Eq. (12.19) in the following way

$$\varepsilon(t) = J_\infty \sigma(t) + \int_0^\infty J'(s)(\sigma(t-s) - \sigma(t)) \, ds \,. \quad (12.20)$$

In accordance to Eq. (12.14), Eq. (12.20) is one-dimensional and the most general representation of a linearly viscoelastic solid.

12.2.6.3 Oscillating Loading

In order to study the material behavior of a linearly viscoelastic solid, a purely harmonically oscillating motion is also suitable besides the previously described standard tests. Such dynamic experiments require an device that produces a deformation sinusoidally variable in time as input and registers the resulting stress as output. Hydraulically and electrically working pulsation devices are here well known. In these experiments a probe is subjected to a sinusoidal deformation with an angular frequency ω and an amplitude ε_0. For linear material behavior the material also reacts with a sinusoidal stress curve of the same frequency, however, stress and strain, are not in phase as indicated in Fig. 12.12.

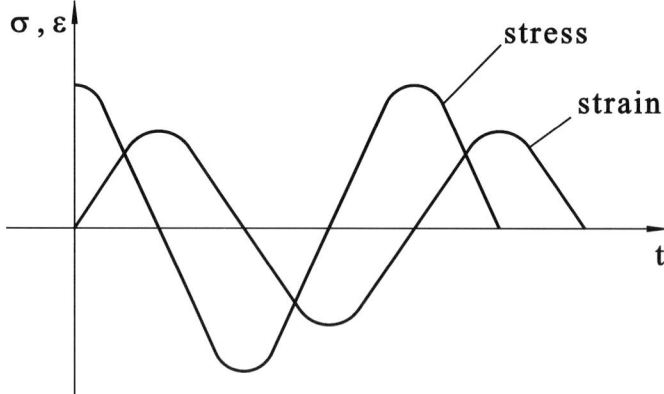

Fig. 12.12 Stress and strain in harmonic loading

Here one works most comfortably with complex quantities for the basic tests, but in each case only the real term or the imaginary term respectively has physical relevance. If one prescribes the strain or strain rate as a harmonic function of time

$$\varepsilon = \varepsilon_0 e^{i\omega t}, \quad \dot{\varepsilon} = i\omega\varepsilon_0 e^{i\omega t} \qquad (12.21)$$

and insert this in Eq. (12.1), one obtains the following relation for the stress

$$\sigma(t) = \hat{E}(\omega)\varepsilon(t) + \frac{\hat{\hat{E}}(\omega)}{\omega}\dot{\varepsilon}(t). \qquad (12.22)$$

$\hat{E}(\omega)$ is called *dynamic stiffness* or *storage modulus* and $\hat{\hat{E}}(\omega)$ is called *dynamic viscosity* or alternatively $\hat{\hat{E}}(\omega)/\omega$ is the *loss modulus*. It can easily be seen that the ratio of loss modulus and storage modulus is the *loss tangent* ($\tan\delta$),

$$\tan\delta = \frac{\hat{\hat{E}}(\omega)}{\hat{E}(\omega)}. \qquad (12.23)$$

The phase angle δ is often called *loss angle*.

The Poynting–Thomson material yields the following relations for the storage modulus and loss modulus which are visualized in Figs. 12.13 and 12.14:

$$\hat{E}(\omega) = E_2 \frac{1 + (E_1/E_2)T^2\omega^2}{1 + T^2\omega^2}, \qquad (12.24)$$

$$\hat{\hat{E}}(\omega) = E_1 \frac{T\omega}{1 + T^2\omega^2}. \qquad (12.25)$$

It can be seen in Fig. 12.13 that with increasing frequency the stiffness will also increase. This behavior is described as *dynamic stiffening*. In Fig. 12.14 one can see that the loss modulus disappears at the frequency $\omega = 0$ and at very high frequencies, i. e. the material behaves purely elastic in case of statics and at very high frequencies respectively. In between there must be a maximum and one asks about the conse-

12.2 Material Description

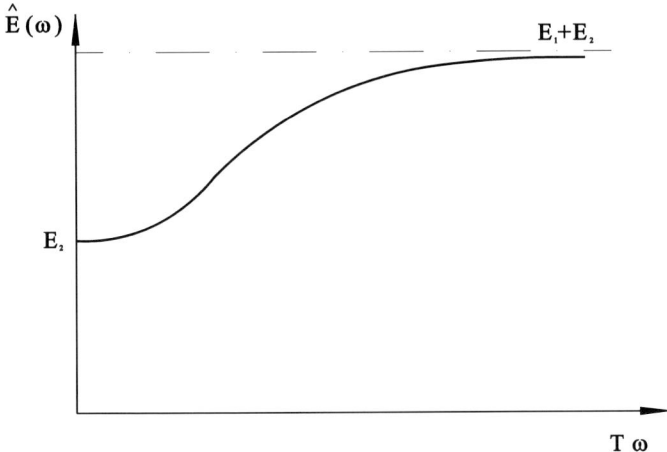

Fig. 12.13 Storage modulus $\hat{E}$ of the Poynting–Thomson material

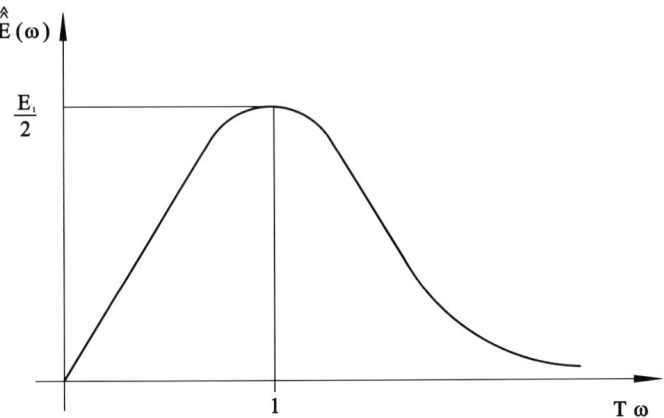

Fig. 12.14 Loss modulus $\hat{\hat{E}}$ of the Poynting–Thomson material

quences. If one considers that the damping is highest at this point, it will be clear that for problems where high damping is requested the balance between characteristic process time ω and relaxation time T is tuned in a way that this point is realized.

In biology these are the intervertebral discs in mammals and in technology these are the automotive door frame seals. Here not only a sealing function is wanted but also a damping function for the sound design of car doors.

If one further makes use of Eq. (12.22) and insert only the real or imaginary terms of Eq. (12.21), the relation between stress and strain can be represented in form of

$$\left(\sigma(t) - \hat{E}(\omega)\varepsilon(t)\right)^2 + \hat{\hat{E}}^2(\omega)\varepsilon^2(t) = \hat{\hat{E}}^2(\omega)\varepsilon_0^2. \tag{12.26}$$

One realizes that this equation describes tilted ellipses in the $\sigma - \varepsilon$ plane that are cycled counter-clockwise (see Fig 12.15). In the limit cases $\omega = 0$ and $\omega \to \infty$

the ellipses degenerate into two straight lines with different slopes. In case of the Poynting–Thomson material this results in the slopes E_2 for $\omega = 0$ and $E_1 + E_2$ for $\omega \to \infty$ respectively.

It should be mentioned that Eq. (12.22) does not describe any material law although the relation between stress and stain is given formally. The indicated relation is only valid for the kinematic constraint that the strain occurs in form of Eq. (12.21). If one asks for the relations derived here and for the general material representation, as given by Eq. (12.14), Eq. (12.21) has to be inserted into Eq. (12.14) in order to again obtain Eq. (12.22). The storage modulus and the loss modulus are given by the following equations:

$$\hat{E}(\omega) = E_\infty + \int_0^\infty E'(s)(\cos(\omega s) - 1)\,ds , \qquad (12.27)$$

$$\hat{\hat{E}}(\omega) = -\int_0^\infty E'(s)\sin(\omega s)\,ds . \qquad (12.28)$$

As one can see, the storage modulus and the loss modulus are not independent of each other. Both are only different expressions of one single material quantity which is the relaxation function. The oscillation test has turned out to be the most important one and the quantities $\hat{E}$, $\hat{\hat{E}}$ have been established as dynamic characteristic values in engineering.

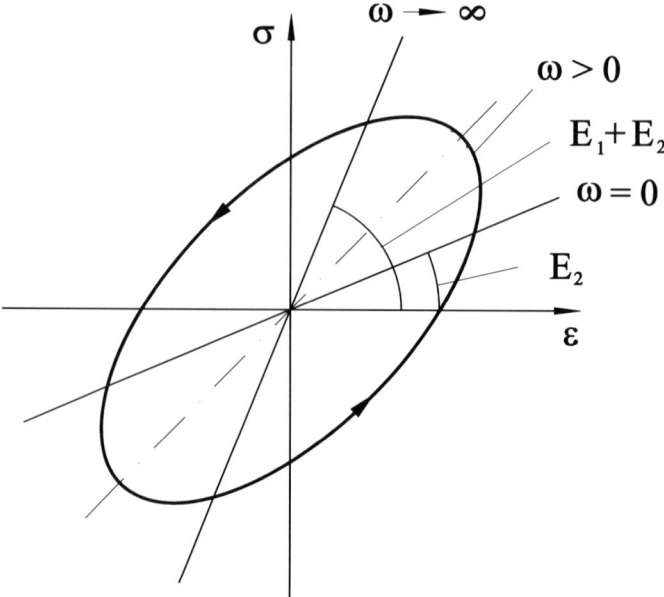

Fig. 12.15 Stress-strain hysteresis of a Poynting–Thomson material

12.2.7 Remarks

The Poynting–Thomson material exhibits a strong stress relaxation and strain retardation behavior as one could see from the discussion of the step test and the creep test respectively. This similar behavior regarding stress and strain is on the one hand due to the symmetry of Eqs. (12.1) or (12.3) and on the other hand due to the fact that stress and strain are connected with each other by a linear differential equation. If one takes into consideration that Eq. (12.3) is the simplest kind of a fully symmetrical differential equation, it will be clear why the Poynting–Thomson material represents the simplest rheological model to describe qualitatively correctly the viscoelastic properties of soft substances. Table 12.1 shows the most important material models which are made from combinations of springs and dashpots.

In Table 12.1, if one considers the Kelvin-Voigt material which is also often used for describing soft materials, one can see from the given stress-strain relation why this material describes creep deformation but does not correctly describe stress relaxation.

It is clear that one can continue extending Table 12.1 by producing many material equations by arbitrarily combining springs and dashpots. However, all material descriptions obtained in this way can be uniformly described by using the material description given by Eq. (12.14). Therefore, such spring dashpot combinations have only a descriptive value. Despite the universality of Eq. (12.14), this material law still has deficiencies, because it is basically linear and it cannot directly be translated to multidimensional problems. Before discussing nonlinear material descriptions for viscoelasticity, some basics of continuum mechanics and of material theory will be provided in the next two sections.

Table 12.1 Simple rheological models

Figure	Name	Type of Material	Material Law
E spring	Hookean material	elastic solid	$\sigma = E\varepsilon$
η dashpot	Newtonian material	viscous fluid	$\sigma = \eta\dot{\varepsilon}$
E – η (series)	Maxwell material	viscoelastic fluid	$\dot{\sigma} + \frac{E}{\eta}\sigma = E\dot{\varepsilon}$
E ∥ η (parallel)	Kelvin–Voigt material	elastic-viscoelastic solid	$\sigma = E\varepsilon + \eta\dot{\varepsilon}$
E_1 ∥ (E_2 – η)	Poynting–Thomson material	viscoelastic solid	$\dot{\sigma} + \frac{\sigma}{T} = (E_1 + E_2)\left(\dot{\varepsilon} + \frac{\varepsilon}{T^*}\right)$; $T = \frac{\eta}{E_2}$; $T^* = T\left(1 + \frac{E_2}{E_1}\right)$
E ∥ η_2 – η_1	Jeffreys–Lether material	viscoelastic fluid	$\dot{\sigma} + \frac{\sigma}{T} = \frac{\eta_1}{1+\eta_1/\eta_2}\left(\ddot{\varepsilon} + \frac{\dot{\varepsilon}}{T^*}\right)$; $T = \frac{\eta_2}{E}\left(1 + \frac{\eta_1}{\eta_2}\right)$; $T^* = \frac{\eta_1}{E}$

12.3 Basics of Continuum Mechanics

12.3.1 Kinematics

As the name continuum mechanics already says, matter is regarded in this theory as a continuum in the three-dimensional Euclidean space E^3. The points forming the continuum are called material points or particles. Essential for the applicability of the theory is that the material points can be identified. The material points of the continuum are referred to as X. A body B is a coherent compact set of material points. The boundary of the point set is referred to as ∂B and is called the surface of the body. The configuration of a body B is a continuous and one-to-one mapping of the material points X of a body B to the position vectors $\mathbf{x} \in E^3$,

$$\mathbf{x} = X(X) \,. \tag{12.29}$$

A continuous sequence of configurations is called the motion of the body:

$$\mathbf{x} = X(X, t) \,. \tag{12.30}$$

1. Reference configuration and current configuration
   The reference configuration of a body B is used to assign a name X to the particle $X \in B$. One conveniently identifies the coordinates of the particle X with the position vector $\mathbf{X}$ that represents the place occupied by the particle X at reference time t_0

$$\mathbf{X} = X(X, t_0) \,. \tag{12.31}$$

The motion then can be described with the relation

$$\mathbf{x} = \mathbf{x}(\mathbf{X}, t, t_0) \,. \tag{12.32}$$

The configuration of a body B at time $t = t_0$ is referred to as reference configuration. The place $\mathbf{x}$ occupied by the particle X at time t is referred to as current configuration. The inverse of Eq. (12.32) is given by the relation.

$$\mathbf{X} = \mathbf{X}(\mathbf{x}, t, t_0) \,. \tag{12.33}$$

Equation (12.33) describes the material point that occupies a place $\mathbf{x}$ at time $t = t_0$. According to the given definitions the following relations are then valid:

$$\mathbf{X} = \mathbf{x}(\mathbf{X}, t_0, t_0) \,, \tag{12.34}$$

$$\mathbf{x} = \mathbf{X}(\mathbf{x}, t, t) \,. \tag{12.35}$$

2. Deformation history
   If a particle $\mathbf{X}$ at time t occupies a place $\mathbf{x}$ and occupied the place ξ at time $t - s$ for $s > 0$, the motion of a particle can be described with the relation

$$\xi = \xi(\mathbf{x}, t - s, t) \,. \tag{12.36}$$

12.3 Basics of Continuum Mechanics

and is referred to as deformation history. The current configuration now acts as the reference configuration. With reference to what has been said above, this image describes the pathline of the particle that now ($s = 0$) occupies the place **x**.

An other important concept in the kinematics of the continuum is the *gradient of deformation*. One defines the *deformation gradient*

$$\mathbf{F}(\mathbf{X}, t, t_0) := \frac{\partial \mathbf{x}(\mathbf{X}, t, t_0)}{\partial \mathbf{X}} \tag{12.37}$$

as well as the relative deformation gradient

$$\mathbf{F}_t(\mathbf{x}, t - s, t) := \frac{\partial \boldsymbol{\xi}(\mathbf{x}, t - s, t)}{\partial \mathbf{x}} . \tag{12.38}$$

The deformation gradients posses the following characteristics: If d**X**, d**x**, d$\boldsymbol{\xi}$ describe the same material line-elements in the reference configuration, the current configuration and the configuration at the time $t - s$, the following relations are valid:

$$\mathbf{F}(\mathbf{X}, t, t_0) \, d\mathbf{X} = d\mathbf{x} , \tag{12.39}$$
$$\mathbf{F}(\mathbf{X}, t - s, t_0) \, d\mathbf{X} = d\boldsymbol{\xi} , \tag{12.40}$$
$$\mathbf{F}_t(\mathbf{x}, t - s, t) \, d\mathbf{x} = d\boldsymbol{\xi} . \tag{12.41}$$

Another calculation instruction results from these relations for the relative deformation gradient

$$\mathbf{F}_t(\mathbf{x}, t - s, t) = \mathbf{F}(\mathbf{X}, t - s, t_0) \mathbf{F}^{-1}(\mathbf{X}, t, t_0) . \tag{12.42}$$

If the deformation gradient and the relative deformation gradient, respectively, do not depend on the material particles **X** and **x**, respectively, one speaks of a homogeneous deformation. For $\det \mathbf{F} = \det \mathbf{F}_t = 1$ the motion is volume preserving (isochoric). Figure 12.16 illustrates once again the different configurational terms and different transformation characteristics of the individual deformation gradients.

The *polar decomposition theorem* is the basis of the definition of deformation tensors:

$$\mathbf{F} = \mathbf{R}\mathbf{U} = \mathbf{V}\mathbf{R} . \tag{12.43}$$

It ensures the unique multiplicative decomposition of a regular tensor **F** into an orthogonal rotation tensor **R**, $\mathbf{R}\mathbf{R}^T = \mathbf{I}$, and the symmetric positive-definite tensors **U** and **V**, respectively. **U** and **V** are called *left Cauchy–Green deformation tensor* and *right Cauchy–Green deformation tensor*, respectively. One defines the right Cauchy–Green deformation tensor,

$$\mathbf{C} = \mathbf{F}^T \mathbf{F} = \mathbf{U}^2 \tag{12.44}$$

the left Cauchy–Green deformation tensor

$$\mathbf{B} = \mathbf{F}\mathbf{F}^T = \mathbf{V}^2 \tag{12.45}$$

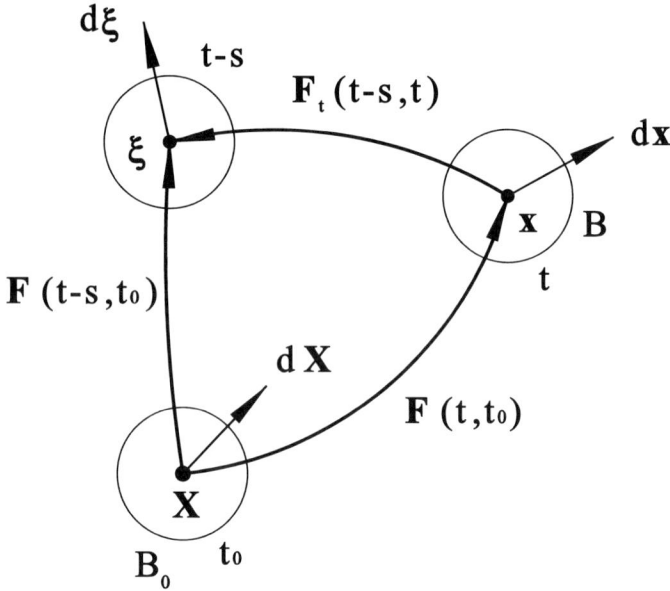

Fig. 12.16 Different configurations of a body

as well as the relative right Cauchy–Green deformation tensor

$$C_t = F_t^T F_t .\tag{12.46}$$

These tensors possess the following characteristics:

$$dX \cdot C\, dX = dx \cdot dx ,\tag{12.47}$$
$$dx \cdot C_t\, dx = d\xi \cdot d\xi ,\tag{12.48}$$
$$dx \cdot B^{-1}\, dx = dX \cdot dX .\tag{12.49}$$

They describe the change of line elements in length and angle during the transition of one configuration to the other, thus justifying the concept of strain tensors. A further quantity used when formulating material descriptions is the *deformation history* **G**. One defines

$$G = C_t - I .\tag{12.50}$$

12.3.2 Stress State

The instantaneous force vector d**k** acting on a material surface element dA can be indicated by two stress vectors:

$$dk = t_0\, dA_0 = t\, dA .\tag{12.51}$$

12.3 Basics of Continuum Mechanics

Here dA and dA_0, respectively, is the same material surface element in the current configuration and in the reference configuration, respectively. If $\mathbf{n}$ and $\mathbf{n}_0$, respectively, is the orientation vector of the surface element in both configurations, then the conservation theorems of momentum and of moment of momentum result in two stress tensors σ and Σ. σ is the Cauchy stress tensor with the characteristics:

$$\mathbf{t} = \sigma\mathbf{n}, \quad \sigma = \sigma^{\mathrm{T}}. \tag{12.52}$$

The stress tensor Σ is called the 1st Piola–Kirchhoff stress tensor or Lagrange stress tensor and has the characteristics:

$$\mathbf{t}_0 = \Sigma\mathbf{n}_0, \quad \Sigma\mathbf{F}^{\mathrm{T}} = \mathbf{F}\Sigma^{\mathrm{T}}. \tag{12.53}$$

Between σ and Σ the following relation exists:

$$\Sigma = (\det \mathbf{F})\sigma\,(\mathbf{F}^{\mathrm{T}})^{-1}. \tag{12.54}$$

12.3.3 Balance Equations

The current configuration and the reference configuration give reason to represent the (local) balance equations in two different forms. If one refers to the current configuration, it holds:

Balance of linear momentum

$$\rho\ddot{\mathbf{x}} = \rho\mathbf{f} + \mathrm{div}\,\sigma \tag{12.55}$$

and balance of mass

$$\dot{\rho} + \rho\,\mathrm{div}\,\dot{\mathbf{x}} = 0. \tag{12.56}$$

In the reference configuration it holds:

Balance of linear momentum

$$\rho_0\ddot{\mathbf{x}} = \rho_0\mathbf{f} + \mathrm{Div}\,\Sigma \tag{12.57}$$

and balance of mass

$$\rho_0 = \rho\det\mathbf{F}. \tag{12.58}$$

In these equations, $\rho\mathbf{f}$ is the external body force and ρ and ρ_0 are densities in the current configuration and reference configuration, respectively. The operation div refers to the material particle $\mathbf{x}$ of the current configuration and the operation Div refers to the material particle $\mathbf{X}$ of the reference configuration. The local balance equations are systems of partial differential equation which hold universally for any material in the interior of the body B. In three-dimensional space the momentum balance are three equations for six unknown stress components and three accelerations. Therefore they must be completed with the material equations which relate stresses with deformation.

We pause for the observation that the balance equations have no unique solution. The balance of the linear momentum in the actual configuration in the absence of body forces and $\ddot{\mathbf{x}} = \mathbf{0}$ is expressed by the homogeneous equation

$$\mathbf{0} = \text{div}\sigma \ . \tag{12.59}$$

Let the body be without any external load on the traction boundary ∂B_σ. These equations have a nontrivial solution ρ not identically zero which is called residual stress or self-equilibrated stress because it is in equilibrium with no loads. Adding Eqs. (12.51) and (12.59) and the traction boundaries shows that $\sigma + \rho$ is a solution besides σ. Residual stresses are "locked-in" within a material and are developed during most common manufacturing processes, for example welding, cold working and grinding. Residual stress can be quite large, and can have serious effects on brittle materials or may cause buckling of thin structures.

In technology the shrinking of wood or concrete may produce residual stress. In biomechanics residual stress must be expected as a consequence of growth or remodeling processes. It has been suggested that the human fingerprints patterns are created because the less stiff basal layer, which separates the outer layer of the epidermis and the inner dermis, growths faster than the two other layers. Compressive residual stress is generated that makes the basal layer buckle thus forming ridges which become visible as fingerprints patterns [30].

The "hidden" residual stresses can be measured indirectly by removing some of the stressed material, say by hole drilling, and by measuring the strain changes due to the stress redistribution in the remaining material. Similar methods are applied in tissue mechanics so that the released residual stress produce split lines in bones or Langer lines in the skin on a cadaver [10]. The latter cleavage lines are of particular interest to the surgeon because an incision made parallel to the lines leads to wider openings and heals with a fines scar compared to an incision across the lines.

12.3.4 Material Equation

There are three basic postulates for the formulation of physically meaningful mechanical material equations:

1. *The principle of determinism:*
   The current stress state depends only on the present and former deformation events, i. e. on the complete deformation history.
2. *The principle of local action:*
   This is an hysteresis principle and requires, in its simplest case, that the stress state on a particle is determined by the history of the deformation gradient. Such a material is called "simple material".

3. *The principle of material objectivity:*
   This postulate requires that the form of a material equation is independent of the reference frame and this has the consequence that only certain dynamic and kinematic variables can be connected in constitutive equations.
   Taking into account these postulates, the material equation for isotropic, homogeneous, simple substances has the following most general form

$$\sigma(t) = f(\mathbf{B}(t)) + \underset{s=0}{\overset{\infty}{\mathcal{F}}} (\mathbf{B}(t), \mathbf{G}(s)) \tag{12.60}$$

with

$$\underset{s=0}{\overset{\infty}{\mathcal{F}}} (\mathbf{B}(t), \mathbf{0}) = \mathbf{0} \quad \text{for all} \quad \mathbf{B}(t). \tag{12.61}$$

$f(\mathbf{B})$ is an isotropic tensor valued tensor function, $\mathcal{F}$ is an isotropic tensor valued functional in $\mathbf{G}$, and $\mathbf{B}$ is an isotropic tensor valued functional of $\mathbf{G}(s)$ with $\mathbf{B}(t)$ as the parameter.

The constitutive Eq. (12.60) expresses that stresses split additively into an elastic term depending on the current deformation condition relatively to the reference configuration and into an hysteresis term depending on the current deformation condition as well as on the complete *deformation history*. In the case of statics, the memory term disappears, i. e. a simple material behaves elastically here.

12.3.5 Constraints

A biopolymer changes its volume only under very high pressure. Whereas small forces already suffice for stretching or shearing it. If one looks at deformation problems of this substance class, the density does generally not change at all. One can describe this material behavior in an idealized way by assuming it to be incompressible. Similarly, much lower forces are required for bending a thin steel wire rather than stretching it. While describing the bending of a wire, one can neglect the mean strain and therefore assume the wire to be inextensible. Incompressibility and inextensibility are examples for geometrical constraints that limit the degrees of freedom of materials. Constrain forces are required to avoid motions that may violate a constraint. Since they do not correspond to any condition taking place, they are kinematically undetermined, i. e. they cannot be determined from the local motion of a material but result from the overall motion and the boundary conditions. Stresses resulting from the deformation have to be amended by a constraint stress tensor $\mathbf{Z}$. In the case of incompressibility, this constraint stress tensor is spherically symmetric and the only component to be determined is the pressure p

$$\mathbf{Z} = -p\mathbf{I}. \tag{12.62}$$

Considering this constraint stress tensor, the constitutive equation for incompressible, isotropic, homogeneous, simple materials assumes the following form

$$\sigma(t) = -p(t)\mathbf{I} + f(\mathbf{B}(t)) + \mathop{\mathcal{F}}_{s=0}^{\infty} (\mathbf{B}(t), \mathbf{G}(s)) \tag{12.63}$$

with the properties of Eq. (12.61).

12.4 Basics of Material Theory

12.4.1 Mathematical Fundamentals

This subsection deals with the auxiliary mathematical tools. These tools cover required definitions and theorems from the representation theory of tensor functions as well as from functional analysis. Firstly, some fundamentals of tensor calculus are reminded of. For every tensor $\mathbf{B}$ a transposed tensor $\mathbf{B}^\mathrm{T}$ is defined by the following rule

$$B_{ik}^\mathrm{T} = B_{ki} . \tag{12.64}$$

In the matrix representation (in Cartesian coordinates) rows and columns of of the transposed tensor $\mathbf{B}^\mathrm{T}$ are exchanged compared with the matrix representation of $\mathbf{B}$. The following relation is valid for all vectors $\mathbf{a}$ and $\mathbf{b}$:

$$\mathbf{b} \cdot (\mathbf{B}\mathbf{a}) = (\mathbf{B}^\mathrm{T} \mathbf{b}) \cdot \mathbf{a} . \tag{12.65}$$

This relation can be considered as a coordinate invariant definition of the transposition. A tensor is called symmetric if $\mathbf{A}^\mathrm{T} = \mathbf{A}$. It is known that one can represent a symmetric tensor in principal axis form, i.e. there is a special Cartesian coordinate system where only the diagonal elements of the matrix representing the tensor are different from zero. These three not disappearing elements $\lambda_1, \lambda_2, \lambda_3$ are the eigenvalues of the tensor which one finds by solving the eigenvalue problem. This problem consists of finding all vectors $\mathbf{a}$ so that $\mathbf{A}\mathbf{a}$ is parallel or anti-parallel to $\mathbf{a}$. This results in identifying the vectors $\mathbf{a}$ and the numbers λ that satisfy the relation $\mathbf{A}\mathbf{a} = \lambda \mathbf{a}$ or

$$(\mathbf{A} - \lambda \mathbf{I})\mathbf{a} = \mathbf{0} . \tag{12.66}$$

The homogeneous system of Eq. (12.66) for the three components of the sought eigenvectors $\mathbf{a}$ only has non-trivial solutions if its determinant disappears

$$\det(\mathbf{A} - \lambda \mathbf{I}) = 0 . \tag{12.67}$$

12.4 Basics of Material Theory

Equation (12.67) is called characteristic polynomial of the tensor **A**. In E^3 this is a cubic equation for the eigenvalues λ which is written in detail as follows

$$\lambda^3 + I_1\lambda^2 + I_2\lambda - I_3 = 0 . \tag{12.68}$$

The three coefficients I_1, I_2, I_3 are algebraic functions of the elements A_{ik} of **A**

$$I_1 = A_{11} + A_{22} + A_{33} = tr\mathbf{A} , \tag{12.69}$$

$$I_2 = \begin{vmatrix} A_{11} & A_{12} \\ A_{21} & A_{22} \end{vmatrix} + \begin{vmatrix} A_{11} & A_{13} \\ A_{31} & A_{33} \end{vmatrix} + \begin{vmatrix} A_{22} & A_{23} \\ A_{32} & A_{33} \end{vmatrix} , \tag{12.70}$$

$$I_3 = \begin{vmatrix} A_{11} & A_{12} & A_{13} \\ A_{21} & A_{22} & A_{23} \\ A_{31} & A_{32} & A_{33} \end{vmatrix} = \det \mathbf{A} . \tag{12.71}$$

These coefficients are called principal invariants of the tensor **A**, since their values are independent of the chosen coordinate system. The linear invariant I_1 is denoted the trace of the tensor **A**.

According to the Cayley-Hamilton theorem any tensor **B**, that maps a vector space onto itself, satisfies its own characteristic polynomial. In the cases considered here, the relevant vector space is the Euclidean vector space E^3 and the theorem is

$$\mathbf{B}^3 - I_1\mathbf{B}^2 + I_2\mathbf{B} - I_3\mathbf{I} = 0 . \tag{12.72}$$

One can use this relation in order to express tensor powers larger than 2 with lower ones. Examples are:

$$\mathbf{B}^3 = I_1\mathbf{B}^2 - I_2\mathbf{B} + I_3\mathbf{I} , \tag{12.73}$$

$$\mathbf{B}^4 = (I_1^2 - I_2)\mathbf{B}^2 + (I_3 - I_2I_1)\mathbf{B} + I_3I_1\mathbf{I} . \tag{12.74}$$

It is clear that one can represent tensor polynomials and analytic tensor functions with this possibility in a strongly simplified form. If one considers e.g. an analytic, tensor valued tensor function $f(\mathbf{B})$ and considers that all higher tensor powers larger than 2 can be expressed with lower ones, the following representation is then plausible

$$f(\mathbf{B}) = \phi_0\mathbf{I} + \phi_1\mathbf{B} + \phi_2\mathbf{B}^2 . \tag{12.75}$$

The coefficients ϕ_i are scalar functions of the principal invariants I_1, I_2, I_3 of **B**. If **B** is regular (i.e. $I_3 \neq 0$) and if one multiplies the Eq. (12.73) with $\mathbf{B}^{-1}$, one gets the following with the help of Eq. (12.75)

$$f(\mathbf{B}) = \omega_0\mathbf{I} + \omega_1\mathbf{B} + \omega_{-1}\mathbf{B}^{-1} , \tag{12.76}$$

$$\omega_0 = \phi_0 - I_2\phi_2 , \quad \omega_1 = \phi_1 + \phi_2 I_1 , \quad \omega_{-1} = \phi_2 I_3 . \tag{12.77}$$

In the case of symmetry of the tensor pre-images and tensors images, one needs in the most general case the knowledge of 6 functions depending on 6 variables for the description of a tensor valued tensor function. The representation (Eqs. (12.75) and (12.76)) shows that one only needs the knowledge of three functions for the description of analytic tensor functions which depend on three variables. The same simplification arises in the context of isotropic tensor functions. A tensor valued tensor function $f(\mathbf{B})$ is called isotropic if the following relation holds for all rotations $\mathbf{Q}$ (with $\mathbf{QQ}^T = \mathbf{I}$):

$$\mathbf{Q} f(\mathbf{B}) \mathbf{Q}^T = f(\mathbf{QBQ}^T) . \tag{12.78}$$

According to an important theorem of representation theory, the representation (Eqs. (12.75) and (12.76)) is also valid for isotropic tensor functions. One easily convinces oneself that every analytical tensor function is isotropic. The representation of an isotropic tensor function that depends on two variables $\tilde{f}(\mathbf{B}, \mathbf{G})$ and is linear in the second argument is needed for the later use. A fourth order tensor $\mathbf{K}$ exists due to the linearity in the argument $\mathbf{G}$, and the function of interest has the following form

$$\tilde{f}(\mathbf{B}, \mathbf{G}) = \mathbf{K}(\mathbf{B}) : \mathbf{G} . \tag{12.79}$$

$\mathbf{K} : \mathbf{G} = K_{ijkl} G_{kl}$ is the double transvection, where according to Einstein one has to sum over double indices ($k, l = 1, 2, 3$). If there is no further information available, one needs, in the case of symmetry of the tensor pre-images and tensor images, $6 \times 6 = 36$ functions that depend on the 6 six elements of $\mathbf{B}$ in order to describe the tensor $\mathbf{K}$. If all considered arguments are isotropic, it holds

$$\begin{aligned}\mathbf{K}(\mathbf{B}) : \mathbf{G} = {}& \mathbf{M}_1(\mathbf{B}) \mathbf{G} + \mathbf{G} \mathbf{M}_1(\mathbf{B}) + tr(\mathbf{M}_2(\mathbf{B}) \mathbf{G}) \mathbf{I} \\ & + tr(\mathbf{M}_3(\mathbf{B}) \mathbf{G}) \mathbf{B} + tr(\mathbf{M}_4(\mathbf{B}) \mathbf{G}) \mathbf{B}^2 .\end{aligned} \tag{12.80}$$

The $\mathbf{M}_i$ are isotropic tensor functions of $\mathbf{B}$ that can be described in the same way as in Eq. (12.75):

$$\mathbf{M}_i(\mathbf{B}) = \phi_{0i} \mathbf{I} + \phi_{1i} \mathbf{B} + \phi_{2i} \mathbf{B}^2 . \tag{12.81}$$

The functions $\phi_{0i}, \phi_{1i}, \phi_{2i}$ depend on the three principal invariants of $\mathbf{B}$. Unlike 36 functions, with assumed symmetry one only needs $3 \times 4 = 12$ functions depending on the principal invariants of $\mathbf{B}$. One needs further tools of functional analysis in order to process further on the hysteresis term of the material laws. One defines the functional derivatives in analogy to ordinary calculus.

Let $\mathcal{F}$ be a functional that maps the Hilbert space $\mathcal{H}$ onto real numbers R. One can then differentiate $\mathcal{F}$ around point $x \in \mathcal{H}$, if there exists a linearly bounded functional $\delta \mathcal{F}$ with the property

$$\mathcal{F}(x+h) = \mathcal{F}(x) + \delta \mathcal{F}(x; h) + o(|h|) . \tag{12.82}$$

12.4 Basics of Material Theory

$o(|h|)$ is a zero function that goes to zero faster than linear. The linearity of the functional refers to the second argument h. The functional is expanded around point x which generally comes in nonlinearly as a parameter. In a Hilbert space $\mathcal{H}$ the possible structure of a linear functional can be clarified by a theorem proved by Friedrich Riesz. According to this theorem, an unique element $y \in \mathcal{H}$ exists to every linearly continuous functional, so that the functional can be represented by a scalar product between y and the argument h.

Riesz representation theorem:
Every linearly continuous functional $\delta \mathcal{F}(h)$ can be represented over a Hilbert space $\mathcal{H}$ in the form

$$\delta \mathcal{F}(h) = \langle y | h \rangle \quad \text{for all} \quad h. \qquad (12.83)$$

$y \in \mathcal{H}$ is uniquely determined by the linear functional and $\langle y | h \rangle$ describes the scalar product.

12.4.2 Necessity of Asymptotic Representations

While considering the general basic postulates of the material theorem, the strategy of the simple material that goes back to Rivlin and Noll only says on which kinematic quantities the Cauchy stress tensor depends. The form of this dependence is not yet fixed except for the isotropy condition, which results from the objectivity. Apart from these quite modest requirements there is no further restriction put on the material laws Eqs. (12.60) and (12.63). As Eqs. (12.60) and (12.63) do not define the type of connection between the stress and the strain and the strain history, it is thus problematic to determine general material laws of such a substance. For this purpose, one would have to record the stress curve for the infinite variety of all possible non-steady-strains and one would have to suggest the construction of the functionals from all the observations. Since this is impossible, the general material laws of real substances remain unknown in the form Eqs. (12.60) and (12.63), and therefore the Eqs. (12.60) and (12.63) are practically useless. This leads to the necessity to put this dependency in well defined terms under the addition of broader conditions and assumptions. From the physical point of view, one will set as a condition for the concretization of material laws that the appearing material constants or functions are measurable in principle. From the technical point of view, one demands that the effort for measuring open quantities and the technical effort for calculating certain deformation problems does not get too big. The possibility of deriving approximations of the general material equation arises with the "principle of fading memory" by adding a fourth physically plausible requirement to the three basic postulates.

12.4.3 Principle of Fading Memory

This principle can be qualitatively formulated as follows: The value of the functional $\mathcal{F}$ depends the lesser on the course of the history of $\mathbf{G}(s)$, the higher s is, in other words the present stress state is influenced lesser by deformation events the further these are behind in time. Fundamental for a quantitative formulation of this postulate is the definition of the norm of a deformation history

$$\|\mathbf{G}(s)\|^2 = \int_0^\infty tr\left(\mathbf{G}\mathbf{G}^\mathrm{T}(s)\right) h^2(s)\,ds \qquad (12.84)$$

as well as the definition of a scalar product between two deformation events. It shall be noted that the tensors of interest are symmetrical

$$\langle \mathbf{G}(s)|\mathbf{H}(s)\rangle = \int_0^\infty tr\left(\mathbf{G}(s)\mathbf{H}^\mathrm{T}(s)\right) h^2(s)\,ds \;. \qquad (12.85)$$

$h(s)$ is a continuous real-valued function that approaches zero "sufficiently fast":

1. An infinitesimal norm of a deformation history is equivalent to a rigid body movement or rest.
2. Two deformation histories are almost equal in the sense of this norm if their progressions are almost equal in a near past. Big differences in the progressions of the distant past are suppressed by the norms.
3. The norm is small if either $tr(\mathbf{G}^2)$ is always small or if the deformation takes place sufficiently slow. During the deceleration of a motion, individual events move further back in time and therefore influence the norm with a smaller weight h.

From the fact, that according to Eq. (12.84) the entirety $\mathcal{H}$ of all square integrable deformation histories forms with definition Eq. (12.85) a complete Euclidean vector space (Hilbert space), one obtains the possibility to formulate the postulate of fading memory as a continuity statement and as a statement for differentiability (smoothness).

In the simplest case, one obtains:

The functional $\mathcal{F}_{s=0}^\infty(\mathbf{G}(s))$ possesses fading memory in terms of the norm Eq. (12.84) if it is continuous in a neighborhood of the zero history ($\mathbf{G}(s) = \mathbf{0}$).

A sharper formulation requires:

The functional $\mathcal{F}_{s=0}^\infty(\mathbf{G}(s))$ possesses fading memory in terms of the norm (12.84) if it is at least differentiable in time in the neighborhood of the zero history ($\mathbf{G}(s) = \mathbf{0}$).

According to the remarks of the last section, every component of the tensor valued functional of the material Eqs. (12.60) and (12.63), respectively, assumes in this

12.4 Basics of Material Theory

case the following form

$$\tilde{\mathcal{F}}_{ij}(\mathbf{B}, \mathbf{G}(s)) \underset{s=0}{\overset{\infty}{=}} \tilde{\mathcal{F}}_{ij}(\mathbf{B}, 0) \underset{s=0}{\overset{\infty}{+}} \langle L_{ij}(\mathbf{B}, s) | \mathbf{G}(s) \rangle + o(\|\mathbf{G}\|)$$

$$= 0 + \int_{s=0}^{\infty} Sp(L_{ij}(\mathbf{B}, s)\mathbf{G}(s))h^2(s)\,ds + \dots \quad (12.86)$$

$$= \int_{s=0}^{\infty} L_{ijkl}(\mathbf{B}, s) G_{kl} h^2\,ds + \dots .$$

If one makes use of the tensor notation again and introduces the abbreviation $\mathbf{K} = h^2 \mathbf{L}$, one obtains the tensor notation

$$\underset{s=0}{\overset{\infty}{\mathcal{F}}}(\mathbf{B}, \mathbf{G}(s)) = \int_{s=0}^{\infty} \mathbf{K}(\mathbf{B}, s) : \mathbf{G}(s)\,ds + o(\|\mathbf{G}(s)\|) . \quad (12.87)$$

The term free of $\mathbf{G}$ in the functional expansion has dropped due to the normalization condition given with Eq. (12.61).

12.4.4 Asymptotic Representation of the Hysteresis Term

The detailed exposition of the following presentation can be found in the literature [19, 45]. These references deal with technical problems, however, it turns out that in the broader sense this deformation class is of great relevance in biology as well.

If one looks at the hysteresis term Eq. (12.87), one sees that due to the many unknown quantities further restrictions are necessary. The deformation class one wants to consider will be asked for and one will adapt the material law to this class. An extremely important class, which is responsible for both technical and biological questions, consists of a static or slowly variable basic deformation superimposed by oscillations of small amplitudes.

In order to include this class, it is sufficient to linearize the hysteresis term physically Eq. (12.87) in $\mathbf{B}$. If one considers that due to the incompressibility

- all spherically symmetric terms will be added to the pressure and
- in the available free energy the deviation of the term $tr\mathbf{B} - 3$ from zero is of quadratic order,

the representation

$$\underset{s=0}{\overset{\infty}{\mathcal{F}}}(\mathbf{B}, \mathbf{G}(s)) = \int_{s=0}^{\infty} (\mu_1(s)\mathbf{G} + \mu_2(s)(\mathbf{GB} + \mathbf{BG}))\,ds \quad (12.88)$$

follows after a lengthier calculation or a little converted

$$\mathcal{F}_{s=0}^{\infty}(\mathbf{B},\mathbf{G}(s)) = \int_{s=0}^{\infty} (\mu'(s)\mathbf{G} + \delta(s)(\mathbf{G}(\mathbf{B}-\mathbf{I}) + (\mathbf{B}-\mathbf{I})\mathbf{G}))\,\mathrm{d}s\,. \quad (12.89)$$

As one can see, only two material functions are left that can be interpreted as follows:

$\mu'(s)$ is the relaxation kernel of the linear elasticity theory,
$\delta(s)$ describes effects of second order.

The material laws of multilinear or finite linear viscoelasticity are generally valid in the case of statics and in the case of dynamics for sufficiently small or for deformation histories which differ little from the static deformation. They hold asymptotically under the following process restrictions:

- the deformation consists of a static or slowly variable basic deformation
- superimposed by oscillations of small amplitude.

A technical example of the mentioned processes is shown in Fig. 12.17. There the progress of the temporal deformation of a motor mounting is shown qualitatively.

The first deformation process is the static compression where the dead load of the engine is decisive. While the engine is running, oscillations of low amplitude and higher frequency will be superimposed on this compression. While driving, oscillations of higher amplitude but lower frequency will also be added.

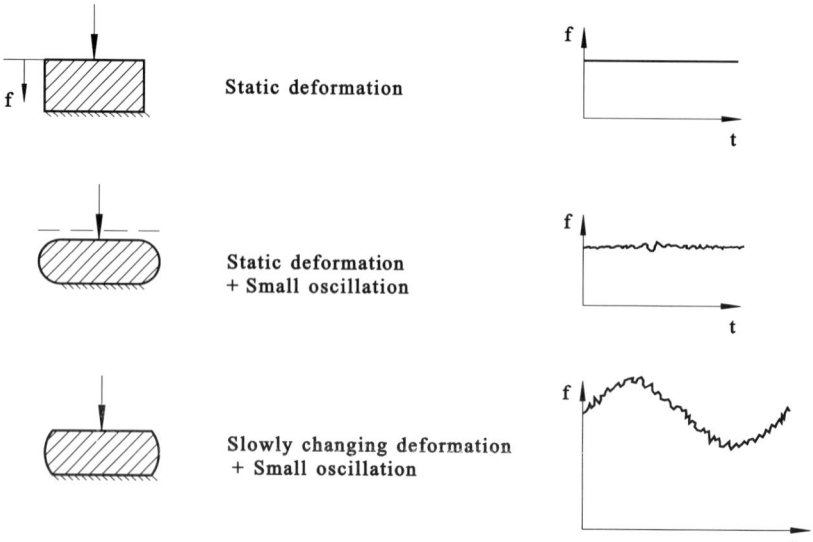

Fig. 12.17 Progress of temporal deformation of a motor mounting

12.4 Basics of Material Theory

The type of deformation considered here is also visible in running humans and in general in mammals. One was successful in simplifying the constitutive model while considering the expected class of deformation. What remains is the examination of the static term.

12.4.5 Examination of the Static Term

With presumed incompressibility and isotropy, one can write the material law given with Eq. (12.75) in the following form

$$\boldsymbol{\sigma} + p\mathbf{I} = \varphi_1 \mathbf{B} + \varphi_2 \mathbf{B}^2 . \tag{12.90}$$

The constraint stress p maintains the incompressibility condition. The functions φ_1 and φ_2 depend on the principal invariants I_1, I_2. The third invariant is $I_3 = \det \mathbf{B} = 1$ and therefore can be neglected. Due to the tensor derivatives used in the following, one introduces the new invariants $\tilde{I}_1$ and $\tilde{I}_2$ with

$$\tilde{I}_1 = I_1 = tr\mathbf{B} , \tag{12.91}$$

$$\tilde{I}_2 = I_1^2 - 2I_2 = tr\mathbf{B}^2 . \tag{12.92}$$

One further requires that Eq. (12.90) may be derived from a potential ψ. The representation

$$\boldsymbol{\sigma} + p\mathbf{I} = \rho_0 \left(\frac{\partial \psi}{\partial \tilde{I}_1} \frac{\partial \tilde{I}_1}{\partial \mathbf{F}} + \frac{\partial \psi}{\partial \tilde{I}_2} \frac{\partial \tilde{I}_2}{\partial \mathbf{F}} \right) \mathbf{F}^T \tag{12.93}$$

follows where

$$\frac{\partial \tilde{I}_1}{\partial \mathbf{F}} \mathbf{F}^T = 2\mathbf{B} ; \quad \frac{\partial \tilde{I}_2}{\partial \mathbf{F}} \mathbf{F}^T = 4\mathbf{B}^2 \tag{12.94}$$

have been used. Since ρ_0 represents the constant density, the scalar function ψ is the free energy referred to the mass. If one compares Eq. (12.93) with Eq. (12.90), one obtains

$$\varphi_1 = 2\rho_0 \frac{\partial \psi}{\partial \tilde{I}_1} ; \quad \varphi_2 = 4\rho_0 \frac{\partial \psi}{\partial \tilde{I}_2} \tag{12.95}$$

and the functions φ_1, φ_2 have to satisfy an integrability condition

$$\frac{\partial \varphi_1}{\partial \tilde{I}_2} = \frac{1}{2} \frac{\partial \varphi_2}{\partial \tilde{I}_1} . \tag{12.96}$$

These integrability conditions are always satisfied, if φ_1 is only a function of $\tilde{I}_1$, and if φ_2 is only a function of $\tilde{I}_2$. Due to this, one makes a constitutive assumption and postulates that

$$\psi = \psi_1(I_1) + \psi_2(I_2) . \tag{12.97}$$

Whether this condition is really satisfied or not, has to be verified by experiments. It seems reasonable to associate the first term with a very soft substance behavior and the second term with the stiffening of the material that occurs at higher strains. However, one has to remind of the phenomenological thermodynamics.

Boundary value and initial boundary value problems formulated properly have a clear solution if the energy ψ is a convex function. The Hessian matrix then must be positive definite, i. e. for all tensors $\mathbf{A}$ it holds

$$\mathbf{A} : \frac{\partial^2 \psi}{\partial \mathbf{F} \partial \mathbf{F}} : \mathbf{A} \geq 0 . \tag{12.98}$$

During large deformations, the convexity can lead to unphysical behavior. However, the existence of solutions can also be shown with the weaker condition that ψ is a polyconvex energy function. This mathematical concept, introduced already in [11] but gets attention only lately. One may here refer to the literature [9, 44]. Polyconvex biomechanics models have been proposed in [2, 26].

12.4.6 Considering Phenomenological Thermodynamics

Since the readership consisting of engineers, biologists and physicians, as a rule is rather acquainted with thermodynamics than with mathematics, the reminder of the thermodynamics turns out considerably shorter than the mathematical part.

The two laws of thermodynamics in integral form are given by the energy theorem

$$\dot{U} + \dot{K} = \dot{Q} + \dot{W} \tag{12.99}$$

and by the entropy inequality

$$\dot{S} \geq \frac{\dot{Q}}{T} . \tag{12.100}$$

The sign $>$ in Eq. (12.100) describes irreversible processes, and the sign $=$ describes reversible processes. The individual names mean

U: Internal energy
K: Kinetic energy
Q: Heat
W: Work of the external forces and body forces
T: Total temperature and
S: Entropy.

12.4 Basics of Material Theory

Equation (12.100) applies to materially closed systems that can only exchange heat with their neighborhood. S is a quantity increasing monotonically in adiabatic systems where the heat exchange is also prevented ($\dot{Q} = 0$). In equilibrium, where all temporal fluctuations come to rest, the entropy S assumes its maximum value. A natural finding is incorporated here that is unfortunately never explicitly mentioned thus causing a lot of discussions and confusions. There is no growth without limits in the *real world*. One has to consider that the second law of thermodynamics with the entropy inequality already contains this information. One has a sequence monotonically increasing in time and *bounded from above*. One concludes logically from this that an upper bound is assumed. The *final state* is reached. Therewith the second law of thermodynamics forms the basis of all extremum principles in nature. If one eliminates $\dot{Q}$ in the energy balance in Eq. (12.99) with the help of Eq. (12.100), one obtains while neglecting the kinetic energy K:

$$\dot{U} \leq T\dot{S} + \dot{W} . \tag{12.101}$$

This relation can be transferred into a mechanically more relevant form

$$(U - TS - W)^{\cdot} \leq S\dot{T} . \tag{12.102}$$

For an isothermal process $\dot{T} = 0$, the left side of Eq. (12.102) assumes a minimum. The two laws given by Eqs. (12.99) and (12.100) are universally valid. They are also valid for materials considered in a uniaxial tension test. In the following, one can neglect the kinetic energy as well as the term resulting from the body forces $\dot{W}$. If one looks at the strain of an incompressible rod in a tension test, as shown in Fig. 12.18, one obtains from both laws in the reversible case the relation named after Gibbs

$$T\,dS = dU - P\,dl . \tag{12.103}$$

With the free energy

$$F = \int_B \rho_0 \psi \, dv = U - TS \tag{12.104}$$

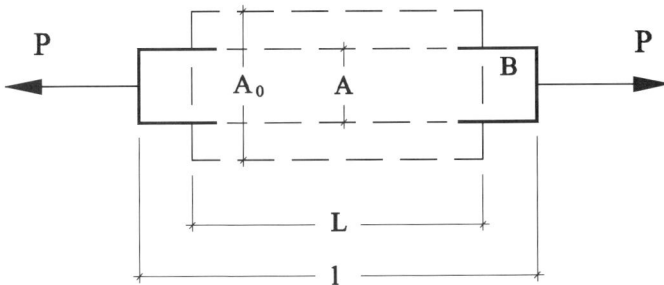

Fig. 12.18 Uniaxial tension test

one can write Eq. (12.103) alternatively as

$$dF = -S\,dT + P\,dl\,. \tag{12.105}$$

As one can see from the right side of equation (12.105), F, S, and P are functions of T and l. From this equation, one further obtains with constant temperature ($dT = 0$) the force P as following

$$P = \frac{\partial U}{\partial l}(l,T) - T\frac{\partial S}{\partial l}(l,T)\,. \tag{12.106}$$

One interprets this equation by saying that P has an energetic term and an entropic term. Both terms can be traced back to the constitutive law since the integrability condition given by Eq. (12.105) implies

$$-\frac{\partial S}{\partial l}(l,T) = \frac{\partial P}{\partial T}(l,T) \tag{12.107}$$

and therefore one obtains

$$P(l,T) = \frac{\partial U}{\partial l}(l,T_0) + \frac{\partial P}{\partial T}(l,T)T\,. \tag{12.108}$$

As one can easily see, this is the tangent to the curve $P(l,T)$ in the case of a fixed l and variable temperature T. The energetic term $\partial U/\partial l(l,T_0)$ is the axis section on the ordinate and the gradient factor is $\tan\alpha = \partial P/\partial T(l,T^*)$. Figure 12.19 illustrates these statements.

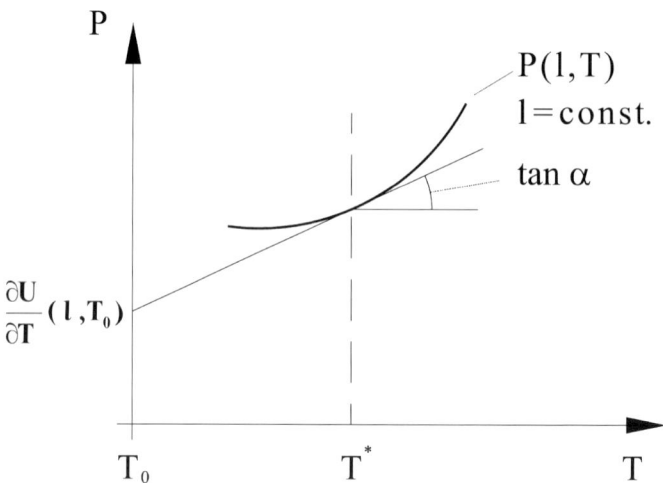

Fig. 12.19 Entropic term and energetic term of the tensile force P

12.4 Basics of Material Theory

Let us note what one gets from this simple calorimetric experiment:

$$P = \frac{\partial U}{\partial l}(l,T) - T\frac{\partial S}{\partial l}(l,T) = \frac{\partial F_2}{\partial l} + \frac{\partial F_1}{\partial l}, \qquad (12.109)$$

$$\frac{\partial F_2}{\partial l} = \frac{\partial U}{\partial l}(l,T), \qquad (12.110)$$

$$\frac{\partial F_1}{\partial l} = -T\frac{\partial S}{\partial l}(l,T). \qquad (12.111)$$

Since Eq. (12.111) is responsible for the rigid range of the force elongation relation therefore for smaller elongations, and Eq. (12.110) describes the change of the internal energy at great elongations, it seems reasonable to connect F_1 with the linear invariants of **B** being I_1 and to connect F_2 with the quadratic invariants of **B** being I_2.

Hypotheses: For F and ψ, respectively, it follows

$$F = F_1(I_1) + F_2(I_2), \qquad (12.112)$$
$$\psi = \psi_1(I_1) + \psi_2(I_2). \qquad (12.113)$$

Let us consider the procedure:

From the calorimetric experiment one obtains the split of the tensile force with

$$P = \underbrace{Th_1(\lambda)}_{\substack{\text{entropy-}\\\text{elastic term}}} + \underbrace{ah_2(\lambda)}_{\substack{\text{energy-}\\\text{elastic term}}}. \qquad (12.114)$$

In the tension test one obtains for an incompressible material:

$$\mathbf{F} = \begin{pmatrix} \lambda & 0 & 0 \\ 0 & \frac{1}{\sqrt{\lambda}} & 0 \\ 0 & 0 & \frac{1}{\sqrt{\lambda}} \end{pmatrix}, \qquad (12.115)$$

$$\mathbf{B} = \begin{pmatrix} \lambda^2 & 0 & 0 \\ 0 & \frac{1}{\lambda} & 0 \\ 0 & 0 & \frac{1}{\lambda} \end{pmatrix}, \qquad (12.116)$$

$$I_1 = \lambda^2 + \frac{2}{\lambda}; \quad I_2 = \lambda^4 + \frac{2}{\lambda^2}, \qquad (12.117)$$

$$\sigma = \begin{pmatrix} \sigma_{11} & 0 & 0 \\ 0 & \sigma_{22} & 0 \\ 0 & 0 & \sigma_{33} \end{pmatrix}, \qquad (12.118)$$

$$\sigma = -p\mathbf{I} + \varphi_1 \mathbf{B} + \varphi_2 \mathbf{B}^2, \qquad (12.119)$$
$$\sigma_{11} = -p + \varphi_1 \lambda^2 + \varphi_2 \lambda^4, \qquad (12.120)$$
$$\sigma_{33} = 0 = \sigma_{22} = -p + \varphi_1 \frac{1}{\lambda} + \varphi_2 \frac{1}{\lambda^2}. \qquad (12.121)$$

If one eliminates the undetermined pressure from Eq. (12.120) with Eq. (12.121), it follows

$$\sigma_{11} = \varphi_1 \left(\lambda^2 - \frac{1}{\lambda} \right) + \varphi_2 \left(\lambda^4 - \frac{1}{\lambda^2} \right) \qquad (12.122)$$

or

$$\sigma_{11} = \left(\lambda^2 - \frac{1}{\lambda} \right) \left(\varphi_1 + \varphi_2 \left(\lambda^2 + \frac{1}{\lambda} \right) \right) . \qquad (12.123)$$

The force is

$$P = \sigma_{11} A . \qquad (12.124)$$

Incompressibility provides that $Al = A_0 L$ and therefore $A = A_0 L/l = A_0 1/\lambda$ due to which the force assumes the following form

$$P = \sigma_{11} \frac{1}{\lambda} A_0 = A_0 \left(\lambda - \frac{1}{\lambda^2} \right) \left(\varphi_1 + \varphi_2 \left(\lambda^2 + \frac{1}{\lambda} \right) \right) \qquad (12.125)$$

If one compares Eq. (12.125) with Eq. (12.114) and takes into account that $\mathbf{B}^2$ in Eq. (12.119) has arisen by differentiating with respect to I_2, the correlation is unique

$$T h_1(\tilde{\lambda}_1) = A_0 \left(\tilde{\lambda}_1 - \frac{1}{\tilde{\lambda}_1^2} \right) \varphi_1 , \qquad (12.126)$$

$$a h_2(\tilde{\lambda}_2) = A_0 \left(\tilde{\lambda}_2 - \frac{1}{\tilde{\lambda}_2^2} \right) \left(\tilde{\lambda}_2^2 + \frac{1}{\tilde{\lambda}_2} \right) \varphi_2 . \qquad (12.127)$$

The invariants $\tilde{\lambda}_1$ and $\tilde{\lambda}_2^2$ are considered as the solutions

$$\tilde{\lambda}_1 = \tilde{\lambda}_1(I_1) ; \quad \tilde{\lambda}_2 = \tilde{\lambda}_2(I_2) \qquad (12.128)$$

of Eq. (12.117). This is equivalent to the cubic equations

$$\tilde{\lambda}_1^3 - I_1 \tilde{\lambda}_1 + 2 = 0 ; \quad \tilde{\lambda}_2^6 - I_2 \tilde{\lambda}_2^2 + 2 = 0 \qquad (12.129)$$

for $\tilde{\lambda}_1$ and $\tilde{\lambda}_2^2$ to be solved with Cardano's formula. The indices 1 and 2 are attached for better distinction. One can convince oneself very easily that $I_1 \geq 3$ is always valid and that the solutions $\tilde{\lambda}_1$ are always real-valued in this case. One root of the first Eq. (12.129) is negative, another one lies between 0 and 1, the third root is greater or equal 1 and thus has to be chosen for the tension test. With the auxiliary quantities

$$r_1 = \sqrt{\frac{I_1}{3}}, \quad \cos \beta_1 = \frac{1}{r_1^3} \qquad (12.130)$$

one obtains the solution

$$\tilde{\lambda}_1 = 2r_1 \cos\left(\frac{\pi - \beta_1}{3}\right). \tag{12.131}$$

The solution of the second equation of Eq. (12.129) provides analogically

$$r_2 = \sqrt{\frac{I_2}{3}}, \quad \cos\beta_2 = \frac{1}{r_2^3}, \tag{12.132}$$

$$\tilde{\lambda}_2 = \sqrt{2r_2 \cos\left(\frac{\pi - \beta_2}{3}\right)}. \tag{12.133}$$

For the uniaxial tension test $\tilde{\lambda}_1 = \tilde{\lambda}_2 = \lambda$.

Even for a very simple material law for elastomers such as the Mooney–Rivlin law the elastic material properties cannot solely be determined with the help of uniaxial tension tests [48]. The split into the entropy elastic term $Th_1(\tilde{\lambda}_1(I_1))$ and the energy elastic term $ah_2(\tilde{\lambda}_2(I_2))$ leads to constitutive laws, for which all material constants can determined economically with one single measurement. In the end, one considers an equibiaxial loading that provides a possibility of testing for verifying previously made assumptions in addition to the tension test with which one determines the constitutive law. It shall be mentioned that analyzing an equibiaxial extension cannot provide any verification of the assumptions. However, falsification is possible. Very sensitive but quite involved characterizations of the nonlinearly elastic behavior in E^2 or E^3 are possible by measuring the strain dependent sound speeds or time-of-flight of longitudinal and transversal acoustic waves [47].

12.4.7 Equibiaxial Loading

This type of deformation possesses an independent right for making material assumptions besides the previously mentioned control possibility. As has already been mentioned in Sect. 12.3.3, biopolymers form different spatial structures. These are mostly laminar structures that are found in the skin, the veins, the bladder etc. For these structures, the equibiaxial loading provides an adapted measuring method. In this test, a thin membrane is inflated. One obtains an only approximately spherical cup since the equibiaxial stretch is disturbed at the clamped boundary. One can obtain more accurate models by solving partial differential equations of the system [18]. Nevertheless, the following simple model already describes the basic behavior. Another problem for the evaluation are unknown residual stresses which always arise in biological tissue due to growth, swelling, or dehydration. An obvious example is the formation of wrinkles in aging tissues as a result of compressive residual stresses.

Figure 12.20 roughly sketches the experimental setup and indicates important parameters of the equipment, deformation and loading.

$\Delta p = p_i - p_a$:   Pressure difference
r: Current radius of curvature of the spherical cup
R: Radius of the flat membrane
h: Maximum deflection
S: Current size of S
$\lambda = S/R$:   Stretch

If one looks at a the deflected membrane in comparison to the not deflected disc, the relations in Fig. 12.21 are valid. The thickness D of the flat membrane has the current value d.

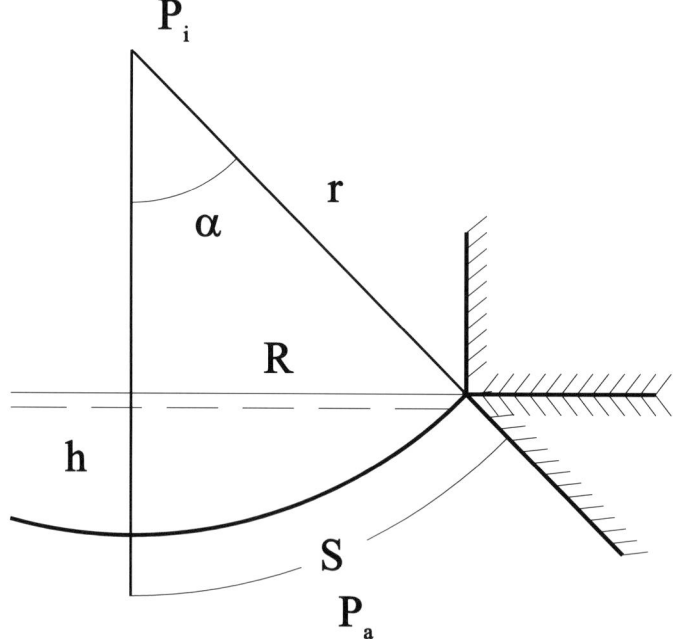

Fig. 12.20 Equibiaxial loading experiment (bulging test)

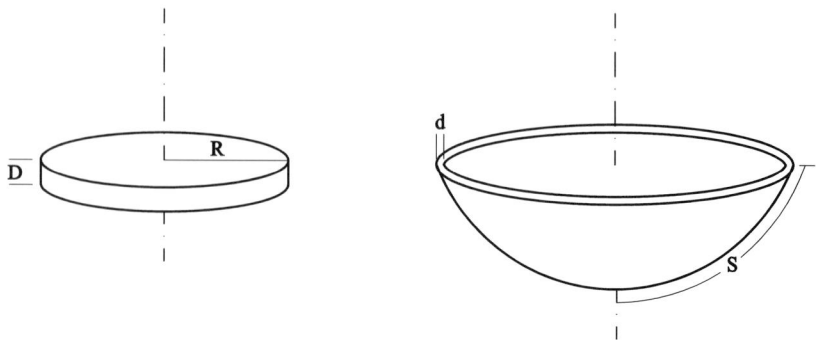

Fig. 12.21 Equibiaxial loading experiment (bulging test)

12.4 Basics of Material Theory

In the case that the membrane is very thin $D/R \ll 1$ the bending component can be neglected compared to the strain component and therefore the pressure jump can be neglected compared to the tangential stress $\sigma_{11} = \sigma_{22}$. For simplicity equibiaxial residual stresses $\rho_{11} = \rho_{22}$ will be assumed. In this case, the equilibrium conditions provide

$$p_i - p_a = \Delta p = 2(\sigma_{11} + \rho_{11}) \frac{d}{r}. \tag{12.134}$$

With the abbreviation $\xi = h/R$ one obtains the kinematics

$$\lambda := \lambda_1 = \lambda_2 = \frac{s}{R} = \frac{r\alpha}{R}, \quad \lambda_3 = \frac{d}{D}, \tag{12.135}$$

$$\cos\alpha = \frac{|R^2 - h^2|}{R^2 + h^2} = \frac{|1 - \xi^2|}{1 + \xi^2}, \tag{12.136}$$

$$\lambda = \frac{\alpha}{\sin\alpha} = \frac{1 + \xi^2}{2\xi} \alpha, \tag{12.137}$$

$$\frac{d}{r} = 8 \frac{D}{R} \frac{\xi^3}{(1+\xi^2)^3} \frac{1}{\alpha^2} = \frac{D}{R} \frac{\alpha}{\lambda^3}, \tag{12.138}$$

As one can see from Eqs. (12.134) and (12.138), the assumption of pure stretching

$$\mathbf{F} = \begin{pmatrix} \lambda & 0 & 0 \\ 0 & \lambda & 0 \\ 0 & 0 & \frac{1}{\lambda^2} \end{pmatrix}, \tag{12.139}$$

$$\mathbf{B} = \begin{pmatrix} \lambda^2 & 0 & 0 \\ 0 & \lambda^2 & 0 \\ 0 & 0 & \frac{1}{\lambda^4} \end{pmatrix}, \tag{12.140}$$

$$I_1 = 2\lambda^2 + \frac{1}{\lambda^4}, \quad I_2 = 2\lambda^4 + \frac{1}{\lambda^8} \tag{12.141}$$

is well approximated for $D/R \ll 1$ and $\xi = h/R \gg 0$. For the stresses one obtains

$$\sigma_{11} = \sigma_{22} = -p + \varphi_1 \lambda^2 + \varphi_2 \lambda^4, \tag{12.142}$$

$$0 = \sigma_{33} = -p + \varphi_1 \frac{1}{\lambda^4} + \varphi_2 \frac{1}{\lambda^8}. \tag{12.143}$$

If one eliminates the constraint stress p in Eq. (12.142) (no causal connection with Δp), one obtains

$$\sigma_{11} = \left(\lambda^2 - \frac{1}{\lambda^4}\right)\left(\varphi_1 + \varphi_2\left(\lambda^2 + \frac{1}{\lambda^4}\right)\right). \tag{12.144}$$

φ_1 and φ_2 are functions of I_1 and I_2 that have been determined before in Sect. 12.4.6. With this experiment, one is able to check the assumptions made in Sect. 12.4.6. The presented experiment is of independent interest for biopolymers. This test should be particularly adequate for questions regarding fatigue strength and delamination. With the Eqs. (12.136), (12.138) and (12.144) and the equilibrium condition given by Eq. (12.134) one obtains the equation of the measured curve

$$\Delta p = \frac{2D}{R\lambda^3}\left(\left(\lambda^2 - \frac{1}{\lambda^4}\right)\left(\varphi_1 + \varphi_2\left(\lambda^2 + \frac{1}{\lambda^4}\right)\right) + \rho_{11}\right)\arccos\left(\frac{1-\xi^2}{1+\xi^2}\right). \tag{12.145}$$

As a typical result for structures that basically deform perpendicular to the load, one recognizes that the material functions φ_1 and φ_2 can only be determined together with the residual stress. In the uniaxial tension test, there are no residual stresses since the problem is statically determined.

Layered biopolymers are made up of several substances such as elastin and collagen fibers that are incorporated in it. Questions regarding delamination or blister of collagen fibers may be examined with the introduced equibiaxial loading experiment superposed by a periodic loading. What is left in the end is the question about concrete constitutive laws which will be discussed in the next section.

12.5 Material Laws for Technical and Biological Polymers

12.5.1 Technical Polymers (Gauß Networks and Kilian Networks)

Kilian makes an interesting suggestion for technical polymers (elastomers) [28, 29]. In analogy to van der Waals' idea he modifies the model of the "ideal conformation gas" as which Eq. (12.146) can be understood. Starting from the ideal equations of state, he obtains the equation describing the real network. If one considers the tension test as shown in Fig. 12.18, one obtains the equation of the ideal conformation gas describing the Gauß (Gauss) network

$$P = \frac{\rho RT}{\lambda_m^2 M_m} A_0 \left(\lambda - \frac{1}{\lambda^2}\right). \tag{12.146}$$

Here ρ is the density, $R = 8.31451\,\mathrm{J\,mol^{-1}\,K^{-1}}$ the (general) molar gas constant, M_m the molecular weight of the stretch invariant basic unit thus $\rho = 1\,\mathrm{g\,cm^{-3}}$ and $M_m = 68.11\,\mathrm{g\,mol^{-1}}$ if natural rubber (C_5H_8) is considered. Further statistical mechanics models of rubber elasticity are discussed in [33, 41].

The quantity λ_m deserves special interest because it can be related to the ultimate elongation of the elastomer. At first one sees that the elastomer becomes stiffer in case of a smaller λ_m. It should be clear that this is connected to the quantity of the cross linking agent. As one can tell by Eq. (12.146), λ_m is already given by the initial slope of the curve at $\lambda = 1$. Therefore it could be shown in Sponagel et al. [46] that this quantity can be measured by a simple Shore hardness test and that it represents the ultimate stretch of the elastomer. So all strains $\lambda > \lambda_m$ are excluded.

12.5 Material Laws for Technical and Biological Polymers

In the representation of Eq. (12.122) it follows $\varphi_2 = 0$ and $\varphi_1 = $ constant. If one compares relation Eq. (12.122) to Eq. (12.146), one obtains

$$\varphi_1 = \frac{\rho RT}{\lambda_m^2 M_m}. \tag{12.147}$$

One sees that φ_1 is completely interpreted on molecular level and the quantity λ_m is determinable in a very simple test [46].

Kilian extends Eq. (12.146) analogically to the van der Waals gas, in which he brings analogous physical effects into play [28, 29]. He postulates the real network behavior in a tension test which we rewrite in the correct van der Waals form

$$\frac{P}{A_0} = \frac{\rho RT}{\lambda_m^2 M_m} \frac{\Lambda}{1 - \frac{\Lambda}{\Lambda_m}} - a\Lambda^2, \tag{12.148}$$

$$\Lambda(\lambda) = \lambda - \frac{1}{\lambda^2}, \quad \Lambda_m = \Lambda(\lambda_m). \tag{12.149}$$

This model is called "Kilian network" in the following. If one compares Eq. (12.148) with the known relation for gases, one sees that the exclusion volume of the network chains takes the position of the exclusion volume of the molecules in the gas. This is the meaning of λ_m. The cohesive forces between the molecules in the gas are described by Kilian with the parameter $a > 0$.

These effects only occur for larger deformations since the elastomer chains must accordingly get close to each other. Rubber tends to crystallize and these energy elastic effects are included here. This circumstance also justifies an extension of the Kilian model by the assumption that this additional term is only described by the invariant I_2.

For the tension test, Eqs. (12.148) and (12.149) can directly be used. For this, one compares relation Eqs. (12.148), (12.149) with Eq. (12.126) and Eq. (12.127). With this unique assignment, one obtains the entropy-elastic term with the Eqs. (12.126) and (12.148)

$$\varphi_1(I_1) = \frac{\rho RT}{\lambda_m^2 M_m} \left(\frac{1}{1 - \frac{\Lambda_1}{\Lambda_m}} \right), \quad \Lambda_1 = \Lambda(\tilde{\lambda}_1(I_1)). \tag{12.150}$$

The energy-elastic term is obtained with the Eqs. (12.127) and (12.149)

$$\varphi_2(I_2) = \frac{-a\Lambda_2}{\tilde{\lambda}_2^2(I_2) + \frac{1}{\tilde{\lambda}_2(I_2)}}, \quad \Lambda_2 = \Lambda(\tilde{\lambda}_2(I_2)). \tag{12.151}$$

Equation (12.148) for the tension test is unchanged because for this test $\tilde{\lambda}_1 = \tilde{\lambda}_2 = \lambda$. Figure 12.22a shows the quasi-static stress-stretch curve of natural rubber at room temperature. The material constants are found by approximation to this tension test.

Of course, it cannot be expected that this assumption of the split into an entropy elastic term (function of I_1) and an energy elastic term (function of I_2) is universally valid. However, it remains to hope that it proves to be useable within an approximation. The biaxial strain tests will be considered for a validation. For the bulging test, the new invariants $\tilde{\lambda}_1(I_1)$ and $\tilde{\lambda}_2(I_2)$ are calculated with I_1 and I_2 taken for the equibiaxial stretch. It is assumed that residual stresses are negligible.

With Eqs. (12.145), (12.150, 12.151), with the identified constants $\lambda_m = 10.77$, $a = 0.17\,\text{N}\,\text{mm}^{-2}$ and with replacing $\rho RT/(\lambda_m^2 M_m)$ by $0.904\,\text{N}\,\text{mm}^{-2}$ for the same natural rubber one obtains a prediction of the bulging test. It is shown together with measurements in Fig. 12.22b. As one can see, the made assumptions about the invariants of **B** seem to lead to useable results. The remaining discrepancies may be addressed to the too simple evaluation that assumes a perfect ball shape of the inflated membrane. In the experiment described in Sect. 12.4.7, the clamped support actually prevents the undisturbed formation of an equibiaxial stress state. A more exact measurement of the stretch λ and of the membrane curvatures may be made with a video-based system in three measuring planes [23].

Examinations of real spatial problems as they occur for rubber springs however prove the assumptions and with known M_m one obtains the statement:

Technical polymers can be well described with two measured parameters λ_m, a and two relaxation functions $\mu'(s)$ and $\delta(s)$.

The importance of this statement will be clear to everybody who knows the complexity and the size of the elastomer processing industry. Figure 12.23 shows that the maximum stretch in tension is larger than in equibiaxial loading and it also shows

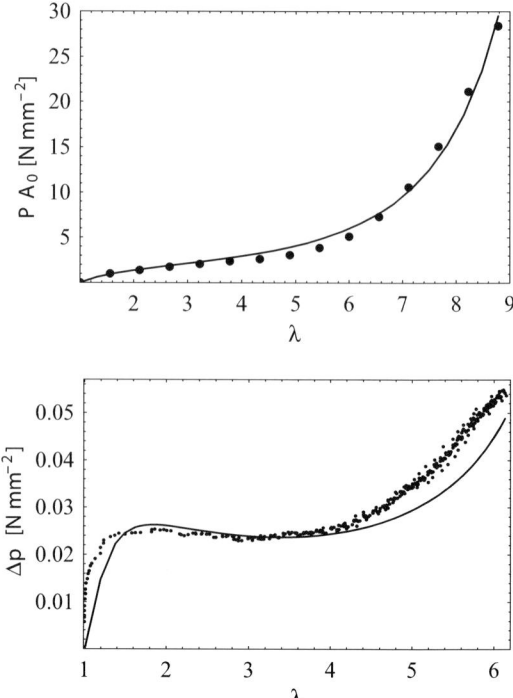

Fig. 12.22 Tests and approximation with the modified Kilian model for natural rubber. **a** Tension test, **b** Bulging test

12.5 Material Laws for Technical and Biological Polymers

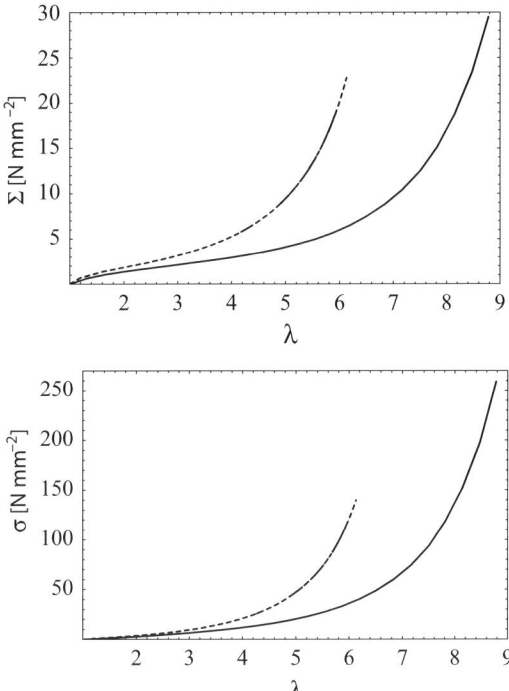

Fig. 12.23 Modified Kilian Model for natural rubber compared in uniaxial tension (—) and in equibiaxial loading (- - -) **a** 1st Piola–Kirchhoff stresses, **b** Cauchy stresses σ

that although rubber is weak in terms of technical stress Σ it is relatively strong in terms of true stress σ.

12.5.2 Biological Polymers (Collagen)

Despite of some similarities with the behavior of technical polymers [20] the situation with biological polymers is much more complex. On the one hand, this stems from the variety of the materials as well as from the structures that can arise and on the other hand from the fact that these substances are in a aqueous environment. From the mechanical point of view, however, it attracts attention that the force-stretch relation in principle proceeds differently than for technical polymers. Technical polymers show in their initial force-stretch relation, no matter whether Gauß or Kilian networks are considered, a basically degressive behavior, whereas this is particularly different for biopolymers. Figure 12.24 shows this difference and immediately makes clear to the engineer, with the area under the different curves, which energies are stored and what this means for a fracture.

In Fig. 12.24b a dotted line is marked, that does not go through the origin $P = 0$ and $\lambda = 1$. This is an remarkable circumstance, because here one sees the tissue prestress P_0 that everyone experiences in his ageing skin. Young skin possesses a big P_0 and old skin a small P_0. As one sees, exactly this prestress has an essential

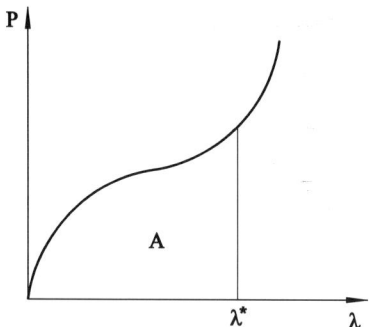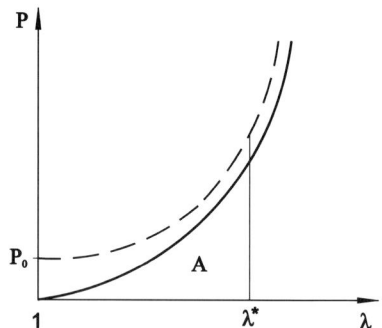

Fig. 12.24 a Kilian network. **b** Biopolymer

influence on biological tissues. What this means for the cosmetics industry, should be clear to everyone.

Yamada [51] presented extensive measurements for biological tissue. Here only some aspects can be illustrated. Fung demonstrated in [15] that uniaxial J-shaped stress-strain curve of many tissues follows an exponential law which could later be extended to more dimensions and to anisotropic materials. For a more complete overview over the extensive state of modern material theories and their applications in the biomechanics of soft tissues one should refer to the literature [10, 21, 22, 25].

Before discussing the molecular mechanisms, that may evoke a J-shaped stress-strain curve as well as the previously mentioned residual prestress, the advantages that are provided by such elastic behavior are considered first. When it turned out from biomechanical examinations that many biological diaphragms and vessel walls, such as arteries, are exceptionally ductile, biomechanics inferred from this that the fracture energy of these materials must be very high. The biomechanics did not find any particular high values when they determined the fracture energy for a large number of animal tissues on the basis of exact measurements. Most fracture energies were between $10^3 \ldots 10^4 \, \text{J/m}^2$. The tough skin of a rat, a worm or the vessel wall of a human artery differs less from metal sheets and rubber based on the fracture energy but rather in terms of the different stress-strain curves.

The material has an insignificantly small shear modulus in the area of small stresses and strains in which the J-curve practically extends horizontally. Therefore, the deformation energy that would become free in case of a fracture cannot be transferred efficiently into the fracture zone. That is a simple but very elegant mechanism in order to increase the safety. Furthermore, the J-curve has other mathematical features than the S-curve so that it does not advantage instabilities in pressurized tubes – aneurysms therefore usually cannot develop in arteries. The situation is quite different for Kilian networks (one thinks of the phase transition during the liquefaction of gas). The advantageous material properties of animal skin are reflected in the traditionally wide application of leather and untanned furs. Before synthetic substances with similar properties were introduced, there was no adequate substitute for leather.

12.5 Material Laws for Technical and Biological Polymers

In order to make clear the connection between the J-shaped course of the stress-strain curve and the toughness of a material, the consequences of another curve shape is discussed for the example of the amnion in eggs. The skin lining the inside of the solid porous eggshells is used for protecting the embryo against moisture and pathogens. The hatching chick which is not especially strong must be able to free itself from its egg. This would be enormously complicated by a very tough amnion. Actually, this skin can tear easily. Up to strains of 22%, the material behavior follows Hooke's law and therefore it represents a special case within the animal tissues. Which molecular mechanism causes this behavior is still unsolved. A extension of covalent bonds by 22% can certainly be ruled out, since the elasticity modulus with $7\,\text{N/mm}^2$ is too small for this.

One finds some good examples for the meaning of J-curves with regard to the tearing of textiles. For a knitted material, one can feel the consequences of the J-shaped stress-strain curves very easily: Sweaters or socks can hardly be torn by pulling at them. Woven materials stretch in accordance to Hooke's law in the two directions of warp and weft. For other directions, diagonal to the weft, the strain behavior corresponds to a J-curve. Most textiles can be torn precisely along a line – however only in direction of the main threads. Tailors frequently make use of this. On the other hand, it is very difficult to tear a woven cloth diagonal to the main threads. Therefore, rips are usually L-shaped in woven clothing because the material tears most easily in the two perpendicular thread directions.

Remarkably, also the human amnion skin, that encloses an embryo, has a J-shaped stress-strain curve that rises however very steeply. This skin shall finally rupture at birth. Within the last two months of pregnancy, it weakens considerably as already shown by Yamada [51]. Probably, this is based on physiological changes that do not directly influence the elasticity of the tissue, but weaken the restrains in the amnion tissue in some way. Biochemical and clinical influences have been discussed in [7].

Also one's skin can become weak and brittle through infections. Exactly this point is of special interest, because here many biologists, biochemists and dermatologists, asks what to do in order to maintain the state of youth. This question is of course not new, but the chances to answer these questions grow with the knowledge of these substance classes in biomechanics.

At the end of these considerations towards biopolymers, the prestress P_0 will be examined because this is actually easy to see. For the following considerations, it is not important whether the side chains are hydrophilic or hydrophobic. If they are hydrophilic, the macromolecule is shrouded in a water drop. In an other case, the environment is water. This is visualized in Fig. 12.25.

If the macromolecule in Fig. 12.25 is elongated it starts to untangle itself causing the entropy S to decrease and causing the interface between water and macromolecule to enlarge. Using the relation given by Eq. (12.108) one obtains:

$$P = \underbrace{\frac{\partial U}{\partial l}(l,T)}_{\text{Change of internal energy due to enlargement of interface}} - \underbrace{T\frac{\partial S}{\partial l}(l,T)}_{\text{Change of entropy}} \qquad (12.152)$$

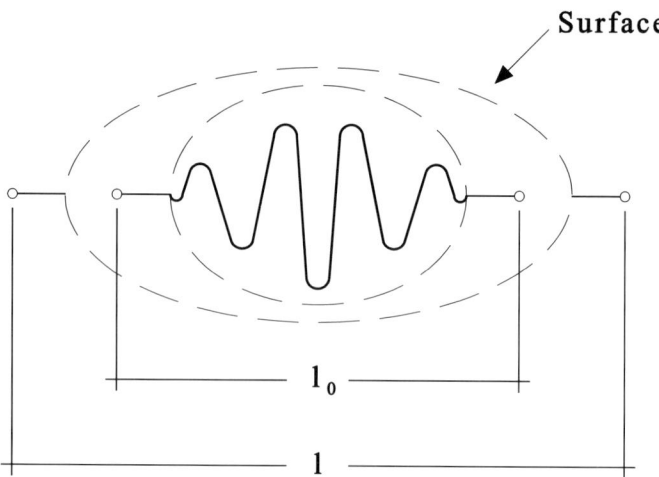

Fig. 12.25 Macromolecule

One sets $U = P_0 l$ and $S = -cl^2/2$. Here the whole internal energy term has been added to the interface energy,

$$P = \underbrace{P_0}_{\substack{\text{Surface} \\ \text{stress}}} + \underbrace{Tc}_{\substack{\text{Stiffness} \\ \text{due to} \\ \text{entropy} \\ \text{change}}} l \qquad (12.153)$$

One further has to make oneself clear that the aqueous milieu keeps the macromolecules in an extended state and therefore it prevents crystallization. So one can realize that the ability to elongate is always given except in exceptional conditions (e. g. dry lips) and that the parameter a as it exists in Kilian networks does not play a significant role here.

Collagen as a central component of human and animal tendons is biomechanically especially interesting. There are two important functions tendons have to fulfill: They are used on the one hand to transfer tensional forces to bones and to other parts of the body for enabling body motion. On the other hand, they act as springs for storing energy.

Muscles in vertebrates are designed biomechanically in a way that they work over certain distances. But if all muscles would be directly attached to bones, this would result in a less efficient construction principle. Anyway, one can hardly imagine that hands with muscles directly attached to phalanges would enable differentiated motions. In reality, muscles situated well above in the arm and connected to the bones of the hand by long thin tendons participate in the motions of the hand. One can easily feel them if one moves the fingers of a hand and simultaneously the forearm with the other hand. Tendons are used for storing deformation energy besides being

12.5 Material Laws for Technical and Biological Polymers

used for the power transmission between muscles and bones – therein they resemble a spring. One just has to observe exactly when deers or horses jump, when cats catch mice or when squirrels and monkeys jump from branch to branch in order to understand that animals are equipped with a variety of very efficient devices for storing deformation energy.

Tendons which mainly consist of parallel collagen fibrils are stiff enough for transferring reliably muscle forces – the maximum strain lies around 8 – 10% which is small by biological standards. Tendons on the other hand can store extremely high strain energies – much more than traditional materials can do.

Pechhold and coworkers were able to interpret several physical properties of polymers with the meander model. In [40] they have therewith described the mechanical properties of myosin, elastin and collagen and have introduced a muscle model in [39]. The mechanics of individual collagen fibers can be described on micro scale with the help of the so called Wormlike-Chain-Model (WCM).

The WCM starts from an arbitrary but continuing curvature of collagen fibers in the unstressed state (Fig. 12.26). The relation between the force P_C applied to a collagen fiber, and the average end-to-end distance r of a collagen fiber is given in [31, 38]:

$$P_C = \frac{kT}{4L_P} \left[4\frac{\lambda_c - 1}{\lambda_m - 1} + \left(1 - \frac{\lambda_c - 1}{\lambda_m - 1}\right)^{-2} - 1 \right]. \qquad (12.154)$$

$k = 1.3806568 \times 10^{-23}$ J/K is the Boltzmann constant, T is the total temperature, L_P describes the persistent length as a measure for the initial stiffness of a collagen fiber, $\lambda_c = r/L_P$ is the stretch of a collagen fiber, and λ_m is the maximum stretch in the completely stretched condition. A continuum model can be derived form the

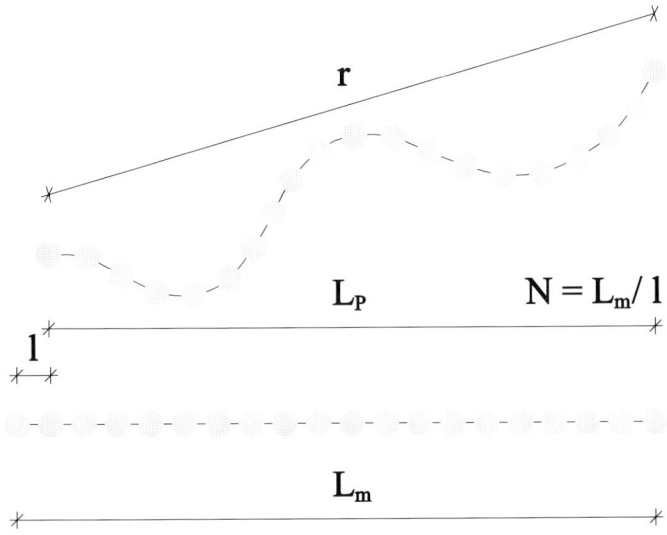

Fig. 12.26 Wormlike-Chain-Model [31]

WCM by building a four-chain-network model and eight-chain-network model in E^2 (e. g. skin, amnion) and E^3, respectively.

One also finds here the exclusion volume corresponding to the Kilian network. What is missing is the term of the internal energy that describes the crystallization. This circumstance could lead to the assumption that the internal energy only depends on I_1.

With biopolymers, however, the necessity for a spatial description is not given as much as for technical polymers. Collagen is build up linearly and fibrous, respectively, and due to this the deformation is predefined. It consists of a simple extension and spatial considerations become irrelevant. It shall be noted that only two free parameters are existing in Eq. (12.154). These are the persistence length L_P and the maximum lengths L_m and λ_m, respectively.

Lambertz examined animal tendons in vibration experiments [32]. The question he examined was connected to the locomotion of animals. Due to the equivalence of this deformation class occurring in nature and the one which has been examined here, he achieved results that agreed well with the material models developed in this article.

The approach is given by

$$\frac{P}{A_0} = E\varepsilon(t) + \beta\varepsilon^2(t) + 2\int_0^\infty \{K_0(s) - 2(K_0(s) - K_1(s)\varepsilon(t))\}\varepsilon_t(s)\,ds \quad (12.155)$$

with

A_0:   initial area of the tendon,
$E = \frac{3}{2}\frac{kT}{A_0 L_P(\lambda_m - 1)}$:   initial rigidity of the fibers,
$\beta = \frac{3}{4}\frac{kT}{A_0(\lambda_m - 1)^2}$:   further rigidity measurement describing the deviation from linear behavior,
$\varepsilon = \lambda - 1$:   strain,
$\left.\begin{array}{l} K_0 = -k_0 e^{-s\lambda_0}, \ k_0, \lambda_0 > 0 \\ K_1 = -k_1 e^{-s\lambda_1}, \ k_1, \lambda_1 > 0 \end{array}\right\}$:   relaxation functions,
$\varepsilon_t(s) = \varepsilon(t-s) - \varepsilon(t)$:   strain history.

The integral free term of the approach corresponds of the expansion of Eq. (12.154) up to the quadratic terms. The hysteresis term is the expansion (Eq. (12.88)), Eq. (12.89) responsible for the considered class of motions.

For input signals of the form

$$\varepsilon(t) = \varepsilon_0 + \varepsilon_1 \sin(\omega t), \quad \varepsilon_1 < \varepsilon_0 \quad (12.156)$$

the evaluations of the vibration experiments yield the results shown in Fig. 12.27.

As one can see, the developed theory also agrees well with the theory, at least for biopolymers. Also in this case the following is sufficient to define the model:

two numbers L_P, λ_m and two relaxation functions $\mu'(s)$, $\delta(s)$.

Here, the biologists, biochemists and physicians see the enormous advantage they attain. Physiological conditions that can be influenced by training, medications etc.

12.6 Volume Change in Biopolymers

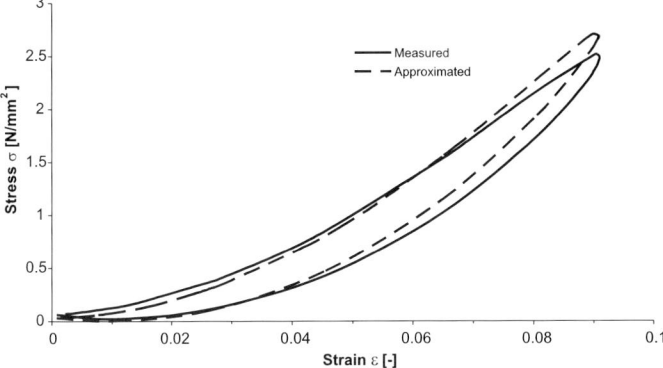

Fig. 12.27 Theoretical and real stress-strain curve of a tendon at 10 Hz [32]

make themselves noticeable in a change of the free parameters. The number of these free parameters has diminished to a minimum and is microscopically interpretable.

Finally, we consider another very interesting feature of biopolymers which leads very much into the proximity of the motoric function of biological substances. This last section deals with the swelling and the shrinking of this substance class.

12.6 Volume Change in Biopolymers

In Sect. 12.2.3, it was mentioned that biopolymers are normally available in the form of a mixture with a liquid. If the liquid is water, one talks about hydrogels and in general one speaks of near valence gels. The liquid acts as a weak solvent and influences the volume of the actual gel as well as the quantity of the valence bonds. A typical example in technology is polyvinyl chloride (PVC) which is available in two trade forms, as non-rigid PVC and as rigid PVC. Well-known examples in medicine and biology are for example skin, hyaline cartilage, hemoglobin, etc.

One knows from the moisture loss of one's lips what it means if these substances lose their moisture. They become rough and tear. But also an active process of "swelling and shrinking" can be observed for biological hydrogels. The network of the biopolymers possesses acid side chains, more exactly carboxyl groups COOH. These ionize in aqueous solutions, so that a COO^- ion remains in the network, while a H^+ ion is released. The latter quickly accumulates at a water molecule and this makes up a H_3O^+ ion which one calls counter ion. Gels swell in aqueous solutions by absorbing the water and they shrink through expression of water. These volume changes can be caused by

- a temperature change,
- an alteration of the ionization degree,
- mechanical loading.

In a typical condition, the gel consists of three particle types: Chain links of the polymer molecules – some ionized –, counter ions, and water molecules. Usually one represents these observed appearances in a (T, υ) diagram with T as the temperature and υ as the degree of shrinking that is defined as the ratio of the number of polymer links to the total number of the particles in a gel. Therefore $\upsilon = 1$ holds, if the gel shrank completely; alternatively $\upsilon < 1$. Figure 12.28 shows a schematic diagram of this type.

In Fig. 12.28a, the dashed line corresponds to a stronger ionization of a polyelectrolytic gel. For lower temperatures, the gel shrank and it only swells a little for an increase in temperature. Only for a transition temperature T_t the process of swelling starts and continues up to values of $\upsilon \ll 1$, so that the volume of the gel can increase strongly. For a further temperature increase, only insignificant swelling takes place. For a subsequent decrease of temperature, the graph is run through backwards. Especially, the essential part of the shrinking procedure also takes place at the temperature T_t. Analyzing Fig. 12.28a, this phenomenon reminds of a phase transition meaning a stability problem. Indeed, one describes swelling and shrinking as a result of a competition of three thermodynamic forces – two of them are entropic and one is energetic. These forces are

- osmotic pressure of counter ions – an entropic expanding force,
- network elasticity – an entropic contracting force,
- molecular interaction. This is an energetically contracting force for the gel.

At the transition temperature, the expanding tendency of the osmotic pressure prevails over the two contracting tendencies. This leads to the abrupt swelling process. In the following, these forces will be discussed in more detail while analyzing the reasons for their appearance.

The electrostatic interaction of the cross-linked ionomers and the counter ions, enforces an electro neutrality of the system at each point. Since the cross-linked ionomers are bonded to the gel, the counter ions consequently cannot leave the gel. On the other hand, the counter ions are practically free particles, and as such, they have the tendency to homogeneously fill the whole system of gel and water-bath. Therefore, they press from the interior against the surface of the gel and expand this through partial entanglement of its long-chain molecules. Water molecules fill

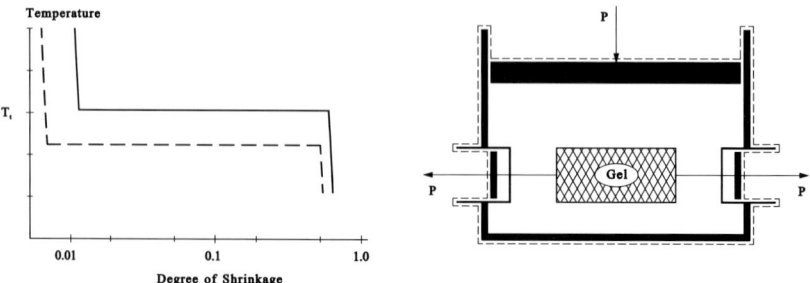

Fig. 12.28 a (T, υ)-diagram for two polyelectrolytic gels. **b** Gel in bath loaded with traction P

12.6 Volume Change in Biopolymers

the area that is created by the entanglement in the gel during the expansion process; these can freely enter and leave the gel. The surface of the gel therefore acts as a semi-permeable wall and the effect of the counter ions can be described as osmotic pressure. The expansion of the gel sets the long-chain molecules into the unlikely untangled condition. They have the tendency to entangle themselves and generate in this way the network elasticity which is an entropic force counteracting the osmotic pressure. Both forces are linearly dependent on the temperature. The network is energetically in the most favorable situation if only members of the network form next neighbor pairs. There is an energy penalty for the formation of a pair made up of a member and a water molecule or any other unequal pair. This effect is largely independent on the temperature and therefore definitely determines the behavior at low temperatures – where the two other forces are small – and leads to shrinking. A physician or biologist describes the sketched system in Fig. 12.28b, which one calls gel plus water-bath, as gel. He imagines, for example, a red blood cell or a intervertebral disk. In order to combine these different ideas, one assumes the water-bath as the volume of water that is needed by the swollen gel in order to homogeneously fill the volume. The sketched figure only has a heuristic value analogously to the spring-damper circuits of Table 12.1. Considering what has been said so far, it becomes clear what is meant with this figure: it describes the degree of homogeneity of a mixture. One imagines the water-bath as being of maximal volume if the gel is a pure polymer and contains no water. In the other case, the whole water reservoir is optimally dissolved in the gel, and the gel fills the whole volume. In this example the swelling process and shrinking process will be described as a function of the temperature.

For the considered system of Fig. 12.28b, one uses the second law of thermodynamics and obtains from Eq. (12.102)

$$(U - TS - W)^{\bullet} \leq S\dot{T} . \tag{12.157}$$

With $P = $ const.

$$\dot{W} = P\dot{l} + p\dot{V} \tag{12.158}$$

it follows

$$(U - TS - Pl)^{\bullet} \leq S\dot{T} + p\dot{V} . \tag{12.159}$$

From this one concludes that for isothermal and isochoric processes ($T = $ const., $V = $ const.) the left quantity in Eq. (12.159) called available free energy A assumes a minimum,

$$A = U - TS - Pl \to \text{Minimum} . \tag{12.160}$$

One must now associate the internal energy and the entropy with the parameters of the system. In order to characterize the system, one introduces the following quantities:

n_1 Number of water molecules,
$n_1^{(1)}$ Number of water molecules in the gel,
$n_1^{(2)}$ Number of water molecules in water bath,
n_2 Number of polymer chains,
xn_2 Number of chain links of molecular size,
n_3 Number of counter ions,
$n' = n_1' + xn_2 + n_3$ Number of particles of molecular size in the gel.

By introducing xn_2 one only has particles of equal size that form the gel: Water molecules, counter ions and chain links. One defines the degree of shrinking v,

$$v = \frac{xn_2}{n'} = \frac{V_0}{V} \qquad (12.161)$$

as the ratio of molecular-sized chain links to the total number. This means macroscopically for an uniaxial tension with the deformation gradient

$$\mathbf{F} = \begin{pmatrix} \lambda & 0 & 0 \\ 0 & \beta & 0 \\ 0 & 0 & \beta \end{pmatrix} \qquad (12.162)$$

and the determinant

$$\det \mathbf{F} = \lambda \beta^2 = \frac{1}{v} . \qquad (12.163)$$

Besides v and λ one introduces the degree of ionization ζ,

$$\zeta = \frac{n_3}{xn_2} . \qquad (12.164)$$

ζ describes the fraction of the ionized chain links. Since the quantities x, n_2, n_3 are constants for a gel, only n_1' and v, respectively, and λ remain as variables in the available free energy A. Therewith one obtains a relation of the type

$$A = A_0 + xn_2 kT \bar{A}(v, \lambda, e, x, \zeta, f, T) \qquad (12.165)$$

with

$$\bar{A} = \frac{1 - (1 + \zeta)v}{v} \ln[1 - (1 + \zeta)v] + \zeta \ln v + \frac{e}{kT}\left(\zeta v - (1 + \zeta)^2 v\right) \\ + \frac{1}{2x}\left(\lambda^2 + \frac{2}{\lambda v} + 2\ln v\right) - \frac{1}{x} f\lambda . \qquad (12.166)$$

In Eq. (12.166) $f = PL/(n_2 kT)$ is the dimensionless force acting on the gel and e is a factor that takes into account the energy alteration, if unequal neighbor pairs are formed. The parameter e has the dimension of kT, from this follows that e/kT and $\bar{A}$ are dimensionless as well. According to the theorem of the minimum of free

12.6 Volume Change in Biopolymers

energy the partial derivatives with respect to υ and λ disappear with fixed values e, x, ζ, f,

$$\frac{\delta \bar{A}}{\delta \upsilon}(\upsilon, \lambda, e, x, \zeta, f, T) = 0, \tag{12.167}$$

$$\frac{\delta \bar{A}}{\delta \lambda}(\upsilon, \lambda, e, x, \zeta, f, T) = 0. \tag{12.168}$$

From this, one obtains the equations

$$\frac{kT}{e} = \frac{-(1+\zeta+\zeta^2)\upsilon^2}{\ln[1-(1+\zeta)\upsilon]+\left(1-\frac{1}{x}\right)\upsilon+\frac{1}{x\lambda}}, \tag{12.169}$$

$$f = \lambda - \frac{1}{\upsilon\lambda^2} \tag{12.170}$$

that describe the dimensionless temperature kT/e and the dimensionless force f as a function of λ and υ. Figures 12.29 and 12.30 describe these relations.

Figure 12.29 seems unsuspicious whereas Fig. 12.30 does not. If one looks more detailed at Fig. 12.30 for a fixed f and if one increases the temperature T, one finds the possibilities sketched in Fig. 12.31.

For low as well as for high temperatures, only one value of υ exists in each case: a high value of υ for the shrunk gel and a low value of υ for the swollen gel. In case of a intermediate temperature three points of intersections exist, and one asks which of them will be realized. About this, the free energy $\bar{A}$ gives information. Because it is easy to understand that the right and left intersection is connected with local minima and the intersection in the middle is connected with a local maximum. If one sketches $\bar{A}$ in dependence on υ for the four temperature steps, the situations shown in Fig. 12.32 follow.

For temperatures between k_1T/e and k_3T/e one sees that two minima can exist, but the right one is always the absolute minimum, curve (2″). Curve (2) in Fig. 12.32

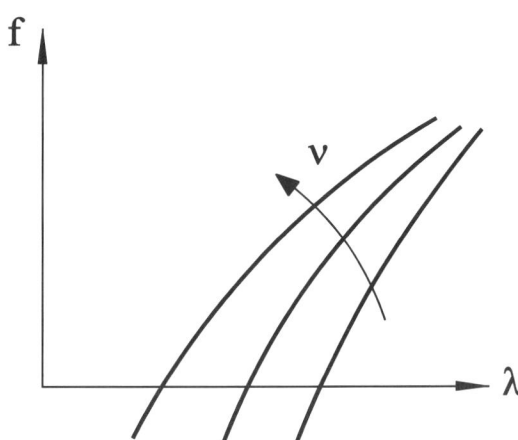

Fig. 12.29 Dimensionless force f as a function of the deformation λ. The parameter is υ

Fig. 12.30 Temperature as a function of the degree of shrinking υ. The parameter is f

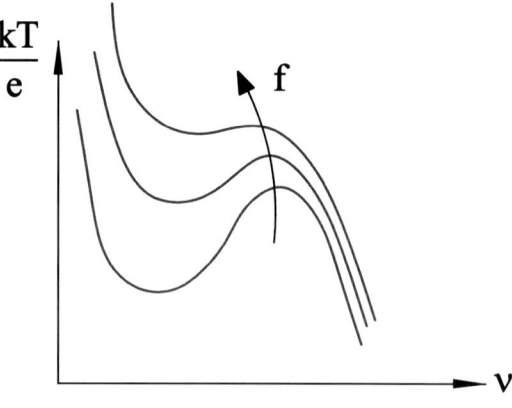

Fig. 12.31 Phase transition of υ for $f = $ const.

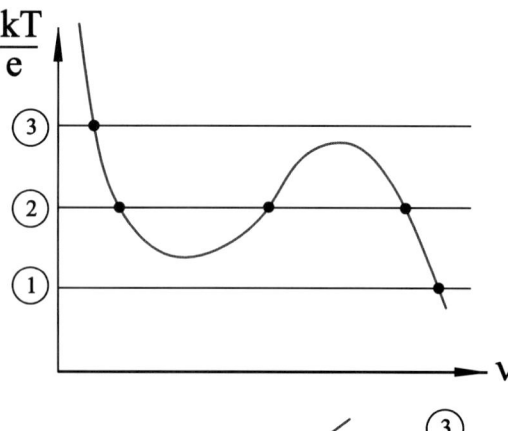

Fig. 12.32 Free energy $\bar{A}$ at different temperatures T

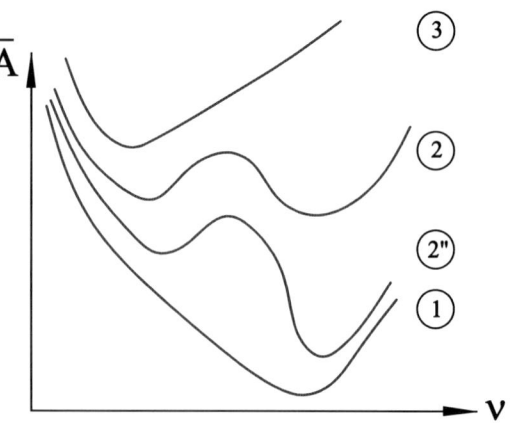

occurs if the intersections lead to areas above and below of equal size in Fig. 12.31. At the corresponding temperature $k_2 T/e$ the level for both minima is equally high and thus this is the temperature that describes the phase transition. For temperatures above this limit temperature, the stability shifts from the right to the left intersection which is connected with a spontaneous swelling. It is obvious that a tension force

12.6 Volume Change in Biopolymers

facilitates the swelling and that compressive loads hinder the swelling. Exact calculations with the relations given by Eqs. (12.169) and (12.170) show this too. For an exact derivation of the free energy one should consult the literature in [13, 37]. This swelling behavior and shrinking behavior of gels will be used in order to interpret two phenomena. On the one hand, the passage behavior of red blood cells through capillaries as a function of the temperature is of great interest [27]. On the other hand, the behavior of hyaline cartilage (intervertebral disk) as a function of stress is of central interest [6].

Phenomenon 1: Red blood cells

One asks oneself, why gas exchange happens in the capillaries and not in the remaining blood vessel system. In the capillary system, high traction forces are exerted on the hemoglobin elastomer which is in the red blood cells. Figure 12.33 shows the deformation problem of the blood cell which becomes longer and thinner.

Through incorporation of free water, the particle becomes softer at constant volume, the network extends and the gas exchange is facilitated or can happen at all. This also would explain why the passage ability is not given below certain critical temperatures, as described by Artmann and coworkers [27, 52].

Phenomenon 2: Hyaline cartilage

One finds this material class in intervertebral disks and as articular cartilage in the diarthroidal joints. It therefore is quite important for the mobility. One asks about the supply mechanisms and waste disposal mechanisms of hyaline cartilage, because it does not possess an individual system of blood vessels. Again one postulates the swelling ability as a supply mechanism – and analogously the shrinking behavior as a waste disposal mechanism – of this material class. This would explain the measured phase transition as it is sketched in Fig. 12.34.

One sees from this experiment that the phase transition is reversible for a constant loading [6]. Exactly this is predicted by the theory. If one believes in this hypothesis, one understands further hypotheses which state that weak, alternating, electric fields build cartilages because a *better supply* could be guaranteed.

Whether one proves these hypotheses or refutes them, it does not play a role from the point of view of mechanics as well as of thermodynamics. Mechanics and thermodynamics deliver under exactly defined, external inputs clear relations between questions (experiments) and statements. They can be extremely helpful for the questions of biology, chemistry and medicine. Alternative models of the intervertebal disc are used in [11].

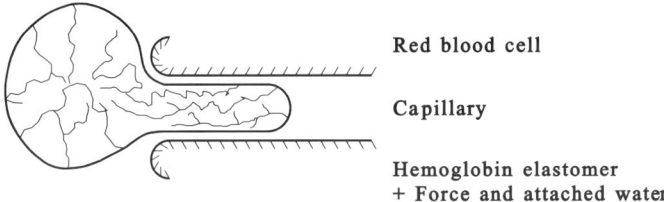

Fig. 12.33 Phase transition of hemoglobin in a red blood cell

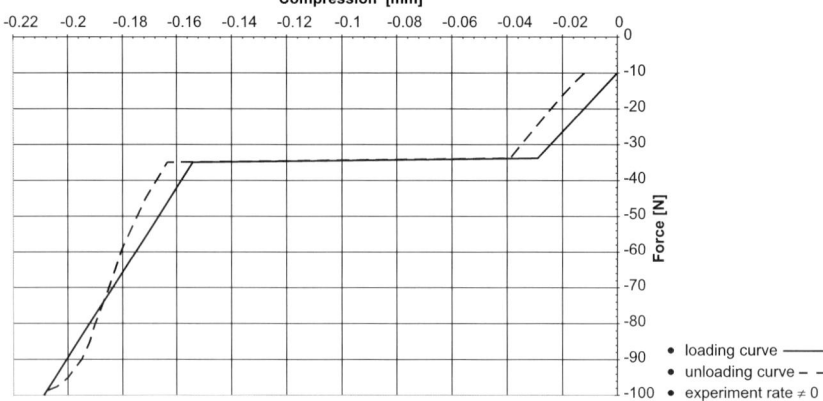

Fig. 12.34 Stress experiment with an intervertebral disk of a cow [6]

12.7 Summary and Outlook

As mentioned in the introduction, the turning to dynamic structures and mechanisms caused a total rethinking in materials engineering. Areas like *biology, medicine* and *technology* are today closely interleaved with each other. High quality biomechanical research is done in large automotive industries, because the product car which is to be developed has to fit to the human being in terms of safety and comfort. Not without reason, renowned German automobile producers operate a rocket sled test room in the pathology centre in Heidelberg. Somewhere, a balance has to be found to the known human substitute "dummy". Furthermore, today many parents of small children use the vibration-controlled comfort of automobiles as a modern cradle. The slow vibration at high amplitudes, as shown in Fig. 12.17, corresponds to the cradle movement of small children. The biological memory reminds one of this movement that incidentally corresponds to a pregnant woman's movement and thus makes driving to a calming and comfortable experience. On the contrary one feels the high frequency vibrations of the motor as disturbing.

But not only technology learns from biology and medicine but also vice versa this connection increasingly gains importance. The biologists, biochemists and physicians depend on observations (experiments). This means an enormous amount of time and effort, and they will do everything in order to reduce this effort. Exactly here, technology, physics and mathematics gives strong guiding.

The tool used in this article was the "phenomenological material theory" borrowing from thermodynamics. On a level that can be understood by the reader, it has been tired to develop the modern theories with their basic experiments. In the area of statics, the phenomenological area has been left and results of statistical thermodynamics have been used instead. The statements were heuristic at many places but not less usable.

The Kilian network is a true pearl for the rubber processing industry. It is also easily expandable to the use of most different fillers like soot or ceramic particles.

12.7 Summary and Outlook

The consideration of the tendon led to similar simplifications and is therefore a big help for questions from biology and medicine.

It has been tried to clarify that the term substance itself is not necessarily interconnected with a material law. Not until the expected deformation as well as the experiment allows the development of a concrete material law from the general formulation [24] demonstrates this approach. The general formulation is practically useless owing to its universality. It has been started with dynamic problems where contact times respectively impact problems become important. This was also the reason why soft highly expansible materials have been considered. Here the fundamental problem of dynamic deformation has been considered and the hysteresis term of the material law has been specified and greatly simplified for the important case of motion that is shown in Fig. 12.17. With the help of thermodynamics as well as statistical mechanics, the statistical part could also be reduced to a great extend. Here, however, the problems are even connected with the substance and with the structure that is trained.

Questions that involve mixtures are of high interest to biology, biochemistry and medicine but could only be considered incompletely in this article due to space limitation. For further reading on continuum biomechanics references [10,21,22,25] are recommended as well as Fung's authoritative book [16].

Lastly, it can be said that the speed of knowledge and development has increased strongly in the last decades and that this trend continues. The researchers as well as the developers will increasingly use basic knowledge and basic methods from neighbor or even distant areas as a result of pressure of time and need for results in order to achieve their goals in a acceptable time.

The trend for specialization as could be observed in the last century seems to turn back and appears to lead to a convergence of natural sciences. The *universal-scholar* will indeed not be recreated, the areas have become too complex for this, but we can imagine very well *universal teams of scholars*. This however makes great demands on the individual team member, because besides technical and social competence, it requires something, which can be called communicative competence. The team member has to combine his/her intellectual world with others without rating and has to create something new for the collective benefit even if this means that he/she has to give up well-loved ideas.

References

1. Ball JM (1977) Convexity conditions and existence theorems in nonlinear elasticity. Archive for Rational Mechanics and Analysis 63:337–403
2. Balzani D, Schroeder J, Gross D, Neff P (2005) Modeling of anisotropic damage in arterial walls based on polyconvex stored energy functions. In: Oñate E, Owen DRJ (eds) VIII International Conference on Computational Plasticity COMPLAS VIII, CIMNE, Barcelona
3. Baumgärtner A (1991) Gummielastizität. In 22. IFF Ferienkurs: Physik der Polymere. Jülich: Forschungszentrum Jülich, 21:1–38
4. Becker E, Bürger W (1975) Kontinuumsmechanik. Teubner, Stuttgart
5. Böhme G (2000) Strömungsmechanik nichtnewtonscher Fluide. Teubner, Stuttgart
6. Braunschweig H (2004) Viskoelastische Stoßprobleme. Diploma thesis, Aachen University of Applied Sciences
7. Bryant-Greenwood GD, Millar LK (2000) Human fetal membranes: their preterm premature rupture. Biol Reprod 63(6):1575–1579
8. Buggisch H, Mazilu P, Weber H (1988) Parameter identification for viscoelastic materials. Rheologica Acta 27(4):363–368
9. Ciarlet PG (1988) Mathematical Elasticity, Volume 1: Three Dimensional Elasticity. Elsevier, North-Holland
10. Cowin SC, Doty SB (2007) Tissue Mechanics. Springer, New York
11. Ehlers W, Karajan N, Markert B (2006) A porous media model describing the inhomogeneous behaviour of the human intervertebral disc. Materialwissenschaft und Werkstofftechnik 37:546–551
12. Eringen AC (1967) Mechanics of Continua. John Wiley, New York
13. Flory PJ (1953) Principles of Polymer Science. Cornell University Press, Ithaca, NY, London
14. Fung YC (1955) An introduction to the theory of aeroelasticity. John Wiley, New York
15. Fung YC (1967) Elasticity of soft tissue in simple elongation. Am J Physiol 213:1532–1544
16. Fung YC (1993) Biomechanics: Mechanical Properties of Living Tissues. 2nd ed., Springer, New York
17. Kassab GS (2004) Fung (Bert) YC (ed) The Father of Modern Biomechanics. Molecular & Cellular Biomechanics 1:5–22
18. Green AE, Adkins JA (1970) Large Elastic Deformation. Oxford University Press, Oxford
19. Haupt P (1971) Viskoelastizität inkompressibler isotroper Stoffe. Dissertation, Universität Berlin
20. Holzapfel GA (2005) Similarities between soft biological tissues and rubberlike materials. In: Austrell P-E, Kari L (eds) Constitutive Models for Rubber IV. A. A. Balkema, Leiden, pp 607–617
21. Holzapfel GA, Ogden RW (eds) (2003) Biomechanics of soft tissue in cardiovascular systems. Springer, Wien
22. Holzapfel GA, Ogden RW (eds) (2003) Mechanics of biological tissue. Springer, Berlin

23. Hsu FPK, Downs J, Liu AMC, Rigamonti D, Humphrey JD (1995) A triplane video-based experimental system for studying axisymmetrically inflated biomembranes. IEEE Trans Biomed Eng 42:442–449
24. Huilgol RR (1972) Continuum Mechanics. John Wiley, New York
25. Humphrey JD (2002) Cardiovascular Solid mechanics: Cells, Tissues, and Organs. Springer, New York
26. Itskov M, Ehret AE, Mavritas D (2006) A polyconvex anisotropic strain–energy function for soft collagenous tissues. Biomech Model Mechanbiol 5:17–26
27. Kelemen C, Chien S, Artmann GM (2001) Temperature transition of human hemoglobin at body temperature: Effects of calcium. Biophysical Journal 80:2622–2630
28. Kilian H-G (1985) An interpretation of the strain-invariants in largely strained networks. Colloid Polymer Sci 263:30–34
29. Kilian HG (1986) Möglichkeiten und Grenzen der Charakterisierung von Elastomeren im Zugversuch. GAK 39(10):548–553
30. Kuecken M (2004) On the Formation of Fingerprints. PhD thesis, University of Arizona
31. Kuhl E, Garikipati K, Arruda EM, Grosh K (2005) Remodeling of biological tissue: Mechanically induced reorientation of a transversely isotropic chain network. Journal of the Mechanics and Physics of Solids 53(7):1552–1573
32. Lambertz D (1985) Mechanische Kenngrößen biologischer Strukturen bei dynamischer Belastung. Diploma thesis, Aachen University of Applied Sciences
33. Landau LD, Lifschitz EM (1965) Elastizitätstheorie. Akademie Verlag, Berlin
34. Mazilu P (1973) On the constitutive law of Boltzmann-Volterra. Rev Roum Math Pures et Appl 18:1067–1069
35. Mazilu P (1985) Die Onsager'schen Reziprozitätsbeziehungen in der Thermodynamik der Boltzmann-Volterra Materialien. Z angew Math Mech 65:137–149
36. Müller I (1973) Thermodynamik. Bertelsmann, Düsseldorf
37. Müller I (2001) Grundzüge der Thermodynamik. Springer, Berlin
38. Ogden RW, Saccomandi G, Sgura I (2006) On worm-like chain models within the three-dimensional continuum mechanics framework. Proc Roy Soc A462:749–768
39. Pechhold W, von Soden W, Kimmich R (1973) The meander model of muscle. Kolloid-Z u Z Polymere 252:829–842
40. Pechhold WR, Gross T, Grossmann HP (1982) Meander model of amorphous polymers. Colloid Polymer Sci 260:378–393
41. Reese S (2000) Thermomechanische Modellierung gummiartiger Polymerstrukturen. Habilitation, Universität Hannover
42. Roy CS (1880–82) The elastic properties of the arterial wall. Journal of Physiology 3:125–159
43. Schajer GS, Steinzig M (2005) Full-field calculation of hole-drilling residual stresses from electronic speckle pattern interferometry data. Experimental Mechanics 45(6):526–532
44. Šilhavý M (1997) The Mechanics and Thermodynamics of Continuous Media. Springer, Berlin
45. Sponagel S (1987) Gummi-Metall-Bauteile. Dissertation, Universität Kaiserslautern
46. Sponagel S, Unger J, Spies KH (2003) Härtebegriff im Zusammenhang mit Vernetzung, Bruchdehnung und Dauerfestigkeit eines Elastomers. Kautschuk Gummi Kunststoffe 56(11):608–613
47. Staat M, Ballmann J (1989) Fundamental aspects of numerical methods for the propagation of multidimensional nonlinear waves in solids. In: Ballmann J, Jeltsch R (eds) Nonlinear Hyperbolic Equations – Theory, Computation Methods, and Applications. Vieweg, Braunschweig, Wiesbaden, pp 574–588
48. Staat M, Ballmann J (1989) Zur Problematik tensorieller Verallgemeinerungen einachsiger nichtlinearer Materialgesetze. Z angew Math Mech 69(2):73–81
49. Timoshenko SP, Goodier JN (1970) Theory of Elasticity. MacGraw-Hill, New York
50. Truesdell C, Noll W (1965) The Nonlinear Field Theories of Mechanics. In: Flügge S (ed) Handbuch der Physik, Vol. III/3. Springer, Berlin, Göttingen, Heidelberg
51. Yamada H (1970) Strength of Biological Materials. Williams & Wilkins, Baltimore

52. Artmann GM, Zerlin KF, Digel I (2008) Hemoglobin senses body temperature. In Artmann GM, Chien S (eds) Bioengineering in Cell and Tissue Research. Springer, Heidelberg, Berlin, pp 415–442

Chapter 13
Modeling Cellular Adaptation to Mechanical Stress

Roland Kaunas

Department of Biomedical Engineering and the Cardiovascular Research Institute, 337 Zachry Engineering Center, Texas A&M University, College Station, USA, rkaunas@bme.tamu.edu

Abstract It is well-accepted that cyclic circumferential stretching of the arterial wall regulates endothelial cell form and function, with the actin cytoskeleton playing a central role. While there has been much progress in characterizing the contribution of the actin cytoskeleton to cellular mechanical properties and to mechanotransduction, it is not well understood how cell mechanical properties influence mechanotransduction. Further, the ability of actin filaments to assemble and disassemble renders the mechanical properties of the actin cytoskeleton highly dynamic. It is suggested herein that endothelial cells can adapt to certain patterns of stretch through directed cytoskeletal remodeling to minimize stretch-induced stress, and this in turn alters the activity of stretch-induced signaling events. A continuum adaptive constitutive model, formulated using mixture theory and motivated by cellular microstructure, is proposed in order to predict the initial stresses developed in stretched adherent cells, and the ensuing microstructural changes which act to minimize intracellular stresses as the actin cytoskeleton remodels. An experimental system is described for testing the model and predictions are made for the time evolution of intracellular stresses, actin organization and mechanotransduction in endothelial cells subjected to different patterns of stretch, which correlate with experimentally observed results.

13.1 Introduction

Situated at the luminal surface of arteries, the endothelium is subjected to both fluid shear stress and mechanical normal stresses. These mechanical factors each contribute to the morphology of endothelial cells (ECs) along the arterial tree. In relatively straight, unbranched arteries, ECs are oriented in a longitudinal direction, which is parallel to the direction of wall fluid shear stress and perpendicular to the direction of circumferential stretch. At branch points, ECs do not have clear patterns of orientation (Nerem et al. 1981). Correlations between local fluid dynamic variables at these branch points and sites of intimal thickening suggest that atherosclerotic plaques tend to occur at sites of low and oscillating wall shear stress (Giddens et al. 1993) and that such shear stress patterns also lead to a lack of EC orientation

(Nerem et al. 1981). The cyclic distension of the vessel wall also influences EC morphology and appears to participate in atherogenesis (Kakisis et al. 2004; Thubrikar and Robicsek 1995). There has been extensive research regarding the roles of fluid shear stress in EC function in both health and disease, however the roles of stretch have not been examined in a comparable manner.

In addition to arteries, other tissues such as bone and skeletal muscle are subjected to compressive and tensile forces, respectively, which direct their form and function. The key question, however, is: What governs the response of a tissue to mechanical forces? In *The Wisdom of the Body* (1932), Walter B. Cannon wrote:

When we consider the extreme instability of our bodily structure, its readiness for disturbance by the slightest application of external forces ... its persistence through so many decades seems almost miraculous. The wonder increases when we realize that the system is open, engaging in free exchange with the outer world, and that the structure itself is not permanent, but is being continuously broken down by the wear and tear of action, and as continuously built up again by processes of repair.

Cannon was referring to the ability of the human body to adapt to its environment, with the external forces ranging from extremes in climate, nutrition, and disease to maintain a state of *homeostasis*. At the tissue level, blood vessels have been proposed to remodel in response to deviations in stress or strain from a homeostatic value [reviewed by Taber (1995)]. The remodeling of tissue requires a response by the resident cells. As described in the 2003 Walter B. Cannon Memorial Lecture entitled "Mechanotransduction and Endothelial Homeostasis: the Wisdom of the Cell", cellular remodeling occurs in response to mechanical loading, which modulates the ensuing signal transduction events (Chien 2006). What remains to be clarified are the mechanisms by which cellular remodeling regulates mechanotransduction.

This chapter describes a new approach toward developing an adaptive constitutive model of adherent cells subjected to mechanical stretch. Such a model would provide a framework by which the accelerating accumulation of mechanotransduction data can be interpreted, thus providing insight into why cells remodel in response to mechanical loading and how this affects mechanotransduction. To illustrate certain concepts, the model is applied to describe how ECs adapt to different modes of stretch. First, however, it is informative to review current techniques for probing the mechanical properties of adherent cells and the constitutive models developed from these measurements.

13.2 A Brief Review of Stretch-Induced Cell Remodeling

13.2.1 Cell Mechanics and Dynamics

The structural and mechanical properties of adherent cells can be largely attributed to their nucleus, plasma membrane, and cytoskeleton. The nucleus is a relatively stiff structure, supported by a network of nuclear lamins on the inner surface of the nuclear membrane. Using micropipette aspiration, Guilak et al. (2000) demonstrated that the nuclei of chondrocytes are ~3 times stiffer and ~2 times more viscous than the cytoplasm. The high viscosity of the nucleus translates into a slow viscoelastic response, so that under short time-scales the nucleus behaves nearly elastically. The nucleus is also very stable, except during cell division when the nuclear membrane completely breaks down and reforms again in the daughter cells.

The plasma membrane is composed mainly of phospholipids, which render the behavior of membrane to be fluid-like. The plasma membrane of red blood cells (the model cell for studies of plasma membrane mechanics) are very resistant to changes in area, with little resistance to in-plane extension or bending (Evans 1989). Thus one of the primary contributions of the plasma membrane to cell mechanics is maintenance of cell surface area with little restriction to changes in cell shape. The plasma membrane does show some viscoelastic behavior, but on a very short timescale (0.1 sec) (Evans 1989).

The mechanics of the cytoplasm is dominated by the aqueous cytosol and the cytoskeleton. The liquid cytosol contributes to the viscous nature of the cytoplasm. The cytoskeleton consists of a fibrous network of actin microfilaments, microtubules, and intermediate filaments. Actin microfilaments are capable of forming various structures through interactions with diverse cross-linking proteins, ranging from lattice-like networks to thick bundles of parallel filaments, termed *stress fibers*. These stress fibers are of particular importance in cell mechanics since they provide the primary force-producing structure in non-muscle cells (Burridge 1981). Stress fibers are typically anchored to the extracellular matrix (ECM) at each end of the cell via focal adhesions, so that myosin-induced translation of filaments in opposing directions generates isometric tension.

Cells are not static entities, as can be observed when following the "milling about" of cells under time-lapse microscopy. Cell locomotion requires the making and breaking of adhesions to the underlying substrate and the remodeling of the cytoskeletal network as a cell moves from one location to another. Actin filaments and microtubules are biopolymers that rapidly polymerize and depolymerize to facilitate the turnover of individual filaments on the timescale of minutes. Intermediate filaments are relatively stable structures, but also disassemble and reassemble during the course of cell shape change and migration.

13.2.2 Actin Cytoskeletal Remodeling in Response to Different Modes of Stretch

Mechanical perturbations induce dramatic remodeling of the cell cytoskeleton. Figure 13.1 illustrates rearrangements of actin stress fibers that may occur when an adherent cell is suddenly subjected to either a tensile or compressive uniaxial stretch. In Fig. 13.1B, a cell expressing GFP-actin was stretched by impaling and stretching the substrate along the direction of the arrow with a micropipette (Kave-

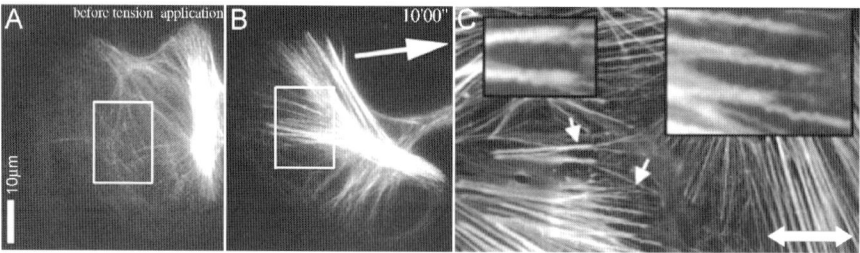

Fig. 13.1 Effects of a step change in strain on actin stress fiber remodeling. **A** and **B**: GFP-actin in a fibroblast before (**A**) and after (**B**) a step in tensile stretch along the direction of the *arrow* in **B**. (reproduced with permission of the Company of Biologists from (Kaverina et al. 2002)). **C**: Stress fibers in cells fixed immediately following a 27% compression in the horizontal direction as indicated by the *double-headed arrow*. The *single-headed arrows* indicate the positions where the image was magnified in the insets (reproduced with permission of John Wiley & Sons from (Costa et al. 2002))

rina et al. 2002). After 10 min, new actin stress fibers were observed, and these new fibers were oriented roughly parallel to the direction of the stretch. The effects of compressive stretch are illustrated in Fig. 13.1C. The elastic substrate was shortened in the direction of the double-headed arrow, then the cells were fixed in formaldehyde and stress fibers were observed using a fluorescent stain (Costa et al. 2002). The stress fibers oriented parallel to the direction of compression were observed to buckle, while stress fiber oriented in the perpendicular direction remained straight. If the cells were not fixed immediately following the compressive strain, the cells would become largely devoid of stress fibers, suggesting that the stress fibers disassemble shortly after buckling. Together, these results implicate tension as a factor which promotes stress fiber formation and stability.

Stress fibers in cells subjected to cyclic uniaxial stretch organize differently from stress fibers in cells subjected to a step-change in uniaxial stretch. Within minutes of initiating cyclic uniaxial stretch, stress fibers begin to orient perpendicular, rather than parallel, to the direction of stretch. An initial attempt to model stress fiber reorientation in response to cyclic stretch was proposed by Wang (2000) in which individual stress fibers were modeled as linearly elastic filaments with strain energy (w) described as:

$$w = \frac{1}{2}k(\delta + L\varepsilon_\mathrm{f})^2, \qquad (13.1)$$

where k is the spring constant, δ is the basal level of strain in the stress fibers of unstretched cell, L is the initial length of the stress fiber and ε_f is the fiber strain due to stretching the substrate upon which the cell is attached. The fiber strain is related to the material strain (ε_{ij}) through Eq. (13.2):

$$\varepsilon_\mathrm{f} = \varepsilon_{11}\cos^2\theta + \varepsilon_{22}\sin^2\theta. \qquad (13.2)$$

13.2 A Brief Review of Stretch-Induced Cell Remodeling

The model further assumes that stress fibers are only stable when the strain energy remains between zero and twice the basal level (i. e., $0 < L\varepsilon_f < \delta$). Under this assumption, it is predicted that stress fibers in cyclically stretched cells will orient toward the direction of minimal perturbation in strain. When there is compression in the substrate in the direction perpendicular to the principal direction of stretch due to a Poisson-type effect (in which case the linearized strains in the x- and y-directions are related by the Poisson ratio ($v = -\varepsilon_{22}/\varepsilon_{11}$) of the substrate through Eq. (13.2), the model predicts stress fibers will orient about an angle $\theta = \cos^{-1}\sqrt{v/(1+v)}$ relative to the direction of stretch. For silicone rubber membranes where $v \sim 0.35$, the resulting angle is approximately 60°. In the case where lateral compression is absent, the model predicts that stress fibers will orient at an angle of 90°. The model also implies that stress fiber orientations will be distributed about this critical angle and that the variance in the distribution decreases with increasing substrate strain.

The predictions from Wang's model match the observed orientation of stress fibers in cells subjected to cyclic uniaxial stretch with (Wang 2000) and without (Kaunas et al. 2005) lateral compression[1]. Kaunas et al. (2005) examined the effects of independently varying stretch magnitude and Rho-induced cell contractility in cyclic uniaxial stretch-induced stress fiber organization, which essentially acts to vary the values of δ and ε_f, respectively, in the strain energy formulation described by Eq. (13.1). As illustrated in Fig. 13.2, stress fibers in unstretched ECs are not oriented in any particular direction (Fig. 13.2A), but become oriented perpendicular the direction of cyclic stretch (Fig. 13.2B). The stress fibers oriented at a 90° angle relative to the direction of stretch since lateral compression was essentially eliminated. Transfecting the cells with an enzyme inhibitor of Rho activity attenuated stress fiber formation in unstretched cells (Fig. 13.2C), but stress fibers were formed once the cells were subjected to cyclic stretch (Fig. 13.2D). Importantly, the stretch-induced stress fibers in the cells treated with the Rho inhibitor were oriented parallel, rather than perpendicular, to the direction of stretch.

The stress fiber organizations observed in Fig. 13.2 appear to obey the restriction on fiber strain energy in a manner similar to that described in Eq. (13.1),

$$w = \frac{1}{2}k(\delta_h + L\varepsilon_f)^2, \tag{13.3}$$

where δ_h is a homeostatic value for fiber strain, with fibers only being stable when the fiber strain energies remain in the range $0 < w < k\delta_h^2$. A plausible value for δ_h may be the basal level of strain in the stress fibers of unstretched cells with normal contractile function. In the case of cells with suppressed Rho activity, the basal fiber strain energy is expected to be zero, so fibers are not formed unless the strain energy is increased through cell stretching. Thus Wang's model describes some important features of stretch-induced stress fiber orientation which should be considered in the development of a constitutive model for stretching of adherent cells.

---

[1] Uniaxial stretch without lateral compression is also known as strip biaxial stretch.

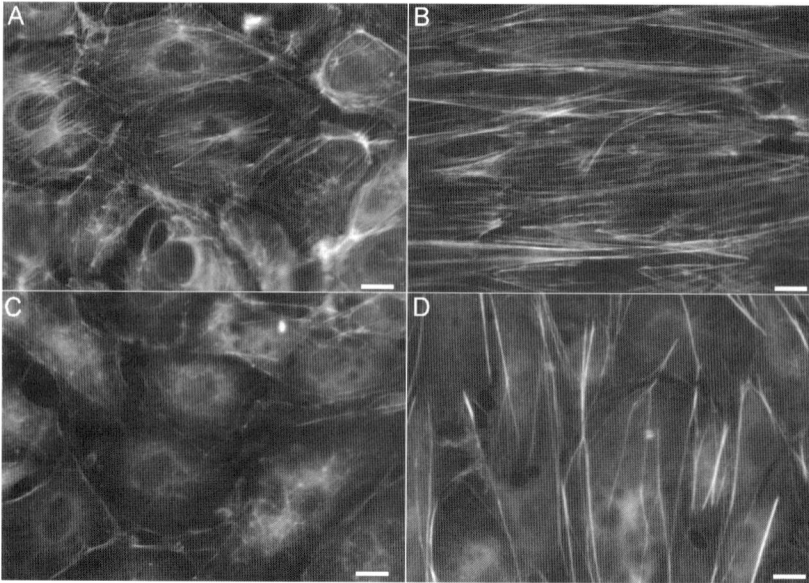

Fig. 13.2 Effects of cyclic uniaxial stretch and Rho-induced contractility on actin fiber organization in endothelial cells. Endothelial cells with normal Rho activity (**A, B**) or with inhibited Rho activity via C3 exoenzyme transfection (**C, D**) were either kept unstretched (**A, C**) or subjected to 10% cyclic uniaxial stretch at a frequency of 1 Hz for 6 hr (**B, D**). The direction of stretch was in the vertical direction of the image. *Scale bar*: 10 μm. (Used with permission from (Kaunas et al. 2005))

13.3 Measurements, Modeling, and Mechanotransduction

Unlike more traditional passive materials, cells are active "materials" whose mechanical properties evolve in response to changes in their mechanical environment. Just as the unique properties of living tissues stimulated new approaches to describe tissue mechanics, new approaches for describing the mechanics of cells need to be developed. Cells, like many tissues, can produce forces via active contraction. These forces can be used by the cell to alter its shape or to apply forces to the surrounding ECM and neighboring cells. An important distinction between tissue and cell biomechanics is the rate at which remodeling occurs – tissues can remodel over a period on the order of days, while cells can disassemble structural components within seconds (Costa et al. 2002). Presently, however, cell remodeling is a phenomenon that is lacking in most mathematical models of cell mechanics (see (Humphrey 2002) for review). Present methods have provided valuable insight indicating the need for adaptive models for stretched cells, but a new approach is necessary which specifically addresses mechanical testing, constitutive modeling and mechanotransduction studies for adherent cells which remodel in response to stretching.

13.3.1 Mechanical Testing of Adherent Cells

To study how cells adapt to stretch, there must first be a way to measure the stresses and strains to which the cell is subjected. A number of techniques have been developed to study the mechanical behavior of cells.

In the atomic force microscopy (AFM) indentation test, a soft cantilever tip locally indents the surface of a cell, and the local stiffness can be obtained through the resulting force-indentation depth relationship. Using AFM, Hofmann and colleagues showed that the local elastic modulus can vary by two orders of magnitude within individual cells, with maximum stiffnesses measured atop actin stress fibers (Hofmann et al. 1997). Further, the stiffness of the cells drops uniformly to the minimum values when actin filaments are depolymerized using cytochalasin B. Thus the actin cytoskeleton appears to be the main contributor to cell stiffness. The time required to obtain such high-resolution maps can be several minutes, however, which limits the temporal resolution of these measurements unless only small areas of a cell are to be interrogated.

In Magnetic Twisting Cytometry (MTC), the cell is probed using a paramagnetic bead coated with an integrin ligand (e.g. RGD-peptide) so as to create a mechanical link between the bead and integrins on the cell surface. A magnetic pole is induced in the bead by a large-magnitude magnetic pulse in either the x- or y-direction. As illustrated in Fig. 13.3A, a torque is generated when an oscillating magnetic field is applied in the z-direction, causing the bead to twist about the axis orthogonal to both the axis of polarization and the z-axis. Since the bead is attached to the cell surface, the torque causes the bead to roll and this can be quantified by phase microscopy and image correlation.

Tension in individual stress fibers are vector quantities oriented parallel to the stress fibers, hence a preferred orientation of the stress fibers can result in anisotropic

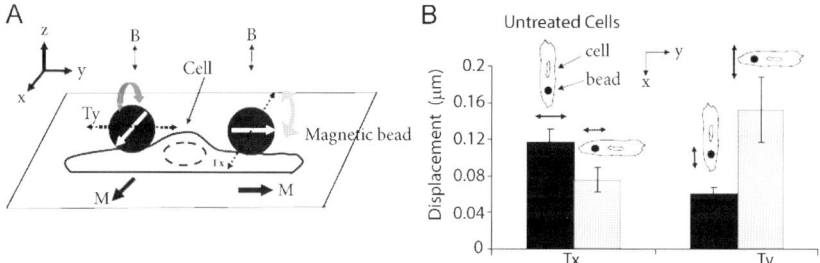

Fig. 13.3 Cell material anisotropy as demonstrated by magnetic twisting cytometry. **A**: Schematic of a torque applied to a cell along two different directions. *Left*: the bead rotated in the x–z plane (*curved double arrow*) in response to T_y (*dashed arrows*) generated by a twisting field B in z after magnetization M in the x-direction. *Right*: the bead rotated in the y–z plane (*curved double arrow*) in response to T_x (*dashed arrows*) generated by a twisting field B in z after magnetization M in the y-direction. **B**: Data from normal untreated smooth muscle cells plated on collagen-1-coated dishes. The cell's long axis was positioned along the x-direction or along the y-direction. *Solid dots* represent a magnetic bead attached to a cell; *double arrows* represent bead rotation directions. These results illustrate that the cell is relatively more stiff along the long axis of the cell (used with permission from (Hu et al. 2004))

cell material properties. Hu and colleagues (Hu et al. 2004) demonstrated such anisotropic properties using MTC. Figure 13.3B illustrates how the direction of force applied relative to the orientation of the cell results in different resistances to magnetic bead displacement. Displacements are larger when the torque is applied about the long axis (resulting in bead translation along the short axis of the cell) than when the same magnitude of torque is applied about the short axis of the cell (resulting in bead translation along the long axis of the cell).

One feature that AFM and MTC have in common is that they interrogate the response of a cell to a localized load. This may not be appropriate when studying the response of a cell to substate stretching, which results in a distributed load via a multitude of cell-matrix adhesions. Measurement of forces produced at the cell-matrix interface can be estimated using a technique originally developed by Dembo and Wang, which they termed *traction microscopy* (Dembo and Wang 1999). In their pioneering study, Dembo and Wang estimated the surface tractions generated by cells pulling on their underlying substrate. Fluorescent markers positioned at the surface of the substrate are used to measure the displacement field. If the material properties of the substrate are known and are elastic, the traction field can be calculated from the displacements based on the Boussinesq solution for the displacement field on the surface of a semi-infinite elastic half-space (Landau and Lifshitz 1986). The Boussinesq solution essentially provides a solution for the deformation field produced by a force applied parallel to the surface of the half-space at a point. Further, the deformation fields from multiple such point forces can be linearly superimposed to give the overall resultant deformation field.

For a cell applying traction forces on an elastic substrate, the point forces are transmitted from the cell via transmembrane integrins locally aggregated into focal adhesions. By assuming traction forces can only localize at regions containing fluorescently-tagged focal adhesion protein (GFP-vimentin), Balaban and colleagues (Balaban et al. 2001) demonstrated that traction forces correlated with the orientation, fluorescent intensity (i. e. vimentin content), and area of the focal adhesions. Importantly, the traction force magnitude correlated with focal adhesion area to indicate cells maintain a constant stress of $5.5 \pm 2\,\mathrm{nN}/\mu\mathrm{m}^2$. Thus, cells appear to regulate the level of stress at focal adhesions at a particular magnitude. Further, since focal adhesion orientation is dependent on the orientation of the attached stress fiber, the orientation of traction forces appear to depend on the orientation of the associated stress fibers.

A fundamental difficulty with traction microscopy is that finding a unique solution of the field of traction forces to reproduce the observed strain field can only be achieved by placing constraints on the solution, such as predefining the locations of the point forces, the matching of displacements, and/or rules on the distribution of forces as a function of position (e. g. smoothing of the field of forces) (Beningo and Wang 2002). Such *a priori* assumptions of the forces may introduce significant errors. To avoid the need for smoothing, Butler and colleagues (Butler et al. 2002) developed a technique to compute the traction field by recasting the relationship between displacements and tractions into Fourier space, which results in a traction field that exactly matches the observed displacements. Using *Fourier Transform Traction*

Microscopy (FTTM), this group was able to measure traction forces with high temporal (40 sec intervals) and spatial resolution (2.7 microns) (Tolic-Norrelykke et al. 2002). Importantly, traction force was demonstrated to correlate with the contractile state of a cell, and the tractions concentrated at the far ends of cellular extension where stress fibers typically adhere to the substrate via integrin linkages.

A more recent technique for measuring cell traction fields was developed using a microfabricated array of closely-spaced vertical elastic posts (Tan et al. 2003). The microprobe array is of sufficiently small dimension that spreading cells can attach to >10 posts and cell adhesions are confined to the upper surfaces of the posts. The posts act as cantilevers and the traction force (F) applied to the post can be estimated from the observed deflection of the post (δ) through

$$F = \frac{3EI}{L^3}\delta, \tag{13.4}$$

where E, I and L are the Young's modulus, moment of inertia and length of the post, respectively. A major advantage of this technique over traction microscopy is that the locations of the force application are known rather than estimated. The downside to the microprobe array is that adhesions cannot form continuously along the basolateral surface of the cell, but can only form atop the microprobes. Currently, the spacing between posts is large enough to produce noticeable effects on the cell morphology, but this may improve as denser arrays are fabricated.

13.3.2 Non-adaptive Constitutive Models of Adherent Cells

Strains in tissues such as arteries can often be measured, or at least estimated, with reasonable accuracy. Complete mechanical measurements cannot be performed on cells *in vivo*, however, so there is a need for accurate constitutive relationships to estimate the stresses produced by a given level of strain in the underlying matrix. A number of constitutive equations have been developed to relate strains to stresses produced by particular methods such as the shear stress generated by MTC. To extrapolate these relationships to estimating the stresses produced by whole cell deformations is questionable. Structural and mechanical anisotropy is also clearly important, but has yet to be incorporated into most constitutive models. That said, it is important to build upon what has already been learned from previous models. We will briefly review some mathematical models that have been proposed for adherent cells and the experimental observations that motivated them.

As described above, actin stress fibers are tension-bearing structures. Microtubules, on the other hand, appear to bear compression rather than tension. This is supported by the observation that microtubules in purified preparations have a persistence length of 5200 μm, yet there is significant curvature of these filaments within intact cells that have diameters of <100 μm (Gittes et al. 1993). In contrast, actin microfilaments in purified solutions have a relatively short persistence length of 17.7 μm, but contractile stress fibers >50 μm in length appear straight. Motivated by these observations, Ingber and colleagues proposed the *tensegrity* model of the cell (Ingber et al. 1981). Introduced by Buckminster Fuller (Fuller 1961), tensegrity is based on a building system for structures with *tensional integrity*, in which the stability of the shape of a structure is maintained by a network of self-equilibrating structural members requiring preexisting stress (or, prestress) to maintain structural integrity. Ordinary elastic materials do not require such prestress to maintain their structural integrity. A key feature of tensegrity structures is that the stiffness of the network is proportional to the level of prestress that it supports (Volokh et al. 2000). This has been verified experimentally in cells by correlating the stiffness measured by MTC with the levels of stress measured by traction microscopy (Wang et al. 2002).

There are two mathematical models of the cytoskeleton based on tensegrity – cable-and-strut models and reticulated networks. In the cable-and-strut model, the prestress in cables are balanced by compressive struts. Actin fibers and microtubules are always represented in these models as the cables and struts, respectively. Intermediate filaments have sometimes been included in these models as nonlinear elastic cables. In reticulated networks, the tension in the tensile elements is balanced externally by attachments to the extracellular matrix and/or by cytoplasmic swelling rather than by compressive struts. A combination of the two models, where some compression is supported by both external and internal supports, indicates that the level of spreading of a cell determines the relative contributions of the extracellular adhesions and microtubules to the compressive forces (Stamenovic 2005). The microtubule network is predicted to have a negligible contribution to cell stiffness in well-spread cells, and a significant contribution in more rounded cells. Thus, the tensegrity-based models are capable of predicting several experimentally observed phenomena in cells. It is worth noting that the results from the cable-and-strut studies are typically based on a very simplified isotropic model composed of six struts with cables connecting the ends of each strut. Considering the contribution of cytoskeletal orientation to mechanical anisotropy, it is desirable to have a more structurally accurate model for correlating cytoskeletal structure with cell mechanical properties.

13.3.3 Adaptive Constitutive Models of Adherent Cells

Most constitutive models of the cell, such as those based on tensegrity, are useful for describing the mechanical response of cell over short time scales where cytoskeletal turnover may not be significant. An accurate description of the cellular mechanical response to a mechanical load over time must also address the remodeling of intracellular structures.

To model the mechanical behavior of the cytoplasm, which is dominated by the cytosol and the cytoskeletal filaments, Humphrey (2007) formulated a continuum model based on the theory of mixtures, first introduced by Truesdell (1965). Here, the cell consists of a mixture of components which can respond separately to changing chemical and mechanical conditions. It is known that different cytoskeletal pro-

13.3 Measurements, Modeling, and Mechanotransduction

teins interact via cross-linking proteins (e. g. plectins) (Svitkina et al. 1996), however there is a lack of detailed data regarding the mechanics of these interactions so these interactions are purposefully avoided in the mixture model.

Consider the following general visco-hyperelastic relation for Cauchy stress (σ_{ij}), which is valid for finite strains:

$$\sigma_{ij} = -p\delta_{ij} + \phi^c 2\mu(\phi^c) D_{ij} + 2 \frac{\partial W}{\partial C_{MN}} F_{iM} F_{jN} . \tag{13.5}$$

Here p is a Lagrangian multiplier, δ_{ij} is the Kronecker delta, D_{ij} is the stretching tensor, F_{ij} is the deformation tensor, and C_{ij} is the right Cauchy–Green tensor. The viscosity (μ) may depend on the mass fraction of the cytosol (ϕ^c) since the depolymerization of the other constituents is expected to increase the apparent viscosity of the cytosol. A major feature of the mixture theory is that the strain energy function (W) can be expressed as the additive contributions of the various constituents.

$$W = \sum_{k=1}^{N} \phi^k W^k \tag{13.6}$$

Eq. (13.6) is subject to the restriction that mass fractions total unity:

$$\phi^c + \sum_{k=1}^{N} \phi^k = 1 , \tag{13.7}$$

where ϕ^k and W^k are the mass fraction and strain energy of each structurally-important constituent k of the cytoplasm. To account for the unique mechanical properties of actin, microtubules and intermediate filaments, a separate strain energy function can be formulated for each filament family.

To address the mechanics of constituent dynamics, Humphrey suggested the following mass balance equation using a hereditary integral to describe the effects of constituent formation and subsequent disappearance:

$$\int_{-\infty}^{t} \frac{\partial \rho^k}{\partial \tau} d\tau = \int_{-\infty}^{0} m^k(\tau) q^k(t-\tau) d\tau + \int_{0}^{t} m^k(\tau) q^k(t-\tau) d\tau , \tag{13.8}$$

where $\frac{\partial \rho^k}{\partial \tau}$ is the fractional change in the mass density of constituent k over the time period $d\tau$. The function $m^k(t)$ describes the rates of production of constituent k at time t. The function $q^k(t-\tau)$ describes the percentage of material produced at time τ which survives to time t. This is reminiscent of the use of hereditary integrals in viscoelasticity.

Taking into account the turnover of a constituent, Humphrey proposed a general constitutive relation of the form (Baek 2006):

$$W^k(t) = \frac{\rho^k(0)}{\rho} Q^k(t) \hat{W}^k \left(C_{n(0)}^k(t) \right) \\ + \int_0^t \frac{m^k(\tau)}{\rho} q^k(t-\tau) \hat{W}^k \left(C_{n(\tau)}^k(t) \right) d\tau , \tag{13.9}$$

where $\rho = \sum \rho^k(t)$ and $\hat{W}^k$ is the energy stored elastically in constituent k, which depends on the right Cauchy–Green tensor $\left(C^k_{n(\tau)}(t)\right)$ of each constituent relative to its individual natural configuration. In particular, $C^k_{n(\tau)}(t) = \left(F^k_{n(\tau)}(t)\right)^T F^k_{n(\tau)}(t)$ where $F^k_{n(\tau)}(t) = \partial x(t)/\partial X^k(\tau)$ and $x^k(t) = x(t)$. $Q^k(t)$ represents the fraction of constituent k that was produced at or before time $t = 0$ and survives to time t with $Q^k(0) = 1$. Clearly, the strain energy for a constituent contributes to the total strain energy only for as long as the constituent exists.

To account for cytoskeletal organization, Humphrey and colleagues proposed a microstructurally-motivated, but phenomenological, relation based on the work of Lanir (1979). In the case of zero constituent turnover:

$$\hat{W}^k = \int_0^{2\pi} \int_{-\pi/2}^{\pi/2} \phi^k R^k(\varphi, \theta) w^k\left(\alpha^k\right) \cos\varphi \, d\varphi \, d\theta \,. \tag{13.10}$$

The function $R^k(\varphi, \theta)$ represents the original distribution of filament orientations in spherical coordinates for the constituent family k. The function $w^k(\alpha^k)$ is a one-dimensional strain energy function for a filament of the constituent k and α^k is its stretch. The nonlinear mechanical behavior of actin fibers led Humphrey and colleagues (Na et al. 2004) to suggest the following form for the one-dimensional energy function:

$$w^k(\alpha^k) = \frac{1}{2} c^k \left[\exp\left(c_1^k \left(\alpha^k - 1\right)^2 \right) - 1 \right], \tag{13.11}$$

where c^k and c_1^k are separate material parameters for each family of constituents k. Humphrey offered an alternative form originally put forth by Lanir (1979):

$$w^k\left(\alpha^k\right) = \frac{1}{2} c_2^k \left(\alpha^k - 1\right)^2 \tag{13.12}$$

Taken together, these equations result in the following constitutive relation to describe the effects of intracellular remodeling:

$$\sigma_{ij}(t) = -p(t)\delta_{ij} + \phi^c(t) 2\tilde{\mu}(\phi^c) D_{ij}(t)$$

$$+ 2 \frac{\partial \sum W^k(t)}{\partial C_{MN}(t)} F_{iM}(t) F_{jN}(t) \tag{13.13a}$$

$$W^k(t) = \frac{\rho^k(0)}{\rho} Q^k(t) \int_0^{2\pi} \int_{-\pi/2}^{\pi/2} R^k(\varphi, \theta) w^k\left(\alpha^k\right) \cos\varphi \, d\varphi \, d\theta$$

$$+ \int_0^t \frac{m^k(\tau)}{\rho} q^k(t - \tau) \left(\int_0^{2\pi} \int_{-\pi/2}^{\pi/2} R^k(\varphi, \theta) w^k\left(\alpha^k\right) \cos\varphi \, d\varphi \, d\theta \right) d\tau \,.$$

$$\tag{13.13b}$$

These equations provide a constitutive framework that account for the turnover of filaments and their orientations. This formulation allows freedom for the choice of equations to describe the kinetics of constituent formation and disappearance. Humphrey neglected to specify a form for these equations to describe intracellular constituent turnover due to lack of necessary kinetic data, but rather suggested, "There is, therefore, a pressing need for rigorous identification of functional forms for survival functions. Candidate phenomenological functions also include differences in Heaviside step functions, Avrami-type equations, and so forth, but again there is a need to consider the biochemical kinetics." (Humphrey 2007). Similarly, there is also a need for identification of functional forms for the formation functions.

Recently, Deshpande and colleagues (Deshpande et al. 2006) developed a biochemo-mechanical model of the cell which, in general terms, is consistent with the framework put forth by Humphrey, but is based on linearized strains. Their model, hereafter referred to as the DME model, is motivated by three processes: (1) an activation signal that stimulates actin polymerization/contraction, (2) a tension-dependent orientation of actin filaments, and (3) a cross-bridge cycling of myosin motor proteins along actin filaments that generates the tension. Based on these phenomena, a coupled model was proposed to describe the evolution of stress magnitude and direction in response to a given impulse activation signal and mechanical boundary conditions.

Given a two-dimensional cell geometry with prescribed boundary conditions, an initially uncontracted cell is activated to begin contraction. The activation signal for stimulating contractile force (e. g. increased concentration of intracellular calcium) is assumed to follow first-order kinetics

$$C = \exp(-t/\tau_a), \qquad (13.14)$$

where τ_a is the time constant for decay of the signal and t is the time measured from the instant the most recent impulse signal was applied. The signal stimulates the activation of actin polymerization/contraction which is characterized by an activation level, $\eta(\theta)$, of actin filaments oriented at an angle θ. The rate equation describing the rate of change in the activation level for fibers of orientation θ is given as:

$$\frac{d\eta(\theta)}{dt} = [1 - \eta(\theta)]\frac{Ck_f}{\tau_a} - \left[1 - \frac{\sigma(\theta)}{\sigma_0(\theta)}\right]\eta(\theta)\frac{k_b}{\tau_a}, \qquad (13.15)$$

where the isometric stress is $\sigma_o(\theta) = \eta(\theta)\sigma_{max}$ and σ_{max} is a constant representing the tensile stress exerted by a stress fiber at maximum activation. The dimensionless rate constants k_f and k_b govern the rates of formation and dissociation of the fibers, respectively. The first term in the right side of Eq. (13.15) represents the rate of increase in activation that depends on the magnitude of the activation signal. The second term represents the rate of decrease in activation and is proportional to the level of activation and the magnitude of stress in the filaments (σ) through a simplified model based on the force-velocity relationship developed by Hill for muscle fibers (1938):

$$\frac{\sigma(\theta)}{\sigma_0(\theta)} = \begin{cases} 0 & \frac{\dot{\varepsilon}}{\dot{\varepsilon}_0} < -\frac{\eta}{k_v} \\ 1 + \frac{k_v}{\eta(\theta)} \frac{\dot{\varepsilon}}{\dot{\varepsilon}_0} & -\frac{\eta}{k_v} \leq \frac{\dot{\varepsilon}}{\dot{\varepsilon}_0} \leq 0 \\ 1 & \frac{\dot{\varepsilon}}{\dot{\varepsilon}_0} > 0 \end{cases}, \quad (13.16)$$

where $\dot{\varepsilon} = d\varepsilon/dt$ is the rate of (linearized) strain of a filament. The non-dimensional constant k_v is the fractional reduction in stress when the strain rate increases by the reference value $\dot{\varepsilon}_0$. The stress in a filament is thus assumed to equal the isometric value when the strain rate is positive. When strain rate is negative, the stress decreases linearly to zero until a critical strain rate ($\dot{\varepsilon} = \dot{\varepsilon}_0 k_v/\eta$) is reached, below which filament stress is zero. Clearly, fiber shortening leads to stress reduction in a fiber, which contributes to fiber deactivation. The fiber strain rate is related to the material strain rate using the affine strain approximation:

$$\dot{\varepsilon}_f = \dot{\varepsilon}_{11} \cos^2\theta + \dot{\varepsilon}_{22} \sin^2\theta + \dot{\varepsilon}_{12} \sin 2\theta. \quad (13.17)$$

The average Cauchy stress generated by the fibers is determined by summing the contributions of all actin filaments:

$$\sigma_{11}^a = \frac{1}{\pi} \int_{-\pi/2}^{\pi/2} \sigma(\theta) \cos^2\theta \, d\theta$$

$$\sigma_{12}^a = \sigma_{21}^a = \frac{1}{\pi} \int_{-\pi/2}^{\pi/2} \frac{\sigma(\theta)}{2} \sin 2\theta \, d\theta \quad (13.18)$$

$$\sigma_{22}^a = \frac{1}{\pi} \int_{-\pi/2}^{\pi/2} \sigma(\theta) \sin^2\theta \, d\theta$$

where $\sigma^a(\theta)$ is the stress contribution from fibers oriented at an angle θ.

The total Cauchy stress is assumed to be dominated by contributions from actin fibers and intermediate filaments. Due to the random organization and relative stability of intermediate filaments, their contribution to the Cauchy stress is assumed to follow an isotropic linear elastic Hooke's law for infinitesimal deformation:

$$W = \frac{E}{2(1+v)} \varepsilon_{ij}\varepsilon_{ij} + \frac{Ev}{(1-2v)(1+v)} \varepsilon_{kk}^2, \quad (13.19)$$

where E and v are the Young's modulus and Poisson's ratio, respectively. The Cauchy stress tensor is derived from the strain energy function as follows:

$$\sigma_{ij}^i = \frac{\partial W}{\partial \varepsilon_{ij}} = \frac{E}{1+v}\varepsilon_{ij} + \frac{Ev}{(1-2v)(1+v)}\varepsilon_{kk}\delta_{ij}. \quad (13.20)$$

From Eqs. (13.18) and (13.20), the total Cauchy stress due to the additive contributions from the actin and intermediate filaments is given as

$$\sigma_{ij}^{\text{total}} = \sigma_{ij}^a + \sigma_{ij}^i. \quad (13.21)$$

The model is capable of describing several experimental findings from cells attached to a microsensor array: i) the dependence of cell force generation on substrate stiffness, ii) the influence of cell shape and boundary conditions on actin fiber organization, and iii) the high concentration of actin stress fibers at focal adhesions (Deshpande et al. 2006).

When considering the DME model to describe the response of a cell to finite strains, the use of linearized strain tensors becomes questionable. The use of linear strain tensors comes from the so-called *small deformation theory* of continuum mechanics, which has as its basic condition the requirement that the displacement gradients and rotations be small compared to unity. Because of the large strains experienced by cells in the artery, it is appropriate to develop a model based on finite elasticity as considered by Humphrey (2007).

13.3.4 Mechanotransduction

An important goal of developing an accurate constitutive model of stretched cells is to understand how a stress or strain applied to a cell will elicit a biochemical response, which in turn will result in a change in cell function. To study the mechanism of mechanotransduction, various devices have been devised to apply defined strains to cells cultured on deformable substrates (see (Brown 2000) for review). Vascular cells transduce stretch into intracellular signals that lead to changes in gene expression (see (Haga et al. 2006; Wang and Thampatty 2006) for review). These biochemical changes then lead to changes in cell function, including cell proliferation, apoptosis, migration, and remodeling. A question often asked in the field of cell mechanobiology is "What does a cell actually respond to?" Is it strain, stress, rate of strain, or some other combination of these mechanical measures? Humphrey suggests that these are merely convenient mathematical concepts rather than measurable, physical quantities, and that "cells cannot respond directly to these continuum metrics or to quantities derived from them – mechanistic models will need to be based on more fundamental quantities, as, for example, inter-atomic forces or conformational changes of the appropriate molecules" (Humphrey 2001). That said, it is still useful to correlate these continuum metrics to mechanotransduction events for the purposes of developing mathematical models relating intracellular stresses and strains generated by substrate stretch to the ensuing biochemical events.

In most studies of mechanotransduction, a population of cells is subjected to a stretch in a well-defined manner for a particular duration of time, after which point the cells are lysed and the intracellular contents collected for biochemical analysis. This results in a 'snapshot in time' of what is happening in the cells. Further, the heterogeneous responses from the cell population are also lost by averaging the responses as a single population-averaged result. When constructing a time course by repeating the experiment after different experimental durations, variability between samples make small changes in biochemical events difficult to detect. Thus, it is imperative that mechanical measurements be collected simultaneously with measurements of biochemical events on a single cell basis. This would provide the data necessary for making accurate correlations between these mechanical and biochemical metrics.

13.4 A New Approach for the Study of the Mechanobiology of Cell Stretching

Despite decades of research in cell mechanobiology, we still do not clearly understand the mechanical properties of adherent cells, nor do we understand the mechanisms by which these cells respond to a given mechanical stimulus to result in biochemical events. The approaches described above have certainly provided valuable insight into this problem, but a new methodology is needed to both study the mechanical properties of cells and relate mechanical inputs to biochemical outcomes. While there are several very well-conceived constitutive relations developed to describe the cell response to very specific types of loading (e. g. magnetic bead twisting), there is a fundamental problem with applying such relationships to the case of stretching a cell. A constitutive relation does not describe a material, rather it describes the response of a material under certain conditions. In the case of a cell attached to a stretched substrate, the constitutive relation must relate the stresses generated as the cell is deformed to the intracellular strains caused by stretching. Due to the dynamic structural properties of cells, a similarly dynamic constitutive model capable of adapting to the mechanical conditions is needed. Such a model may provide new insight into the mechanisms involved in mechanotransduction.

Once armed with an appropriate form for the model, the material parameters are determined by probing the cell with a controlled stretch, estimating tractions at the cell surface, and solving the associated initial-boundary value problem. Since the structure of the cell is dynamic and geometrically complex, the ability to perform several measurements simultaneously (e. g. traction microscopy or microprobe arrays) is preferable to measurements at a single point (e. g. AFM or MTC). Such distributed measurements can immediately identify spatial information such as material anisotropy with less potential for ambiguous interpretation of the data.

To relate continuum metrics to stretch-induced biochemical events, the method of mechanical probing of the cell must allow the simultaneous collection of biochemical measurements. The mechanical measurements are already technically difficult, thus collection of additional biochemical measurements for correlation with the mechanical measurements is often performed in separate experiments. As with all scientific experiments, one wishes to minimize extraneous factors that introduce additional uncertainty to the measurements. Thus, the ideal situation is to be able to perform all the necessary measurements on the same cell at the same time.

The remainder of this chapter is dedicated to proposing an approach which, by overcoming limitations from previous approaches, provides the necessary framework for developing a constitutive model of cells to predict how structural and mechanical changes affect mechanotransduction.

13.4.1 An Adaptive, Microstructure-based Constitutive Model for Stretched Cells

The Constrained Mixture model proposed by Humphrey (2007) provides the basis for an adaptive constitutive model of adherent cells subjected to two-dimensional stretching of the underlying matrix. As an initial approximation, the only structural constituents that will be considered are actin filaments. Further, the k-th component of the actin cytoskeleton is defined as the family of actin filaments which are formed at the same instant in time and are oriented in the same direction. Thus, actin filament formed in different directions and/or formed at different times will be treated mathematically as separate components of the cytoskeleton. Additionally, the filaments are all assumed to be located in the two-dimensional plane adjacent to the basolateral surface of the cell.

To implement the model, specific forms of the constituent formation and survival functions must be defined that describe the adaptation response of the cytoskeletal filaments to stretching. Motivated by chemical kinetics and the observation that stress fibers form in response to stretch (see Figs. 13.1B and 13.2D), the following equation relating $m^k(\theta)$ to the availability of actin monomers and the axial stretch in a fiber is proposed:

$$m^k(\theta,t) = \begin{cases} 0 & \dot{\alpha}^k(\theta,t) < \dfrac{-m_0\alpha_0}{m_1} \\ \phi^g\left[m_0\alpha_0 + m_1\dot{\alpha}^k(\theta,t)\right] & \dot{\alpha}^k(\theta,t) \geq \dfrac{-m_0\alpha_0}{m_1} \end{cases} \quad (13.22)$$

where m_0 and m_1 are proportionality constants and ϕ^g is the mass fraction of G-actin. Assuming no change in the expression of actin in response to stretch, the mass balance in the cytoplasm at any given time is defined as $\phi^c + \phi^g + \sum \phi_f^k = 1$. The rate of stretch in the θ-direction ($\dot{\alpha}^k(\theta,t)$) is the rate the new fiber would be stretched if it formed at time t. The motivation for making $m^k(\theta,t)$ proportional to the fiber deposition stretch (α_0) and $\dot{\alpha}^k(\theta,t)$ is that fiber formation is assumed to stem from the gathering together of a loose network of actin filaments into a parallel bundle (Fig. 13.4). This gathering process can occur by myosin-driven filament translation or by stretching of the actin filament network.

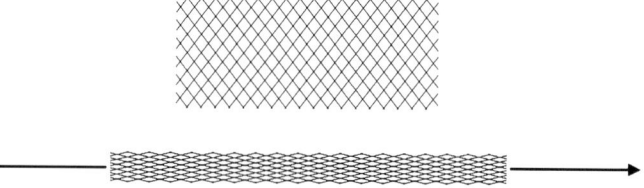

Fig. 13.4 Formation of an actin stress fiber by drawing together of a loose actin filament network. *Top*: In the absence of an applied stress, the actin cytoskeleton is illustrated as a loose network of skew filaments. *Bottom*: Upon applying a uniaxial stress, the network elongates and collapses into a tight bundle

These mechanisms for fiber formation are supported by the observations that stress fibers assemble in response to contractility (Chrzanowska-Wodnicka and Burridge 1996) and cyclic stretch (Kaunas et al. 2005). It has been shown that actin fibers are pre-stretched to a value of 1.2 to 1.3 under normal, unperturbed conditions (Costa et al. 2002; Deguchi et al. 2005). In Eq. (13.21), the level of pre-stretch is represented α_0, which is assumed to be proportional to the basal level of myosin II activity.

The form of the survival function $q^k(t-\tau,\theta)$ should be chosen based on observations of actin fiber turnover. Unfortunately, such data are not readily available so a phenomenological approach must be taken. An initial approximation for the form of the survival function might be an exponential decay with a constant half-life (cf. Humphrey 2007). Motivated by the concept of homeostasis as well as the study by Wang (Wang 2000), it is expected that the rate of disappearance of an actin fiber is proportional to the deviation of fiber stretch from a homeostatic value (α_h), in which case an exponential decay with a half-life dependent on $|\alpha^k(t)-\alpha_h|$ may be more appropriate:

$$q^k(t-\tau,\theta) = \exp\left\{-(t-\tau)q_0\left[1+q_1\left[\frac{\alpha^k(t-\tau,\theta)-\alpha_h}{\alpha_h}\right]^2\right]\right\}, \quad (13.23)$$

where q_0 and q_1 are constants relating the sensitivity of the half-life to the deviation of fiber stretch from the homeostatic value. An initial estimate for α_h can be taken from the preferred pre-stretch of ~1.2 to 1.3 of stress fibers in unstretched ECs (Costa et al. 2002; Deguchi et al. 2005).

Obviously, assuming the actin cytoskeleton can completely describe the mechanical properties of adherent cells is an oversimplification. For the purposes of illustration of the basic concepts of the model, such a simplification is sufficient. The contributions from the nucleus, plasma membrane, microtubules and intermediate filaments may improve the quantitative accuracy of the model, but are not expected to change the general predictions of the present model.

Characteristics of the substrate can also affect actin fiber properties. For instance, the amount of stress fibers in cells cultured on soft substrates is proportional to substrate stiffness (Yeung et al. 2005), which suggests that for stress fibers to form, the matrix must be sufficiently stiff to support the tension. Similarly, the adhesive strength of focal adhesions must also be capable of withstanding tensions developed in stress fibers in order to maintain stable actin fibers (DiMilla et al. 1991). These phenomena should be incorporated into the formation and/or survival equations once appropriate equations are determined. The predictive value of these forms of these equations must then be verified by running simulations under different stretch conditions.

13.4.2 Biaxial Loading Traction Microscopy

As stated above, a device is needed that interrogates the stress/strain relationship for cells deformed on stretched matrices. Data collected from such experiments would provide the necessary information to develop a constitutive relation for stretched cells. Motivated by the aforementioned traction microscopy methodology, the present design is of a biomechanical culture system capable of subjecting adherent cells to diverse biaxial stress/strain culture conditions and tests while permitting live microscopic imaging. Figure 13.5 illustrates a testing platform for stretching cells on a cruciform specimen of uniform thickness in biaxial loading. The hydrogel is coupled to the device via a solid porous material into which the hydrogel intercalates, thus providing a large surface area for binding and the application of biaxial loading. The bars constrain the stress (and strain) in the hydrogel locally, but these local influences become negligible a sufficient distance from the bars (Mönch and Galster 1963). The hydrogel is stretched symmetrically along two orthogonal axes using four computer-controlled stepper motors. This arrangement allows control of multiple parameters including the strain magnitude, strain rate, and temporal waveform along each axis.

The stretching platform is mounted on the stage of an upright microscope with the hydrogel centered under a water-dipping objective. Axisymmetric stretching is

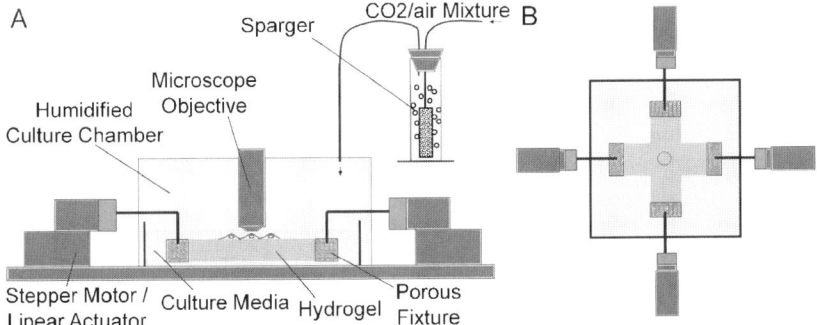

Fig. 13.5 Schema for the proposed adherent cell mechanical testing system. **A**: *Front view* of the system showing the stepper motors/linear actuators, the hydrogel culture system, CO_2/air mixture humidifier, and the microscope objective. **B**: *Top view* showing the configuration of the stepper motors/linear actuators for axisymmetric deformation of the hydrogel. The location of the objective is indicated as a *circle* in the center of the hydrogel

necessary to keep a cell within the field-of-view of the microscope during stretching. The entire hydrogel culture chamber is perfused with a mixture of 5%CO_2/95% air to maintain the culture media at pH 7.4 using a bicarbonate buffer. To minimize evaporation, the gas is humidified by bubbling through warm, sterile water contained in a sparger. The microscope stage and stretch device temperature are controlled at 37 °C.

Hydrogel stretching is controlled by specifying the displacements of the hydrogel fixtures using the computer-controlled stepper motors. The size of the steps must be adequately small to provide a smooth displacement waveform at the highest applied strain rates. Even in the absence of cell traction forces, the strain applied locally to the cell of interest may differ from the average strain for the entire hydrogel, so the local strain must be measured using the fiducial markers in the hydrogel which provide real-time feedback to the computer using a high-speed interface. For each image capture, a search algorithm based on pixel intensity values locates the pixel coordinates for the centroid of each marker. The software then computes the stretch ratios by comparing the current marker positions, given by pixel coordinates, to the reference positions using bilinear isoparametric interpolation (Humphrey et al. 1990). Keeping the surface of the hydrogel in focus as the hydrogel is stretched will be challenging, so it may be necessary to stop the stretch periodically to refocus for each image.

The existing methodology of traction microscopy must be modified to measure traction forces produced by deforming a cell during stretching. As illustrated in Fig. 13.6, a minimum of four images of the markers on the surface of the hydrogel must be recorded, rather than the two needed for traction microscopy of unstretched cells. After the cell adheres and spreads on the hydrogel surface, the first image will be of the contracted cell along with the surrounding markers in the unstretched state (Fig. 13.6A). Next, the substrate, and hence the cell, is stretched and the new positions of the markers and the cell border are recorded (Fig. 13.6B). The cell is then removed to eliminate traction forces generated by the cell so that the traction-free positions of the markers can be recorded in both the stretched (Fig. 13.6C) and unstretched (Fig. 13.6D) states. Cell removal can be accomplished by exchanging the culture media for 2% trypsin-EDTA in phosphate buffered saline and gently flushing the gel with a syringe to remove loosely-attached cells from the hydrogel.

From these four images, the marker displacements due to cell contractile forces alone are determined by comparing images represented by Fig. 13.6A and D. Next, the marker displacements due to combined effects of contractility and cell deformation on traction force are measured by comparing the images represented by Fig. 13.6B and C. The total traction at a particular location on the stretched substrate is the vectorial addition of the tractions due to cell contraction and cell deformation. To estimate traction fields over time, images recorded at different times are compared to the reference images (Fig. 13.6C and D).

When developing a constitutive model to describe the behavior of a material, it is desirable to make a model which is valid over a wide range of conditions (e. g. Navier–Stokes equation for Newtonian fluids) rather than simply capable of describing the material response to a very limited range of conditions (e. g. response

13.4 A New Approach for the Study of the Mechanobiology of Cell Stretching

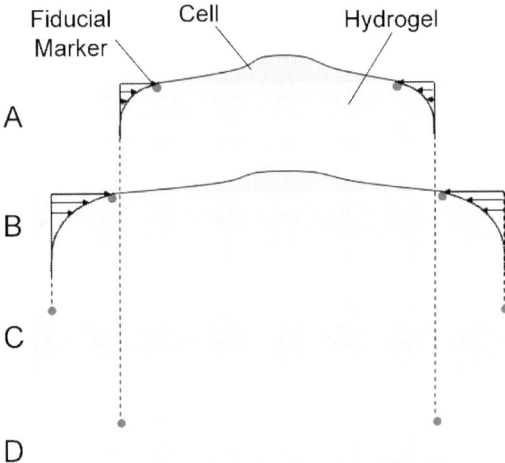

Fig. 13.6 Traction force-induced hydrogel deformations and marker positions are illustrated at different steps of an experiment. For simplicity, only two markers are shown. **A**: A contracting cell exerts tractions on the underlying substrate, thus pulling the markers towards the cell. The deformation due to tractions applied to the hydrogel surface decays with depth into the hydrogel. **B**: Stretching the substrate increases the traction magnitudes, and hence increases the surface deformation of the hydrogel. **C**: Removing the cell allows the substrate to relax and the markers to move to their traction-free positions in the stretched hydrogel. **D**: Removing substrate stretch allows the markers to move to their traction-free positions in the unstretched hydrogel

to a point load on the surface of the cell). With the proposed device, a range of test conditions should be employed to explore the mechanical properties of the cell under study. This would involve varying the pattern of stretch (e. g. uniaxial vs. equibiaxial), the magnitude of strain, and the strain-rates.

One drawback to the proposed system is that forces within the cell itself are not measured directly. Relationships between force and extent of elongation of isolated stress fibers (Deguchi et al. 2006) and even individual actin filaments (Kishino and Yanagida 1988; Liu and Pollack 2002) have been measured. The preexisting tension in actin bundles extracted from intact cells have been estimated by stretching the extracted fibers to the original lengths and measuring the tensions (Deguchi et al. 2005). Laser nanoscissors have recently been used to sever individual stress fibers to measure the mechanics of fiber retraction as well as the associated changes in the traction field (Kumar et al. 2006). Such experiments need to be interpreted within the framework of a constitutive model in order to isolate the contributions of individual components of the entire intracellular structure.

Since the model is based on cytoskeletal organization, it would be very useful to image cytoskeletal organization using fluorescently-labeled proteins simultaneous with the traction microscopy measurements. Variants of Green Fluorescent Protein, or newer alternatives such as DsRed, conjugated to a structural protein of interest can be transfected into the cells to be expressed and incorporated into the specific intracellular structures [e. g. EYFP-actin into stress fibers (Kumar et al. 2006)]. Such fluorescently-tagged proteins are also useful fiducial markers for measuring intracellular strains. For the purposes of producing discrete markers along cytoskeletal fibers, it is better to express low, rather than high, levels of fluorescently-tagged cytoskeletal monomers – a technique commonly referred to as Fluorescent Speckle Microscopy (Danuser and Waterman-Storer 2003). The resulting variations in the fluorescence intensity along a fiber allow the measurement of fiber translation, and potentially local fiber strain. Alternatively, proteins (e.g. α-actinin) or small organelles (e.g. mitochondria) which bind along the lengths of cytoskeletal fibers can be used as fiducial markers (Hu et al. 2003; Peterson et al. 2004). These approaches may also provide some indications of cytoskeletal dynamics, however a more quantitative approach is preferable. For example, actin filament turnover can be measured using Fluorescent Recovery After Photobleaching (FRAP) or Photoactivated Fluorescence (PAF) with the data interpreted using a diffusion model to isolate the effects of G-actin diffusion and polymerization/depolymerization kinetics (Tardy et al. 1995).

There is evidence that arterial ECs in vivo do not contain the relatively high density of stress fibers observed in ECs in cell culture. It remains to be determined if cells without stress fibers, such as those cultured on very soft hydrogels, can still have anisotropic material properties. Actin filaments which are not bundled into stress fibers could still be organized such that the overall orientation may be aligned to result in anisotropic material properties, however such structures are too small to be resolved by fluorescent microscopy and other imaging techniques would be necessary (i.e. electron microscopy).

13.4.3 Mechanotransduction Experiments

While it is generally accepted that mechanical forces can be transduced into biochemical signals, the mechanisms by which this occurs remain to be elucidated. There is evidence indicating that mechanical forces contribute to conformational changes in signaling proteins to alter their signaling activity. A well-studied mechanism for regulation of protein activity is through conformational changes between auto-inhibited and active (non-inhibited) states. As illustrated in Fig. 13.7, this can occur through the binding of a regulatory protein which destabilizes the intramolecular binding of the active domain of the protein to the inhibitory domain. This requires that the binding of the regulatory protein provides sufficient energy to overcome the energy barrier necessary to dissociate the bond between the inhibitory and catalytic domains. It is possible that mechanical energy transmitted from a stretched matrix into the cell could provide the energy necessary to overcome such energy barriers to activate proteins.

The protein vinculin represents a particularly plausible mechanosensitive protein whose activity appears to be regulated by conformation changes (Bakolitsa et al. 2004). Vinculin localizes to the cytoplasmic side of focal adhesions, as well as adherens junctions which join cells to cells. The molecular structure of vinculin can

13.4 A New Approach for the Study of the Mechanobiology of Cell Stretching

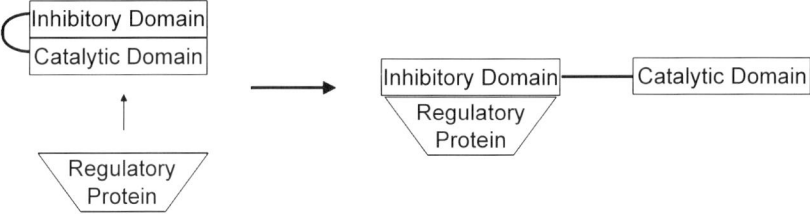

Fig. 13.7 A mechanism for protein activation by disrupting the binding of the inhibitory domain of a protein to the catalytic domain. In the inactive state, the inhibitory domain binds to the catalytic domain of the protein. This intramolecular binding is disrupted upon binding of a regulatory protein to the inhibitory domain, allowing the catalytic domain to interact with a substrate protein

be generalized as consisting of a 'head' region of roughly 850 amino acids and a 'neck' region of about 200 amino acids. Intramolecular binding of these two regions blocks the ability of vinculin to interact with its binding partners within cell junction complexes. Mechanical forces applied to vinculin could potentially disrupt this intramolecular bond to allow vinculin interactions with other protein binding partners.

So how might mechanical stretch affect the conformation of a protein? Sheetz and colleagues sought to answer this question by measuring biochemical changes in response to stretching the intact cytoskeleton of adherent cells treated with a detergent to remove the cell plasma membrane (Sawada and Sheetz 2002). By removing the plasma membrane, the effects of transmembrane ion currents were removed. The binding affinities of various signaling proteins was increased in the stretched cytoskeletons as compared to static cytoskeletons, suggesting that mechanotransduction occurs through conformational changes in cytoskeletal proteins generated by force-dependent deformations.

An important end result of mechanotransduction is regulation of gene expression, and cyclic stretch has been demonstrated to regulate the expression of numerous genes. Gene expression is regulated by upstream signaling pathways, which lead to activation of transcription factors that directly interact with the promoter elements of a gene (Haga et al. 2006). Several such pathways have been identified for regulating mechanosensitive genes, including NF-κB and the Mitogen-Activated Protein Kinase (MAPKs) family signaling pathways. One member of the MAPK family, JNK, has been shown to be differentially activated by cyclic uniaxial and equibiaxial stretch in ECs (Kaunas et al. 2006). Importantly, stretch-induced JNK activation subsides as the stress fibers within the cells align perpendicular to the direction of stretch. This suggests that cells may perceive mechanical stretch differently depending on the orientation of stress fibers relative to the direction of stretch. The activation of other proteins (Hornberger et al. 2005) and the expression of genes (Park et al. 2004) have also been found to depend on the mode by which cells are stretched.

Typically gene expression is measured by collecting mRNA from cells after subjecting the cells to mechanical stimulation and quantifying expression of a single gene by Northern blot or a large set of genes by cDNA microarray analysis. Using micropatterned elastic substrate to control cell and stress fiber alignment, Kurpinski and co-workers (2006) used microarray to show that expression profile of mesenchymal stem cells subjected to cyclic uniaxial stretch depended on the orientation of the cells relative to the direction of stretch. This only provides information about the expression pattern of genes at a particular time point for the entire population of cells in the sample. If one is interested in temporal changes in gene expression then these assays must be repeated for each time point using different samples of cells. It would be very useful to be able to correlate gene expression with the mechanical stress/strain state of individual cells. Since the mechanical stress/strain state can change significantly over time, a technique for monitoring gene expression in individual cells over time is necessary. This can be performed using reporter genes encoding for fluorescently-labeled proteins such as variants of GFP conjugated to the promoter sequence of the gene of interest (Thompson et al. 2004). The expression pattern of multiple reporter genes encoding fluorescent proteins of different colors could be simultaneously monitored assuming each emission wavelength is sufficiently resolved, thus allowing the comparison of expression of different genes in the same cell under identical mechanical conditions.

There is a need to develop models by which the dynamic changes of cell structure and mechanical quantities are related to biochemical outcomes such as protein activation and gene expression. The time course of stretch-induced activation of JNK in ECs correlates with stress fiber orientation perpendicular to the direction of stretch (Kaunas et al. 2006), which is predicted to result in a decrease in the contribution of cell deformation on overall mechanical stress in the cell. To correlate the activation of a protein, which is a scalar value, a scalar representation of the level of stress in the cell is necessary. One candidate is the total energy transmitted by the cell to elastic distortion of the substrate (Butler et al. 2002):

$$U = \frac{1}{2} \sum_{i=1}^{N} \vec{T}_i(x, y) \cdot \vec{u}_i(x, y), \quad (13.24)$$

where U is strain energy, $\vec{T}_i(x, y)$ is a traction force applied at position (x, y) and $\vec{u}_i(x, y)$ is the displacement of the hydrogel at that position. Studies correlating the magnitudes of stretch with protein activity or gene expression suggest that increasing the stretch magnitude leads to an increase in the biochemical signal up to a saturating value (Li et al. 1998). Such saturation of the biochemical response could be expressed as a sigmoidal function of U:

$$r_a(U) = \frac{r_{max}}{1 + e^{-A_r(U - U_{1/2})}}, \quad (13.25)$$

where r_a is the magnitude of the response (e. g. protein activity or level of gene expression), which reaches a maximum value (r_{max}) at saturating levels of substrate strain energy. The constant A_r describes the sensitivity of the biochemical response to the substrate strain energy and $U_{1/2}$ is the magnitude of U for half-maximal activation of the biochemical signal.

It remains to be determined if U or some other scalar representation of the state of stress in the cell best correlates with the stimulation of biochemical events in stretched cells, in which case another scalar variable could be used in place of U in Eq. (13.24). A biochemical basis for relating mechanical stress (or strain) to JNK activation comes from the observation that mechanical strain stimulates conformational activation of integrins and that stretch-induced JNK activation requires the formation of new integrin-mediated adhesions (Katsumi et al. 2005). It is postulated that tension in actin fibers is transmitted to integrins at focal adhesion, which stimulates turnover of the stress fibers and associated focal adhesions.

To correlate cell stress and/or strain with the level of protein activation or gene expression, it would be very useful to measure these biochemical changes simultaneous to the traction force measurements in real time. Biosensors using Fluorescence Resonance Energy Transfer (FRET) as a readout for conformational changes related to protein activation have been successfully used to measure spatial and temporal changes in the activation of proteins in response to mechanical stimuli (Wang et al. 2005). Similarly, GFP reporter gene expression can be used to track gene expression in live cells. The levels of fluorescence from these protein biosensors or reporter genes can be followed for individual cells over time and correlated with various measures of the state of stress/strain in the cell. The capability to resolve spatially-varying activation of proteins within individual cells using FRET biosensors raises the possibility that local stress/strain concentration in a cell can be correlated with local biochemical activities. Clearly there is potentially much knowledge to be gained by combining intracellular biosensor technologies with an adaptive mechanical model of adherent cells.

13.5 Illustrative Examples

13.5.1 Step Stretch: Effect of Ramp Rate on the Cell Stress and Fiber Organization

Let us consider the response of an adherent cell to a ramp in pure uniaxial strain (a.k.a. strip biaxial stretch) to a final stretch ratio λ_f in the x_1-direction while maintaining the stretch in the x_2-direction constant at unity:

$$\lambda_1 = \begin{cases} \dfrac{\lambda_f}{t_f} t & 0 \le t \le t_f \\ \lambda_f & t > t_f \end{cases} \quad (13.26)$$

$$\lambda_2 = 1 ,$$

where $\lambda_i = L_i/L_{i0}$ is the change in substrate stretch in the x_i-direction for $i = 1, 2$. For very fast ramp rates the response will be similar to a step-change in λ_1. Prior to the ramp in stretch, the axial fiber strain is assumed to have an initial value α_0

representing the deposition stretch corresponding to the contractility-induced pre-existing stretch in the fiber. As illustrated in Fig. 13.8, the average axial fiber stretch ratio in fibers oriented in the x_1-direction will initially increase sharply from α_0 to $\alpha_0 \lambda_f$.

Now consider what happens after the initial application of stretch. From Eq. (13.21), it is clear that immediately following the stretch the model predicts an increase in the rate of formation of fibers oriented toward the direction of stretch, however these newly formed fibers will have a natural configuration based on the deformed state of the substrate rather than the original state, and hence the axial strain in these fibers will equal to α_0. From Eq. (13.22), there is an expected simultaneous increase in the rate of disappearance of the stretched fibers assuming that the new axial fiber strain has increased above the homeostatic strain value. The net result is that the average axial fiber strain $\langle \alpha \rangle$ "relaxes" back to α_0 with an exponential decay characterized by the half-life of the stretched fibers (Fig. 13.8). Meanwhile, there is no change in axial fiber stretch in the x_2-direction. Since fiber stress is a nonlinear, but monotonically increasing, function of fiber stretch, the net result is an increase in stress in the direction of substrate stretch which relaxes back to the original stress magnitude. The apparent stress relaxation gives the overall appearance that the cell material properties are viscoelastic, however this is actually due to fiber turnover rather than viscoelastic properties of the fibers themselves (cf. Humphrey 2007).

With the increase in traction forces applied by the cell to the substrate during the period of elevated stress, there would be a transient increase in the substrate strain energy (Eq. (13.23)). Equation (13.24) then predicts an increase in the activity of stress-dependent proteins, followed by a decline back to the basal level of activity as the strain energy returns to basal level.

Now let us consider what happens when the ramp rate is much slower than the rate of fiber turnover. Under these conditions, the fibers will disappear and reform in the current material configuration such that the strain contributed by axial deformation is negligible at all times ($\alpha^k \approx \alpha_0$). Thus, there is not expected to be a significant increase in traction force applied to the substrate, and hence there is not an increase in stress-dependent biochemical activity. This is an important prediction since this suggests that cells have a "short memory" of the substrate configuration.

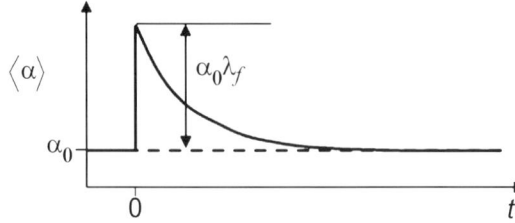

Fig. 13.8 Actin fiber stretch response in response to a step in strip biaxial stretch. The change in fiber axial stretch ratio is shown for a fiber oriented in the x_1-direction (*solid curve*) and x_2-direction (*dashed curve*)

13.5.2 Cyclic Stretch: Effect of Frequency on the Cell Stress, Fiber Organization, and Mechanotransduction

Let's next consider the response to cyclic uniaxial stretch (a.k.a. cyclic strip biaxial stretch):

$$\lambda_1 = 1 + (\lambda_f - 1)\left(\frac{1}{2} - \frac{1}{2}\cos\omega t\right) \tag{13.27}$$

$$\lambda_2 = 1$$

where ω and λ_f are the frequency and amplitude of stretch, respectively. A physiologically relevant cycle period for arterial cells is the heart rate (1 Hz), the period of which is expected to be significantly lower than the half-life of a normal actin fiber. Now when a new fiber is formed, the reference configuration is based on whatever the configuration is at that point in the stretch cycle. Since new fibers can be deposited at any time during a cycle, the mean axial fiber stretch approaches α_0 as the original fibers are replaced by new fibers. As illustrated in Fig. 13.9, the time-averaged axial fiber stretch will be equal to that of the unstretched fibers oriented in the x_2-direction, hence the average rate of fiber formation will be equal in all directions. The axial stretch in these fibers oriented in the x_1-direction will deviate from the homeostatic value for most of a cycle, however, leading to an increased rate of disappearance of fibers oriented toward the direction of stretch. The net result is a net accumulation of fibers oriented perpendicular to the direction of stretch, the extent of which increases with increasing stretch amplitude, as has been observed experimentally (Kaunas et al. 2005).

In the time period immediately following the onset of cyclic stretch, U, and hence r_a, are expected to increase initially due to traction forces generated by cell deformation. As the actin fibers becomes preferentially oriented pendicular to the direction of stretch, U and r_a are expected to return to near basal levels due to a decrease in traction forces generated by fiber distension. These model predictions on adaptations in actin fiber orientation and biochemical activity closely follow the observed perpendicular orientations of stress fibers and transient activation of JNK in endothelial cells subjected to cyclic uniaxial stretch (Kaunas et al. 2006). When the frequency of stretch is very small, the model prediction are similar to the case of

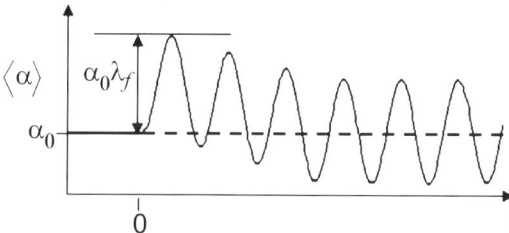

Fig. 13.9 Fiber stretch in response to cyclic strip biaxial stretch. The expected axial fiber stretch ratio is shown for fibers oriented in the x_1-direction (*solid curve*) and x_2-direction (*dashed curve*)

a slow ramp in stretch – the fibers will disappear and reform in the current material configuration such that the strain contributed by axial deformation is negligible at all times ($\alpha^k \approx \alpha_0$), and thus the tractions and associated stress-dependent protein activities will not increase.

In the case of cyclic equibiaxial stretch, the orientation of a fiber does not affect the fiber stretch since the substrate stretch is isotropic. It is readily apparent that the model predicts a sustained increase in average axial fiber stretch since the fibers cannot reorient in a direction perpendicular to substrate stretch. The resulting sustained increase in U and r_a suggests that the mechanism for sustained activation of JNK in endothelial cells subjected to cyclic equibiaxial stretch is the inability of stress fibers to adapt to equibiaxial stretch (Kaunas et al. 2006).

13.6 Closure

To quote Charles Darwin: "It is not the strongest of the species that survives, nor the most intelligent, but rather the one most responsive to change." In the face of environmental stresses, species must adapt to survive. It seems the same concept holds true at the subcellular level – only when the cytoskeletal network is able to adapt to changes in mechanical stresses can the cell attain mechanical homeostasis. Cells regulate the growth and remodeling of the surrounding tissue in response to macroscopic mechanical stresses, suggesting the adaptation process is connected across many orders of length scale.

Homeostasis also involves maintaining a balance in the cell's biochemistry. A daunting task in the field of signal transduction is to make sense of how interconnected signaling pathways interact to regulate gene expression and cell function in the face of extracellular perturbations. The temporal and spatial activation patterns of signaling molecules may dictate how activation of a common signaling pathway can result in different patterns of gene expression and cell responses. Add to this time-dependent and spatially-varying stresses and strains generated in a cell in response to mechanical loading of a tissue – there is clearly a pressing need for an accurate constitutive model of adherent cells to be able to interpret the results of the mounting accumulation of biochemical data in the field of mechanotransduction.

While the approach outlined in this chapter were motivated by a need to understand the effects of stretch on EC function, the concepts can also be applied to other cell types. This methodology is particularly well-suited for studying the mechanobiology of cells (epithelial and endothelial) that normally reside on the surface of a tissue. A challenging next step is to develop a similar approach for studying the mechanobiology of cells residing within a tissue. To measure traction forces for a cell in a three-dimensional matrix, the elastic substrate needed for performing traction microscopy would also need to support the culture of cells inside of the substrate. Ideally a natural matrix such as collagen would be used, yet the material properties of collagen are complex and depend on many variables, including remodeling by the cells. To this end, mixture theory has been applied toward describing the time-dependent, adaptive material properties of tissues (reviewed in Humphrey 2007).

Acknowledgements I would like to thank Drs. Jay Humphrey and Jimmy Moore Jr. for their thoughtful critiques of this chapter.

References

1. Bakolitsa CD, Cohen M, Bankston LA, Bobkov AA, Cadwell GW, Jennings L, Critchley DR, Craig SW, Liddington RC (2004) Structural basis for vinculin activation at sites of cell adhesion. Nature 430:583–6
2. Baek S, Rajagopal KR, Humphrey JD (2006). A theoretical model of enlarging intracranial fusiform aneurysms. J Biomech Eng 128:142–149
3. Balaban NQ, Schwarz US, Riveline D, Goichberg P, Tzur G, Sabanay I, Mahalu D, Safran S, Bershadsky A, Addadi L, Geiger B. (2001). Force and focal adhesion assembly: a close relationship studied using elastic micropatterned substrates. Nat Cell Biol 3:466–72
4. Beningo KA, Wang YL (2002). Flexible substrata for the detection of cellular traction forces. Trends Cell Biol 12:79–84
5. Brown TD (2000) Techniques for mechanical stimulation of cells in vitro: a review. J Biomech 33:3–14
6. Burridge K (1981) Are stress fibres contractile? Nature 294:691–2
7. Butler JP, Tolic-Norrelykke IM, Fabry B, Fredberg JJ (2002) Traction fields, moments, and strain energy that cells exert on their surroundings. Am J Physiol Cell Physiol 282:C595–605
8. Cannon WB (1932) The Wisdom of the Body. W.W. Norton & Co., New York
9. Chien S (2006) Mechanotransduction and endothelial cell homeostasis: the wisdom of the cell. Am J Physiol Heart Circ Physiol 292:H1209–24
10. Chrzanowska-Wodnicka M, Burridge K (1996) Rho-stimulated contractility drives the formation of stress fibers and focal adhesions. J Cell Biol 133:1403–15
11. Costa KD, Hucker WJ, Yin FC (2002) Buckling of actin stress fibers: a new wrinkle in the cytoskeletal tapestry. Cell Motil Cytoskeleton 52:266–74
12. Danuser G, Waterman-Storer CM (2003) Quantitative fluorescent speckle microscopy: where it came from and where it is going. J Microsc 211:191–207
13. Deguchi S, Ohashi T, Sato M (2005) Evaluation of tension in actin bundle of endothelial cells based on preexisting strain and tensile properties measurements. Mol Cell Biomech 2:125–33
14. Deguchi S, Ohashi T, Sato M (2006) Tensile properties of single stress fibers isolated from cultured vascular smooth muscle cells. J Biomech 39:2603–10
15. Dembo M, Wang YL (1999) Stresses at the cell-to-substrate interface during locomotion of fibroblasts. Biophys J 76:2307–16
16. Deshpande VS, McMeeking RM, Evans AG (2006) A bio-chemo-mechanical model for cell contractility. Proc Natl Acad Sci USA 103:14015–20
17. DiMilla PA, Barbee K, Lauffenburger DA (1991) Mathematical model for the effects of adhesion and mechanics on cell migration speed. Biophys J 60:15–37
18. Evans EA (1989) Structure and deformation properties of red blood cells: concepts and quantitative methods. Methods Enzymol 173:3–35
19. Fuller B (1961) Tensegrity. Portfolio Artnews Annual 4:112–127

20. Giddens DP, Zarins CK, Glagov S (1993) The role of fluid mechanics in the localization and detection of atherosclerosis. J Biomech Eng 115:588–94
21. Gittes F, Mickey B, Nettleton J, Howard J (1993) Flexural rigidity of microtubules and actin filaments measured from thermal fluctuations in shape. J Cell Biol 120:923–34
22. Guilak F, Tedrow JR, Burgkart R (2000) Viscoelastic properties of the cell nucleus. Biochem Biophys Res Commun 269:781–6
23. Haga JH, Li YS, Chien S (2006) Molecular basis of the effects of mechanical stretch on vascular smooth muscle cells. J Biomech 40:947–60
24. Hill AV (1938) The heat of shortening and the dynamic constants of muscle. Proc R Soc Lond Ser B 126:136–95
25. Hofmann UG, Rotsch C, Parak WJ, Radmacher M (1997) Investigating the cytoskeleton of chicken cardiocytes with the atomic force microscope. J Struct Biol 119:84–91
26. Hornberger TA, Armstrong DD, Koh TJ, Burkholder TJ, Esser KA (2005) Intracellular signaling specificity in response to uniaxial vs. multiaxial stretch: implications for mechanotransduction. Am J Physiol Cell Physiol 288:C185–94
27. Hu S, Chen J, Fabry B, Numaguchi Y, Gouldstone A, Ingber DE, Fredberg JJ, Butler JP, Wang N (2003) Intracellular stress tomography reveals stress focusing and structural anisotropy in cytoskeleton of living cells. Am J Physiol Cell Physiol 285:C1082–90
28. Hu S, Eberhard L, Chen J, Love JC, Butler JP, Fredberg JJ, Whitesides GM, Wang N (2004) Mechanical anisotropy of adherent cells probed by a three-dimensional magnetic twisting device. Am J Physiol Cell Physiol 287:C1184–91
29. Humphrey JD (2001) Stress, strain, and mechanotransduction in cells. J Biomech Eng 123:638–41
30. Humphrey JD (2002) On mechanical modeling of dynamic changes in the structure and properties of adherent cells. Math Mech Sol 7:521–39
31. Humphrey JD (2007) Need for a continuum biochemomechanical theory of soft tissue and cell growth and remodeling. In: Holzapfel GA, Ogden RW (eds) Biomechanical Modeling at the Molecular, Cellular, and Tissue Levels. Springer, Berlin, 2007 (in press)
32. Humphrey JD, Strumpf RK, Yin FC (1990) Determination of a constitutive relation for passive myocardium: I. A new functional form. J Biomech Eng 112:333–9
33. Ingber DE, Madri JA, Jamieson JD (1981) Role of basal lamina in neoplastic disorganization of tissue architecture. Proc Natl Acad Sci USA 78:3901–5
34. Kakisis JD, Liapis CD, Sumpio BE (2004) Effects of cyclic strain on vascular cells. Endothelium 11:17–28
35. Katsumi A, Naoe T, Matsushita T, Kaibuchi K, Schwartz MA (2005) Integrin activation and matrix binding mediate cellular responses to mechanical stretch. J Biol Chem 280:16546–9
36. Kaunas R, Nguyen P, Usami S, Chien S (2005) Cooperative effects of Rho and mechanical stretch on stress fiber organization. Proc Natl Acad Sci USA 102:15895–900
37. Kaunas R, Usami S, Chien S (2006) Regulation of stretch-induced JNK activation by stress fiber orientation. Cell Signal 18:1924–31
38. Kaverina I, Krylyshkina O, Beningo K, Anderson K, Wang YL, Small JV (2002) Tensile stress stimulates microtubule outgrowth in living cells. J Cell Sci 115:2283–91
39. Kishino A, Yanagida T (1988) Force measurements by micromanipulation of a single actin filament by glass needles. Nature 334:74–6
40. Kumar S, Maxwell IZ, Heisterkamp A, Polte TR, Lele TP, Salanga M, Mazur E, Ingber DE (2006) Viscoelastic retraction of single living stress fibers and its impact on cell shape, cytoskeletal organization, and extracellular matrix mechanics. Biophys J 90:3762–73
41. Kurpinski K, Chu J, Hashi C, Li S (2006) Anisotropic mechanosensing by mesenchymal stem cells. Proc Natl Acad Sci USA 103:16095–100
42. Landau LD, Lifshitz EM (1986) Theory of Elasticity. Pergamon, Oxford
43. Lanir Y (1979) A structural theory for the homogeneous biaxial stress-strain relationships in flat collagenous tissues. J Biomech 12:423–36
44. Li Q, Muragaki Y, Hatamura I, Ueno H, Ooshima A (1998) Stretch-induced collagen synthesis in cultured smooth muscle cells from rabbit aortic media and a possible involvement of angiotensin II and transforming growth factor-beta. J Vasc Res 35:93–103

45. Liu X, Pollack GH (2002) Mechanics of F-actin characterized with microfabricated cantilevers. Biophys J 83:2705–15
46. Mönch E, Galster D (1963) A method for producing a defined uniform biaxial tensile stress field. Brit J Appl Phys 14:810–2
47. Na S, Sun Z, Meininger GA, Humphrey JD (2004) On atomic force microscopy and the constitutive behavior of living cells. Biomech Model Mechanobiol 3:75–84
48. Nerem RM, Levesque MJ, Cornhill JF (1981) Vascular endothelial morphology as an indicator of the pattern of blood flow. J Biomech Eng 103:172–6
49. Park JS, Chu JS, Cheng C, Chen F, Chen D, Li S (2004) Differential effects of equiaxial and uniaxial strain on mesenchymal stem cells. Biotechnol Bioeng 88:359–68
50. Peterson LJ, Rajfur Z, Maddox AS, Freel CD, Chen Y, Edlund M, Otey C, Burridge K (2004) Simultaneous stretching and contraction of stress fibers in vivo. Mol Biol Cell 15:3497–508
51. Sawada Y, Sheetz MP (2002) Force transduction by Triton cytoskeletons. J Cell Biol 156:609–15
52. Stamenovic D (2005) Microtubules may harden or soften cells, depending of the extent of cell distension. J Biomech 38:1728–32
53. Svitkina TM, Verkhovsky AB, Borisy GG (1996) Plectin sidearms mediate interaction of intermediate filaments with microtubules and other components of the cytoskeleton. J Cell Biol 135:991–1007
54. Taber LA (1995) Biomechanics of growth, remodeling, and morphogenesis. Appl Mech Rev 48,487–545
55. Tan JL, Tien J, Pirone DM, Gray DS, Bhadriraju K, Chen CS (2003) Cells lying on a bed of microneedles: an approach to isolate mechanical force. Proc Natl Acad Sci USA 100:1484–9
56. Tardy Y, McGrath JL, Hartwig JH, Dewey CF (1995) Interpreting photoactivated fluorescence microscopy measurements of steady-state actin dynamics. Biophys J 69:1674–82
57. Thompson DM, King KR, Wieder KJ, Toner M, Yarmush ML, Jayaraman A (2004) Dynamic gene expression profiling using a microfabricated living cell array. Anal Chem 76:4098–103
58. Thubrikar MJ, Robicsek F (1995) Pressure-induced arterial wall stress and atherosclerosis. Ann Thorac Surg 59:1594–603
59. Tolic-Norrelykke IM, Butler JP, Chen J, Wang N (2002) Spatial and temporal traction response in human airway smooth muscle cells. Am J Physiol Cell Physiol 283:C1254–66
60. Truesdell C, Noll W (1965) The non-linear field theories of mechanics. In: Flugge S (ed) Handbuch der Physik, Vol III/3. Springer, Berlin Heidelberg New York
61. Volokh KY, Vilnay O, Belsky M (2000) Tensegrity architecture explains linear stiffening and predicts softening of living cells. J Biomech 33:1543–9
62. Wang JH (2000) Substrate deformation determines actin cytoskeleton reorganization: A mathematical modeling and experimental study. J Theor Biol 202:33–41
63. Wang JH, Thampatty BP (2006) An introductory review of cell mechanobiology. Biomech Model Mechanobiol 5:1–16
64. Wang N, Tolic-Norrelykke IM, Chen J, Mijailovich SM, Butler JP, Fredberg JJ, Stamenovic D (2002) Cell prestress. I. Stiffness and prestress are closely associated in adherent contractile cells. Am J Physiol Cell Physiol 282:C606–16
65. Wang Y, Botvinick EL, Zhao Y, Berns MW, Usami S, Tsien RY, Chien S (2005) Visualizing the mechanical activation of Src. Nature 434:1040–5
66. Yeung T, Georges PC, Flanagan LA, Marg B, Ortiz M, Funaki M, Zahir N, Ming W, Weaver V, Janmey PA (2005) Effects of substrate stiffness on cell morphology, cytoskeletal structure, and adhesion. Cell Motil Cytoskeleton 60:24–34

Chapter 14
How Strong is the Beating of Cardiac Myocytes? – The CellDrum Solution

Jürgen Trzewik[1], Peter Linder[1] and Kay F. Zerlin[2]

[1] Aachen University of Applied Sciences, Cellular Engineering, Ginsterweg 1, 52428 Jülich, Germany
linder@fh-aachen.de
[2] Research Center Jülich, PTJ, Wilhelm-Johnen-Straße, 52428 Jülich, Germany

Abstract Widespread diseases like hernia formation or pelvic floor malfunction are directly related to a reduced biomechanical functionality of the concerned soft tissue. These exemplary diseases, but also dysfunction of the cardiovascular system are caused by processes, which are based on changes of biomechanical properties at the cellular level. The systems for the in vitro evaluation of biomechanical material properties, as described in the literature, are adopted from standard material test methods mainly focusing on an uniaxial loading. But uniaxial loading is rarely seen in nature since most load cases are two or three dimensional. This work describes a new measurement principle, which allows the cultivation of cell monolayers or thin tissue composites under biaxial load conditions and mechanical evaluation at the same time.

The new cell cultivation module, termed CellDrum, was developed for that purpose. The Celldrum consists of a thin, biocompatible silicon membrane attached to a cylindrical well ($\varnothing$ 16 mm). The CellDrum membrane can be populated with cells grown in monolayer structures or serves as a sealing boundary layer for thin film cell-matrix constructs anchored with the Cell Drum's wall.

The measurement system described in this work offers the opportunity to analyze the mechanical properties of cell constructs over a long period of time under defined mechanical boundary. It is the first system which offers the novel and unique possibility to cultivate synchronously beating cardiomyocytes and evaluate the contractile behaviour together with the monitored beating frequency.

14.1 Introduction

A person's health is determined by the biomechanical integrity and performance of tissue and organs. Widespread diseases like hernia formation or pelvic floor malfunction are directly related to a reduced biomechanical functionality of the concerned soft tissue. These exemplary diseases, but also dysfunction of the cardiovascular system, are caused by processes, which are based on changes of biomechanical properties at the cellular level. Fundamental research in this topic on humans is limited for practical and ethical reasons, especially the evaluation of active substances.

As compared to the tremendous efforts that have been made in the past to analyze biological samples for their biochemical cellular processes, only little is known so far about technologies acquiring information on cellular processes related to mechanical properties of living cells. One reason for this situation was linked to the absence of suitable technologies analyzing mechanical properties of living cells in a natural environment (Trzewik et al. 2004).

The popular use of complete cardiomyocyte monolayer cultures, even when cultivated on flexible silicon matrices (Komuro et al. 1990) has a major drawback so far: the major function in the heart, namely loaded isometric contraction could not be measured. In this chapter, we will present a method allowing us to undertake these kind of measurements

In recent years, evidence has been growing about important roles of mechanical forces in regulating the behaviour of single cells and their communities (Chicurel et al. 1998; Galbraith and Sheetz 1998; Geiger et al. 2001). Force on cells can be either external (e.g., resulting from blood flow or traction of other cells) or internal. In animal cells, internal forces are mostly generated by the actin cytoskeleton and transmitted to the extracellular matrix (ECM) through cell-matrix adhesion proteins. For stationary animal cells cultured on flat substrates, the most prominent type of cell-matrix interactions are focal adhesions (FAs) (Burridge and Chrzanowska-Wodnicka 1996; Geiger and Bershadsky 2001). FAs are large supramolecular protein assemblies, consisting of a submembrane plaque with more than 50 different proteins (including vinculin and paxillin) and a transmembrane part provided by receptors of the integrin family.

Cells sense changes in their mechanical environment and promote in return alterations and adaptations in tissue structure and function. In a feedback process, mechanical stimuli regulate fundamental processes as cell division and differentiation and therefore play a crucial role in tissue growth and regulation. Addressing the loading conditions of test samples is important. Not only biomechanical engineers but also biologists must consider protein and gene expressions studies (Huang et al. 2002; Langholz et al. 1995) of tissue constructs and their relation to mechanical load conditions. The activity of a cell is regulated, in part, by changes in the mechanical environment in which it resides (Eastwood et al. 1998; Wakatsuki et al. 2002).

The cellular impact on tissue tension plays an important role in numerous physiological and pathological processes. An important examples is wound healing (Tejero-Trujeque 2001; Banes et al. 2001) and connective tissue homeostasis (Banes et al. 2001; Brown et al. 1998e; Tomasek et al. 2002). Furthermore another important part of tissue forces research are cardiac muscle contraction (Eschenhagen et al. 1997; Langendorff 1895) and morphogenesis (Benjamin and Hillen 2003). More obviously, mechanical loading is of substantial importance to the development, function and repair of all tissues in the musculoskeletal system (Henderson and Carter 2002; Payumo et al. 2002; Vandeburgh 1992), including bone, ligament, tendon, skeletal muscle, intervertebral disc and meniscus. Vice versa, information

14.1 Introduction

upon these cellular processes can be derived from the mechanical properties of living cells. In particular, measurements of mechanical properties of cell constructs would provide a valuable insight into various cellular processes.

Devices and methods reported previously were able to monitor mechanical properties of tissue constructs or individual cells, respectively. All so far proposed experimental approaches have very complicated setups in common. This limits the scientific benefit to the investigator's experimental skills and experience. It would be unreasonable to assume that any of the described methods would have the potential to be upscaled for pharmaceutically relevant, high-throughput screening methods. Nevertheless, another profound reason for developing a new analysis device for cellular components is related to the absence of well defined and biologically relevant boundary conditions. Formerly adherent cell aspirated into a micropipette (Sato et al. 1987; Theret et al. 1988) do obviously not mimic any in vivo situation but provide complicated boundary conditions. Also, cells introducing wrinkles into stress free soft substrata (Harris et al. 1980) or the ability of cultured fibroblasts to reorganize and contract free floating three dimensional collagen I gels (Bell et al. 1979; Grinnell et al. 1999) is not related to any in vivo situation. However, such unstrained gels must be considered as mechanically completely different models to study mechanically regulated cellular processes.

Rectangular gels suffer from "necking" due to the stress variations imposed during uniaxial force measurements. The uniaxial loading also introduces a physiologically unknown parallel aligning in response to the applied force. It may be concluded, that the object to be investigated is actively modified by the experimental setup. Another drawback of many concepts is that they do not consider the specific needs of a sterile, cell culture compatible experimental setup. However, this is quite essential since the interpretation of any experimental result is critically related to a long term observation (over weeks) of all relevant parameters. The cellular response is definitely determined by the time course of the cultivation. Furthermore, the impact of contamination by fungi, bacterial or other microorganism would also influence cellular responses of test specimen and may result in the misinterpretation of experimental data.

As a general remark it must be mentioned that cellular and tissue biophysical forces are very small. Only little disturbances e. g. temperature and pressure changes, humidity or surface tension on boundary layers can make measurements irreproducible as can different culture media and devices. This may lead to misinterpretations of measurement values.

Since the fundamental work by Langendorff (1895) accomplished more than 100 years ago the isolated perfused heart became a widely accepted experimental model in cardiovascular research. Cultures of dissociated embryonic or neonatal cardiac myocytes, respectively, have been used in studies on the pharmacology of cardiovascular physiology for more than 50 years (Cavanaugh and Cavanaugh, 1957). Various approaches were made using these cells as models to investigate myocyte related characteristics like ion channel function (Iijima et al. 1984; Morales et al. 1996), cardiac hypertrophy (Fink et al. 2000) and the contractility of cardiac structures (Brady et al. 1979; Lin et al. 2001; Palmer et al. 1996; Tasche et al. 1999). The patch-clamp technique (Hamill and Sakmann 1981) became a well accepted tool for analysing electrical properties like action potentials of isolated cardiac myocytes. Multiple approaches and principles were published using systems measuring forces exerted by single myocytes as the isometric force-related displacement of glass cantilevers (Tarr et al. 1979), micro-electromechanical systems (MEMS) (Lin et al. 2001), cells attached to capacitive force transducers (Bluhm et al. 1995) and ultra sensitive cantilever designs (Tasche et al. 1999) with a low compliance. Those methods lacked easy-to-use handling procedures. Major problems occurred when the force-detection system was not fully submergible. In those cases, some parts must pass the liquid-air interface. The surface tension of the meniscus around the connecting element through the interface is about $10\,\mu N$. The forces generated by single myocytes, however, may be one to two orders of magnitude lower than this (Brady 1991). Attempts to use complete cardiomyocyte monolayer cultures on flexible silicon matrices (Komuro et al. 1990) failed because the major functional parameter of the heart, loaded isometric contraction, could not be measured.

14.2 The CellDrum Technique

Besides technical problems, the systems for the in vitro evaluation of biomechanical material properties, as previously described in the literature, are adopted from standard material test methods mainly focusing on an uniaxial loading. But uniaxial loading is rarely seen in nature because most load cases are two or three dimensional. The CellDrum technique is a new measurement principle, which allows the cultivation of cell monolayers or thin tissue composites under biaxial load conditions and mechanical evaluation at the same time.

The CellDrum principle was developed to assess biomechanical properties of various cell types at in vitro conditions. A CellDrum (Fig. 14.1) consist of a plastic

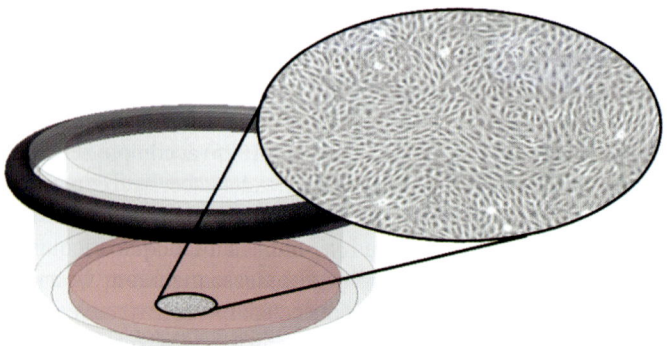

Fig. 14.1 Schematic of the standard CellDrum with a cell populated silicon membrane

cylinder which is sealed on one end with a thin, typically 1 μm thick, biocompatible silicon membrane. A rubber ring is used to fix the CellDrum to the experimental setup. The membrane allows cell attachment and proliferation at in vitro cell culture conditions. Two different experimental setups based on the CellDrum principle had to be developed to meet the different needs of standard monolayer cell cultures on the one hand and of tissue equivalents on the other hand.

14.2.1 The Measurement Principle

The basic concept, for the validation of the biomechanical properties of cells, was to monitor the stress-strain relationship of the cell-membrane composites. This basic system was applied for endothelial and fibroblast cells proliferated on Cell-Drum membranes. Furthermore it was applied to self-contracting cardiomyocytes in monolayer cultures and cardiomyocytes embedded in collagen I matrices reassembling a heart tissue equivalent. The concept of tissue equivalents was also realized for fibroblast populated collagen I matrices.

The relative displacement of silicon membranes attached to cylindrical wells (diameter 16 mm) was measured with non contact displacement sensors at a resolution in the μm-range. The CellDrum membrane is populated with cells grown as monolayer structures or it served as a sealing boundary for thin film cell-matrix constructs anchored to the CellDrum's wall. A highly sensitive laser triangulation sensor and a custom made image orientated CCD sensor (CMS) can be used. The relative membrane-cell composite displacement is the characteristic variable for various experimental setup variants.

In this chapter we want to focus on a special modification of the general load-displacement measurement system (Fig. 14.2):

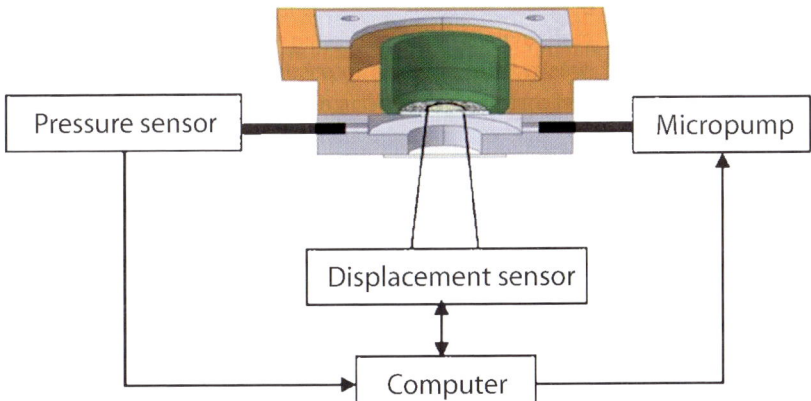

Fig. 14.2 Schematic of the system with the measurement chamber (*orange*) and an embedded CellDrum (*green*). The appropriate computer controlled sensors and actuators are indicated schematically

14.2.2 The Steady State Measurement, a Useful Modification

Steady state measurements function is especially useful for the evaluation of contracting cardiomyocytes. The system is similar to the load deflection setup, but the membrane displacement was kept constant during the experiment and pressure changes were analyzed. The pressure sensor offered a higher sensitivity than the displacement measurement alone, which was especially useful in the investigation on cardiomyocytes. Variations in the displacement of the membrane caused by contracting cardiomyocytes are almost undetectable in an optical setup. But this contraction causes pressure variation in a pressure range of several Pa, which is detectable by a highly sensitive pressure sensor. This pressure signal can be separated from mechanical and electrical noise due to the repetitive nature of the cell contraction. This separation is accomplished after the application of a software integrated filter algorithm.

But this method can not be used to analyze stress variations introduced by the slow contraction processes of non-muscle cells. Due to the slow and non repetitive nature of the related pressure changes it is not possible to subtract cell signals from signals related to electrical and mechanical background noise.

The dynamic measurement is the strongest diversification of the basic CellDrum concept. Complete membranes consisting of tissue equivalent cell-collagen structures were analyzed by exciting the membrane tissue compound with a brief air pressure pulse. The resulting resonance oscillation was monitored by a laser-based deflection sensor. Frequency and damping were analyzed revealing information on mechanical properties of the tissue construct. This system is very tolerant of most effects causing errors in all other described systems. Furthermore it offers the opportunity to perform a high number of measurements in a short period of time. The system is perfectly suited for high throughput screening evaluations.

14.3 Preparation of Samples

14.3.1 Methods of Cell-Membrane Connection

CellDrum membranes were manufactured using 184 Sylgard™ polydimethylsiloxane (Dow Corning, Michigan, USA) known as a silicon rubber. Individual batches were manufactured for typical membrane thickness varying from 1 μm to 10 μm. The CellDrum is now commercially available at Cell&Tissue Technology Corporation (Juelich, Germany). The polydimethylsiloxane (PDMS) based CellDrum membranes show different, but more adequate properties of the biological environment as compared to surfaces and substrates usually used in cell culture technologies. The CellDrum membranes are permeable for gases, flexible (soft) and coatable with ECM proteins. The control of extra cellular matrix (ECM) organization and cell adhesion are critical for the attachment of cells. Adequate procedures were therefore developed to exhibit relevant matrix proteins at the CellDrum surface. Other aspects

14.3 Preparation of Samples

of biocompatibility are related to the segregation of harmful molecules (as for example monomers) from the materials, leading to necrotic or apoptotic cell death, respectively. Various investigations were accomplished to exactly reveal these features and properties.

Cell-Tak™ is described to promote an extremely strong binding between cell and substrata (Miyazaki et al. 2000). A strong force transmission between cell and CellDrum membrane was desired in our monolayer experiments, and therefore Cell-Tak™ was seen to be the best choice for membrane coating. Cell-Tak™ Cell adhesive is a formulation of polyphenolic proteins extracted from Mytilus edulis (marine mussel). The adhesive is described to be biocompatible. Cell proliferation and viability tests for BAEC cells grown on precoated CellDrum membranes, which were not exposed to strain variations during culturing supported that assertion.

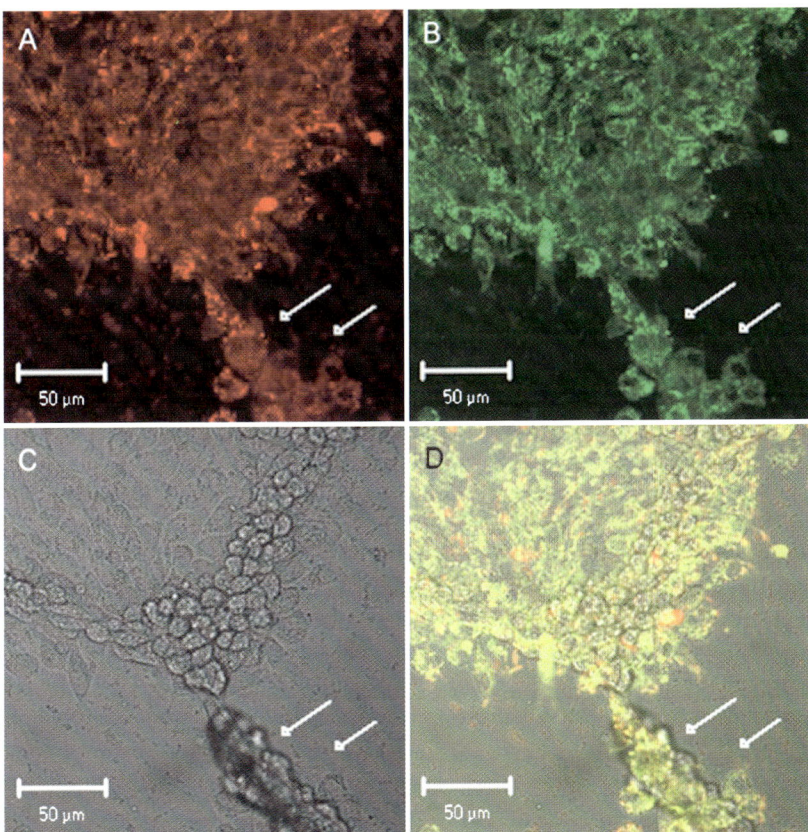

Fig. 14.3 Mitocaputre stained BAEC's cultivated on a Cell-tak coated CellDrum after exposition to mechanical strain ($\varepsilon = 0.5\%$) for 2 h: Clusters of cells separated (*white arrows*) from the monolayer and remained only in lose contact with the monolayer. Nonapoptotic cells display a *red* staining (**A**) indicating an active metabolism within their mitochondria. *Green* staining (**B**) displays the control image for a successful total staining. The superimposed control image **D** indicates untruthful image areas in *red*, but apoptotic cells in *green* and nonapoptotic cells in *yellow*. Image C in brightfield mode displays the disrupted BAEC monolayer on top of the CellDrum

However, the replacement of fibronectin with Cell-Tak™ as a membrane coating led to ablation and destruction of cells after mechanical stretching. The response was completely different from identical experiments carried out with fibronectin coated membranes. After exposure to mechanical stress adherent cells (in particular BAECs) grown on Cell-Tak™ coated surfaces, lost contact to their substrate and reshaped into a round structure. This is an indicator of cell death. The underlying mechanism of this phenomenon was unclear, but nevertheless important to identify, in regard of a meaningful experimental data analysis.

The apoptosis analysis of BAEC's grown on fibronectin or Cell-Tak™ coated CellDrum membranes which were exposed to mechanical stress did not show any differences in the appearance of apoptotic cells (Figs. 14.3; 14.4). It seem that the ablation of cells due to mechanical impact at different surface adhesion sites is not due to an apoptosis inducing process.

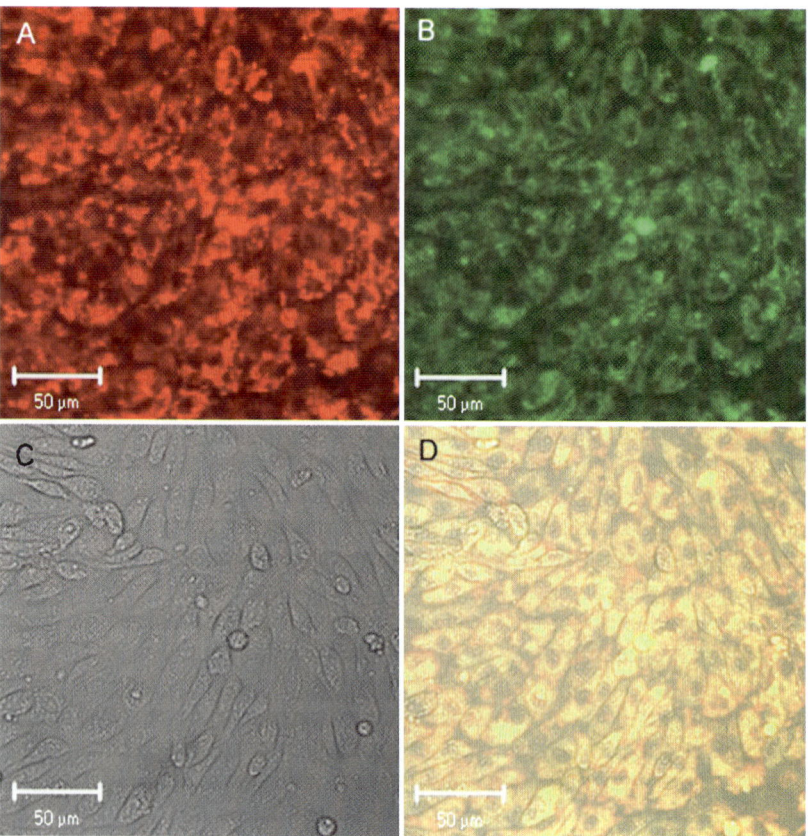

Fig. 14.4 Mitocaputre™ stained BAEC's cultivated on a fibronectin coated CellDrum after expo-sition to mechanical strain ($\varepsilon = 0.5\%$) for 2 h: Nonapoptotic cells display a *red* staining (**A**) in-dicating a active metabolism within their mitochondria. *Green* staining (**B**) displays the control image for a successful total staining. The superimposed control image **D** indicates untruthful areas in *red*, but apoptotic cells in *green* and nonapoptotic cells in *yellow*. Image **C** in brightfield mode displays the intact BAEC monolayer on top of the CellDrum

14.3.2 Visualization of Location and Orientation of the Cells Inside the Gel

A fibroblast seeded collagen matrix was used as model (Bell et al. 1979; Trzewik 2004) to study the ability of fibroblasts to reorganize and contract collagen matrices in vitro. Appropriate biological protocols and engineering tools to monitor cell orientation were developed, for example with live 3T3 NIH fibroblasts:

The tissue constructs were prepared as described above. After gelation, this resulted in a laterally fixed gel (Velcro® enhanced CellDrum), approximately 1 mm in thickness. A uniaxial mechanical loading chamber (Fig. 14.5) was used to prepare freely floating but uniaxially strained collagen gels.

After adding culture medium, all samples were incubated at 37 °C and 5% CO_2 in humidified incubator. Culture media were changed daily. After gelation, cells adhered to the collagen fibers and started to reorganize within the matrix. The gel's plane surface represents the x–y-plane. In order to visualize location and orientation of the cells inside the gel, a new procedure based on confocal imaging was introduced (Trzewik et al. 2004). 3D confocal images indicate that fibroblasts in freely-floating gels were randomly distributed. The orientation of biaxially fixed cells (Fig. 14.6) was randomly distributed in the x–y-plane, but there was no tendency to grow perpendicularly along the z-direction, out of the plane of tension.

A different situation occurs in collagen gels, which were strained by 5% after gelation. The cells were elongated and oriented in the direction of the externally applied stress (Fig. 14.7), but did not spread out of the x–y-plane, which is again parallel to the plane of tension.

The use of fluorescent stained live cells in combination with a confocal microscope enabled acquiring depth encoded information on the orientation of cells in

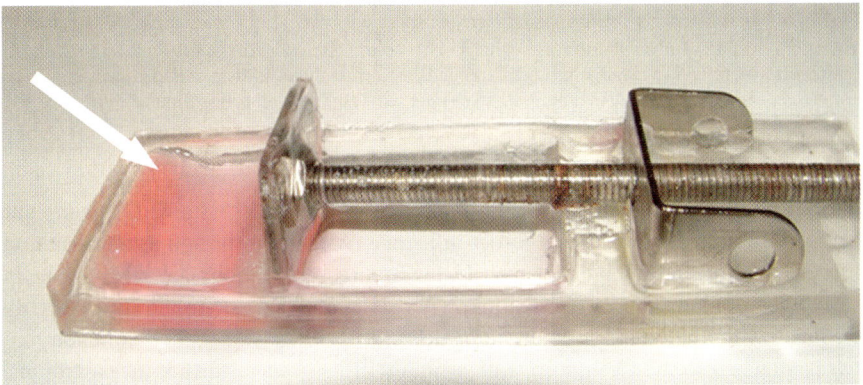

Fig. 14.5 Uniaxial loading device. The FCM (*pink*) is connected to parallel walls via Velcro® rods (*white arrow*), where the left wall was fixed and the right wall can be pulled to the right. The FCM was strained by at maximum 5% after gelation (3T3 fibroblasts were labelled with cytoplasmic fluorescent dye and added to the gel at a final concentration of 1×10^5 cells/ml)

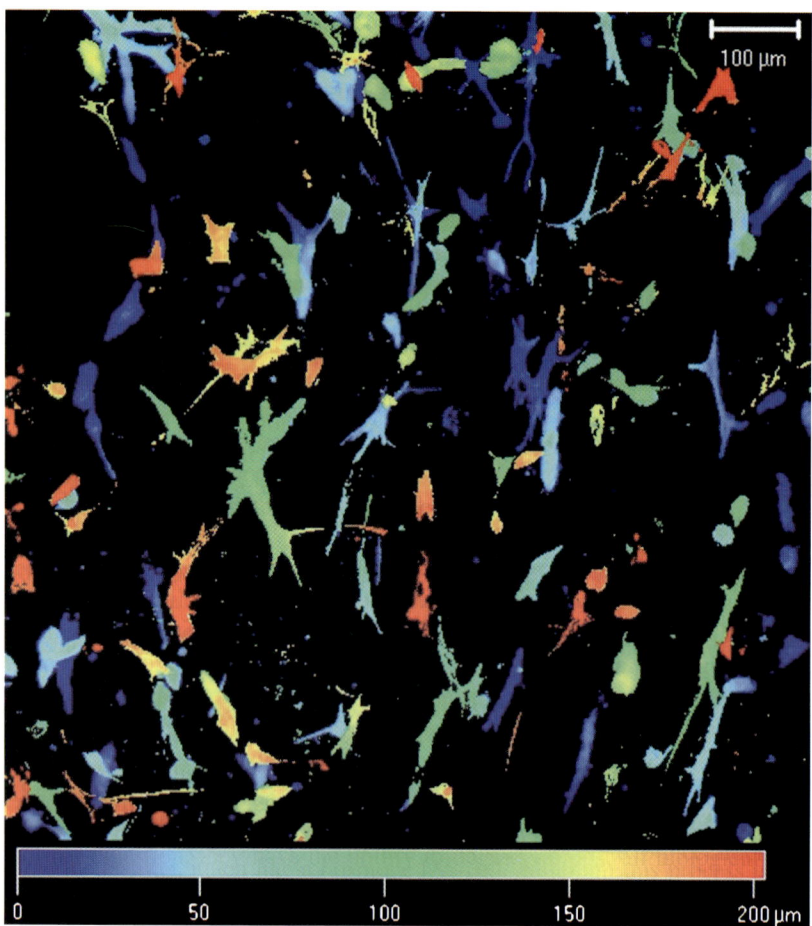

Fig. 14.6 Depth encoded 3-dimensional distribution of 3T3 fibroblasts dyed with Cell-Tracker™. The gel's surface represents the x–y-plane. The *colours* encode the location of the cells in z-direction (depth). *Dark red* represents top cells and *dark blue* bottom cells, respectively (see *color bar*). The Image displays a biaxially fixed gel with random cell orientation

three dimensions. The experimental approach of uniaxially strained, rectangular cell seeded collagen gels has a direct influence on the distribution and orientation of cells within the collagen matrix. Uniaxial stress conditions are rarely seen in vivo, except for structures like tendon, whereas biaxial stress conditions are highly frequent in the body as for example in skin. In biaxially cell seeded collagen matrixes attached to the CellDrum wall, which will be used in further investigations, it appeared reasonable to assume an isotropic distribution of cell orientation in the x–y plane. Since there were no measurable cellular protrusions into z-directions and therefore no major traction forces, which would thin out the membrane, we assumed an almost constant collagen thickness during the time of cultivation.

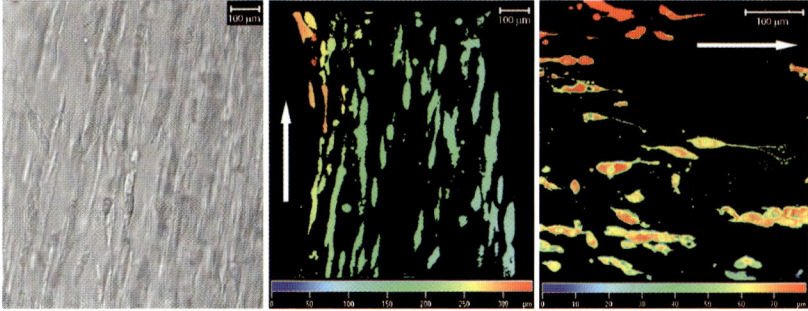

Fig. 14.7 Depth encoded 3-dimensional distribution of Celltracker™ dyed 3T3 cells in a uniaxially strained matrix (*middle*), compared to a brightfield image of cells in the setting (*left*). The *right image* displays an uniaxially strained gel at higher magnification and higher dissection resolution. Individual protrusions became visible, growing in parallel to the principal strain directions. The direction of strain is indicated by *white arrows*

14.3.3 Example Cardiac Myocytes

The system consists of the CellDrum with its typical circular, thin, flexible and transparent silicon membrane. Two types of experiments were carried out:

1. cardiomyocyte monolayer contractility and
2. 3D tissue construct contractility measurements.

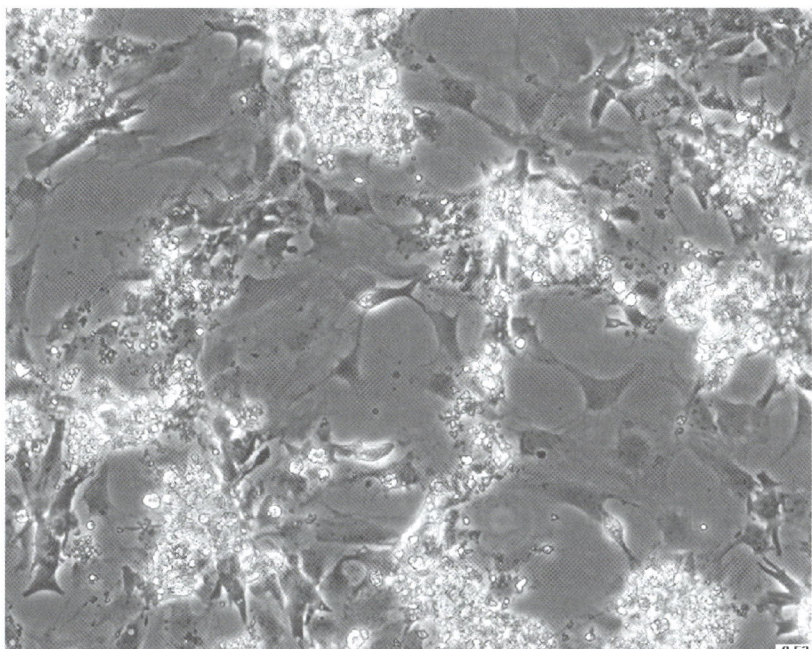

Fig. 14.8 Neonatal rat cardiomyocytes grown at a subconfluent state

Sample preparation: 1) Spontaneously contracting cardiomyocyte monolayers were grown on top of a fibronectin-coated silicon CellDrum membrane (Fig. 14.8).

Cell-populated collagen gels were poured onto the CellDrum forming a 3D tissue construct after gelation. Mesh-like anchors assured the attachment of the tissue construct to the CellDrum wall. An experimental setup was used in which membrane and cells were bulged outwards by air pressure. A laser-based deflection sensor measured the centre deflection as a function of applied pressure Fig. 14.2. The setup allowed presetting the stress level to the cell constructs externally. This is required when simulating in vivo situations like different levels of hypertrophy. The relevant forces of spontaneously contracting hybrid cardiomyocyte-collagen films, or monolayers respectively, were measured as oscillating differential pressure change between the pressurised measurement chamber and the surrounding environment. Neonatal rat cardiomyocytes were seeded at an initial concentration of 2×10^4 cells/μm^3 in between two layers of collagen anchored to the CellDrum wall or directly onto the CellDrum membrane pre-coated with fibronectin (30 μg/ml), respectively (culture time 2 days). The membrane deflection was preset and managed automatically by software-controlled adjustment of chamber pressure.

14.3.4 Beating of Cardiac Myocytes

In this study we present an experimental system allowing direct simultaneous mechanical tension and frequency measurements of cardiomyocyte monolayers as well as of thin 3D cardiomyocyte tissue constructs. Monolayers were grown on top of fibronectin-coated silicone CellDrum membranes. In 3D tissue constructs, the cardiomyocytes were grown within a collagen gel of 0.5 mm thickness, which was fixed circumferentially to the CellDrum body.

Mechanical load is an important growth regulator in the developing heart (Cooper, 1987) and the orientation and alignment of neonatal rat cardiomyocytes is stress sensitive (Fink et al. 2000; Terracio et al. 1988). This is why it was necessary to develop the CellDrum technology with its biaxial stress-strain distribution and very defined mechanical boundary conditions. In both setups, cells were exposed to strain in two directions, radially and circumferentially. This is similar to the biaxial loading, which we observe in real heart tissue. Thus, from a biomechanical point of view the system is more advantageous then previous setups based on uniaxial approaches (Eschenhagen et al. 1997; Zimmermann et al. 2000). However, achieving experimental data on mechanical tensions that are at least one order of magnitude lower than regular surface tension of water needed much engineering effort. To our knowledge, the CellDrum application so far is the only one allowing simultaneous frequency and mechanical tension measurements at biaxial mechanical loading conditions. Potentially, it will be useful in studies aiming at hypertension and heart failure (Trzewik 2006; Cooper 1987; Fink et al. 2000; Komuro et al. 1990). This is because the residual tension in the 3D tissue construct can be adjusted to any level below the burst-tension of the construct. Residual tension sounds more like a technical term. However, it is directly related to the pressure existing in the measurement chamber (Fig. 14.2) underneath the 3D construct mimicking blood pressure (Trzewik et al. 2002; Trzewik 2006; Trzewik et al. 2004).

A direct mechanical tension measurement was possible with both cardiomyocyte monolayers (Fig. 14.8) as well as with 3D tissue constructs. We first analysed mechanical tension induced by cardiomyocyte monolayers. Microscopic images showed clearly visible and synchronously beating cells. Original tension signals are shown in Figs. 14.9 and 14.10A, left.

The signal was very noisy and sinusoidal rhythms, as expected from the contracting monolayer, were hardly visible. When Triton X-100 was added to the cells in order to kill them the signal changed visibly (Fig. 14.10A, right). After signal analysis (Fig 14.10 B) and Fast Fourier Transformation (Fig. 14.10C) it became obvious, that cardiomyocyte contraction was present in the left curves of Fig. 14.10. Whereas in the right curves there was only noise indicating that contraction had diminished because of cell death. This experiment proved experimentally that the rhythmic contraction originated from synchronously beating cardiomyocytes and was not an artefact. The beating rates of 3.3 Hz found in cardiomyocyte monolayers and of 4.4 Hz equal 198 beats/min and 264 beats/min, respectively Fig. 14.10, were in good agreement with published data (Jongsma et al. 1983). Single isolated neonatal rat heart cells beat slowly and irregularly. An average beating interval is in the range of seconds. The coefficient of variation is over 40%. It was shown before that slowness and irregularity of beating are intrinsic properties of the cells and are not caused by dissociation damage or lack of conditioning factors in the culture medium (Jongsma et al. 1983). As soon as intracellular cell contacts are established by letting the cultures grow for a sufficient time or by plating them at high cell densities

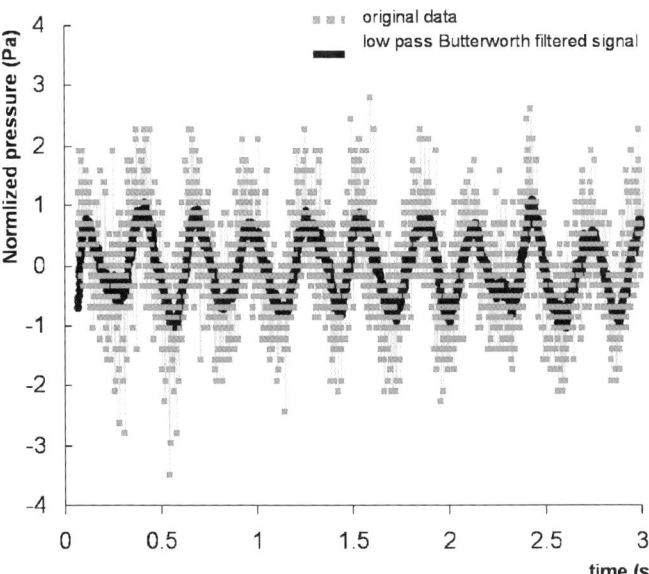

Fig. 14.9 Rhythmic contraction of a cardiomyocyte monolayer

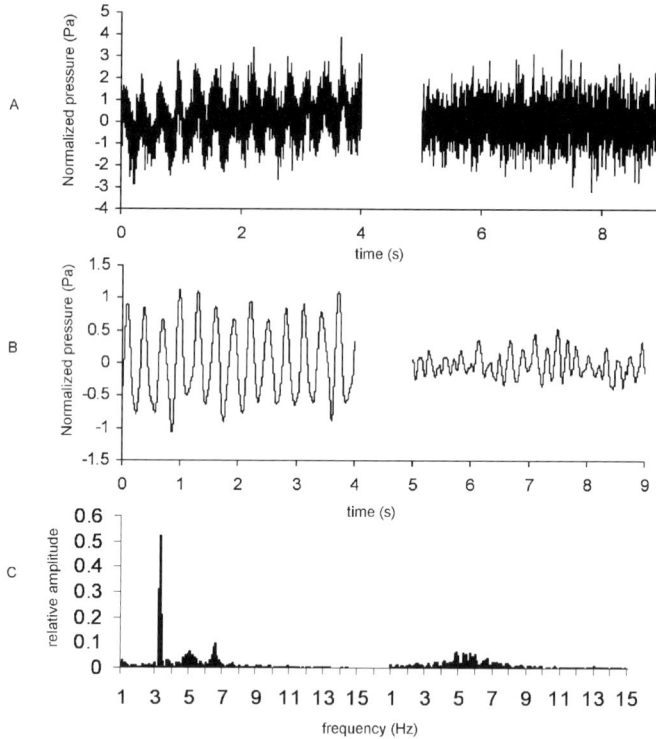

Fig. 14.10 Pulsation characteristics of myocardial cells before (*left*) and after treatment with 10% Triton X-100: **A** measured pressure curve; **B** the pressure curve filtered by a Butterworth filter kernel **C** frequency domain

both interval duration and irregularity decreased. The 3D cell-collagen hybrid structure chosen to establish the 3D tissue construct mimics physiological conditions much better then monolayers do. This is because in 3D constructs there are not only end-to-end (intercalated discs) but also side-to-side (desmosomes, gap junctions and tight junctions) contacts. These histological features better resemble the architecture of the myocardium and seem to maintain a higher degree of cellular differentiation than monolayers (Eschenhagen et al. 1997).

Drugs were chosen with known effects to heart tissue in order to verify the system (Fig. 14.11). Noradrenalin enhanced the beating rate whereas beta-blocker slowed it down. However, Fig. 14.11 shows at the same time, that there is more to say about drug action. This is for example the time course of drug action including delay times and slopes after drug administration. In the future, this may give valuable additional information on drug action.

The mechanical tension in monolayers was $1.62 \pm 0.17\,\mu N/mm^2$, whereas in 3D tissue constructs it was only $1.28 \pm 0.13\,\mu N/mm^2$. This may be related to the elastic properties of the collagen matrix and to the (unknown) cell density in both setups. By principle, it was possible to analyse the cell density since the silicone

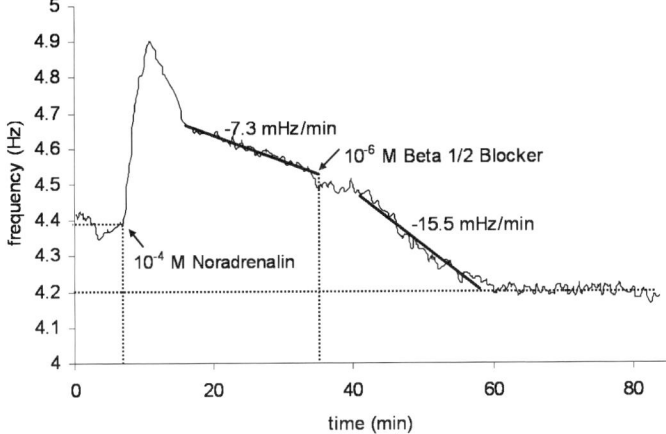

Fig. 14.11 Effect of noradrenalin and Beta 1/2 blocker on contraction frequency of myocardial cells

membrane and the tissue construct are permeable for light and confocal microscopy can be applied for cell counting. In future studies this should be considered since it will allow measuring the average tension per individual cell within a 3D construct (Trzewik 2006; Artmann 2005).

14.3.5 Medical Application in Future

As described previously (Trzewik et al. 2004) the CellDrum system provides an isotropic stress-strain distribution within the 3D tissue construct mimicking the biomechanical situation in real heart tissue. The application of the CellDrum technology offers many possibilities for in vitro investigations of cardiac function (Trzewik 2006; Artmann 2005). The setup can be easily turned into a high-throughput secondary drug screening system and might become a very useful tool in drug and basic research on hypertension and cardiac failure.

In general:

The major prospects for all described methods lie in a development toward a high throughput screening system. The use of circular samples constructs is becoming standard in many biotechnological screening assays where standard plates for 96 or 384 samples are common. In this regard, the CellDrum technology satisfies all necessary requirements to be embedded in an industry standard. Drug screening assays could be performed using either cell monolayers or collagen-based tissue equivalents. Both primary cell cultures and continuous cell lines can serve as cell source for the experiments. Continuous cell lines may be used as a standard system, since they can be subcultivated without any further variations of their phenotype. For example, continuous cardiomyocyte line HL-1 is constructed and is under evaluation at the moment. This may lead to a significant reduction in animal experiments together with an increased efficiency in sample handling and throughput. The use of a cardiomyocyte-based high throughput system is very promising but nevertheless it may need plenty of engineering time and resources. Many of this difficulties in advancing the steady-state method towards a high throughput system will not occur in the dynamic method. The dynamic method is highly resistant to disturbances introduced by the systems environment, so that any noise or variation in the boundary conditions can easily be identified and filtered out by subsequent software analysis. Furthermore, the system provides sufficient number of measurements per time together with an efficient sample handling. The stress strain-analysis demands engineering resourses, but it may be worth applying because it offers unique insights in important tissue parameters like residual tension and the tissues' elastic modulus. It can be easily up-scaled to larger sample sizes in sample geometry and may serve as a qualifying tool for skin tissue equivalents to be implanted in patients.

The system can be also easily adapted to such an important medical application as the determination of mechanical properties of an amniotic sack, whose mechanical parameters are seemingly responsible for many cases of premature birth. The system may also provide access to a deeper understanding of widespread diseases which are related to connective tissue dysfunction like Hernia formation or female pelvic floor prolaps.

Acknowledgements This chapter includes parts of the master thesis of Dipl.-Ing. Peter Linder and parts of the PhD-thesis of Dipl.-Ing. Jürgen Trzewik. It reviews experimental studies which were carried out at the laboratories of Prof. Dr. Gerhard M. Artmann, Dept. of Cell Biophysics, University of Applied Sciences Aachen, Jülich. Parts of both studies have been published before (Linder et al. 2006).

References

1. Artmann GM (2005) Keynote Lecture: Contemporary Bioengineering in Cell & Tissue Research. 2nd World Conference for Regenerative Medicine, Leipzig, Germany
2. Banes AJ, Lee G, Graff R, Otey C, Archambault J, Tsuzaki M, Elfervig M, Qi J (2001) Mechanical forces and signaling in connective tissue cells: cellular mechanisms of detection, transduction, and responses to mechanical deformation. Current Opinions in Orthopaedics 12:389–396
3. Benjamin M, Hillen B (2003) Mechanical influences on cells, tissues and organs – Mechanical Morphogenesis. Eur J Morphol 41:3–7
4. Bell E, Ivarsson B, Merrill C (1979) Production of a tissue-like structure by contraction of collagen lattices by human fibroblasts of different proliferative potential in vitro. Proc Natl Acad Sci USA 76:1274–1278
5. Bluhm WF, McCulloch AD, Lew WY (1995) Active force in rabbit ventricular myocytes. J Biomech 28(9):1119–1122
6. Brady AJ (1991) Mechanical properties of isolated cardiac myocytes. Physiol Rev 71(2):413–428
7. Brady AJ, Tan ST, Ricchiuti NV (1979) Contractile force measured in unskinned isolated adult rat heart fibres. Nature 13, 282(5740):728–729
8. Burridge K, Chrzanowska-Wodnicka M (1996) Focal adhesions, contractility, and signaling. Annu Rev Cell Dev Biol 12:463–518
9. Cavanaugh MW, Cavanaugh DJ (1957) Studies on the pharmacology of tissue cultures. I. The action of quinidine on cultures of dissociated chick embryo heart cells. Arch Int Pharmacodyn Ther 110(1):43–55
10. Cooper G (1987) Cardiocyte adaptation to chronically altered load. Annu Rev Physiol 49:501–518
11. Chicurel ME, Chen LS, Ingber DE (1998) Cellular control lies in the balance of forces. Curr Opin Cell Biol 10:232–239
12. Eschenhagen T, Fink C, Remmers U, Scholz H, Wattchow J, Weil J (1997) Three-dimensional reconstitution of embryonic cardiomyocytes in a collagen matrix: a new heart muscle model system. FASEB J 11:683–694
13. Eastwood M, Mudera VC, McGrouther DA, Brown RA (1998) Effect of precise mechanical loading on fibroblast populated collagen lattices: morphological changes. Cell Motil Cytoskeleton 40:13–21
14. Fink C, Ergun S, Kralisch D, Remmers U, Weil J, Eschenhagen T (2000) Chronic stretch of engineered heart tissue induces hypertrophy and functional improvement. FASEB J 14:669–679
15. Galbraith CG, Sheetz MP (1997) A micromachined device provides a new bend on fibroblast traction forces. Proc Natl Acad Sci USA 94:9114–9118

16. Geiger B, Bershadsky A (2001) Assembly and mechanosensory function of focal contacts. Curr Opin Cell Biol 13:584–592
17. Grinnell F, Ho CH, Lin YC, Skuta G (1999) Differences in the regulation of fibroblast contraction of floating versus stressed collagen matrices. J Biol Chem 274:918–923
18. Harris AK, Wild P, Stopak D (1980) Silicone rubber substrata: a new wrinkle in the study of cell locomotion. Science 208:177–179
19. Hamill OP, Sakmann B (1981) Multiple conductance states of single acetylcholine receptor channels in embryonic muscle cells. Nature 294:462–464
20. Henderson JH, Carter DR (2002) Mechanical induction in limb morphogenesis: the role of growth-generated strains and pressures. Bone 31:645–653
21. Huang W, Sher YP, Peck K, Fung YC (2002b) Matching gene activity with physiological functions. Proc Natl Acad Sci USA 99:2603–2608
22. Iijima T, Yanagisawa T, Taira N (1984) Increase in the slow inward current by intracellularly applied nifedipine and nicardipine in single ventricular cells of the guinea-pig heart. J Mol Cell Cardiol 16:1173–1177
23. Jongsma HJ, Tsjernina L, de Bruijne J (1983) The establishment of regular beating in populations of pacemaker heart cells. A study with tissue-cultured rat heart cells. J Mol Cell Cardiol 15:123–133
24. Morales E, Cole WC, Remillard CV, Leblane N (1996) Block of large conductance Ca(2+)-activated K^+ channels in rabbit vascular myocytes by internal Mg^{2+} and Na^+. J Physiol 15(495 Pt 3):701–716
25. Komuro I, Kaida T, Shibazaki Y, Kurabayashi M, Katoh Y, Hoh E (1990) Stretching cardiac myocytes stimulates protooncogene expression. J Biol Chem 265:3595–3598
26. Langendorff O (1895) Untersuchungen am überlebenden Säugetierherzen. Pügers Arch ges Physiologie 61:291–332
27. Langholz O, Rockel D, Mauch C, Kozlowska E, Bank I, Krieg T, Eckes B (1995) Collagen and collagenase gene expression in three-dimensional collagen lattices are differentially regulated by alpha 1 beta 1 and alpha 2 beta 1 integrins. J Cell Biol 131:1903–1915
28. Lin G, Palmer RE, Pister KS, Roos KP (2001) Miniature heart cell force transducer system implemented in MEMS technology. IEEE Trans Biomed Eng 48:996–1006
29. Linder P, Trzewik J, Rüffer M, Artmann-Temiz A, Digel I, Kurz R, Rothermel A, Robitzki A, Artmann GM (2006) Contractile tension and beating rates of self-exciting monolayers and 3D-tissue-constructs of neonatal rat cardiomyocytes. Cardiovascular research FF:111–111, under submission
30. Palmer RE, Brady AJ, Roos KP (1996) Mechanical measurements from isolated cardiac myocytes using a pipette attachment system. Am J Physiol 270:C697–C704
31. Payumo FC, Kim HD, Sherling MA, Smith LP, Powell C, Wang X, Keeping HS, Valentini RF, Vandenburgh HH (2002) Tissue engineering skeletal muscle for orthopaedic applications. Clin Orthop, pp S228–S242
32. Sato M, Levesque MJ, Nerem RM (1987) An application of the micropipette technique to the measurement of the mechanical properties of cultured bovine aortic endothelial cells. J Biomech Eng 109:27–34
33. Tasche C, Meyhofer E, Brenner B (1999) A force transducer for measuring mechanical properties of single cardiac myocytes. Am J Physiol 277:H2400–H2408
34. Tarr M, Trank JW, Leiffer P, Shepherd N (1979) Sarcomere length-resting tension relation in single frog atrial cardiac cells. Circ Res 45:554–559
35. Tejero-Trujeque R (2001) How do fibroblasts interact with the extracellular matrix in wound contraction? J Wound Care 10:237–242
36. Terracio L, Miller B, Borg TK (1988) Effects of cyclic mechanical stimulation of the cellular components of the heart: in vitro. In Vitro Cell Dev Biol 24:53–58
37. Theret DP, Levesque MJ, Sato M, Nerem RM, Wheeler LT (1988) The application of a homogeneous half-space model in the analysis of endothelial cell micropipette measurements. J Biomech Eng 110:190–199
38. Tomasek JJ, Gabbiani G, Hinz B, Chaponnier C, Brown RA (2002) Myofibroblasts and mechano-regulation of connective tissue remodelling. Nat Rev Mol Cell Biol 3:349–363

39. Trzewik J (2007) Experimental analysis of biaxial mechanical tension in cell monolayers and cultured three-dimensional tissues. The CellDrum Technology. PhD Thesis, University of Applied Sciences Aachen and Technical University Ilmenau
40. Trzewik J, Artmann-Temiz A, Linder P, Demirci T, Digel I, Artmann GM (2004) Evaluation of lateral mechanical tension in thin-film tissue constructs. Ann Biomed Eng 32:1243–1251
41. Trzewik J, Ates M, Artmann GM (2002) A novel method to quantify mechanical tension in cell monolayers. Biomed Tech (Berl) 47(1 Pt 1):379–381
42. Vandenburgh HH (1992) Mechanical forces and their second messengers in stimulating cell growth in vitro. Am J Physiol 262:R350–R355
43. Zimmermann WH, Fink C, Kralisch D, Remmers U, Weil J, Eschenhagen T (2000) Three-dimensional engineered heart tissue from neonatal rat cardiac myocytes. Biotechnol Bioeng 68(1):106–114

Chapter 15
Mechanical Homeostasis of Cardiovascular Tissue

Ghassan S. Kassab

Department of Biomedical Engineering, Surgery, Cellular and Integrative Physiology; Indiana University Purdue University Indianapolis, IN 46202, gkassab@iupui.edu

Abstract There is ample evidence in the literature that the cardiovascular (CV) system is in a state of mechanical homeostasis; i. e., the variation of stresses and strains in the CV is thought to be relatively small. Several hypotheses regarding the homeostatic state of stresses and strains have been proposed for the CV system such as: 1) uniform wall shear stress; 2) uniform mean circumferential stress and strain; 3) uniform transmural stress and strain hypothesis; and 4) biaxial (circumferential and axial) stress and strain hypothesis. The objective of this review is to critically evaluate the various hypotheses in lieu of experimental data and analytical models. We shall outline the mechanical stimuli that are most closely regulated in the CV system. In the process, we will consider the evidence for these hypotheses in the normal CV system, in flow-overload, pressure-overload, during development and in atherogenesis. The implications of these hypotheses on mechano-transduction and on vascular growth and remodeling will be highlighted.

15.1 Introduction

The structure-function relation is one of the oldest paradigms in biology and medicine. One premise of the structure-function relation is the notion of homeostasis and the major impetus in the field is motivated by the desire to understand function and physiology and subsequently patho-physiology. More than a century ago (1850–1860s), Claude Bernard developed the idea that the living system is in a state of internal chemical equilibrium (Bernard 1957). In 1929, WB Cannon coined the word "homeostasis" to describe such a state (Cannon 1973). The Greek word *homoios* means like or resembling whereas *stasis* means position. Hence, homeostasis in an organism means the maintenance of a steady state by coordinated physiological processes. Although the mechanical state of a living body is variable, the variation of stresses and strains is thought to be relatively small. Hence, there exists a mechanical homeostasis of the living tissues. A mechanical perturbation of the system generally leads to some adaptation to restore mechanical homeostasis. Pathology may be viewed as an adaptation process gone awry. Hence, an understanding of mechanical homeostasis and adaptation is important to clarify the borderline between physiology and patho-physiology.

There is ample evidence in the literature that the cardiovascular (CV) system is in a state of mechanical homeostasis. The definition of mechanical homeostasis is the regulation of CV wall mechanics in the long term; i. e., chronic regulation. Remodeling, hypertrophy, atrophy, hyperplasia, etc. are all chronic responses which could regulate long-term mechanics. Several hypotheses regarding the homeostatic states of stress and strain have been proposed for the CV system; e. g., 1) the uniform shear rate or shear stress hypothesis; 2) the uniform circumferential stress and strain hypothesis; 3) the transmural stress and strain hypothesis; and 4) the biaxial (circumferential and axial) stress and strain hypothesis. We shall review these various hypotheses in lieu of experimental data and theoretical predictions in Sects. 15.2, 15.3 and 15.4. In Sect. 15.5, we will consider the effect of perturbation of mechanical homeostasis in CV dysfunction processes. Finally, we shall discuss the implications of these hypotheses on mechano-transduction and on vascular growth and remodeling.

15.2 Shear Stress and Scaling Laws of Vascular System

15.2.1 Variation of Shear Stress

The CV system is constantly loaded by blood pressure and blood flow which induce circumferential, axial and radial stresses and strains in the vessel wall and shear stress on the endothelial cells, respectively. It is well accepted that there exists a "homeostatic" state of stress in the CV system (Kassab and Navia 2006). Kamiya et al. (1984) have previously examined the variations of wall shear stress (WSS) in the CV system. They collected diameter and velocity data from the literature and estimated the WSS throughout the arterial system. The WSS is proportional to the volumetric flow rate and viscosity of blood and inversely proportional to the cube of vessel diameter assuming a laminar, incompressible, Newtonian flow through a rigid cylindrical vessel (Kassab and Fung 1995).

Kamiya et al. (1984) reported that the WSS varies by about a factor of 2 (approximately $10-20\,\mathrm{dyne/cm^2}$) in the arterial tree (aorta to pre-capillary arterioles). Hence, they proposed the existence of homeostatic state of WSS in the vascular system given the relatively small variation of WSS relative to the large variation in diameters (4 orders of magnitude) and flows (8 orders of magnitude).

15.2.2 Design of Vascular System: Minimum Energy Hypothesis

In contemplating the design of the vascular system, Murray (1926) proposed a compromise between the frictional and metabolic cost expressed as a cost function. The formulation of the minimum energy hypothesis led to the well known Murray's law which states that flow rate is proportional to the cube of the vessel diameter. This relation implies that a 10% increase in diameter will cause a 33% increase in flow through the vessel.

Zhou, Kassab and Molloi, ZKM model (Zhou et al. 1999; Kassab 2007) generalized the "minimum energy hypothesis" to an entire coronary arterial tree. In the process, a vessel segment was defined as a stem and the entire tree distal to the stem was defined as a crown. Subsequently, they derived a flow-diameter relationship similar to Murray's law with the important difference that the exponent was not fixed at a value of 3.0 but depends on the resistance of the vascular system of interest. In addition, the ZKM model predicts new scaling design laws that relate the diameter and length; and the volume and length of vessels through power-law relations. These relationships have been validated for the coronary arterial trees in the following four ways: 1) a hemodynamic analysis of coronary arterial blood flow based on detailed anatomical data yields these relationships over the entire arterial network (Kassab et al. 1997; Zhou et al. 1999; Kassab 2007), 2) a generalization of Murray's cost function and conservation of energy predict the same result over the entire coronary arterial tree (Zhou et al. 1999; Kassab 2007), 3) in vivo data on coronary segment flow, tree length and volume using digital subtraction angiography (Zhou et al. 2002) verified the relationships for vessels proximal to 0.5 mm in diameter as observed in an angiogram, and 4) a generalization of metabolic dissipation term to include passive and active vessel wall as well as blood (Liu and Kassab 2007). Recently, Kassab (2006) has shown that these scaling relations extend beyond the coronaries (Kassab 2007) to pulmonary vessels, vessels of various skeletal muscles, mesentery, omentum and conjunctiva vessels in various species ranging from rat to human.

15.2.3 Mechanism for Murray's Minimum Energy Hypothesis

Numerous other theoretical attempts have been made to explain the design of vascular trees based on the principle of minimum blood volume, minimum lumen surface area and minimum drag force on the vessel wall (Kamiya and Togawa 1972; Zamir 1976a). Any proposed principle or hypothesis must provide a mechanism for the implementation of the principle. One of the advantages of the minimum energy hypothesis is that it provides a mechanism for the local implementation of minimum energy. It has been previously shown that the global requirement of "minimum energy" is consistent with the local condition of "uniform shear" on the vessel endothelium (Kassab and Fung 1995; LaBarbera 1990; Zamir 1976b). Hence, through the regulation of vessel caliber in response to changes in WSS, the endothelial cells are responsible for the local enforcement of the global minimum energy. This hypothesis will be explored more critically below.

15.2.4 Uniform Shear Hypothesis

The "uniform shear" hypothesis implies Murray's law with an exponent of 3.0 (Kassab and Fung, 1995). This is certainly not the case for many vascular trees as reported by Kassab (2006). Other experimental data negates the notion of uniformity of WSS throughout the vascular system. In the mesentery, it has been found the WSS is amplified in the microcirculation (vessels <100 μm in diameter) increasing from approximately 10 dynes/cm^2 to 60 dynes/cm^2 towards the capillary vessels (Lipowsky and Zweifach 1974; Lipowsky et al. 1978). A similar conclusion was made by Pries et al. (1995) on the mesenteric microcirculation. They showed that although the variation in WSS is significant, it is pressure-dependent; i.e., the arteriole, capillary and venule shear stress variation reduce to a single curve when examined as a function of pressure which led to their pressure-shear hypothesis. In a different organ, Stepp et al. (1999) measured the velocity of blood and diameter of epicardial vessels (50–450 μm in diameter) in the dog heart. They computed the WSS and found it to be heterogeneously distributed where arterioles < 160 μm in diameter have increased WSS relative to the larger vessels. The variation of WSS in the baseline and adenosine dilated vessels was approximately 10–30 dynes/cm^2 and 10–40 dynes/cm^2 in the 50–450 μm diameter range, respectively. Hence, the narrow variation of shear stress (10–20 dynes/cm^2) reported by Kamiya et al. (1984) was a consequence of the approximate database on the mean diameter and velocity which were not measured simultaneously on the same vessels.

The uniform shear hypothesis has other difficulties when applied to the microcirculation. Rodbard (1975) and Hacking et al. (1996) showed that any network with more than one flow pathway is unstable and would eventually reduce to a single flow pathway if the uniform shear hypothesis holds. Furthermore, this hypothesis does not apply to the venous system where the diameters are larger for similar flows resulting in significantly lower shear stresses.

15.3 Stress and Strain

15.3.1 Tension

The blood pressure is primarily opposed by the forces of elastin, collagen and smooth muscle cells that are oriented to form well defined layers (see review in Rhodin 1979). Thick elastin bands form concentric lamellae while finer elastin fibers form networks between lamellae. The collagen fibers are distributed circumferentially in the interstices. In a comparative study of aorta from various species ranging from mouse to pig, Wolinsky and Glagov (1967) found that the total number of medial lamellar units is proportional to the aortic diameter. Hence, despite a large variation in aortic diameter, the average tension per lamellar unit of an aortic media was fairly constant.

15.3.2 Definitions of Stress and Strain

Laplace's equation states that the mean circumferential stress in the vessel wall is directly proportional to the blood pressure, P, and inversely proportional to the vessel thickness-to-radius ratio. This relation applies to a cylindrical vessel under static equilibrium where the stress represents the force per unit area. For a thin walled vessel, the circumferential stress is twice the axial stress in the absence of any axial pre-stretch and the radial stress is equal to one half of the pressure. The radial stress obviously has a smaller magnitude than the circumferential and axial stresses for a given pressure. Also, unlike the circumferential and axial stresses which are tensile, the radial stress is compressive.

The circumferential deformation or stretch of an artery may be described by the mid-wall circumferential stretch ratio, λ_θ ($\lambda_\theta = c/C$); where c refers to the mid-wall circumference of the vessel in the loaded state ($c = 2\pi r_m$ where r_m is the midwall radius) and C refers to the corresponding mid-wall circumference in the zero-stress state. Similarly, the axial deformation or strain is given by the axial stretch ratio, λ_z ($\lambda_z = L/L_0$); where L and L_0 are the axial lengths in the loaded and zero-stress states, respectively. The axial stretch ratio can be measured by placing markers along the length of the vessel and measuring the length change from the loaded to the zero-stress state.

15.3.3 Spatial Variation of Circumferential Stress and Strain in the CV System

Guo and Kassab (2004) determined the distribution of circumferential stress and strain along the aorta and throughout the coronary arterial tree to examine the variations of stress and strain. They showed that the Green strain (circumference of the artery at physiological loading relative to the zero-stress state) and stress varied from 0.2–0.7 (with lower strain in smaller vessels) and 10 – 150 kPa, respectively, along the aorta and the entire LAD arterial tree (more than three orders of magnitude difference in vessel diameters). The relative uniformity of strain from the proximal aorta to a 10 μm arteriole relative to stress implies that there is a homeostasis of strain in the CV system which has important implications for mechanotransduction and for vascular growth and remodeling as discussed below.

It should be noted, however, that the average circumferential wall stress reported does not correspond to the cell stress. In larger vessels where the average wall stress is higher (∼150 kPa), it is generally assumed that the cells are bearing a small fraction of the load because of the abundance of the load-bearing extracellular matrix (ECM). If we assume a constrained mixture model (Gleason and Humphrey 2005b) where the all components undergo the same strain but the stresses are additive, this implies that the cell stress is much less than the average wall stress. In smaller vessels where the average wall stress is significantly lower (∼10 kPa), the cell stress would be closer to the average wall stress because there is much less ECM. Mathematical models are needed to compute the cellular stresses in small and large vessels to clarify the micromechanics of vessel wall.

The stress and strain can be modulated by vasoactivity of smooth muscle cells. Since the data discussed above were obtained in a vasodilated state, it represents the largest variation of stress and strain. In the small arteries and arterioles, smooth muscle contraction can significantly reduce wall strain and stress. The effect on stress is more profound, however, since the radius is decreased while the wall thickness is increased. Although the effect on strain is expected to be smaller, the tone in the smaller vessels is expected to increase the strain over the vasodilated measurements which will further narrow the range of variation in deformation. Hence, tone may be a critical element in maintaining mechanical homeostasis as it can regulate circumferential strain and stress as well as WSS.

For the heart, finite element models of the heart have shown that there is substantial regional heterogeneity of ventricular mechanics even under normal conditions (Costa and McCulloch 1994). The fiber strain distribution, however, is remarkably uniform despite significant variations in fiber stress. One mechanism for this fiber strain uniformity is the torsional deformation that results from the helical fiber orientations and the anisotropy of myocardium (Guccione et al. 1991). The homeostasis of strain in the CV system (heart and blood vessels) is remarkable.

15.3.4 Dynamic Variations of Stress and Strain

The variations described above are spatial but averaged over the cardiac cycle. The temporal variations of mechanical stresses and strains have a significant impact on vascular physiology and pathophysiology. Temporal gradient of WSS or oscillatory shear index (OSI) has been shown to induce significant proliferation of endothelial cells (Bao et al. 2001) and correlate with site of atherosclerosis (Ku et al. 1985). Similarly strain rate has been shown to differentially regulate mitogen activated protein kinase (MAPK) molecules in endothelial cells (Azuma et al. 2000).

15.4 Intramural Stress and Strain

15.4.1 Residual Circumferential Strain

Prior to 1983, it was believed that stress concentration occurred at the intima of the blood vessel and the subendocardium of ventricle, to the extent that the circumferential stress at the inner wall was much higher than that at the outer wall (Choung and Fung 1983). The stress concentration at the inner wall was a direct consequence of the assumption that the unloaded (zero transmural pressure) blood vessel or ventricle is at the zero-stress state. Simultaneously and independently, Fung (1983) and Vaishnav and Vossoughi (1983) challenged the starting assumption that the unloaded blood vessel is at the zero-stress state. A radial cut of a blood vessel ring relieves the residual stress and strain and changes the no-load circular geometry into an open sector (Fung 1990). The open sector represents the zero-stress state which was quantified by the opening angle. The finding of circumferential residual stress removed the concept of stress concentration at the inner wall of the vessel in the *in vivo* state (Choung and Fung 1983). The circumferential residual strain led to the "transmural uniform stress" hypothesis proposed by Fung (1983); i. e., the circumferential stresses at the inner and outer wall are nearly equal.

15.4.2 Axial Pre-stretch

In vivo, the change in vessel length in the CV cycle is negligible as compared to the pulsation of diameter. The length is constrained by vessel branches and surrounding tissue. More importantly, the vessel is also pre-stretched axially (Dobrin et al. 1990). The existence of pre-stretch and axial tethering was documented much earlier than circumferential residual strain (Bergel 1961; Fuchs 1900; Hesses 1926; McDonald 1974; Patel and Fry 1966; Patel and Vaishnav 1972; Guo and Kassab 2003). The axial pre-stretch is typically characterized by the axial stretch ratio, λ_z, which is the ratio of the axial length of the vessel *in situ* to that *in vitro*.

In a recent study, Zhang et al. (2005) considered the effect of axial stretch on axial and circumferential stresses and strains. For the coronary artery, an axial stretch ratio of 1.5 results in identical circumferential and axial stresses and strains. This is interesting since the axial pre-stretch ratio for the left anterior descending (LAD) artery is about 1.4 (Lu et al. 2003). These results indicate that the circumferential and axial wall stresses and strains in the LAD artery become more similar as the axial stretch ratio increases. In other words, under the same physiological pressure, a more isotropic state of stress and strain may be obtained by pre-stretching the vessel.

15.4.3 Radial Tissue Constraint

The mechanical properties of blood vessels depend not only on the micro-structural components of the vessel wall such as collagen and elastin fibers, smooth muscle cells, and ground substances but also on the properties of neighboring tissue. All blood vessels receive some perivascular support from the surrounding tissue. Hamza et al. (2003) determined the effect of surrounding tissue on the mechanical properties of blood vessels. They considered an epicardial LAD artery surrounded by the passive myocardium. The pressure-cross sectional area (CSA) relation was measured for the LAD artery *in situ* and *in vitro* using digital subtraction angiography. They found that, at pressure of 100 mm Hg, the CSA *in situ* is 34% smaller than that at the *in vitro* state. This corresponds to a 19% decrease in diameter due to the surrounding tissue constraint. Hence, the coronary arteries are radially constrained by the surrounding tissue and myocardium (Hamza et al. 2003). The obvious question is what are the mechanical implications of these observations? A finite element analysis shows that even a small external compression (10%) of the myocardium on the LAD artery causes a large reduction in circumferential stress and tends towards a biaxially isotropic (circumferential and axial) stress and strain state (Zhang et al. 2005).

15.5 Perturbation of Mechanical Homeostasis

Stress and strain are intimately related to tissue function, growth and remodeling. Hence, a thorough understanding of the homeostatic state of stress and strain state in the normal vessel wall can be used as a physiological reference state. Changes in the mechanical loading and hence in the internal stresses and strains lead to an adaptive process in structure, material properties and possibly function. The following is a brief review.

15.5.1 Flow Reversal

Flow reversal represents a perturbation from the forward homeostatic streamlined flow. Numerous methods have been used to document flow reversal in the CV system: *In vivo* (Moore and Ku 1994; Oshinski et al. 1995; Bogren et al. 1997; Krams et al. 1997), *in vitro* (Moore and Ku 1994; Gharib and Beizai 2003) and *in silico* (Marques et al. 2003). The conclusion of the various methods has been that transient flow reversal does occur in the normal CV system near bifurcations, along curved vessels and in the abdominal aorta (infrarenal) below the renal artery (Shaaban and Duerinckx 2000). These regions have been of particular interest because of their predilection to atherosclerosis. Indeed, the connection between flow or WSS and atherogenesis has been investigated for the past four decades. Abnormally high shear stresses and WSS spatial gradients can cause endothelial damage (Fry 1968; Herrman et al. 1994; Zeindler et al. 1989), whereas the location of atherosclerotic plaques is related to regions of low and oscillatory WSS (Caro et al. 1969; Gidden et al. 1993; Barakat et al. 1999).

Nitric oxide (NO) production increases in proportion to the magnitude of flow and is known to be athero-protective (Mochizuki et al. 1999; Traub and Berk 1998; Li and Forstermann 2000). What was not previously known, however, is the effect of flow reversal on NO bioavailability. Lu and Kassab (2004) have recently shown that [NO] is significantly reduced in reverse flow and that the reduction is mediated through an increase in superoxide (O_2^-) production during flow reversal. This is a demonstration of a chemical imbalance (NO vs. O_2^-) during perturbation of homeostatic flow.

15.5.2 *Flow-Reduction and Flow-Overload*

The uniformity of WSS is restored when blood flow is perturbed such as in flow-overload (Kamiya and Togawa 1980) or in flow-reduction (Langille and O'Donnel 1986). Blood vessels can accommodate such regulation through a change in vessel diameter at two levels: acutely through vasoactive mechanisms (flow-dependent constriction or dilation), and chronically by adjusting vascular caliber (Kamiya and Togawa 1980; Zarins et al. 1987; Glagov et al. 1987; Lu et al. 2002).

The acute response is limited by the degree of vasodilation governed by smooth muscle cells of the media. If the acute response is not sufficient to restore WSS, a chronic response ensues. Remodeling of the arterial wall in response to changes in flow (increase or decrease) occurs over weeks to months and the net result is a tendency to restore WSS to the baseline level. Chronic flow increase results in an enlarged lumen (Kamiya and Togawa 1980), whereas reduced flow induces intimal thickening and a decrease in lumen diameter (Langille and O'Donnel 1986). These steady state adaptations are under the command of the endothelium such that the removal of the endothelium prevents the response (Sumpio et al. 1987). The regulatory mechanism under dynamic conditions has also been previously investigated using an aortocaval fistula in the rat (Driss et al. 1997). Phasic hemodynamic measurements and aortic wall dimension were measured using ultrasound. It was concluded that systolic and mean WSS were normalized after 2 months while the diastolic shear was not.

It has been postulated that there exists a threshold shear stimulus below which the vessel remodeling fails to normalize the WSS (Brownlee and Langille 1991). Brownlee and Langille (1991) observed that the carotid arterial diameter of mature rabbits was insensitive to smaller (60%) increase in blood flow after 2 months and that WSS remained significantly elevated. They also found that, in contrast to the adult vessels, vessels of weanling rabbits exhibited sufficient diameter enlargement subsequent to blood flow increases to normalize the WSS, even though the blood flow increases were more modest (47%). Miyashiro et al. (1997) have also reported that flow-induced vascular remodeling is age-dependent in the rat carotid artery. Kassab et al. (2002) have shown that the remodeling of the lumen of the trunk of the LAD artery is consistent with the constant shear hypothesis, for flow increases of 18–52% in young swine.

In the heart, Grossman (1980) proposed that the increase in myocardial wall thickness normalizes the systolic circumferential wall stress in pressure-overload. Nguyen et al (1993) proposed that end diastolic (ED) stress rather than end systolic stress is normalized in pressure-overload hypertrophy and hence is the more likely stimulus. Similarly, ED stress was proposed to stimulate growth in volume-overload hypertrophy (Florenzano and Glantz 1987; Grossman 1980). In those studies, however, strain was not measured. Emery and Omens (1997) measured strain in their arteriovenous fistula rat model and reported that midwall strains at ED pressure were normalized, but ventricular wall stresses remained substantially elevated in conjunction with an order of magnitude increase in stiffness 6 weeks after volume hypertrophy. Hence, ED fiber strain is a likely stimulus for remodeling (Omens 1998).

Time-varying dimension changes (strain rate, volume rate, etc.) have been proposed as possible candidates for cardiac hypertrophy in volume overload (Holmes 2004).

15.5.3 Pressure-Overload

The "uniform tension" hypothesis described above was generalized into a "uniform stress" hypothesis as the principle that dictates the remodeling of arteries in hypertension (Matsumoto and Hayashi 1994; Vaishnav et al. 1990; Wolinsky 1971, 1972). It has been observed that the wall thickness-to-radius ratio increases in proportion to the increase in pressure such that the circumferential wall stress is restored after some period of growth and remodeling. The increase in the ratio of wall thickness-to-radius ratio occurs by an increase in muscle mass of the media or by decrease in the lumen without medial hypertrophy. Hence, this hypothesis presupposes a homeostatic state of stress.

In addition to the "uniform circumferential stress" hypothesis, Kassab and colleagues have recently found that the circumferential strain (computed in reference to the zero-stress state) responds faster and recovers more quickly than the circumferential stress in flow-overload of an arterial–venous (a–v) fistula (Lu et al. 2001) as shown in Fig. 15.1. Figure 15.2 shows similar data in a model of pressure-overload by obstruction of the common bile duct (Dang et al. 2004). Hence, the circumferential strain appears to be biologically regulated. The observed growth and remodeling processes that return the stresses and strains to some "set values" support the hypothesis of a "homeostatic" state of these parameters but with very different time constants.

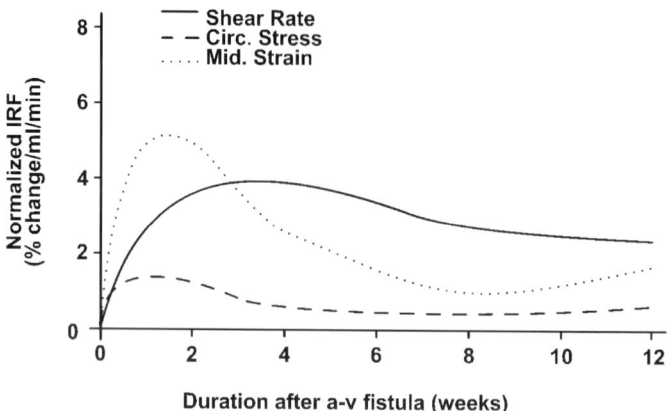

Fig. 15.1 Indicial Response Functions (IRFs) of shear rate, mean circumferential stress and midwall strain normalized to the respective initial values. The IRF represents the change of a particular feature of a blood vessel; e. g., shear rate, mean circumferential stress or midwall strain, in response to a unit step change of input variable after the arterial-venous (a–v) fistula; e. g., flow. Figure reproduced from Lu et al. 2002 by permission

15.5 Perturbation of Mechanical Homeostasis

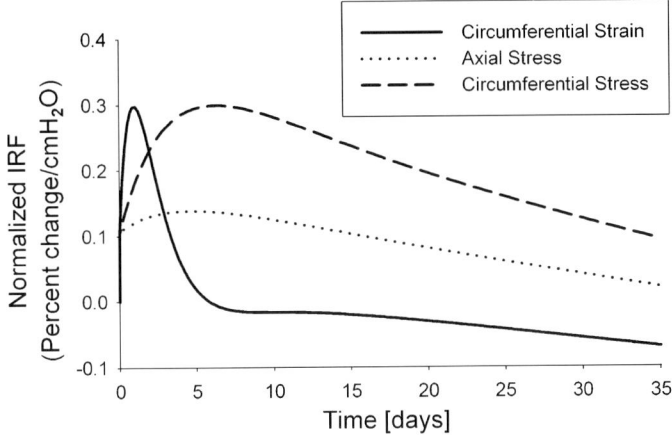

Fig. 15.2 Indicial Response Functions (IRFs) of axial, circumferential stress and strain normalized to the respective initial values. The IRF represents the change of a particular feature of a blood vessel; e.g., axial, circumferential stress and strain, in response to a unit step change of input variable; e.g., pressure. Figure reproduced from Dang et al. 2005 by permission

15.5.4 Simultaneous Hypertension and Flow-Overload

A change in flow often accompanies a change in pressure. A question arises as to which principles does vessel remodeling obeys when both blood flow and pressure are changed. To address this question, Kassab et al. (2002) examined the LAD artery in a supra-valvular aortic stenosis (SVAS) model which experiences both an increase in flow and pressure. They found that the WSS is normalized along the entire length of the LAD artery. The circumferential stress, however, was found to be elevated in the proximal LAD artery (approximately 23% greater than the control) where the flow increase was fairly large. In the distal LAD artery, where the flow increase was smaller, the circumferential stress was nearly normalized (within 6% of the control value). The change in circumferential stress depends on the change in the wall thickness-to-radius ratio, which was found to decrease in the proximal but increase in the distal LAD artery. The non-uniformity in remodeling was due to the differential hemodynamic conditions; i.e., an increase in pressure and flow in the proximal while primarily an increase in pressure in the distal artery.

In conclusion, it appears that the constant WSS hypothesis takes precedence in the SVAS model. Indeed, in models of flow-overload where the vessel dilates to normalize the WSS, the circumferential stress and strain will increase since the wall thickness is unchanged (Lu et al. 2001). A number of studies have reported an increase in circumferential stress despite the normalization of WSS (Masuda et al. 1989; Girerd et al. 1996). Girerd et al. (1996) have reported a 51% increase in circumferential stress despite normalization of WSS in an a–v fistula of patients with end-stage renal disease. In a study using an a–v fistula in rats, Lu et al. (2001) found that the circumferential stress and strain normalize after a period of 4 and 8 weeks, respectively (Fig. 15.1). It is possible that longer duration of increased pressure and flow may produce different results and that the circumferential stress in the LAD artery may normalize beyond the 5-week period.

15.5.5 Changes in Axial Stretch

Several studies have examined the effects of axial pre-stretch on cellular tissue growth and remodeling in vivo and ex vivo (Han et al. 2003; Jackson et al. 2002; Clerin et al. 2003). It was concluded that the growth restores the axial strain while maintaining similar material properties. In a computational study, Gleason and Humphrey (2005a) used a mathematical growth and remodeling model to predict similar findings. It is interesting to note that the axial retraction is small in the young and increases with postnatal growth and development as the vessels are stretched by body growth (Dobrin et al. 1975; Huang et al. 2006) as described below. It is also interesting to note that the axial prestretch decreases with aging (Kassab, unpublished results).

15.5.6 Postnatal Growth and Development

At the time of birth, the major hemodynamic changes include an increase of cardiac output (9 fold) and an increase in systemic blood pressure (2–3 fold) (Huang et al. 2006; Heymann et al. 1981; Wiesmann et al. 2000). We have recently explored the principle of mechanical homeostasis in the postnatal model of mouse aorta (Huang et al. 2006; Guo and Kassab 2003). Our data show a rapid (3–4 folds) increase of the circumferential Cauchy stress in the first 30 days of age. The change in stress at 10 weeks in comparison to 30 days is significant. Hence, the circumferential stress continues to increase. The circumferential elastic modulus of the mouse aorta increases gradually from the time of birth and reaches a relatively constant value after 2 weeks of age. These results suggest that the aorta becomes stiffer with development, with stiffness of abdominal > thoracic aorta. The modulus also continues to increase from 30 days to 10 weeks, although the change is not statistically significant. The strain, however, is unchanged at 10 weeks relative to 30 days (Fig. 15.3). Indeed, the circumferential Green strain of the mouse aorta reaches the adult value at the age of 2 weeks as shown in Fig. 15.3. The tendency to reach a homeostatic strain is apparent.

The WSS at birth is significantly higher than physiological values and decreases linearly during development. This is certainly a strong stimulus for the remodeling of the lumen of the aorta. Interestingly, the WSS is essentially normalized at 30 days. In conclusion, it appears that the circumferential strain and WSS normalize faster than the circumferential stress. It should be noted, however, that it is difficult to compare fluid-induced WSS with pressure-induced circumferential stress (five

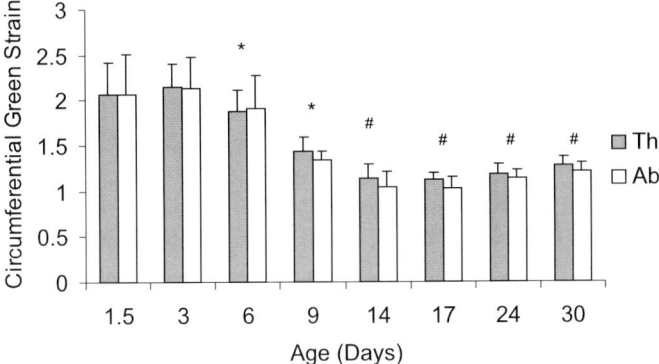

Fig. 15.3 The temporal changes of circumferential Green strain after birth. Th and Ab represent thoracic and abdominal aorta, respectively. "*" denotes significant difference when comparing different segments along the aorta. "#" denotes significant difference when comparing thoracic and abdominal aorta. Figure reproduced from Huang et al. 2006 by permission. The strain is unchanged at 10 weeks (Guo and Kassab 2003) relative to 30 days (mean of 1.3 ± 0.26 and 1.2 ± 0.11, respectively)

orders of magnitude difference). Does mechanotransduction proceed through the same molecular pathways? Do the cells sense these huge differences in magnitude in the same way? Is there a normalizing factor from one to the other? These and many other issues remain to be elucidated.

15.5.7 Implications for Atherosclerosis

There is no doubt that the intramural stress and strain have bearing on the predilection of blood vessel for atherosclerosis. A perturbation of mechanical homeostasis leads to a vascular response which results in structural and mechanical remodeling of vessel wall. An interesting example of the relation between the mechanical status of the vessel wall and atherosclerosis is as follows: the epicardial arteries develop atherosclerosis while the intramural arteries, surrounded by myocardium, do not (Kenyon 1979). Atherosclerotic changes involving the epicardial portion of the coronary artery stop where the artery enters the myocardium. The cyclically contracting heart compresses the intramural arteries during the cardiac cycle. The external compression of the intramural arteries may affect both the fluid mechanics of the blood flow in the vessel lumen and the solid mechanics of the vessel wall. An additional example of vessels that are devoid of atherosclerosis due to the mechanical influence of the surrounding tissue is the vertebral artery. Angiograms of the vessel reveal alternating pattern of disease where the portion surrounded by bone is disease free, while the inter-vertebral segments are prone to atherosclerosis (Meyer and Naujokat 1964).

Zhang et al. (2005) simulated the fluid-solid interaction, to determine the flow and stress fields in the vessel lumen and wall, respectively, for vessels with and without the influence of surrounding tissue. On the fluid mechanics side, the external compression may have a "wash out" effect on the blood flow; i. e., reduce the transit time of blood flow and hence reduce the transport time of low-density lipoprotein across the vessel wall. On the solid mechanics side, the vessel wall stress may be decreased because the surrounding tissue bears some of the stress and strain. The simulation revealed that the changes of flow velocity and WSS, in response to cyclical external loading, appear less important than the circumferential stress and strain reduction in the vessel wall under the proposed boundary conditions. These results have important implications since high stresses and strains can induce growth, remodeling and atherosclerosis; and hence a reduction of stress and strain may be athero-protective.

15.6 Limitations, Implications and Future Directions

Stress and strain are mathematical concepts based on the continuum model developed to describe the force intensity (force per area) and the deformation (change of length) in the vessel wall, respectively. As such, there are many ways to define stress and strain which are simply mathematical concepts that are intended to describe a physical quantity. Hence, biological cells may not "sense" strain or stress directly but rather may be stimulated by more fundamental quantities such as conformational changes or molecular forces. Since it is difficult to measure the later quantities directly, the measurements of stress and strain as quantities averaged over a continuum scale are very useful.

15.6.1 What is the Stimulus for Mechanotransduction?

It has not been possible to establish whether blood vessels respond to stress or strain because it is difficult to experimentally separate the two mechanical stimuli. Stress and strain are intimately related through the material properties of the vessel wall (constitutive relation) and hence it is impossible to change one without affecting the other. It is likely that strain or possibly strain rate is the stimulus for mechanotransduction based on the premise that forces transmitted via individual proteins (integrins) causes conformational changes that alter their biological activity. The force transmission may occur either at the site of cell adhesion or within the stress-bearing members of the cytoskeleton (Ingber 2003). The altered equilibrium state can subsequently initiate a biochemical signaling cascade or produce a local structural change. Since conformational change of a molecule or enzyme is inherently related to deformation, it is very likely that chemical kinetics is affected by deformation. Alternatively, it is possible that specific elements in smooth muscle cells experience deformations that are proportional to stress and would thus act as sensors of stress. Much work remains to be done in this area.

There is significant data in the literature that supports the existence of stretch or deformation-activated ion channels in the vascular system (see review in Sachs 1988). Specifically, stretch-sensitive calcium ion channels have been identified and studied in endothelial cells (Hoyer et al. 1997; Murase et al. 2001; Naruse et al. 1998; Naruse and Sokabe 1993). Experiments that use micropipettes to stretch an endothelial cell membrane show the degree of stretch or deformation is related to the opening of trans-membrane cation channels. The major effect of the activation of these mechano-sensitive ion channels is the influx of calcium and hence the depolarization of the cell. Moreover, there is a causal relation between MAPK molecules of endothelial cells and both shear stress and cyclic strain (Azuma et al. 2000; Tseng et al. 1995).

15.6.2 Strain Homeostasis

For a given blood vessel, there appears a "set point" for the WSS that is regulated by the endothelium (Kassab et al. 2002). Murray's law implies that the set point is exactly the same throughout the vasculature. Since Murray's law does not hold throughout the vasculature, the set point cannot be the same for all blood vessels. This raises a very interesting issue which requires a change in the current paradigm. It may be that the shear stress is not the "sensed" element. The fluid WSS provides the loading on the endothelial cells which causes axial tension and deformation in those cells. Fung and Liu (1993) have previously shown that the tension in the endothelial cell is proportional to the WSS. The higher shear stress in the microcirculation suggests a larger tension on those endothelial cells. The tension will cause some level of deformation depending on the mechanical properties (stiffness) of the cell. Suppose that the endothelial cells regulate deformation rather than shear stress and pressure. A set point of strain for the endothelium may exist throughout the entire vascular system provided that the endothelial cells of micro-vessels are stiffer than those of larger vessels. The changes in stiffness may be locally controlled by cell volume regulation. Future studies are needed to elucidate whether a strain-dependent set point can be locally controlled by endothelial cell volume regulation to adjust the stiffness to the tension imposed by the shear stress.

15.6.3 Existence of Gauge Length?

If strain is the "sensed" element, there must be some gauge length relative to which deformation causes a response. Although the precise gauge length is unknown, it is clear that many enzymes, receptors and ionic channels are sensitive to conformational or length changes. A possible gauge length may relate to the zero-stress which serves as the reference length for measurement of strain or deformation. Fung (1990) has previously proposed the lengths of the vessel sector in the zero-stress state are a sensitive index of growth and remodeling. Specifically, he proposed that that the non-uniformity (inner relative to outer wall) of growth and resorption can alter the zero-stress state. Fung and Liu showed that hypertension induces growth of intimal circumference that exceeds that of the adventitia (Fung and Liu 1989, 1992; Liu and Fung 1989). Consequently, the vessel sector in the zero-stress state shows an outward bend and hence an increase in the opening angle. Conversely, Lu et al. (2001) showed that flow-overload induces growth of adventitial circumferential length that exceeds that of intima. Hence, the vessel sector bends inward and decreases the opening angle. The connection between these length changes and a possible gauge length requires future investigations.

15.6.4 Other Possibilities

The pressure is pulsatile *in vivo* and hence the stress and strain vary throughout the cardiac cycle. The foregoing discussion of mechanical homeostasis is based on the mean values. It may be, however, that the pulsatile features are important and that the rates of deformation (shear rate, strain rate, etc.) have physiological significance. Furthermore, it may be that other quantities are the control variables that relate to stress and strain. For example, local energy consumption (or power as energy rate) may be expressed as the product of stress and strain (or strain rate) and may be hypothesized as a possible mechanotransducer.

15.6.5 Summary and Future Directions

There is no doubt that the CV function is under homeostatic regulation. This involves neural, humoral and mechanical regulation. There is significant evidence that the internal mechanical factors are narrowly bounded and carefully regulated such that a perturbation of mechanical loading causes adaptive responses to restore mechanical homeostasis. Here, we explore the evidence for mechanical homeostasis and its implications on mechanotransduction in cells and how cells maintain function under altered mechanical loading conditions. Models and experiments to uncover the mechanisms of these phenomena will be valuable for determining how different pathophysiologies could arise with dysregulation of the process.

Acknowledgements This research was supported in part by the National Institute of Health-National Heart, Lung, and Blood Institute Grant 2 R01 HL55554-7, by the National Science Foundation grant #9978199 and by the American Heart Association 0140036N.

References

1. Azuma N, Duzgun SA, Ikeda M, Kito H, Akasaka N, Sasajima T, Sumpio BE (2000) Endothelial cell response to different mechanical forces. J Vasc Surg 32(4):789–794
2. Bao X, Lu C, Frangos JA (2001) Mechanism of temporal gradients in shear-induced ERK1/2 activation and proliferation in endothelial cells. Am J Physiol Heart Circ Physiol 281:H22–H29
3. Barakat AI, Leaver EV, Pappone PA, Davies PF (1999) A flow-activated chloride-selective membrane current in vascular endothelial cells. Circ Res 85(9):820–828
4. Ben Driss A, Benessiano J, Poitevin P, Levy BI, Michel JB (1997) Arterial expansive remodeling induced by high flow rates. Am J Physiol 272(2 Pt 2):H851–H858
5. Bergel DH (1961) The static elastic properties of the arterial wall. J Physiol 156:445–457
6. Bernard C (1957) An introduction to the study of experimental medicine. New York: Dover Publications
7. Bogren HG, Mohiaddin RH, Kilner PJ, Jimenez-Borreguero LJ, Yang GZ, Firmin DN (1997) Blood flow patterns in the thoracic aorta studied with three-directional MR velocity mapping: The effects of age and coronary artery disease. J Magn Reson Imaging 7(5):784–793
8. Brownlee RD, Langille BL (1991) Arterial adaptations to altered blood flow. Can J Physiol Pharmacol 69(7):978–983
9. Buga GM, Gold ME, Fukuto JM, Ignarro LJ (1991) Shear stress-induced release of nitric oxide from endothelial cells grown on beads. Hypertension Dallas 17:187–193
10. Cannon WB (1929) Organization for physiological homeostasis. Physiol Rev 9(3):399–431
11. Caro CG, Fitz-Gerald JM, Schroter RC (1969) Arterial wall shear stress and distribution of early atheroma in man. Nature 223(211):1159–1161
12. Chuong CT, Fung YC (1983) Three-dimensional stress distribution in arteries. J Biomech Eng 105(3):268–274
13. Clerin V, Nichol JW, Petko M, Myung RJ, Gaynor JW, Gooch KJ (2003) Tissue engineering of arteries by directed remodeling of intact arterial segments. Tissue Engineering 9:461–472
14. Costa KD, McCulloch AD (1994) Relationship between regional geometry and mechanics in a three-dimensional finite element model of the left ventricle. Adv Bioeng 28:11–12
15. Dang Q, Gregersen H, Duch B, Kassab GS (2004) Indicial response of growth and remodeling of common bile duct post obstruction. Am J Physiol Gastrointest Liver Physiol 286(3):G420–G427
16. Dobrin PB, Canfield T, Sinha S (1975) Development of longitudinal retraction of carotid arteries in neonatal dogs. Experientia 31(11):1295–1296
17. Dobrin PB, Schwarcz TH, Mrkvicka R (1990) Longitudinal retractive force in pressurized dog and human arteries. J Surg Res 48(2):116–120
18. Dobrin PB (1986) Biaxial anisotropy of dog carotid artery: estimation of circumferential elastic modulus. J Biomech 19(5):351–358

19. Emery JL, Omens JH (1997) Mechanical regulation of myocardial growth during volume-overload hypertrophy in the rat. Am J Physiol 273 (Heart Circ Physiol 42):H1198–H1204
20. Florenzano F, Glantz S (1987) Left ventricular mechanical adaptation to chronic aortic regurgitation in intact dogs. Am J Physiol 252 (Heart Circ Physiol 21):H969–H984
21. Forstermann U, Pollocki JS, Schmidt HH, Heller M, Murad F (1991) Calmodulin-dependent endothelium-derived relaxing factor synthase activity is present in the particulate and cystolic fractions of bovine aortic endothelial cells. Proc Natl Acad Sci USA 88:1788–1792
22. Frobert O, Gregersen H, Bjerre J, Bagger JP, Kassab GS (1998) Relation between the zero-stress state and the branching orders of the porcine left coronary arterial tree. Am J Physiol 275 (Heart Circ Physiol 44):H2283–H2290
23. Fry DL (1968) Acute vascular endothelial changes associated with increased blood velocity gradients. Circ Res 22(2):165–197
24. Fuchs RF (1900) Zur Physiologie und Wachstummsmechanik des blutgefassystems. Arch Ges 28:7
25. Fung YC, Liu SQ (1993) Elementary mechanics of the endothelium of blood vessels. J Biomech Eng 115:1–12
26. Fung YC, Liu SQ (1989) Change of residual strains in arteries due to hypertrophy caused by aortic constriction. Circ Res 65(5):1340–1349
27. Fung YC, Liu SQ (1991) Changes of zero-stress state of rat pulmonary arteries in hypoxic hypertension. J Appl Physiol 70(6):2455–2470
28. Fung YC (1990) Biomechanics: Motion, flow, stress, and growth. New York, NY, Springer-Verlag
29. Fung YC (1983) What principle governs the stress distribution in living organs? In: Fung YC, Fukada E, Junjian W, eds. Biomechanics in China, Japan and USA. Beijing, China, Science, pp 1–13
30. Gharib M, Beizai M (2003) On the correlation between negative near-wall shear stress in human aorta and various stages of congestive heart failure. Ann Biomed Eng 31:678–685
31. Gibbons GH, Dzau VJ (1994) The emerging concept of vascular remodeling. NEJM 330:1431–1438
32. Girerd X, London G, Boutouyrie P, Mourad JJ, Safar M, Laurent S (1996) Remodeling of the radial artery in response to a chronic increase in shear stress. Hypertension 27(3 Pt 2):799–803
33. Glagov S, Weisenberg E, Zarins CK, Stankunavicius R, Kolettis GJ (1987) Compensatory enlargement of human atherosclerotic coronary arteries. N Engl J Med 316(22):1371–1375
34. Gleason RL, Humphrey JD (2005a) Effects of a sustained extension on arterial growth and remodeling: a theoretical study. J Biomech 38:1255–1261
35. Gleason RL, Humphrey JD (2005b) A 2D constrained mixture model for arterial adaptations to large changes in flow, pressure and axial stretch. Math Med Biol 22(4):347–69
36. Grossman W (1980) Cardiac hypertrophy: useful adaptation or pathological process? Am J Med 69:576–583
37. Guccione J, McCulloch A, Waldman L (1991) Passive material properties of intact ventricular myocardium determined from a cylindrical model. ASME, J Biomech Eng 113:42–55
38. Guo X, Kassab GS (2004) Distribution of stress and strain along the porcine aorta and coronary arterial tree. Am J Physiol (Heart Circ Physiol) 286(6):H2361–H2368
39. Guo X, Kassab GS (2003) The Variation of Mechanical Properties along the Length of the Aorta in the C57BL/6 Mouse. Am J Physiol 285(6):H2614–H2622
40. Hacking WJG, Van Bavel E, Spaan JAE (1996) Shear stress is not sufficient to control growth of vascular networks: a model study. Am J Physiol 270:H364–H375
41. Hamza LH, Dang Q, Lu X, Mian A, Mollio S, Kassab GS (2003) Effect of passive myocardium on the compliance of porcine coronary arteries. Am J Physiol Heart Circ Physio 285(2):H653–H660
42. Han HC, Ku DN, Vito RP (2003) Arterial wall adaptation under elevated longitudinal stretch in organ culture. Ann Biomed Eng 31(4):403–411

43. Herrmann RA, Malinauskas RA, Truskey GA (1994) Characterization of sites of elevated low density lipoprotein at the intercostals celiac and iliac branches of the rabbit aorta, Arterioscler. Thromb Vasc Biol 14:313–323
44. Hesses M (1926) Über die pathologischen Veranderungen der Arterien der oberen extremitat. Virchows Arch Path Anat Physiol 261:225–252
45. Heymann MA, Iwamoto HS, Rudolph AM (1981) Factors Affecting Changes in the Neonatal Systemic Circulation. Annu Rev Physiol 43:371–383
46. Holmes JW (2004) Candidate mechanical stimuli for hypertrophy during volume overload. J Appl Physiol 97:1453–1460
47. Hoyer J, R Kohler,Distler A (1997) Mechanosensitive cation channels in aortic endothelium of normotensive and hypertensive rats. Hypertension 30(1 Pt 1):112–9
48. Huang Y, Guo X, Kassab GS (2006) Axial Non-uniformity of Geometric and Mechanical Properties of Mouse Aorta Increase during Postnatal Growth. Am J Physiol 290:H657–H664
49. Humphrey JD, Kang T, Sakarda P, Anjanappa M (1993) Computer-aided vascular experimentation: a new electromechanical test system. Ann Biomed Eng 21(1):33–43
50. Ingber DE (2003) Mechanobiology and diseases of mechanotransduction. Ann Med 35:564–577
51. Jackson ZS, Gotlieb AI, Langille BL (2002) Wall tissue remodeling regulates longitudinal tension in arteries. Circulation Research 90:918–925
52. Kamiya A, Togawa T (1972) Optimal branching structure of the vascular tree. Bull Math Biophys 34:431–438
53. Kamiya A, Bukhari R, Togawa T (1984) Adaptive regulation of wall shear stress optimizing vascular tree function. Bull Math Biol 46(1):127–137
54. Kamiya A, Togawa T (1980) Adaptive regulation of wall shear stress to flow change in the canine carotid artery. Am J Physiol 239:H14–H21
55. Kassab GS, Berkley J, Fung YC (1997) Analysis of pig's coronary arterial blood flow with detailed anatomical data. Ann Biomed Eng 25(1):204–217
56. Kassab GS, Fung YC (1995) The pattern of coronary arteriolar bifurcations and the uniform shear hypothesis. Ann Biomed Eng 23(1):13–20
57. Kassab GS, Gregersen H, Nielsen SL, Liu X, Tanko LB, Falk E (2002) Remodeling of the left anterior descending artery a porcine model of supravalvular aortic stenosis. J Hypertension 20(12):2429–2437
58. Kassab GS (2006) Design Laws of Vascular Trees: Of Form and Function. Am J Physiol 290:H894–H903
59. Kassab GS (2007) Design of Coronary Circulation: The Minimum Energy Hypothesis. Computer Methods in Applied Mechanics and Engineering 196:3033–3042
60. Kassab GS, Navia JA (2006) Biomechanical Considerations in the Design of Graft: The Homeostasis Hypothesis. Annu Rev Biomed Eng 8:499–535
61. Kenyon D (1979) Anatomical and physiological characteristics of arteries. In: Wolf S, Werthessen NT, Eds. Dynamics of arterial flow. New York, NY, Plenum 115:48
62. Krams R, Wentzel JJ, Oomen JAF, Vinke R, Schuurbiers JCH, de Feyter PJ, Serruys PW, Slager CJ (1997) Evaluation of endothelial shear stress and 3D reconstruction from angiography and IVUs (ANGUS) with computational fluid dynamics. Arterioscler Thromb Vasc Biol 17(10):2061–2065
63. Ku DN, Gidden DP, Zarins CK, Glagov S (1985) Pulsatile flow and atherosclerosis in the human carotid bifurcation, Arteriosclerosis 5:293–302
64. La Barbera M (1990) Principles of design of fluid transport systems in biology. Science 249:992–1000
65. Langille BL, O'Donnell F (1986) Reductions in arterial diameter produced by chronic decreases in blood flow are endothelium-dependent. Science 231(4736):405–407
66. Li H, Forstermann U (2000) Nitric oxide in the pathogenesis of vascular disease. J Pathol; 190(3):244–254
67. Lipowsky HH, Zweifach BW (1974) Network analysis of microcirculation of cat mesentery. Microvasc Res 7:73–83

68. Lipowsky HH, Kovalacheck S, Zweifach BW (1978) The distribution of blood rheological parameters in the microvasculaure of cat mesentery. Circ Res 43:738–749
69. Liu SQ, Fung YC (1989) Relationship between hypertension, hypertrophy, and opening angle of zero-stress state of arteries following aortic constriction. J Biomech Eng 111(4):325–335
70. Liu Y, Kassab GS (2007) Metabolic Dissipation in Vascular Trees. Am J Physiol 292(3):1336–9
71. Lu X, Kassab GS (2004) Nitric oxide is significantly reduced in ex vivo porcine arteries during reverse flow because of increased superoxide production. J Physiol 561(2):575–582
72. Lu X, Yang J, Zhao JB, Gregersen H, Kassab GS (2003) Shear modulus of coronary arteries: Contribution of media and adventitia. Am J Physiol Heart Circ Physiol 285(5):H1966–H1975
73. Lu X, Zhao JB, Wang GR, Gregersen H, Kassab GS (2001) Remodeling of the zero-stress state of the femoral artery in response to flow-overload. Am J Physiol Heart Circ Physiol 280(4):H1547–H1559
74. Marques PF, Oliveira MEC, Franca AS, Pinotti M (2003) Modeling and simulation of pulsatile blood flow with a physiologic wave pattern. Artif Organs 27(5):478–485
75. Masuda H, Bassionny H, Glagov S, Zarins CK (1989) Artery wall restructuring in response to increase flow. Surg Forum 40:285–286
76. Matsumoto T, Hayashi K (1994) Mechanical and dimensional adaptation of rat aorta to hypertension. J Biomech Eng 116(3):278–283
77. McDonald DA (1974) Blood flow in arteries. Baltimore, MD: William Wilkins
78. Meyer WW (1964) Über die rhythmische Lokalisation der atherosklerotischen Herde im cervikalen Abschnitt der vertebralen Arterie. Beitr Path Anat 130:24–39
79. Miyashiro JK, Poppa V, Berk BC (1997) Flow-induced vascular remodeling in the rat carotid artery diminishes with age. Circ Res 81(3):311–319
80. Mochizuki S, Goto M, Chiba Y, Ogasawara Y, Kajiya F (1999) Flow dependence and time constant of the change in nitric oxide concentration measured in the vascular media. Med Biol Eng Comput 37(4):497–503
81. Moore JE, Ku DN (1994) Pulsatile velocity measurements in a model of the human abdominal aorta under resting conditions. J Biomech Eng 116(3):337–346
82. Murase K, Naruse K, Kimura A, Okumura K, T Hayakawa, Sokabe M (2001) Protamine auguments stretch induced calcium increase in vascular endothelium. British Journal of Pharmacology 134(7):1403–1410
83. Murray CD (1926) The physiological principle of minimum work. I. The vascular system and the cost of blood volume, Proc Natl Acad Sci USA 12:207–214
84. Naruse K, Sokabe M (1993) Involvement of stretch-activated ion channels in Ca2+ mobilization to mechanical stretch in endothelial cells. Am J Physiol 264(4 Pt 1):C1037–1044
85. Naruse K, Sai X, Yokoyama N, Sokabe M (1998) Uni-axial cyclic stretch induces c-src activatin and translocation in human endothelial cells via SA channel activation. Febs Letters 441(1):111–5
86. Nguyen TN, Chagas ACP, Glantz SA (1993) Left ventricular adaptation to gradual renovascular hypertension in dogs. Am J Physiol 265 (Heart Circ Physiol 24):H22–H38
87. Omens JH (1998) Stress and strain as regulators of myocardial growth. Prog Biophys Mol Bio 69:559–572
88. Oshinski JN, Ku DN, Mukundan S, Loth F, Pettigrew RI (1995) Determination of wall shear stress in the aorta with the use of MR phase velocity mapping. J Magn Reson Imaging 5(6):640–7
89. Patel DJ, Fry DL (1966) Longitudinal tethering of arteries in dogs. Circ Res 19(6):1011–1021
90. Patel DJ, Vaishnav RN (1972) The rheology of large blood vessels. In: Bergel DH, ed. Cardiovascular Fluid Dynamics. Vol. 2. New York, NY: Academic Press, pp 1–64
91. Pries AR, Secomb TW, Gaehtgens P (1995) Design principles of vascular beds. Circ Res 77(5):1017–23
92. Rhodin JAG (1979) Architecture of the vessel wall. In: Berne RM, ed. Handbook of Physiology, Sect. 2, Vol. 2, American Physiology Society

93. Rodbard S (1975) Vascular caliber. Cardiology 60:4–49
94. Sachs F (1988) Mechanical transduction in biological systems. CRC Crit Rev Biomed Eng 16:141–169
95. Satcher Dewey RCF Jr, Hartwig JH (1997) Mechanical remodeling of the endothelial surface and actin cytoskeleton induced by fluid flow. Microcirculation, 4:439–453
96. Shaaban AM, Duerinckx AJ (2000) Wall shear stress and early atherosclerosis: a review. AJR Am J Roentgenol 174(6):1657–1665
97. Stepp DW, Nishikawa Y, Chilian WM (1999) Regulation of shear stress in the canine coronary microcirculation. Circ Res 100:1555–1561
98. Sumpio BE, Banes AJ, Buckley M, Johnson G Jr (1988) Alterations in aortic endothelial cell morphology and cytoskeletal protein synthesis during cyclic tensional deformation. J Vasc Surg 7(1):130–138
99. Traub O, Berk BC (1998) Laminar shear stress: mechanisms by which endothelial cells transducer an atheroprotective force. Arteriosc Thromb Vasc Bio 18(5):677–685
100. Tseng H, TE Peterson, BC Berk (1995) Fluid shear stress stimulates mitogen-activated protein kinase in endothelial cells. Circ Res 77(5):869–78
101. Vaishnav RN, Vossoughi J, Patel DJ, Cothran LN, Coleman BR, Ison-Franklin EL (1990) Effect of hypertension on elasticity and geometry of aortic tissue from dogs. J Biomech Eng 112(1):70–74
102. Vaishnav RN, Vossoughi J (1983) Estimation of residual strains in aortic segments. In: Hall CW, ed. Biomedical Engineering II Recent Developments. New York, NY, Pergamon Press, pp 330–333
103. Weizsacker HW, Lambert H, Pascale K (1983) Analysis of the passive mechanical properties of rat carotid arteries. J Biomech 16(9):703–715
104. Wiesmann F, Ruff J, Hiller KH, Rommel E, Haase A, Neubauer S (2000) Developmental changes of cardiac function and mass assessed with MRI in neonatal,juvenile,and adult mice. Am J Physiol Heart Circ Physiol 278:H652–H657
105. Wolinsky H (1971) Effects of hypertension and its reversal on the thoracic aorta of male and female rats. Morphological and chemical studies. Circ Res 28(6):622–637
106. Wolinsky H (1972) Long-term effects of hypertension on the rat aortic wall and their relation to concurrent aging changes. Morphological and chemical studies. Circ Res 30(3):301–309
107. Wolinsky H, S Glagov (1967) A lamellar unit of aortic medial structure and function in mammals. Circ Res 20:99–111
108. Zamir M (1976a) Shear forces and blood vessel radii in the cardiovascular system. J Gen Physiol 69:213–222
109. Zamir M (1976b) The role of shear forces in arterial branching. J Gen Physiol 67(2):213–222
110. Zarins CK, Zatina MA, Giddens DP, Ku DN, Glagov S (1987) Shear stress regulation of artery lumen diameter in experimental atherogenesis. J Vasc Surg 5(3):413–420
111. Zeindler CM, Kratky RG, Roach MR (1989) Quantitative measurements of early atherosclerotic lesions on rabbit aortae from vascular casts, Atherosclerosis 76:245–255
112. Zhang W, Herrera C, Atluri SN, Kassab GS (2004) Effect of surrounding tissue on vessel fluid and solid mechanics. J Biomech Eng 126(6):760–769
113. Zhang W, Herrera C, Atluri SN, Kassab GS (2005) The Effect of Longitudinal Pre-Stretch and Radial Constraint on the Stress Distribution in the Vessel Wall: A New Hypothesis. MCB (Mechanics & Chemistry of Biosystems) 2(1):41–52
114. Zhou Y, Kassab GS, Molloi S (2002) In vivo validation of the design rules of the coronary arteries and their application in the assessment of diffuse disease. Phys Med Bio47(6):977–993
115. Zhou Y, Kassab GS, Molloi S (1999) On the design of the coronary arterial tree: A generalization of Murray's law. Phys Med Biol 44(12):2929–2945
116. Ziegler T, Nerem RM (1994) Effect of flow on the process of endothelial cell division. Arteriosclerosis and Thrombosis 14:636–643

Chapter 16
The Role of Macromolecules in Stabilization and De-Stabilization of Biofluids

Björn Neu[1] and Herbert J. Meiselman[2]

[1]Division of Bioengineering, School of Chemical and Biomedical Engineering, Nanyang Technological University, Singapore
[2]Department of Physiology and Biophysics, Keck School of Medicine, University of Southern California, Los Angeles, USA

> Generally speaking, it may be said that a reduction of the suspension stability of the blood is one of the most common general reactions of the organism in disease, perhaps the most common.
>
> Robin Fahraeus, 1929.

Abstract Interactions between biological particles, including red blood cells (RBC), white blood cells, platelets and endothelial cells lining blood vessels, continue to be of both basic science and clinical interest. Numerous studies have detailed several "lock-key" mechanisms (e.g., antigen-antibody) leading to cell–cell attraction and adhesion, and specific molecules and binding sites have been identified. Conversely, the non-specific interactions of macromolecules with various cell types have received less attention. The material below focuses on the interactions between red blood cells as influenced by the presence of macromolecules in the suspending phase. RBC are the most numerous cells in blood, have an important biological function (i.e., transport of oxygen), and enhanced attractive interaction between RBC increases low shear blood viscosity and adversely affects blood flow in small vessels. Depletion of macromolecules near the RBC surface is dealt with in detail since this phenomenon appears to be the most likely mechanism for RBC aggregation. Stabilization of RBC systems via small polymers or co-valent attachment of polymers to the membrane surface is also considered. We believe that increased attention to the biophysical aspects of cell–cell interactions will yield important new information that may lead to improved disease diagnosis, patient care and advances in cellular- and bio-engineering.

16.1 Introduction

Blood is often referred to as a highly specialized tissue circulating through the vascular system, with many functions from oxygen delivery and immunological functions to the regulation of body temperature. Alterations of the flow properties of blood can lead to many problems including tissue dysfunction or ischemia. Consequently, the regulation of these flow properties is of paramount importance for the functioning of the vascular system. Thus, studying the mechanisms regulating the flow-properties of bio-fluids and what factors eventually lead to their de-stabilization is not only of basic scientific interest but also of great importance for understanding of the patho-pathology of diseases which have been associated with vascular problems. Identifying and quantifying these factors and mechanisms should help in the development of therapeutic and diagnostic agents and should also be invaluable in other areas like the development of biomaterials for either in vivo or in vitro applications.

16.1.1 Macromolecules as a Determinant of Blood Cell Interaction

Crucial for the understanding of blood flow is a detailed understanding of the interactions of all the cellular components. Cell interaction and cell adhesion are regulated by multiple forces ranging from specific or lock and key forces to non specific forces like electrostatic or van der Waals interaction [1]. In many cases as in blood, one also has to consider the presence of macromolecules or plasma proteins. These molecules not only affect the physicochemical properties of the plasma (e. g., viscosity) but, in addition, regulate the interactions between many cellular components in normal and pathological conditions. For example, in case of agglutination of cells or particles in the circulation, antibodies or immunoglobulins specific to certain antigens on the surface of these cells (e. g., red blood cells) can link them together causing them to agglutinate. Other examples where cell–cell interactions are orchestrated by plasma proteins include platelet or leukocyte adhesion to the endothelium, platelet aggregation in thrombosis and haemostatic plug formation.

The above examples for cell interactions induced by macromolecules have one thing in common: the respective molecule acts as a ligand or linkage between adjacent cells by specific or non-specific binding. An example where the presence of macromolecules is of crucial importance yet the mechanism seems quite different is the aggregation of RBC. At stasis or during slow flow, RBC in plasma or in solutions containing large polymers (e. g., dextran with a molecular weight larger than 40 kDa) aggregate into characteristic linear arrays (i. e., rouleaux), then form secondary three-dimensional branched structures. In normal, non-pathological blood, the aggregates break up when subjected to relatively low shear rates (e. g., $20-40 \text{ s}^{-1}$), suggesting weak attractive forces. However, these attractive forces can become much stronger in pathological conditions or if RBC are suspended in suspensions containing certain concentrations of larger polymers like dextran or poly (ethylene glycol) [2]. Note that the presence of large polymers or proteins is required for RBC aggregation: RBC washed and re-suspended in protein or polymer-free saline do not aggregate. Further, there is now general agreement regarding the correlations between elevated levels of fibrinogen or other large plasma proteins and enhanced RBC aggregation, and the effects of molecular mass and concentration for neutral polymers such as dextran [3].

16.1.2 The Impact of Red Blood Cell Aggregation on Blood Flow

Reversible RBC aggregation is the major determinant of the non-Newtonian (i.e., shear thinning) flow behavior of blood and other RBC suspensions, with both low shear viscosity and the degree of non-Newtonian behavior increased with enhanced aggregation. Tube flow of blood and other RBC suspensions is also affected by RBC aggregation: enhanced aggregation leads to the formation of a cell-poor marginal layer and impaired oxygen diffusion/release. Likewise, in vivo and clinical studies have documented the marked adverse effects of increased RBC aggregation (Fig. 16.1). It has been shown that peripheral vascular resistance is elevated, especially in cases of limited vascular vasodilatory reserve, and microcirculatory flow dynamics are markedly altered (i.e., blunted velocity profiles, RBC maldistribution, plasma skimming). Numerous clinical reports have demonstrated greater aggregation and elevated low shear blood viscosity in several clinical states (e.g., diabetes, myocardial infarction, nephrotic syndrome, sepsis, stroke, vascular disease). Eventually, such intense aggregation might lead to hindrance or even blockage of blood flow in small vessels, and thus ultimately to reduced tissue perfusion and ischemia.

There is also strong epidemiological evidence indicating that RBC aggregation is a major risk factor for cardiovascular disease [4]. Apart from these direct consequences, RBC aggregation may also influence the functioning of other cells. For example, it has been shown that RBC aggregation is partly responsible for the margination of white blood cells towards the vessel wall and their interaction with the endothelium [5]. This, in turn, might enhance inflammatory conditions and the expression of inflammatory agents (i.e., fibrinogen, immunoglobulins) which are known to further enhance RBC aggregation. Consequently, the reversible aggregation of RBC has been of basic scientific and clinical interest for

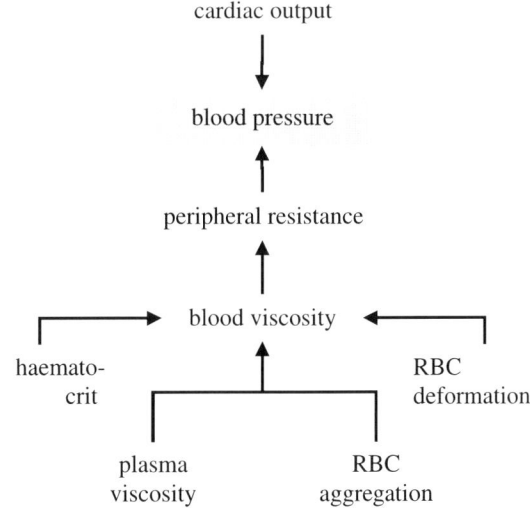

Fig. 16.1 Schematic illustration of the impact of RBC on the functioning of the vascular system. Blood viscosity is determined by the haematocrit, plasma viscosity, RBC aggregation and the deformability of RBC and is one of the major determinants of peripheral resistance, which together with the cardiac output, determine arterial blood pressure

decades [6–17]. However, the specific mechanisms involved in RBC aggregation and thus the role of macromolecules in RBC-RBC interaction have not yet been fully elucidated [18–23].

16.2 The Effects of Macromolecules on the Stability of Colloids

When macromolecules are introduced into colloidal suspensions destabilization or flocculation can occur via two distinct mechanisms depending on the interaction between these molecules and the colloidal particles. If this interaction is attractive and the polymer concentration is not sufficient to yield full coverage of the surface, polymer chains can adsorb onto adjacent surfaces leading to so-called bridging flocculation. However with increasing polymer concentration full coverage of single colloid particles can occur, eventually leading to steric stabilization of the suspension [24]. On the other hand, if the colloid-polymer interaction is repulsive or vanishes, polymer adsorption is not favored [25]. In this case, the concentration of the polymer segment vanishes at the surface of the colloid and increases with the distance from the surface as illustrated in Fig. 16.2.

The driving force behind this depletion effect is as follows: if a polymer molecule close to a surface experiences a loss in conformational entropy, and if such a loss is not compensated by adsorption energy, a depletion layer develops. If two colloid particles approach one another, the volume of the depletion zone is decreased and hence solvent is displaced from the depletion zone into the bulk phase (Fig. 16.3). Thus, the free energy of the system is lowered since the solvent is transferred from a region of higher chemical potential to one of lower chemical potential leading to depletion aggregation or flocculation [26–29]. In the past, several reports have dealt with the experimental and theoretical aspects of depletion aggregation as applied to the general field of colloid chemistry [26–29]. Polymer depletion as a mechanism in bio-fluids has received much less attention. However, recent reports have demonstrated that, in the same way depletion layers develop at colloid surfaces, one could expect the existence of macromolecular depletion layers at biological interfaces.

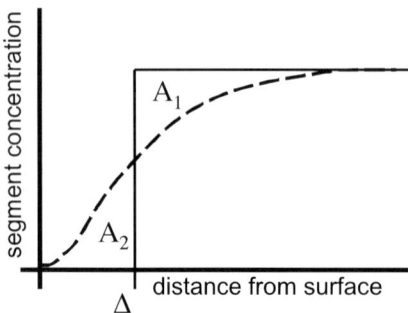

Fig. 16.2 Polymer segment concentration (*dashed line*) at a non-adsorbing interface. The conformational entropy restriction close to the surface is not compensated by an adsorption energy leading to a depletion layer thickness Δ, which is defined by the equality of A_1 and A_2 [24]

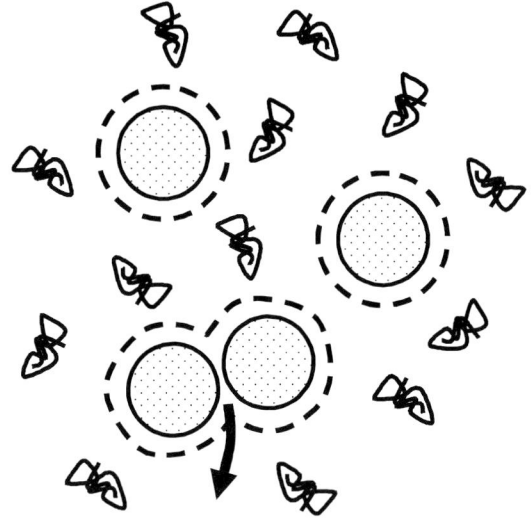

Fig. 16.3 Illustration of the polymer poor region around colloid particles and the depletion interaction via solvent displacement from the depletion zone into the bulk phase as two colloids approach

16.3 Macromolecular Depletion at Biological Interfaces

At present, there are two co-existing "models" for RBC aggregation: bridging and depletion. In the bridging model, red cell aggregation is proposed to occur when the bridging forces due to the adsorption of macromolecules onto adjacent cell surfaces exceed disaggregating forces due to electrostatic repulsion, membrane strain and mechanical shearing [3, 30–34]. This model seems to be similar to other cell interactions like agglutination, with the only difference being that the proposed adsorption energy of the macromolecules is much smaller in order to be consistent with the relative weakness of these forces. In contrast, the depletion model proposes quite the opposite. In this model RBC aggregation occurs as a result of a lower localized protein or polymer concentration near the cell surface compared to the suspending medium (i. e., relative depletion near the cell surface). This exclusion of macromolecules near the cell surface leads to an osmotic gradient and thus depletion interaction [19]. As with the bridging model, disaggregation forces are electrostatic repulsion, membrane strain and mechanical shearing.

As described in the previous section, the preferential exclusion of macromolecules leading to depletion flocculation has received considerable attention by colloid scientists during the past decades. This, in turn, makes it surprising that this phenomenon has been almost ignored when considering the stability of cell suspensions. In part, this lack of attention is due to the previous and current problems associated with deciding if a depletion layer has occurred at cell surfaces. One reason is the small extension of such depletion layers. Depletion layers are in the same size range as the hydrated size of the depleted macromolecule, meaning for most plasma proteins known to induce aggregation, it is only just a few nanometers; the thickness of the RBC glycocalyx has been determined to also be a few nanometers. Thus, it is not only impossible to detect such a layer with direct optical observations but, in addition, it is also quite challenging to distinguish between weak absorption or a depletion effect since the latter might also involve some intermixing (i.e., penetration) of the macromolecule and the glycocalyx. So even though past reports have described surface adsorption of dextran, and specific binding mechanism between fibrinogen and RBC have been outlined [6, 30, 31, 35], such results should be interpreted with great care. In fact, an extensive review of literature values by Janzen and Brooks [36] has detailed likely technical artifacts (e.g., trapped fluid between RBC) and thus the extremely wide range of reported data for fibrinogen and dextran binding.

Several studies have investigated the structure and extension of depletion layers using various techniques [28, 37–39]. One method, which is also applicable to macromolecular depletion at biological interfaces, is to investigate the depletion layer by means of electrophoresis. In solutions of neutral soluble polymers that give rise to depletion layers, particles have an unexpectedly high mobility [20, 34, 40, 41]. This effect is due to the reduced viscosity near the particle surface due to the depletion effect [22]: electro-osmotic flow decreases rapidly outside the electric double layer, so if the double layer thickness is comparable to or larger than that of the depletion layer, the influence of suspending phase viscosity is reduced (Fig. 16.4).

Figure 16.5 presents results from a study which was directed toward validating the existence of the depletion layer by employing measurements of unit-gravity cell sedimentation and of cell mobility for RBC in various polymer solutions [20]. These studies thus tested the hypothesis that regardless of polymer molecular weight, the effects of suspending medium viscosity on sedimentation could be predicted via the Stokes Equation, whereas EPM would become less sensitive to medium viscosity with increasing molecular weight (i.e., with increasing depletion layer thick-

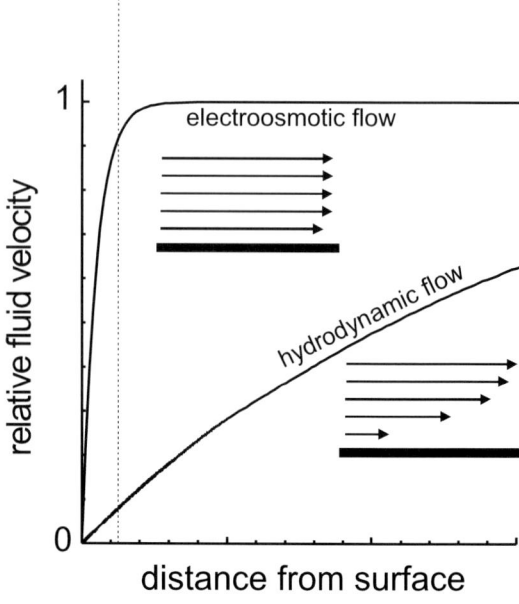

Fig. 16.4 Schematic diagram of velocity profiles along a surface. In the case of electroosmosis the mechanical energy dissipation occurs within the range of the Debye length (*dotted line*). Thus, for a small enough Debye length, the energy dissipation in electroosmotic flow occurs over a shorter range compared to that of a hydrodynamic flow

16.3 Macromolecular Depletion at Biological Interfaces

ness). As shown in Fig. 16.5, the Stokes Equation is applicable regardless of polymer molecular weight, whereas the electrophoretic mobility follows the expected inverse relation only for small polymers (i. e., 10.5 kDa dextran) with RBC electrophoretic mobility essentially independent of medium viscosity for large polymers (i. e., 519 kDa dextran). The latter point is of particular interest in that it demonstrates the existence of the depletion layer and its dependence on molecular size; dextran 10.5 kDa has a hydrated radius of about 3 nm compared to about 20 nm for the 519 kDa dextran [20].

Consequently, particle electrophoresis has been used extensively to study macromolecules at biological interfaces. Via changing the ionic strength of the suspending medium it is even possible to probe the extent of the depletion layer. Variation of ionic strength varies the thickness of the double layer, and thus measurement of mobility at different ionic strengths allows the estimation of depletion layer thickness [38]. As long as the double layer is significantly thinner than the depletion layer, the mobility will be higher than predicted based upon media viscosity. With increasing double layer thickness the influence of bulk viscosity increases and when the double layer thickness becomes significantly greater than the depletion layer, the measured mobilities or ζ-potentials are unaffected by viscosity in the depletion region. Figure 16.6 shows an example of such an approach, in which the ratio of glutaraldehyde-fixed RBC mobility in polymer free solution and in solutions containing dextran of various molecular weights are plotted versus suspending phase viscosity [19]. According to the Smoluchowski Equations, we would expect these mobility ratios to fall exactly on the inverse viscosity ratios which are shown as dashed lines. Instead, the ratio is well below the expected value for smaller Debye lengths indicating a significantly lower viscosity close to the RBC surface. Using this approach it has been confirmed that the thickness of such depletion layers is in the same range as expected for the hydrated size of these polymers, thereby agreeing with the concept of polymer depletion near the RBC surface and lending strong support to a "depletion model" mechanism for reversible RBC aggregation [19, 39, 42, 43].

It should also be noted that some studies have evaluated depletion of proteins and polyelectrolytes at RBC surfaces by means of particle electrophoresis. However, the details of such experiments as well as interpretation of the data are not as straight forward as with neutral polymers. For charged polyelectrolytes and hence also for proteins, it is also necessary to consider electrostatic forces which can also affect depletion or adsorption [24, 44]. In addition, forces between the particle surface and the polyelectrolyte differ with ionic strength, and thus adsorption and depletion can vary with the extent of the double layer as well as the confirmation of the charged macromolecule. For example, for polymers and surfaces of the same sign, only slight or no adsorption is often observed at low ionic strength, whereas with increasing ionic strength, the adsorbed amount increases. This behavior is due to the electrostatic screening, which increases with increasing salt concentration; in the case of pure electro-adsorption (i. e., when the surface charges and the polymer charges have opposite signs), the opposite effects of ionic strength are observed [24, 44].

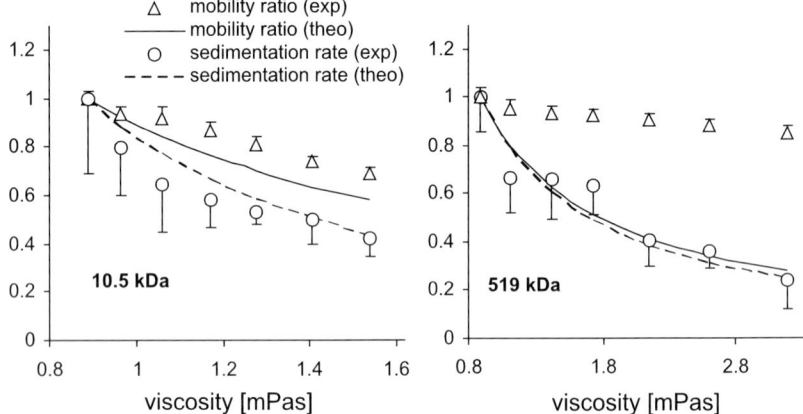

Fig. 16.5 Experimental and predicted values of red blood cell mobility and sedimentation rate versus suspending phase viscosity for red blood cells in various solutions of 10.5 kDa and 519 kDa dextran. All mobility and sedimentation data are expressed relative to values obtained for cells in dextran-free buffer, and thus have a value of unity at a suspending medium viscosity of 0.89 mPas. Symbols are: $\triangle$ = experimental mobility ratio; $\circ$ = experimental sedimentation ratio; *solid line* = predicted mobility; *dashed line* = predicted sedimentation. Data are mean ±SD (modified from [20])

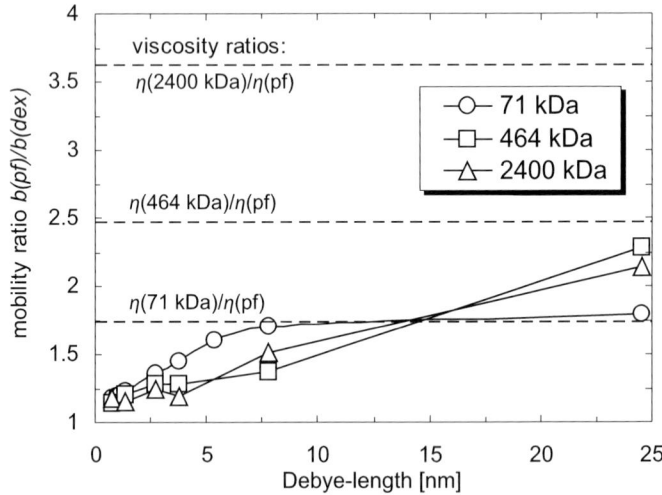

Fig. 16.6 Ratio of the mobility for RBC in polymer free solution (pf) and in solutions containing 2 g/dl of dextran having three different molecular weights as well as the inverse viscosity ratio, i. e., the expected mobility ratio after Smoluchowski's equation as a function of the Debye length (modified from [19])

One consequence of the above-mentioned phenomena is that it is not possible to determine the thickness of the depletion layer by simply changing the ionic strength of the suspending media as is utilized for neutral polymers such as dextran. Further, if only measuring at a single ionic strength, it is sometimes not possible to decide if apparent changes of the particle mobility should be attributed to depletion or adsorption. Thus, even though protein and polyelectrolyte depletion can be detected with particle electrophoresis, it is also clear that for the investigation of plasma proteins at bio-interfaces it is still necessary to develop other techniques in order to obtain quantitative results.

16.4 Cell–Cell Interactions Mediated by Macromolecular Depletion

In order to calculate surface affinities between RBC suspended in polymer solutions, it is first necessary to define the nature of the cell–cell interaction. The exterior RBC surface, termed the glycocalyx, consists of a complex layer of proteins and glycoproteins and bears a net negative charge that is primarily due to ionized sialic acid groups [45]. In the theoretical model employed herein, only depletion and electrostatic interactions have to be considered. Owing to the high electrostatic repulsion, cell–cell distances at which minimal interaction energy (i. e., maximal surface affinity) occurs are always greater than twice the thickness of the cell's glycocalyx. Thus, steric interactions between glycocalyx on adjacent RBC can be neglected. Further, calculated total interaction energies are in the order of 1 – 10 µJ/m^2, whereas for cell separations greater than twice the glycocalyx thickness van der Waals interactions are in the range of 10^{-2} µJ/m^2 [46], and thus can also be neglected.

16.4.1 Depletion Interaction Energy

Examination of the energetics of depletion layers requires distinguishing between so-called "hard" and "soft or hairy" surfaces. Hard surfaces are considered to be smooth and do not allow polymer penetration into the surface, whereas soft surfaces, such as the RBC glycocalyx, are characterized by a layer of attached macromolecules that can be penetrated in part or entirely by the free polymer in solution [28, 47]. Figure 16.7 presents a stylized representation of polymer concentrations adjacent to a cell or particle with a soft surface.

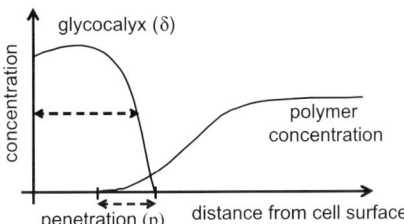

Fig. 16.7 Concentration-distance profile near a surface having a glycocalyx

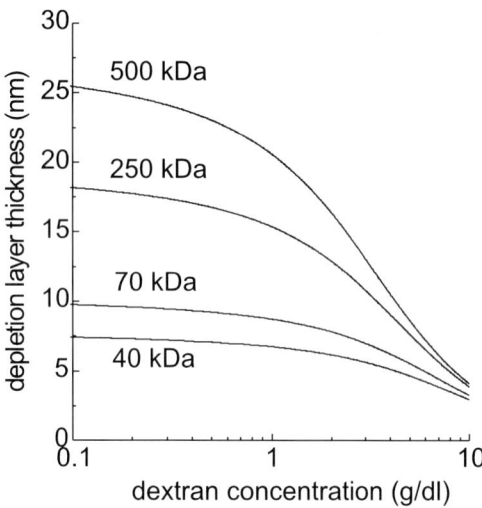

Fig. 16.8 Effects of polymer concentration on the depletion layer thickness for different molecular weight fractions of dextran

One way of estimating the depletion interaction energy w_D is by assuming a step profile for the free polymer [23, 28, 48]. Given a depletion layer thickness Δ and a separation distance of d between adjacent surfaces, w_D is than given by

$$w_D = -2\Pi\left(\Delta - \frac{d}{2} + \delta - p\right), \quad (16.1)$$

where $(d/2 - \delta + p) < \Delta$ and equals zero for $(d/2 - \delta + p) > \Delta$; Π denotes the osmotic pressure.

To evaluate the depletion interaction with Eq. (16.1) it is also necessary to estimate the depletion layer thickness (Fig. 16.8). Several theoretical approaches have been suggested [49]. An approach introduced by Vincent [27] is based upon calculation of the equilibrium between the compressional or elastic free energy and the osmotic force experienced by polymer chains at a non-absorbing surface and yields:

$$\Delta = -\frac{1}{2}\frac{\Pi}{D} + \frac{1}{2}\sqrt{\left(\frac{\Pi^2}{D}\right) + 4\Delta_0^2}, \quad (16.2)$$

where Δ_0 is the depletion thickness for vanishing polymer concentration and is equal to $1.4 \cdot R_g$, with R_g being the polymer's radius of gyration [27]. The parameter D is a function of the bulk polymer concentration (c):

$$D = \frac{k_B T}{\Delta_0^2}\left(c\frac{N_A}{M}\right)^{\frac{2}{3}}, \quad (16.3)$$

where k_B and N_A are the Boltzmann constant and Avogadro number.

16.4.2 Depletion at Soft Surfaces

The next step is to consider the penetration depth p of the free polymer into the attached layer. Intuitively, this penetration should depend on the polymer type, concentration and molecular size, and would be expected to be larger for small molecules and to increase with increasing polymer concentration due to increasing osmotic pressure. One possibility is to calculate p by assuming that penetration proceeds until the local osmotic pressure developed in the attached layer is balanced by the osmotic pressure of the bulk solution [28]. It is also possible to consider that the attached polymers collapse under the osmotic pressure of the bulk polymer [47]. However, it is difficult to accurately apply such a model to RBC in polymer or protein solutions, since too little is known about the physicochemical properties of the glycocalyx, and in particular, about the interaction between the glycocalyx and different polymers or proteins. However, one possibility is to use an exponential approximation for the concentration dependence of the penetration depth:

$$p = \delta \left(1 - e^{-\frac{c}{c_p}}\right), \tag{16.4}$$

where c_p is the penetration constant of the polymer in solution (i.e., when c_p equals c, p is 63% of δ). In this approach δ is assumed to be independent of bulk polymer concentration, and therefore p is essentially a linear function of c at low concentrations (relative to c_p) and asymptotically approaches δ at high concentrations.

16.4.3 Electrostatic Forces

As mentioned above, the strong electrostatic repulsion between adjacent cells permits forces other than depletion and electrostatic to be neglected in this simple model [23, 50]. The electrostatic free energy of two cells can be calculated by considering an isothermal charging process. For $d \geq 2\delta$ the electrostatic repulsion is than given by [23]:

$$w_E = \frac{\sigma^2}{\delta^2 \varepsilon \varepsilon_0 \kappa^3} \sinh(\kappa \delta) \left(e^{\kappa \delta - \kappa d} - e^{-\kappa d}\right), \tag{16.5}$$

where ε, ε_0, κ and σ are the relative and absolute permitivities, the inverse of the Debye-length and the surface charge density (i.e., charge per surface area). Finally, the total interaction energy (w_T) as a function of the cell–cell separation is than given by the sum of the electrostatic interaction energy (w_E) and the depletion interaction energy (w_D) as plotted in Fig. 16.9.

Figure 16.10 summarizes the major aspects of the model for several molecular weight dextran fractions. The results shown are consistent with experimental measurements of RBC aggregation in dextran solutions [32, 51–53], in that both

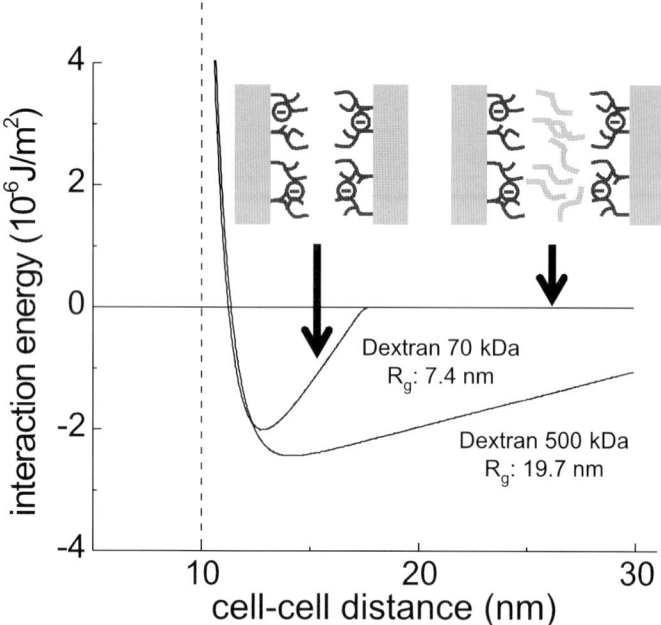

Fig. 16.9 Total interaction energy between approaching RBC as a function of cell–cell distance (d: 5 nm, p: 5 nm, c: 1 g/dl). As the cells approach there is basically no depletion interaction as long as polymer remains in the depletion zone. However, as the cells come closer they are attracted by depletion interaction until electrostatic repulsion limits their approach

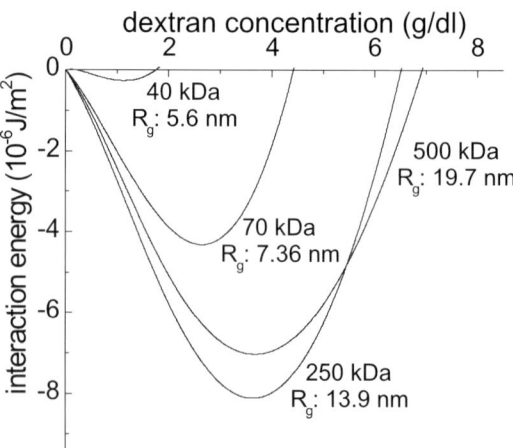

Fig. 16.10 Effects of bulk phase polymer concentration (c) on the total energy (w_T) for RBC suspended in various solutions of dextran (δ: 5 nm, p: 5 nm)

16.4 Cell–Cell Interactions Mediated by Macromolecular Depletion

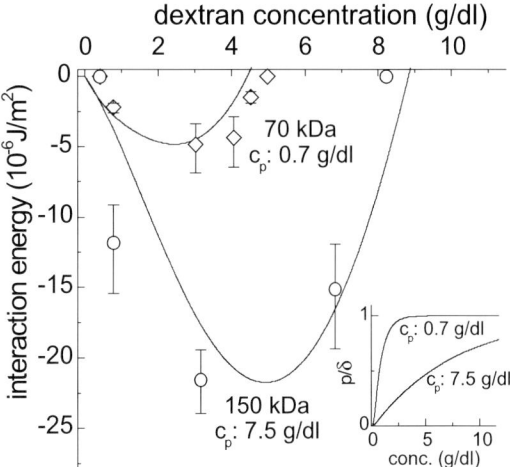

Fig. 16.11 Comparison between calculated (*solid lines*) and experimental values (*data points*) of total interaction energy; c_p is the penetration constant (modified from [50])

a lower molecular weight threshold for aggregation and biphasic responses to increasing polymer concentration have been experimentally demonstrated. Note also that the agreement between theoretical predictions and experimental findings provides a basis for exploring the effects of altered glycocalyx properties on red cell aggregability.

It is also important to note that the calculated energies are in the expected range, in that experimental measurements of the energies leading to aggregation are usually just a few $\mu J/m^2$ depending on the method employed. By employing a micropipette technique Buxbaum et al. [54] determined cell–cell surface affinities for normal human red blood cells in various dextran solutions, with their results indicating biphasic affinity-concentration relations (Fig. 16.11). In order to quantitatively compare their experimental findings with the depletion model described herein, the empirical approach for penetration described above was employed (Eq. (16.4)). In this approach p increases with increasing polymer concentration (small diagram in Fig. 16.11) gradually reaching complete penetration ($p = \delta$). The only variable is the penetration constant c_p. The theoretical results are also shown in Fig. 16.11. The shapes of the calculated $w_T - c$ relations and the upper polymer concentrations at which w_T declines to zero are in excellent agreement with the experimental data. In addition, the values of c_p required for achieving equality between the calculated and experimental peak w_T levels are consistent with the expected effect of polymer molecular mass on glycocalyx penetration: the lower value for the smaller dextran indicates greater penetration ability for smaller dextran molecules.

16.4.4 Cell Surface Properties as a Determinant of Depletion Interaction

It is important to understand that cell surface properties can also have a dramatic impact on the depletion layer thickness and thus can also play a crucial role in the destabilization of blood. An example of how rather small changes in RBC surface properties can markedly affect cell–cell interaction is the impact of cell age on the tendency to form aggregates. During their 120-day life span, human RBC undergo several physicochemical changes, including an increased tendency to aggregate in plasma or polymer solutions [55]. Even though this phenomenon has been investigated for decades, and it has been shown that the differences are not related to altered cell volume, deformability or surface IgG levels, a rational explanation for these age-associated differences in RBC aggregation has been lacking until recently.

Based on the model presented herein, one possible reason for this increased aggregation for older RBC would be an age dependent loss of surface charges and thus less repulsion between the cells. Another possibility would be that the depletion layer thickness increases as the cells age. Electrophoretic mobility studies of age-separated RBC have revealed an interesting pattern: no significant difference of electrophoretic mobility for cells in protein-free or polymer-free media, yet a small ($\approx 5\%$) but significant greater mobility for older cells in plasma, serum or dextran solutions [56]. This age-dependent difference is most likely due to a reduced local viscosity at the cell surface, and thus an increased depletion layer at the surface of the old cells. There are basically two possibilities for this depletion layer increase: decreasing polymer penetration or decreasing glycocalyx thickness with cell age. Both could both explain the lower viscosities at the cell surface and thus higher mobilities in polymer solution. Theoretical calculations show that a decrease of 15% for either p or δ would lead to a 5% increase in mobility, thus agreeing with experimental mobility data, and to a 70–80% increase of the cell surface affinity which would explain the differences in aggregation.

In overview, the model for depletion-mediated aggregation [40] provides a rational framework for explaining the observed equal mobility values in buffer, the different values in polymer solutions, and the marked increases of polymer-induced aggregation for older, age-separated RBC. That is, compared to younger cells, older cells may have either a slightly thinner glycocalyx or a slightly less polymer penetration into their glycocalyx, and thus greater polymer depletion and larger osmotic forces favoring aggregation [50].

16.5 Stabilization of Bio-Fluids via Macromolecules

Studies of RBC aggregation primarily utilize one or more polymers or proteins dissolved in an isotonic buffer and determine the effects of such parameters as molecular weight and concentration. Literature reports in this area are numerous, and include general observations for the plasma protein fibrinogen [57] and the neutral polysaccharide dextran [32], promotion and inhibition of aggregation in binary mixtures of poly(ethylene glycol) and dextran [58], thermodynamic and experimental analyses of non-specific adsorption and its relation to aggregation [36], inhibition of polymer-induced aggregation by a poloxamer [59, 60] and evaluation of relationships between polymer hydrated size and inhibition or promotion of aggregation [61]. The rheological effects of the covalent attachment of polymers to the cell membrane have more recently become of interest, with these studies mostly focused on promotion and/or inhibition of RBC aggregation.

The ability of a covalently attached polymer to stabilize a RBC suspension via a reduction of aggregation was first reported by Armstrong and co-workers [62], and followed from earlier reports indicating that the addition of a poly(ethylene glycol)-containing block copolymer, termed a poloxamer, reduced aggregation [59, 63]. Since the poloxamer molecule consists of two hydrophilic poly(ethylene glycol) (PEG) chains connected by a hydrophobic poly(propylene glycol) (PPG) core and is 80% PEG, these authors reasoned that attachment of PEG to the RBC membrane surface would also reduce aggregation. A simple method to covalently bind PEG to the RBC was developed, in which monomethoxy-PEG of 5 kDa molecular mass activated with cyanuric chloride was added to a suspension of washed RBC in buffer, gently mixed for at room temperature for one hour, then the RBC washed and resuspended in autologous plasma. This process yielded RBC with normal morphology and deformability, and with 5 kDa PEG covalently attached to cell membrane, presumably via lysine groups. Rheological testing of PEG-coated cells in plasma indicated a 93% reduction in the extent of RBC aggregation at stasis and a 75% reduction of low shear blood viscosity; viscosity-shear rate relations for the suspensions were nearly Newtonian [62].

Two studies have been conducted in order to characterize the regions near the glycocalyx of PEG coated RBC [64,65]. Both of these studies employed micro electrophoretic methods to determine RBC electrophoretic mobility (EPM), and in each study, cells were coated with linear or 8-armed PEGs of various molecular weights (i.e., 3.35, 18.5, 35.0 and 35.9 kDa) using different polymer/RBC ratios during attachment. Salient aspects of these studies include: 1) PEG coating markedly decreases EPM by up to 80% for both linear and branched PEGs, with the effect increasing with increasing polymer/RBC ratios; 2) The ability of PEGs to reduce EPM is a function of molecular size and geometry, with the effectiveness ranked as branched 35.9 > linear 18.5 > linear 3.35 > linear 35.0; 3) The calculated thickness and hydrodynamic friction of the PEG layer are also functions of molecular size, with the thickness about 6–8 nm for 3.35 kDa PEG and 20 nm for 20 kDa PEG and the friction about $0.2\,\text{nm}^{-1}$ and $0.04\,\text{nm}^{-1}$ for the same PEGs. Note that since PEG is neutral and its covalent attachment only removes positively-charges amino groups on the cell membrane, an increase of EPM (i.e., more negative surface charge), rather that the observed decrease, would be expected. Note also that the EPM results are consistent with the decrease of RBC aggregation for PEG coated cells, and thus lend support to the depletion-mediated aggregation model discussed below.

PEG coating of RBC has been used as a tool to explore possible mechanisms involved in RBC aggregation [66], specifically the so-called bridging theory. In this approach it is assumed that aggregating molecules are large enough to bridge the gap between adjacent RBC at their closest distance of approach, and that they bind to the surface of each RBC with sufficient affinity to overcome electrostatic forces tending to separate the cells [67]. To evaluate this approach, 35 kDa PEG, a potent aggregator of RBC, was covalently attached to RBC, and thus formed one-half of the bridge between cells. The coated RBC were then suspended in an isotonic, 0.5% solution of 35 kDa PEG and their aggregation behavior tested. The bridging theory would predict enhanced aggregation for RBC coated with 35 kDa PEG and suspended in a solution of the same PEG inasmuch as one-half of the PEG bridge was already formed. In fact, the *opposite* was observed: PEG coated cells exhibited no aggregation in the PEG solution whereas control cells were strongly aggregated [66]. These results thus call into question the bridging approach, at least for commonly used non-ionic polymers such as PEG, dextran and polyvinylpyrrolidone [55].

16.6 Destabilization of Bio-Fluids via Macromolecular Binding

The studies of PEG coated RBC described above have a common characteristic: aggregation was decreased to various degrees subsequent to the attachment of PEG to the cell surface [66]. However, in many applications it is of interest to increase RBC aggregation and destabilize the suspension without the addition of polymer to the suspending medium, and hence without altering the physicochemical properties of the medium. To that end, Armstrong et al. [68] evaluated the rheological effects of covalent attachment of nonionic poly(ethylene glycol)-containing block copolymers, termed poloxamers, to the cell surface. These block copolymers exhibit a critical micellization temperature (CMT) above which there is a phase transition from predominately single, fully hydrated copolymer chains to micelle-like structures; the CMT is a function of molecular mass and concentration [69]. Therefore, for poloxamer coated cells at temperatures above the CMT, the formation of micelles made up of polymers from different cells would be expected to cause cell aggregation.

In their studies [68], reactive derivatives of poloxamers containing 80% PEG and having molecular masses of 8.4, 11.4, 13.0 and 14.6 kDa were covalently attached to the cell surface; the attachment was done at temperatures below their CMT in order to avoid the formation of permanent covalently cross-linked aggregates. Coated cells were re-suspended in native, unaltered autologous plasma and tested via viscometry. As anticipated, both poloxamer molecular mass and measurement temperature affected rheological behavior: 1) At both 25 and 37 °C, cells coated with the 8.4 kDa poloxamer (CMT = 48 °C) showed nearly Newtonian behavior and markedly reduced low shear viscosity; 2) At 25 °C, cells coated with the 11.4 kDa (CMT = 37 °C) had flow behavior similar to control cells, indicating a similar degree of RBC aggregation in the control and treated samples. However, at 37 °C, low shear viscosity was significantly above control, indicating an enhancement of aggregation; 3) At both 25 and 37 °C, cells coated with either 13.0 kDa (CMT = 30 °C) or 14.6 kDa (CMT = 23 °C) showed greatly increased low shear viscosity that was 4-fold higher than control.

The poloxamer results presented above [68] provide an unique approach for inhibiting or enhancing RBC aggregation for cells suspended in unaltered plasma, and thus avoid addition of water-soluble polymers to the suspending medium. In addition to being of scientific interest, this approach now allows carrying out studies in which medium physicochemical properties and RBC aggregation can be independently varied. For example, it should now be possible to evaluate the effects of aggregation and/or medium viscosity on low shear in vitro rheological behavior of RBC suspensions; to date such studies do not appear to exist. Studies of the specific effects of RBC aggregation and/or plasma viscosity on in vivo blood flow are now also possible, inasmuch as infusion of an aggregating polymer into the circulation is not necessary; at least one report using the approach has been published [70].

16.7 Conclusion & Outlook

As outlined in this article, macromolecules can stabilize or destabilize biological fluids depending on their properties and their interactions with the particles or cells in suspension. We have emphasized macromolecular depletion as a mechanism for cell–cell interactions and its effects on RBC aggregation and hence on the flow properties of blood. While we recognize that some caution should be exercise regarding the validity of this concept for RBC interactions, evidence supporting depletion as the main mechanism has been growing in the last few years (e. g., [19, 23]).

One of the primary reasons for seeking a better understanding of the mechanism for RBC aggregation, and thus the impact of macromolecules on the flow properties of blood, arises since large alterations of RBC aggregation occur due to either cellular changes or alterations of plasma protein levels. Age-separated human RBC exhibit marked differences in aggregability, with older, denser cells having 3-fold greater affinity than younger, less dense cells in both plasma and polymer solutions [23]. We also know that in some clinical states or diseases (e. g., diabetes, myocardial infarction, sickle cell disease), RBC surface properties that regulate depletion interaction are altered, resulting in significant increases in the tendency of RBC to form aggregates. Additionally, significant variations have been observed when comparing the aggregability of RBC from healthy donors [2]. The reasons for such large differences in healthy donors, as well as their significance, still need to be investigated: it is not even clear if these changes are merely an epi-phenomena or a precursor of certain diseases.

A better understanding of the role of macromolecules in bio-fluids may have physiological and biomedical implications. As mentioned above, enhanced RBC aggregation and aggregability have been observed in pathological conditions. Our recent understanding of red cell aggregation suggests that these differences are due to altered depletion forces between cells, with changes of both cellular and plasma properties responsible for the greater attractive forces [2,50]. Developing techniques to modify cellular and/or plasma properties in order to normalize RBC aggregation would clearly be of value, and would most likely lead to improved clinical care. Some encouraging progress has already been achieved in this area (i. e., decreasing both RBC aggregation and aggregability in diabetes [71]), indicating the potential merit of this approach. While the above discussion has dealt with RBC–RBC interactions, it seems quite reasonable to believe that the general observations for RBC are also applicable to other cells in suspension (e. g., platelet aggregation) and to the interaction of cells with vascular endothelium (e. g., thrombus formation). Finally, we believe that it may be useful to utilize altered depletion forces for biotechnological applications such as cell separation or diagnostics, and to consider such forces when evaluating the biocompatibility of natural or man-made biomaterials.

References

1. Sackmann E, Bruinsma RF (2002) Cell adhesion as wetting transition? Chem Phys Chem 3:262–269
2. Rampling MW, Meiselman HJ, Neu B, Baskurt OK (2004) Influence of cell-specific factors on red blood cell aggregation. Biorheology 41:91–112
3. Chien S, Lang LA (1987) Physicochemical basis and clinical implications of red cell aggregation. Clin Hemorheol 7:71–91
4. Lowe GDO (1995) Fibrinogen and Cardiovascular-Disease – Historical Introduction. European Heart Journal 16:2–5
5. Yedgar S, Koshkaryev A, Barshtein G (2002) The red blood cell in vascular occlusion. Pathophysiology of Haemostasis & Thrombosis 32:263–268
6. Chien S, Simchon S, Abbot RE, Jan KM (1977) Surface adsorption of dextrans on human red cell membrane. J Colloid Interface Sci 62:461–470
7. Lowe GDO (1988) Clinical Blood Rheology. Boca Raton, Florida: CRC Press, p 554
8. Stoltz JF, Singh M, Riha P (1999) Hemorheology in Practice. Amsterdam: IOS Press
9. Meiselman HJ, Baskurt OK, Sowemimo-Coker SO, Wenby RB (1999) Cell electrophoresis studies relevant to red blood cell aggregation. Biorheology 36:427–432
10. Lim B, Bascom PAJ, Cobbold RSC (1997) Simulation of red blood cell aggregation in smear flow. Biorheology 34:423–441
11. Kounov NB, Petrov VG (1999) Determination of erythrocyte aggregation. Math Biosci 157:345–356
12. Hovav T, Yedgar S, Manny N, Barshtein G (1999) Alteration of red cell aggregability and shape during blood storage. Transfusion 39:277–281
13. Holley L, Woodland N, Hung WT, Cordatos K, Reuben A (1999) Influence of fibrinogen and haematocrit on erythrocyte sedimentation kinetics. Biorheology 36:287–297
14. Cloutier G, Qin Z (1997) Ultrasound backscattering from non-aggregating and aggregating erythrocytes – A review. Biorheology 34:443–470
15. Sennaoui A, Boynard M, Pautou C (1997) Characterization of red blood cell aggregate formation using an analytical model of the ultrasonic backscattering coefficient. IEEE Trans Biomed Eng 44:585–591
16. Evans EA, Parsegian VA (1983) Energetics of membrane deformation and adhesion in cell and vesicle aggregation. Ann NY Acad Sci 416:13–33
17. Evans E, Buxbaum K (1981) Affinity of red blood cell membrane for particle surfaces measured by the extent of particle encapsulation. Biophys J 34:1–12
18. Armstrong JK, Meiselman HJ, Wenby RB, Fisher TC (2001) Modulation of red blood cell aggregation and blood viscosity by the covalent attachment of Pluronic copolymers. Biorheology 38:239–247
19. Bäumler H, Donath E, Krabi A, Knippel W, Budde A, Kiesewetter H (1996) Electrophoresis of human red blood cells and platelets. Evidence for depletion of dextran. Biorheology 33:333–351

20. Neu B, Meiselman HJ (2001) Sedimentation and electrophoretic mobility behavior of human red blood cells in various dextran solutions. Langmuir 17:7973–7975
21. Van Oss CJ, Arnold K, Coakley WT (1990) Depletion flocculation and depletion stabilization of erythrocytes. Cell Biophys 17:1–10
22. Bäumler H, Donath E (1987) Does dextran indeed significantly increase the surface potential of human red blood cells? Studia Biophysica 120:113–122
23. Neu B, Meiselman HJ (2002) Depletion-mediated red blood cell aggregation in polymer solutions. Biophys J 83:2482–2490
24. Fleer GJ, Cohen Stuart MA, Scheutjens JHMH, Cosgrove T, Vincent B (1993) Polymers at interfaces. London, Chapman & Hall, p 502
25. Asakura S, Oosawa F (1954) On interaction bewteen two bodies immersed in a solution of macromolecules. J Chem Phys 22:1255–1256
26. Jenkins P, Vincent B (1996) Depletion flocculation of nonaqueous dispersions containing binary mixtures of nonadsorbing polymers. Evidence for Nonequilibrium effects. Langmuir 12:3107–3113
27. Vincent B (1990) The calculation of depletion layer thickness as a function of bulk polymer concentration. Colloids and Surfaces 50:241–249
28. Vincent B, Edwards J, Emmett S, Jones A (1986) Depletion flocculation in dispersions of sterically-stabilised particles ("soft spheres"). Colloids and Surfaces 18:261–281
29. Feign RI, Napper DH (1980) Depletion stabilization and depletion flocculation. J Colloid Interface Sci 75:525–541
30. Brooks DE (1973) The effect of neutral polymers on the electrokinetic potential of cells and other charged particles. J Colloid and Interface Sci 43:700–713
31. Brooks DE (1988) Mechanism of red cell aggregation. In: Platt D (ed) Blood Cells, Rheology and Aging. Springer Verlag, Berlin, pp 158–162
32. Chien S (1975) Biophysical behavior of red cells in suspensions. In: Surgenor DM (ed) The Red Blood Cell. Academic Press. New York, pp 1031–1133
33. Chien S, Dellenback RJ, Usami S, Burton DA, Gustavson PF, Magazinovic V (1973) Blood volume, hemodynamic, and metabolic changes in hemorrhagic shock in normal and splenectomized dogs. Am J Physiol 225:866–879
34. Snabre P, Mills P (1985) Effect of Dextran Polymer on Glycalyx Structure and Cell Electrophoretic Mobility. Colloid Polym Sci 263:494–500
35. Lominadze D, Dean WL (2002) Involvement of fibrinogen specific binding in erythrocyte aggregation. Febs Letters 517:41–44
36. Janzen J, Brooks DE (1991) A critical reevaluation of the nonspecific adsorption of plasma proteins and dextrans to erythrocytes and the role of these in rouleaux formation. In Interfacial Phenomena in Biological Systems. M Bender, editor. Marcel Dekker, New York, 193–250
37. Cowell C, Liinon R, Vincent B (1978) Reversible Flocculation of Sterically-Stabilized Dispersions. Journal of the Chemical Society-Faraday Transactions I 74:337–347
38. Donath E, Kuzmin P, Krabi A, Voigt A (1993) Electrokinetics of Structured Interfaces with Polymer Depletion – a Theoretical-Study. Colloid Polym. Sci 271:930–939
39. Krabi A, Donath E (1994) Polymer Depletion Layers as Measured by Electrophoresis. Colloid Surf A-Physicochem Eng Asp 92:175–182
40. Brooks DE, Seaman GVF (1973) Effect of Neutral Polymers on Electrokinetic Potential of Cells and Other Charged-Particles. 1. Models for Zeta Potential Increase. Journal of Colloid and Interface Science 43:670–686
41. Brooks DE (1973). Effect of Neutral Polymers on Electrokinetic Potential of Cells and Other Charged-Particles. 2. Model for Effect of Adsorbed Polymer on Diffuse Double-Layer. Journal of Colloid and Interface Science 43:687–699
42. Donath E, Krabi A, Nirschl M, Shilov VM, Zharkikh MI, Vincent B (1997) Stokes friction coefficient of spherical particles in the presence of polymer depletion layers – Analytical and numerical calculations, comparison with experimental data. J Chem Soc–Faraday Trans 93:115–119
43. Donath E, Pratsch L, Bäumler H, Voigt A, Taeger M (1989) Macromolecule Depletion at Membranes. Studia Biophysica 130:117–122

References

44. Bohmer MR, Evers OA, Scheutjens J (1990) Weak Polyelectrolytes between 2 Surfaces – Adsorption and Stabilization. Macromolecules 23:2288–2301
45. Seaman GVF (1975) Electrokinetic behavior of red cells. In: Surgenor DM (ed) The Red Blood Cell. Academic Press, New York, pp 1135–1229
46. Lerche D (1984) Electrostatic fixed charge distribution in the RBC-glycocalyx and their influence upon the total free interaction energy. Biorheology 21:477–492
47. Jones A, Vincent B (1989) Depletion Flocculation in Dispersions of Sterically-Stabilized Particles. 2. Modifications to Theory and Further-Studies. Colloids and Surfaces 42:113–138
48. Fleer GJ, Scheutjens JHMH, Vincent B (1984) The stability of dispersions of hard spherical particles in the presence of nonadsorbing polymer. In Polymer adsorption and dispersion stability. E. D. Goddard andB. Vincent, editors. ACS. Washington DC, pp 245–263
49. Jenkins P, Snowden M (1996) Depletion flocculation in colloidal dispersions. Advances in Colloid and Interface Science 68:57–96
50. Neu B, Sowemimo S-Coker, Meiselman H (2003) Cell–Cell Affinity of Senescent Human Erythrocytes. Biophys J 85:75–84
51. Baskurt OK, Farley RA, Meiselman HJ (1997) Erythrocyte aggregation tendency and cellular properties in horse, human, and rat: a comparative study. Am J Physiol 273:H2604–2612
52. Chien S, Jan KM (1973) Ultrastructural basis of the mechanism of rouleaux formation. Microvasc Res 5:155–166
53. Nash GB, Wenby RB, Sowemimo-Coker SO, Meiselman HJ (1987) Influence of cellular properties on red cell aggregation. Clin. Hemorheol. Microcirc 7:93–108
54. Buxbaum K, Evans E, Brooks DE (1982) Quantitation of surface affinities of red blood cells in dextran solutions and plasma. Biochemistry 21:3235–3239
55. Meiselman HJ (1993) Red-blood-cell role in RBC aggregation. Clin. Hemorheol Microcirc 13:575–592
56. Sowemimo-Coker SO, Whittingstall P, Pietsch L, Bauersachs RM, Wenby RB, Meiselman HJ (1989) Effects of cellular factors on the aggregation behavior of human, rat and bovine erythrocytes. Clin. Hemorheol. Microcirc 9:723–737
57. Rampling MW, Whittingstall P (1986) A comparison of five methods for estimating red cell aggregation. Klinische Wochenschrift 64:1084–1088
58. Neu B, Armstrong JK, Fisher TC, Meiselman HJ (2001) Aggregation of human RBC in binary dextran-PEG polymer mixtures. Biorheology 38:53–68
59. Carter C, Fisher TC, Hamai H, Johnson CS, Meiselman HJ, Nash BG, Stuart J (1992) Hemorheological effects of a nonionic copolymer surfactant (poloxamer 188). Clin Hemorheol 12:109–120
60. Toth K, Wenby RB, Meiselman HJ (2000) Inhibition of polymer-induced red blood cell aggregation by poloxamer 188. Biorheology 37:301–312
61. Armstrong JK, Wenby RB, Meiselman HJ, Fisher TC (2004) The hydrodynamic radii of macromolecules and their effect on red blood cell aggregation. Biophys J 87:4259–4270
62. Armstrong JK, Meiselman HJ, Fisher TC (1997) Covalent binding of poly(ethylene glycol) (PEG) to the surface of red blood cells inhibits aggregation and reduces low shear blood viscosity. Am J Hematol 56:26–28
63. Toth K, Wenby R, Meiselman HJ (2000) Inhibition of polymer-induced red blood cell aggregationby poloxamer 188. Biorheology 37:301–312
64. Neu B, Armstrong JK, Fisher TC, Bäumler H, Meiselman HJ (2001) Electrophoretic mobility of human red blood cells coated with poly(ethylene)glycol. Biorheology 38:389–403
65. Neu B, Armstrong JK, Fisher TC, Meiselman HJ (2003) Surface characterization of poly(ethylene glycol) coated human red blood cells by particle electrophoresis. Biorheology 40:477–487
66. Armstrong JK, Meiselman HJ, Fisher TC (1999) Evidence against macromolecular "bridging" as the mechanism of red blood cell aggregation induced by nonionic polymers. Biorheology 36:433–437
67. Maeda N, Shiga T (1999) Inhibition and acceleration of erythrocyte aggregation induced by small macromolecules. Biochim Biophys Acta 843:128–136

68. Armstrong JK, Meiselman HJ, Wenby R, Fisher TC (2005) Modulation of red blood cell aggregation and blood viscosity by the covalent attachment of Pluronic copolymers. Biorheology 38:239–247
69. Alexandridis P, Hatton TA (2006) PEG-PPO-PEG block copolymer surfactants in aqueous solutions and at interfaces–thermodynamics, structures, dynamics and modeling. Colloids and Surfaces 96:1–46
70. Yalcin O, Aydin F, Ulker P, Uyuklu M, Gungor F, Armstrong JK, Meiselman HJ, Baskurt OK (2006) Effects of red blood cell aggregation on myocardial hematocrit gradient using two approaches to increase aggregation. Am J Physiol-Heart Circul Physiol 290:H765–H771
71. Chong-Martinez B, Buchanan TA, Wenby R, Meiselman HJ (2003) Decreased red blood cell aggregation subsequent to improved glycemic control in type 2 diabetes mellitus. Diabetic Med 20:301–306

Chapter 17
Hemoglobin Senses Body Temperature

Gerhard M. Artmann, Kay F. Zerlin* and Ilya Digel

Aachen University of Applied Sciences, Institute of Bioengineering,
artmann@fh-aachen.de;
* PTJ Jülich, Research Centre Jülich GmbH, Germany

Abstract The chapter addresses the question whether hemoglobin (protein) can "sense" body temperature. When human red blood cells (RBCs) were aspirated into 1.3 µm pipettes ($\Delta P = -2.3$ kPa), a transition from blocking the pipette below $T_c = 36.3 \pm 0.3\,°C$ to passing it above T_c occurred (passage transition). With a 1.1 µm pipette no passage was seen and RBC volume measurements were possible. With increasing temperature RBCs lost volume significantly faster below than above $T_c = 36.4 \pm 0.7\,°C$ (RBC volume transition). Colloid osmotic pressure (COP) measurements of RBCs in plasma ($25\,°C \leq T \leq 39.5\,°C$) showed a turning point at $T_c = 37.1 \pm 0.2\,°C$ above which the COP rapidly decreased (COP transition). In NMR T1 relaxation time measurements the T1 of RBCs in plasma changed from a linear ($r = 0.99$) increment of T1 below $T_c = 37 \pm 1\,°C$ at a rate of 0.023 s/K, into a parallel to the temperature axis above this point (RBC T1 transition). In conclusion: during micropipette aspiration, an amorphous gel forms in the spherical trail of the aspirated RBC, consisting of mostly hemoglobin and water. At T_c a fluidization of the gel occurs and non-covalent bonds (Van-der-Waals bonds) break down due to thermal energy enabling cell passage. The passage, the volume, the COP, and the RBC T1 transitions all happen at distinct T_c close to body temperature. We suggest a transition gel to liquid to be a common mechanism of these phenomena. T_c may mark the set point of a species' normal body temperature which might be inscribed in the primary structure of a species' hemoglobin and possibly in other proteins. The concepts of non-linearity and phase transitions in protein-water systems might bring novel exciting aspects into cell biology.

17.1 Instead of an Introduction

A book chapter is not a scientific paper and should leave space for extra thoughts, remarks and even episodes science sometimes goes through. One of those extras was an unlucky experiment, which we performed in 1994, at the University of San Diego, California, Dept. Bioengineering, being on the very first sabbatical leave. We were aspirating human red blood cells (RBCs) into very narrow micropipettes

(1.3 μm) where they were stuck. After having been blown out, RBCs showed a different shape than before; discocytes had turned into echinocytes (Bessis and Lessin 1970). Then, after a while, the echinocytic shape recovered spontaneously into discocytes (Figs. 17.1–17.4).

The basic working hypothesis of those experiments was that isotropic membrane tension (Evans et al. 1976) exerted to the cell membrane during aspiration induced a lipid translocation (i. e. 'lipid flow') from the inside to the outside of the lipid bilayer (Artmann et al. 1997). After releasing this tension by blowing the aspirated cell back into the suspending buffer, those 'extra lipid molecules' in the outer layer that had been translocated during pipetting remained trapped. As consequence, in the outer membrane those trapped extra lipid molecules caused an area excess as compared to the inner leaflet causing an outward bending of the lipid bilayer (spicules). This resulted in the echinocytic shape (Fig. 17.4). This shape change was transitory because of the activity of phospholipid translocases (Devaux 1992; Soupene and Kuypers 2006; Ikeda et al. 2006, Fischer 2004).

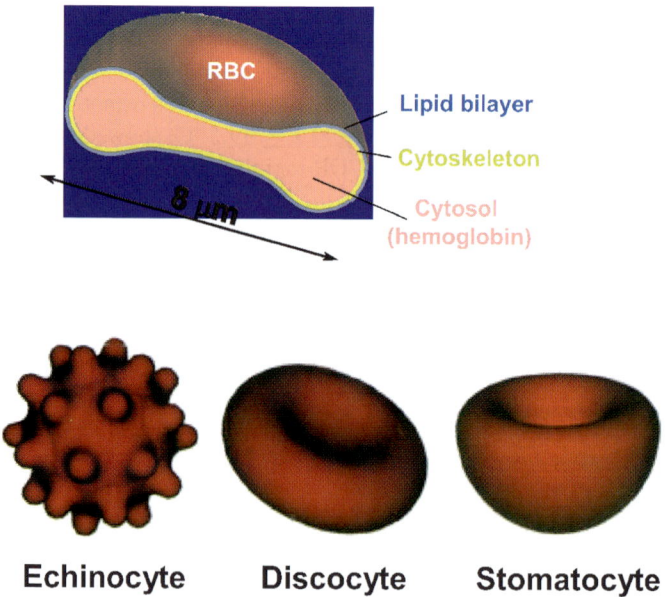

Fig. 17.1 Human RBCs. The regular RBC shape is discocytic (*middle*). However, there is a continuous shape change possible turning discocytes into exchinocytes and stomatocytes (Fig. 17.2), respectively. Echinocytes can be formed by adding foreign molecules to the outer lipid bilayer half of the cell. These can be drugs or lipids. Drugs must be negatively charged at physiological pH to insert into the outer leaflet. Lipids enter the outer leaflet as well because of their lipophilicity (Claessens et al. 2007). In both cases, an outward bending moment in the lipid bilayer causes echinocytosis and RBC membrane spicules. Stomatocytes are caused by positively charged drugs causing an inward bending due to insertion into the inner leaflet and/or due to drug interaction with transmembrane proteins (Chen et al. 2003; Deuticke 1968; Grebe and Zuckermann 1990; Artmann et al. 1996; Schmid Schoenbein et al. 1986 a and b)

17.1 Instead of an Introduction

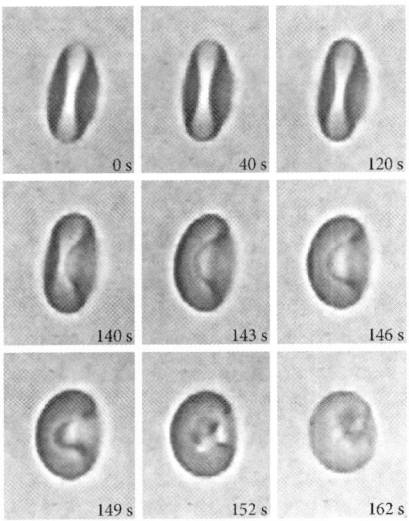

Fig. 17.2 Example for the action of the cationic drug chlorpromazine on the plasma membrane of a RBC. Human RBC treated with chlorpromazine (2.5 µM) (CPZ) at time zero (*inserts*) and room temperature (Artmann et al. 1996). CPZ gradually transformes a native RBCs (in upright position) into a sphero-stomatocyte, a process taking place within about 3 min. Chlorpromazine exhibits specific and unspecific effects in other cell types as well (Bastianetto et al. 2006; Hueck et al. 2000)

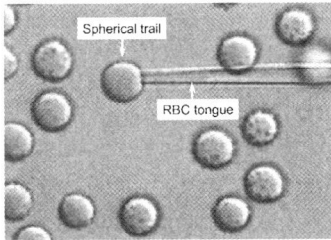

Fig. 17.3 Human RBCs. A RBC was aspirated into a micropipette at an inner diameter of 1.1 µm and at −2.3 kPa aspiration pressure. In the spherical trail hemoglobin and hemoglobin bound cell water assembled as a gel not permitting RBC passage through the pipette (Artmann et al. 1997, 1998; Cribier et al. 1993)

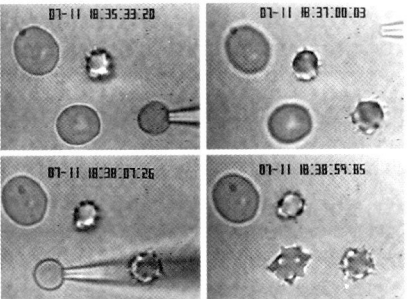

Fig. 17.4 Human RBCs in a micropipette experiment. One RBC at a time was aspirated into the pipette and kept there for 10 s. Afterwards it was blown out and the next RBC was aspirated. Clearly, the pipette induced echinocytic shape can be seen. After releasing the cell from the pipette its shape recovered spontaneously because of the activity of phospholipidtranslocases (Artmann et al. 1997; Devaux 1992)

Three classes of lipid translocases (Ikeda et al. 2006), P-type ATPases (Pierre and Xie 2006), ABC transporters (Wenzel et al. 2007), and scramblases (Sahu et al. 2007), are known to be involved in the regulation of lipid asymmetry, which is required for the mechanical stability of the membrane and for vesicular transport. In general, concerning many eukaryotic cells, local or global changes in lipid asymmetry are important for cell cycle progression, apoptosis, and platelet coagulation (Ikeda et al. 2006).

In our experiments, the translocase activity reversed the mechanically induced shape change in micropipette experiments within the order of minutes. Since the activities of those 'working' proteins depend on temperature (Artmann et al. 1997; Cribier et al. 1993), we wanted to measure the shape recovery time at 37 °C where enzymes should work most efficiently. It was after midnight, at about one o'clock, when we began the experiment. We pulled the cells into the pipette at 36 °C. They formed a nice spherical trail and a long tongue inside the pipette (Fig. 17.3). After blowing the cell out and waiting until the RBC shape had recovered we raised the buffer temperature to 37 °C and repeated the experiment with the same cell. To our great surprise the cell entered and – *passed* the pipette with only some seconds time delay. This caused us troubles because we needed a sufficient amount of cells being stuck in the pipette, changing their shape after blow-out and recovering at a better rate than at 36 °C. However, nothing like this happened. We tried other cells. Only ten out of over hundred did what we 'wanted' them to do – being stuck in the pipette. We took those cells; acquired the average recovery time and adjusted the temperature to 38 °C. To our greater surprise, the RBCs did not even visibly stuck in the pipette anymore but passed fast. At 39 °C, the pipette was no obstacle any longer for a red blood cell to pass. There was no chance of getting shape recovery data of red blood cells with that size of inner pipette diameter. We left the lab and saw daylight. Another day in beautiful California began with a marine layer covering the coastline in the morning and a clear blue sky from eleven o'clock on. Many days have passed since then and uncounted lab hours were spent trying to understand what happened to the red blood cell at 37 °C. We knew, by that time we had discovered an effect, a strange effect, which only was visible at certain conditions. The discovery was that a red blood cell when aspirated into a pipette of a diameter of 1.3 μm blocks the pipette at 36 °C and passes it at 37 °C. Thus, we discovered a transition from blockage to passage with a midpoint at 36.4 ± 0.3 °C as we found later (Artmann et al. 1998). However, what we did not know was that this effect would attract our attention for more than a decade.

17.2 Physiological Aspects of Thermoregulation in the Body

The body temperature and activity of animals depend on the temperature of their environment. Simply spoken, there exist two different methods nature uses to adjust a species' body temperature. Animals can be either poikilothermal or homeothermal

17.2 Physiological Aspects of Thermoregulation in the Body

(warm-blooded). However, there is no sharp line between the two animal groups. The activity of poikilothermal animals is directly affected by the environmental temperature (like for most reptiles). Homeothermal animals (like mammals) can better regulate their body temperature keeping it relatively constant over a broad range of environmental temperatures. The thermoregulation can be behavioural (warm up in the sun e. g.) and physiological.

The physiology of body temperature is an old topic of research, a good example for this is the report of Simpson and Galbraith published in 1905 (Simpson and Galbraith 1905). Up to the present, many aspects of the physiology of thermoregulation and body temperature have been published giving an excellent overview to this topic as well as many examples (Penzlin 1977; Whittow 1971).

A review of Ross and Christiano (Ross and Christiano 2006) deals with physiological aspects of skin and bones. Not surprisingly, skin and the formation of sweat are major elements of the physiological thermoregulation. Blood and the vascular system are further elements of thermoregulation. Blood executes many vital functions such as transport of gases, heat, regulation of water and electrolyte homeostasis (Schmidt et al. 2005). Thermoregulation however, is nowadays not only a topic of physiology.

Biochemists are studying its particular regulatory aspects like proteins and signalling pathways involved. Mozo described in 2005 (Mozo et al. 2005) the role of a specific protein group, called UCPs, in mammals and birds' thermoregulation. UCP1 is an uncoupling protein of the mitochondrial transporter family. It creates a proton leak at the mitochondrial membrane uncoupled to ATP synthesis. Because of this the electrochemical gradient is lowered, the respiratory chain increases its activity and, as a result, acid reserves are lowered. Simply spoken, "heat" is produced. Further examples were reviewed by Silva in 2006 (Silva 2006) who described in detail known thermogenic mechanisms and their hormonal regulation. According to this review, homeothermic species use several mechanisms to produce heat based on the regulation of the thyroid hormone (TH) and the sympathetic nervous system (Silva 2006). Body temperature and its regulation remains an important element of scientific interest especially in medical research as for example to combat fever, to understand enzyme activities or to reveal further important links between blood and thermoregulation, respectively.

Mechanisms of biological thermosensation received relatively little attention from physiologists for many years. Recent advances in thermal physiology research have disclosed temperature-sensitive ion channels belonging to the "transient receptor potential" (TRP) family in the peripheral sensory neurons and in the brain. Among them, the TRPV3 protein is one of the well-studied, whose role as a "molecular thermometer" has been recently proved using knock-out mice. A link between temperature slope and ion permeability of many TRP proteins has been established. Nevertheless, the particular mechanism of temperature-induced signalling remains unknown and stays out of the scope of most studies published so far (Oberwinkler 2007; Tominaga 2004; Nomoto et al. 2004).

17.3 Red Blood Cells

Knowing the complexity of the field we were about to move in that was one thing. The strange effect of RBC passage through narrow pipettes, which we had observed, was another one. With a background as biophysicists as well as bioengineers and having looked at red cells for now more then 25 years, we perceived the investigation of the details of this transition as our due.

Blood is a suspension with about 45% content of solid components consisting of three major cell types; red blood cells (RBC), thrombocytes and leukocytes. The quantitatively biggest compartment is RBCs with about 4.8 to 5.3 million cells per µL blood, which is ten times more than the number of thrombocytes, and thousand times more than the one of leukocytes. Normal human RBCs have a biconcave shape (Figs. 17.1, 17.2) with the diameter of about 7.5 µm and maximum thickness of about 2 µm. RBCs have a lipid bilayer membrane. At the intracellular face of the membrane, a net-like two-dimensional protein structure, the cytoskeleton, is linked to the membrane and adds shear elasticity and mechanical strength to the RBC's properties (Artmann 1995; Evans et al. 1976; Evans 1983; Hochmuth et al. 1979; Hochmuth and Waugh 1987; Hochmuth 1993). The cytoskeleton mostly made of spektrin enables RBCs keeping their structure and integrity in the circulation. The RBC cytoplasm contains a very high concentration of dissolved hemoglobin (330 mg/ml). RBC's have to pass the smallest capillaries having a diameter of about 3 to 8 µm where the gas exchange takes place (Chien 1981). Consequently, RBCs during their life span, have to go several million times through an opening smaller than their own diameter. This is why RBCs need to be extremely flexible (Artmann 1995; Chien et al. 1978; Chien 1981; Fischer 1989 and 2004; Shi et al. 1998; Temiz et al. 2000).

17.4 Temperature Transition in RBC Passage Through Micropipettes

The micropipette aspiration technique was used to study biophysical properties of cells like membrane shear modulus and relaxation time (Artmann et al. 1997; Chien et al. 1978; Engström and Meiselman 1997; Evans 1983; Hochmuth 1993; Mohandas and Chasis 1993). A sudden change in the passage behaviour of red blood cells at a critical temperature T_c when cells entered the micropipettes was repeatedly reported in our previous works (Artmann et al. 1998; Kelemen et al. 2001).

Red blood cells were aspirated with a pressure of -2.3 kPa into micropipettes having diameters around 1.3 µm (Fig. 17.3). Experiments were performed in standard phosphate buffer systems in a microscope chamber allowing adjustment of suspension temperature. Beginning at 25 °C, the passage behaviour of individual RBCs was systematically investigated (about 50 cells per temperature step). Below a critical temperature, $T_c = 36.4 \pm 0.3$ °C all aspirated RBCs blocked the pipette's entrance. At T_c the cells began to pass and at 37 °C all cells passed. At higher temperatures, passage became so fast that the RBC hardly stopped when mastering the passage through a pipette, which was about six times smaller than its diameter. Thus a temperature transition was born of red blood cells taking place at human body temperature (from now on termed 'micropipette passage transition') (Fig. 17.6).

Discussing those observations with colleagues, and this is the advantage of writing a book chapter instead of a paper, we received different responses. Some said "cannot be" and mentioned an "almost inextensible RBC plasma membrane (references were cited), which would not allow...", some avowed frankly "I don't believe you" and one highly regarded scientist said carefully "how interesting" (Chien et al. 1978). The most interesting smile, however, ran trough someone's face when we told him the finding. His words were "never heard about it" (Kassab 2004). By that time, we were no students anymore but when "the father of modern biomechanics" mumbled "never heard about it" then you should repeat your experiment. To make sure that not only our talent and motivation made the observation possible, we designed the practical course "Temperature Transition of Red Blood Cells in Micropipette Experiments" for our Bachelor students in Cell Biophysics. Only some students failed to prove the existence of the micropipette passage transition.

17.5 The Molecular Mechanism of the Micropipette Passage Transition

Why do RBCs flow freely through a 1.3 μm micropipette above a critical temperature and why do they block the same pipette below? This question was the initial question of a whole series of micropipette experiments. Firstly, we performed optical density measurements using an interference light filter at a wavelength of 415 nm, where hemoglobin (Hb) has its maximum of absorbance (Soiret band). This enabled us visualizing microscopically the distribution of hemoglobin within an individual aspirated RBC (Fig. 17.5). At 36 °C (Fig. 17.5 above) the spherical RBC trail was large in diameter and the tongue short. Only one degree higher we observed the opposite. Waiting just 10 s more after Fig. 17.5 below was taken we would have observed that the RBC had completely passed the pipette, without membrane fracture. The spherical trail of an aspirated RBC below T_c contained as much as 45–50 g/dL Hb (physiological value 33 g/dL). A side effect of this observation was that at 36 °C the RBC tongue was empty of Hb whereas at 37 °C it was slightly filled suggesting an ongoing Hb redistribution process preceding the passage (Fig. 17.5).

Thus, from the beginning we were suspicious about the role of hemoglobin in this transition. The key observation, however, is shown in Fig. 17.6. Whole RBCs showed a very sharp "step function"-like transition from blocking the pipette to freely passing the same pipette at $T_c = 36.4 \pm 0.3$ °C at which 50% of RBC passed and 50% did not pass (Fig. 17.6). Just to consider ALL options, there was a chance, that cell passage was enabled by temperature dependent RBC *membrane* properties. To study this, RBC "ghosts", cellular membranes after cytosol removal were used (Artmann et al. 1998; Friederichs et al. 1992). Using the same settings, a temperature transition was observed, however, at about 28.3 ± 2.3 °C. This temperature was clearly far away from $T_c = 36.4$ °C observed with intact RBCs. This particular tran-

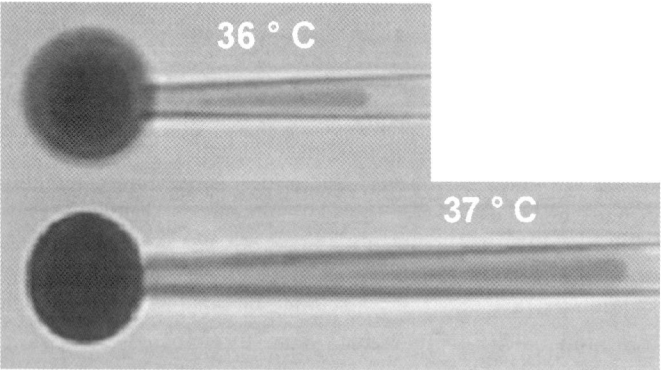

Fig. 17.5 above: Aspirated RBC at 36 °C (1.3μ inner pipette diameter), −2.3 kPa aspiration pressure (spherical trail is a little blurred). Clearly visible is the tongue inside the pipette. **below:** The same RBC as above. However, the spherical trail is smaller in diameter and tongue length has almost doubled

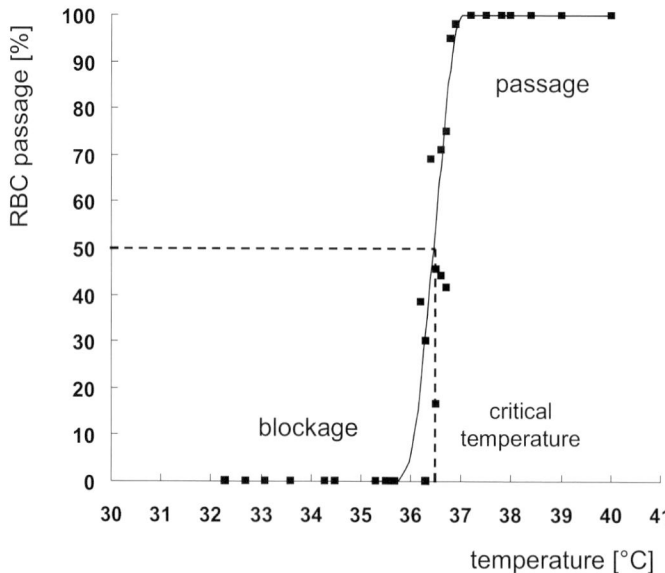

Fig. 17.6 Result of a micropipette aspiration experiment. The red blood cell passage curve was obtained with human RBCs. The figure shows the percentage of passing RBCs through a 1.3 mm micropipette at −2.3 kPa aspiration pressure as a function of temperature. The critical temperature T_c was 36.4 ± 0.3 °C (50% passage). RBCs passed above and blocked below this temperature. Above 37 °C, all RBC passed the micropipette and the speed of RBC passages increased drastically with temperature (Artmann et al. 1998)

sition at 28.3 °C we interpreted as a sudden elastomeric expansion (phase transition) of the extremely strained cytoskeletal spectrin network (Discher et al. 1994; Pollack 2001 a and b). Some linear scientists were sceptical about that idea (Lee et al. 1999). However, we did not pursue our hypothesis further since we were already on the Hb track. This is why the mechanically induced cytoskeletal phase transition remains a suggestion until today worthy to put some more theoretical and experimental effort in. Moreover, non-linear elastomeric phase transitions of the RBC cytoskeleton must be modelled theoretically and proven experimentally.

17.6 Hemoglobin Viscosity Transition

After having observed that in the spherical trail of an aspirated RBC highly concentrated Hb assembled forming a protein plug below T_c the next experiments were logical to do, namely, temperature dependent Hb solution viscosity measurements. Viscosity data of Hb solutions were obtained in the early eighties of last century. Veldkamp and Votano reported (Veldkamp and Votano 1980) the temperature dependency of macromolecular interactions in dilute and concentrated Hb solutions. These data showed a transition in the conformation of hemoglobin at approximately 22 °C, independently of the Hb concentration. At temperatures above 36 °C, the diffusion coefficient remained virtually independent of temperature. Additionally, they reported on an increase of the particle radius beginning at about 22 °C. In other studies on viscosity of Hb solution performed earlier, our good friend Herb Meiselman did not find anything surprising although he went up to 45 g/dL (Muller et al. 1992).

Sometimes it is good to hold on some strong finding since times can be hard and doubts may overwhelm you. We were lucky enough throughout the scientific story told here, to have found such a strong, undeniable, unforgettable and beautiful cellular effect taking place at a temperature that is unique (Fig. 17.6). This is why we always knew that it was worth looking in the strangest corners to find something at T_c. As for example, if you plan on viscosity studies with Hb solutions – for what reason should you mix unphysiological concentrations of 50 g/dl Hb? The ones who saw how much stuff even 33 g Hb was and what little amount instead 100 mL water appeared to be – what reason would they have to mix these components making an super unphysiological 'marmalade' to perform viscosity studies with? We had a reason – this made the difference and the reason was based on a much earlier experiment that went almost wrong.

Micropipette aspiration experiments with RBCs allowed us assuming a rapidly changing viscosity of the cytosol of RBCs (Artmann et al. 1998; Kelemen et al. 2001). Kelemen et al. discovered in 2001 a clear, sudden viscosity transition of highly concentrated Hb solutions at $T_c = 36 \pm 1$ °C. This viscosity drop was seen only at and above 45 g/dL concentration where two Hb molecules are separated by only ten times the diameter of a water molecule (Kelemen et al. 2001). The transition was NOT seen at physiological concentrations (Fig. 17.7). As a mechanism, it was proposed that the high-to-low viscosity transition at T_c was mediated by a partial release of hemoglobin-bound water. Increasing temperature may release bound protein water to an extent that at T_c the Hb suspension viscosity breaks down. The conclusions from these experiments were that the temperature transition of viscosity was 1) not limited to intra cellular Hb, 2) visible only at high Hb-concentrations that were far out of the physiological range, and, last but not least, 3) it was hemoglobin that caused the transition and NOT the RBC membrane.

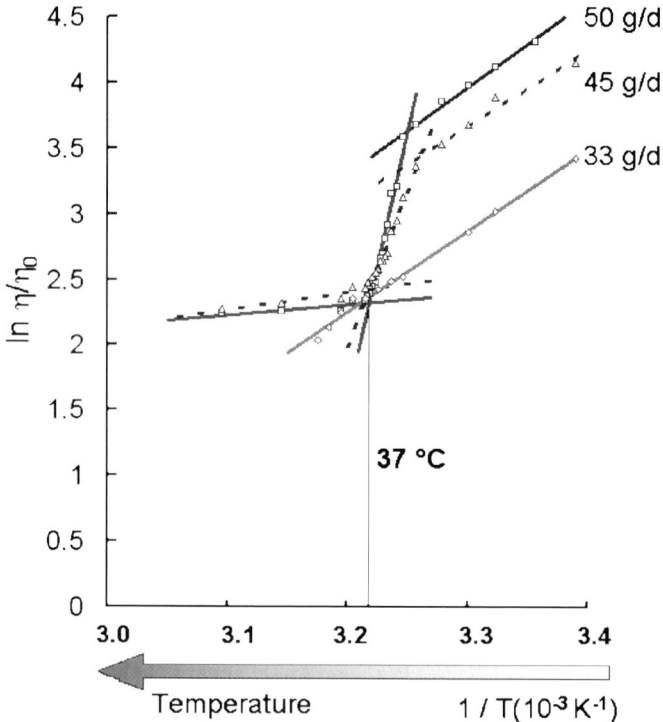

Fig. 17.7 Arrhenius plot of the relative viscosity (vs. phosphate buffer) of highly concentrated hemoglobin solutions in phosphate buffer (Kelemen et al. 2001)

17.7 Circular Dichroism Transition in Diluted Hb Solutions

When there was a macroscopic effect, there must be a molecular reason. We assumed very small structural changes of the Hb molecule as reason for the molecular effect. They have to be small, because Hb is one of the best-known proteins ever! If there was a big and obvious effect, it would have been found already! Here was some lesson to learn for the mainstreamers among us. It may seem that red blood cells are out and stem cells are in; it may look as if Hb is out and G proteins are in; we may feel the need to follow keywords down the road and sometimes we MUST follow (because of funding). However, once we have the freedom, why not reconsider a potato field that belongs to a very nice farmer named Max Perutz (we bow humbly before the Creation!) who said 'We have already harvested. If you still can find a potato, it is yours'. Thus, having in mind our discovery (Fig. 17.5) we went on (Blow 2002).

A common method for the analysis of general protein structures is circular dichroism spectroscopy. This is a chiroptical measurement where the chirality of molecules is quantified. If polarised light is send through a solution containing chiral molecules, the plane of its polarization is changed. Light absorption is dependent

17.7 Circular Dichroism Transition in Diluted Hb Solutions

on the molecular extinction coefficient, ε. In optically active substances, it differs for right-handed (εR) and left-handed (εL) polarised light, respectively. The difference $\Delta\varepsilon$ (mdegr) defines θ_λ, the ellipticity, which is wavelength dependent. Equation (17.1) gives the definition of the ellipticity (Fasman 1996).

$$\tan\theta = \frac{\varepsilon_r - \varepsilon_l}{\varepsilon_r + \varepsilon_l}. \qquad (17.1)$$

By combining with Beer's law the ellipticity can be written as:

$$\tan\theta_\lambda = \ln 10 \cdot \frac{180°}{4\pi} \cdot (\varepsilon_L - \varepsilon_R) \cdot c \cdot d, \qquad (17.2)$$

where c is the concentration of the optically active compound and d is a length of the optical path.

In a CD-spectrum obtained by a wavelength scan, information on the content of secondary protein structures (helix, parallel and antiparallel beta-sheet, beta-turn and random coil) is contained. Using proper deconvolution software the percentage of secondary structures in a sample can be estimated. At a wavelength of 222 nm, the unfolding of α-helices is most visible in the spectrum (Bohm et al. 1992).

In 2004, Artmann et al. reported a circular dichroism spectroscopy study on structural changes during thermal denaturation of Hb between 25 °C and 60 °C (Artmann et al. 2004). Wavelength scans from 190 nm to 260 nm were performed at temperature steps of one degree and ellipticity data were taken at 222 nm. Results demonstrated a non-linear dependency of the ellipticity vs. temperature. As expected, the α-helix content decreased with increasing temperature indicating thermal protein denaturation (Kinderlerer et al. 1970, 1973). However, as opposed to Kindelerers findings, the plot of the relative ellipticity, F_{obs}, at 222 nm against temperature showed a S-shaped curve showing a change point at 37 °C. Relative ellipticity was calculated $F_{obs} = [E_{obs}(T) - E60]/[E25 - E60]$, where $E_{obs}(T)$ was the ellipticity at 222nm at temperature T, $E60$ the ellipticity at 60 °C, and $E25$ the ellipticity at 25 °C. In other words, at 37 °C the 'speed' (change of ellipticity/temperature change) of denaturation was highest (Fig. 17.8). The obtained critical temperature was $T_c = 37.2 \pm 0.6$ °C (Artmann et al. 2004). Further experiments showed no influence of pH of the solution and oxygenation/deoxygenation of the Hb sample on the effect. HbS of sickle cell anemia did show the same effect as well (Fig. 17.8). Thus, the s-shaped denaturing curve of Hb around body temperature was highly conserved. It was important to note that this denaturing seen in CD spectra was reversible at least up to 39 °C. Significant result of this work was that the structural transition temperature was remarkably close to the critical temperature seen in the micropipette aspiration experiments with human RBCs (Fig. 17.6).

Circular dichroism data were encouraging, because they gave us hints on the existence of a structural transition at a critical temperature close to body temperature. However, the used Hb solutions were highly diluted and therefore far away from Hb-concentrations seen in the red blood cell. As we learned in the paragraphs on Hb suspension viscosity (Fig. 17.7) and on the micropipette passage transition, very

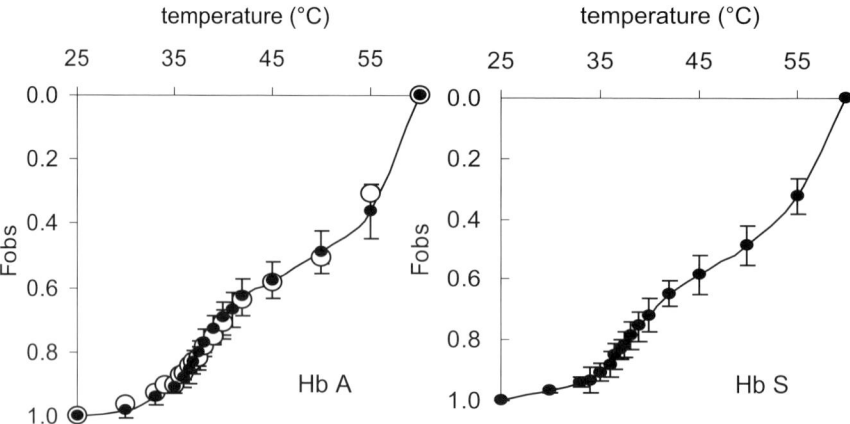

Fig. 17.8 CD spectroscopy; fractional changes (F_{obs}) in the ellipticity at 222 nm with temperature for human hemoglobin A (*left*) and hemoglobin S (*right*, Hb from sickle cell anemia). Filled circles represent the oxygenated state of the hemoglobins, open circles the deoxygenated state. The error bars represent the standard deviation of the respective fractional changes. The deoxygenated state of hemoglobin S was not studied because of the molecular precipitation occurring in the deoxygenated state (Artmann et al. 2004)

high Hb concentrations were necessary to appreciate a strong transition effect. Thus, we returned to experiments with whole red blood cells and even to experiments with RBCs suspended in their own (autologous) plasma.

17.8 A RBC Volume Transition Revealed with Micropipette Studies

When an RBC enters a narrow micropipette, the cytosol is sheared and an isotropic RBC membrane tension builds up until steady state of forces is reached (Evans 1983; Hochmuth et al. 1979; Chien et al. 1978). Cellular bulk water is squeezed out across the membrane and the RBC lost volume. As good channel candidates for water release we considered band 3 and band 4.5 proteins, channel-forming integral proteins (CHIP), and aquaporin 1 (AQP1) (Fairbanks et al. 1971; Farinas et al. 1993; Jay 1996; Poschl et al. 2003; Walz et al. 1997; Yokoyama et al. 1978). Since a short term increase of the aspiration pressure up to -12 kPa did not show any further RBC volume decrement, the Hb molecules were assumed to be in closest possible contact to each other leaving only traces of bulk water inside the RBC, if at all. The remaining cell water must therefore be highly bound (Dencher et al. 2000) to mostly Hb. Thus, a Hb-water-gel had formed in the RBC's spherical trail (Fig. 17.3) which did not allow RBC passages through the pipette below T_c (Artmann et al. 1998).

Single RBC volume determinations with micropipettes were carried out with established methods (Engström and Meiselman 1996, 1997). The inner pipette diam-

17.8 A RBC Volume Transition Revealed with Micropipette Studies

eter used when we discovered the micropipette temperature transition (Fig. 17.6) of 1.3 μm was changed to 1.1 μm representing a diameter where no passage of RBCs was possible. Using such a pipette, a RBC was pulled into the pipette at 34.5 °C. The final aspiration pressure $\Delta P = -2.3$ kPa was built up slowly until a steady state was reached i.e. when the RBC tongue length no longer showed any visible changes (Fig. 17.3). Thus, microscopic images of aspirated RBCs were acquired at steady state conditions in the absence of cytosolic shear stress. From these images, geometrical parameters were determined and the RBC volume was calculated (Artmann et al. 2007).

During RBC aspiration tongue formation was observed to occur in two phases, 1) a fast initial entrance which typically lasted less then a second and 2) a subsequent slow creeping phase during a couple of seconds where the tongue length changed less then 10%. A steady state of tongue length was considered to be reached, when an aspiration pressure of −12 kPa was applied for a short time and no further tongue length increment was observed. Figure 17.9 shows the RBC volume vs. temperature curve as obtained from 12 single RBCs. The average RBC volume data showed a distinct and significant (Dikta et al. 2006) change in slope at the critical temperature $T_c = 36.4 \pm 0.7$ ($N = 12$). With increased temperature, the RBC volume below T_c decreased at a rate of 7.8 fL/K. Beginning at T_c and above, the rate decreased by only 3 fL/K (Fig. 17.9). The slope between 34.5 °C and T_c and the one between T_c and 39.5 °C were significantly negative. Additionally, the slope between 34.5 °C and T_c was significantly more negative than the slope between T_c and 39.5 °C. Calculation of the Hb radii corresponded to a water layer thinning rate per one Hb molecule of 0.084 nm/K below T_c and of 0.052 nm/K above, respectively.

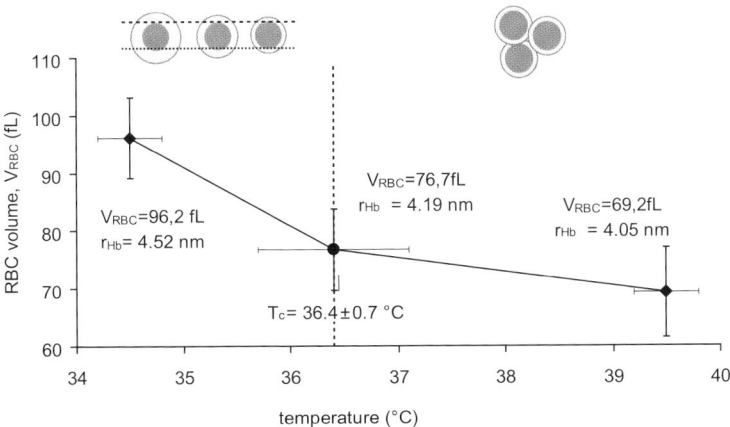

Fig. 17.9 RBC volume vs. temperature in micropipette studies. RBC volume decreased with increasing temperature but at two significantly different slopes below and above $T_c = 36.4 \pm 0.7$ °C, respectively. Below T_c RBCs lost volume at a rate of 7.8 fL^3/K, and above at 3 fL/K. The *inserts* show schematic hemoglobin molecules; *dark*: hemoglobin, *light*: bound water. On the *left*: the water shell thins out with rising temperature but Hb does not expand due to hydrostatic pressure inside the RBC trail. *Right*: Hb aggregates develop

When globular proteins denature, their diameter increases with temperature. Light scattering data (Digel et al. 2006; Zerlin et al. 2007) as well as preliminary X-ray diffraction studies (Büldt et al. 2007d) indeed, showed enhancing hydrodynamic radii and increasing lattice constants with increased temperatures until T_c was reached (in preliminary X-ray experiments on hemoglobin lattice constants, it was interesting that at T_c Hb crystals disintegrated (Büldt et al. 2007). However, Hb molecules within the spherical trail of an aspirated RBC (Fig. 17.5) are subjected to an appreciably high hydrostatic pressure and Hb molecules are therefore mechanically constrained from expanding. That is why their volume vs. temperature course is significantly different from freely suspended Hb molecules. In other words, aspiration compresses the Hb molecules mechanically due to the two-dimensional, isotropic elastic area tension in the red cell membrane during aspiration (Kharakoz and Sarvazyan 1993). This is followed by an enhanced cell internal hydrostatic pressure acting against Hb's thermal unfolding (Artmann et al. 1997; Evans et al. 1976; Staat 2007). Secondly, the Hb molecule continuously "melted off" bound water and as a result the Hb's diameter lowers rapidly. This is what causes RBC volume loss below T_c (the term "melting" used here is just a more descriptive word for a shift of the equilibrium between Hb-bound water and bulk water toward bulk water).

At T_c the Hb water shell was thinned out in a way that the remaining shell thickness gave rise to destabilize the Hb-water complex, leading to an overall entropy change followed by the on-set of Hb aggregation (Digel et al. 2006; Zerlin et al. 2007). Above T_c, Hb aggregation determined the RBC volume loss observed in micropipette experiments according to vant Hoff's law (Artmann et al. 2007) (Fig. 17.9).

17.9 Micropipette Passage Transition in D_2O Buffer

However, this does not explain why, at T_c, RBCs did set on passing almost in a step function (Artmann et al. 1998; Kelemen et al. 2001) (Fig. 17.6). It was evident that a change must take place at T_c which must be linked to the Hb structure and to Hb-bound water (Artmann et al. 2004; Digel et al. 2006). Micropipette experiments between 34.5 °C and 39.5 °C showed that the RBC volume decreased with increasing temperature but at two significantly different slopes below and above $T_c = 36.4 \pm 0.7$ °C, respectively. T_c was only 0.1 degree higher as observed earlier. With a set of small formulae developed for estimating the Hb molecular volume inside an RBC during aspiration, we determined Hb molecule radii, rHb, (Fig. 17.9). Results on rHb fitted to previously obtained light scattering data (Digel et al. 2006; Zerlin et al. 2007). In terms of developing bioengineering, this approach may have marked a step forward in terms of estimating the Hb molecule radius from micropipette experiments under hydrostatic pressure (Dadarlat and Post 2006; Kharakoz and Sarvazyan 1993; Staat 2007). The major findings were that RBCs lost volume with increasing temperature significantly more readily below than above

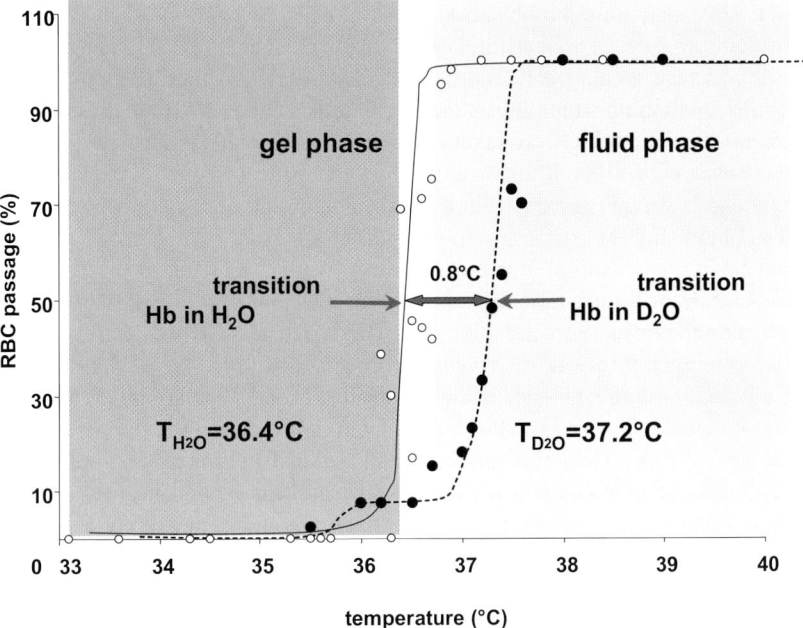

Fig. 17.10 Micropipette passage experiments (see Fig. 17.6). The red blood cell passage curves were obtained with human RBCs. *Left curve*: RBCs were suspended in a H₂O based buffer (*open circles*), *Right curve*: RBCs were suspended in a D₂O based buffer (*filled circles*). The T_c measured with H₂O based buffer differed from the one with D₂O by 0.8 °C. This indicated that stronger molecular bonds exerted by deuterium replacing hydrogen caused a shift in the critical temperature by plus 0.8 °C. *Thus, the passage phenomenon was related to protein bound water*

a critical temperature, T_c. We concluded that at T_c, a structural change must have occurred in Hb, handling bound and free water in two distinctly different ways below and above T_c, respectively (Kelemen et al. 2001). In micropipette experiments performed with a heavy water buffer (Stadler et al. 2007; Büldt 2007) and human RBCs, the temperature transition was observed at a 0.8 degree higher temperature than with a regular water buffer (Fig. 17.10). We attributed this shift of T_c to stronger hydrogen bonds and higher hydration shell stability, inhibiting the passage transition at body temperature. Because of stronger bonds a tiny higher amount of thermal energy is needed to enable the transition. It appears at a 0.8 °C higher temperature (Cioni and Strambini 2002; Niimura et al. 2004; Büldt et al. 2007). Thus, water has to be the key to our data interpretation.

17.10 NMR T1 Relaxation Time Transition of RBCs in Autologous Plasma

Performing red blood cell and whole blood experiments as presented in the next paragraphs may appear to indicate a return to an old-fashioned approach to cell biology. However, some biophysical effects in cells can only be seen when cells and proteins are kept in their natural environment (Eisenberg 2003; Zheng and Pollack 2003; Lucas et al. 1991; Todd III and Mollitt 1994). Furthermore, if water and hydrogen binding strength was the key to look at, then NMR studies on spin – lattices interactions with the relaxation time T1 as parameter was the method of choice (Finnie et al. 1986; Zefirova et al. 1991). T1 reflects the magnetic interaction strength of a hydrogen nucleus with its environment (lattice) and the entropy of the sample (Li et al. 1996).

Two NMR T1 relaxation time measurements were carried out in parallel 1) with blood plasma containing 83% RBCs (RBC-in-plasma-sample) and 2) with blood plasma alone (plasma-only-sample). With both samples, a two-phasic behavior of T1 vs. temperature was observed, showing a kink at a $T_c = 37 \pm 1\,°C$. The RBC-in-plasma-sample showed a change from a steady, linear ($r = 0.99$) increment of T1 below T_c at a rate of 0.023 s/K into a parallel to the temperature axis above (Fig. 17.11A). The temperature dependency of T1 of the plasma-only-sample showed converse characteristics, thus, no temperature dependency below 37 °C, a turning point at 37 °C and followed by a decrease with temperature at a modest slope of 3.8×10^{-3} above (Fig. 17.11B). The T1 of the plasma-only-sample were three to five times higher as compared to the RBC-in-plasma-sample.

The cytosol of an individual RBC comprises three major phases: Hb, Hb-bound water and osmotically active free (bulk) water. Using whole blood at high hematocrit ensured that the sample contained as much Hb in its natural environment as possible. At high sample hematocrit only small amounts of free/confined water outside RBCs in the suspending blood plasma remained. At a hematocrit of 83% chosen in NMR T1 experiments and with only blood plasma suspending the RBCs, the blood sample

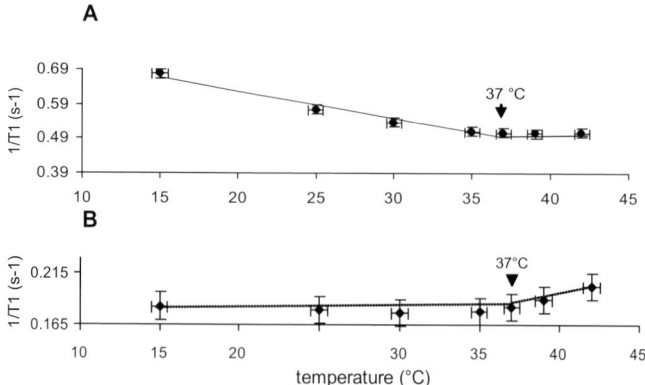

Fig. 17.11 A: NMR T1 relaxation time measurements for blood plasma containing 83% RBCs (RBC-in-plasma-sample). The T1 relaxation time increased with temperature at a rate of about 0.023 s/K below 37 °C. Above 37 °C, it remained constant. **B**: NMR T1 relaxation time measurements for blood plasma alone (plasma-only-sample). The T1 change over temperature in the plasma sample showed converse characteristics, no change below 37 °C and an increase above 37 °C (note the different scales 1/T1 used in the two graphs)

17.10 NMR T1 Relaxation Time Transition of RBCs in Autologous Plasma

consisted on average of a total of 26.4 g/dl Hb. Thus, much of the T1 signal must be attributed to hydrogen in the Hb molecules as well as to Hb-bound water.

We observed a non-linearity in the T1 temperature dependency (Fig. 17.11A). Below T_c, and with increasing temperature, T1 rose linearly which reflected an enhancing sample entropy (Cameron et al. 1988; Zefirova et al. 1991). A new development was observed at $T_c = 37\,°C$ and above. T1 did not change any longer with temperature but remained constant. Thus at T_c, the sample entropy had changed from a continuous increase below to a constant T1 above T_c reflecting no entropy change with temperature above T_c. A sample entropy change should be accompanied by a change in the specific heat at constant pressure suggesting as principle mechanism of the transition a phase transition of second order (Goldenfeld 1992). However, we did not see such change in differential scanning calorimetry experiments (Digel et al. 2007; Michnik et al. 2005b).

How can the slope change in Fig. 17.11A be explained? The thermally expanding (mechanically unconstrained) Hb molecule destabilizes; the hydrodynamic radius and the molecular entropy increase continuously with temperature below T_c. Hb molecules loose bound water (Fig. 17.12) stabilizing Hb's structure until T_c was reached. At T_c Hb condenses (Fig. 17.8), LOSES entropy and Hb molecules aggregate (Fig. 17.12). Thus, there is less freedom of movements of Hb's hydrogen atoms

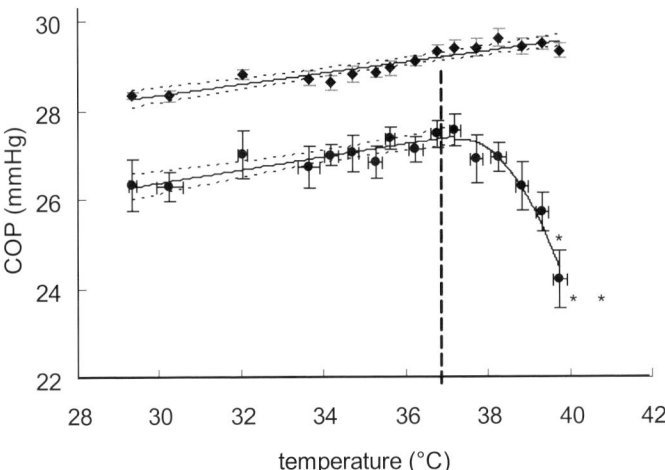

Fig. 17.12 For colloid osmotic pressure measurements samples were injected into the sample chamber of a WESCOR colloid osmometer which was placed into a heating unit. After temperature equilibrium was reached at 39 °C, the COP was taken. At the same temperature, a plasma sample of the same donor was measured. Then the unit temperature was decreased and a new temperature equilibrium was established. Measurements were repeated until the temperature range from 39 °C to 29 °C was investigated (16 donors). COP data are presented as average ±1 SEM. *Dotted lines* show the 95% confidence interval. The RBC-in-plasma-sample (HKT 77.6 ± 5.3%) showed a 2 mm Hg lower COD below T_c and a linear increment with temperature which went in parallel with the plasma-only-sample. Beginning at $T_c = 37\,°C$, the COP of the RBC-in-plasma sample dropped significantly

above T_c then below. On the other hand, bound water turns into bulk water, which is released to the plasma fraction outside the RBCs for colloid osmotic reasons (Figs. 17.9 and 17.12). This fraction gains entropy. The two entropy contributions may sum up in a way that total entropy remains constant as T1 does between 37 °C and 42 °C (Fig. 17.11A). These NMR experiments were carried out with RBCs in plasma in order to remain below the set point of irreversible Hb denaturation. However, there must be another change point of T1 vs. temperature at around 42 °C when irreversible denaturing of Hb sets on. Ongoing neutron scattering experiments seem to confirm this. In brief: Body temperature of many species correlates to a slight entropy phase transition of Hb (Fig. 17.16). In terms of enthalpy changes at T_c we may deal with a zero-sum game. We must NOT observe a change in the sample's specific heat although there was a change point in T1 at T_c (Artmann et al. 2007, Stadler et al. 2007, Digel et al. 2007; Michnik et al. 2005).

T1 of the plasma-only-sample did show a turning point at 37 °C as well (Fig. 17.11B). However, the slope of T1 vs. temperature showed an opposite tendency: below 37 °C, it was constant, and at T_c and above it decreased. Since blood plasma mostly consists of water and human serum albumin, T1 changes must be attributed to mostly human albumin. Thus, below body temperature the total plasma entropy remained unchanged, whereas above 37 °C it gradually decreased. The latter may be due to a beginning moderate thermal denaturation of human albumin from T_c on.

17.11 Colloid Osmotic Pressure Transition of RBC Suspended in Plasma

The colloid osmotic pressure (COP) of blood has been studied early (Roche et al. 1932). However, the philosophy of the set of experiments discussed here was rather different. We wanted to see whether there was a critical temperature visible in the COP of RBCs suspended in autologous plasma. The idea was as follows: if inside RBCs Hb molecules above a postulated critical temperature would set on aggregating upon heating, this would result in a decrease of the number of particles in the cytosol and, consequently, in a drop of the cytosolic COP. Thus, an imbalance between the cytosolic and the (outside) plasma COP would appear and the cytoplasmatic water would move outwards to re establish equilibrium of both COPs. Consequently the COP of the plasma compartment would drop setting on at the postulated T_c.

In contrast to micropipette experiments, COP measurements were gathered both at shear stress-free conditions within the RBC cytosol and at a cell internal hydrostatic pressure unmodified by external mechanical forces on RBC membranes. It was found that 1) the COP of the plasma-only-sample at 29 °C fits to the physiological COP, 2) it increased with temperature as predictable by van't Hoff's law, and 3) it did not show any transition (change in slope) at any temperature. However, the RBC-in-plasma-sample 1) exhibited a 2 mm Hg lower COP, and 2) showed a turning point at a critical temperature, $T_c = 37.1 \pm 0.2\,°C$ above which the COP decreased with temperature.

Now we apply our hypothesis from above for an explanation. As for the RBC-in-plasma-sample, two sample volume compartments must be considered, 1) the volume occupied by RBCs, on average $77.6 \pm 5.3\%$, and 2) the remaining plasma volume at 22.4%. The COP of this sample should be identical to the plasma-only-sample since RBCs are too big to contribute to the sample COP. Below 37 °C, the COP was instead 2 mm Hg smaller. This difference most likely resulted from plasma albumin molecules binding to RBC membranes which in turn would thin-out the remaining plasma. In order to understand why the COP of the RBC-in-plasma-samples dropped above T_c, we need to remember that the COP is proportional to the particle number in the RBC cytosol and not to its total protein content. When Hb molecules inside RBCs aggregate, the particle number and at the same time the intracellular COP decrease. The balance between the RBC internal and the RBC external COP is shifted. Consequently, cell water moves outwards, diluting the outside plasma until the two COPs are rebalanced. Due to this extra water derived form the RBCs cytosol above T_c the COP of whole RBC-in-plasma-sample decreases.

17.12 The Temperature Transition Effect so Far

In a discussion with Prof. Shu Chien, our good friend, he asked "Why is the transition that sharp in pipette and viscosity experiments and why not in all other experiments in your studies"? – We may have come closer to answer this question.

In many experiments on the temperature dependency of Hb properties using light scattering, CD spectroscopy (Artmann et al. 2004; Digel et al. 2006; Zerlin et al. 2007), micropipette volume calculations (Fig. 17.9) or Hb viscosity measurements (Kelemen et al. 2001) a transition temperature, T_c, was found. However, the transition was never as sharp as in the micropipette passage (Fig. 17.6) as well as in viscosity experiments of highly concentrated Hb solutions (Kelemen et al. 2001). This indicates the possibility of an unknown mechanism making that particular transition significantly sharper. We suggest that T_c indicated the temperature at which a phase transition of second order occurred (Goldenfeld 1992; Heller and Hofer 1975; Landau 2007). The transition would take place inside the spherical trail of the aspirated RBC where the Hb concentration was between 45 g/dL and 50 g/dL (Kelemen et al. 2001). Below T_c, a disordered 'amorphous polymer' was formed consisting of mostly Hb plus bound water (gel-like). Below T_c, sample enthalpy, entropy and volume are continuous functions of temperature. Near the transition temperature, T_c, the Hb-gel would "soften" and set on flowing under mechanical cytosolic shearing exerted when an RBC enters the micropipette (Fig. 17.3). Above T_c the spherical trail's cytosol (Fig. 17.9) would exhibit properties approaching those of an ordinary fluid (Artmann et al. 1998; Kelemen et al. 2001), although still more viscous than low-molecular weight liquids. Thus, the transition might constitute a gel transition known to occur in proteins at low temperatures (Vitkup et al. 2000). There was a slope change in T1 vs. temperature from a steady T1 increase below T_c to a constant T1 above supporting the idea of a glass-like transition. The turning point at T_c was not sharp, because we must consider that the RBC-in-plasma-sample was not ideally consisting of only one kind of protein where next neighbours were of the same kind. Other proteins like RBC membrane proteins as well as even lipids were in close vicinity to hemoglobin which affect an individual Hb's T_c depending on its individual environment (spin lattice interaction) (Finnie et al. 1986).

In summary, nature might have arranged a plot like this: below T_c both Hb and water increase their entropy and therefore T1 rises. At T_c a change point in T1 occurs (glass-like transition) so that T1 turns parallel to the temperature axis. From T_c on, the entropy changes for water alone and for Hb alone, respectively, develop into two opposite directions keeping the *total* sample entropy constant with temperature – T1 remains in parallel to the T-axis. The sum enthalpy of the RBC-in-plasma-sample may not change, although entropy did. Physically this would be a rare but not impossible event. The parallelism of T1 to the temperature axis, however cannot be maintained. At the protein denaturing temperature 42.6 °C irreversibly protein denaturation (Schmidt et al. 2005) would rise the sample entropy again. Thus, between T_c and the protein denaturing temperature, a plateau in the Hb-water entropy vs. temperature (i.e. T1 relaxation time vs. temperature) had formed inscribing surprisingly well the temperature range of pyrexia in humans.

17.13 Strange coevals – *Ornithorhynchus anatinus* and *Tachyglossus aculeatus*

Because of the experimental evidence in human Hb one particular question emerged: Was it possible, that Hbs of species having body temperatures different from humans would show a transition as well, and if so, would the critical temperature correlate with the species' body temperature?

The answer is "yes and yes".

A study by Digel et al. (2006) focused on Hb of two exceptional species, the duck-billed platypus (*Ornithorhynchus anatinus*, body temperature $T = 31 - 33$ °C) (Fig. 17.13), and the echidna (*Tachyglossus aculeatus*, body temperature $T = 32 - 33$ °C). CD-experiments with Hb of both animals showed s-shaped thermal denaturation curves similar to those of human Hb, however with the critical temperature between $32 - 33$ °C which was close to their body temperature. CD-spectroscopy results were later confirmed by dynamic light scattering measurements (DLS). DLS temperature scans of both Hbs showed a strong increase of the hydrodynamic radius (Fig. 17.14) at a specific transition temperature, T_c, between 32 and 33 °C (Digel et al. 2006; Zerlin et al. 2007).

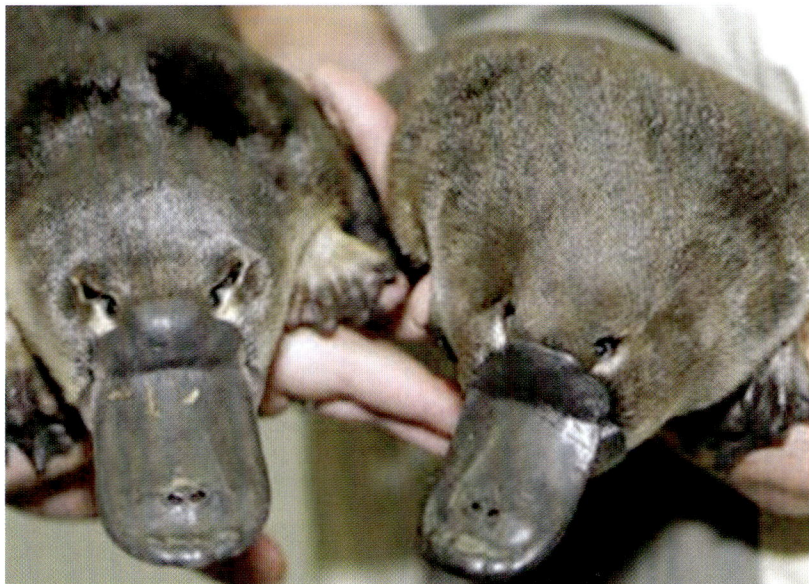

Fig. 17.13 The platypus is among nature's most unlikely animals. In fact, the first scientists to examine a specimen believed they were the victims of a hoax. The animal is best described as a hodgepodge of more familiar species: the duck (bill and webbed feet), beaver (tail), and otter (body and fur). Males are also poisonous. They have sharp stingers on the heels of their rear feet and can use them to deliver a strong toxic blow to any foe. Platypuses hunt underwater. Folds of skin cover their eyes and ears to prevent water from entering, and the nostrils close with a watertight seal. On land, the webbing on their feet retracts to expose individual nails and allow the creatures to run. The platypus is one of only two mammals (the echidna is the other) that lay eggs. A mother typically produces one or two eggs and keeps them warm by holding them between her body and her tail. The eggs hatch in about ten days, but platypus infants are the size of lima beans and totally helpless. Females nurse their young for three to four months until the babies can swim on their own (Photograph from http://dsc.discovery.com)

17.14 Hb Temperature Transition of Species with Body Temperatures Different from 37 °C

Body temperature of most homeothermal animals is in a range between 33 °C and 41 °C. Zerlin et al. (2007) reported a study with hemoglobins of fourteen species with different body temperatures (Table 17.1). At this point, we would like to make some remarks that never would appear in a paper. Logistically, this study was complex. It took almost a year officially obtaining platypus and echidna blood samples from Australia. Additionally, blood samples were from various sources and experiments were carried out in different laboratories 10,000 km away from each other. Furthermore, when there was a DLS instrument available in the one lab there was no CD instrument available and vice versa. And we were lucky enough to find both instruments in the same lab then the veterinarian of the closest zoo might not have been that compliant.

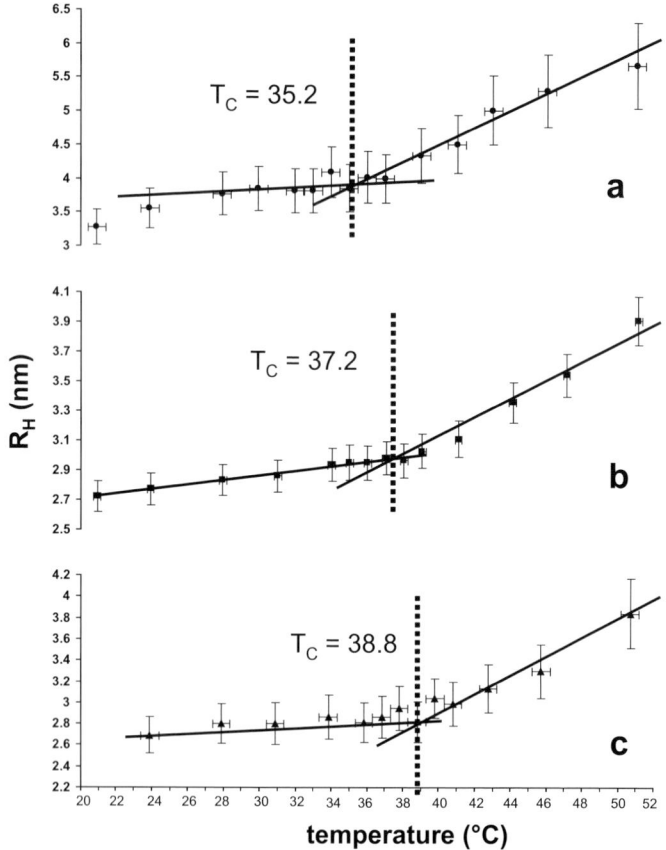

Fig. 17.14 Hydrodynamic Hb molecule radius (DLS data) vs. temperature. **a** *Ornithorhynchus anatinus* ($T_B = 33.0 \pm 1.0$), **b** *Homo sapiens sapiens* ($T_B = 36.6 \pm 0.2$), **c** *Bos taurus taurus* ($T_B = 38.6 \pm 0.3$). T_c indicates the observed transition temperature (Zerlin et al. 2007)

Dynamic light scattering experiments: Diffusion coefficients and hydrodynamic radii of six hemoglobins were analyzed between 21 °C and 51 °C (Figs. 17.14, 17.15).

Table 17.2 shows average hydrodynamic radii (R_H), calculated coefficients of diffusion (D), the animals' body temperatures (T_B) and the structural transition temperatures (T_c). In DLS experiments, the R_H of Hb molecules increased linearly with temperature.

At the critical transition temperature, T_c, the slope of the curves suddenly rose. The slope of the coefficient of diffusion vs. temperature was negative but showed a change point at T_c. Below T_c, the slope was small whereas above it was significantly more negative than below.

Higher hydrodynamic radii and corresponding lower coefficients of diffusion were interpreted as the result of Hb aggregation above T_c (Zerlin et al. 2007).

CD experiments: In CD experiments, the temperature dependencies of the ellipticity at 222 nm from hemoglobins of in total ten different species. When plotting

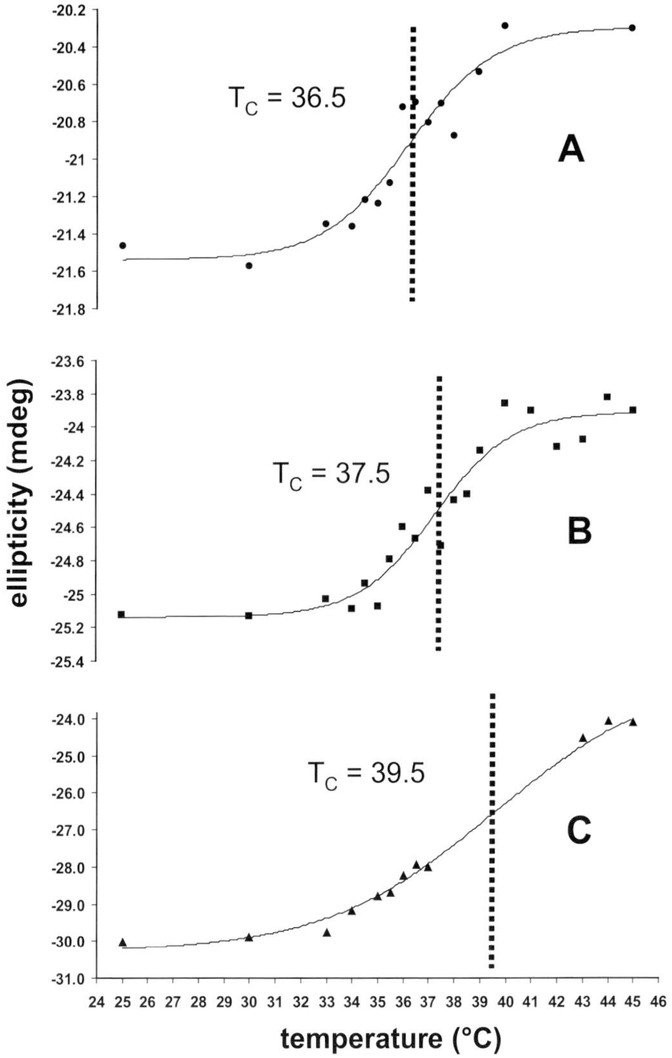

17.14 Hb Temperature Transition of Species with Body Temperatures Different from 37 °C   437

Fig. 17.15 Original plot of three circular dichroism experiments with hemoglobin from **A** *Catagonus wagneri* ($T_B = 36.17 \pm 0.5$), **B** *Camelus bactrianus* ($T_B = 38.0 \pm 0.5$), and **C** *Eurypyga helias* ($T_B = 41.06 \pm 0.5$). Midpoint temperatures of the fitted sigmoidal curves were defined as structural transition temperature (Artmann et al. 2004; Digel et al. 2006; Zerlin et al. 2007)

the ellipticity vs. temperature two curvature changes were seen since curves were s-shaped (with increasing temperature: low slope, high slope, low slope). The critical temperature, T_c, was taken as the mid-point temperature between the lower and the upper change point.

CD and DLS experiments carried out in parallel with identical Hb samples showed that the T_c obtained with DLS and the T_c from CD measurements were

Table 17.1 Listing of species and their body temperatures. Hemoglobin of these species were analyzed with CD spectroscopy as well as with Dynamic Light Scattering (Zerlin et al. 2007)

Species	Latin Name	T_B (°C)
Platypus	Ornithorhynchus anatinus	33.0
Echidna	Tachyglossidae	33.0
Coala	Phascolarctos cinereus	35.2
Chacoan peccary	Catagonus wagneri	36.2
Human	Homo sapiens sapiens	36.7
Dusty pademelon	Thylogale	37.0
Fishing cat	Felis viverrina	37.6
Felis serval	Felis silvestris forma catus	37.7
Bactrian camel	Camelus bactrianus	38.0
Cow	Bos taurus taurus	38.6
Pig	Sus scrofa domestica	39.7
Chicken	Gallus gallus domestica	41.0
Greater sunbittern	Eurypyga helias	41.1
Spotted nutchracker	Nucifraga caryocatactes	42.2

Table 17.2 Results of DLS experiments with hemoglobin of different species (Zerlin et al. 2007)

Species	Body-temperature (°C)	Structural transition temperature (°C)	R_H 25 °C (nm)	D 25 °C $(m^2 sec^{-1}) \cdot 10^{-11}$
Tachyglossus aculeatus	33.0 ± 1.0	34.9 ± 1.3	8.82 ± 0.23	2.76 ± 0.01
Ornithorhynchus anatinus	33.0 ± 1.0	33.4 ± 2.1	3.49 ± 0.25	6.98 ± 0.02
Homo sapiens sapiens	36.6 ± 0.2	36.4 ± 0.8	3.59 ± 0.09	6.79 ± 0.02
Bos taurus taurus	38.6 ± 0.3	38.1 ± 1.9	3.83 ± 0.16	6.36 ± 0.02
Sus scrofa domesticus	39.7 ± 0.5	40.1 ± 1.3	12.70 ± 4.38*	1.92 ± 0.01
Gallus gallus domesticus	41.0 ± 0.5	41.8 ± 1.9	4.43 ± 0.28	5.50 ± 0.02

the same within experimental errors. The temperature transitions occurred in hemoglobins from ALL species investigated. They were correlated linearly ($R^2 = 0.97$) with the species' body temperature at a slope of 0.81. Transitions always happened and were always close to the species' body temperature (Zerlin et al. 2006). Thus, normal body temperature marks a unique position on the temperature scale (Fig. 17.16).

17.15 Molecular Structural Mechanism of the Temperature Transitions

Based on two-dimensional infrared spectroscopy a two-step model of human Hb unfolding was proposed (Yan et al. 2002, 2003). An initial structural perturbation occurs between 30 °C and 44 °C followed by an initial thermal unfolding between

17.15 Molecular Structural Mechanism of the Temperature Transitions

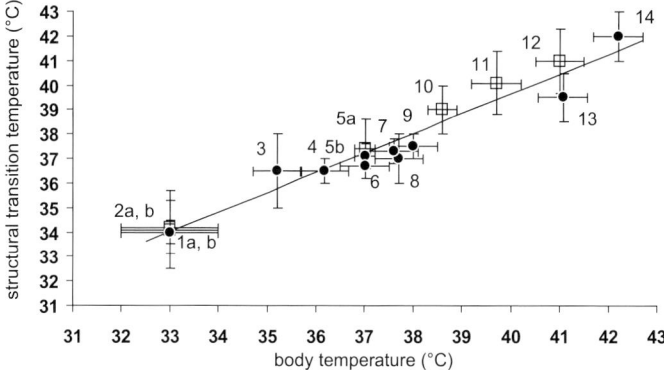

Fig. 17.16 Correlation between the structural transition temperatures of hemoglobins from various species and the species' body temperature. *Black circles*: Circular dichroism data, *open squares*: DLS data. The *line* represents a linear regression fit with a slope of 0.81 and a coefficient of correlation of 0.97. Numbering: *1a* and *1b* Ornithorhynchus anatinus (platypus), *2a* and *2b* Tachyglossus aculeatus (echidna), *3*. Phascolarctos cinereus (coala), *4*. Catagonus wagneri (chacoan peccary), *5a* and *5b* Homo sapiens sapiens (human), *6*. Thylogale sp. (dusty pademelon), *7*. Felis viverrina (fishing cat), *8*. Felis silvestris forma catus (felis serval), *9*. Camelus bactrianus (bactrian camel), *10*. Bos taurus taurus (cow), *11*. Sus scrofa domestica (pig), *12*. Gallus gallus domesticus (chicken), *13*. Eurypyga helias (greater sunnbittern), *14*. Nucrifraga caryocatactes (spotted nutcracker). (*a* and *b* at same number indicate that both DLS and CD T_cs were obtained.) (Zerlin et al. 2007)

44 °C and 54 °C. The second step is irreversible and induces thermal aggregation between 54 and 70 °C. Figure 17.17 shows a thermal denaturation curve of Homo sapiens sapiens Hb (CD data). The structural transition discussed here occurred in the initial structural perturbation stage.

The studied hemoglobins show very similar molecular properties e. g. molecular weight, amino acid chain length of the subunits and α-helical content (Zerlin et al. 2007). All in online databases available amino acid sequences of hemoglobins (α- and β-subunits) we studied were aligned with the ClustalX (1.83) software (Fig. 17.18). The "in silico" predicted α-helical parts of the subunits were marked grey. The total averaged similarity of the amino acid units were calculated. In both subunits, the first two α-helical parts of the sequences (amino acid numbers 7–32) were interestingly of low similarity! Both parts are located at external parts of the subunit (solvent exposed). In contrast, the other parts of the subunits, in particular middle parts (located inside the Hb with contacts to the heme group) show high similarities between 70 to 100%. A first careful conclusion can be drawn based on the alignment study in Fig. 17.18. A sequence part with low conserved sequence similarity but highly conserved structural elements (α-helix) might be a very promising candidate to be in charge of determining the critical transition temperature of a species. In other words, nature may use protein primary structure in order to adjust where body temperature has to be set on the temperature scale. Alternatively,

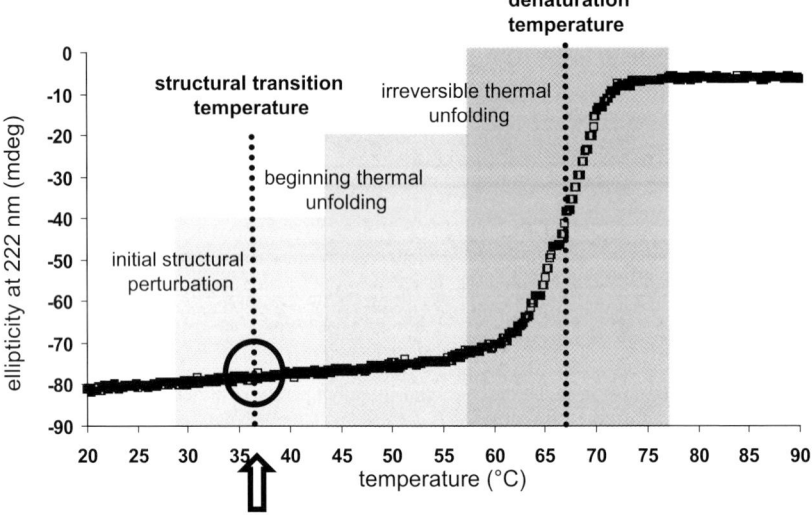

Fig. 17.17 Complete denaturation curve of Homo sapiens sapiens hemoglobin obtained from CD temperature scans. The ellipticity at 222 nm was normalized. The *red area* shows the region of the described structural perturbation stage, the *grey area* represents the initial thermal unfolding stage and the *dark grey area* the stage of irreversible thermal unfolding. The *black circle* marked with an *arrow* indicates the temperature range where the structural transition was seen. It was in the perturbation stage near body temperature. The denaturation temperature was 67.1 °C (Zerlin et al. 2007)

it can be speculated that the hemoglobin structure adapted itself evolutionarily to a particular body temperature.

17.16 Physics Meets Physiology

What would now be the links of physics to physiology? Our answer is that firstly, T_c marks the set-point of a species body temperature (Digel et al. 2006; Zerlin et al. 2007). Secondly, T_c represents the low-end point of a temperature range above which the entropy of a RBC-in-plasma sample does not change with temperature (Fig. 17.11A) (Artmann et al. 2004, 2007). This temperature range ends when thermal protein denaturing at 42.6 °C occurs enhancing sample entropy drastically (Stadler et al. 2007). We have known this temperature range since long – it is the fever or pyrexia range. In this pyrexia zone depending on the actual stage of cell internal Hb aggregation cell water moves in and out of the RBC depending on the direction of temperature change due to RBC internal Hb aggregation (Fig. 17.12). The latter might contribute to blood homeostasis during fever. The pyrexia zone in humans ends physiologically at a temperature, $T_{ID} = 42.6$ °C, where proteins denature irreversibly. Thus, the pyrexia zone might from now on be defined physically

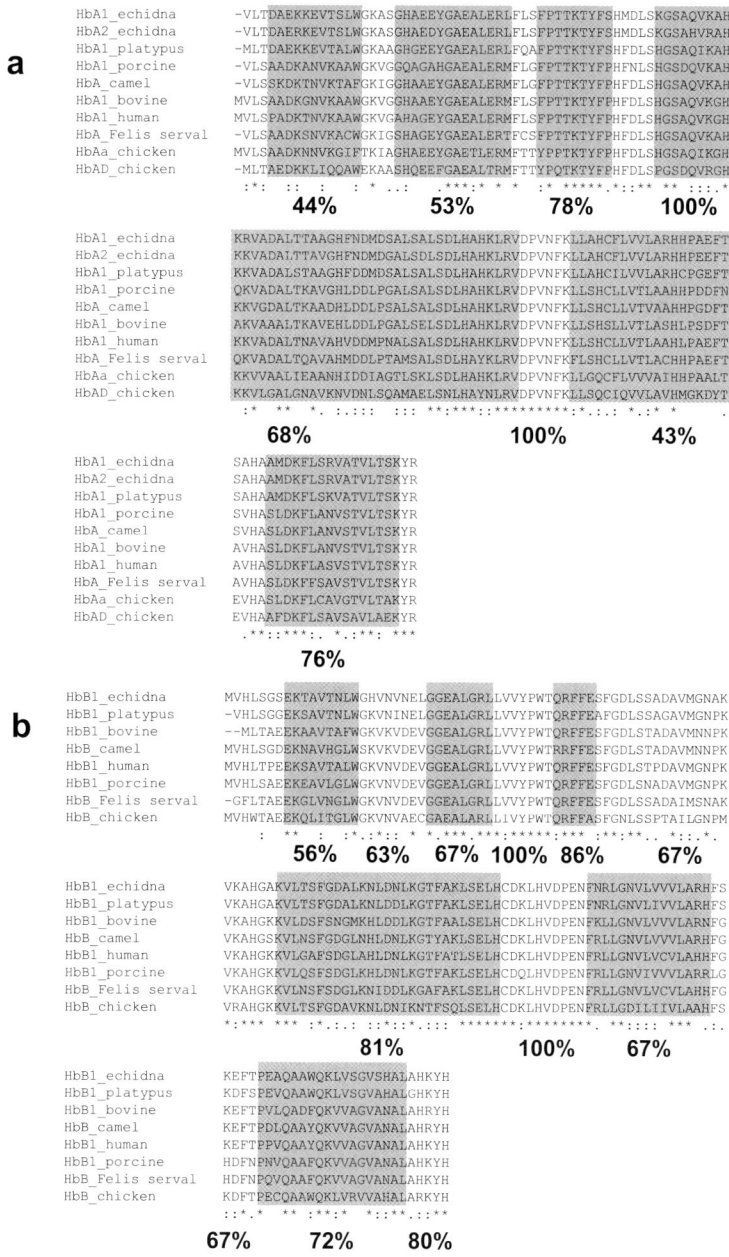

Fig. 17.18 Amino acid sequence alignment of hemoglobin subunits α and β of all species whose sequences were available online (NCBI database). **a**: hemoglobin α-subunits, **b**: hemoglobin β-subunits. The *grey background* indicates predicted α-helical contents of the subunits. For each section an amino acid sequence similarity between all hemoglobins is given (* identical amino acids, : similar amino acids, . less similar amino acids)

by two distinct temperatures, the lower one marked by T_c and the upper one marked by T_{ID}. We human beings as part of nature might exist at a temperature range where the total entropy of RBC in blood plasma "takes a breath" remaining for a couple of degree Celsius almost unchanged with rising temperature. From now on, body temperature and the range of pyrexia of many species might have a physical address which would be the primary structure of some proteins encoding the transition temperature, T_c, i.e. body temperature on the one hand and encoding the beginning of irreversible denaturing at T_{ID}, where entropy would set on rising again with increasing temperature.

The claim here is not that Hb *regulates* body temperature. Hb is only one out of possibly many proteins encoding body temperature. Evidently, new trends on cell biology using completely different approaches from those used by molecular and cell biologists will become more and more important in our attempt to understand *life*.

Acknowledgements This article is dedicated to Ludwig Artmann who died on July 21, 2001 on a beautiful summer day in a village in central Germany, on a day during which we performed experiments to complete this study in San Diego, California. Ludwig Artmann was a man who encouraged us to be strong and to study hard doesn't matter what were his costs.

The work was mostly performed with our lab's research spare money. Dr. Zerlin was financed for two years by a grant from the Ministry of Innovation, Science, Research and Technology of the State of North Rhine-Westphalia to G.M. Artmann and by the Centre of Competence in Bioengineering at Juelich, Germany. The chapter contains results of vitally important experiments conducted by Dipl. Ing. Carsten Meixner, Mrs. Dipl.-Ing. Nicole Kasischke and Mrs. Silvia Engels. We thank Jeff Turnage, San Diego Zoological Society, San Diego, USA, for supplying animal blood samples. We are very grateful to Dr. Ira Tremmel from Max-Planck-Institut für biophysikalische Chemie in Göttingen for patience and perseverance in securing valuable monotremal blood samples for the experiments. We thank Prof. Shu Chien and Prof. Y.C. Fung, Whittaker Institute for Bioengineering (UCSD), San Diego, for interesting and helpful discussions. Finally, we thank our colleagues Prof. G. Büldt, Prof. J. Zaccai, Cand. PhD A. Stadler and PD Dr. Fitter, Research Centre Jülich, ILL Grenoble, France, for their support.

References

1. Artmann,G.M. 1995. Microscopic photometric quantification of stiffness and relaxation time of red blood cells in a flow chamber. *Biorheology* 32:553-570.
2. Artmann,G.M., L.Burns, J.M.Canaves, A.Temiz-Artmann, G.W.Schmid-Schonbein, S.Chien, and C.Maggakis-Kelemen. 2004. Circular dichroism spectra of human hemoglobin reveal a reversible structural transition at body temperature. *Eur. Biophys. J.* 33:490-496.
3. Artmann, G. M., I.Digel, K.F.Zerlin, Ch.Maggakis-Kelemen, Pt.Linder, D.Porst, P.Kayser, G.Dikta, and A.Temiz Artmann. 2007. Hemoglobin senses body temperature. 2007. Unpublished Work, submitted to *Eur. Biophys. J.*
4. Artmann,G.M., C.Kelemen, D.Porst, G.Buldt, and S.Chien. 1998. Temperature transitions of protein properties in human red blood cells. *Biophys. J.* 75:3179-3183.
5. Artmann,G.M., A.Li, J.Ziemer, G.Schneider, and U.Sahm. 1996. A photometric method to analyze induced erythrocyte shape changes. *Biorheology* 33:251-265.
6. Artmann,G.M., K.L.Sung, T.Horn, D.Whittemore, G.Norwich, and S.Chien. 1997. Micropipette aspiration of human erythrocytes induces echinocytes via membrane phospholipid translocation. *Biophys. J.* 72:1434-1441.
7. Bastianetto,S., M.Danik, F.Mennicken, S.Williams, and R.Quirion. 2006. Prototypical antipsychotic drugs protect hippocampal neuronal cultures against cell death induced by growth medium deprivation. *BMC. Neurosci.* 7:28.
8. Bessis,M. and L.S.Lessin. 1970. The discocyte-echinocyte equilibrium of the normal and pathologic red cell. *Blood* 36:399-403.
9. Blow,D. 2002. Max Perutz (1914-2002). *Q. Rev. Biophys.* 35:201-204.
10. Bohm,G., R.Muhr, and R.Jaenicke. 1992. Quantitative analysis of protein far UV circular dichroism spectra by neural networks. *Protein Eng* 5:191-195.
11. Buldt, G., Artmann, G. M., Zaccai, G., Digel, I., and Stadler, A. M. Differences of water and D2O binding to proteins; temperature dependent DSC measurements and lattice constants of hemoglobin crystals. 7-5-2007. Personal Communication
12. Cameron,I.L., V.A.Ord, and G.D.Fullerton. 1988. Water of hydration in the intra- and extracellular environment of human erythrocytes. *Biochem. Cell Biol.* 66:1186-1199.
13. Chen,J.Y., L.S.Brunauer, F.C.Chu, C.M.Helsel, M.M.Gedde, and W.H.Huestis. 2003. Selective amphipathic nature of chlorpromazine binding to plasma membrane bilayers. *Biochim. Biophys. Acta* 1616:95-105.
14. Chien,S. 1981. Filterability and other methods of approaching red cell deformability. Determinants of blood viscosity and red cell deformability. *Scand. J. Clin. Lab Invest Suppl* 156:7-12.
15. Chien,S., K.L.Sung, R.Skalak, S.Usami, and A.Tozeren. 1978. Theoretical and experimental studies on viscoelastic properties of erythrocyte membrane. *Biophys. J.* 24:463-487.
16. Cioni,P. and G.B.Strambini. 2002. Effect of heavy water on protein flexibility and lattice constants of protein crystals. *Biophys. J.* 82:3246-3253.

17. Claessens,M.M., F.A.Leermakers, F.A.Hoekstra, and M.A.Stuart. 2007. Opposing effects of cation binding and hydration on the bending rigidity of anionic lipid bilayers. *J. Phys. Chem. B* 111:7127-7132.
18. Cribier,S., J.Sainte-Marie, and P.F.Devaux. 1993. Quantitative comparison between aminophospholipid translocase activity in human erythrocytes and in K562 cells. *Biochim. Biophys. Acta* 1148:85-90.
19. Dadarlat,V.M. and C.B.Post. 2006. Decomposition of protein experimental compressibility into intrinsic and hydration shell contributions. *Biophys. J.* 91:4544-4554.
20. Dencher,N.A., H.J.Sass, and G.Buldt. 2000. Water and bacteriorhodopsin: structure, dynamics, and function. *Biochim. Biophys. Acta* 1460:192-203.
21. Deuticke,B. 1968. Transformation and restoration of biconcave shape of human erythrocytes induced by amphiphilic agents and changes of ionic environment. *Biochim. Biophys. Acta* 163:494-500.
22. Devaux,P.F. 1992. Protein involvement in transmembrane lipid asymmetry. *Annu. Rev. Biophys. Biomol. Struct.* 21:417-439.
23. Digel, I, Engels, S, Hoffmann, B, Porst, D., and Artmann, G. M. Differential Scanning Calorimetry and micropipette experiments of human Red Blood Cells. 7-8-0007. Unpublished Work
24. Digel,I., C.Maggakis-Kelemen, K.F.Zerlin, P.Linder, N.Kasischke, P.Kayser, D.Porst, A.A.Temiz, and G.M.Artmann. 2006. Body temperature-related structural transitions of monotremal and human hemoglobin. *Biophys. J.* 91:3014-3021.
25. Dikta,G., M.Kvesic, and C.Schmidt. 2006. Bootstrap Approximations in Model Checks for Binary Data. *Journal of the American Statistical Association* 101:521-530.
26. Discher,D.E., N.Mohandas, and E.A.Evans. 1994. Molecular maps of red cell deformation: hidden elasticity and in situ connectivity. *Science* 266:1032-1035.
27. Eisenberg,H. 2003. Adair was right in his time. *Eur. Biophys. J.* 32:406-411.
28. Engstrom,K.G. and H.J.Meiselman. 1996. Effects of pressure on red blood cell geometry during micropipette aspiration. *Cytometry* 23:22-27.
29. Engstrom,K.G. and H.J.Meiselman. 1997. Combined use of micropipette aspiration and perifusion for studying red blood cell volume regulation. *Cytometry* 27:345-352.
30. Evans,E.A. 1983. Bending elastic modulus of red blood cell membrane derived from buckling instability in micropipet aspiration tests. *Biophys. J.* 43:27-30.
31. Evans,E.A., R.Waugh, and L.Melnik. 1976. Elastic area compressibility modulus of red cell membrane. *Biophys. J.* 16:585-595.
32. Fairbanks,G., T.L.Steck, and D.F.Wallach. 1971. Electrophoretic analysis of the major polypeptides of the human erythrocyte membrane. *Biochemistry* 10:2606-2617.
33. Farinas,J., A.N.Van Hoek, L.B.Shi, C.Erickson, and A.S.Verkman. 1993. Nonpolar environment of tryptophans in erythrocyte water channel CHIP28 determined by fluorescence quenching. *Biochemistry* 32:11857-11864.
34. Fasman,G.D. 1996. Circular Dichroism and the Conformational Analysis of Biomolecules. Plenum Press, New York.
35. Finnie,M., G.D.Fullerton, and I.L.Cameron. 1986. Molecular masking and unmasking of the paramagnetic effect of iron on the proton spin-lattice (T1) relaxation time in blood and blood clots. *Magn Reson. Imaging* 4:305-310.
36. Fischer,T.M. 1989. Erythrocyte deformation under shear flow. *Blood* 73:1074-1075.
37. Fischer,T.M. 2004. Shape memory of human red blood cells. *Biophys. J.* 86:3304-3313.
38. Friederichs,E., R.A.Farley, and H.J.Meiselman. 1992. Influence of calcium permeabilization and membrane-attached hemoglobin on erythrocyte deformability. *Am. J. Hematol.* 41:170-177.
39. Goldenfeld,N. 1992. Lectures on Phase Transitions and the Renormalization Group.
40. Grebe,R. and M.J.Zuckermann. 1990. Erythrocyte shape simulation by numerical optimization. *Biorheology* 27:735-746.
41. Heller,K.B. and M.Hofer. 1975. Temperature dependence of the energy-linked monosaccharide transport across the cell membrane of Rhodotorula gracilis. *J. Membr. Biol.* 21:261-271.

References

42. Hochmuth,R.M. 1993. Measuring the mechanical properties of individual human blood cells. *J. Biomech. Eng* 115:515-519.
43. Hochmuth,R.M. and R.E.Waugh. 1987. Erythrocyte membrane elasticity and viscosity. *Annu. Rev. Physiol* 49:209-219.
44. Hochmuth,R.M., P.R.Worthy, and E.A.Evans. 1979. Red cell extensional recovery and the determination of membrane viscosity. *Biophys. J.* 26:101-114.
45. Hueck,I.S., H.G.Hollweg, G.W.Schmid-Schonbein, and G.M.Artmann. 2000. Chlorpromazine modulates the morphological macro- and microstructure of endothelial cells. *Am. J. Physiol Cell Physiol* 278:C873-C878.
46. Ikeda,M., A.Kihara, and Y.Igarashi. 2006. Lipid asymmetry of the eukaryotic plasma membrane: functions and related enzymes. *Biol. Pharm. Bull.* 29:1542-1546.
47. Jay,D.G. 1996. Role of band 3 in homeostasis and cell shape. *Cell* 86:853-854.
48. Kassab,G.S. 2004. Y.C. "Bert" Fung: the father of modern biomechanics. *Mech. Chem. Biosyst.* 1:5-22.
49. Kelemen,C., S.Chien, and G.M.Artmann. 2001. Temperature transition of human hemoglobin at body temperature: effects of calcium. *Biophys. J.* 80:2622-2630.
50. Kharakoz,D.P. and A.P.Sarvazyan. 1993. Hydrational and intrinsic compressibilities of globular proteins. *Biopolymers* 33:11-26.
51. Kinderlerer,J., H.Lehmann, and K.F.Tipton. 1970. Thermal denaturation of human haemoglobins. *Biochem. J.* 119:66-67.
52. Kinderlerer,J., H.Lehmann, and K.F.Tipton. 1973. The thermal denaturation of human oxyhaemoglobins A, A2, C and S. *Biochem. J.* 135:805-814.
53. Landau,L.D.a.L.E.M. 2007. Statistical Physics Part 1. Pergamon.
54. Lee,J.C., D.T.Wong, and D.E.Discher. 1999. Direct measures of large, anisotropic strains in deformation of the erythrocyte cytoskeleton. *Biophys. J.* 77:853-864.
55. Li,Z., S.Raychaudhuri, and A.J.Wand. 1996. Insights into the local residual entropy of proteins provided by NMR relaxation. *Protein Sci.* 5:2647-2650.
56. Lucas,C.E., A.M.Ledgerwood, W.J.Rachwal, D.Grabow, and J.M.Saxe. 1991. Colloid oncotic pressure and body water dynamics in septic and injured patients. *J. Trauma* 31:927-931.
57. Michnik,A., Z.Drzazga, A.Kluczewska, and K.Michalik. 2005. Differential scanning microcalorimetry study of the thermal denaturation of haemoglobin. *Biophys. Chem.* 118:93-101.
58. Mohandas,N. and J.A.Chasis. 1993. Red blood cell deformability, membrane material properties and shape: regulation by transmembrane, skeletal and cytosolic proteins and lipids. *Semin. Hematol.* 30:171-192.
59. Mozo,J., Y.Emre, F.Bouillaud, D.Ricquier, and F.Criscuolo. 2005. Thermoregulation: what role for UCPs in mammals and birds? *Biosci. Rep.* 25:227-249.
60. Muller,G.H., H.Schmid-Schonbein, and H.J.Meiselman. 1992. Development of viscoelasticity in heated hemoglobin solutions. *Biorheology* 29:203-216.
61. Niimura,N., T.Chatake, K.Kurihara, and M.Maeda. 2004. Hydrogen and hydration in proteins. *Cell Biochem. Biophys.* 40:351-369.
62. Nomoto,S., M.Shibata, M.Iriki, and W.Riedel. 2004. Role of afferent pathways of heat and cold in body temperature regulation. *Int. J. Biometeorol.* 49:67-85.
63. Oberwinkler,J. 2007. TRPM3, a biophysical enigma? *Biochem. Soc. Trans.* 35:89-90.
64. Penzlin,H. 1977. Lehrbuch der Tierphysiologie. *Gustav Fischer Verlag*, Stuttgart New York.
65. Pierre,S.V. and Z.Xie. 2006. The Na,K-ATPase receptor complex: its organization and membership. *Cell Biochem. Biophys.* 46:303-316.
66. Pollack,G.H. 2001a. Cells, Gels and the Engines of life. Ebner and Sons, Seattle.
67. Pollack,G.H. 2001b. Is the cell a gel–and why does it matter? *Jpn. J. Physiol* 51:649-660.
68. Poschl,J.M., C.Leray, P.Ruef, J.P.Cazenave, and O.Linderkamp. 2003. Endotoxin binding to erythrocyte membrane and erythrocyte deformability in human sepsis and in vitro. *Crit Care Med.* 31:924-928.
69. Roche,J., A.Roche, G.S.Adair, and M.E.Adair. 1932. The osmotic pressure of globin. *Biochem. J.* 26:1811-1828.

70. Ross,F.P. and A.M.Christiano. 2006. Nothing but skin and bone. *J. Clin. Invest* 116:1140-1149.
71. Sahu,S.K., S.N.Gummadi, N.Manoj, and G.K.Aradhyam. 2007. Phospholipid scramblases: an overview. *Arch. Biochem. Biophys.* 462:103-114.
72. Schmid-Schonbein,H., H.Heidtmann, and R.Grebe. 1986a. Spectrin, red cell shape and deformability. I. Membrane curvature in genetic spectrin deficiency. *Blut* 52:131-147.
73. Schmid-Schonbein,H., H.Heidtmann, and R.Grebe. 1986b. Spectrin, red cell shape and deformability. II. The antagonistic action of spectrin and sialic acid residues in determining membrane curvature in genetic spectrin deficiency in mice. *Blut* 52:149-164.
74. Schmidt,R., F.Lang, and G.Thews. 2005. Physiologie des Menschen mit Pathophysiologie. Springer Verlag.
75. Shi,Y.D., G.Artmann, R.Agosti, and E.Longhini. 1998. A modified Casson equation to characterize blood rheology for hypertension. *Clin. Hemorheol. Microcirc.* 19:115-127.
76. Silva,J.E. 2006. Thermogenic mechanisms and their hormonal regulation. *Physiol Rev.* 86:435-464.
77. Simpson,S. and J.J.Galbraith. 1905. Observations on the normal temperatures of the monkey and its diurnal variation, and on the effects of changes in the daily routine in this variation. *Transaction of the royal society of Edinburgh* 45:65-104.
78. Soupene,E. and F.A.Kuypers. 2006. Identification of an erythroid ATP-dependent aminophospholipid transporter. *Br. J. Haematol.* 133:436-438.
79. Staat, M. Biomechanical considerations on Red Blood Cell aspiration into narrow micropipettes. 6-6-2007. 6-6-2007. Personal Communication
80. Stadler, A. M., Zerlin, K. F., Digel, I., Artmann, G. M., Embs, J. P., Buldt, G., and Zaccai, G. Dynamics of hemoglobin and water in human red blood cells and concentrated hemoglobin solutions. Abstract 128. 14-7-2007. London, European Biophysics Congress. 14-7-2007. Conference Proceeding
81. Temiz,A., O.K.Baskurt, C.Pekcetin, F.Kandemir, and A.Gure. 2000. Leukocyte activation, oxidant stress and red blood cell properties after acute, exhausting exercise in rats. *Clin. Hemorheol. Microcirc.* 22:253-259.
82. Todd,J.C., III and D.L.Mollitt. 1994. Sepsis-induced alterations in the erythrocyte membrane. *Am. Surg.* 60:954-957.
83. Tominaga,M. 2004. [Molecular mechanisms of thermosensation]. *Nippon Yakurigaku Zasshi* 124:219-227.
84. Tremmel,I.G., E.Weis, and G.D.Farquhar. 2005. The influence of protein-protein interactions on the organization of proteins within thylakoid membranes. *Biophys. J.* 88:2650-2660.
85. Veldkamp,W.B. and J.R.Votano. 1980. Temperature dependence of macromolecular interactions in dilute and concentrated hemoglobin solutions. *Biopolymers* 19:111-124.
86. Vitkup,D., D.Ringe, G.A.Petsko, and M.Karplus. 2000. Solvent mobility and the protein 'glass' transition. *Nat. Struct. Biol.* 7:34-38.
87. Walz,T., T.Hirai, K.Murata, J.B.Heymann, K.Mitsuoka, Y.Fujiyoshi, B.L.Smith, P.Agre, and A.Engel. 1997. The three-dimensional structure of aquaporin-1. *Nature* 387:624-627.
88. Wenzel,J.J., A.Piehler, and W.E.Kaminski. 2007. ABC A-subclass proteins: gatekeepers of cellular phospho- and sphingolipid transport. *Front Biosci.* 12:3177-3193.
89. Whittow,C. 1971. Comparative physiology of thermoregulation. *Academic Press*, New York and London.
90. Yan,Y.B., Q.Wang, H.W.He, X.Y.Hu, R.Q.Zhang, and H.M.Zhou. 2003. Two-dimensional infrared correlation spectroscopy study of sequential events in the heat-induced unfolding and aggregation process of myoglobin. *Biophys J* 85:1959-1967.
91. Yan,Y.B., R.Q.Zhang, and H.M.Zhou. 2002. Biphasic reductive unfolding of ribonuclease A is temperature dependent. *Eur J Biochem.* 269:5314-5322.
92. Yokoyama,K., T.Terao, and T.Osawa. 1978. Membrane receptors of human erythrocytes for bacterial lipopolysaccharide (LPS). *Jpn. J. Exp. Med.* 48:511-517.
93. Zefirova,T.P., A.N.Glebov, E.N.Gur'ev, R.S.Mavliautdinov, and O.I.Tarasov. 1991. [Nuclear magnetic relaxation of aqueous solutions of proteins, plasma, erythrocytes, and blood]. *Biull. Eksp. Biol. Med.* 112:378-381.

94. Zerlin, K.F., N.Kasischke, I.Digel, C.Maggakis-Kelemen, A.A.Temiz, D.Porst, P.Kayser, P.Linder, and G.M.Artmann. 2007. Structural transition temperature of hemoglobins correlates with species' body temperature. *Eur Biophys J* 37:1-10.
95. Zheng, J.M. and G.H.Pollack. 2003. Long-range forces extending from polymer-gel surfaces. *Phys. Rev. E. Stat. Nonlin. Soft. Matter Phys.* 68:031408.

Part V
Bioengineering in Clinical Applications

Chapter 18
Nitric Oxide in the Vascular System: Meet a Challenge

Stefanie Keymel, Malte Kelm, Petra Kleinbongard

Department of Medicine, Medical Clinic I, University Hospital RWTH Aachen, Pauwelsstraße 30, 52074 Aachen, Germany,
PKleinbongard@ukaachen.de

Abstract After a small introduction in the topic and relative young history of nitric oxide (NO) we would like to illustrate a few aspects of the wide field in NO research in biomedicine. The theoretical background of NO metabolism is spiked with numerous examples of methodologies from bioassays to biochemistry. Furthermore, first steps in the realization of new concepts in NO research in microcirculation are developed.

18.1 Nitric Oxide: NO

First, we give a short overview of known functions of NO in humans. NO is a gas which controls endless functions in the body. It is important to mention that new NO-dependent processes are discovered daily. Scientific journals operating in the biomedical, biological and medical field are flooded with new insights to the biological activity and potential clinical approaches of NO. Each revelation underlines the already resumed functions of NO in controlling blood circulation, and regulating brain, lung, liver, kidney, stomach, gut, genital activity as well as other organs. It causes penile erections by dilating blood vessels, and controls the action of almost every orifice from swallowing to defecation. The immune system uses NO in fighting viral, bacterial and parasitic infections, and tumours. NO mediates messages between nerve cells and is associated with the processes of learning, memory, sleeping, feeling pain, and, probably, depression. It is a mediator in inflammation and rheumatism. In summary, NO participates in nearly every regulating cascade [1, 2].

18.2 NO in Vascular Biology

Although very short living, NO has emerged to one of the key mediators within the cardiovascular system with an essential role in physiological and pathological processes. About ten years ago, one could hardly find any information in physiology textbooks about the action and importance of NO. Now, information about NO is widely spread among research and teaching.

The most extensively characterized effect of NO in vascular biology is its vasodilatory effect on blood vessels. By this, NO becomes part of the complex regulatory mechanisms of blood pressure and blood flow in humans. Besides the myogenic and metabolic responses, NO acts with different other vasoactive substances such as angiotensin II, vasopressin, prostanoids and natriuretic peptides and plays a central role in the local regulation of vessel tone and blood flow. NO is synthesized in endothelial cells from the amino acid L-arginine whereby the reaction is accelerated by the constitutively expressed endothelial nitric oxide synthase (eNOS) [3–5]. The highly membrane permeable product NO can quickly reach the vascular smooth muscle cells where it activates the soluble guanylate cyclase leading to an increased intracellular cyclic GMP formation and relaxation of the smooth muscle cells [6]. An activation of the eNOS and an increased production of NO occurs by different mechanisms. Under physiological conditions, one of the main mechanisms is the increase in intracellular calcium concentration in the endothelial cells by diverse substances, e. g. acetylcholine or bradykinin. The activity of eNOS is also modified by shear stress, whereas an increase of the shear stress on the vascular wall leads to a stimulation of eNOS with a consecutive local vasodilation (Fig. 18.1) [7].

The vasodilating effect of NO is not only the best known effect, but also the first effect described. NO generating agents were started to use about 150 years ago with the purpose to improve the myocardial blood flow without any knowledge regarding the active principal. The beneficial effect of nitroglycerine was first described in 1875 in "The Lancet". Irony of fate is that Alfred Nobel had to ingest nitroglycerine to improve his heart function during aging under the synonym Trinitrin. The molecular mechanism was discovered step by step: In the early 20th century, scientists worked on *in-vitro* actions of nitrate containing compounds although little progress

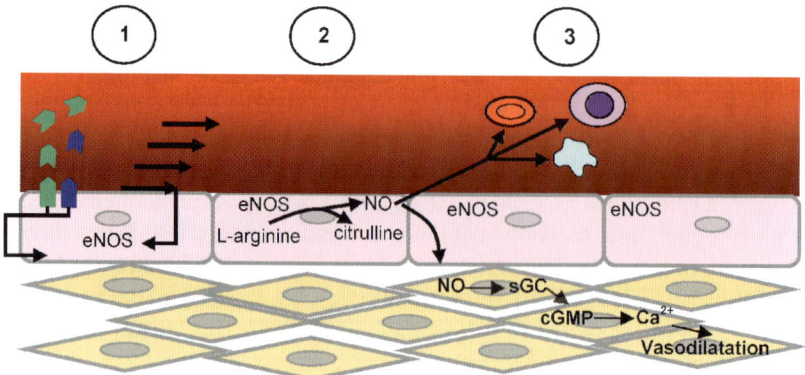

Fig. 18.1 eNOS signalling. (*1*) Endothelial NOS can be stimulated by various mechanisms, for example increased shear stress or receptor activation by vasoactive substances. (*2*) NO is produced in endothelial cells from the substrate L-arginine. (*3*) Produced NO can diffuse either into the vascular wall to induce a relaxation of smooth muscle cells or to the luminal side to interact with blood cells

was made towards understanding the cellular mode of action. Ferid Murad discovered the release of NO from nitroglycerine and its action on vascular smooth muscle in 1977 [8]. Robert Furchgott and John Zawadski recognized the importance of the endothelium in acetylcholine induced vasorelaxation (1980) and Louis Ignarro and Salvador Moncada identified endothelial derived relaxing factor (EDRF) as NO (1987) [9–11]. In 1989, the biosynthesis of NO was discovered and 1998 the scientists Furchgott, Ignarro and Murad received the Nobel Prize for their pioneer work on NO. Interestingly, glycerol trinitrate remains the agents of choice for relieving angina symptoms; other organic esters and inorganic nitrates are also used, but the rapid action of nitroglycerine and its established efficacy makes it the mainstay of angina pectoris relief [12].

The exogenously applied or endothelium-derived NO can not only diffuse to the abluminal side of the endothelial cells, but also to the luminal side. Within the blood NO can exert multiple functions: it reduces platelet activation and platelet aggregation, it reduces leukocyte adhesion and influences red blood cell deformability (Fig. 18.1) [13–15].

According to the diverse functions in vascular biology, a dysbalance in the L-arginine-NO-pathway plays a key role in the pathogenesis of atherogenesis [16]. Patients with an endothelial dysfunction, an early stage in the development of atherosclerosis, already show a reduced production and properties of the vasodilating mediator NO and a disturbed blood cell function which promotes the degenerative changes in the vessel walls [17]. As atherosclerotic disease is the leading cause of death in the developed countries, the interest in the identification of main mechanisms, the early diagnosis and therapy of cardiovascular diseases is very high [18].

18.3 Key Questions

What are the exact mechanisms of NO signalling? Is it possible to detect the very short lived molecule? What are the physiological and pathophysiological effects of NO? Where medicine meets biotechnology and bioengineering!

18.4 Assessment of NO Mediated Vasoactivity

The *ex-vivo* detection of vasoactivity or vasodilating agents is an essential feature of cardiovascular research. Consequently, bioassays for analysing the mechanisms of vasodilation are essential.

Extracted aortic rings of mice, rats or rabbit survive in an organ bath filled with buffer. This is a traditional experimental setup that has been widely used to investigate the physiology and pharmacology of *ex-vivo* tissue preparations of arterial rings. The aortic segments are cut into ring segments. The tissue is then suspended in a temperature controlled and with a gas diffuser equipped chamber – the 'organ bath' – (37 °C, bubbled with 95% O_2 and 5% CO_2). After an equilibration period, a normalization technique is often applied to set the vascular ring segments at a pressure comparable to that at the *in-vivo* situation. Briefly, each arterial segment is stretched up in progressive steps to determine the individual length tension curve. The contractile function is recorded with either a force (isometric) or displacement (isotonic) transducer. The transducer is connected to a preamplifier (usually a bridge amplifier) which is connected to the data acquisition system.

Many extensive efforts to get further insights in the mechanism of vasodilatation under physiological and pathophysiological conditions are solved with this assay. This assay has the potential to differentiate between endothelial and non-endothelial vasodilation. Therefore, the endothelium of some aortic ring segments can be denuded by gently rubbing the intimal surface with a thin polyethylene tube. The endothelium-dependent vasodilatation can be ruled out. The NO-mediated contraction and relaxation is clearly abolished because the source of NO production is destroyed. Furthermore, there is clear evidence for the regulation mechanisms of NO dependent vasodilatation. For example: Nitroglycerine induces a vasorelaxation in denuded aortic segments, but acetylcholine does not relax the segments without endothelial cells. This was the first hint for a functional eNOS regulated by an acetylcholine receptor [19, 20].

The aorta ring assay overcomes the gap between *in-vivo* and *in-vitro* assays. The vasodilating effect of NO can not only be observed during the *in-vitro* or *ex-vivo* setting of vascular ring experiments. It has already been transferred to *in-vivo* investigation of humans as it is the fundamental principle of the assessment of endothelium-dependent vasodilation in the coronary or peripheral circulation. In general, endothelium-dependent vasodilation needs a stimulus for endothelial cells to release NO. This can be different vasoactive substances or an accelerated flow in the examined vessel which both increase eNOS activity. The intracoronary infusion of acetylcholine has long been used for testing endothelium-dependent vasodilation of the coronary arteries [21]. In healthy vessels, the infusion of acetylcholine results in a NO-mediated dilation of the target vessels. However, patients with a disturbance of endothelial cell function due to atherosclerotic lesions show a reduced vasorelaxing effect or even a paradoxical vasoconstriction response to acetylcholine [22]. This diagnostic tool is still considered as the gold standard for the assessment of the endothelial function, but it is limited because of its invasive nature.

The non-invasive measurement of flow-mediated dilation (FMD) of conduit arteries has emerged to a widely used method for testing endothelial. The main advances of this method are the non-invasive stimulation of eNOS by increasing the flow in the peripheral artery and the non-invasive detection of the vessel reaction by high resolution ultrasound [23]. For the measurement of the FMD the patient is lying in a supine position and the diameter of the brachial artery is quantified just above the antecubital fossa by high resolution ultrasound. After the basal measurement, a cuff is placed around the forearm and is inflated to a suprasystolic pressure level for 5 minutes which results in a hypoxic vasodilation of the forearm vasculature. When the cuff is deflated reactive hyperemia occurs and causes an increased blood flow in the brachial artery to accommodate the dilated vessels in the forearm vasculature. The increased blood flow and the increased shear stress on the vascular wall of the brachial artery stimulates eNOS and the production of NO [24]. One minute after cuff release the vasodilating effect can be quantified as an increased vessel diameter in the ultrasound scan. Interestingly, the FMD is not only reduced in patients with atherosclerotic disease, for instance coronary artery disease, but already in patients with cardiovascular risk factors [25, 26]. As atherosclerosis is not a local process, but a systemic disease, the reduced FMD of the brachial artery reflects a systemic endothelial dysfunction and is associated with an increased risk for cardiovascular events, for example myocardial infarction or stroke. Therefore, the NO-dependent FMD can be used for the early detection of patients with high risk for cardiovascular disease and allows early interventions before atherosclerotic lesions occur.

Beside the functional assessment of endothelial function, structural alterations of the vessel wall can be monitored by high resolution ultrasound as well. For this purpose the intima-media thickness of the arterial wall is measured whereby very small changes can be detected [27]. The thickening of the arterial intima-media complex can be explained by the formation of (pre-)atherosclerotic deposits with a cellular and a matrix component and contributes to the disturbance of the endothelial function.

18.5 From the *In-Vivo* and *Ex-Vivo* Detection of NO Effects to Biochemical Assessment of NO

Shortly, after the discovery of NO as a vasodilator it seemed to be unimaginable to detect this highly reactive gas in the blood stream. Why? NO is a small, diffusible, highly reactive free radical with a short half-life, and is present in low concentrations. Therefore, real time detection of NO is extremely difficult [28]. Even after some years of NO research, it is still a challenge to optimize useful methods for the detection of NO formation [29]. Up to now there are many different ways of NO detection whereas direct and indirect techniques are used for the measurement. Direct NO detection methods often use the chemical behaviour of the radical. Reactants are used that possess a strong linkage to NO. This stoichiometric behaviour allows a concentration-dependent detection of NO. Products of these reactions are characterised by an uncomplicated detection. The fast reaction of NO with oxyhemoglobin [30] or the reaction with fluorescent dyes like DAF (4-amino-5-methylamino-2'7'-difluorofluorescein) [31] as well as an agent like DETC (diethyldithiocarbamate) [32] are used. Indirect techniques for NO detection use products or metabolites of the NO pathway. One of these methods is the arginine-citrulline-assay. It is based on the enzymatic conversion of the tritium labelled arginine (L-[$3H$]-arginine) to NO and L-[$3H$]-citrulline. Afterwards, the radioactive labelled citrulline is detectable [33]. All described assays are highly specialized experimental setups for the NO detection. But, they can not be applied to *ex-vivo* detection of *in-vivo* NO formation. These assays need special characteristics: It should work a) without prior *in-vivo* reactions or with prior applied reactants, and b) it should be highly sensitive. Additionally, the desired *ex-vivo* marker for NO should be c) as specific as possible for the enzymatic NO formation.

18.6 On the Road to a Potential Sensitive Marker for NO Formation: Is Nitrite a Candidate?

To understand the difficulties in selecting the right NO metabolite and the best detection method the *in-vivo* metabolism of NO should be explained more detailed: Enzymatically produced NO is released abluminally and into the vascular lumen. Because of the rapid distribution and metabolism of NO a direct reliable quantification of its basal production *ex-vivo* has been rather challenging. The majority of intravascular NO is inactivated by its reaction with hemoglobin to form nitrate. Nitrate concentrations are influenced by a variety of NO-synthase (NOS) independent factors, including dietary nitrate intake, formation of saliva, bacterial nitrate synthesis within the bowel, denitrifying liver enzymes, inhalation of atmospheric gaseous nitrogen oxides, and renal function. Due to these factors and the high background level, small changes in plasma nitrate concentrations may not sensitively reflect acute changes in NOS activity. Apart from the predominant metabolism of NO to nitrate a minor portion of NO undergoes conversion to nitrosated and nitrosylated species and nitrite, respectively [34]. Formation of plasma nitros(yl)ated NO-species (RXNO: the sum of *S*-nitrosothiols, *N*-nitrosamines, iron-nitrosyl species) may act to conserve and transport NO. Whatever mechanisms may turn out to best describe how the effects of NO are conserved in the circulation, it is clear that besides the local actions it elicits NO can also be transported throughout the body to function in a paracrine fashion, much like a hormone [35–38]. Although this process is likely to be limited and the oxidation to nitrite is assumed to be the major pathway of the mentioned reactions beside the oxidation to nitrate (Fig. 18.2). The rapid nitrite metabolism appears to be a sensitive marker with an estimated half life of nitrite in the range of minutes in protein containing samples. The half life allows the detection with high sensitivity. Therefore, nitrite detection was performed in biological samples like blood which were rather challenging [39].

Detection systems are available to measure nitrite in the mM range in aqueous samples. These methods include the well known Griess reaction and the high performance liquid chromatography (HPLC). These approaches led to the first results of nitrite detection in human plasma. In the Griess reaction, nitrite chemically reacts with a diazotizing reagent, sulfanilamide, in acidic media to form a diazonium salt. This intermediate then reacts with a coupling reagent, *N*-naphtyl-ethylenediamine, to form a stable azo compound. The intense pink colour of the product can be detected at its peak maximum absorption of 540 nm. The absorption is detected photometrically [40, 41]. HPLC is a useful separation based analytical technique for the determination of nitrite. An injector, two-piston pumps, a guard column as well as an analytical column and a detector built the basis of this chromatographic system. The analytes permeate through the analytical column at different rates due to differences

18.6 On the Road to a Potential Sensitive Marker for NO Formation: Is Nitrite a Candidate? 457

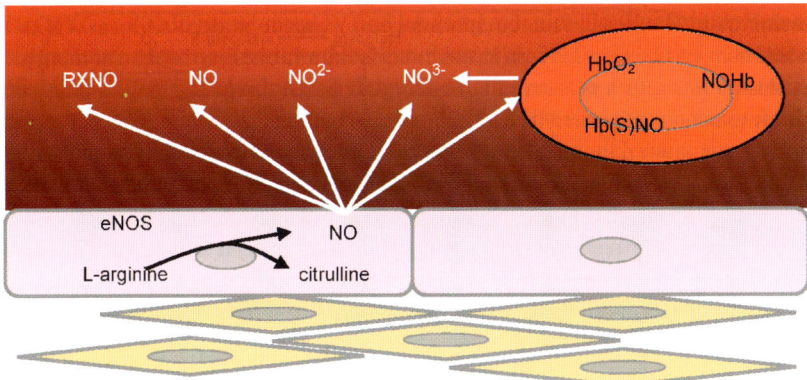

Fig. 18.2 Metabolites of the L-arginine-NO-pathway. Oxidative products of NO in human plasma are nitrite and nitrite which represent the major reaction products. Plasma NO can also bind to different proteins (RXNO). Within the RBC, NO reacts with the heme groups of deoxyhemoglobin to form iron-nitrosyl hemoglobin (NOHb) and with the cysteine at position 93 of the beta chain of hemoglobin to form *S*-nitrosohemoglobin (SNOHb)

in their partitioning behaviour between the mobile liquid phase and the stationary phase. Nitrite absorbs UV light at 200–220 nm. Therefore, a direct detection after separation by using an anion exchanger and an UV detector set at 210 or 214 nm is possible [42]. Up to date, more nitrite detection methods are available, but irrespective of this and the extremely improved methodologies, enormous differences in nitrite levels were detected in human plasma. This can be explained with the variations in blood sampling and sample processing as well as with the methodological problems inherent to the analytical procedures. Some methods do not possess the sensitivity necessary to measure nitrite precisely. In addition, the analysis might be affected by proteins, varying redox conditions, and trace contamination with nitrite during sample processing [39].

To get an impression of the applicability of nitrite to be a marker for NO synthesis three distinct analytical methods with adequate sensitivity were used to detect nanomolar levels of plasma nitrite in different mammalian species in spite of subtle differences in sample preparation. The gas-phase chemiluminescence, a method that depends on the reductive conversion of nitrite to NO and subsequent detection of the liberated NO by its reaction with ozone, an HPLC technique which employs ion chromatography to separate nitrite from other plasma constituents, online reduction and subsequent postcolumn derivatization with the Griess reagent, and flow injection analysis in combination with the Griess reagent, a method that employs colorimetric reaction with nitrite without prior separation from other plasma constituents was applied. All three used methods should measure similar plasma nitrite levels, leading to the presumption that these methods are sensitive enough to detect nitrite in biological fluids [43]. After evaluating a highly sensitive method we looked for a characteristic of nitrite to be a marker for NOS activity. Therefore, eNOS in the human forearm vasculature was stimulated which results in acute changes in plasma nitrite concentration. Data from experimental studies underlined this result, showing that 70–90% of the circulating plasma nitrite is derived from eNOS. By inhibition of eNOS activity with a selective eNOS inhibitor it could be demonstrated that plasma nitrite is a sensitive marker for eNOS activity [43]. Besides many other systems to detect NOS activity or NO directly or indirectly we established a system with a high potential in examining NOS activity in experimental setups and in clinical studies, as well. First results showed a decreased nitrite concentration in human volunteers with increasing numbers of cardiovascular risk factors (smoking, hypertension, hypercholesterolemia, obesity) [44]. In the future, it will be very interesting to monitor the efficacy of therapeutic interventions influencing endothelial function and the NO metabolism in clinical trials.

18.7 More Information About NO Interactions in the Blood

Since NO is released not only abluminally to exert its effects on cells of the vascular wall, but also into the vessel lumen, a significant part of the NO produced by the endothelium is believed to come into direct contact with blood. The fate of this fraction of NO is thought to be dictated largely by its interaction with erythrocytic hemoglobin. RBC are believed to be a major sink for NO by virtue of the rapid dioxygenation reaction of NO with oxyhemoglobin to form methemoglobin and nitrate the second–order rate constant of which approaches $3-4 \times 10^7 \, M^{-1} \, s^{-1}$ [45]. Although this reaction has appreciated widespread recognition as the major inactivation pathway of NO *in-vivo*, recent results obtained in humans suggested that this may not be the sole route under all conditions [46]. Modelling analyses showed that endothelium produced NO reacted as rapidly with blood as it does with free hemoglobin [47, 48]. Of particular importance in this context is the finding that the reaction rate of NO with oxyhemoglobin within erythrocytes is limited by its diffusion into the cell and occurs ∼1000 times slower compared to the reaction with free oxyhemoglobin [47]. Consumption of plasma NO by RBC is reduced by increased flow, an unstirred plasma layer surrounding the RBC, the cell-free zone near the vascular wall, and a reduced diffusion rate over the cellular membrane [49–51]. Alternatively, NO may bind to the heme group of deoxyhemoglobin to form NOHb [52]. A third possibility is the reaction of NO, or a higher oxidation product, such as NO_2 or N_2O_3, with cysteine-93 of the β-globin chains (β-Cys^{93}) of hemoglobin, leading to formation of an S-nitrosated derivative of oxyhemoglobin (SNOHb) (Fig. 18.2) [53]. SNOHb has been suggested to participate in the regulation of blood flow [54]. This result turns the role of RBC in the NO metabolism. The inactive scavenger becomes to be a "transporter" of bioactive NO.

Other indications for an active role of RBC in NO metabolism are findings describing a reaction of nitrite with hemoglobin of RBC. For a long time nitrite has been considered as an end product of NO reaction with oxygen and without intrinsic vasodilatory activity. However, recent studies have revealed a striking effect of nitrite infusions on forearm and systemic blood flow. Infusions of supraphysiological and near physiological concentrations of nitrite into the brachial arteries of healthy volunteers significantly increased blood flow and were associated with the formation of NO-hemoglobin-compounds. The verification of the chemical reaction between nitrite and deoxyhemoglobin to form NO supports the idea of a novel function for nitrite in the regulation of vasodilation [55]. This raises the question, whether the application of nitrite may open new avenues for therapeutical approaches [56, 57].

18.8 Intravascular Sources of NO

NOS isoforms are described in all blood cells [58, 59]. But, taken together, a constitutive NOS activity under resting conditions is not extensively characterized. Impact of NO on blood cell function like leukocyte adhesion or platelet aggregation is well described [13, 14]. Preliminary – and likewise not uniformly – data pointed towards the possibility that RBC might carry NOS protein. Some diverging results on RBC and the potential NOS activity are under investigation. Some groups postulated either a basal or total inactive NOS isoform or an origin by a nonenzymatic NO synthesis within RBC [60].

This view could be enlarged by demonstrating that RBC constitutively synthesize NO. This NO formation is modulated by supplementation of eNOS substrate in RBC. Using biochemical analyses of NO and bioassays for NOS activity, unequivocal evidence that RBC constitutively carry an active NOS could be provided. Thus RBC not simply scavenges NO, but instead represent an important source of vascular NO formation. RBC NOS resembles a variety of specific regulatory pathways of eNOS, in that it is stereospecifically stimulated by the substrate L-arginine, it is sensitive to common NOS inhibitors, and its regulation depends on the intracellular calcium level and the phosphorylation at serine 1177 regulated by the PI3K. Erythrocytes carry important enzymes of the L-arginine metabolism, such as arginase degrading the eNOS substrate, dimethylarginine dimethylaminohydrolase (DDAH), an enzyme metabolising endogenous NOS inhibitors, and cationic amino acid transporters. Although admittedly speculative, RBC might fine tune their NO production via control of substrate availability [61].

The RBC NOS may substantially increase the local NO concentration at the immediate vicinity of the outer membrane thus contributing to an intrinsic barrier preventing consumption of NO derived from other sources than RBC themselves. Alternatively, intrinsic NO formation may alter the electromechanical properties of the RBC membrane such as proteins and lipoproteins preventing consumption of NO by RBC.

18.9 The Potential Relevance of RBC NOS Activity

After the discovery and characterization of the RBC NOS the main question that arises: What is the relevance of RBC NOS?

It is supposed that RBC NOS is most important for the function and perfusion in the microcirculatory bed. The microcirculation is defined as vessels with a diameter of less than 150 μm. This definition includes arterioles (diameter 10 – 100 μm), capillaries (4 – 10 μm) and venules (10 – 100 μm). The main mechanisms of blood flow regulation in the microcirculatory bed are the active tone of the arterioles on the one hand and the hemorheological properties on the other hand. In larger vessels with a diameter of more than 30 μm the active tone and the vessel resistance, respectively, are considered as the major determinant of the total resistance. But with decreasing vessel diameter the contribution of hemorheology to the total resistance increases. Rheology is the study of the deformation and flow of matter under the influence of an applied stress. Corresponding to this definition, hemorheology deals with the flow properties of blood. It can be distinguished between plasmatic factors such as protein content and blood cell properties, for example RBC deformability, RBC aggregation, platelet aggregation and platelet and leukocyte adhesion. It is known that several of the mentioned blood cell functions are influenced by NO. Beside the major role of NO, the regulation of the vascular tone, NO influences RBC deformability, platelet aggregation and adhesion of leukocytes to the endothelium. NO in the cardiovascular system mainly comes from the constitutively expressed endothelial NOS. But from the larger to the smaller vessels, the expression of the endothelial NOS in the vascular wall is more and more reduced [62].

RBC deformability is an excellent example of the relevance of RBC NOS. RBC passing through the capillaries need to be flexible as their average diameter of 7.6 μm is larger than the diameter of the smallest capillaries. RBC deformability is not only modulated by NO produced by the constitutively expressed endothelial NOS or by NO-releasing drugs, but by the RBC-derived NO. To test the hypothesis that RBC modulate their deformability via endogenous NO synthesis, erythrocytes were subjected to an assay which indexes the degree of cell flexibility in relation to a flow rate of cells through a filter system. Stimulation of RBC NOS increased flow rate and thus RBC deformability. Following NOS inhibition, or addition of the NO scavenger oxyhemoglobin, erythrocytes became more rigid preventing the adequate passage of red blood cells. In the microcirculation, where endothelial NOS expression level is low, RBC may regulate their deformability by their "own" NOS allowing RBC to react on the special structural, functional and metabolic circumstances during the passage through the capillary bed [61]. The mechanisms of NO influencing RBC deformability are still unknown. There are different methods measuring RBC deformability, for example the filtration method [63], the ektacytometry [64] or the micropipette aspiration [65, 66] (Fig. 18.3). As the described methods detect different elements of RBC deformability, comparing measurements may help to understand NO action on RBC. In addition, *in-vitro* experiments showed as well, that the activity of RBC NOS affects platelet aggregation. Stimulation of RBC NOS by the substrate L-arginine results in a reduction of platelet aggregation whereas inhibition of RBC NOS by a competitive inhibitor increases platelet aggregation. This is believed to be an additional hint of the relevance of RBC NOS [61].

Until today, it is not possible to exclusively influence RBC NOS activity. Thereby, it can only be speculated about the importance of RBC NOS *in-vivo*. It is suggested,

18.9 The Potential Relevance of RBC NOS Activity

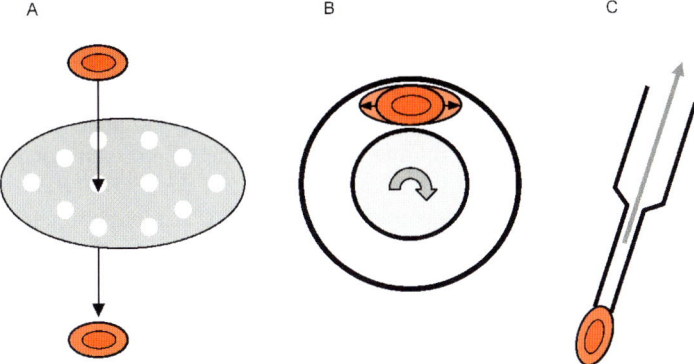

Fig. 18.3 Measurement of RBC deformability. Different methods detect different aspects of RBC deformability. The filtration of RBC forces the cells to fold and squeeze through the pores (**A**). When measuring RBC deformability by ektacytometry RBC have to align and to elongate in dependence of the applied shear stress (**B**). The micropipette aspiration is the only method that can distinguish between the cytosolic and membranous component of RBC deformability (**C**)

that it may be most important in the microcirculation with its very small capillaries. Two arguments underline this suggestion: 1. Endothelial NOS expression in the microcirculation is low. Therefore, NO derived from RBC NOS contributes to a main part to the local NO production. 2. Due to the structural properties and blood flow profile in the microcirculatory bed, blood cells are in close contact allowing direct interaction between NO-producing RBC and other blood cells or the microvascular endothelium.

Throughout the literature there are interesting links of a potential interaction of RBC NOS to diverse relevant *in-vivo* aspects. NO determines survival in transgenic mice over-expressing erythropoietin. In these animals, systemic inhibition of NO synthesis was associated with occlusive RBC accumulation in terminal arterioles and death of all animals within hours [67]. This result points to the relevance of RBC deformability for the passage of blood through the microvasculature. NO donors affect membrane fluidity and also the deformability of RBC [15, 68, 69]. The impairment of RBC deformability via scavenging of RBC-derived NO may explain the deleterious effects of blood substitutes containing free hemoglobin and also the microcirculatory damage observed in sickle cell patients during hemolytic crises. NO bioavailability is crucial in patients with sickle cell disease and modifications of the circulating NO pool by inhaled NO or L-arginine exerts beneficial effects in acute vaso-occlusive crises and pulmonary hypertension [70–72]. In general, RBC-derived NO may be important in diseases associated with altered hemorheologic features such as thalassemia, spherocytosis, thrombocytosis and malaria. NO is beneficial in *falciparum* malaria by inhibiting adherence of infected RBC to the vascular wall [73]. Further investigations are needed to strengthen the relevance of RBC NOS. Does RBC NOS also influence RBC aggregation and RBC adhesion to the endothelium? Does RBC NOS affect platelet activation or adhesion? What are the mechanisms of action on the cellular level?

18.10 Outlook

The first step for a better understanding of the importance of RBC NOS *in-vivo* is an improvement of the visualization techniques of the microcirculatory bed. Diverse diagnostic tools in the clinical routine which examine the coronary or peripheral microcirculation are developed. An example for the assessment of the coronary microcirculation is the quantitative coronary angiography. An intracoronary Doppler wire can be placed through a catheter in distal segments of the coronary vascular tree. On the tip of the Doppler wire probe a piezoelectric crystal is mounted which is the key element of this technique. This technique gives information about the

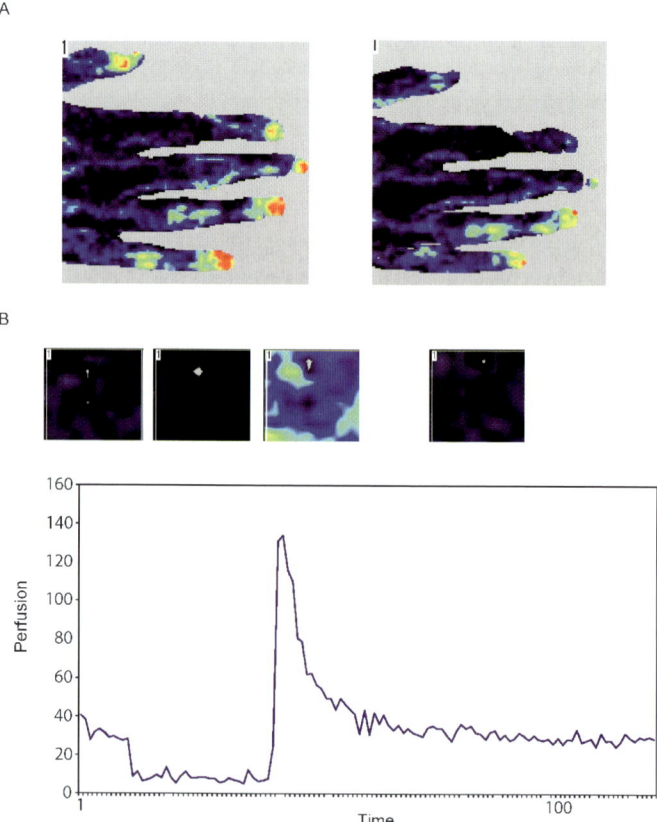

Fig. 18.4 Original registration of Laser Doppler Perfusion Imaging (LDPI). **A** LDPI allows the scanning of cutaneous microcirculation of extended areas. This technique can be applied for the cold pressure test of the hand. The *upper panel* shows an original registration of the basal perfusion, the *lower panel* the perfusion of the same hand after 30 seconds of ice water. **B** In addition, LDPI can serve for the detection of the microcirculatory dynamics. The *upper panel* shows the field of registration, the *lower panel* shows the changes in perfusion in the setting of a reactive hyperemia (*1*: basal perfusion, *2*: perfusion during arterial occlusion, *3*: Perfusion after release of occlusion, *4*: Return to baseline perfusion)

18.10 Outlook

blood flow velocity and the flow profile in the smaller coronary vessels. In addition, the functional capacity can be measured as the coronary flow reserve after the coronary application of a vasodilator, for instance adenosine or nitroglycerine. Due to the invasive nature of the coronary angiography, intensive work is currently done on the improvement of existing imaging techniques, for example contrast-enhanced magnetic resonance tomography. In the peripheral circulation, the microcirculation can be examined non-invasively by the Laser Doppler technique (Fig. 18.4). The principle of this method is the reflection of the laser light by moving red blood cells. In comparison to the emitted light, the reflected light shows a shift in its frequency which is linearly dependent on the RBC velocity. But all mentioned methods have a common restriction: they do show neither the vessels nor the blood cells directly.

For human investigations there are two methods that can visualize vessels on the capillary level: the capillaroscopy and Orthogonal Polarized Spectral (OPS) imaging. Capillaroscopy or capillary microscopy is usually applied on the finger nailfold. The native technique allows the evaluation of the morphology and the capillary den-

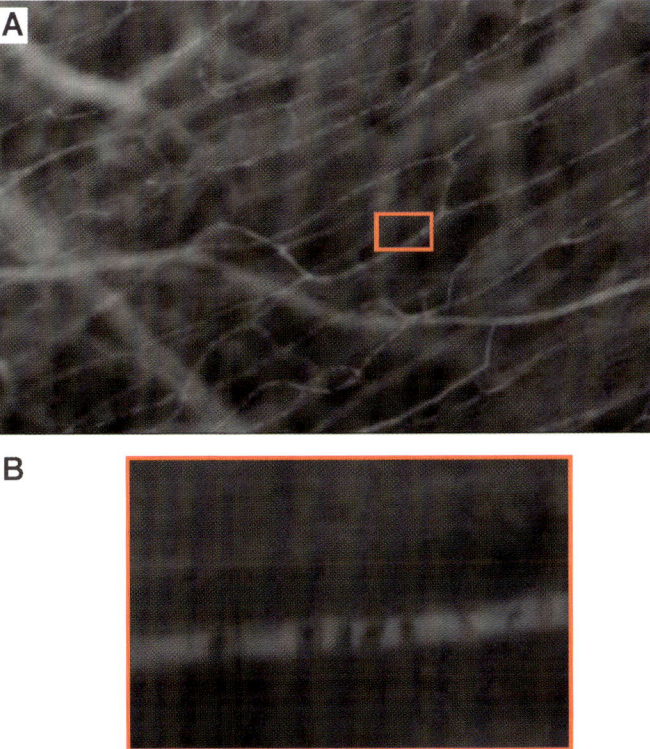

Fig. 18.5 Videomicroscopic detection of the vascular structure and blood cells in the mouse model of the dorsal skinfold chamber. This technique allows the analysis of the morphology and function of the microcirculatory bed (**A**) and the real time detection of circulating blood cells (**B**). The *red rectangle* indicates the location of the highly magnified section in **B**

sity. After injection of a fluorescent dye blood flow velocity can be measured and blood flow distribution patterns can be analyzed. In addition, the capillary pressure can be examined by the cannulation of nailfold capillaries using very thin glass micropipettes [74]. OPS imaging is – simplified – a portable videomicroscope. Emitted polarized light is absorbed by hemoglobin and reflected by the background. Therefore, RBC appear dark in the resulting image. OPS can be applied on tissues covered by a thin epithelial layer, for example the sublingual mucosa [75].

Not in human, but in animal studies, high resolution microscopy is on its way in visualizing microcirculation on the cellular level. Different animal models have long been used for intravital microscopy of the microcirculatory bed, for example the rat or mouse mesentery, the cremaster muscle, the ear of the hairless mouse and the hamster or mouse dorsal skinfold chamber [76]. With the technical progress in real time image acquisition and analysis the investigations of the microcirculation on the single cell have started (Fig. 18.5). The systematic real time assessment of hemorheological parameters and the evaluation RBC NOS *in-vivo* can follow.

References

1. Pryor WA (1998) Nitric oxide. General introduction. Free Radic Biol Med 25:383
2. Kuo PC, Schroeder RA (1995) The emerging multifaceted roles of nitric oxide. Ann Surg 221:220–235
3. Moncada S, Higgs A (1993) The L-arginine-nitric oxide pathway. N Engl J Med 329:2002–2012
4. Alderton WK, Cooper CE, Knowles RG (2001) Nitric oxide synthases: structure, function and inhibition. Biochem J 357:593–615
5. Förstermann U, Kleinert H (1995) Nitric oxide synthase: expression and expressional control of the three isoforms. Naunyn Schmiedebergs Arch Pharmacol 352:351–364
6. Ignarro LJ, Kadowitz PJ (1985) Pharmacological and physiological role of cGMP in vascular smooth muscle relaxation. Annu Rev Pharmacol Toxicol 25:171–191
7. Moncada S (1997) Nitric oxide in the vasculature: physiology and pathophysiology. Ann N Y Acad Sci 811:60–69
8. Arnold WP, Mittal CK, Katsuki S, Murad F (1977) Nitric oxide activates guanylate cyclase and increases guanosine $3':5'$-cyclic monophosphate levels in various tissue preparations. Proc Natl Acad Sci USA 74:3203–3207
9. Furchgott RF, Zawadzki JV (1980) The obligatory role of endothelial cells in the relaxation of arterial smooth muscle by acetylcholine. Nature 288:373–376
10. Ignarro LJ, Buga GM, Wood KS, Byrns RE, Chaudhuri G (1987) Endothelium-derived relaxing factor produced and released from artery and vein is nitric oxide. Proc Natl Acad Sci USA 84:9265–9269
11. Palmer RMJ, Ferrige AG, Moncada S (1987) Nitric oxide release accounts for the biological activity of endothelium-derived relaxing factor. Nature 327:524–526
12. Ignarro LJ, Cirino G, Casini A, Napoli C (1999) Nitric oxide as a signaling molecule in the vascular system: An overview. J Cardiovasc Pharmacol 34:879–886
13. Radomski MW, Palmer RM, Moncada S (1990) An L-arginine/nitric oxide pathway present in human platelets regulates aggregation. Proc Natl Acad Sci USA 87:5193–5197
14. Radomski MW, Palmer RMJ, Moncada S (1987) Endogenous nitric oxide inhibits human platelet adhesion to vascular endothelium. Lancet 7:1057–1058
15. Starzyk D, Korbut R, Gryglewski RJ (1999) Effects of nitric oxide and prostacyclin on deformability and aggregability of red blood cells of rats ex vivo and in vitro. J Physiol Pharmacol 50:629–637
16. Arnal J-F, Dinh-Xuan A-T, Pueyo M, Darblade B, Rami J (1999) Endothelium-derived nitric oxide and vascular physiology and pathology. Cell Mol Life Sci 55:1078–1087
17. Kawashima S, Yokoyama M (2004) Dysfunction of endothelial nitric oxide synthase and atherosclerosis. Arterioscler Thromb Vasc Biol 24:998–1005
18. Vallance P, Chan N (2001) Endothelial function and nitric oxide: clinical relevance. Heart 85:342–350

19. Rapoport RM, Murad F (1983) Agonist-induced endothelium-dependent relaxation in rat thoracic aorta may be mediated through cGMP. Circ Res 52:352–357
20. van de Voorde J, Leusen I (1986) Endothelium-dependent and independent relaxation of aortic rings from hypertensive rats. Am J Physiol Heart Circ Physiol 250:H711–H717
21. Horio Y, Yasue H, Rokutanda M et al. (1986) Effects of intracoronary injection of acetylcholine on coronary arterial diameter. Am J Cardiol 57:984–989
22. Takase B, Hamabe A, Satomura K et al. (2006) Comparable prognostic value of vasodilator response to acetylcholine in brachial and coronary arteries for predicting long-term cardiovascular events in suspected coronary artery disease. Circulation 70:49–56
23. Coretti MC, Anderson TJ, Benjamin EJ et al. (2002) Guidelines for the ultrasound assessment of endothelial-dependent flow-mediated vasodilation of the brachial artery. J Am Coll Cardiol 39:257–265
24. Joannides R, Haefeli WE, Linder L et al. (1995) Nitric oxide is responsible for flow-dependent dilatation of human peripheral conduit arteries in vivo. Circulation 91:1314–1319
25. Celermajer DS, Sorensen KE, Bull C, Robinson J, Deanfield JE (1994) Endothelium-dependent dilation in the systemic arteries of asymptomatic subjects relates to coronary risk factors and their interaction. J Am Coll Cardiol 24:1468–1474
26. Kelm M (2002) Flow-mediated dilatation in human circulation: diagnostic and therapeutic aspects. Am J Physiol Heart Circ Physiol 282:H1–H5
27. Lorenz MW, Markus HS, Bots ML, Rosvall M, Sitzer M (2007) Prediction of clinical cardiovascular events with carotid intima-media thickness: a systematic review and meta-analysis. Circulation 115:459–467
28. Kelm M, Strauer BE (2002) Microvascular dysfunction and myocardial ischaemia in hypertensive patients. Excerpta Medica 12:4–7
29. Tarpey MM, Fridovich I (2001) Methods of detection of vascular reactive species Nitric Oxide, Superoxide, Hydrogen Peroxide, and Peroxynitrite. Circ Res 89:224–236
30. Kelm M, Dahmann R, Wink D, Feelisch M (1997) The nitric oxide-superoxide assay: insights into the biological chemistry of the NO/O_2-interaction. J Biol Chem 272:9922–9932
31. Kojima H, Nakatsubo N, Kikuchi K et al. (1998) Detection and imaging of nitric oxide with novel fluorescent indicators: diaminofluoresceins. Anal Chem 70:2446–2453
32. Porasuphatana S, Weaver J, Budzichowski TA, Tsai P, Rosen GM (2001) Differential effect of buffer on the spin trapping of nitric oxide by iron chelates. Anal Biochem 298:50–56
33. Heger J, Gödecke A, Flögel U et al. (2002) Cardiac-specific overexpression of inducible nitric oxide synthase does not result in severe cardiac dysfunction. Circ Res 90:93–99
34. Kelm M (1999) Nitric oxide metabolism and breakdown. Biochim Biophys Acta 1411:273–289
35. Stamler JS, Jaraki O, Osborne J et al. (1992) Nitric oxide circulates in mammalian plasma primarily as S-nitroso adduct of serum albumin. Proc Natl Acad Sci USA 89:7674–7677
36. Ignarro LJ, Lippton H, Edwards JC et al. (1981) Mechanism of vascular smooth muscle relaxation by organic nitrates, nitrites, nitroprusside and nitric oxide: evidence for the involvement of S-nitrosothiols as active intermediates. J Pharmacol Exp Ther 218:739–749
37. Rassaf T, Preik M, Kleinbongard P et al. (2002) Evidence for in vivo transport of bioactive nitric oxide in human plasma. J Clin Invest 109:1241–1248
38. Rassaf T, Kleinbongard P, Preik M et al. (2002) Plasma nitrosothiols contribute to the systemic vasodilator effects of intravenously applied NO: experimental and clinical study on the fate of NO in human blood. Circ Res 91:470–477
39. Grau M, Hendgen-Cotta U, Brouzos P et al. (2007) Recent methodological advances in the analysis of nitrite in the human circulation: Nitrite as a biochemical parameter of the L-arginine/NO pathway. Journal of Chromatography B 851:106–123
40. Griess JP (1879) Bemerkungen zu der Abhandlung der HH. Wesely und Benedikt "Über einige Azoverbindungen". Ber Deutsch Chem Ges 12:426–428
41. Kleinbongard P, Rassaf T, Dejam A, Kerber S, Kelm M (2002) Griess method for nitrite measurement of aqueous and protein-containing sample. Methods Enzymol 359:158–168

References

42. Preik-Steinhoff H, Kelm M (1996) Determination of nitrite in human blood by combination of a specific sample preparation with high performance anion-exchange chromatography and electrochemical determination. J Chromatogr B Biomed Appl 685:348–352
43. Kleinbongard P, Dejam A, Lauer T et al. (2003) Plasma nitrite reflects constitutive nitric oxide synthase activity in mammals. Free Radic Biol Med 35:790–796
44. Kleinbongard P, Dejam A, Lauer T et al. (2006) Plasma nitrite concentrations reflect the degree of endothelial dysfunction in humans. Free Radic Biol Med 40:295–302
45. Doyle MP, Hoekstra JW (1981) Oxidation of nitrogen oxides by bound dioxygen in hemoproteins. J Inorg Biochem 14:351–358
46. McMahon T, Moon RE, Luschinger BP et al. (2002) Nitric oxide in the human respiratory cycle. Nat Med 8:711–717
47. Joshi MS, Ferguson TB Jr, Han TH et al. (2002) Nitric oxide is consumed, rather than conserved, by reaction with oxyhemoglobin under physiological conditions. Proc Natl Acad Sci USA 99:10341–10346
48. Lancaster JR Jr (1997) A tutorial on the diffusibility and reactivity of free nitric oxide. Nitric Oxide 1:18–30
49. Liao JC, Hein TW, Vaughn MW, Huang K-T, Kuo L (1999) Intravascular flow decreases erythrocyte consumption of nitric oxide. Proc Natl Acad Sci USA 96:8757–8761
50. Liu X, Miller MJ, Joshi MS et al. (1998) Diffusion-limited reaction of free nitric oxide with erythrocytes. J Biol Chem 273:18709–18713
51. Vaughn MW, Huang KT, Kuo L, Liao JC (2000) Erythrocytes possess an intrinsic barrier to nitric oxide consumption. J Biol Chem 275:2342–2348
52. Wennmalm A, Benthin G, Edlund A et al. (1993) Metabolism and excretion of nitric oxide in humans. An experimental and clinical study. Circ Res 73:1121–1127
53. Stamler JS, Jia L, Eu JP et al. (1997) Blood flow regulation by S-nitrosohemoglobin in the physiological oxygen gradient. Science 276:2034–2037
54. McMahon TJ, Stone AE, Bonaventura J, Singel DJ, Stamler JS (2000) Functional coupling of oxygen binding and vasoactivity in S-nitrosohemoglobin. J Biol Chem 275:16738–16745
55. Gladwin MT (2005) Nitrite as an intrinsic signaling molecule. Nature Chemical Biology 1:245–246
56. Dejam A, Hunter CJ, Schechter AN, Gladwin MT (2004) Emerging role of nitrite in human biology. Blood Cells Mol Dis 32:423–429
57. Dejam A, Hunter CJ, Pelletier MM et al. (2005) Erythrocytes are the Major Intravascular Storage Sites of Nitrite in Human Blood. Blood 106:734–739
58. Hickey M, Granger D (2001) Inducible nitric oxide synthase (iNOS) and regulation of leucocyte/endothelial cell interactions: studies in iNOS-deficient mice. Acta Physiol Scand 173:119–126
59. Freedman JE, Sauter R, Battinelli EM et al. (1999) Deficient platelet-derived nitric oxide and enhanced hemostasis in mice lacking the NOSIII gene. Circ Res 84:1416–1421
60. Metha JL, Metha P, Li D (2000) Nitric oxide synthase in adult red blood cells: vestige of an earlier age or a biologically active enzyme? J Lab Clin Med 135:430–431
61. Kleinbongard P, Schulz R, Rassaf T et al. (2006) Red blood cells express a functional endothelial nitric oxide synthase. Blood 107:2943–2951
62. Teichert A-M, Miller TL, Tai SC et al. (2000) In vivo expression profile of an endothelial nitric oxide synthase promoter-reporter transgene. Am J Physiol Heart Circ Physiol 278:H1352–H1361
63. Reid HL, Barnes AJ, Lock PJ, Dormandy JA, Dormandy TL (1976) Technical methods. A simple method for measuring erythrocyte deformability. J Clin Path 29:855–858
64. Dobbe JG, Streekstra GJ, Hardeman MR, Ince C, Grimbergen CA (2002) Measurement of the distribution of red blood cell deformability using an automated rheoscope. Cytometry 50:313–325
65. Maggakis-Kelemen C, Biselli M, Artmann GM (2002) Determination of the elastic shear modulus of cultured human red blood cells. Biomed Tech (Berl) 47:106–109

66. Artmann GM, Paul Sung K-L, Horn T et al. (1997) Micropipette Aspiration of Human Erythrocytes Induces Echinocytes via Membrane Phospholipid Translocation. Biophys J 72:1434–1441
67. Ruschitzka FT, Wenger RH, Stallmach T et al. (2000) Nitric oxide prevents cardiovascular disease and determines survival in polyglobulic mice overexpressing erythropoietin. Proc Natl Acad Sci USA 97:11609–11613
68. Tsuda K, Kimura K, Nishio I, Masuyama Y (2000) Nitric oxide improves membrane fluidity of erythrocytes in essential hypertension: an electron paramagnetic resonance investigation. Biochem Biophys Res Commun 275:946–954
69. Bor-Kucukatay M, Wenby RB, Meiselman HJ, Baskurt OK (2003) Effects of nitric oxide on red blood cell deformability. Am J Physiol Heart Circ Physiol 284:H1577–H1584
70. Reiter CD, Wang X, Tanus-Santos JE et al. (2002) Cell – free hemoglobin limits nitric oxide bioavailability in sickle – cell disease. Nat Med 8:1383–1389
71. Weiner DL, Hibberd PL, Betit P et al. (2003) Preliminary assessment of inhaled nitric oxide for acute vaso-occlusive crisis in pediatric patients with sickle cell disease. JAMA 289:1136–1142
72. Morris CR, Kuypers FA, Larkin S et al. (2000) Arginine therapy: a novel strategy to induce nitric oxide production in sickle cell disease. Br J Haematol 111:498–500
73. Serirom S, Raharjo WH, Chotivanich K et al. (2003) Anti-adhesive effect of nitric oxide on Plasmodium falciparum cytoadherence under flow. Am J Pathol 162:1651–1660
74. Shore AC (2000) Capillaroscopy and the measurement of capillary pressure. Br J Clin Pharmacol 50:501–513
75. Ince C (2005) The microcirculation is the motor of sepsis. Crit Care 9:S13–S19
76. Menger MD, Laschke MW, Amon M et al. (2003) Experimental models to study microcirculatory dysfunction in muscle ischemia-reperfusion and osteomyocutaneous flap transfer. Langenbecks Arch Surg 388:281–290

Chapter 19
Vascular Endothelial Responses to Disturbed Flow: Pathologic Implications for Atherosclerosis

Jeng-Jiann Chiu[1], Shunichi Usami[2], Shu Chien[2,3]

[1] Division of Medical Engineering Research, National Health Research Institutes, Miaoli 350, Taiwan, ROC,
jjchiu@nhri.org.tw
[2] Department of Bioengineering and Whitaker Institute of Biomedical Engineering, University of California, San Diego, La Jolla, CA 92093-0412
[3] Department of Medicine, University of California, San Diego, La Jolla, CA 92093-0427

Abstract Atherosclerosis is prone to develop at branches and bends of the arterial tree, where laminar blood flow is disturbed by recirculation, with a non-uniform and irregular distribution of wall shear stress. Vascular endothelial cells (ECs) form an interface between the flowing blood and the vessel wall, and are exposed to blood flow-induced shear stress. Recent evidence suggests that laminar blood flow and sustained high shear stress modulate the expression of EC genes and proteins that function to protect against atherosclerosis, whereas disturbed flow and the associated oscillatory and low shear stress up-regulate pro-atherosclerotic genes and proteins that promote development of atherosclerosis. Understanding of the effects of disturbed flow on ECs not only provides mechanistic insights into the role of complex flow patterns in the pathogenesis of atherosclerosis, but also helps to define the differences between quiescent (non-atherogenic) and activated (atherogenic) ECs, which may lead to the discovery and identification of new therapeutic strategies. In this chapter, we summarize the current experimental and theoretical knowledge on the effects of disturbed flow on ECs, in terms of their signal transduction, gene expression, structure, and function. Our purpose is to provide the basic information on the effects of disturbed flow on ECs that is necessary to understand the etiology of lesion development in the disturbed flow-regions of the arterial tree.

19.1 Introduction

Atherosclerosis, which is responsible for most cardiovascular-related morbidity and mortality, develops preferentially in regions of the arterial tree where non-uniform and irregular distribution of wall shear stress is generated by complex patterns of blood flow. It has been recognized that hemodynamic characteristics determine the location of lesions and contribute to the pathogenesis of atherosclerosis. Vascular endothelial cells (ECs), which form an interface between the flowing blood and the vessel wall, occupy an unique location directly exposing to blood flow-induced shear stress. Recent evidence suggests that laminar blood flow and sustained high shear stress in the straight part of the artery down-regulate atherogenesis-related genes [e. g., monocyte chemotactic protein-1 (MCP-1)] and up-regulate antioxidant and growth-arrest genes in ECs. In contrast, disturbed flow observed at branch points of the arterial tree and the associated oscillatory and low shear stress cause sustained activation of MCP-1 expression in ECs and enhance monocyte infiltration into the arterial wall. These findings provide a cellular and molecular basis for the explanation of the preferential localization of atherosclerotic lesions at regions of disturbed flow, such as the arterial branch points and curvatures. Understanding of the effects of disturbed flow on EC signaling, gene expression, structure, and function not only provides the molecular and cellular bases for the role of complex flow patterns in the pathogenesis of atherosclerosis, but also helps to define the differences between quiescent (non-atherogenic) and activated (atherogenic) ECs at the molecular and gene expression levels, which may consequently lead to the discovery and identification of novel atherogenesis-related genes and new therapeutic strategies. In this chapter, we summarize the current experimental and theoretical knowledge on the effects of disturbed flow on ECs, in terms of their signal transduction, gene expression, structure, and function. The discussion focuses mainly on vascular endothelial responses to disturbed flow observed at branch points and curvatures and simulated in the vertical step-flow channel. Our purpose is to provide the basic information on the effects of disturbed flow on ECs that is necessary to understand the etiology of lesion development in the disturbed flow-regions of the arterial tree.

19.2 Endothelial Dysfunction is a Marker of Atherosclerotic Risk

Atherosclerosis is responsible for more than 50% of all mortality in the USA, Europe and some Asian countries. It is a principal contributor to the pathogenesis of heart attack and stroke. The lesions proceed through a series of pathological stages, including intimal thickening, fatty streaks, fibrous plaques, and the formation of complicated plaques, and involve components including dysfunction of vascular ECs and their interactions with circulating white blood cells (WBCs), platelets, and lipids, the extracellular matrix (ECM), and activated smooth muscle cells (SMCs) (Ross 1993; Lusis 2000; Libby 2002; Libby et al. 2002). ECs form an interface between the flowing blood and the vessel wall, and are subjected to various chemical and mechanical stimuli (Ross 1993; Lusis 2000). In addition to serving as a permeability barrier, ECs perform many important functions, such as the production, secretion, and metabolism of biochemical substances, the control of contractility of the underlying SMCs, and the recruitment of circulating WBCs into the vessel wall. Dysfunction of ECs induced by unfavorable chemical and mechanical stimuli has been shown to be a critical pathogenic risk factor for atherosclerosis

(Gimbrone et al. 2000) and its complications (Kinlay and Ganz 1997; Ross 1999; Bonetti et al. 2003). The pathophysiological consequences of endothelial dysfunction include the impairment of EC-dependent vasodilation, increased permeability to macromolecules such as lipoproteins, increased expression of pro-inflammatory agents, enhanced recruitment and accumulation of monocytes in the intima as foam cells, altered regulation in growth and survival of vascular cells (e. g., decreased EC regeneration and increased SMC proliferation and migration), and altered hemostatic/fibrinolytic balances (enhanced thrombin generation, platelet aggregation and adhesion, and fibrin deposition) (Lerman and Burnett 1992; Gimbrone et al. 2000; Bonetti et al. 2003). Biochemical and cellular factors that can cause endothelial dysfunction especially relevant to atherogenesis include cytokines and chemokines, growth factors, and hormones; infection by bacteria, viruses, and other pathogens; chronic exposure to hyperhomocysteinemia and/or hyperlipidemia; and accumulation of oxidized low density lipoprotein (oxLDL) and their components (e. g., lysophosphatidylcholine) within the vessel wall (Gimbrone et al. 2000). In addition, various types of hemodynamic forces can also modulate the expression of pathophysiologically relevant genes in ECs and the consequent modulation of their structure and function (Resnick and Gimbrone 1995; Gimbrone et al. 1997, 2000). The possibility that hemodynamic forces act as atherogenic stimuli for endothelial dysfunction provides a rational explanation for the long-standing observation that the earliest lesions of atherosclerosis characteristically develop in a non-random pattern, the geometry of which correlates with branch points and curvatures, i. e., regions with disturbed flow (Cornhill and Roach 1976; Glagov et al. 1988; Gimbrone et al. 2000).

19.3 Correlation Between Lesion Locations and Disturbed Flow Regions of the Arterial Tree

Blood vessels are constantly exposed to various types of hemodynamic forces, including fluid shear stress, cyclic stretch, and hydrostatic pressure, induced by the pulsatile blood flow and pressure. ECs bear primarily the wall shear stress, the component of frictional forces arising from the blood flow and acting parallel to the vessel luminal surface (Davies 1995; Fung 1997; Li et al. 2005). The magnitude of shear stress can be estimated in straight vessels by using the Poiseuille's law (Fung 1997; Malek et al. 1999), which defines that shear stress is proportional to the viscosity of blood and inversely proportional to the third power of the internal radius of vessel (Zamir 1976; Kamiya et al. 1984; LaBarbera 1990; Hishikawa et al. 1995). Experimental measurements using different methods have shown that the magnitudes of shear stress range from 1 to 6 dynes/cm^2 in the venous system and from 10 to 70 dynes/cm^2 in the arterial network (Nerem et al. 1998; Malek et al. 1999).

There has been a long-standing observation that atherosclerotic lesions develop near the arterial branches and curvatures, where the local flow is disturbed (e. g., non-uniform and irregular oscillation and recirculation) (Caro et al. 1971; Bharadvaj et al. 1982; Zarins et al. 1983; Motomiya and Karino 1984; Asakura and Karino 1990; Nerem 1992; Malek et al. 1999). Conversely, the processes of lesion formation are generally absent in the linear regions of vasculature, which are typically associated with less complex and more uniform flows. The atherosclerotic lesions are preferentially located at the outer walls of bifurcations and regions of flow recirculation (Malek et al. 1999). In these geometrically predisposed regions, fluid shear stress on the vessel wall is significantly lower in magnitude and exhibits directional changes with flow separation and reattachment. Direct measurements and fluid mechanics analyses of models of these lesion-prone areas revealed that shear stress are on the order of $\pm 4\,\mathrm{dynes/cm^2}$ in these areas with a very low net magnitude, in contrast to the values of $>12\,\mathrm{dynes/cm^2}$ in the lesion-free areas (Zarins et al. 1983; Jou et al. 1996; Malek et al. 1999).

The co-localization of atherosclerotic lesions with regions of disturbed flow with low and oscillatory shear stress has been shown throughout the arterial systems, including carotid bifurcation (Zarin et al. 1983; Motomiya and Karino 1984; Gnasso et al. 1997), aortic arch (Topper and Gimbrone 1999), coronary arteries (Asakura and Karino 1990; Friedman et al. 1993), and infrarenal and femoral arterial trees (Pedersen et al. 1997). For example, examination of pathological sections following carotid endarterectomy showed greatest thickness of plaque in the outer wall of carotid sinus, where recirculation flow occurs with the existence of low and oscillatory shear stress (Zarins et al. 1987; Malek et al. 1999). Using echo-Doppler ultrasound to measure the wall shear stress in carotid arteries in young patients, Gnasso et al. (1996) found an inverse relationship between the intima/media thickness ratio and local wall shear stress. They further found that the lesion-affected human carotid arteries exhibited significantly lower wall shear stress than disease-free controls (Gnasso et al. 1997). Ku et al. (1985) showed that intimal thickening correlates well with the oscillatory shear index in carotid bifurcation. High-speed cinematography and flow dynamic analysis using microparticles in postmortem coronary arteries showed good correlations of intimal thickening with regions of bifurcation having disturbed flow and low shear stress (Asakura and Karino 1990); in contrast, the flow-dividers and inner walls with high shear stress showed a lack of development of lesions. Measuring the atherosclerotic lesions in patients with coronary artery disease using serial quantitative coronary angiography, Gibson et al. (1993) found that the progression rates of lesions correlated inversely with the magnitudes of shear stress under the control of systemic risk factors. The co-localization of atherosclerotic lesions with disturbed and low shear regions has been further established in human abdominal aortas both at autopsy (Pedersen et al. 1997) and with noninvasive magnetic resonance phase velocity mapping (Oyre et al. 1997; Oshinski et al. 1995). En-face examination of the luminal surface of human thoracic aortas showed dilated intercellular clefts, irregular endothelial morphology with denuded regions covered with platelets and WBCs, and accumulation of subendothelial macrophages and lymphocytes; these changes were seen in the outer walls of the bifurcations, but not in the inner walls or flow divider (Kolpakov et al. 1996). Similar localization of atherosclerotic lesions has been found in studies on the aorta (Bassiouny et al. 1994; Sawchuk et al. 1994) and carotid arteries (Beere et al. 1992) of experimental animals. These non-random patterns of early atherosclerotic lesions, which are seen in both experimental animals and humans, suggest that complex flow pattern and the associated low and oscillatory shear stress are critical factors in the initiation and progression of atherosclerosis.

19.4 *In Vitro* Studies on the Effects of Disturbed Flow on ECs

19.4.1 *In Vitro* Models for Studying the Effects of Disturbed Flow on ECs

Several commonly used *in vitro* models have been developed for studying the effects of fluid shear stress on ECs. These models include parallel-plate flow chamber and cone-and-plate viscometer (Dewey et al. 1981; Usami et al. 1993). To simulate some features of disturbed flow patterns near arterial branches and bends (e. g., flow separation and reattachment, recirculation, and non-uniform shear stress distribution) (Ku et al. 1985), several in vitro models, including tubular sudden expansion (Karino and Goldsmith 1979; Pritchard et al. 1995; Hinds et al. 2001) and backward-facing step (i. e., vertical-step) (DePaola et al. 1992, 1999; Truskey et al. 1995; Chiu et al. 1998; Barber et al. 1998; Skilbeck et al. 2001), have been established to produce a laminar eddy (Macagno and Hung 1967, 1970). An example of a typical vertical-step flow model is shown in Fig. 19.1 (Chiu et al. 1998, 2003; Chen et al. 2006). The vertical-step flow channel is composed of two parts with different channel heights along its length. The glass slide with EC monolayers and the gaskets were fastened between a polycarbonate base plate and a stainless plate, using vacuum suction or

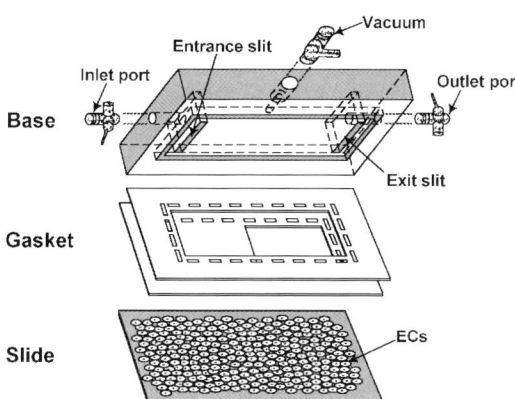

Fig. 19.1 Diagram showing the parallel-plate flow chamber for vertical-step flow. The polycarbonate base plate, the gasket, and the glass slide with EC monolayer are held together by a vacuum suction applied at the perimeter of the slide via port, forming a channel with variable depth. Cultured medium enters at inlet port through entrance slit into the channel, and exits through exit slit and outlet port

clamping. The channel width (w) is 10 mm. The entrance region has a height (h) of 0.25 mm and a length of 15 mm; the main channel has a height (H) of 0.5 mm and a length of 45 mm (Fig. 19.2A). When a designated rate of flow is imposed into the channel, a well-defined laminar recirculation eddy without transition to turbulence is created immediately downstream of the step; this is followed by a region of flow reattachment, and fnally a unidirectional laminar flow is re-established further downstream. The experimental flow patterns can be visualized by phase microscopy

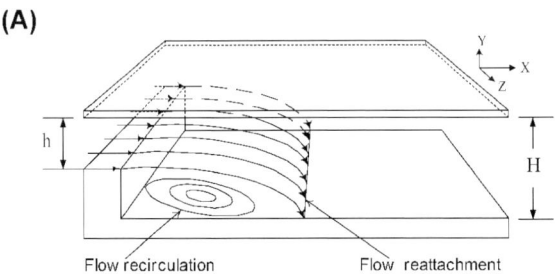

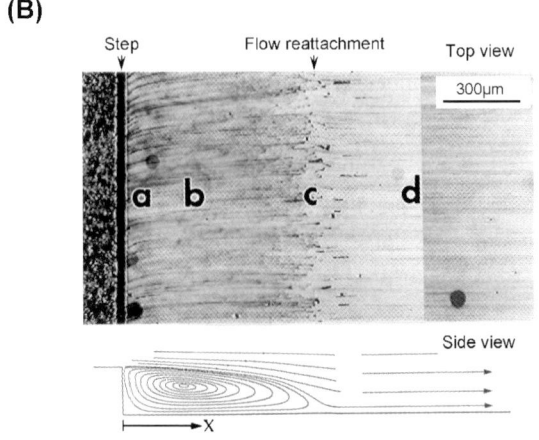

Fig. 19.2 Visualization of flow patterns created in vertical-step flow channel. **A** Schematic diagram of the flow channel and test section. **B** *Upper part*: Phase-contrast photomicrograph (*top view*) of experimental flow patterns in the vertical-step flow channel. Flow is from *left* to *right* and is made visible with the marker particles. Flow separation occurs in the region distal to the step, forming four specific flow areas: *a*, the stagnant flow area; *b*, the center of the recirculation eddy; *c*, the reattachment flow area; *d*, the fully developed flow area. From on-line microscopic observations, the particles transported from the bulk flow along the curved streamlines with decreasing velocities towards the wall near the reattachment point (area *c*). Some of the particles moved in a retrograde direction (upstream) towards the step in the eddy, while others moved forward to rejoin the mainstream with increasing velocities. In the recirculation eddy, the particles moved upstream from the reattachment point c with increasing velocities and then decelerated when close to the wall of the step (area *a*). The particles were carried away from the surface of chamber by upward curved streamlines as they approached the step. *Lower part*: Schematic drawing of the side view of the streamlines in the vertical step flow deduced from the top view photograph. Reproduced with permission from Chiu et al. (2003)

19.4 In Vitro Studies on the Effects of Disturbed Flow on ECs

using marker particles (Fig. 19.2B). Position "*a*" is the flow stagnation area; "*b*" is the area below the center of the recirculation eddy; "*c*" is the reattachment flow area; and "*d*" is located in the fully developed flow area. The flow separation region (i.e., areas *a*, *b*, and *c*) is designated as disturbed flow and the restored unidirectional region downstream (i.e., area *d*) represents undisturbed laminar flow. Thus, this *in vitro* model of flow disturbance permits the study of ECs located in different flow environments in the same monolayer.

The wall shear stress created in the vertical-step flow can be characterized by measurements using micron-resolution particle image velocimetry (μPIV) and by computational simulation (Chiu et al. 1998, 2003). Figure 19.3A shows an example of μPIV-measured near-wall velocity components and contours parallel to the surface of the channel in the vicinity of the reattachment point [at a Reynolds number (based on the inlet flow rate and hydraulic diameter) of 100]. The wall shear stress distributions were determined from the measured velocity gradients at the wall or by computational simulation (Fig. 19.3B) (Chiu et al. 2003). It is noted that large shear stress gradients exist in the regions of flow disturbance, particularly at the sites near the step and the reattachment point, where the mean shear stress values are very low ($0 \sim 0.5$ dynes/cm^2). The shear stress below the center of the recirculation eddy in disturbed flow is relatively high (~ 6 dynes/cm^2), in agreement with the findings of Truskey et al. (1995). This high level of shear stress is apparently due to the high velocity gradient of the recirculation eddy. The shear stress distribution in the fully developed flow area (*d*) is nearly constant (~ 7 dynes/cm^2). These results in the fully developed area are consistent with those obtained for a uniform Poiseuille

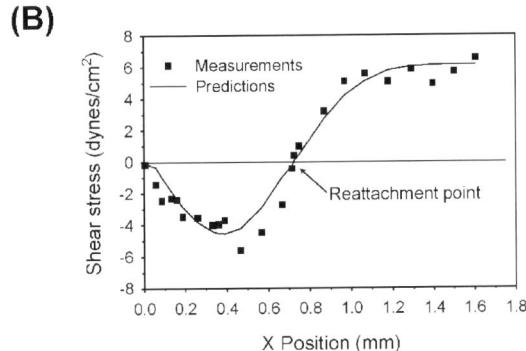

Fig. 19.3 Measurement of wall shear stress in the vertical-step flow channel. **A** A representative plot of the μPIV-measured near-wall velocity components and contours parallel to the surface of the channel in the vicinity of the reattachment point. **B** Comparison of computed and measured shear stress distributions along the axisymmetric centerline of the flow channel. Reproduced with permission from Chiu et al. (2003)

flow, $\tau = 6Q\mu/(wH^2)$, where Q is the flow rate, μ is the dynamic viscosity, w is the channel width, and H is the channel height.

19.4.2 Effects of Disturbed Flow on EC Morphology, Cytoskeletal Organization, and Junctional Proteins

ECs change their shape in a flow pattern- and shear stress-dependent manner. An example of morphological changes of cultured ECs induced by exposure to a disturbed flow in vertical-step flow channel is shown in Fig. 19.4 (at a Reynolds number of 243). A pronounced cell alignment with the flow direction can be observed in the fully developed flow area d (unidirectional laminar flow) at a relatively high level of shear stress (21 dynes/cm^2). Most cells in areas a and c are rounded in shape, as both areas have similar relatively static flow environments with a very low shear stress ($\sim$0.5 dynes/cm^2); however, the cells in area b tend to be partly elongated and aligned with the local flow direction due to the relatively high shear stress level (22 dynes/cm^2) below the center of the eddy. These results on cell shape and orientation are comparable with those reported by the *in vitro* studies on steady laminar flow (Dewey et al. 1981; Eskin et al. 1984; Levesque and Nerem 1985) and disturbed flow (Davies et al. 1992; DePaola et al. 1992; Truskey et al. 1995), as well as *in vivo* observations in the primate thoracic aorta (Davies et al. 1997, 2001), which showed that ECs are aligned with the longitudinal axis of the vessel in a straight segment with no branches, but they transform to a polygonal morphology in regions around branches, where the shear stress is estimated to move back and forth across the boundary region during each cardiac cycle. These results are also in agreement with the *in vivo* findings by Nerem et al. (1981) on the rabbit aortic intercostal ostia that marked changes in EC morphology were found in the regions proximal and distal to ostia as well as around flow dividers, where the flow is disturbed and the shear stress gradient is high, whereas cells on the aorta are aligned with the flow direction. These findings suggest that EC morphology and orientation may act as a natural marker or indicator of the detailed features of blood flow.

The distribution and organization of EC cytoskeleton can also be altered by changes in flow patterns and shear stress distribution (Franke et al. 1984; Ives

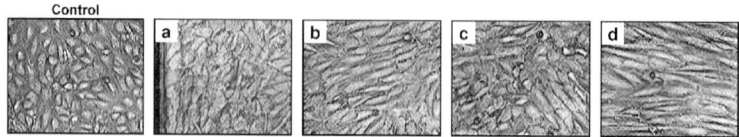

Fig. 19.4 The morphological changes induced in the confluent EC monolayer by exposure to vertical-step flow. **a** stagnant flow area; **b** below center of the recirculation eddy; **c** reattachment flow area; **d** fully developed flow area. The direction of the main flow is *left* to *right*. 0 h sample (control) was photographed from the same experimental monolayer before shearing. Modified from Chiu et al. (1998)

et al. 1986; Kim et al. 1989; Vyalov et al. 1996; Galbraith et al. 1998; Helmke et al. 2000; Lee and Gotlieb 2003). When ECs are subjected to shear flow in the vertical-step flow channel, one can observe their responses to both laminar and disturbed flows in the same chamber. In the fully developed flow area d (with shear stress of 21 dynes/cm^2 at Reynolds number of 243) the ECs display very long, well-organized, parallel actin stress fibers aligned with the flow direction in the central regions of the cells (Fig. 19.5A), and the formation of bright F-actin bundles at cell peripheries is prominent. In the disturbed flow areas (areas a and c) where shear stress is low (~ 0.5 dynes/cm^2) and there is flow reversal, the actin filaments tend to localize mainly at the periphery of the cells, and there are occasional cells that have a complete or partial loss of peripheral F-actin microfilaments, together with the formation of random and short actin bundles. It is notable that the distributions of actin filaments are similar in areas b and d of the vertical-step flow; both areas have a relatively higher shear stress with a clear directionality (21–22 dynes/cm^2), albeit the directions are opposite (forward in b and backward in d). These results are in agreement with those reported by others *in vivo* that low hemodynamic shear favors prominent peripheral actin microfilaments (Vyalov et al. 1996) and high hemodynamic shear (Langille et al. 1991) promotes an increase in central actin stress fibers. In general, the microtubules are distributed relatively uniformly throughout most cell regions, except for their accumulation adjacent to the nucleus (Fig. 19.5B). Mostly, the microtubules display a linear organization and an alignment with the flow direction in the fully developed flow areas (d), whereas they show a randomly organized pattern within the cells in the disturbed flow areas (areas a and c).

Different flow patterns and associated shear stress have also been shown to regulate the expression and distribution of intercellular junctional proteins in ECs. ECs mainly express gap junctional proteins connexin (Cx) 40, 37, and 43, among which Cx43 is the most prominent connexin *in vitro* (Bruzzone et al. 1993). Gabriels and Paul (1998) reported that Cx43 immunoreactivity is present at specific vascular branch points in rat aorta, but absent in samples of the ventral aortic wall distant from branch points, suggesting an association of Cx43 protein expression with complex flow conditions and spatial shear stress gradients. By using a vertical-step flow

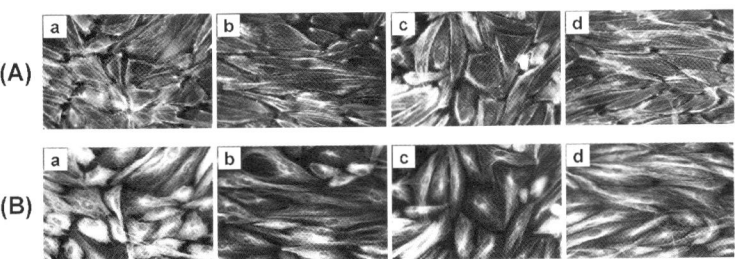

Fig. 19.5 Fluorescence photographs of F-actin filaments (**A**) and microtubules (**B**) after exposure to vertical-step flow for 24 h. **a** stagnant flow area; **b** below center of the recirculation eddy; **c** reattachment flow area; **d** fully developed flow area. The direction of the main flow is *left* to *right*. Magnification: 400×. Modified from Chiu et al. (1998)

device, DePaola et al. (1999) demonstrated that flow disturbance and spatial shear stress gradients in regions of disturbed flow can induce the expression of Cx43, with sustained disruption of its punctuate distribution at cell borders and inhibition of intercellular communication (demonstrated by dye transfer experiment), as compared to the fully developed flow with constant shear stress. Dai et al. (2004) showed that the Cx43 expression is up-regulated by athero-prone waveform (mainly disturbed flow) of shear stress, whereas elevated Cx37 and Cx40 expression levels are associated with the athero-protective waveform (mainly laminar flow). These observations in Cx patterns induced by athero-prone and athero-protective waveforms of shear stress are in agreement with the *in vivo* findings by Kwak et al. (2002), who showed that Cx43 is detectable in the endothelium covering the shoulder of the plaques in human carotid arteries, whereas Cx37 and Cx40 are present in the endothelium of nondiseased arteries and not detectable in the endothelium convering the plaques. These results suggest a prominent role for complex hemodynamic flow conditions in determining regional differences in gap junction expression and cell–cell communication that may contribute to vascular pathological changes in regions of flow disturbance.

19.4.3 Effects of Disturbed Flow on EC Proliferation and Migration

There is considerable evidence that the growth status of ECs can be regulated by the type of flow patterns and shear stresses to which they are exposed (for review see Li et al. 2005). Earlier studies have shown that laminar shear stress causes a dose-related reduction of EC proliferative rate (Levesque et al. 1990). It has also been shown that ECs subjected to a long duration of laminar flow at sufficiently high shear stresses have a lower rate of DNA synthesis than ECs under static condition (Akimoto et al. 2000). Laminar flow has been shown to reduce the number of cells entering the cell cycle, with the majority of cells being arrested in the G_0 or G_1 phase (Akimoto et al. 2000; Lin et al. 2000). ECs at branch points, where disturbed flow predominates, appear to age faster (Cooke 2003). In human iliac arteries, ECs at the branch points have shorter telomeres, consistent with a focal acceleration of senescence (Chang and Harley 1995). ECs exposed to an *in vitro* model of disturbed flow in a vertical-step flow channel turn over more rapidly than those kept in static condition (Davies et al. 1986; DePaola et al. 1992; Tardy et al. 1997). The rate of DNA synthesis is significantly higher in ECs in the vicinity of the reattachment flow area than that in the laminar flow area (Chiu et al. 1998). Li et al. (2005) showed that the enhancement of EC proliferation in the flow reattachment area is accompanied by a sustained activation of extracellular-signal regulated kinase (ERK) and that this disturbed flow-induced ERK activation can be abolished by the MEK/ERK inhibitor PD98095. Moreover, disturbed flow may favor EC proliferation through the release of p21 suppression of cyclin-dependent kinase (CDK) activity *via* G_0/G_1-S transition, as a result of the low shear stress (Akimoto et al. 2000; Davies 2000). These results support the notion that laminar flow, with a sufficiently high shear stress and a definitive direction, serves a protective role by reducing EC turnover to prevent the occurrence of lesions at the straight parts of arteries, whereas disturbed flow, with a low and oscillatory shear stress and a high shear stress gradient, promotes EC turnover that contributes to the formation and progression of atherosclerotic lesions at branch points of the arterial tree.

Studies using the flow channel *in vitro* (Tardy et al. 1997; Hsu et al. 2001; Li et al. 2005) and canine carotid-femoral grafts *in vivo* (Sprague et al. 1997; Wu et al. 1995) have shown that EC migration in wound healing is significantly enhanced by laminar shear stress, whereas disturbed flow has less effect on wound healing. The shear-induced directional migration may be related to the shear regulation of lamellipodia protrusion and focal adhesions remodeling in the direction of flow (Davies et al. 1994; Li et al. 2002, 2005). Such shear-induced protrusion of lamellipodia and remodeling of focal adhesions and actin cytoskeleton in ECs have been shown to be regulated by Rho family small GTPases (Davies 1993; Li et al. 1997, 2002; Hsu et al. 2001; Hu et al. 2002; Wojciak-Stothard and Ridley 2003; for review see Chien et al. 2005). In addition, shear stress enhances EC-ECM interaction, as demonstrated by measuring the traction forces exerted by ECs on a deformable substrate (Li et al. 2005). Different flow patterns and associated shear stresses may exert differential effects on EC-ECM interactions, which may consequently result in a differential modulation in EC migration.

19.4.4 Effects of Disturbed Flow on EC Permeability

The intercellular junction of the normal endothelium, especially that in the straight part of the arterial tree, does not allow the passage of macromolecules such as lipoproteins. The atherosclerosis-prone areas such as branch points, however, have been shown to have an increase in macromolecular permeability in studies on experimental animals (Bell et al. 1974; Stemerman et al. 1986). The arterial wall in regions prone to atherosclerosis have increased uptake of fluorescent, radiolabeled and chromogenic tracers used to represent serum lipoprotein particles (Weinberg 2004). In normal rabbit aorta, these lesion-prone branch sites have a permeability that is about four times that of the lesion-resistant non-branch areas (Schwenke and Carew 1989). In the porcine proximal external iliac artery, where there is a complex blood flow pattern, endothelial permeability to albumin-tagged Evans blue dye decreased with increased time-averaged shear stress in the physiological range; in contrast, the permeability increased under oscillatory shear conditions with little time-averaged shear stress (Himburg et al. 2004). By exposing EC monolayers to disturbed flow in a vertical-step flow channel, Phelps and DePaola (2000) demonstrated that the transendothelial transport of dextran (molecular weight 70,000) in the vicinity of flow reattachment was significantly higher than that in the fully developed flow area. These *in vivo* and *in vitro* experiments suggest that endothelial barrier function and macromolecular permeability can be regulated by flow disturbance and spatial variations in shear stress, which may contribute to the increased macromolecular permeability found in the atherosclerosis-prone areas such as branch points.

The mechanisms by which flow disturbance and spatial shear stress gradients enhance macromolecular permeability in disturbed flow created in the vertical-step flow channel *in vitro* or near the arterial branch points *in vivo* are likely to be multifactorial. The "cell turnover-leaky junction hypothesis", which was proposed by Weinbaum et al. (1985) and supported by several *in vivo* and *in vitro* experiments (Lin et al. 1989; Chuang et al. 1990; Huang et al. 1992; Chen et al. 1995), suggest that the intercellular junctions of ECs become widened and leaky to allow passage of macromolecules such as lipoproteins as cell turnover rate is increased. Systematic topographic mapping of the thoracic aorta of the rabbit has shown that the regional distribution of EC mitosis and macromolecular permeability can be correlated with high shear stress gradients. Since ECs in regions of disturbed flow, particularly in the area of flow reattachment, have higher turnover rates in comparison to the cells under undirectional laminar flow (DePaola et al. 1992, 1999; Chiu et al. 1998), it is very likely that the higher turnover rate of rounded ECs in the vicinity of flow separation in arteries *in vivo* contributes to the local increase in macromolecular permeability. In addition, disturbed flow may also directly influence endothelial integrity, and thus permeability, as a result of modulations of paracellular gaps and cytoskeletal organization (Albelda et al. 1988; Casnocha et al. 1989). Local differences in flow patterns and shear forces among neighboring cells in the disturbed flow region may widen the intercellular junctions and increase macromolecule transport. This is consistent with the findings by Phelps and DePaola (2000) that shear stress gradients can regulate intercellular junctional structure; specifically normal Cx43 gap junctions distributed at the cell periphery are severely disassembled in regions of disturbed flow, whereas much less disruption of the normal junctional pattern was observed in regions of fully developed flow (DePaola et al. 1999). Moreover, the expression of VE-cadherin, an important intercellular adhesion molecule that plays a role in mediating endothelial integrity and permeability (Dejana 1996; Corada et al. 1999), at EC borders was found to be significantly higher in the descending thoracic aorta and abdominal aorta, where the pulsatile flow has a strong net forward component, than in the aortic arch and the post-stenotic dilatation site beyond an experimental constriction, where the flow near the wall is complex and reciprocating with little net flow (Miao et al. 2005). VE-cadherin staining of the ECs in the flow channel showed that prolonged exposure of ECs to pulsatile flow (12 ± 4 dynes/cm^2 at 1 Hz) results in a coutinuous staining of VE-cadherin at cell borders, whereas the reciprocating flow (0.5 ± 4 dynes/cm^2 at 1 Hz) results in a discontinuous staining (Miao et al. 2005). The ECs in regions of flow reattachment with low shear stress magnitude and high shear stress gradients may also have a complete or partial loss of peripheral F-actin microfilaments in the periphery (Chiu et al. 1998). These redistributions of gap junctions and cytoskeletal proteins in regions of flow disturbance may contribute to intercellular widening, and consequently enhanced endothelial macromolecular permeability.

19.4.5 Effects of Disturbed Flow on EC Signaling and Gene Expression

ECs respond to mechanical stimuli by converting them into intracellular signals that ultimately regulate downstream gene expression and cellular function. There have been extensive studies on the effects of laminar shear stress on the signaling and gene expression in ECs (for review see Traub and Berk 1998; Chien et al. 1998; Chien 2003, 2007; Gimbrone et al. 2000; Davies et al. 2005; Li et al. 2005). However, the mechanisms that signal the mechanical stimuli induced by disturbed flow and regulate the downstream gene expression and cellular function have not been fully clarified. *In vitro* studies using parallel-plate flow chamber or cone-and-plate viscometer have identified a number of pathophysiologically relevant genes, such as platelet-derived growth factor (PDGF)-BB (Hsieh et al. 1991; Resnick et al. 1993; Khachigian et al. 1995), intercellular adhesion molecule-1 (ICAM-1) (Nagel et al. 1994; Tsuboi et al. 1995; Sampath et al. 1995), and MCP-1 (Shyy et al. 1994, 1995), whose expression is modulated by different types of shear stress (for review see Chien et al. 1998; Chien 2003, 2007; Li et al. 2005). The induction of genes in ECs exposed to shear stress is believed to involve the activation of various signaling molecules, including protein kinases, calcium influx, inositol trisphosphate, NO, cGMP, G proteins, reactive oxygen species (ROS), and a variety of transcription factors, including nuclear factor-κB (NF-κB), early growth response factor-1 (Egr-1), and activator protein-1 (AP-1), which contains c-fos and c-Jun (Curran and Franza 1988). Using a model of vertical-step flow and a quantitative image analysis technique, Nagel et al. (1999) demonstrated that ECs subjected to disturbed flow with spatial variations of shear stress exhibit increased levels of localized NF-κB, Egr-1, c-Jun, and c-fos in their nuclei, as compared with the cells exposed to unidirectional laminar flow or maintained under static conditions. This differential regulation of transcription factor expression by disturbed *vs.* laminar flows indicates that regional differences in blood flow patterns *in vivo*, particularly the occurrence of spatial shear stress gradients, may represent important local modulators for EC signaling and gene expression at anatomic sites predisposed for atherosclerotic development (Nagel et al. 1999).

It is difficult to analyze the effects of disturbed flow created in vertical-step flow channel on gene expression in ECs, because changes of very local hemodynamic environments occur in the channel with length scales of tens to hundreds of microns and the cell numbers in different flow areas (i. e., flow separation, recirculation, and reattachment) are sparse. A precise *in-situ* assessment of ECs in different flow areas in the vertical-step flow channel is needed for comparisons of the gene expression profiles induced by different flow patterns (especially laminar flow *vs.* disturbed flow). The application of the microdissection or micropipette technique may be useful for obtaining small numbers of ECs from regions with different flow patterns in the vertical-step flow channel. To evaluate the degree of heterogeneity between gene expression in individual ECs within different flow regions, an antisense RNA (aRNA) amplication method (Eberwine et al. 1992; Davies et al. 2001) was applied to individual ECs following their isolation by microdissection from disturbed and laminar flow sites. Transcription profiles of individual ECs isolated from disturbed flow region exhibited more expression heterogeneity than those from laminar flow region. Flow disturbance *in vitro* induced differential expression at the single gene level, as illustrated for the expression of Cx43 (Davies et al. 2001). It is very likely that within highly heterogeneous populations of ECs located in disturbed flow regions, the expression of pro-atherosclerotic genes may occur within the range of expression profiles induced by the local hemodynamics, which may consequently contribute to the focal initiation of the atherosclerotic lesions (Davies et al. 2001).

There have been several reports showing the regulatory effects of disturbed flow (created in a vertical-step flow channel) on the expression of pathophysiologically relevant genes or proteins in ECs. For example, Chiu et al. (2003) examined the long-term effects of disturbed flow on the expression of adhesion molecules ICAM-1, vascular adhesion molecule-1 (VCAM-1), and E-selectin on ECs using immunofluorescence staining with antibodies against these proteins. The exposure of ECs to vertical-step flow for 24 h leads to fluorescence-positive staining for ICAM-1 and E-selectin expressions, but not for VCAM-1. Histochemical studies on the activation of sterol regulatory element binding proteins (SREBPs), which cause increases in the expressions of LDL receptor, cholesterol synthase, and fatty acid synthase, in ECs have shown that, in contrast to the transient SREBP activation by laminar flow, ECs in the disturbed flow region in a vertical-step flow channel exhibit a sustained activation of SREBP and SRE-mediated gene expression, and hence enhanced LDL uptake and lipid synthesis (Liu et al. 2002). Dunzendorfer et al. (2004) showed that laminar flow applied to ECs results in a down-regulation of gene and protein expression of toll-like receptor 2 (TLR2), which has been found to be augmented in activated ECs of atherosclerotic lesions (Edfeldt et al. 2002; Bjorkbacka et al. 2004; Michelsen et al. 2004). Disturbed flow applied to ECs in a step flow channel retains the responsiveness of ECs to TLR agonists such as lipopolysaccharide and TNF-α.

Several *in vitro* studies using genomic approaches have identified a number of pathophysiologically relevant genes in ECs that are regulated by different types of shear stresses, including high and low laminar shear stresses, turbulent shear stress, and disturbed flow (Garcia-Cardena et al. 2001; McCormick et al. 2001; Chen et al. 2001; Wasserman et al. 2002; Brooks et al. 2002). The results from these studies have suggested that high-shear laminar flow modulates EC gene expressions and functions that protect against atherogenesis, whereas disturbed flow up-regulates pro-atherosclerotic genes or proteins that promote development of atherosclerosis. To more realistically assess the modulation of gene expression profiles in ECs induced by mechanical forces that are actually present in the atherosclerosis-susceptible and atherosclerosis-resistant regions of human arteries, Dai et al. (2004) have analyzed the flow patterns present in the human carotid bifurcation, using three-dimensional computational fluid dynamic analyses based on the actual geometries and flow profiles measured by magnetic resonance imaging (MRI) and ultrasound. Their results showed that athero-prone and athero-protective waveforms of shear stress differentially regulate EC gene expression, with up-regulation by athero-prone waveform of a number of genes encoding pro-inflammatory [e.g.,

IL-8, chemokine receptor 4 (CXCR4), and tumor necrosis factor receptor superfamily, member 21 (TNFRSF21)] and angiogenic functions [e. g., connective tissue growth factor (CTGF)] and those implicated in atherogenesis [e. g., thrombospondin 1 (THBS1) and matrix metalloproteinase 1 (MMP1)] (Oemar et al. 1997; Galis and Khatri 2002; Stenina et al. 2003). In contrast, exposure to the athero-protective waveform of shear stress increased the expression of sets of genes that prevent development of atherosclerosis [such as krüppel-like factor 2 (KLF2) and those related to the NO pathway, including endothelial NO synthase (eNOS) and guanylate cyclase 1 α_3 (GUCY1A3)] (Dekker et al. 2002; SenBanerjee et al. 2004; Wang et al. 2006). This approach using actual profiles present in the atherosclerosis-susceptible and resistant regions of human arteries facilitates the identification of genes that may have direct pathophysiological relevance to the atherosclerotic disease process *in vivo*.

19.5 *In Vivo* Studies on the Effects of Disturbed Flow on ECs

In contrast to the many *in vitro* studies on the effects of disturbed flow on ECs, there have been limited numbers of *in vivo* investigations on the EC responses to disturbed flow, mainly due to the lack of appropriate *in vivo* model that can generate well-controlled flow patterns and the difficulty of determine the molecular and cellular responses *in vivo*. There is a critical need of *in vivo* studies of cellular responses to disturbed flow to substantiate the *in vitro* findings of the effects of disturbed flow on ECs. Hutchison (1991) investigated the effect of different flow patterns on EC morphology *in vivo* by introducing graded stenoses (40–64% diameter reduction) of dog common carotid arteries and examining the post-stenotic velocity field with transcutaneous pulsed Doppler velocimetry. In this study, ECs are found to be aligned and elongated in the direction of flow five diameters upstream of the stenosis throat, where the flow is laminar with a relatively higher shear stress. The cells are maximally rounded immediately downstream to the stenosis throat and gradually returned to upstream elongation by five diameters downstream. Flow measurements show the existence of a region of flow separation and low-velocity recirculation between the stenosis and the downstream laminar flow area; thus, the downstream cell rounding is associated with flow separation and low shear stress value. Gabriels and Paul (1998) examined the relationship between altered shear stress and Cx43 expression *in vivo* by inducing a flow disturbance in a segment of abdominal aorta by coarctation. Within eight days, a strong but local up-regulation of Cx43 is observed at the leading edge of the coarctation, suggesting that Cx43 expression can be modulated by changes in hemodynamic pattern *in vivo*. Miao et al. (2005) introduced a local stenosis in the rat abdominal aorta with a U-shaped titanium clip and demonstrated that VE-cadherin was highly expressed at EC borders in the abdominal aorta at sites without stenosis, where the blood flow is laminar. In contrast, there was little or no VE-cadherin expression in the EC borders in the post-stenotic dilatation sites, where the flow is disturbed. With the same experimental model, Wang et al. (2006) further showed that KLF2 was highly expressed in

ECs at laminar flow, but there was virtually no KLF2 expression in the ECs at the post-stenotic sites, where the flow pattern is disturbed. These results indicate that the EC structure and gene and protein expressions can be modulated by different flow patterns *in vivo* in a manner similar to that observed *in vitro*.

Cheng et al. (2005) developed an *in vivo* model that can create a well-defined disturbed flow and variations in shear stress to investigate their effects on the expression and intracellular distribution of eNOS. A cylinder with a tapered lumen was placed around the carotid artery of rabbits and transgenic mice, in which eNOS expression and distribution can be monitored by fusing a protein with green fluorescent protein (GFP) (van Haperen et al. 2003). A region of low shear stress and a region of flow disturbance with oscillatory shear stress were created upstream and downstream of the device, respectively, and flow dynamics was measured by a Doppler probe. The conical cylinder lumen induces a progressive stenosis of the vessel and a gradual increase in shear stress in the stenotic zone. Strong induction of eNOS was found in the EC membrane and Golgi complex in the high shear stress region, as compared with the other regions and with the untreated control carotid arteries. Both the low and oscillatory shear stress regions showed a 3-fold decrease in eNOS mRNA expression, as compared with the control vessel (undisturbed shear stress). By using *en face* immunostaining on carotid arteries with antibodies against phospho-eNOS, they further demonstrated that the levels of eNOS phosphorylation in ECs in the high shear stress region were significantly higher than the cells in regions of low and undisturbed shear stress (6-fold) and oscillatory shear stress (2-fold). This study represents the first report to show the responses of mRNA and protein expressions *in vivo* to disturbed flow with spatial variations of shear stress.

Porat et al. (2004) showed that the expression of EC-specific receptor *tie1* in the arterial tree in *tie1-lacZ* transgenic mice is enahnced in vascular bifurcations and branch points, i.e., regions with atherogenic disturbed flow. In aortic valves, *tie1* promoter is exclusively expressed in ECs lining the inner aspect of the cup-shaped cusps, but not in ECs at the outer aspect of the same leaflet. This asymmetrical pattern of expression suggests that hemodynamic patterns play a significant role in regulating *tie1* induction. Flow disturbance induced by the surgical interposition of a vein into an artery led to an induction of *tie1* expression in the region immediately downstream of the artery/vein junction, where the lumen enlargement due to exposure to arterial pressure provides a geometry of post-stenotic dilatation. This *tie1* induction by flow disturbance *in vivo* was confirmed by *in vitro* study using a step flow channel, where the up-regulation of *tie1* promoter activity is seen only in ECs residing in the region of flow separation and recirculation downstream to the step.

To discriminate the gene expression profiles of ECs between sites of disturbed flow and laminar flow *in vivo*, Passerini et al. (2004) conducted DNA microarray study on fresh ECs isolated from regions of inner aortic arch (areas exposed to disturbed flow) and descending thoracic aorta (areas exposed to laminar flow) of normal adult pigs and subjected the results to biological pathways analysis. In disturbed flow regions, there is the up-regulation of several inflammatory cytokines and receptors, as well as elements of the NF-κB system, which reflects a pro-inflammatory phenotype. Athero-protective profiles are also seen in disturbed flow regions, notably an enhanced antioxidative gene expression. Thus, the gene expression profile in disturbed flow regions may reflect a delicate balance between athogenic and athero-proctive genes; the action of risk factors may tip the balance to initiate atherogenesis. The study by Passerini et al. (2004) has provided the first public database of regional EC gene expression in normal animal.

19.6 Summary and Conclusions

Atherosclerosis, although clearly associated with several systemic risk factors (e. g., hyperlipidemia, hypertension, smoking, obesity, and diabetes), has a preferential localization pattern at the outer edges of blood vessel bifurcations and at points of blood flow recirculation and stasis (Packham et al. 1967; Schwenke and Carew 1988, 1989; Wissler 1995; VanderLaan 2004). In these predisposed locations, fluid shear stress on the vessel wall is significantly lower in magnitude and exhibits changes and flow separation with high gradients; these features are absent in the straight part of the vascular tree which are generally spared from atherosclerosis (Glagov et al. 1988; Ku et al. 1985; Friedman et al. 1987). This striking correlation between regional hemodynamics and atherosclerosis has motivated many studies that have attempted to define a mechanistic role for hemodynamic factors, especiallay disturbed flow and associated variations of shear stress, in the pathogenesis of atherosclerosis (Chien et al. 1998; Chien 2003, 2007; Gimbrone et al. 2000; Davies 2000; Berk et al. 2001; Li et al. 2005).

As an interface between the flowing blood and the vessel wall, EC monolayer is directly exposed to changes in blood flow and associated shear stress, and its dysfunction has been well recognized as a critical early event in atherosclerosis. In this chapter, we have summarized the current *in vitro* and *in vivo* studies on the effects of disturbed flow and the associated variations of shear stress on ECs, in terms of their signal transduction, gene expression, structure, and functions (for summary see Table 19.1), which have been implicated in the initiation and progression of atherosclerotic lesions. In relatively straight segments of an artery with a laminar flow and a physiological level of shear stress (10 to 70 dynes/cm^2), ECs are aligned and elongated in the direction of flow. In contrast, in areas of disturbed flow, the ECs are more polygonal in appearance without a clear orientation, and there are redistributions of cytoskeletal organization and intercellular junctional proteins (e. g., Cx and VE-cadherin). These morphological changes are accompanied by increases in cell turnover rate and DNA synthesis and a sustained activation of SREBP, which may consequently contribute to the increases in permeability to macromolecules such as lipoproteins, as well as lipid uptake and synthesis at the branch points.

Laminar shear stress in a physiological range activates signaling pathways that induce endothelial elaboration of a number of vasoactive factors that promote vasodiation and suppress the proliferation and migration of SMCs (e. g., increased expression of eNOS and production of NO) and inhibit adherence of circulating blood elements (e. g., reduced expression of adhesion molecules and chemotac-

Table 19.1 Summary of effects of different flow patterns and associated shear stresses on EC and vascular biology

Laminar flow/high shear stress	Disturbed flow/low and oscillatory shear stress
Vasodilation	Vasoconstriction
Lower EC turnover	Higher EC turnover
Lower macromolecular permeability and LDL uptake	Higher macromolecular permeability and LDL uptake
Lower DNA synthesis	Higher DNA synthesis
Elongated and aligned phenotype	Polygonal phenotype
Lower expression of adhesion molecule, inflammatory, and chemokine genes	Higher expression of adhesion molecule, inflammatory, and chemokine genes
Higher expression of antioxidant genes	Lower expression of antioxidant genes
Inhibition of WBC adhesion and platelet aggregation	Promotion of WBC adhesion and platelet aggregation
Reduced oxidative stress	Sustained elevation of oxidative stress
Reduced SMC activation	Increased VSM activation
Promotion of endothelization	Reduced endothelial repair

tic proteins such as VCAM-1 and MCP-1). Several atheroprotective genes, such as antioxidant, anti-inflammatory, anti-coagulant, and anti-apoptotic genes, are up-regulated by sustained laminar shear stress with a clear direction. In contrast, disturbed flow with a low and oscillating shear stress and a high shear stress gradient elicits factors that impair endothelium-mediated vasodilation and induce SMC activation (e. g., reduced expression of eNOS and production of NO) and increase adhesion of circulating blood elements (e. g., increased expressions of ICAM-1, E-selecitin, MCP-1, and NF-κB). Transcriptional profiling of ECs in different flow regimes reveals that ECs exposed to disturbed flow with a low and oscillatory shear stress have significantly higher levels of a number of pathophysiologically relevant genes whose products may serve pro-inflammatory, pro-coagulant, proliferative, and pro-apoptotic functions, and hence promote atherosclerosis. All these findings indicate that laminar shear stress in a physiological range maintains vascular homeostasis and plays protective roles against atherosclerosis, whereas alterations of EC biology by disturbed flow with low and oscillatory shear stress would predispose these arterial regions to atherogenesis (McLenachan et al. 1990; Topper and Gimbrone 1999; Chien et al. 1998; Chien 2003, 2007; de Nigris et al. 2003; Li et al. 2005).

Atherogenesis involves interactions of multiple factors, including a complex array of circulating blood cells and plasma components, their interactions with the cells and matrix proteins of the arterial wall, and the effects of flow patterns on mass transfer. Investigations on EC responses to disturbed flow will provide important information concerning alterations in vascular signaling, gene expression, and function in the disease-prone regions. Combination of these results with data obtained from ECs impaired from systemic risk factors such as smoking, hyperlipidemia, hyperglycemia, hypertension, obesity, and diabetes will help to distinguish the activated or dysfunctional ECs from healthy or quiescent ECs. The consequences of dy-

19.6 Summary and Conclusions

namic interaction between disturbed flow/oscillatory shear stress-induced changes of EC biology and systemic risk factors will not only enhance our understanding of the mechanisms of atherogenesis, but also the success of therapeutic modalities used to modify disease progression and clinical outcomes. A major challenge in this field is to integrate a large body of data on EC responses to disturbed flow at the signaling and gene expression levels to identify useful biomarkers for atherosclerosis and to discover novel molecular targets, thus facilitating the development of new therapeutic strategies for this pathological change that underlies many cardiovascular diseases.

Acknowledgements This work was supported by National Health Research Institutes (Taiwan) Grant ME-096-PP-06 (to Jeng-Jiann Chiu); National Science Council (Taiwan) NRPGM Grant 96-3112-B-400-009 (to Jeng-Jiann Chiu); and National Heart, Lung, and Blood Institute Grants HL064382 and HL080518 (to Shu Chien). The authors would like to acknowledge the valuable help by Ms. Pei-Ling Lee, Mr. Sheng-Chieh Lien, and Ms. Yi-Ting Yeh during the preparation of the manuscript.

References

1. Akimoto S, Mitsumata M, Sasaguri T, Yoshida Y (2000) Laminar shear stress inhibits vascular endothelial cell proliferation by inducing cyclin-dependent kinase inhibitor p21(Sdi1/Cip1/Waf1). Circ Res 86:185–190
2. Albelda SM, Sampson PM, Haselton FR, McNiff JM, Mueller SN, Williams SK, Fishman AP, Levine EM (1988) Permeability characteristics of cultured endothelial cell monolayers. J Appl Physiol 64:308–322
3. Asakura T, Karino T (1990) Flow patterns and spatial distribution of atherosclerotic lesions in human coronary arteries. Circ Res 66:1045–1066
4. Barber KM, Pinero A, Truskey GA (1998) Effects of recirculating flow on U-937 cell adhesion to human umbilical vein endothelial cells. Am J Physiol 275:H591–H599
5. Bassiouny HS, Zarins CK, Kadowaki MH, Glagov S (1994) Hemodynamic stress and experimental aortoiliac atherosclerosis. J Vasc Surg 19:426–434
6. Beere PA, Glagov S, Zarins CK (1992) Experimental atherosclerosis at the carotid bifurcation of the cynomolgus monkey. Localization, compensatory enlargement, and the sparing effect of lowered heart rate. Arterioscler Thromb 12:1245–1253
7. Bell FP, Adamson IL, Schwartz CJ (1974) Aortic endothelial permeability to albumin: focal and regional patterns of uptake and transmural distribution of 131I-albumin in the young pig. Exp Mol Pathol 20:57–68
8. Berk BC, Abe JI, Min W, Surapisitchat J, Yan C (2001) Endothelial atheroprotective and anti-inflammatory mechanisms. Ann N Y Acad Sci 947:93–109; discussion 109–11
9. Bharadvaj BK, Mabon RF, Giddens DP (1982) Steady flow in a model of the human carotid bifurcation. Part I-flow visualization. J Biomech 15:349–362
10. Bjorkbacka H, Kunjathoor VV, Moore KJ, Koehn S, Ordija CM, Lee MA, Means T, Halmen K, Luster AD, Golenbock DT, Freeman MW (2004) Reduced atherosclerosis in MyD88-null mice links elevated serum cholesterol levels to activation of innate immunity signaling pathways. Nat Med 10:416–421
11. Bonetti PO, Lerman LO, Lerman A (2003) Endothelial dysfunction: a marker of atherosclerotic risk. Arterioscler Thromb Vasc Biol 23:168–175
12. Brooks AR, Lelkes PI, Rubanyi GM (2002) Gene expression profiling of human aortic endothelial cells exposed to disturbed flow and steady laminar flow. Physiol Genomics 9:27–41
13. Bruzzone R, Haefliger JA, Gimlich RL, Paul DL (1993) Connexin40, a component of gap junctions in vascular endothelium, is restricted in its ability to interact with other connexins. Mol Biol Cell 4:7–20
14. Caro CG, Fitz-Gerald JM, Schroter RC (1971) Atheroma and arterial wall shear. Observation, correlation and proposal of a shear dependent mass transfer mechanism for atherogenesis. Proc R Soc Lond B Biol Sci 177:109–159
15. Casnocha SA, Eskin SG, Hall ER, McIntire LV (1989) Permeability of human endothelial monolayers: effect of vasoactive agonists and cAMP. J Appl Physiol 67:1997–2005

16. Chang E, Harley CB (1995) Telomere length and replicative aging in human vascular tissues. Proc Natl Acad Sci USA 92:11190–11194
17. Chen BP, Li YS, Zhao Y, Chen KD, Li S, Lao J, Yuan S, Shyy JY, Chien S (2001) DNA microarray analysis of gene expression in endothelial cells in response to 24-h shear stress. Physiol Genomics 7:55–63
18. Chen CN, Chang SF, Lee PL, Chang K, Chen LJ, Usami S, Chien S, Chiu JJ (2006) Neutrophils, lymphocytes, and monocytes exhibit diverse behaviors in transendothelial and subendothelial migrations under coculture with smooth muscle cells in disturbed flow. Blood 107:1933–1942
19. Chen YL, Jan KM, Lin HS, Chien S (1995) Ultrastructural studies on macromolecular permeability in relation to endothelial cell turnover. Atherosclerosis 118:89–104
20. Cheng C, van Haperen R, de Waard M, van Damme LC, Tempel D, Hanemaaijer L, van Cappellen GW, Bos J, Slager CJ, Duncker DJ, van der Steen AF, de Crom R, Krams R (2005) Shear stress affects the intracellular distribution of eNOS: direct demonstration by a novel in vivo technique. Blood 106:3691–3698
21. Chien S (2003) Molecular and mechanical bases of focal lipid accumulation in arterial wall. Prog Biophys Mol Biol 83:131–151
22. Chien S, Li S, Shiu YT, Li YS (2005) Molecular basis of mechanical modulation of endothelial cell migration. Front Siosci 10:1985–2000
23. Chien S (2007) Mechanotransduction and endothelial cell homeostasis: the wisdom of the cell. Am J Physiol Heart Circ Physiol 292, H1209–H1224
24. Chien S, Li S, Shyy YJ (1998) Effects of mechanical forces on signal transduction and gene expression in endothelial cells. Hypertension 31, 162–169
25. Chiu JJ, Chen CN, Lee PL, Yang CT, Chuang HS, Chien S, Usami S (2003) Analysis of the effect of disturbed flow on monocytic adhesion to endothelial cells. J Biomech 36:1883–1895
26. Chiu JJ, Wang DL, Chien S, Skalak R, Usami S (1998) Effects of disturbed flow on endothelial cells. J Biomech Eng 120:2–8
27. Chuang PT, Cheng HJ, Lin SJ, Jan KM, Lee MM, Chien S (1990) Macromolecular transport across arterial and venous endothelium in rats. Studies with Evans blue-albumin and horseradish peroxidase. Arteriosclerosis 10:188–197
28. Cooke JP (2003) Flow NO, and atherogenesis. Proc Natl Acad Sci USA 100:768–770
29. Corada M, Mariotti M, Thurston G, Smith K, Kunkel R, Brockhaus M, Lampugnani MG, Martin-Padura I, Stoppacciaro A, Ruco L, McDonald DM, Ward PA, Dejana E (1999) Vascular endothelial-cadherin is an important determinant of microvascular integrity in vivo. Proc Natl Acad Sci USA 96:9815–9820
30. Cornhill JF, Roach MR (1976) A quantitative study of the localization of atherosclerotic lesions in the rabbit aorta. Atherosclerosis 23:489–501
31. Curran T, Franza BR Jr (1988) Fos and Jun: the AP-1 connection. Cell 55:395–397
32. Dai G, Kaazempur-Mofrad MR, Natarajan S, Zhang Y, Vaughn S, Blackman BR, Kamm RD, Garcia-Cardena G, Gimbrone MA Jr (2004) Distinct endothelial phenotypes evoked by arterial waveforms derived from atherosclerosis-susceptible and -resistant regions of human vasculature. Proc Natl Acad Sci USA 101:14871–14876
33. Davies PF (1993) Endothelium as a signal transduction interface for flow forces: cell surface dynamics. Thromb Haemost 70:124–128
34. Davies PF (1995) Flow-mediated endothelial mechanotransduction. Physiol Rev 75:519–560
35. Davies PF (2000) Spatial hemodynamics, the endothelium, and focal atherogenesis: a cell cycle link? Circ Res 86:114–116
36. Davies PF, Barbee KA, Volin MV, Robotewskyj A, Chen J, Joseph L, Griem ML, Wernick MN, Jacobs E, Polacek DC, dePaola N, Barakat AI (1997) Spatial relationships in early signaling events of flow-mediated endothelial mechanotransduction. Annu Rev Physiol 59:527–549
37. Davies PF, Remuzzi A, Gordon EJ, Dewey CF Jr, Gimbrone MA Jr (1986) Turbulent fluid shear stress induces vascular endothelial cell turnover in vitro. Proc Natl Acad Sci USA 83:2114–2117

38. Davies PF, Robotewskyj A, Griem ML (1994) Quantitative studies of endothelial cell adhesion. Directional remodeling of focal adhesion sites in response to flow forces. J Clin Invest 93:2031–2038
39. Davies PF, Robotewskyj A, Griem ML, Dull RO, Polacek DC (1992) Hemodynamic forces and vascular cell communication in arteries. Arch Pathol Lab Med 116:1301–1306
40. Davies PF, Shi C, Depaola N, Helmke BP, Polacek DC (2001) Hemodynamics and the focal origin of atherosclerosis: a spatial approach to endothelial structure, gene expression, and function. Ann N Y Acad Sci 947:7–16; discussion 16–17
41. Davies PF, Spaan JA, Krams R (2005) Shear stress biology of the endothelium. Ann Biomed Eng 33:1714–1718
42. de Nigris F, Lerman LO, Ignarro SW, Sica G, Lerman A, Palinski W, Ignarro LJ, Napoli C (2003) Beneficial effects of antioxidants and L-arginine on oxidation-sensitive gene expression and endothelial NO synthase activity at sites of disturbed shear stress. Proc Natl Acad Sci USA 100:1420–1425
43. Dejana E (1996) Endothelial adherens junctions: implications in the control of vascular permeability and angiogenesis. J Clin Invest 98:1949–1953
44. Dekker RJ, van Soest S, Fontijn RD, Salamanca S, de Groot PG, VanBavel E, Pannekoek H, Horrevoets AJ (2002) Prolonged fluid shear stress induces a distinct set of endothelial cell genes, most specifically lung Kruppel-like factor (KLF2). Blood 100:1689–1698
45. DePaola N, Davies PF, Pritchard WF Jr, Florez L, Harbeck N, Polacek DC (1999) Spatial and temporal regulation of gap junction connexin43 in vascular endothelial cells exposed to controlled disturbed flows in vitro. Proc Natl Acad Sci USA 96:3154–3159
46. DePaola N, Gimbrone MA Jr, Davies PF, Dewey CF Jr (1992) Vascular endothelium responds to fluid shear stress gradients. Arterioscler Thromb 12:1254–1257
47. Dewey CF Jr, Bussolari SR, Gimbrone MA Jr, Davies PF (1981) The dynamic response of vascular endothelial cells to fluid shear stress. J Biomech Eng 103:177–185
48. Dunzendorfer S, Lee HK, Tobias PS (2004) Flow-dependent regulation of endothelial Toll-like receptor 2 expression through inhibition of SP1 activity. Circ Res 95:684–691
49. Eberwine J, Yeh H, Miyashiro K, Cao Y, Nair S, Finnell R, Zettel M, Coleman P (1992) Analysis of gene expression in single live neurons. Proc Natl Acad Sci USA 89:3010–3014
50. Edfeldt K, Swedenborg J, Hansson GK, Yan ZQ (2002) Expression of toll-like receptors in human atherosclerotic lesions: a possible pathway for plaque activation. Circulation 105:1158–1161
51. Eskin SG, Ives CL, McIntire LV, Navarro LT (1984) Response of cultured endothelial cells to steady flow. Microvasc Res 28:87–94
52. Franke RP, Grafe M, Schnittler H, Seiffge D, Mittermayer C, Drenckhahn D (1984) Induction of human vascular endothelial stress fibres by fluid shear stress. Nature 307:648–649
53. Friedman MH, Bargeron CB, Deters OJ, Hutchins GM, Mark FF (1987) Correlation between wall shear and intimal thickness at a coronary artery branch. Atherosclerosis 68:27–33
54. Friedman MH, Brinkman AM, Qin JJ, Seed WA (1993) Relation between coronary artery geometry and the distribution of early sudanophilic lesions. Atherosclerosis 98:193–199
55. Fung YC (1997) Biomechanics: Circulation. New York, NY: Springer
56. Gabriels JE, Paul DL (1998) Connexin43 is highly localized to sites of disturbed flow in rat aortic endothelium but connexin37 and connexin40 are more uniformly distributed. Circ Res 83:636–643
57. Galbraith CG, Skalak R, Chien S (1998) Shear stress induces spatial reorganization of the endothelial cell cytoskeleton. Cell Motil Cytoskeleton 40:317–330
58. Galis ZS, Khatri JJ (2002) Matrix metalloproteinases in vascular remodeling and atherogenesis: the good, the bad, and the ugly. Circ Res 90:251–262
59. Garcia-Cardena G, Comander J, Anderson KR, Blackman BR, Gimbrone MA Jr (2001) Biomechanical activation of vascular endothelium as a determinant of its functional phenotype. Proc Natl Acad Sci USA 98:4478–4485
60. Gibson CM, Diaz L, Kandarpa K, Sacks FM, Pasternak RC, Sandor T, Feldman C, Stone PH (1993) Relation of vessel wall shear stress to atherosclerosis progression in human coronary arteries. Arterioscler Thromb 13:310–315

61. Gimbrone MA Jr, Nagel T, Topper JN (1997) Biomechanical activation: an emerging paradigm in endothelial adhesion biology. J Clin Invest 99:1809–1813
62. Gimbrone MA, Jr, Topper JN, Nagel T, Anderson KR, Garcia-Cardena G (2000) Endothelial dysfunction, hemodynamic forces, and atherogenesis. Ann N Y Acad Sci 902:230–239; discussion 239–240
63. Glagov S, Zarins C, Giddens DP, Ku DN (1988) Hemodynamics and atherosclerosis. Insights and perspectives gained from studies of human arteries. Arch Pathol Lab Med 112:1018–1031
64. Gnasso A, Carallo C, Irace C, Spagnuolo V, De Novara G, Mattioli PL, Pujia A (1996) Association between intima-media thickness and wall shear stress in common carotid arteries in healthy male subjects. Circulation 94:3257–3262
65. Gnasso A, Irace C, Carallo C, De Franceschi MS, Motti C, Mattioli PL, Pujia A (1997) In vivo association between low wall shear stress and plaque in subjects with asymmetrical carotid atherosclerosis. Stroke 28:993–998
66. Helmke BP, Goldman RD, Davies PF (2000) Rapid displacement of vimentin intermediate filaments in living endothelial cells exposed to flow. Circ Res 86:745–752
67. Himburg HA, Grzybowski DM, Hazel AL, LaMack JA, Li XM, Friedman MH (2004) Spatial comparison between wall shear stress measures and porcine arterial endothelial permeability. Am J Physiol Heart Circ Physiol 286:H1916–H1922
68. Hinds MT, Park YJ, Jones SA, Giddens DP, Alevriadou BR (2001) Local hemodynamics affect monocytic cell adhesion to a three-dimensional flow model coated with E-selectin. J Biomech 34:95–103
69. Hishikawa K, Nakaki T, Marumo T, Suzuki H, Kato R, Saruta T (1995) Pressure enhances endothelin-1 release from cultured human endothelial cells. Hypertension 25:449–452
70. Hsieh HJ, Li NQ, Frangos JA (1991) Shear stress increases endothelial platelet-derived growth factor mRNA levels. Am J Physiol 260:H642–H646
71. Hsu PP, Li S, Li YS, Usami S, Ratcliffe A, Wang X, Chien S (2001) Effects of flow patterns on endothelial cell migration into a zone of mechanical denudation. Biochem Biophys Res Commun 285:751–759
72. Hu YL, Li S, Miao H, Tsou TC, del Pozo MA, Chien S (2002) Roles of microtubule dynamics and small GTPase Rac in endothelial cell migration and lamellipodium formation under flow. J Vasc Res 39:465–476
73. Huang AL, Jan KM, Chien S (1992) Role of intercellular junctions in the passage of horseradish peroxidase across aortic endothelium. Lab Invest 67:201–209
74. Hutchison KJ (1991) Endothelial cell morphology around graded stenoses of the dog common carotid artery. Blood Vessels 28:396–406
75. Ives CL, Eskin SG, McIntire LV (1986) Mechanical effects on endothelial cell morphology: in vitro assessment. In Vitro Cell Dev Biol 22:500–507
76. Jou LD, van Tyen R, Berger SA, Saloner D (1996) Calculation of the magnetization distribution for fluid flow in curved vessels. Magn Reson Med 35:577–584
77. Kamiya A, Bukhari R, Togawa T (1984) Adaptive regulation of wall shear stress optimizing vascular tree function. Bull Math Biol 46:127–137
78. Karino T, Goldsmith HL (1979) Adhesion of human platelets to collagen on the walls distal to a tubular expansion. Microvasc Res 17:238–262
79. Khachigian LM, Resnick N, Gimbrone MA Jr, Collins T (1995) Nuclear factor-kappa B interacts functionally with the platelet-derived growth factor B-chain shear-stress response element in vascular endothelial cells exposed to fluid shear stress. J Clin Invest 96:1169–1175
80. Kim DW, Langille BL, Wong MK, Gotlieb AI (1989) Patterns of endothelial microfilament distribution in the rabbit aorta in situ. Circ Res 64:21–31
81. Kinlay S, Ganz P (1997) Role of endothelial dysfunction in coronary artery disease and implications for therapy. Am J Cardiol 80:11I-16I
82. Kolpakov V, Polishchuk R, Bannykh S, Rekhter M, Solovjev P, Romanov Y, Tararak E, Antonov A, Mironov A (1996) Atherosclerosis-prone branch regions in human aorta: microarchitecture and cell composition of intima. Atherosclerosis 122:173–189

83. Ku DN, Giddens DP, Zarins CK, Glagov S (1985) Pulsatile flow and atherosclerosis in the human carotid bifurcation. Positive correlation between plaque location and low oscillating shear stress. Arteriosclerosis 5:293–302
84. Kwak BR, Mulhaupt F, Veillard N, Gros DB, Mach F (2002) Altered pattern of vascular connexin expression in atherosclerotic plaques. Arterioscler Thromb Vasc Biol 22:225–230
85. LaBarbera M (1990) Principles of design of fluid transport systems in zoology. Science 249:992–1000
86. Langille BL, Graham JJ, Kim D, Gotlieb AI (1991) Dynamics of shear-induced redistribution of F-actin in endothelial cells in vivo. Arterioscler Thromb 11:1814–1820
87. Lee TY, Gotlieb AI (2003) Microfilaments and microtubules maintain endothelial integrity. Microsc Res Tech 60:115–127
88. Lerman A, Burnett JC Jr (1992) Intact and altered endothelium in regulation of vasomotion. Circulation 86:III12–III19
89. Levesque MJ, Nerem RM (1985) The elongation and orientation of cultured endothelial cells in response to shear stress. J Biomech Eng 107:341–347
90. Levesque MJ, Nerem RM, Sprague EA (1990) Vascular endothelial cell proliferation in culture and the influence of flow. Biomaterials 11:702–707
91. Li S, Butler P, Wang Y, Hu Y, Han DC, Usami S, Guan JL, Chien S (2002) The role of the dynamics of focal adhesion kinase in the mechanotaxis of endothelial cells. Proc Natl Acad Sci USA 99:3546–3551
92. Li S, Kim M, Hu YL, Jalali S, Schlaepfer DD, Hunter T, Chien S, Shyy JY (1997) Fluid shear stress activation of focal adhesion kinase. Linking to mitogen-activated protein kinases. J Biol Chem 272:30455–30462
93. Li YS, Haga JH, Chien S (2005) Molecular basis of the effects of shear stress on vascular endothelial cells. J Biomech 38:1949–1971
94. Libby P (2002) Inflammation in atherosclerosis. Nature 420:868–874
95. Libby P, Ridker PM, Maseri A (2002) Inflammation and atherosclerosis. Circulation 105:1135–1143
96. Lin K, Hsu PP, Chen BP, Yuan S, Usami S, Shyy JY, Li YS, Chien S (2000) Molecular mechanism of endothelial growth arrest by laminar shear stress. Proc Natl Acad Sci USA 97:9385–9389
97. Lin SJ, Jan KM, Weinbaum S, Chien S (1989) Transendothelial transport of low density lipoprotein in association with cell mitosis in rat aorta. Arteriosclerosis 9:230–236
98. Liu Y, Chen BP, Lu M, Zhu Y, Stemerman MB, Chien S, Shyy JY (2002) Shear stress activation of SREBP1 in endothelial cells is mediated by integrins. Arterioscler Thromb Vasc Biol 22:76–81
99. Lusis AJ (2000) Atherosclerosis. Nature 407:233–241
100. Macagno EQ, Hung TK (1967) Computational and experimental study of a captive annular eddy. J Fluid Mech 28:43–64
101. Macagno EQ, Hung TK (1970) Computational study of accelerated flow in a two-dimensional conduit expansion. J Hydraulic Res 8:41–64
102. Malek AM, Alper SL, Izumo S (1999) Hemodynamic shear stress and its role in atherosclerosis. JAMA 282:2035–2042
103. McCormick SM, Eskin SG, McIntire LV, Teng CL, Lu CM, Russell CG, Chittur KK (2001) DNA microarray reveals changes in gene expression of shear stressed human umbilical vein endothelial cells. Proc Natl Acad Sci USA 98:8955–8960
104. McLenachan JM, Vita J, Fish DR, Treasure CB, Cox DA, Ganz P, Selwyn AP (1990) Early evidence of endothelial vasodilator dysfunction at coronary branch points. Circulation 82:1169–1173
105. Miao H, Hu YL, Shiu YT, Yuan S, Zhao Y, Kaunas R, Wang Y, Jin G, Usami S, Chien S (2005) Effects of flow patterns on the localization and expression of VE-cadherin at vascular endothelial cell junctions: in vivo and in vitro investigations. J Vasc Res 42:77–89
106. Michelsen KS, Wong MH, Shah PK, Zhang W, Yano J, Doherty TM, Akira S, Rajavashisth TB, Arditi M (2004) Lack of Toll-like receptor 4 or myeloid differentiation factor 88 reduces atherosclerosis and alters plaque phenotype in mice deficient in apolipoprotein E. Proc Natl Acad Sci USA 101:10679–10684

107. Motomiya M, Karino T (1984) Flow patterns in the human carotid artery bifurcation. Stroke 15:50–56
108. Nagel T, Resnick N, Atkinson WJ, Dewey CF Jr, Gimbrone MA Jr (1994) Shear stress selectively upregulates intercellular adhesion molecule-1 expression in cultured human vascular endothelial cells. J Clin Invest 94:885–891
109. Nagel T, Resnick N, Dewey CF Jr, Gimbrone MA Jr (1999) Vascular endothelial cells respond to spatial gradients in fluid shear stress by enhanced activation of transcription factors. Arterioscler Thromb Vasc Biol 19:1825–1834
110. Nerem RM (1992) Vascular fluid mechanics, the arterial wall, and atherosclerosis. J Biomech Eng 114:274–282
111. Nerem RM, Alexander RW, Chappell DC, Medford RM, Varner SE, Taylor WR (1998) The study of the influence of flow on vascular endothelial biology. Am J Med Sci 316:169–175
112. Nerem RM, Levesque MJ, Cornhill JF (1981) Vascular endothelial morphology as an indicator of the pattern of blood flow. J Biomech Eng 103:172–176
113. Oemar BS, Werner A, Garnier JM, Do DD, Godoy N, Nauck M, Marz W, Rupp J, Pech M, Luscher TF (1997) Human connective tissue growth factor is expressed in advanced atherosclerotic lesions. Circulation 95:831–839
114. Oshinski JN, Ku DN, Mukundan S Jr, Loth F, Pettigrew RI (1995) Determination of wall shear stress in the aorta with the use of MR phase velocity mapping. J Magn Reson Imaging 5:640–647
115. Oyre S, Pedersen EM, Ringgaard S, Boesiger P, Paaske WP (1997) In vivo wall shear stress measured by magnetic resonance velocity mapping in the normal human abdominal aorta. Eur J Vasc Endovasc Surg 13:263–271
116. Packham N, Sheil AG, Loewenthal J (1967) Aortic aneurysms: a review with reports of 62 cases. Med J Aust 2:833–836
117. Passerini AG, Polacek DC, Shi C, Francesco NM, Manduchi E, Grant GR, Pritchard WF, Powell S, Chang GY, Stoeckert CJ Jr, Davies PF (2004) Coexisting proinflammatory and antioxidative endothelial transcription profiles in a disturbed flow region of the adult porcine aorta. Proc Natl Acad Sci USA 101:2482–2487
118. Pedersen EM, Agerbaek M, Kristensen IB, Yoganathan AP (1997) Wall shear stress and early atherosclerotic lesions in the abdominal aorta in young adults. Eur J Vasc Endovasc Surg 13:443–451
119. Phelps JE, DePaola N (2000) Spatial variations in endothelial barrier function in disturbed flows in vitro. Am J Physiol Heart Circ Physiol 278:H469–H476
120. Porat RM, Grunewald M, Globerman A, Itin A, Barshtein G, Alhonen L, Alitalo K, Keshet E (2004) Specific induction of tie1 promoter by disturbed flow in atherosclerosis-prone vascular niches and flow-obstructing pathologies. Circ Res 94:394–401
121. Pritchard WF, Davies PF, Derafshi Z, Polacek DC, Tsao R, Dull RO, Jones SA, Giddens DP (1995) Effects of wall shear stress and fluid recirculation on the localization of circulating monocytes in a three-dimensional flow model. J Biomech 28:1459–1469
122. Resnick N, Collins T, Atkinson W, Bonthron DT, Dewey CF Jr., Gimbrone MA Jr (1993) Platelet-derived growth factor B chain promoter contains a cis-acting fluid shear-stress-responsive element. Proc Natl Acad Sci USA 90:4591–4595
123. Resnick N, Gimbrone MA Jr (1995) Hemodynamic forces are complex regulators of endothelial gene expression. FASEB J 9:874–882
124. Ross R (1993) The pathogenesis of atherosclerosis: a perspective for the (1990s) Nature 362:801–809
125. Ross R (1999) Atherosclerosis-an inflammatory disease. N Engl J Med 340:115–126
126. Sampath R, Kukielka GL, Smith CW, Eskin SG, McIntire LV (1995) Shear stress-mediated changes in the expression of leukocyte adhesion receptors on human umbilical vein endothelial cells in vitro. Ann Biomed Eng 23:247–256
127. Sawchuk AP, Unthank JL, Davis TE, Dalsing MC (1994) A prospective, in vivo study of the relationship between blood flow hemodynamics and atherosclerosis in a hyperlipidemic swine model. J Vasc Surg 19:58–63; discussion 63–54

128. Schwenke DC, Carew TE (1988) Quantification in vivo of increased LDL content and rate of LDL degradation in normal rabbit aorta occurring at sites susceptible to early atherosclerotic lesions. Circ Res 62:699–710
129. Schwenke DC, Carew TE (1989) Initiation of atherosclerotic lesions in cholesterol-fed rabbits. I. Focal increases in arterial LDL concentration precede development of fatty streak lesions. Arteriosclerosis 9:895–907
130. SenBanerjee S, Lin Z, Atkins GB, Greif DM, Rao RM, Kumar A, Feinberg MW, Chen Z, Simon DI, Luscinskas FW, Michel TM, Gimbrone M.A. Jr, Garcia-Cardena G, Jain MK (2004) KLF2 Is a novel transcriptional regulator of endothelial proinflammatory activation. J Exp Med 199:1305–1315
131. Shyy YJ, Hsieh HJ, Usami S, Chien S (1994) Fluid shear stress induces a biphasic response of human monocyte chemotactic protein 1 gene expression in vascular endothelium. Proc Natl Acad Sci USA 91:4678–4682
132. Shyy JY, Lin MC, Han J, Lu Y, Petrime M, Chien S (1995) The cis-acting phorbol ester "12-O-tetradecanoylphorbol 13-acetate"-responsive element is involved in shear stress-induced monocyte chemotactic protein 1 gene expression. Proc Natl Acad Sci USA 92:8069–8073
133. Skilbeck C, Westwood SM, Walker PG, David T, Nash GB (2001) Dependence of adhesive behavior of neutrophils on local fluid dynamics in a region with recirculating flow. Biorheology 38:213–227
134. Sprague EA, Luo J, Palmaz JC (1997) Human aortic endothelial cell migration onto stent surfaces under static and flow conditions. J Vasc Interv Radiol 8:83–92
135. Stemerman MB, Morrel EM, Burke KR, Colton CK, Smith KA, Lees RS (1986) Local variation in arterial wall permeability to low density lipoprotein in normal rabbit aorta. Arteriosclerosis 6:64–69
136. Stenina OI, Krukovets I, Wang K, Zhou Z, Forudi F, Penn MS, Topol EJ, Plow EF (2003) Increased expression of thrombospondin-1 in vessel wall of diabetic Zucker rat. Circulation 107:3209–3215
137. Tardy Y, Resnick N, Nagel T, Gimbrone MA Jr, Dewey CF Jr (1997) Shear stress gradients remodel endothelial monolayers in vitro via a cell proliferation-migration-loss cycle. Arterioscler Thromb Vasc Biol 17:3102–3106
138. Topper JN, Gimbrone MA Jr (1999) Blood flow and vascular gene expression: fluid shear stress as a modulator of endothelial phenotype. Mol Med Today 5:40–46
139. Traub O, Berk BC (1998) Laminar shear stress: mechanisms by which endothelial cells transduce an atheroprotective force. Arterioscler Thromb Vasc Biol 18:677–685
140. Truskey GA, Barber KM, Robey TC, Olivier LA, Combs MP (1995) Characterization of a sudden expansion flow chamber to study the response of endothelium to flow recirculation. J Biomech Eng 117:203–210
141. Tsuboi H, Ando J, Korenaga R, Takada Y, Kamiya A (1995) Flow stimulates ICAM-1 expression time and shear stress dependently in cultured human endothelial cells. Biochem Biophys Res Commun 206:988–996
142. Usami S, Chen HH, Zhao Y, Chien S, Skalak R (1993) Design and construction of a linear shear stress flow chamber. Ann Biomed Eng 21:77–83
143. van Haperen R, Cheng C, Mees BM, van Deel E, de Waard M, van Damme LC, van Gent T, van Aken T, Krams R, Duncker DJ, de Crom R (2003) Functional expression of endothelial nitric oxide synthase fused to green fluorescent protein in transgenic mice. Am J Pathol 163:1677–1686
144. VanderLaan PA, Reardon CA, Getz GS (2004) Site specificity of atherosclerosis: site-selective responses to atherosclerotic modulators. Arterioscler Thromb Vasc Biol 24:12–22
145. Vyalov S, Langille BL, Gotlieb AI (1996) Decreased blood flow rate disrupts endothelial repair in vivo. Am J Pathol 149:2107–2118
146. Wang N, Miao H, Li YS, Zhang P, Haga JH, Hu Y, Young A, Yuan S, Nguyen P, Wu CC, Chien S (2006) Shear stress regulation of Kruppel-like factor 2 expression is flow pattern-specific. Biochem Biophys Res Commun 341:1244–1251

147. Wasserman SM, Mehraban F, Komuves LG, Yang RB, Tomlinson JE, Zhang Y, Spriggs F, Topper JN (2002) Gene expression profile of human endothelial cells exposed to sustained fluid shear stress. Physiol Genomics 12:13–23
148. Weinbaum S, Tzeghai G, Ganatos P, Pfeffer R, Chien S (1985) Effect of cell turnover and leaky junctions on arterial macromolecular transport. Am J Physiol 248:H945–H960
149. Weinberg PD (2004) Rate-limiting steps in the development of atherosclerosis: the response-to-influx theory. J Vasc Res 41:1–17
150. Wissler RW (1995) An overview of the quantitative influence of several risk factors on progression of atherosclerosis in young people in the United States. Pathobiological Determinants of Atherosclerosis in Youth (PDAY) Research Group. Am J Med Sci 310(1):S29–S36
151. Wojciak-Stothard B, Ridley AJ (2003) Shear stress-induced endothelial cell polarization is mediated by Rho and Rac but not Cdc42 or PI 3-kinases. J Cell Biol 161:429–439
152. Wu MH, Kouchi Y, Onuki Y, Shi Q, Yoshida H, Kaplan S, Viggers RF, Ghali R, Sauvage LR (1995) Effect of differential shear stress on platelet aggregation, surface thrombosis, and endothelialization of bilateral carotid-femoral grafts in the dog. J Vasc Surg 22:382–390; discussion 390–382
153. Zamir M (1976) The role of shear forces in arterial branching. J Gen Physiol 67:213–222
154. Zarins CK, Giddens DP, Bharadvaj BK, Sottiurai VS, Mabon RF, Glagov S (1983) Carotid bifurcation atherosclerosis. Quantitative correlation of plaque localization with flow velocity profiles and wall shear stress. Circ Res 53:502–514
155. Zarins CK, Zatina MA, Giddens DP, Ku DN, Glagov S (1987) Shear stress regulation of artery lumen diameter in experimental atherogenesis. J Vasc Surg 5:413–420

Chapter 20
Why is Sepsis an Ongoing Clinical Challenge? Lipopolysaccharide Effects on Red Blood Cell Volume

Aysegül Temiz Artmann and Peter Kayser

Medical & Molecular Medicine, Aachen University of Applied Sciences, Ginsterweg 1, 52428 Juelich, Germany and Centre of Competence in Bioengineering, Aachen University of Applied Sciences, Ginsterweg 1, 52428 Juelich, Germany, a.artmann@fh-aachen.de

Abstract Sepsis is a systemic inflammatory reaction response to infection (SIRS). It is difficult to diagnose and hard to treat, but the incidence in patients steadily increasing. Sepsis is called "severe" if any organ fails; whereas septic shock involves a failure of the cardiovascular system. The mortality rate for sepsis is 20–30%, rising to over 50–90% in cases of septic shock. Death may occur early in the course of disease due to irreversible shock, or later due to development of multiple organ dysfunctions. Hypotension is caused by myocardial depression, pathological vasodilatation and extravasations of circulating volume due to widespread capillary leak. The endotoxin "*lipopolysaccharide*" (LPS) is the primary gram-negative bacterial product responsible for sepsis. LPS causes a shock-like state with hypotension and organ dysfunction when injected into animals and has predictable effects in healthy humans. A single controlled intravenous bolus of *E. Coli* endotoxin evokes many of the responses like malaise, headaches, nausea and vomiting, and clinical signs such as tachycardia, tachypnea, leukocytosis and hypotension that characterise septic shock.

Red blood cells (RBC) affect blood flow and circulation in the micro and macrovasculature depending on their membrane, cytosolic components and their mechanical parameters. Most *in-vivo* studies conducted applying microorganisms like *Escherichia Coli* to experimental animals, showed deleterious effects of the infection on RBC. RBC deformability decrements and aggregation increments were reported. In contrary, our *in-vitro* experiments adding LPS to whole blood could not confirm these findings. The RBC deformability remained unchanged and RBC lost their capability of forming rouleaux represents RBC aggregation, ($p < 0.05$). The effect on RBC aggregation was observed visually (differential interference contrast images), by RBC sedimentation rate and with the Laser Optical Rotational Cell Analyzer. Surprisingly, the loss of RBC aggregation was determined immediately after LPS was added at concentrations of $1-10\,\mu g/mL$ to whole blood. However, RBC aggregation was reestablished after LPS containing plasma was exchanged with LPS free plasma. Additionally, in plasma containing LPS, RBC appeared slightly swollen. The measured RBC diameter increased significantly ($p < 0.05$). The possible volume change of RBC was also followed by micropipette aspiration experiments. The RBC tongue length found significantly decreased after LPS application ($p < 0.05$) that represented the possible RBC volume increase. These results took our attention to plasma proteins and colloid osmotic pressure. *In-vitro* LPS induced blood showed a significant immediate colloidal osmotic pressure decrease ($p < 0.05$).

As septic shock occurs from volume depletion where the volume from circulation is mostly lost because of the increase of endothelium permeability, possible RBC water influx might constitute another reason for rapid volume depletion. The *in-vitro* results are difficult to prove in *in-vivo* systems because of strongly controlled homeostasis. The new hypothesis may open a new discussion on LPS-induced sepsis and septic shock, potentially being followed by a new therapeutic approach.

20.1 Introduction

A review of the literature on sepsis leads to the following conclusions: sepsis is difficult to diagnose and treat although the number of cases observed is steadily increasing. The incidence of this condition has increased by over 300% in the last twenty years. The mortality rate of patients with septic shock is 50–90% and is currently the 10th leading cause of death [1].

Sepsis is characterized by systemic activation of the inflammatory and coagulation cascades during microbial infection. As the inflammatory response to the infection is uncontrolled, it may cause unsuppressed inhibition of fibrinolysis, thus promoting thrombosis. Subsequently, the progression of sepsis leads to hypotension, microvascular dysfunction and disseminated intravascular coagulation. Hypotension is caused by myocardial depression, pathological vasodilatation and extravasations of the circulating volume of blood due to widespread capillary leak [2]. Septic shock may follow sepsis, resulting in multiple organ dysfunction syndrome and death. The condition is defined as severe sepsis if more than one organ fails, and septic shock if there is a failure of the cardiovascular system. The synthesized mediators may cause circulatory collapse and pan-endothelial injury that may increase microvascular permeability everywhere in the body. Briefly the host reaction to inflammation can be clinically manifested by the systemic inflammatory response syndrome (SIRS) progressing – in response to infection – to severe sepsis, septic shock, adult respiratory distress syndrome (ARDS) and multiple organ dysfunction syndrome. Secondary infection may subsequently occur in addition to organ failure.

The endotoxin "*lipopolysaccharide*" (LPS) is the primary gram-negative bacterial product responsible for many sepsis cases which is a principal initiator in the pathogenesis of sepsis. LPS causes a shock-like state with hypotension and organ dysfunction when injected into animals and has predictable effects in healthy humans. The history of *lipopolysaccharides*, recently documented by Rietschel and Westphal [3], began in the 18th century. They searched for the fever and disease-producing substance that was associated with unhygienic conditions. It was referred to as the pyrogenic material, putrid poison, or toxin. By 1872 the possible role for living organisms in human disease allowed Klebs, a German bacteriologist, to attribute the majority of military war deaths to a pyrogenic substance from micro organisms which he called "*Microsporon septicum*". Two years later, a Danish pathologist Panum reported a non-volatile, heat-resistant, water-soluble, pyrogenic toxin obtained from putrid matter. In 1892, Pfeiffer reported that the agent of cholera, *Vibrio cholerae*, produces a pyrogenic, non-secreted toxin that is heat-stabile, in addition to a secreted, heat-labile toxin. He called it endotoxin, a term still used today for the *lipopolysaccharides* that were later found to constitute endotoxins. Techniques to extract and to prepare endotoxins which were sufficiently pure for use in structural studies were slow to develop. In the 1930s and 1940s, fairly pure preparations were reported to consist of a *polysaccharide*, a lipid part, and a small amount of protein, and received the name *lipopolysaccharide* [3]. In an infected host, small amounts of LPS can be beneficial, as they can be used to stimulate the immune system in order to shrink tumours; however larger amounts induce high fever, increase heart rate, and lead to septic shock and death by lung and kidney failure, intravascular coagulation, and systemic inflammatory response. The enterobacterial LPSs (especially those of *Escherichia coli* and *Salmonella enterica serovar typhimurium*) were most thoroughly studied.

Gram negative septic shock comprises 50% of total cases of sepsis. Endotoxic shock is ascribed to lipid A attached to the LPS core *polysaccharide* common to most gram negative bacteria [4].

20.2 Physiopathological Events During Sepsis

Mononuclear phagocytes recognise microbial structures such as endotoxins by CD14 receptors. The CD14 receptor has no intracellular domain, and thus needs the help of adjacent Toll-like receptors. Toll-like receptors are able to mediate the signal into the cell, resulting in activation of nuclear factor κB, a transcription factor, which promotes the synthesis and release of a variety of pro- and anti-inflammatory cytokines by activated mononuclear phagocytes.

20.3 Markers in Clinical Diagnosis of Sepsis

The markers of systemic inflammation are very important to identify patients at risk of severe sepsis or septic shock at an early stage of the disease. Systemic inflammation is characterised by synthesized proinflammatory cytokines (tumour necrosis factor, interleukin-1, and interleukin-8). These cytokines and vasoactive mediators are some of the markers to diagnose sepsis. The most common current approach in the diagnosis of sepsis and the most beneficial therapy is to check not a single marker but rather to provide the necessary information. The set of markers reflecting pro and anti-inflammatory activation profiles are mostly used. Tumour necrosis factor α and interleukin-1 are associated with the early pathophysiological events of septic shock. Procalcitonin (PCT) is an innovative and highly specific marker used for the diagnosis of clinically relevant bacterial infections and sepsis. Procalcitonin supports early diagnosis and clinical decision-making which could direct a timely effective therapy at the right time and prove cost-effective for critically ill patients.

20.4 Microcirculation and Sepsis

Sepsis is associated with haemodynamic alterations and microcirculatory disturbances. Early studies suggest that the microcirculation is a major site of endotoxin attack during septic shock. More recent studies examined [5] possible impacts of endotoxins on the microcirculation. Many changes take place, such as the slowing of capillary blood flow due to depressed perfusion pressure as a result of systemic pressure decrease and local arteriolar constriction. Observations suggest that the microcirculation is shut off early in severe sepsis, allowing the effects of hypoperfusion and widespread capillary dilation to ultimately occur [5].

Decreased capillary blood flow during shock is a result of the failure in the normal passage of cellular elements, including erythrocytes and neutrophils. This defect occurs, in part, because of decreased perfusion pressure, decreased deformability of red and white blood cells, constricted arterioles, circulating obstructive fragments (including hemoglobin), and the plugging of microvessels with "sludge". Other factors include the adherence of cells to capillary and venular epithelial membranes creating increased resistance to flow, loss of fluid through abnormal transcapillary exchange, differential vascular resistance changes between various beds (e. g., intestinal vs. muscle), and the relative absence of regulatory neurohumoral control of small vessel segments of the circulation. During sepsis/septic shock, endothelial cells are reported to modulate vascular tone, control local blood flow, influence the rate of leakage of fluids and plasma proteins into tissues, modulate the accumulation and extravasation of white cells into tissues, and influence white cell activation. As a result of the predominance of many destructive factors, a subsequent round of tissue damage may occur. Because of prolonged capillary vascular stasis, deficient flow, and factors released from injured cells, the microcirculation becomes a site for uncontrolled bacterial growth enhanced by sustained hypoxemia, acidosis and toxaemia. These events may combine to contribute to the loss of normal cell integrity and death of the host.

20.5 Therapy

Despite active treatment, the death rate of septic shock patients is very high. Losses of intravascular integrity [6], and increased capillary membrane permeability are the known causes of septic shock that conceals hypovolemic shock by internal fluid loss. Aggressive fluid therapy is very important during septic shock therapy. Septic patients with impending circulatory collapse often need five or more litres per day of fluid to support their blood pressure although the fluid which is given is not held in circulation and lost to the third space permanently [7]. Crystalloids or colloids may be used to this end, although crystalloids are preferred. Physicians often favour plasma therapies rather than replacing the volume only with isotonic salt solutions. Critically ill patients with sepsis may have strong hypoalbuminemia but albumin supplementation to profoundly hypoalbuminemic septic patients has been shown to have no clinically significant effect in reducing microvascular permeability [8].

20.6 Activated Protein C

Recombinant activated protein C (rhAPC) is a new treatment for patients with severe sepsis [20] and recently, placebo-controlled randomized clinical trials demonstrated the efficacy and safety of rhAPC for severe sepsis. As compared to a placebo, a 4-day infusion of recombinant APC leads to a reduction in the occurrence of death by 19.4% and an absolute reduction in the risk of death by 6.1% ($p < 0.005$) [21]. Protein C is a glycoprotein which occurs in blood plasma.

Activated protein C is an important natural anticoagulant, formed from protein C by the action of the thrombin-thrombomodulin (TM) complex on the endothelial cell surface. It inhibits tumour necrosis factor-α production and attenuates different deleterious events induced by LPS, causing the significant reduction of mortality in patients with severe sepsis [22]. Activated protein C effects on inflammatory responses, especially leukocyte endothelium responses, and ischemia/reperfusion induced organ injuries during sepsis to severe sepsis and septic shock cases have been studied thoroughly [22–27]. In one of the studies, the administration of APC decreased pulmonary vascular injury and hypotension as well as coagulation abnormalities by inhibiting production of the tumor necrosis factor-α in rats given endotoxin [25]. In this study, it was concluded that the observations strongly suggested that APC plays an important role in the regulation of inflammation as well as coagulation. The anti-inflammatory and anticoagulant properties of APC are useful in treating patients with severe sepsis.

20.7 Red Blood Cell Behaviour During Sepsis

Red blood cells affect blood flow and circulation in the micro and macrovasculature depending on their membrane, cytosolic components and their mechanical parameters. Reactive oxygen species are continuously generated in the biological system during infection, and play an important role in a variety of processes which are strongly toxic for the organism as they can attack and modify a wide variety of biological molecules [9]. If the quantities of reactive oxygen species are so high that the organism cannot metabolise or inactivate them, surrounding tissues are negatively affected and damaged [10]. Red blood cells have a well defined biconcave shape allowing them to pass through narrow capillaries where they undergo large, reversible, nonlinear elastic deformation. They represent a relatively simple cellular model system, compared to other biological cells. Recent studies have indicated that RBC which are in close contact with activated leukocytes can be damaged, at least in part by oxidative mechanisms, resulting in structural and functional alterations [11–13]. Most *in-vivo* studies conducted applying microorganisms like *Escherichia coli* to experimental animals, showed deleterious effects of the infection on RBC [14]. This decreased flexibility may be responsible, in part, for the microcirculatory abnormalities accompanying sepsis [15]. RBC deformability decrements and aggregation increments were reported although RBC aggregation remains controversial during sepsis, septic shock, and toxic shock [14, 16, 17].

20.8 New Perspective

In our *in-vitro* experiments, where LPS was added to whole blood, no confirmation could be made concerning the increase of RBC aggregation; RBC deformability remained unchanged and RBC lost their capability of forming rouleaux (aggregation) (Fig. 20.1). Blood flow is strongly influenced by erythrocyte flow properties; alterations of the cellular or membrane characteristics of RBC can affect haemodynamics. RBC aggregability is one of the properties which plays a major role in

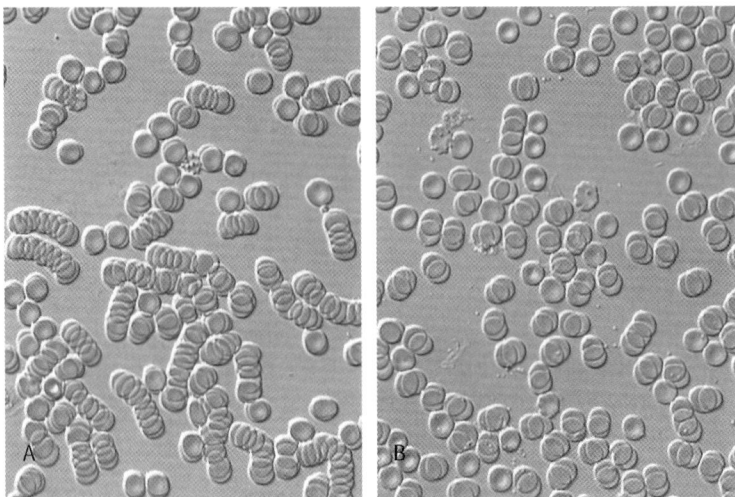

Fig. 20.1 Micro photographs of RBC Aggregation before (**A**) and immediately after (**B**) adding 10 µL PBS (**A**) or 10 µL LPS stock solution (0.1 µg LPS/1 µL LPS stock solution) to 1000 µL whole blood (**B**). The images were chosen from areas where the most RBC appeared. The slight RBC volume changes after LPS are present, although not visible to the naked eye

20.8 New Perspective

haemodynamics. Determinant factors of RBC aggregation include the number of RBC, RBC shape (contact area), RBC deformability, surface charge, concentration of macromolecules (plasma proteins), size of macromolecules, shear rate in blood flow, and the viscosity of the media. The effect on RBC aggregation was observed microscopically (Fig. 20.1) and with the Laser Optical Rotational Cell Analyzer (Fig. 20.2). Surprisingly, the loss of RBC aggregation was determined immediately after LPS was added at a concentration of 1 μg/mL to whole blood (Fig. 20.2). However, RBC aggregation was re-established after LPS-containing plasma was exchanged with LPS free plasma. Additionally, in plasma-containing LPS, RBC ap-

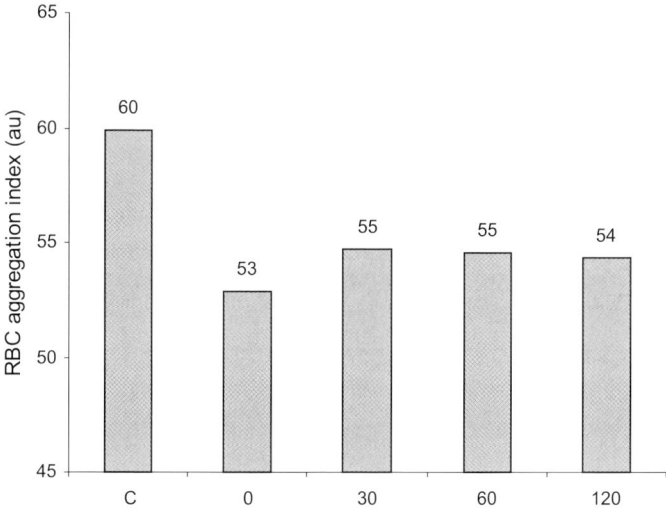

Fig. 20.2 Typical RBC aggregation time course ($n = 1$) over two hours at room temperature. At time zero, 1 μg LPS/mL whole blood was added. As seen here, the anti-aggregative LPS effect appeared immediately after LPS was added to the sample

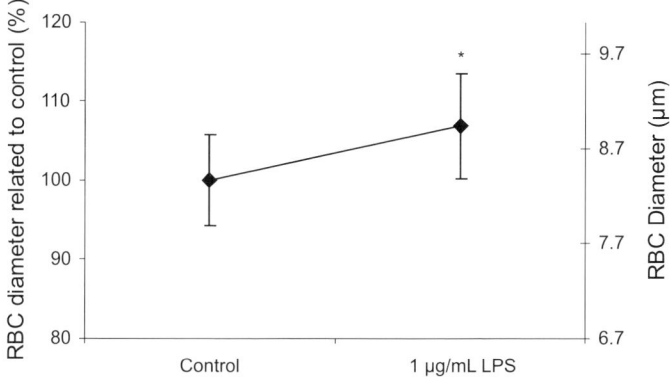

Fig. 20.3 LPS-induced RBC diameter increment due to adding 1 μg LPS/ml whole blood after 2 h of incubation at 37 °C ($n = 50$). The RBC diameter shown here is a projected diameter

peared slightly swollen. The RBC diameter of LPS added blood was found to have been significantly increased to $p < 0.05$ compared to the control by evaluating microscopic images and by the Micropipette Aspiration Technique results (Figs. 20.3, 20.4 and 20.5). The whole blood colloid osmotic pressure was significantly lowered ($p < 0.05$) *in-vitro*. Therefore, possible water influx can cause the RBC volume increase (Fig. 20.6). Together with the observed immediate RBC effects in the

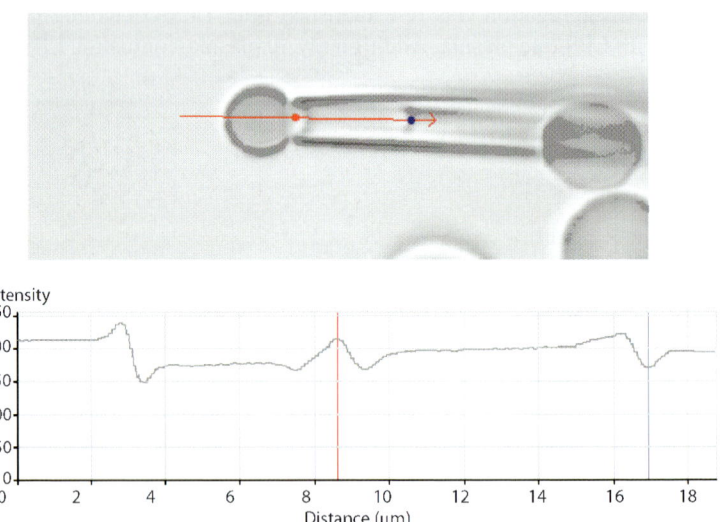

Fig. 20.4 The same micropipette (borosilicate glass) was always used. The negative pressure was constant (26 cm H_2O column). The distance between the *arrows* shows the tongue length of RBC

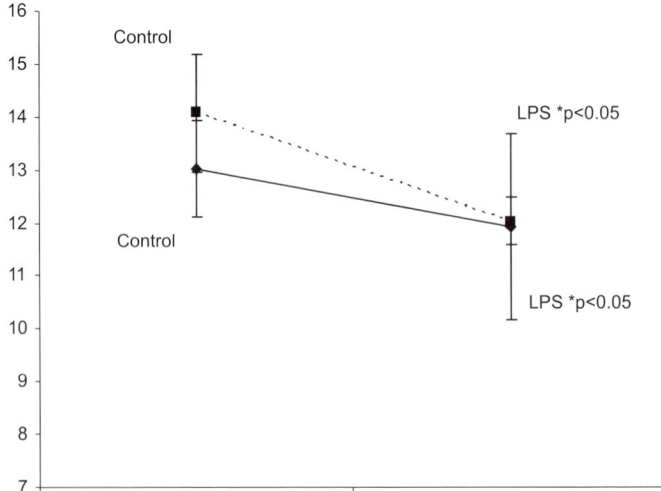

Fig. 20.5 RBC tongue length (μm) before ($n = 30$/donor; 5 donors) and immediately after (*solid line*) adding 1 μg LPS/ml whole blood or after 2 h incubation at 37 °C with 1 μg LPS/ml whole blood ($n = 3$, 15 RBC/donor), respectively (error bars +/− 1 SD)

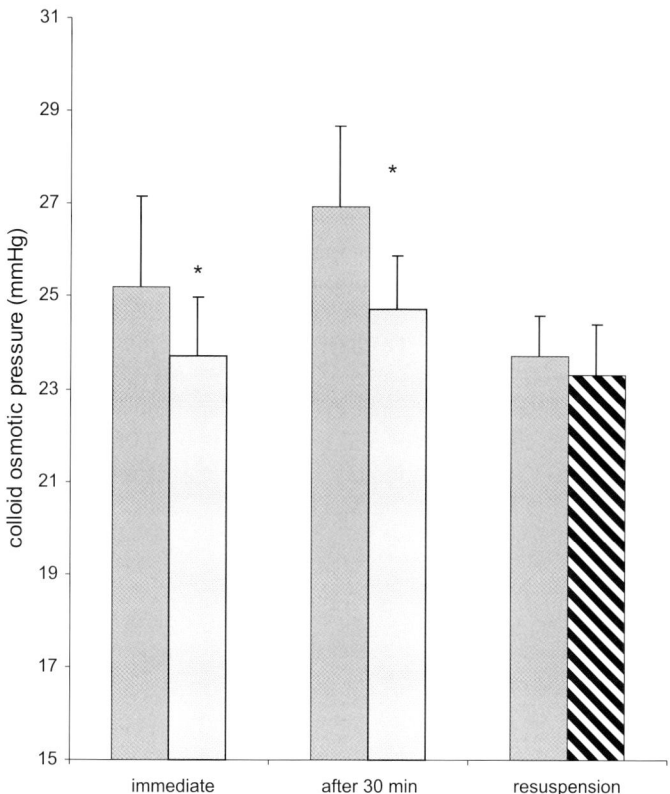

Fig. 20.6 Colloid osmotic pressure (COP) of whole blood at room temperature (average of 5 donors). *Left columns*: immediately after LPS was added to whole blood, the COP decreased significantly by 6%. *Middle columns*: 30 minutes after 1 μg LPS/ml whole blood, a COP decrease of 8% was observed. *Right columns*: LPS-containing plasma was removed and replaced by LPS-free plasma. The observed volume changes recovered to normal (* $p < 0.05$, error bars $+/-1$ SD)

presence of LPS, a biophysical mechanism for the effects can be established. The possible water influx into the RBCs further negatively influences the circulation insufficiency.

The LPSs resisted structural characterization by their amphipathic nature. Their strong tendency to form aggregates made it difficult to determine their molecular weight. The discovery that the lipid region of LPSs could be separated after weak acid hydrolysis helped to establish its general structure comprising of two or three regions: a lipid region, a core, and a third region in bacteria-yielding smooth colonies [18]. In the early 1980s, the structures of both lipid A and the core were established. The regions moving from the bacterial membrane toward the outside are: a lipid (called lipid A) linked to Kdo of the core *oligosaccharide*, itself linked to the second *glycoside* if any, consisting of a sequence of repetitive subunits named O-chain or O-specific antigens. The endotoxins were later recognized as the matrix of the external leaflet of the bacterial outer membrane that form the major component (45%) occupying 75% of the surface of the bacterium. *E. coli* is estimated to have 106 molecules per cell [19]. The *polysaccharide* moiety is directed outwards from the bacterium surface, extending up to 10 nm from the surface. Although not secreted by the cells, small amounts of the LPS are released when cell division occurs and larger amounts are released by bacteria killed by antibiotics, phagocytosis, the complement complex, or through treatment with divalent cation chelators.

The water movement towards RBC is strongly controlled by the osmolarity of the environment. As the rapid LPS anti-aggregant effect on slightly swollen RBCs was observed in our experiments, colloid osmotic pressure (COP) of whole blood at room temperature (average of 5 donors) was determined. Immediately after LPS was added to whole blood, the COP decreased significantly by 6%. Colloid osmotic pressure decrease 30 minutes after 1 μg/mL LPS in plasma was 8%. Following the removal of LPS-containing plasma and replacement by LPS-free plasma, the observed volume changes recovered to normal ($p < 0.05$). The RBC tongue length measurements (30 RBC randomly chosen in any sample) undertaken with micropipette aspiration technique confirmed the MCV changes determined by microphotographs.

References

1. Sun D, Aikawa N (1999) The natural history of the systemic inflammatory response syndrome and the evaluation of SIRS criteria as a predictor of severity in patients hospitalized through emergency services. Keio J Med 48:28–37
2. Shapiro N, Howell MD, Bates DW et al. (2006) The association of sepsis syndrome and organ dysfunction with mortality in emergency department patients with suspected infection. Ann Emerg Med 48:583–90, 590
3. Rietschel ETh, Westphal O (1999) In Endotoxins in Health and Disease. Brade H, Opal SM, Vogel SN, Morrison DC, editors. History of Endotoxins, Marcel Dekker, New York, Ref Type, Generic, pp 1–30
4. Gransden WR, Eykyn SJ, Phillips I, Rowe B (1990) Bacteremia due to Escherichia coli: a study of 861 episodes. Rev Infect Dis 12:1008–1018
5. Hinshaw LB (1996) Sepsis/septic shock: participation of the microcirculation: an abbreviated review. Crit Care Med 24:1072–1078
6. Langheinrich AC, Ritman EL (2006) Quantitative imaging of microvascular permeability in a rat model of lipopolysaccharide-induced sepsis: evaluation using cryostatic micro-computed tomography. Invest Radiol. 41:645–650
7. Sparrow A, Willis F (2004) Management of septic shock in childhood. Emerg Med Australas 16:125–134
8. Margarson MP, Soni NC (2002) Effects of albumin supplementation on microvascular permeability in septic patients. J Appl Physiol 92:2139–2145
9. Southorn PA, Powis G (1988) Free radicals in medicine. II. Involvement in human disease. Mayo Clin Proc 63:390–408
10. Ho JS, Buchweitz JP, Roth RA, Ganey PE (1996) Identification of factors from rat neutrophils responsible for cytotoxicity to isolated hepatocytes. J Leukoc Biol 59:716–724
11. Temiz A, Baskurt OK, Pekcetin C, Kandemir F, Gure A (2000) Leukocyte activation, oxidant stress and red blood cell properties after acute, exhausting exercise in rats. Clin Hemorheol Microcirc 22:253–259
12. Rohn TT, Hinds TR, Vincenzi FF (1995) Inhibition by activated neutrophils of the Ca_{2+} pump ATPase of intact red blood cells. Free Radic Biol Med 18:655–667
13. Baskurt OK, Temiz A, Meiselman HJ (1998) Effect of superoxide anions on red blood cell rheologic properties. Free Radic Biol Med 24:102–110
14. Baskurt OK, Temiz A, Meiselman HJ (1997) Red blood cell aggregation in experimental sepsis. J Lab Clin Med 130:183–190
15. Todd JC, III, Poulos ND, Davidson LW, Mollitt DL (1993) Role of the leukocyte in endotoxin-induced alterations of the red cell membrane. Second place winner of the Conrad Jobst Award in the Gold Medal paper competition. Am Surg 59:9–12
16. Voerman HJ, Fonk T, Thijs LG (1989) Changes in hemorheology in patients with sepsis or septic shock. Circ Shock 29:219–227

17. Alt E, Amann-Vesti BR, Madl C, Funk G, Koppensteiner R (2004) Platelet aggregation and blood rheology in severe sepsis/septic shock: relation to the Sepsis-related Organ Failure Assessment (SOFA) score. Clin Hemorheol Microcirc 30:107–115
18. Osborn MJ (1963) Studies on the Gram-Negative Cell Wall. I. Evidence for the Role OF 2-. Proc Natl Acad Sci USA 50:499–506
19. Raetz CR, Whitfield C (2002) Lipopolysaccharide endotoxins. Annu Rev Biochem 71:635–700
20. Green C, Dinnes J, Takeda A et al. (2005) Clinical effectiveness and cost-effectiveness of drotrecogin alfa (activated) (Xigris) for the treatment of severe sepsis in adults: a systematic review and economic evaluation. Health Technol Assess 9:1–iv
21. Liaw PC, Esmon CT, Kahnamoui K et al. (2004) Patients with severe sepsis vary markedly in their ability to generate activated protein C. Blood 104:3958–3964
22. Yuksel M, Okajima K, Uchiba M, Horiuchi S, Okabe H (2002) Activated protein C inhibits lipopolysaccharide-induced tumor necrosis factor-alpha production by inhibiting activation of both nuclear factor-kappa B and activator protein-1 in human monocytes. Thromb Haemost 88:267–273
23. Murakami K, Okajima K, Uchiba M et al. (1996) Activated protein C attenuates endotoxin-induced pulmonary vascular injury by inhibiting activated leukocytes in rats. Blood 87:642–647
24. Hirose K, Okajima K, Taoka Y et al. (2000) Activated protein C reduces the ischemia/reperfusion-induced spinal cord injury in rats by inhibiting neutrophil activation. Ann Surg 232:272–280
25. Okajima K (2004) Regulation of inflammatory responses by activated protein C: the molecular mechanism(s) and therapeutic implications. Clin Chem Lab Med 42:132–141
26. Pirat B, Muderrisoglu H, Unal MT et al. (2007) Recombinant human-activated protein C inhibits cardiomyocyte apoptosis in a rat model of myocardial ischemia-reperfusion. Coron Artery Dis 18:61–66
27. Finigan JH, Dudek SM, Singleton PA et al. (2005) Activated protein C mediates novel lung endothelial barrier enhancement: role of sphingosine 1-phosphate receptor transactivation. J Biol Chem 280:17286–17293

Chapter 21
Bioengineering of Inflammation and Cell Activation: Autodigestion in Shock

Alexander H. Penn[1], Erik B. Kistler[2], Geert W. Schmid-Schönbein[1]

[1]Department of Bioengineering, The Whitaker Institute of Biomedical Engineering, Powell-Focht Bioengineering Hall, Room 284, 9500 Gilman Drive, University of California San Diego, La Jolla, CA 92093-0412
gwss@bioeng.ucsd.edu
[2]Department of Anesthesiology, University of California San Diego, La Jolla, CA

Death sits in the intestines

Hippocrates (460–370 BC)

Abstract Increasing evidence suggests that most acute and chronic cardiovascular diseases and tumors are associated with an inflammatory cascade. The inflammation is accompanied by activation of cells in the circulation and fundamental changes in the mechanics of the microcirculation, expression of pro-inflammatory genes and downregulation of anti-inflammatory genes, attachment of leukocytes to the endothelium, elevated permeability of the endothelium, thrombosis, mast cell degranulation, apoptosis, macrophage infiltration and growth factor release. Eventually there is resolution of the inflammation with formation of new tissue. But in disease, the inflammatory cascade does not come to a complete resolution, indicating that a basic mechanism persists which triggers the inflammation. The trigger mechanism(s) that may stimulate the inflammatory cascade is currently unknown and the target of our research.

We present here a basic series of studies that was designed to explore trigger mechanisms for inflammation in shock and multi-organ failure, an important clinical problem that is associated with high mortality. We traced the source of the inflammatory mediators to the powerful digestive enzymes in the intestine. Synthesized in the pancreas as part of normal digestion, they have the ability to degrade almost all biological tissues and molecules. In the lumen of the intestine, digestive enzymes are fully activated and auto-digestion of the intestine is prevented by compartmentalization in the lumen of the intestine facilitated by the mucosal epithelial barrier. Under conditions of intestinal ischemia or epithelial stimulation, however, the mucosal barrier becomes permeable to pancreatic enzymes allowing their entry into the wall of the intestine. The process leads to *auto-digestion* of matrix proteins and tissue cells in the intestinal wall and production of a previously undescribed class of inflammatory mediators. These inflammatory mediators are released into the central circulation, lymphatics and the peritoneal cavity. The observations are in line with the hypothesis that multi-organ failure in shock may be due an auto-digestion process by pancreatic digestive enzymes. Our analysis leads to new opportunities; for example, inhibition of pancreatic enzymes in the lumen of the intestine, serves to attenuate the inflammation in several forms of shock.

21.1 Introduction

In the past decades a large number of important diseases have been shown to be associated with the inflammatory cascade. Tell-tale signatures and markers of the inflammatory cascade were initially demonstrated in chronic inflammatory diseases, e. g. rheumatoid arthritis and gastritis, but then also in experimental forms of ischemia of the heart, brain, kidneys, intestine and many other organs. It was shown that atherosclerosis, diabetes, chronic diseases like retinopathy, venous disease or dementia, and more recently also cancer and arterial hypertension have inflammatory markers [1]. These initial suggestions from experimental studies and small scale clinical studies were recently examined in larger clinical trials with selected markers for inflammation and in a variety of patient populations. Many of these trials reveal such high levels of correlation with clinical indices of inflammation that the measurement of selected inflammatory markers, such as c-Reactive Protein levels, has now been incorporated into clinical practice as a triage measure. An increasing number of diseases has been shown to be associated with inflammation; in fact at this stage we like to ask the question, which human disease is *not* associated with inflammation?

The inflammatory cascade serves over a life-time as a repair mechanism after an injury designed to initially eliminate and then restore new tissue, even if it is only a scar tissue without its original functional cell types. The many steps of the inflammatory cascade can be executed in all tissue and they are designed to lead to the resolution of inflammation at critical stages of life, e. g. after birth or trauma.

But in many of the diseases cited above, the markers for inflammation persist for prolonged periods of time without resolution. This evidence indicates that the root cause of inflammation *remains present*. It is our objective to identify the trigger mechanisms that maintain the inflammation in selected diseases. Their identification opens the door to prevention and optimization of medical treatment, quality of life, and longevity.

In the following we will discuss a specific case, *physiological shock with multi-organ failure*. In shock some of the strongest inflammatory signals are encountered, and there are many experimental models to investigate mechanisms. Multi-organ failure is associated with extraordinary high levels of mortality and therefore in term of its medical significance it is a problem second to none. It is a frontier for bioengineering research with a need for engineering analysis.

21.2 Inflammation in Shock and Multi-Organ Failure

Mortality in shock, regardless of the causative agent or action (e.g. hypovolemic, sepsis, cardiogenic, neurogenic, radiation-induced shock) has long been a poorly understood problem. Original theories, especially with regards to hemorrhagic shock, postulated the existence of 'toxic' mediators that interfered with the *homeostasis of the body* in the shock state, e.g. in form of bacterial or endotoxin sources. While this idea has come in and out of fashion throughout the last century, it is now generally accepted that inflammatory mediators, or 'activators' of the cardiovascular and inflammatory systems are important and possibly modifiable components of the shock paradigm. After activation, fundamental biomechanical properties of cells in the circulation are altered with profound impact on the microcirculation.

It was first noticed by Barroso-Aranda et al. [2,3] that otherwise matched animals that had higher circulating levels of activated leukocytes in their circulation during hemorrhagic shock had higher resultant mortality. Leukocyte activation could be demonstrated by use of a test (by nitro-blue tetrazolium reduction) that allows quantitative evaluation of the level of superoxide production by these cells. This test can be directly applied to fresh, *unseparated* blood samples, to maximize direct insight into the state of activation *in vivo*. Furthermore, those animals that displayed higher baseline levels of activated leukocytes pre-shock also had increased mortality secondary to the shock state. These results sparked a search in our laboratory for inflammatory factors in plasma that might initiate or otherwise contribute to the early manifestations of shock. Although many inflammatory mediators are thought to be involved in shock, particularly septic shock, antagonists to different putative mediators have been tried repeatedly with limited clinical success [4–6, 8]. Among these target inflammatory mediators were endotoxin, cytokines, complement and several others. The difficulty with targeting cytokines and other inflammatory mediators is that they appear downstream from the precipitating event and thus act in a parallel fashion; attempting to block one mediator may be thus largely an exercise with contradictory consequences. In fact, cytokines may be part of the repair mechanisms that are inherent in the inflammatory cascade and therefore a reflection of the attempt to repair tissue and come to a resolution of inflammation. Blockade of cytokines may impair repair more than it blocks progression of the inflammation in shock.

The main issue is that these early interventions were oblivious to the actual source of the inflammation, an issue that is especially pressing in the case of shock due to the severity of the inflammatory signals. A critical experiment to demonstrate that removal of a suspected inflammatory mediator from a shock plasma sample *in vitro* leads to actual reduction of the inflammatory activity in the same sample is missing in the shock literature. We were in the past not certain which are the important and active inflammatory mediators in shock plasma or lymph fluid.

Thus we were most interested in elucidating the upstream trigger events in the pathogenesis of shock with the hope to develop an approach of interrupting the inflammatory cascade at its genesis.

21.3 The Pancreas as a Source of Cellular Activating Factors and the Role of Serine Proteases

Since we had found evidence of significant levels of inflammatory mediators in the plasma of animals very early (within 40 minutes) in the shock state, we hypothesized that it was possible that these activating factors were preformed and might simply be released, rather than synthesized *de novo*.

To test this concept and identify more closely the organ source of inflammatory mediators, we collected homogenates from small intestine, spleen, heart, liver, kidney, adrenals, and pancreas and assessed their ability to produce inflammatory mediators in vitro. Visceral organs from the rat were harvested, homogenized and incubated at 38 °C for two hours in buffer to maximize any enzyme activity present and then added to naïve isolated human neutrophils. Cell activation by organ homogenates was tested by microscopic examination of neutrophil pseudopod production and oxygen free radical formation (lucigenin chemiluminescence production and nitroblue tetrazolium reduction). Of all the organ homogenates examined, the pancreas has by far the greatest ability to produce inflammatory mediators (Fig. 21.1). This ability of the pancreas to activate cardiovascular cells, including neutrophils and endothelium *in vitro* was replicated in the pig, suggesting that the effects of pancreatic homogenates on cells are a species-independent phenomenon [9]. These homogenates affect cells in the entire cardiovascular system, including the microcirculation, circulating neutrophils as well as endothelium, increasing neutrophil cell adhesion and oxygen free radical production in *in vivo* mesentery prepa-

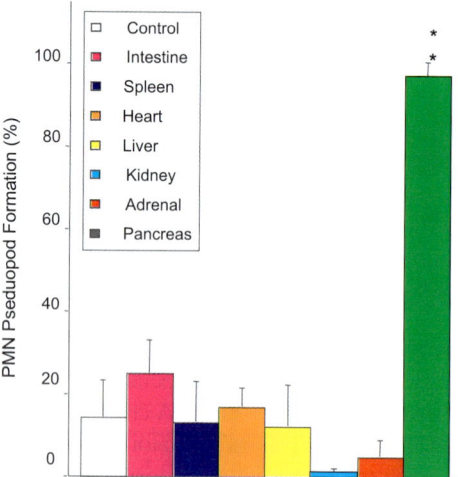

Fig. 21.1 Incubation of human PMNs with tissue homogenates from different organs in the rat for 10 min (and then fixed with gluteraldehyde) demonstrate high levels of cellular activation in cells treated with pancreatic homogenates but not from other organs. These results were duplicated using other methods of defining cell activation, including NBT formation and lucigenin chemiluminescence production. ∗∗$p < 0.01$ compared to controls. There are no significant differences between levels of cellular activation between other tissue homogenates and controls

rations. Circulating factors from pancreatic homogenates produce inflammation and are rapidly lethal when injected into naïve control animals, both whole molecular weight and low-molecular weight (filtered at 3 kD) homogenate fractions [9].

These findings are in line with Lefer's studies of the "myocardial depression factor", a low molecular weight factor that circulates in shock and depresses cardiac contractility, and appears to be of a pancreatic origin as well [10]. Thus the inflammatory factors that produce activation of circulating cells and endothelium may also be depressants of the cardiac myocytes. In fact, our collective experience at this stage indicates that the inflammatory mediators generated in shock depress just about all cell functions tested to date.

The key difference between the tissue homogenates derived from pancreas and other organs appears to be the preponderance of digestive (serine) proteases in the pancreas that are able to degrade tissue. Serine proteases appear to be necessary in order to produce inflammatory mediators in both the pancreas and in other tissues. This can be shown by the result that pancreatic tissue initially treated with a serine protease inhibitor (e. g. Phenylmethylsulphonyl fluoride, PMSF) and then homogenized is unable to produce inflammatory mediators, but if the protease inhibitor is added *after* homogenization there is no decrement in inflammatory activity [11]. While serine proteases can be pro-inflammatory by themselves we encountered in the organ homogenates also inflammatory mediators of lower molecular weight that may be responsible for the shock state. The presence of inflammatory mediators other than serine proteases can also be demonstrated directly by putting tissue homogenates through low molecular weight (<3 kDa) filters. Such fractions continue to retain inflammatory activity in the filtered low molecular weight homogenate [2, 13].

Although the pancreas possesses enzymes necessary to produce inflammatory mediators from tissues, the target tissue that produces the mediators need not necessarily be pancreatic in origin. Indeed, incubation of other organ tissues, such as intestine and liver, with low concentrations of pancreatic homogenates results in the production of inflammatory mediators by these tissues [9, 14]. The production of inflammatory mediators by these tissues when incubated with purified digestive enzymes (either trypsin or chymotrypsin) further reinforces the idea that it is the proteolytic degradation of tissue that results in the release of inflammatory mediators [9].

All in all, these *in vitro* studies served to identify a major source of inflammatory mediators in shock and gave birth to the hypothesis of *auto-digestion* by one's own pancreatic digestive enzymes.

21.4 Blockade on Pancreatic Digestive Enzymes in the Lumen of the Intestine

To test the idea that these *in vitro* factors might have some *in vivo* relevance, we subjected animals to splanchnic arterial occlusion shock, a severe shock model that targets the intestine in particular by occlusion of the celiac and superior mesenteric arteries [15]. Intravenous pretreatment with a serine protease inhibitor (6-amidino-2-naphtyl *p*-guanidinobenzoate dimethanesulfate, Futhan) led to a mild decrease in mortality, suggesting that there may be some leakage of pancreatic proteases into the systemic circulation (which may have also been secondary to inhibition of other activated inflammatory systems) but was otherwise ineffective in preventing the deleterious sequelae of the shock [15].

Repeating these experiments, however, with circulating protease inhibitors in the *lumen of the intestine* during shock completely abolished the high levels of inflammation seen by control animals in splanchnic arterial occlusion shock [16]. In these experiments, the animal's proximal duodenum and terminal ileum were connected to an in-line peristaltic perfusion pump that perfused the small bowel with protease inhibitor or control solution during the shock period. Gut lumen circulating serine protease inhibitors were effective in abolishing the shock, while control animals or those perfused with xanthine oxidase inhibitors (another putative inflammatory agent) displayed no such improvement (Fig. 21.2). In addition to decreased mortality, these experiments showed decreased levels of inflammatory mediators, decreased lung edema, and decreased cellular activation compared with controls. These results were achieved with a variety of serine protease inhibitors [17–19]. These findings have been replicated in endotoxic as well as hemorrhagic shock models, indicating that a gut-derived inflammatory process may be more universal in nature [20, 21].

Doucet et al. [22] confirmed the importance of enteral protease in the pathophysiology of hemorrhagic shock when he infused Futhan into the lumen of the intestine of pigs, whose digestive tracts are similar to our own. Animals with intestinal blockade of digestive enzymes need less transfusion fluids to maintain blood pressure, exhibit decreased inflammation in the lungs, and display decreased injury to the liver and small intestine.

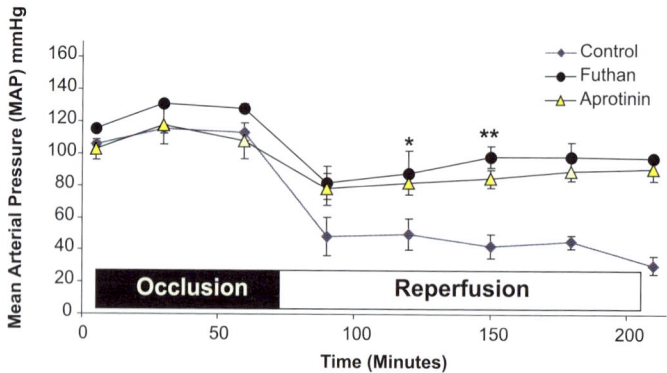

Fig. 21.2 Mean arterial blood pressure (MAP) during (Occlusion) and after reperfusion (Reperfusion) of the splanchnic arteries. MAP is significantly greater in both Futhan ($n = 5$) and Aprotinin ($n = 5$) groups compared to control shocked animals ($n = 5$). $^*p < 0.05$, $^{**}p < 0.01$ compared to controls. There are no significant differences in MAP between Sham SAO shock animals (not shown) and Futhan and Aprotinin-treated animals by 120 min and 180 min, respectively

21.5 What Mechanisms Prevent Auto-digestion?

Under normal physiological conditions, the intestine as the main digestive organ is the single recipient of pancreatic digestive enzymes via the pancreatic duct. These enzymes are fully activated and present in high concentrations as part of the digestive process, capable of breaking down into molecular building blocks all proteins, lipids, carbohydrates and nucleic acids that make up the food supply. From this point of view it is an organ like no other.

In fact, we have to ask the question, what mechanisms facilitate digestion of intestine that is part of a meal, without digestion of one's own intestine. The answer to this question may involve two mechanisms, (a) the use of a barrier that serves to compartmentalize the high concentrations of digestive enzymes and prevent escape from the lumen of the intestine into the wall and into the subepithelial space, and (b) perhaps the presence of digestive enzyme inhibitors in the tissue compartments that surround the lumen of the intestine. To date, we have found limited evidence for significant pancreatic enzyme blockade in the wall of the intestine, sufficient to deal with the high concentrations of the pancreatic enzyme concentrations in the lumen of the intestine. The preponderance of the evidence points towards the mucosal epithelium as a barrier for entry of digestive enzymes into the wall of the intestine, i. e. a central role of the epithelial brush border.

21.6 Triggers of Shock Increase Intestinal Wall Permeability

Shock states are associated with elevated permeability of the brush border [23], a situation that permits entry of digestive enzymes into the space between and underneath the mucosal epithelium and initiation of the auto-digestion process [24]. Entry of digestive enzymes into the interstitial space of intestinal microvilli results in their destruction to the point of complete breakdown of the mucosal barrier, exposing the intestinal submucosa and muscularis to the full concentrations of pancreatic digestive enzymes.

In trauma, sepsis, burns, and other insults, the onset of tissue injury results in decreased blood flow to the gut region. Ischemia causes a rapid increase in permeability of the intestinal mucosal layer [25–29]. Reperfusion injury such as after vascular surgery will also result in increased intestinal permeability, although not necessarily solely due to an ischemic mechanism [30]. Triggers of shock, through a diversity of mechanisms, all result in decreased barrier properties of the intestine allowing first small molecular weight and then even macromolecules in the lumen to enter the circulation, the lymphatics and the peritoneum through the wall of the intestine.

Entry of proteases from the lumen of the intestine or already bound to the intestinal wall through the intestinal wall may render the gut susceptible to permeability changes. Light microscopy of the intestinal lumen shows clear degradation of the microvilli during shock; microvilli of protease inhibitor-treated animals show intact mucosa comparable to that of control subjects [16]. To verify that proteases can enter the intestinal wall during ischemia, Rosario et al. showed using in-situ zymography that protease activity is indeed increased in the intestinal wall after ischemia [24]. It is easy to generate conditions that allow entry of digestive enzymes into the intestinal wall.

In addition to the destructive action of the digestive enzymes under shock conditions, a second injury mechanism appears to be present as well that affects other organs, even if they are innocent bystanders not involved in the initial insult that leads to the breakdown of the brush border permeability. Powerful inflammatory mediators may be formed during cleavage of the wall [9]. Indeed, increased levels of inflammatory mediators have been measured in the lymph of shocked animals, portal vein blood, and peritoneal fluid [16, 25, 31].

21.7 Intestine as Source of Inflammatory Mediators in Shock

It has long been hypothesized that the intestine could be a major source of mediators that, upon introduction to the central circulation following reperfusion of the intestinal blood supply, might lead to multiple organ failure [32–37]. Evans et al. [38] performed one of the most direct studies showing that the intestine is involved in the pathogenesis of shock. After excising 90% of the small intestine, they showed a four-fold increase in the survival of animals subjected to endotoxic shock.

Some of the important early manifestations of shock are the hemorrhagic, necrotic lesions that form at various points along the intestine. This necrosis has been experimentally linked directly to ischemia of the intestine and other shock triggers [32, 33, 37, 39, 40, 42]. In a modification of the above study, Evans et al removed the ileum and jejunum from the gastrointestinal tract by anastomosing the proximal jejunum and terminal ileum to re-establish intestinal continuity and joining the ends of the excised jejunum and ileum to form a loop with blood supply carefully preserved. A week after surgery, they induced endotoxic shock. They saw both an increase in survival and a reduction in hemorrhagic necrosis [40]. This protection may have been due to prevention of new digestive proteases from the pancreas entering the loop and inactivation of existing digestive proteases over the course of the experiment.

21.8 Characterization of Protease-Derived Shock Factors

A key issue is to determine the nature of the factors obtained from protease digestion of intestinal wall homogenates and their mode of operation. An important aspect came to light with the use of tissue digests derived from the small intestine. High-resolution light microscopy of neutrophils exposed to such protease digests over

21.8 Characterization of Protease-Derived Shock Factors

a broader time interval than used when looking solely for pseudopod formation reveals that the protrusions from the cell surface grow with time and resemble less pseudopods and instead more membrane blebs (Fig. 21.3, [43]). Blebs are fluid filled detachments of the plasma membrane from the underlying cytoskeleton and may grow to a size that is larger than the original size of the cell. Both leaflets of the phospholipid bilayer may separate from the cytoskeleton or one leaflet may separate from the other. We observe both types of membrane separation and a full range of bleb sizes. Fluorescent labeling of actin filaments serve to confirm that the protrusions contain no filamentous actin and thus are membrane blebs [44]. Thus there is a need to determine what other effects besides activation the inflammatory factors may have on neutrophils.

Since blebs are indicative of cell damage, we measured cell death by means of a life/death indicator (propidium iodide uptake) in neutrophils. The cells were exposed to intestinal wall homogenate digested for three hours by each of the three

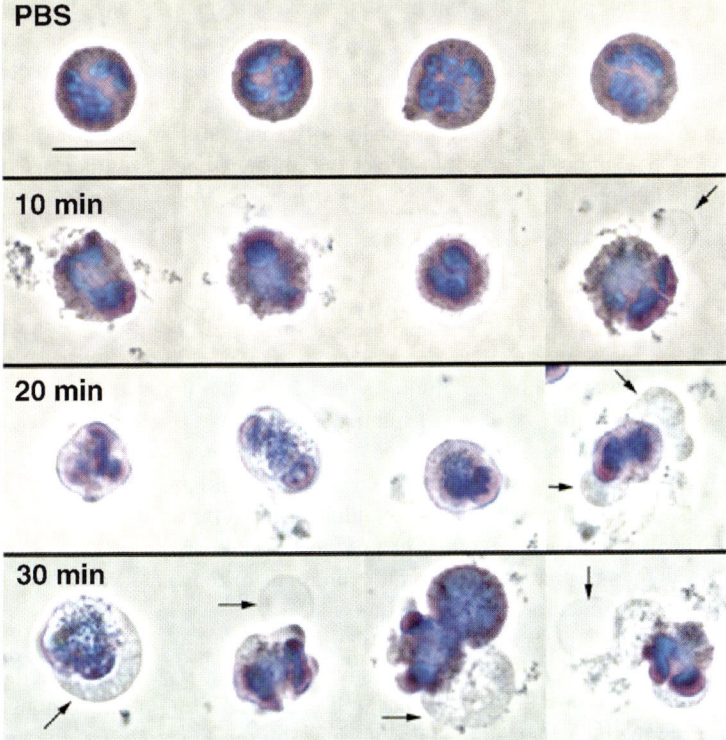

Fig. 21.3 Four representative examples of neutrophils after exposure to buffer (*top row*) or intestinal homogenate digested with chymotrypsin for 10 (2nd row), 20 (3rd row), or 30 minutes (*bottom row*). The cells were fixed in 1% glutaraldehyde and stained with crystal violet in 3% acetic acid. Blebs, round fluid-filled protrusions of the cell membrane from the underlying cytoskeleton that do not appear to readily stain with the crystal violet, are clearly visible (*arrows*) after as little as 10 minutes with the digested homogenate. *Scale at upper left* = 10 microns. From [43]

main proteases secreted by the pancreas into the intestinal lumen: trypsin, chymotrypsin, and elastase. Though the rate of cell death and the time of onset of death varied considerably between proteases and between animals, all but two of the digested homogenates killed a majority of the neutrophils within 80 min, and even those last two homogenates had begun the process of killing neutrophils. Some homogenates even began to kill cells in less than 10 minutes, indicating the presence of a powerful cytotoxic factor. Given its rapid rate, the cell death under these conditions is probably due to a form of necrosis as opposed to apoptosis.

To verify that cytotoxicity could be achieved with the proteases present in the intestinal lumen, we homogenized the enteral content and digested the intestinal homogenates with it. The digests generated with luminal fluid, with its full complement of pancreatic protease activity, generated highly cytotoxic samples. The cytotoxicity was greater than the cytotoxicity generated by the use of individual proteases (e. g. trypsin, chymotrypsin, elastase) [43]. What was surprising was that the luminal fluid on its own was sometimes cytotoxic, though rarely to the same extent as the intestinal homogenate digested with luminal fluid, suggesting that the luminal fluid may contain, besides the pancreatic digestive enzymes, cytotoxic substrate as well.

To link these findings to a traditional *in-vivo* model of shock, we subjected rats to intestinal ischemia by splanchnic artery occlusion for three hours before collecting their small intestines. The homogenates of shock rats, but not of control rats, were cytotoxic, supporting our hypothesis that during shock cytotoxic factors are produced in the intestine [43].

21.9 Cytotoxic Factors Derived from the Intestine

The finding that cell death in addition to cell activation may result from exposure to intestinal digests may have major implications for investigation of shock mechanisms. The evidence supports the hypothesis that the intestine serves as a major source for shock's global inflammation. Cytotoxic factors produced in the intestine may be responsible for the majority of multi-organ dysfunction and cell death observed in the rest of the circulation. The products of necrosis are themselves inflammatory [32, 45], and upon reperfusion the inflammatory mediators will be released into the lymph, circulation, and peritoneum.

Inflammatory mediators generated in the intestine and released into its venules are absorbed under normal circumstances by the liver, as shown by a dramatic reduction of inflammatory activity from the high levels in the portal venous blood coming from the intestine and the hepatic vein. But this absorption of the liver under shock conditions is limited in time, especially under the condition in which major parts of the intestine are ischemic and a source of inflammatory mediators. At a time, when the absorption in the liver becomes deficient, and inflammatory mediators leak into the central circulation the onset of multi-organ failure can be observed [16].

Another common condition during shock is *Disseminated Intravascular Coagulation* (DIC) [46]. This pathology is observed following the uncontrolled activation of clotting factors and fibrinolytic enzymes throughout small blood vessels. Fibrin is deposited, platelets and clotting factors are consumed, and fibrin degradation products inhibit fibrin polymerization, resulting in tissue necrosis and bleeding. Coagulation normally occurs on the surface of platelets after scrambblases in the plasma membrane flip aminophospholipids (phosphotidyl serine and phosphotidyl ethanolamine) from the inner leaflet to the outer leaflet giving coagulation factors a location to assemble [47]. It has been proposed that exposure to debris from necrotic tissue can cause DIC because of the exposed aminophospholipids [48]. Therefore, products of necrosis in the intestine that become flushed into the circulation upon reperfusion in shock represent a possible cause of DIC.

21.10 Removal or Blockade of Intestinal Cytotoxic Mediators

If cytotoxic factors in ischemic intestines contribute to the pathogenesis of shock then it is of interest to determine ways of interfering with their production or removing them from the circulation. We serendipitously uncovered a method for removal of inflammatory factors *in vitro*. In an attempt to reduce debris in the intestinal homogenates we passed homogenates through glass fiber. If filtration was performed before protease digestion, it had little effect on the homogenate's cytotoxicity. If, however, homogenates were filtered after digestion, cytotoxic activity in the intestinal homogenates could be largely reduced [43]. Filtration by size exclusion filters

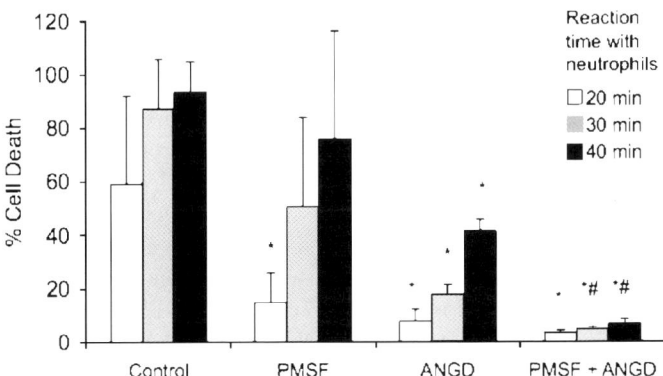

Fig. 21.4 Neutrophil cell death caused by exposure for 20, 30, or 40 min to intestinal wall homogenates digested by filtered luminal fluid for 3 hours at 37 °C. Filtered luminal fluid was incubated with PMSF and/or ANGD or buffer to inhibit digestive enzyme activity in the luminal proteases prior to incubation with intestinal wall (final concentrations of inhibitors were 1 mM for PMSF and 2.31 μM [1.25 mg/ml] for ANGD). Bars are mean ± SD: $N = 4$ animals. *: significant vs. controls. #: significant vs. PMSF (Tukey-Kramer HSD significance test). From [43]

was less effective at reducing cytotoxicity even though the hydrophobic glass fiber filters have larger pore sizes.

An *in vivo* protocol for the use of this technique to remove cytotoxic or inflammatory factors once formed remains to be determined. In the interim, we are interested in developing a method to prevent these factors from forming in the first place.

Our current approach is to block pancreatic proteases in the lumen of the intestine. This raises the question of what protease inhibitors are most appropriate. PMSF, a serine enzyme inhibitor reduces protease activity in luminal fluid by approximately half [43]. Ethylenediaminetetraacetic acid and E-64, inhibitors of metalloproteases and cysteine proteases, respectively, have no effect on pancreatic protease activity. Instead, a combination of Futhan, which is not effective on its own at completely inhibiting the proteases, and PMSF provide dramatic inhibition of proteases in the luminal fluid of the intestine. This combination of inhibitors then successfully prevents cell death from intestinal homogenates incubated with luminal fluid that contains the full spectrum of pancreatic enzymes (Fig. 21.4, [43]). Although PMSF cannot be used *in vivo* due to its toxicity, this observation suggests future development of an inhibitor-based intervention against cell death.

21.11 Conclusions

We traced the source of the inflammatory mediators in several forms of shock to the powerful digestive enzymes in the intestine. Synthesized in the pancreas, as part of normal digestion, they have the ability to degrade almost all biological tissues and molecules. In the lumen of the intestine, digestive enzymes are fully activated and self-digestion of the intestine is prevented by compartmentalization in the lumen of the intestine facilitated by the mucosal epithelial barrier. Under conditions of intestinal ischemia or epithelial stimulation, however, the mucosal barrier becomes permeable to pancreatic enzymes allowing their entry into the wall of the intestine. The process leads to *auto-digestion* of matrix proteins and tissue cells in the intestinal wall and production of a previously unidentified class of inflammatory mediators. We show that most of these are tissue degradation products generated by pancreatic digestive enzymes. These inflammatory mediators are released into the central circulation, lymphatics and the peritoneal cavity. The observations are in line with the hypothesis that multi-organ failure in shock may be due an auto-digestion process by pancreatic digestive enzymes. Our analysis leads to new opportunities; for example, inhibition of pancreatic enzymes in the lumen of the intestine, serves to attenuate the inflammation in several forms of shock. The procedure may have clinical utility.

The analysis we have presented so far covers just the very first step of any bioengineering analysis, i. e. identification of the players involved. A key event is the transport of digestive enzymes in the lumen of the intestine and the barrier prop-

21.11 Conclusions

erties of the mucosal brush border. Our elementary and qualitative understanding of digestive enzyme transport in the intestine needs to be replaced by quantitative models that take into consideration enzymatic activity and the interaction of digestive products with the cells in the intestine and in the circulation. This will be an important contribution by bioengineering to one of the most challenging clinical problems.

Acknowledgements Supported by USPHS Grant NHLBI HL67825 and Program Project Grant HL-43024. The authors thank Dr. Tony Hugli of the Torrey Pines Institute for Molecular Sciences and Dr. David Hoyt of the University of California Irvine for many inspiring discussions and suggestions to this work.

References

1. Schmid-Schönbein GW (2006) Analysis of inflammation. Annu Rev Biomed Eng 8:93–131
2. Barroso-Aranda J, Zweifach BW, Mathison JC et al. (1995) Neutrophil activation, tumor necrosis factor, and survival after endotoxic and hemorrhagic shock. Journal of Cardiovascular Pharmacology 25 Suppl 2(7):S23–S29
3. Barroso-Aranda J, Chavez-Chavez R, Schmid-Schönbein GW (1991) Spontaneous neutrophil activation and the outcome of hemorrhagic shock in rabbits. Circ Shock 36:185–190
4. Abraham E, Anzueto A, Gutierrez G et al. (1998) Double-blind randomised controlled trial of monoclonal antibody to human tumour necrosis factor in treatment of septic shock. NORASEPT II Study Group. Lancet. Mar 28 351(9107):929–933
5. Angus DC, Birmingham MC, Balk RA et al. (2000) E5 murine monoclonal antiendotoxin antibody in gram-negative sepsis: a randomized controlled trial. E5 Study Investigators. Jama, Apr 5 283(13):1723–1730
6. Butty VL, Roux-Lombard P, Garbino J et al. (2003) Anti-inflammatory response after infusion of p55 soluble tumor necrosis factor receptor fusion protein for severe sepsis. Eur Cytokine Netw, Jan–Mar 14(1):15–19
7. Opal SM, Fisher CJ Jr, Dhainaut JF et al. (1997) Confirmatory interleukin-1 receptor antagonist trial in severe sepsis: a phase III, randomized, double-blind, placebo-controlled, multicenter trial. The Interleukin-1 Receptor Antagonist Sepsis Investigator Group. Crit Care Med, Jul 25(7):1115–1124
8. Vincent JL, Spapen H, Bakker J et al. (2000) Phase II multicenter clinical study of the platelet-activating factor receptor antagonist BB-882 in the treatment of sepsis. Crit Care Med, Mar 28(3):638–642
9. Kistler EB, Hugli TE, Schmid-Schönbein GW (2000) The pancreas as a source of cardiovascular cell activating factors. Microcirculation, Jun 7(3):183–192
10. Lefer AM (1987) Interaction between myocardial depressant factor and vasoactive mediators with ischemia and shock. Am J Physiol, Feb 252(2 Pt 2):R193–R205
11. Kistler EB (1998) Humoral Mechanisms of Cellular Activation in Ischemic Shock. [PhD]. La Jolla, California: Department of Bioengineering, University of California San Diego
12. Kistler E, Brondeau A, Pfeiffer P et al. (1997) In vitro chemiluminescence measurements of plasma superoxide production by pancreatic activating factors. Annals of Biomedical Engineering 25(Supplement):S–56
13. Kistler E, Lefer A, Hugli T et al. (1997) In vitro neutrophil activating factor present in procine pancreas. Annals of Biomedical Engineering 25(Supplement):S–56
14. Waldo SW, Rosario HS, Penn AH et al. (2003) Pancreatic digestive enzymes are potent generators of mediators for leukocyte activation and mortality. Shock, Aug 20(2):138–143
15. Kistler EB, Lefer AM, Hugli TE et al. (2000) Plasma activation during splanchnic arterial occlusion shock. Shock, Jul 14(1):30–34

16. Mitsuoka H, Kistler EB, Schmid-Schönbein GW (2000) Generation of in vivo activating factors in the ischemic intestine by pancreatic enzymes. Proceedings of the National Academy of Sciences of the United States of America 97(4):1772–1777
17. Mitsuoka H, Kistler EB, Schmid-Schönbein GW (2002) Protease inhibition in the intestinal lumen: attenuation of systemic inflammation and early indicators of multiple organ failure in shock. Shock, Mar 17(3):205–209
18. Mitsuoka H, Schmid-Schönbein GW (2000) Mechanisms for blockade of in vivo activator production in the ischemic intestine and multi-organ failure. Shock, Nov 14(5):522–527
19. Schimmeyer SM (2005) Intraintestinal Enzyme Inhibition in Splanchnic Arterial Occlusion Shock [Master Thesis]. La Jolla California: Department of Bioengineering, University of California San Diego
20. Fitzal F, Delano FA, Young C et al. (2003) Pancreatic enzymes sustain systemic inflammation after an initial endotoxin challenge. Surgery, Sep 134(3):446–456
21. Fitzal F, Delano FA, Young C et al. (2002) Pancreatic protease inhibition during shock attenuates cell activation and peripheral inflammation. J Vasc Res, Jul–Aug 39(4):320–329
22. Doucet JJ, Hoyt DB, Coimbra R et al. (2004) Inhibition of enteral enzymes by enteroclysis with nafamostat mesilate reduces neutrophil activation and transfusion requirements after hemorrhagic shock. J Trauma, Mar 56(3):501–510; discussion 510–501
23. Rollwagen FM, Li YY, Pacheco ND et al. (2000) Microvascular effects of oral interleukin-6 on ischemia/reperfusion in the murine small intestine. Am J Pathol, Apr 156(4):1177–1182
24. Rosario HS, Waldo SW, Becker SA et al. (2004) Pancreatic trypsin increases matrix metalloproteinase-9 accumulation and activation during acute intestinal ischemia-reperfusion in the rat. Am J Pathol, May 164(5):1707–1716
25. Deitch EA (1990) Intestinal permeability is increased in burn patients shortly after injury. Surgery 107(4):411–416
26. Kanwar S, Windsor AC, Welsh F et al. (2000) Lack of correlation between failure of gut barrier function and septic complications after major upper gastrointestinal surgery. Ann Surg 231(1):88–95
27. Roumen RM, Hendriks T, Wevers RA et al. (1993) Intestinal permeability after severe trauma and hemorrhagic shock is increased without relation to septic complications. Arch Surg 128(4):453–457
28. Welsh FK, Farmery SM, MacLennan K et al. (1998) Gut barrier function in malnourished patients. Gut 42(3):396–401
29. Yu P, Martin CM (2000) Increased gut permeability and bacterial translocation in Pseudomonas pneumonia-induced sepsis. Crit Care Med 28(7):2573–2577
30. Roumen RM, van der Vliet JA, Wevers RA et al. (1993) Intestinal permeability is increased after major vascular surgery. J Vasc Surg 17(4):734–737
31. Ishimaru K, Mitsuoka H, Unno N et al. (2004) Pancreatic proteases and inflammatory mediators in peritoneal fluid during splanchnic arterial occlusion and reperfusion. Shock, Nov 22(5):467–471
32. Bounous G (1967) Role of the intestinal contents in the pathophysiology of acute intestinal ischemia. American Journal of Surgery 114(3):368–375
33. Bounous G (1969) "Tryptic enteritis": its role in the pathogenesis of stress ulcer and shock. Canadian Journal of Surgery 12(4):397–409
34. Bounous G, Brown R, Mulder D et al. (1965) Abolition of 'Trypstic Enteritis' in the Shocked Dog. Arch Surg 91(Sept):371–375
35. Mainous MR, Ertel W, Chaudry IH et al. (1995) The gut: a cytokine-generating organ in systemic inflammation? Shock. Sep 4(3):193–199
36. Moore EE, Moore FA, Franciose RJ et al. (1994) The postischemic gut serves as a priming bed for circulating neutrophils that provoke multiple organ failure. J Trauma, Dec 37(6):881–887
37. Smith EE, Crowell JW, Moran CJ et al. (1967) Intestinal fluid loss in dogs during irreversible hemorrhagic shock. Surgery, Gynecology, and Obstetrics 125:45–48
38. Evans WE, Darin JC (1966) Effect of enterectomy in endotoxin shock. Surgery, Nov 60(5):1026–1029

References

39. Bounous G, Menard D, de Medicis E (1977) Role of pancreatic proteases in the pathogenesis of ischemic enteropathy. Gastroenterology 73:102–108
40. Evans WE, Shore RT, Carey LC et al. (1967) Effect of intestinal exclusion in Escherichia coli endotoxin shock. Arch Surg 95(3):511–516
41. Freiman DG (1965) Hemorrhagic necrosis of the gastrointestinal tract. Circulation 32(3):329–331
42. Manohar M, Tyagi RP (1973) Experimental intestinal ischemia shock in dogs. American Journal of Physiology 225(4):887–892
43. Penn AH, Hugli TE, Schmid-Schönbein GW (2007) Pancreatic Enzymes Generate Cytotoxic Mediators In The Intestine. Shock, Mar 27(3):296–304
44. Penn AH (2005) Digestive enzymes in the generation of cytotoxic mediators during shock
45. Fadok VA, Bratton DL, Guthrie L et al. (2001) Differential effects of apoptotic versus lysed cells on macrophage production of cytokines: role of proteases. J Immunol, Jun 1 166(11):6847–6854
46. Hardy JF, De Moerloose P, Samama M (2004) Massive transfusion and coagulopathy: pathophysiology and implications for clinical management. Can J Anaesth. Apr 51(4):293–310
47. Solum NO (1999) Procoagulant expression in platelets and defects leading to clinical disorders. Arterioscler Thromb Vasc Biol. Dec 19(12):2841–2846
48. Hardaway RM (2000) A review of septic shock. Am Surg Jan 66(1):22–29

Chapter 22
Percutaneous Vertebroplasty: A Review of Two Intraoperative Complications

Christianne Vant[1], Manfred Staat[2], Gamal Baroud[1]

[1] Laboratoire de biomécanique, Université de Sherbrooke, Sherbrooke, Canada, Gamal.Baroud@USherbrooke.ca
[2] Institute of Bioengineering and Biomechanics Laboratory, Aachen University of Applied Sciences, Campus Jülich, 52428 Jülich, Germany

Abstract Vertebroplasty is an interventional radiology procedure used to treat vertebral compression fractures. It shows promising results but has two major drawbacks: the excessive injection pressure and the risk for extraosseus cement leakage. To examine each of these problems, vertebroplasty is divided into intra- and extravertebral components. The excessive injection pressure is an extravertebral problem because 95% of the injection pressure is extravertebral. Extant solutions are cement delivery devices and lowering the cement viscosity. Additionally, the cannula can be redesigned to lower the friction and injection pressure. The extraosseus cement leakage is an intravertebral problem. Literature recommends delaying the injection to increase the control over the infiltration of cement; this hypothesis is tested using a synthetic model. The bone cement has complex rheological properties, which are modified by altering the cement preparation and composition; these may also play a role in controlling cement leakage. Combining the findings, conflicting demands on the cement viscosity are clearly demonstrated; the extravertebral component demands low-viscosity cement while the intravertebral component demands high-viscosity cement. The challenge is therefore to develop biomaterials, techniques and/or devices that can control the conflicting demands on cement viscosity.

22.1 Introduction

Osteoporosis is a skeletal disease that causes gradual loss of bone minerals in trabecular bone thereby decreasing bone density and strength and making it more susceptible to fractures. Since older adults compose the fastest growing population segment in industrialized nations, osteoporosis is expected to escalate. It is diagnosed using dual energy X-ray absorptiometry (DEXA) to measure the bone mineral density relative to a general population (T-score) and to a representative subset of the population (Z-score), by age and gender.

In the United States, osteoporosis instigates 1.5 million fractures annually, with almost half – 700,000 – being vertebral compression fractures (VCF) [1]. Complications include acute pain syndrome, reduced spinal strength, deformity, and loss of independence. The current annual cost of osteoporosis-related fractures in the United States is at least US$ 5–$ 10 billion, with the prediction of an escalation to over US$ 60 billion within a few decades [1].

Traditional treatments include bed rest, pain medication and external bracing yet offer no cure [2], demonstrating the need for an improved treatment method. Ideally, this will increase quality of life by liberating patients from pain and allowing them to independently conduct activities of daily living.

The interventional radiology procedure of percutaneous vertebroplasty offers the potential of fulfilling this need. Promising clinical results have led to its increasing role in the treatment of VCF; an estimated 100,000 patients have already benefited from its therapeutic effects.

22.2 Vertebroplasty: Minimally Invasive and Cost-Effective Solution

Vertebroplasty is a minimally invasive, cost-effective procedure (about C$ 300 per vertebra [3]) whereby bone cement is injected through a cannula under real-time image guidance. It is increasingly used to stabilize vertebrae weakened by osteoporosis or cancer. The bone cement is injected transpedicularly into the vertebra and flows through the trabecular bone structure and, upon hardening, offering mechanical reinforcement [4–7]. Its main benefit is the reduced pain experienced by the patient, with up to 90% of patients experiencing pain relief in the first 24 hours [8–14].

Regrettably, two important complications exist: the excessive pressure required to inject the cement into the vertebra; and the risk of extraosseus cement leakage.

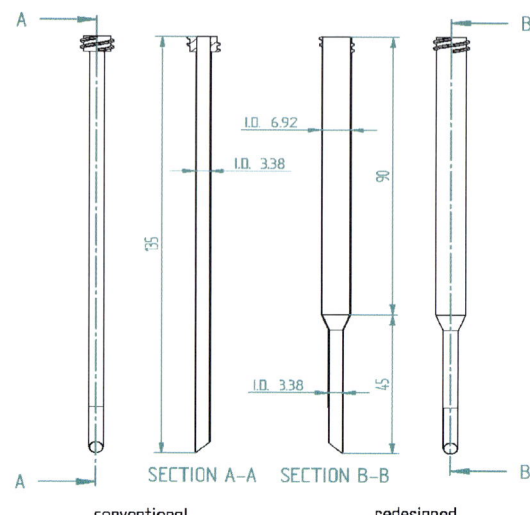

Fig. 22.1 The conventional and redesigned cannulae are similar in overall design; both have a variable output surface and a luer lock for syringe attachment. The conventional cannula has a constant diameter while the redesigned cannula has two – the diameter of the proximal section is twice that of the distal section

The pressure required for injection can reach upwards of 2–3 MPa, which is well above normal human capabilities [15–17]. If such high pressures are reached during injection, the procedure may have to be aborted and result in insufficient filling. Equally important, is that the clinician's concentration is shifted to applying enough force to inject the cement rather than focusing on monitoring the procedure. There are two main research directions focused on lowering the required injection pressure: the design of mechanisms to ease the physical pressure required by the clinician; and lowering the cement's viscosity by altering, for example, the liquid-to-powder ratio (LPR).

Cement leakage is a considerable risk because it is uncontrollable and unpredictable, occurring in up to 73% of vertebroplasty procedures [15, 18–24]. Most cases are clinically irrelevant [25] but all have the potential to result in spinal cord compression, embolism or death [3, 14, 26–29]. The improvement of visualization techniques contributes a large portion of research pertaining to the *detection* of cement leakage. Two common techniques are: (1) a vertebrogram, created to detect potential leakage paths by injecting contrast agent into the vertebra prior to the cement injection; and (2) fluoroscopy, to highlight the cement flow during the procedure by adding radiopacifier to the cement. The latter method is limited because it requires altering the cement's composition and because it is reactive – that is, the clinician must respond to visual clues. Another research direction is the biomechanical *prevention* of leakage by gaining a thorough understanding of the forces underlying cement flow.

This article will focus on the intraoperative biomechanics of vertebroplasty with the goal of identifying the ideal conditions for a safe cement injection. The problem will be viewed systematically by dividing the injection into two components: intra- and extra-vertebral [30]. This comprises the theoretical separation of the injection pressure into *infiltration* and *delivery* components, respectively [31, 32]. The findings will then be combined to arrive at an optimal solution.

The biomechanical perspective on leakage is meant to provide insight into complimentary methods of reducing or preventing leakage. Biomechanics will ideally lead to a method of guiding the flow within a vertebral body, thus reducing the risk of leakage. Visualization will ultimately continue to play a substantial role in ensuring safety, even with a guided flow of cement. When combined, these methods will give rise to a safe, reliable vertebroplasty procedure.

22.3 Extravertebral Biomechanics: Excessive Delivery Pressure

The extravertebral biomechanics are concerned with external aspects of injection including delivery pressure, which must overcome friction in the cannula, and cement preparation.

The infiltration and delivery pressures were compared in intact and damaged vertebrae, using a valve to open and close a fenestration and acquire both measurements from one vertebra. In this investigation, Baroud et al. [30] demonstrated that approx-

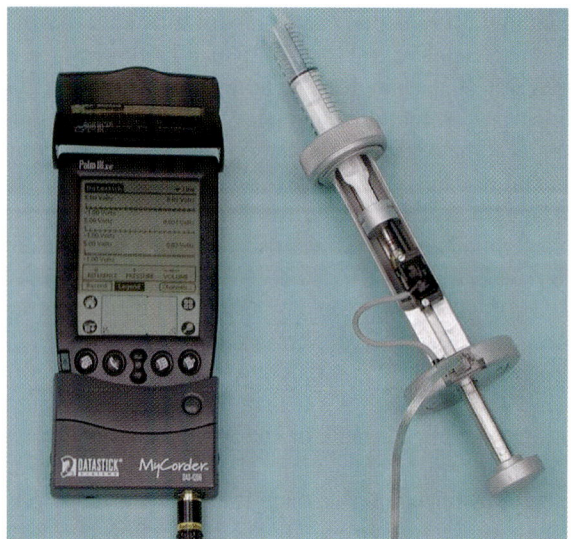

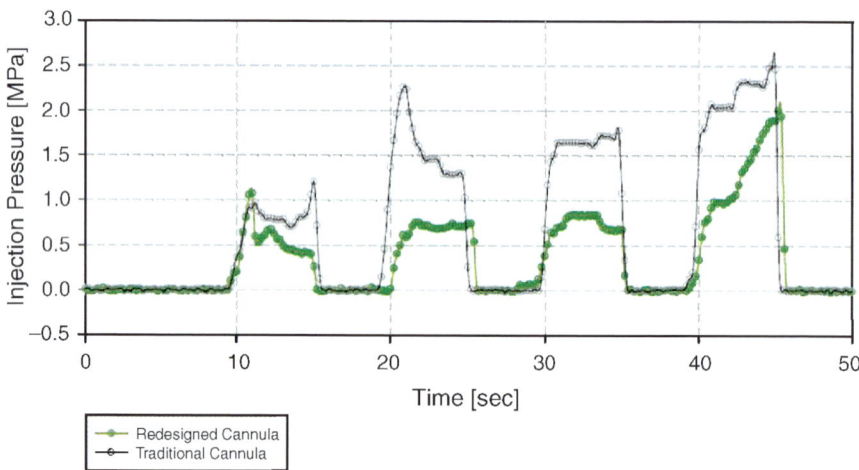

Fig. 22.2 The injection pressures and volumes for a sample injection during cadaveric setting using the conventional cannula design. The injection pressures were measured clinically using the portable device shown. Laboratory results show a decrease in injection pressure of 60% by using the redesigned cannula

imately 95% of the injection pressure is extravertebral, and that only the remaining 5% of the injection pressure is intravertebral. From this, the extravertebral delivery component is identified as the bottleneck of the excessive injection pressure.

There are two existing methods of overcoming the excessive delivery pressure. The first is to increase the pressure applied through the use of a cement delivery device; and the second is to lower the cement viscosity to ease the injection. There are currently at least seven devices available on the market to increase the pressure applied to the cement (e.g. V-max mixing and delivery system, EZFlow cement delivery system, Osteoject), many of which employ a screw-driven mechanism. The drawbacks are that they increase the risk of phase separation, where the liquid and powder separate under pressure, and the risk of cement leakage, resulting from the limited tactile feedback with these devices.

Building on the identification of the extravertebral pressure as the bottleneck of injection, Baroud et al. [33] redesigned an 8-gauge cannula that features two distinct sections as shown in Fig. 22.1. The geometry of the distal third enters the pedicles and is therefore governed by the anatomical limits and remains unchanged. The diameter of the proximal two-thirds, which passes through soft tissue, is approximately twice that of the distal section. A theoretical model predicted a 67% reduction in delivery pressure using these geometrical configurations. Laboratory studies showed that the redesigned cannula reduced the injection pressure by 60%, while *ex vivo* testing demonstrated a 44% reduction in injection pressure. Some results from Baroud et al. [33] are shown in Fig. 22.2. The key benefits are that it eases cement injection by reducing the delivery pressure, it can be easily integrated into the existing procedure, and it is cost-effective.

There are also a number of methods of reducing cement viscosity. It is common clinical practice to increase the liquid-to-powder ratio during mixing; but the increased amount of monomer may cause toxicity, and it decreases the setting times for calcium phosphate (CaP) and calcium sulfate (CaS) cements. Another method is to cool at least one component of the cement to delay polymerization and keep the viscosity low for an extended period of time. The drawbacks of these methods are similar to those of the pressure devices, albeit for different reasons. Phase separation is due to the non-uniformity of the dough formed from cooled components and cement leakage is simply due to the lower viscosity.

A further method of reducing the cement viscosity is by altering the mixing method, that is, using oscillations to enhance the mixing. Oscillation not only significantly lowers cement viscosity, but also has the benefits of significantly increasing the handling time, injectability and reproducibility of the cements. Unlike the other methods, this neither impedes the process of cement polymerization nor increases the risk of phase separation or toxicity. Rather, it improves the cohesiveness and uniformity of the cement, which appears to play a role in reducing the risk of leakage. However, as in other methods, lower viscosity may increase the risk of leakage [31]. The exact interplay between cement cohesiveness, viscosity and risk of leakage remains to be examined.

22.4 Intravertebral Biomechanics: Risk of Extravasation

The intravertebral biomechanics describe the flow of cement within the vertebra; they depend on the properties of the cement and the trabecular structure. Conversely, they are not concerned with resolving the limitation of excessive injection pressure; proposed methods of reducing internal pressure – venting the shell [27] or creating a cavity in the trabecular bone [34] – have been deemed fruitless, as internal pressure comprises such a small portion of the overall injection pressure. In other words, intravertebral biomechanics are primarily concerned with reducing the risk of leakage.

Leakage is the most serious complication that is preventing vertebroplasty from becoming a *routine* procedure [35]. It is attributed to irregularities in the vertebral body, such as blood vessels and vertebral compression fractures [36]. These irregularities provide a directional preference for the flow and create a non-uniform infiltration where the cement spreads into projections like the fingers of a glove [35, 36]. The key to controlling leakage is to guide the filling; that is, to ensure uniform infiltration overcomes the directionality from the vertebra's irregularities.

Kyphoplasty offers one option for guiding the flow; a balloon is inflated inside the vertebra creating a cavity into which cement is injected. This comparatively expensive procedure (US$ 3000 per vertebra) has not demonstrated significant improvements over vertebroplasty, although claims have been made that it reduces injection pressure, increases leakage and restores vertebral height. Although this presents one alternative, it is not the focus of this review.

A delayed injection is an oft-recommended option, which allows the cement to polymerize further prior to injection making it easier to control; the more viscous cement has a lower Reynolds' number, indicating stable and laminar flow. The hypothesis therefore follows that the high-viscosity cement overcomes the directional forces in the vertebra, thus creating a uniform pattern of infiltration.

This hypothesis must be tested using a synthetic model because of the tremendous variability among biological samples. Within this model the properties affecting the intravertebral flow of cement can be determined and understood, which is vital to guiding the flow and controlling leakage.

A synthetic leakage model considering only the infiltration component of the injection was created with the underlying requirements that it be physically accurate, kinematically uniform and biomechanically representative. Morphological and flow properties similar to osteoporotic vertebral bone were captured in custom-made cylindrical porous aluminum foams [ERG Aluminum and Aerospace, California]. Specifically, the open porosity of the foam was measured to be $91.1\% \pm 0.6$ using a micro-CT (Fig. 22.3) and the permeability was measured as $8.47 \pm 0.26 \times 10^{-8}\,\mathrm{m}^2$ using Darcy's flow protocol; these fit well in the osteoporotic range of bone [37].

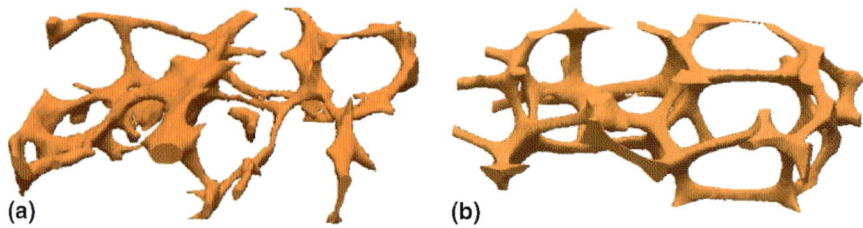

Fig. 22.3 Micro-CT images of the trabecular bone (**a**) and customized trabecular bone substitute (**b**)

22.5 Injectable Biomaterials

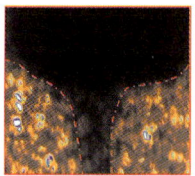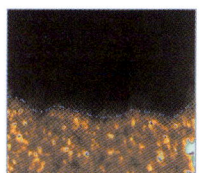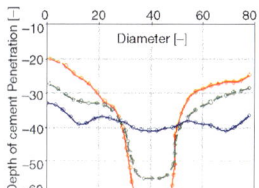

Fig. 22.4 From *left* to *right*, the radiographs **a**, **b** and **c** depict the cement distribution patterns for the cases of immediate extravasation, moderate extravasation and no leakage, respectively. The *black area* represents the distribution of the cement from superior to inferior. From *left* to *right*, the cement was injected 3.5, 8 and 12 minutes following cement mixing, respectively. The *graph* on the *right* depicts the digitalization that was performed to yield numeric values for the amount of leakage

The porous foam samples were placed in polycarbonate syringes [20 mL Medallion, Merit Medical, Utah], which contained the flow and gave rise to a uniform cement infiltration. A 3 mm-diameter longitudinal 'blood vessel' was created in the model to establish a path of least resistance to direct cement flow during infiltration. Finally, the cylindrical samples were saturated with a hydrogel [Jell-0, Kraft Canada] to replicate the complete filling of the trabecular bone cavities with blood and bone marrow and to reproduce the intravertebral pressure.

Using the synthetic model, appropriate methods for guiding the flow of cement within the vertebra can be determined. Preliminary testing was conducted using a constant flow protocol. The results show three distinct leakage/infiltration patterns (Fig. 22.4) identified with respect to elapsed time from cement mixing: (1) *immediate extravasation*, where the cement recognized the path of least resistance immediately and resulted in a large amount of cement leakage; (2) *moderate extravasation*, where a gradual delay in the time of leakage was observed and a smaller amount of cement leaked; and (3) *no leakage*, where leakage was completely suppressed and the infiltration was considered safe. These three infiltration patterns confirm the hypothesis that delaying the injection increases the uniformity of filling.

22.5 Injectable Biomaterials

Cement behaviour is very complex; rheological properties other than the viscosity affect the leakage phenomenon. It is hypothesized that – in addition to a low Reynolds number – the yield stress, the viscoelastic properties and/or the cohesiveness of the cement are key determinants of the leakage effect.

In a recent review of injectable bone cements, Lewis [38] outlined the 18 desirable properties of bone cements; the most important of which were: easy injectability, high radiopacity, and a constant viscosity during polymerization, i.e. between mixing and delivery.

Bohner and Baroud [39] define injectability as the ability of a paste to remain homogeneous during its injection; under pressure, this phase separation is known as filter pressing. This is the most common problem with calcium phosphate (CaP) cements, there are, however, certain additives that can improve their injectability. For example, polysaccharides are a workability-improvement agent, which increase injectability, ultimate compressive strength and fracture toughness of the polymerized cement.

Until recently, radiopacifier was manually added to acrylic cements because there were no FDA-approved cements with the required 10–50 wt% radiopacifier. Although the addition of radiopacifier alters the cement properties, Carrodeguas et al. [40] found the clinical addition of "sufficient" mass of radiopacifier to acrylic cements to be "adequate".

Determining the values of the viscosity of polymerizing cement is crucial to understanding the fluid dynamics of the flow through the cannula and the trabecular bone. The study of the cement's rheological properties, such as the complex viscosity, could lead to the optimization of the cement's composition and the cannula design, as well as determining the optimum time and viscosity of injection; all of which lead to minimizing the risk of extravasation.

Revisiting the synthetic leakage model described above, it can be noted that the complex viscosity measurements of the cement were taken for each injection time, as a measure of the ongoing polymerization. The cement's complex viscosity over time progressed in a sigmoidal relationship. The first and last sections had steep slopes, demonstrating the drastic increase in viscosity relative to time, and the middle section joined them linearly, with a gradual increase in viscosity.

Gisep et al. [41] compared the behaviours of five cements; two calcium phosphate (CaP) and three acrylic cements were injected through a cannula. The injection force required for the CaP cements was fairly constant over time, allowing for controlled injections. Adding hyaluronic acid to chronOS Inject may have altered its behaviour to that of acrylic cement; that is, it had an approximately linear force increase instead of the characteristic exponential force increase of CaP cements. Consequently, all of the chronOS Inject was delivered before reaching an injection force of 200 N, while only 36% of the other CaP cement was delivered before reaching this force. The acrylic cements were more consistent: 54–88% of all cements were delivered at injection forces below 200 N.

22.6 Discussion

Cement injection, or vertebroplasty, is a procedure that has far-reaching and wide-ranging socio-economic implications. In this paper, a combination of clinical, experimental and theoretical research is used to analyze the biomechanics of cement injection with the goal of integrating the various findings. From this, a broader understanding of the factors that affect cement injection should be attained.

Upon concatenation of the "delivery" and "infiltration" components, the competing demands placed on bone cements are evident. Although acrylic and CaP cements present different challenges for cement injection, both of them require low

22.6 Discussion

viscosities to facilitate delivery through the cannula. Conversely, the bone cements must be high-viscosity to uniformly infiltrate the bone and decrease the risk of leakage. Furthermore, the force analysis reveals that approximately 95% of the injection pressure is required to overcome the friction in the cannula, and that only 5% of the injection pressure is required to infiltrate the bone cavities.

In summary, the most interesting global findings are that "delivery" comprises most of the injection pressure and demands cements of low viscosity, whereas "infiltration" comprises very little of the injection pressure but demands cements of high viscosity. The challenge is therefore to develop biomaterials, techniques and/or devices that can control the conflicting demands on cement viscosity. Since the low viscosity and high-pressure requirements of delivery conflict with the requirements of infiltration, solutions that do not alter viscosity are preferred.

A further examination of the findings presented suggests that a window of safe infiltration can be established. The window begins when leakage is eliminated in the synthetic model and ends when the cement viscosity undergoes a rapid increase to achieve complete polymerization. This window describes the period of time during which the cement is viscous enough to safely infiltrate the trabecular structure but before it is completely polymerized.

These results are corroborated by independent *in vivo* observations. By controlling the time at which injection begins, and thereby the cement viscosity, the clinical incidence of leakage can potentially be significantly reduced in a porous medium with irregularities, such as vertebrae afflicted by osteoporosis or cancer. The difficulty is that although the working time of the cement is about 17 minutes, it is not manually injectable with a standard syringe and cannula after about 7 minutes, which is about the same time that 'no leakage' occurs. In this situation, the redesigned cannula discussed above may play an important role in cement injection.

It is important to note that a window of safe infiltration based on cement viscosity does not address the problem of clinically controlling cement viscosity. Viscosity varies with mixing time, injection time, liquid-to-powder ratio and clinician preference. Some clinicians visually screen the cement by inspecting small samples of cement injected into air to determine whether the cement has reached an appropriate consistency for a safe injection. Evidently, this method of screening viscosity depends on the clinician's experience and preference. In response to this problem, some manufacturers have attempted to introduce a viscometer to the vertebroplasty technique.

A future use of the synthetic model is as an aid for rapid *in vitro* screening of bone cements, as suggested by Lewis [38]. To accomplish this, these synthetic models will require a set of international standards describing the injection protocol. At this time, bone cements could be easily tested and compared using various rheological and mechanical properties.

These biomechanical research methods are not competing with visualization research; they aim to compliment each other by attacking the problem from two different angles. When combined, the two methods will produce a safe and reliable vertebroplasty procedure.

References

1. Watts NB, Harris ST, Genant HK (2001) Treatment of painful osteoporotic vertebral fractures with percutaneous vertebroplasty or kyphoplasty. Osteoporos Int 12(6):429–437
2. Leidig G, Minne HW, Sauer P et al. (1990) A study of complaints and their relation to vertebral destruction in patients with osteoporosis. Bone Mineral 8(3):217–229
3. Canadian Coordinating Office for Health Technology Assessment (CCOHTA) (2002) Percutaneous vertebroplasty: a bone cement procedure for spinal pain relief. Iss Emerging Health Tech 31:1–4
4. Baroud G, Nemes J, Ferguson S, Steffen T (2003) Material changes in osteoporotic human cancellous bone following infiltration with acrylic bone cement for a vertebral cement augmentation. Comp Methods Biomech Biomed Eng 6(2):133–139
5. Belkoff SM, Maroney M, Fenton DC, Mathis JM (1999) An in vitro biomechanical evaluation of bone cements used in percutaneous vertebroplasty. Bone 25(2):23S–26S
6. Heini PF, Berlemann U, Kaufmann M, Lippuner K, Frankhauser C, van Landuyt P Augmentation of mechanical properties in osteoporotic vertebral bones – a biomechanical investigation of vertebroplasty efficacy with different bone cements. Eur Spin J 2001; 10(2):164–171
7. Wilson DR, Myers ER, Mathis JM et al. (2000) Effect of augmentation on the mechanics of vertebral wedge fractures. Spine 25(2):158–165
8. Deramond H, Depriseter C, Galibert P et al. (1998) Percutaneous vertebroplasty with polymethylmethacrylate: technique, indications, and results. Radiol Clin North Am 36(3):533–546
9. Fourney DR, Schomer DF, Nader R et al. (2003) Percutaneous vertebroplasty and kyphoplasty for painful vertebral body in cancer patients. J Neurosurg (Spine 1) 98:21–30
10. Heini PF, Walchli B, Berlemann U (2000) Percutaneous transpedicular vertebroplasty with PMMA: operative technique and early results. A prospective study for the treatment of osteoporotic compression fractures. Eur Spine J 9(5):445–450
11. Jarvik J, Kallmes D, Mirza SK (2003) Vertebroplasty: learning more, but not enough. Spine 28(14):1487–1489
12. Jensen ME, Evans AJ, Mathis JM, Kallmes DF, Cloft HJ, Dion JE (1997) Percutaneous polymethylmethacrylate vertebroplasty in the treatment of osteoporotic vertebral body compression fractures: technical aspects. Am J Neuroradiol 18:1897–1904
13. Jensen ME, Kallmes DF (2002) Percutaneous vertebroplasty in the treatment of malignant spine disease. Cancer J 8(2):194–206
14. Mathis JM, Barr JD, Belkoff SM, Barr MS, Jensen ME, Deramond H (2001) Percutaneous vertebroplasty: a developing standard of care for vertebral compression fractures. AJNR Am J Neuroradiol 22:373–381
15. Baroud G, Bohner M, Heini P, Steffen T (2004) Injection biomechanics of bone cements used in vertebroplasty. Biomed Mater Eng 14(4):487–504
16. Baroud G, Heini P, Bohner M, Ferguson S, Steffen T (2003) Drop in pressure at injection and infiltration in vertebroplasty. Presented at: 13th Interdisciplinary Research Conference on Biomaterials (GRIBOI 2003). Baltimore MD

17. Baroud G, Beckman L, Heini P, Ferguson S, Steffen T (2003) Clinical and laboratory analysis of the pressure at injection in a vertebroplasty. Presented at: Annual Meeting of the International Society of the Study of the Lumber Spine. Vancouver, Canada
18. Cortet B, Cotten A, Boutry N et al. (1997) Percutaneous vertebroplasty in patients with osteolytic metastases or multiple myeloma. Rev Rhum Engl Ed 64(3):177–183
19. Cortet B, Cotten A, Boutry N et al. (1999) Percutaneous vertebroplasty in the treatment of osteoporotic vertebral compression fractures: an open prospective study. J Rheumatol 26(10):2222–2228
20. Cotten A, Dewarte F, Cortet B et al. (1996) Percutaneous vertebroplasty for osteolytic metastases and myeloma: effects of the percentage of lesion filling and the leakage of methyl methacrylate at clinical follow-up. Radiology 200:525–530
21. Cyteval C, Sarrabere MP, Roux JO et al. (1999) Acute osteoporotic vertebral collapse: open study on percutaneous injection of acrylic surgical cement in 20 patients. AJR Am J Roentgenol. 173:1685–1690
22. Lin JT, Lane JM (2002) Nonmedical management of osteoporosis. Curr Opin Rheumatol 14(4):441–446
23. Shapiro S, Abel T, Purvines S (2003) Surgical removal of epidural and intradural polymethylmethacrylate extravasation complicating percutaneous vertébroplastie for an osteoporotic lumbar compression fracture. Case report. J Neurosurg 98(1):90–92
24. Weill A, Chiras J, Simon JM, Rose M, Sola-Martinez T, Enkaoua E (1996) Spinal metastases: indications for and results of percutaneous injection of acrylic surgical cement. Radiology 199:241–247
25. Jensen ME, Dion JE (2000) Percutaneous vertebroplasty in the treatment of osteoporotic compression fractures. Neuroimaging Clin North Am 10(3):547–568
26. Aebli N, Krebs J, Davis G, Walton M, Williams MJA (2003) Theis, J-C. Fat embolism and acute hypotension during vertebroplasty. Spine 27(5):460–466
27. Aebli N, Krebs J, Schwenke D, Davis G, Theis J-C (2003) Cardiovascular changes during multiple vertebroplasty with and without vent-hole: an experimental study in sheep. Spine 28(14):1504–1511
28. Padovani B, Kasriel O, Brunner P, Peretti-Viton P (2001) Pulmonary embolism caused by acrylic cement: a rare complication of percutaneous vertebroplasty. AJNR Am J Neuroradiol 20(3):375–377
29. Phillips FM, Todd Wetzel F, Lieberman I, Campbell-Hupp M (2002) An in vivo comparison of the potential for extravertebral cement leak after vertebroplasty and kyphoplasty. Spine 27(19):2173–2178
30. Baroud G, Vant C, Giannitsios D, Bohner M, Steffen T (2005) Effect of vertebral shell on injection pressure in vertebroplasty. Spine 30(1):68–74
31. Baroud G, Matsushita C, Samara M, Beckman L, Steffen T (2004) Influence of oscillatory mixing on the injectability of three acrylic and two calcium-phosphate bone cements for vertebroplasty. J Biomed Mater Res Part B: Appl Biomater 68B:105–111
32. Baroud G, Bohner M, Heini P, Steffen T (2004) Biomechanics of cement injection for vertebroplasty. In: 5th Combined Meeting of the Orthopaedic Research Societies of the USA, Canada, Japan and Europe. Banff, Canada. Abstract No 147
33. Baroud G, Steffen T (2005) A new cannula to ease cement injection during vertebroplasty. Eur Spine J 14(5):474–479
34. Garfin S, Yuan HA, Reiley MA (2001) New technologies in spine: kyphoplasty and vertebroplasty for treatment of painful osteoporotic compression fractures. Spine 26(14):1511–1515
35. Fourney DR, Schomer DF, Nader R et al. (2003) Percutaneous vertebroplasty and kyphoplasty for painful vertebral body fractures in cancer patients. J Neurosurg (Spine 1) 98:21–30
36. Bohner M, Gasser B, Baroud G, Heini P (2003) Theoretical and experimental model to describe the injection of a polymethylmethacrylate cement into a porous structure. Biomaterials 24:2721–2730
37. Baroud G, Falk R, Crookshank M, Sponagel S, Steffen T (2004) Experimental and theoretical investigation of directional permeability of human vertebral cancellous bone for cement infiltration. J Biomech 37(2):189–196

38. Lewis G (2006) Injectable bone cements for use in vertebroplasty and kyphoplasty: state-of-the-art review. J Biomed Mater Res B Appl Biomater 76B(2):456–468
39. Bohner M, Baroud G (2005) Injectability of calcium phosphate pastes. Biomater 26(13):1553–1563
40. Carrodeguas RG, Lasa BV, del Barrio JSN (2004) Injectable acrylic bone cements for vertebroplasty with improved properties. J Biomed Mater Res B Appl Biomater 68B(1):94–104
41. Gisep A, Curtis R, Hanni M, Suhm N (2006) Augmentation of implant purchase with bone cements: an in vitro study of injectability and dough distribution. J Biomed Mater Res B Appl Biomater 77B(1):114–119

Part VI
Plant and Microbial Bioengineering

Chapter 23
Molecular Crowding: A Way to Deal with Crowding in Photosynthetic Membranes

Ira G. Tremmel

Max-Planck-Institut für biophysikalische Chemie, Computational Biomolecular Dynamics Group, Am Fassberg 11, 37077 Göttingen, Germany, ira@germanynet.de

Abstract In the last decades our view of biological systems has changed dramatically. One reason is an increasing awareness of molecular crowding in virtually all living cells. An example for a crowded system is photosynthesis. At the first glance, for many years the riddle of photosynthesis and the involved flow of electrons seemed to be solved since long ago. Nearly all involved proteins were known as well as most mechanisms of electron transfer within them. Between the photosynthetic proteins electrons were assumed to be transported via free diffusion of electron carriers. However, the diffusion of these carriers within the photosynthetic membrane may be strongly influenced by molecular crowding, which might nearly completely restrict it. Nevertheless, effects of molecular crowding are only sparsely investigated in the available literature although they show again that "the whole is more than the sum of its parts" (Aristotle). Even if all single components of a process are known, this does not mean that their interplay is really understood. Apart from diffusion many other important parameters determining the metabolism in a cell or within a membrane, like e. g. reaction equilibria, aggregation, self organisation or reaction rates, are also influenced by molecular crowding. Hence, molecular crowding is an important but underestimated phenomenon that is worth to be investigated in more detail already because of its omnipresence.

23.1 In the Crowd

Cells and biological membranes are packed with large molecules, especially proteins, nucleic acids and complex sugars. Macromolecules are present as soluble species and/or structural arrays at total concentrations of up to several hundred grams per litre in essentially all physiological compartments [73]. In photosynthetic membranes (thylakoids), e. g., an area of more than 70% is covered by proteins [44, 87] (Fig. 23.1, left). Although local composition varies widely between different systems, it is evident that most macromolecular reactions and processes *in vivo* take place in environments in which macromolecules occupy a considerable

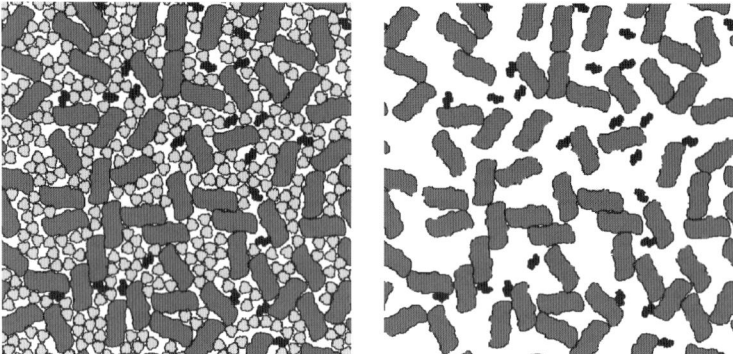

Fig. 23.1 Photosynthetic membranes (thylakoids) as crowded systems. *Left*: Integral proteins located within the thylakoids, randomly distributed. Altogether more than 70% of the membrane are covered by membrane spanning proteins. The light harvesting complexes (*light grey*) are abundant but they are not directly participating in electron transport. *Right*: Only those proteins are shown that are involved in photosynthetic electron transport. *Dark grey*: Photosystem II, *black*: cytochrome *bf* complexes

fraction of the total volume or area. Yet most biochemical studies of macromolecular properties *in vitro* are carried out in dilute solutions in which the total concentration of macromolecules rarely surpasses 1 g/l [24, 70].

Typically the concentration of each defined molecular species in a cell is low (sub micromolar range). However, the combined density of macromolecular components can reach very high values. Such media are termed 'crowded' or 'volume-occupied' rather than 'concentrated', because no single species of macromolecule is necessarily present at a high concentration.

The term 'macromolecular crowding' connotes the non-specific influence of steric repulsions on specific reactions and processes that occur in highly volume-occupied media [73]. The concept of 'non-specific interaction' is widely misunderstood. Many if not most researchers still regard such interactions as an artefact of a particular experimental system that interferes with the acquisition of meaningful data [61]. Strategies such as extrapolation of results to zero macromolecular concentration are devised for the reduction or elimination of the influence of non-specific interactions on a test reaction. Although such procedures may be appropriate in certain specific experimental situations, they do not necessarily provide results that are more meaningful in a biological context. On the contrary, significant non-specific interaction is an unavoidable consequence of crowding and may have a huge influence on the process considered. Thus, extreme caution should be exerted in the extrapolation of experimental results (acquired in dilute solutions) to function in the intact cell or membrane.

Examples for processes strongly influenced by molecular crowding are:

- reaction equilibria;
- aggregation of molecules;
- diffusion processes;
- reaction rates.

23.2 Macromolecular Crowding

The influence of high fractional volume occupancy on the rates and equilibria of macromolecular reactions taking place in crowded solutions or membranes has been recognised since the 1960s, but the biochemical and biophysical implications of these effects have only begun to be appreciated by the wider community of bio-physiological and biomedical researchers within the last 10 years or so. Since the beginning of this century, several minireviews on the subject of macromolecular crowding have appeared [18, 24, 33, 60, 63].

An example for a crowded membrane are the thylakoids where photosynthesis takes place. Here, membrane spanning proteins occupy more than 70% of the membrane area [44, 87]. However, diffusion between these membrane spanning proteins is necessary for photosynthetic electron transport. Furthermore, many reaction mechanisms in photosynthesis are not understood up to date and investigations accounting for molecular crowding are sparse [66, 87, 88]. Yet, photosynthesis takes place in a crowded membrane and it is to be expected that molecular crowding may severely influence photosynthetic electron flow (Fig. 23.1). Thus, albeit photosynthesis is arguably the most important biochemical pathway known, an important factor influencing and maybe controlling it seems to have been severely neglected in photosynthetic research during the last years.

This chapter aims at illustrating effects of molecular crowding on photosynthetic electron transport as an example of a reaction, taking place in a crowded membrane. A way to deal with crowding in photosynthetic membranes is introduced.

In the next section (Sect. 23.2), an introduction into molecular crowding will be given followed by Sect. 23.3 on the basic concepts of photosynthetic electron transport. Eventually, Sect. 23.4 shows the results of an investigation of photosynthetic electron transport dealing with effects of molecular crowding in photosynthetic membranes.

23.2 Macromolecular Crowding

Molecular crowding is more accurately termed the excluded volume effect, because the mutual impenetrability of all molecules is its most basic characteristic [24]. This non-specific steric repulsion is always present, regardless of any other attractive or repulsive interactions that might occur between the molecules. In fact, its ubiquity may cause steric repulsion to be taken for granted and their consequences to be overlooked.

23.2.1 Excluded Volume

The mutual impenetrability of solute molecules, due to the Pauli exclusion, is arguably the most fundamental intermolecular interaction. Because of steric repulsion, no part of any two molecules can be in the same place at the same time. That part of the total volume where the centre of mass of a test particle cannot be placed at a particular instant is called the excluded volume. Alternatively, the part of the total volume that may be occupied is the available volume (Fig. 23.2).

How much of the intracellular or intramembrane area is unavailable to other macromolecules depends upon the numbers, sizes, and shapes of all the molecules present in each compartment. If the test molecule is very small compared to the background molecules the available volume is nearly identical to the total unoccupied volume (Fig. 23.2, left). In contrast, if the size of the test particle is comparable to that of the background molecules, the available volume is considerably smaller (Fig. 23.2, right). For example, in a solution containing 30% by volume of identical globular molecules, less the 1% of the remaining volume is available to an additional molecule of equal size [25]. However, much more of the volume is available to smaller molecules.

In order to illustrate the concept of excluded volume it is possible to employ a macroscopic analogy [33]. Let us consider a beaker filled to the brim with peas. Assuming perfect spherical shapes the randomly close-packed peas occupy about 65% of the volume of the beaker [85], leaving about 35% of the volume in the interstices between the peas. Even though the volume between the peas is 'empty', it is impossible to add a single pea. The interstitial volume is *excluded* to the peas or in other words the *available* volume (i. e. the total volume minus the excluded volume) for peas has become zero. However, the interstitial volume although not available for peas is available for particles that are sufficiently smaller than peas, such as e. g. grains of sand. If we pour sand into the beaker, it will fill in the interstices between the peas. Due to geometrical restraints, however, it will only fill about 65% of the interstices, leaving smaller interstices between the grains. Thus, the grains of sand

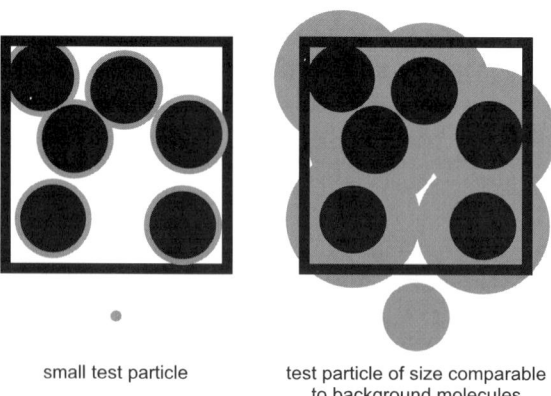

Fig. 23.2 Example of excluded volume considering spherical particles. The regions that are available to the centre of a test particle (*white area*) are restricted to those that are further from any other particle than the sum of the radii of the test particle and pre-existing solutes. *Left*: For a test particle much smaller than the background molecules only a small part of the volume in the box is excluded (*black* and *grey area*). *Right*: For a test particle of similar size to the background molecules nearly the whole volume is excluded and a test molecule can hardly be added to the volume

will fill $\frac{35}{100} \cdot \frac{65}{100}$, i.e. about 23% of the total volume of the beaker, leaving more than 10% of the beaker empty. Whereas this volume is available for even smaller particles it is excluded for the larger peas and grains of sand.

23.2.2 Reaction Equilibria and Aggregation

In freshman chemistry we are taught that the reactivity of a solute is proportional to its concentration, or number of molecules of solute per *total volume*. In fact, this is only strictly true in the highly dilute limit. In a crowded system solutes are spatially constrained by steric repulsion to the available volume, thus, the effective (local) concentration (i. e. the number of molecules per *available volume*) is different from the overall concentration. Here, the reactivity of a solute species is determined by the local concentration.

23.2.2.1 Effects of Crowding on Reaction Equilibria

The influence of volume exclusion on the thermodynamics of chemical equilibria in crowded media becomes evident upon examination of a few simple but fundamental relationships [57, 58]. Let us consider a generalised reversible reaction in solution:

$$r_1 R_1 + r_2 R_2 + \cdots \rightleftharpoons p_1 P_1 + p_2 P_2 + \ldots , \qquad (23.1)$$

where r_i is the stoichiometric coefficient of reactant species R_i and p_i is the stoichiometric coefficient of product species P_i. The equilibrium molar concentrations of reactants are related by the following relationship (law of mass action):

$$K = \frac{c_{P_1}^{p_1} c_{P_2}^{p_2} \cdots}{c_{R_1}^{r_1} c_{R_2}^{r_2} \cdots} , \qquad (23.2)$$

with the equilibrium constant K and c_i the concentration of species i. Considering the available volume a particle can occupy in highly crowded media, however, defines an effective concentration, which can be much higher than the overall concentration in a crowded medium. In a solution of macromolecules interacting exclusively via steric repulsion there exists an extremely simple relationship between the effective concentration a_i and the overall concentration c_i of each molecule [52],

$$\frac{v_{\text{tot}}}{v_{a,i}} = \frac{a_i}{c_i} \equiv \gamma_i , \qquad (23.3)$$

where γ_i is termed the *activity coefficient* of species i, v_{tot} is the the total volume, and $v_{a,i}$ is the volume available to species i.

Thermodynamics teaches that equilibrium constants are generally expressed in terms of equilibrium activities rather than overall concentrations. According to

Eqs. (23.2) and (23.3), the equilibrium constant K may be rewritten in terms of activities:

$$K = \frac{a_{P_1}^{p_1} a_{P_2}^{p_2} \cdots}{a_{R_1}^{r_1} a_{R_2}^{r_2} \cdots} \cdot \frac{\gamma_{R_1}^{r_1} \gamma_{R_2}^{r_2} \cdots}{\gamma_{P_1}^{p_1} \gamma_{P_2}^{p_2} \cdots}. \tag{23.4}$$

Thus, K actually denotes an apparent equilibrium constant, which is composition-dependent under crowding conditions. K is related to the ideal equilibrium constant K^0 in the highly diluted limit according to

$$K = K^0 \Gamma \tag{23.5}$$

with the non-ideality factor
$$\Gamma = \frac{\gamma_{R_1}^{r_1} \gamma_{R_2}^{r_2} \cdots}{\gamma_{P_1}^{p_1} \gamma_{P_2}^{p_2} \cdots}. \tag{23.6}$$

In dilute media the non-ideality factor equals one, and thus the apparent equilibrium constant K is identical to the ideal, highly diluted equilibrium constant K^0. In a crowded medium, however, the effects of crowding on reaction equilibria and reaction rates are highly non-linear with respect to the sizes and concentrations of the interacting molecules. In highly crowded media, the activity coefficient of each macrosolute – dilute as much as concentrated – may deviate from unity by as much as several orders of magnitude, with potentially major impact on reaction equilibria and rates in these solutions [18, 24, 33, 61, 73].

23.2.2.2 Aggregation and Self Organisation

The effect of molecular crowding on available volume, and thus on the non-ideality factor Γ, is sensitive to the relative sizes (compare Fig. 23.2) and shapes of all molecules present. The extent to which a particular molecular species excludes volume to other particles generally increases with the ratio of surface to volume of that species. The larger the excluded volume (and hence the smaller, the available volume) the higher is the contribution of steric repulsion to reduced entropy and increased free energy. Thus, clearly, one of the ways in which the system can reduce its free energy is to maximise the available volume (see e. g. [24, 62, 63, 73]).

A fundamental chemical consequence of macromolecular crowding is therefore the facilitation of processes that lead to a decrease in excluded volume, which may be accomplished in two ways. The first is by favouring compact conformations over extended conformations of flexible macromolecules. The second is by favouring macromolecular associations.

Equilibrium constants of macromolecular associations can be increased by as much as two or three orders of magnitude, depending on the relative sizes and shapes of the reactants and products, and on the size of the background molecules. For example, the equilibrium constant for the association of a spherical monomer of molecular weight 40,000 into a homodimer was found to be 8–40-fold larger (depending on the specific volume of the protein) if the protein is expressed in *E. coli*, compared to its value in an uncrowded solution [96]. For a tetramer, the shift is in the range $10^3 - 10^5$.

23.2 Macromolecular Crowding

Another example is the spontaneous alignment and bundling of self-assembled filaments, a phenomenon that has particular significance for cytoskeletal organisation. For such large and elongated macromolecules, orientational ordering can also change the available volume, as illustrated by J. Herzfeld [36], who showed that, under certain circumstances, crowding provides a non-specific force towards various forms of spatial ordering of highly anisotropic macrosolutes.

These striking effects arise from the increase in available volume for the solute molecules obtained when macromolecules bind to one another. The configurational entropy of each macromolecular species decreases, and its contribution to the total free energy of the solution increases. However, the binding is favoured because the increase of the available volume increases the entropy of the solute molecules and thus, the free energy of the whole system decreases.

Further examples changing the available volume are the formation of oligomeric structures as fibrin, collagen, and multienzyme complexes in metabolic pathways, and the formation of nonfunctional aggregates such as bacterial inclusion bodies or the amyloid deposits (plaques) in some neurodegenerative diseases. In extreme cases, like in the matrix of mitochondria, macromolecules are said to exist and function in a quasi-crystalline state. Not only the binding or aggregation of macromolecules to one another change the available volume. Other processes causing a change in available volume include the folding of newly synthesised polypeptide chains into compact functional proteins, folding of nucleic-acid chains into more compact shapes or the unfolding of proteins induced by stress. This situation may represent a prerequisite for the self-organisation characteristic of all living systems; thermodynamically, the process is entropy-driven due to water exclusion, similar to the hydrophobic effect and protein folding.

23.2.3 Reaction Rates

During recent decades it has gradually become recognised that crowding can considerably alter the reactivity of individual macromolecules, both qualitatively and quantitatively. The effects of crowding on reactivity of macromolecules is also implicated as one of the ways in which cells sense and respond to changes in their overall volume, induced by osmotic alterations caused by transport and metabolism [16, 24, 48, 64]. However, the ramifications of crowding are only now becoming more generally understood. One such ramification is that it is not clear what are the rate laws governing reactions occurring *in vivo*.

The classical Michaelis–Menten formalism for enzyme reactions [56, 80] is based on mass-action laws. However, the law of mass action relies on strict assumptions concerning, for instance, the characteristics of the reaction medium, which must be dilute, perfectly-mixed, and homogeneous. Certainly, many of these assumptions fail in the case of biological reactions in living cells or bio-membranes. In particular, the law of mass action might become invalid in the case of diffusion limited reactions [8, 29].

The rate of slow reactions is ordinarily limited by the rate with which reactants pass over a free energy barrier identified as a transition state. For slow reactions, this rate is sufficiently low that the transition state may be treated as if it were in equilibrium with reactants and products. For fast reactions, however, the rate of association is limited by the rate with which reactants encounter each other. This rate is dominated by translational diffusion, which decreases monotonically with increasing crowding due to the presence of an increasing number of obstacles for diffusion [53, 97]. As a consequence an inhomogeneous distribution of reactants and products may develop and the rate of any biochemical process that is diffusion limited will be reduced by crowding.

In recent years, biochemists have been using various computational frameworks to extract rate laws or empirical reaction equations from numerical simulations. Lattice-gas automata simulations of simple reactions in a particular class of heterogeneous media (low dimensional and fractal media) by Kopelman [45, 46] have shown that assuming an unbounded (or very large) reaction space in the asymptotic limit $t \to \infty$, the rate coefficient k is not a constant but rather has the time-dependent form

$$k(t) \propto t^{-(1-p)}, \qquad (23.7)$$

where p is a non-dimensional index quantifying deviations from the classical law of mass action. This nonconventional type of kinetics has been referred to as 'fractal' kinetics. For well stirred reactions occurring in a homogeneous three-dimensional space, $p = 1$, and the law of mass action is valid.

Fractal kinetics arise from a spatial self organisation of the reactants induced by the properties of diffusion. In the case of the reaction $A + B \to C$, the self organisation can even lead to spontaneous segregation of the reactants into A-only and B-only regions, a phenomenon called the Zeldovich effect [8, 67, 86].

Altogether the effects of crowding on biochemical reaction rates are complex and poorly understood. On one hand crowding reduces diffusion (see e. g. references [11, 25, 87]) and, hence, the rate of encounter. On the other hand crowding increases the thermodynamic activities of the reactants. The precise net result of these opposing effects depends upon the precise nature of each reaction and each crowded system.

23.2.4 Evidence for Consequences of Crowding in Living Systems

Experimental and theoretical work has demonstrated substantial (order-of-magnitude) effects of crowding on a broad range of biophysical, biochemical, and physiological processes, including e. g. nucleic acid and protein conformation and stability, protein-protein and protein-DNA association equilibria and kinetics (including protein crystallisation, protein fibre formation and bundling), catalytic activity of enzymes and cell volume regulation [24, 61, 97]. In Table 23.1 a few sample observations are summarised.

23.2 Macromolecular Crowding

Table 23.1 Some sample reports of experimentally observed crowding effects on macromolecular reactions

Observation	Affected Parameter	Reference
In *in vitro* studies of DNA replication and transcription, it has been found necessary to add crowding agents to extracts to obtain properties akin to those seen *in vivo*.		[97]
The activity coefficient of fibrinogen is tenfold higher in a solution of 80 g/l bovine serum albumin compared to normal buffer. This concentration of crowding agent is close to the total protein concentration in blood plasma, thus, the activity of fibrinogen in its natural environment may be an order of magnitude larger than in buffers commonly used *in vitro*.	activity coefficients, rate constants	[71]
The effective concentration of hemoglobin exceeds the actual concentration by a factor of > 10 at 200 g/l and a factor approaching 100 at 300 g/l. (For reference, the concentration of hemoglobin in a normal red blood cell typically exceeds 300 g/l.) This finding illustrates the strongly non-linear effect of molecular crowding on the effective concentration. The effect can be quantitatively described by a rigid hard sphere model.	activity coefficients, rate constants	[30, 75]
Monomers of the bacterial cell division protein FstZ associate with one another in the presence of crowding agents in a manner quantitatively described by a hard sphere model describing the effects of excluded volume.	association equilibria	[72]
Spontaneous alignment and bundling of self-assembled filaments, a phenomenon that has particular significance for cytoskeletal organisation, was found to be induced by crowding.	self assembly, association equilibria	[36]
Significant effects on the rate of formation of amyloid fibres of α-synclein, a process that is implicated in the pathogenesis of Parkinson's disease were observed.	self assembly, association equilibria	[90]
The thermal stability of α-lactumin is increased by 25 – 30% when encapsulated in a silica matrix. Earlier studies demonstrated that the addition of high concentrations of heat stable proteins stabilises other proteins against denaturation by heat or ethanol.	self assembly, association equilibria	[21, 60]
Enzymes that catalyse proteolysis in dilute solution can catalyse peptide synthesis in sufficiently crowded solutions.	reaction equilibria, rate constants	[81]
The simulated *E. coli* phosphotransferase system (PTS) corresponds to its real counterpart when macromolecular crowding is assumed both, to increase the association rate constants and to decrease the dissociation rate constants of the PTS complexes.	reaction equilibria, rate constants	[74]

Crowding is not confined to cellular interiors, but also occurs in the extracellular matrix of tissues such as cartilage; even blood plasma contains ≈ 80 g/l of protein, a concentration sufficient to cause significant crowding effects.

In the following, photosynthetic electron transport as an example for a process taking place in a crowded membrane will be examined in more detail.

23.3 Photosynthesis in a Crowded Environment

Photosynthesis is one of the most important processes for life on earth; nearly all life depends on it. It is a complex process, comprised of many coordinated biochemical reactions occurring in higher plants, algae, some bacteria, and some protists. Photosynthesising organisms are collectively referred to as photoautotrophs. On one hand photosynthesis fixates carbondioxide, a process that gained increasing attention during the last years in the context of the green-house effect. On the other hand photosynthesis delivers oxygen, necessary for respiration. Photosynthesis uses the energy of light to produce glucose. Simplified, the overall reaction of this process can be written as

$$6CO_2 + 6H_2O + \text{light energy} \rightarrow C_6H_{12}O_6 + 6O_2$$
$$\text{carbon dioxide} + \text{water} + \text{light energy} \rightarrow \text{glucose} + \text{oxygen}$$

Photosynthesis occurs in two stages. In the first phase, also called the light reactions, the energy of light is captured and used to produce high-energy molecules like ATP and NADPH via photosynthetic electron transport. During the second phase, the light-independent reactions (also called the Calvin Cycle), the high-energy molecules are used to capture carbon dioxide (CO_2) and produce the precursors of glucose. Both stages, the Calvin Cycle located in the matrix of the chloroplasts (stroma), as well as the light reactions proceeding within the photosynthetic membrane (thylakoids), are taking place in an environment where molecular crowding may play an important role.

The focus of the work presented here will be on the light reactions taking place within the thylakoids. In this Section the basic concepts of photosynthetic electron transport in a crowded membrane and resulting problems will be outlined. First, the basic principles of photosynthesis will be introduced. This will be followed by a paragraph placing special emphasis on diffusion processes in a crowded membrane.

23.3.1 Basic Principles of Photosynthesis in Higher Plants

A long tradition of photosynthetic research has provided us with profound knowledge of the components involved in photosynthesis. However, in many studies, photosynthetic electron transport has been investigated in artificial systems, using artificial electron donors and/or acceptors. Often, only single components were examined. Thus, single electron transfer steps are well known while the interplay of the involved components is hardly understood. As shown in Fig. 23.1, the photosynthetic membrane, or thylakoid, is stuffed with membrane spanning proteins – certainly a crowded system. In a crowded system, however, the complex organisation of proteins and the interaction of the single components becomes especially important.

23.3.1.1 Components Involved in Photosynthetic Electron Transport

A first step of the conversion of light energy into chemical energy is the absorption of a photon by light-harvesting complexes (LHC) and the subsequent energy transfer to the reaction centre of a photosystem (PS). Two types of photosystems exist; PS I, which is associated with LHC I, and PS II, for which the cross section of light absorption is increased by LHC II. The excitation of a reaction centre drives a charge separation followed by the transfer of an electron to acceptors with a higher (i.e. less negative) redox potential, thus involving several electron transfer steps (see Fig. 23.3).

After the charge separation in the reaction centre of PS II, the electron is transferred via several internal acceptors to the terminal acceptor of PS II, the plastoquinone (PQ) located at the Q_B binding site. Q_B is a two electron gate. After receiving a second electron the reduced PQ leaves the Q_B binding pocket as PQH_2. This way electrons from the huge membrane spanning PS II are taken up by the small lipophilic, mobile PQH_2 and are transported within the thylakoids to another huge membrane spanning complex; cytochrome bf (cyt *bf*). On the donor side of PS II the oxidised reaction centre is eventually reduced by oxidising water via the oxygen evolving complex.

cyt *bf* has two binding sites; Q_o, where PQH_2 is oxidised and Q_r, where PQ is reduced. After binding of the reduced PQ (PQH_2) at the Q_o binding site of cyt *bf*, one electron from the PQH_2 is transferred via cyt *bf* to another mobile carrier, plastocyanin. Thus a semiquinol is left at the Q_o pocket. The second electron is stored within cyt *bf*, and the oxidised PQ leaves the Q_o site [37]. The sequence is repeated when another PQH_2 binds at the Q_o binding site. After the second cycle the cyt *bf* contains two electrons, which are transferred to PQ at the Q_r binding site. Eventually, the uptake of two protons from the chloroplast matrix, the stroma side, leads to the formation of PQH_2, which can leave the Q_r binding site and enter the plastoquinol pool[1].

Plastocyanin leaves cyt *bf* and diffuses within the watery lumen space to PS I. Similar to PS II, PS I absorbs light or excitons from LHC, which is followed by a charge separation at the reaction centre. The separated electron is then transferred via the mobile electron carrier ferredoxin to the terminal acceptor $NADP^+$, and PS I is re-reduced via plastocyanin.

---

[1] This sequence of reactions is called the Q-cycle [65].

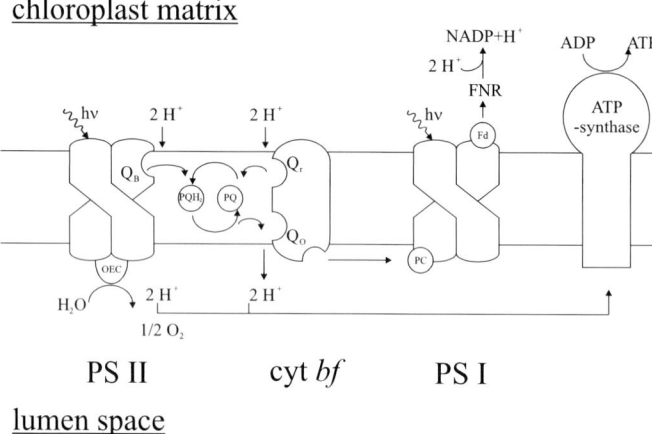

Fig. 23.3 Schematic diagram of intermolecular electron transport in higher plants. Abbreviations: OEC: oxygen evolving complex, Q_B: electron quinone-acceptor of PS II, PQ: plastoquinone, PQH_2: plastoquinol (i. e. reduced plastoquinone), Q_o and Q_r: oxidising and reducing binding site on cyt *bf*, respectively, PC: plastocyanin, Fd: ferredoxin, FNR: ferredoxin-NADP-reductase

The reactions at the PS I and PS II can be considered as fast in relation to the plastoquinol oxidation at the cyt *bf* complex, which is considered to be the slowest step in the electron transport chain [32].

In the course of electron transport from water to $NADP^+$ protons are translocated from the chloroplast stroma to the lumen space and a pH-gradient across the thylakoid membrane is built up (Fig. 23.3). Energy set free by protons leaving the lumenal space along the gradient is converted to chemical energy, ATP, via the ATP-synthase according to the chemiosmotic theory of Mitchell [35, 65]. Between the initial donor, H_2O, and the final electron acceptor, $NADP^+$, is a potential difference of more than 1.2 V. This energy is provided by the absorption of light (similar to a photo cell).

23.3.1.2 Thylakoid Architecture

The site of photosynthetic electron transport, the thylakoid membrane network consists of a single membrane enclosing the lumen space. It is highly structured (cf. Fig. 23.4) and unstacked regions (stroma thylakoids) can be distinguished from stacked regions (granal thylakoids) [1, 2].

A typical grana stack is composed of two to about twenty grana discs depending on growth conditions (for more details see references [7, 82]). Grana consist of a central core with non-exposed membranes forming the grana stack, a peripheral domain that consists of the margins, and two end membranes [1]. In contrast to the non-exposed grana core membranes, the margins, the end membranes, and the stroma lamellae are stroma-exposed. These regions are distinguished from each

23.3 Photosynthesis in a Crowded Environment

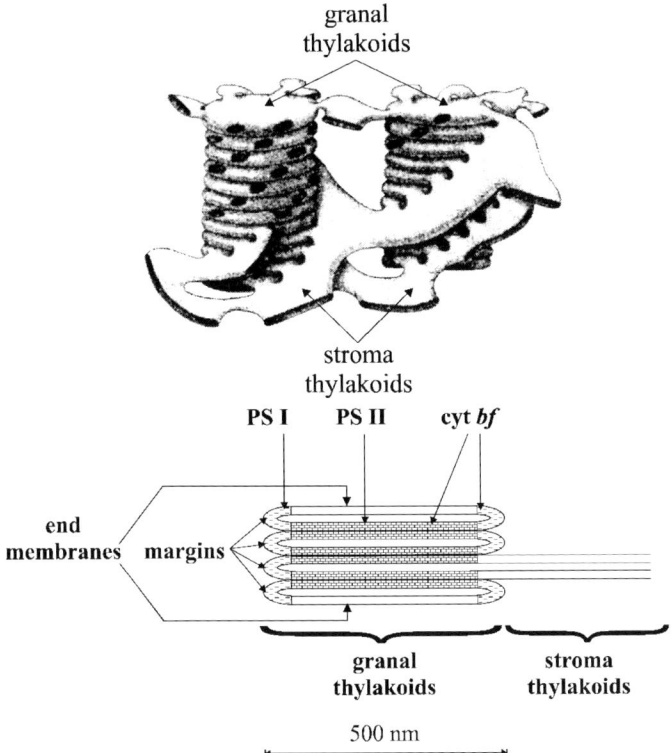

Fig. 23.4 Model of the thylakoid membrane. *Top*: Reprinted from Staehelin and van der Staay [82], copyright (1996) with kind permission from Springer. *Bottom*: Schematic diagram illustrating the different domains of the granal thylakoid membrane according to Wollenberger and co-workers [94]. Only PS II types involved in linear electron transport are shown. LHC and ATP-synthase not shown

other by their biochemical composition and thus their function (cf. Fig. 23.4 and Table 23.2).

The grana core membranes are enriched in PS II centres [6] and contain the highest concentration of cyt *bf* complexes compared to the other thylakoid domains [1]. Both plastoquinone and plastocyanin occur in the grana core [66] whereas PS I cannot be found [50] (Fig. 23.4).

Table 23.2 Distribution of photosynthetic components in the different regions of the thylakoids

Component	Grana	Stroma	Reference
PS II	85%	15%	[82]
PS I	68%	32%	[3]
cyt *bf*	85%	15%	[3]
plastoquinone	77%	22%	[17]

Albertson proposed that linear electron transport is carried out between the PS II centres in the grana core and PS I centres in the margins. The organisation of PS II and PS I in different domains could avoid wasteful 'spill over' between the two photosystems [5]. As a consequence, long-range electron transport by the diffusible electron carriers (plastoquinone and plastocyanin) over up to 250 nm (the radius of a granum) is necessary [1].

23.3.2 The Problem of Fast Electron Transport in Thylakoids

Plastoquinol, which transfers electrons from PS II to cyt bf, has a high diffusion coefficient. However, one has to take into account that the membrane in which it is diffusing is crowded with proteins that may act as obstacles (see e.g. references [42, 51, 77]). First evidence for this arose from estimations of the diffusion coefficient for PQ in phosphatidylcholine lipid vesicles [11]. The diffusion coefficient D is about a factor of ten lower if 20% of the membrane are occupied by proteins. More recent measurements in thylakoids using fluorescence quenching of pyrene yielded a coefficient between $0.1 - 3 \times 10^{-9}$ cm^2/s which is 100 times slower than the coefficient in artificial lipid vesicles without proteins ($D = 3.5 \times 10^{-7}$ cm^2/s) [10]. According to Joliot, Lavergne and Béal [41] also PQ diffusion within thylakoids is impeded. Evidence for this arises from measurements of the apparent equilibrium constant between Q_A and PQ. Values of 1–5 [41] are obtained, which is much lower than that expected by considering the midpoint potential (≥ 70). This can be explained by a further compartmentation of the grana into different quinone diffusion domains. Thus, the accumulation of Q_A^- does not express a global equilibrium, but rather indicates total reduction in micro-domains where the PQ/Q_A ratio is small [41]. This effect is described in more detail in Sect. 23.2.2.1.

Following these findings Joliot, Lavergne, and co-workers developed the microdomain theory. According to this theory the crowding of the membrane by integral proteins results in a network of barriers to diffusion [40,41,49]. This theory is based on a percolation effect[2]. In this case the lateral diffusion coefficient (D) may be distance dependent $D(r)$ [77]. The further a particle diffuses, the higher the probability of encountering an obstacle and hence the more tortous the path. Furthermore, the diffusion coefficient is dependent on the concentration of obstacles (c), thus leading to $D(r, c)$ (see Fig. 23.5).

Above a certain threshold for the concentration of obstacles (percolation threshold c_p) $D(r, c_p)$ approaches zero at some finite r. The physical or biological meaning

---

[2] Historically, percolation theory goes back to Flory [27] and to Stockmayer [84] who used it to describe how small branching molecules form larger and larger macromolecules if more and more chemical bonds are formed between the original molecules. However, usually, the start of percolation theory is associated with a 1957 publication of Broadbent and Hammersley that introduced the name and dealt with it in a more mathematical way [15]. Today there is a large variety of applications of percolation theory as for example animal migration, bush fires, evaluating the distribution of oil or gas inside porous rocks in oil reservoirs, diffusion processes etc. For an overview on diffusion in membranes see Almeida and Vaz [4].

23.3 Photosynthesis in a Crowded Environment 557

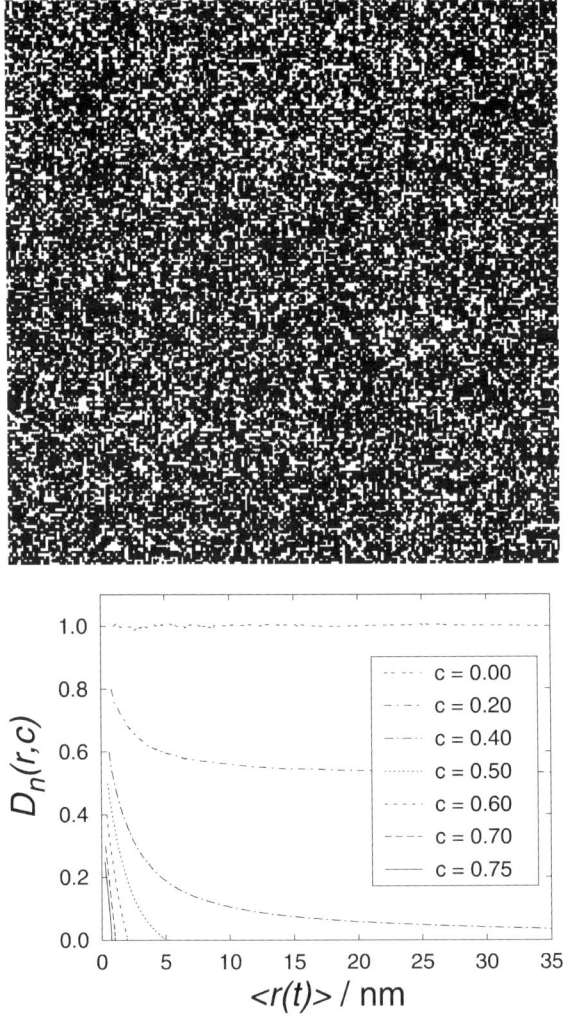

Fig. 23.5 Random distribution of point obstacles on a square lattice, 200×200 nm. The occupied area fraction is 0.6. *Below*: Distance dependence of the normalised diffusion coefficient $D_n(r, c)$ of a point tracer on a square lattice in the presence of immobile point obstacles. *Curves* are for different occupied area fractions. 1000 tracers were used, and 500 different initial distributions of the obstacles. The slight deviation of the diffusion coefficient $D_n(0)$ from one reflects statistical errors. Another indication of the statistical error is the occasional discontinuity in the curves, representing the transition between separate runs. For $c = 0$ for 100 data points $D_n(0) = 1.0044 \pm 0.0042$. No specific interaction between obstacles was assumed. Figure reprinted from reference [87], copyright (2003), with kind permission from Elsevier

of is that, if there are too many obstacles in the diffusion space, the diffusing particle will get 'caught' between the obstacles so that it cannot leave a certain area, thus leading to a diffusion coefficient of zero outside this area ($D(r_{\text{outside}}, c_p) = 0$) but much higher within the area $D(r_{\text{inside}}, c_p) \geq 0$. Thus, micro-domains are regions in the thylakoids within which PQ can diffuse more or less freely but which it cannot leave on a short time scale[3]. The micro-domains are bounded by membrane proteins acting as obstacles to PQ diffusion. It follows that only PS II and cyt *bf* complexes localized in the same micro-domain are connected on a fast time scale.

23.4 Crowding Effects in Photosynthetic Membranes

As discussed in the previous section, the area occupation in thylakoids due to membrane spanning proteins is very high. We may ask whether the threshold conditions that lead to the formation of diffusion domains are indeed met in thylakoids and whether such restricted diffusion is of physiological importance. To shed light onto these questions a random walk simulation approach was developed in [87], taking into account crowding in thylakoids.

23.4.1 Restricted PQ Diffusion in Crowded Thylakoids

As can be seen in Fig. 23.5 for point obstacles each occupying exactly one site on the lattice the percolation threshold is between 0.4 and 0.5 [87]. This value is below the occupied area fraction as found in thylakoids. However, the exact value of the percolation threshold depends strongly not only on the obstacle concentration but also on the shape of the obstacles. Furthermore, for the estimation of the percolation threshold to a value between 0.4 and 0.5 a homogeneous distribution of the obstacles and no obstacle–obstacle interaction were assumed. This probably does not hold for the proteins in the thylakoid membrane. The interplay of the percolation threshold with several factors that may influence thylakoid architecture, like e. g. obstacle shape and size, boundary lipids, protein–protein interaction or obstacle mobility, is analysed in the following.

Shape and size of obstacles: First, it was checked whether the percolation threshold is exceeded in thylakoids if the geometry of photosynthetic proteins is taken into account. Figure 23.6 illustrates the model of the thylakoid membrane used in the simulation. It depicts the shape of the photosynthetic proteins derived from structural analysis of the integral photosynthetic proteins (PS II with tightly bound LHC II [34], cyt *bf* [14], and LHC II [47]) and their random arrangement in the membrane. Different protein densities were investigated and plastoquinone diffusion in such an arrangement was examined [87].

The dependence of the normalised diffusion coefficient $D_n(r, c)$ on the protein concentration and the distance travelled is shown. In accordance with percolation theory the diffusion coefficient in the absence of obstacles $D(0)$ was constant and

---

[3] It has to be taken into account that proteins are not completely immobile as assumed for the determination of the percolation threshold. The membrane proteins are undergoing Brownian motion and are thus moving albeit very slowly compared to the small PQ.

23.4 Crowding Effects in Photosynthetic Membranes 559

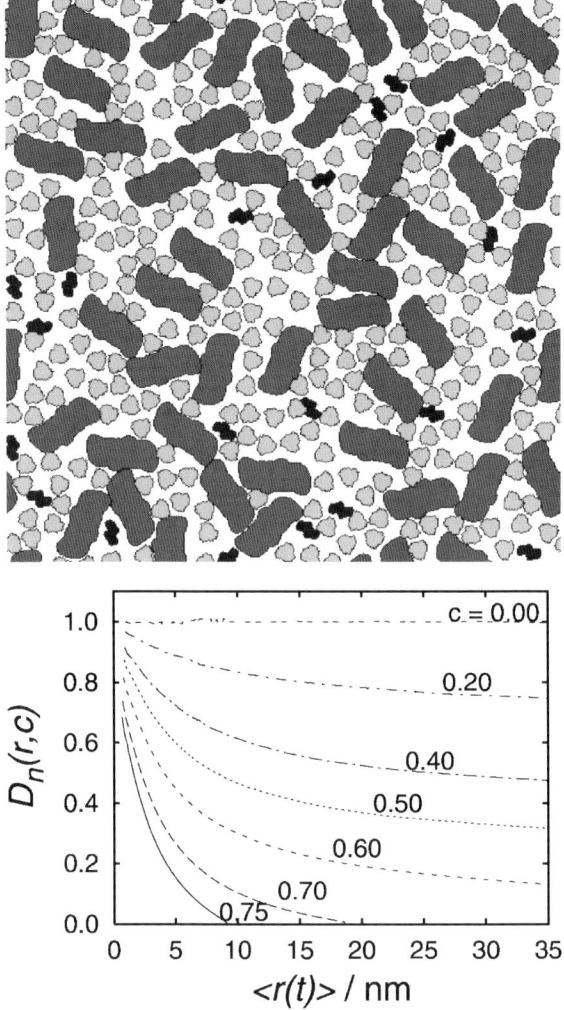

Fig. 23.6 Random distribution of photosynthetic proteins on a square lattice, 200 × 200 nm. *Dark grey*: Photosystem II dimers, *black*: cyt *bf* dimers, *light grey*: free light-harvesting trimers. The stoichiometry of the complexes PS II:cyt *bf*:LHC II is 2.6:1:14.1. The occupied area fraction is 0.6. *Below*: Distance dependence of the normalised diffusion coefficient $D_n(r,c)$ of a PQ in the presence of immobile obstacles exhibiting the shape of photosynthetic proteins. *Curves* are for different occupied area fractions. Conditions as in Fig. 23.5. Figure reprinted from reference [87], copyright (2003), with kind permission from Elsevier

independent of the travelled distance. For diffusion in the presence of obstacles it can be clearly seen that the diffusion coefficient became distance dependent and decreased with increasing distance travelled (r). As expected from percolation theory this effect became more significant for higher area occupation. Finally, the percolation threshold c_p was reached at an occupied area fraction between 0.6 and 0.7, which is significantly higher than for point obstacles.

Below the percolation threshold PQ may travel over a fairly large distance during the time course of photosynthetic electron transport (turnover time $\approx 10-15$ ms) [87]. Figure 23.7 shows how tracers (plastoquinol) spread out from a randomly chosen site (e. g. a binding site on PS II) within a few ms. The occupied area fraction was chosen to be 0.6, which is close to domain formation but still below the percolation threshold. PQ diffusion under these conditions is very fast compared to the time for photosynthetic electron transport. Therefore, even if PQ diffusion is severely slowed, as long as it is not trapped in a diffusion domain, PQ could migrate within a few milliseconds over a large area and visit a large number of cytochrome *bf* complexes.

Beyond the percolation threshold, however, the tracers were trapped within diffusion domains. Interestingly, for the protein shapes investigated (immobile, and without specific protein–protein interactions) this is around the value found in thylakoids (see above and [43, 87]). Under the given conditions, free PQ exchange can probably not occur for distances over more than 20 nm for an occupied area fraction of 0.7 (9.4 nm for a fraction of 0.75). The question now is whether the distance within which PQ can diffuse quickly is sufficient for efficient electron transfer from PS II to cyt *bf*.

The average distance (centre to centre) from one cyt *bf* complex to the closest photosystem II was roughly 15 nm at protein densities of 60–75% (line in Fig. 23.8), whereas diffusion properties changed drastically in this range of area occupation (columns in Fig. 23.8) [87]. Thus, between an area occupation between 0.6 and 0.75 three different scenarios were found:

$c = 0.6$: free (maybe slowed) diffusion of PQ,
$c = 0.7$: PQ is trapped but in average for each cyt *bf* more than one PS II is in close enough vicinity for electron transfer,
$c = 0.75$: cyt *bf* may become isolated from electrons delivered by PS II.

The presented results demonstrate the huge effect that a slightly increased area occupation has if the area covered with proteins is near the percolation threshold, thus illustrating how important it would be to know the exact area occupation in thylakoids. However, only crude estimates are available up to date.

The results shown so far demonstrate that the shape of obstacles may severely influence the percolation threshold. For point obstacles the percolation threshold

Fig. 23.7 The distance plastoquinone may travel on a millisecond time scale if the occupied area fraction is 0.6 (i. e. below the percolation threshold). The area is 300×300 nm. 1000 tracers were placed at a randomly chosen Q_B-binding site at PS II (*arrow*) and their journey was recorded. The *grey scale* represents the frequency of visits of a certain lattice site. *Numbers* indicate how many times (on average) a tracer has occupied a lattice point drawn in the corresponding shade. For example, the *black spots* show the sites that a tracer has occupied between 10 and 19 times. Figure reprinted in modified form from reference [87], copyright (2003), with kind permission from Elsevier

23.4 Crowding Effects in Photosynthetic Membranes 561

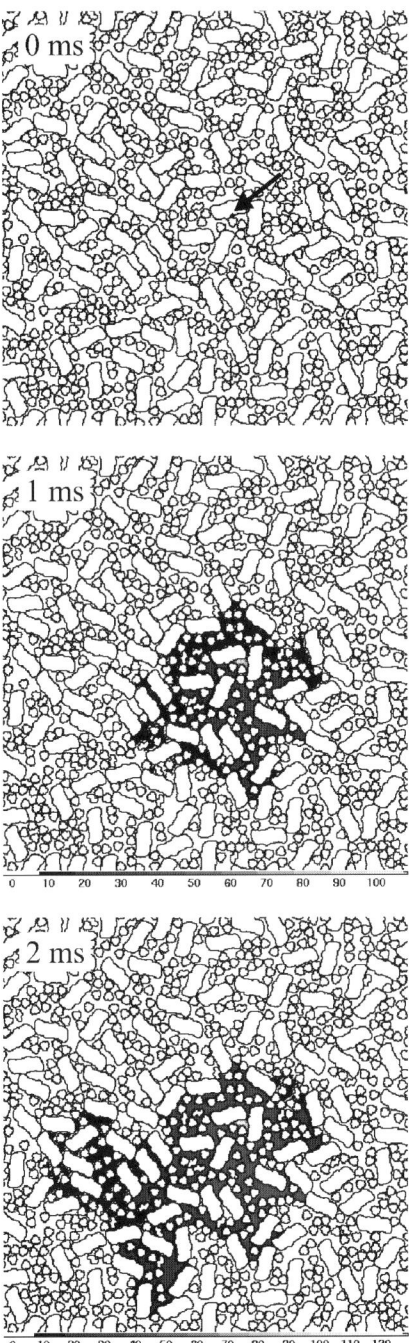

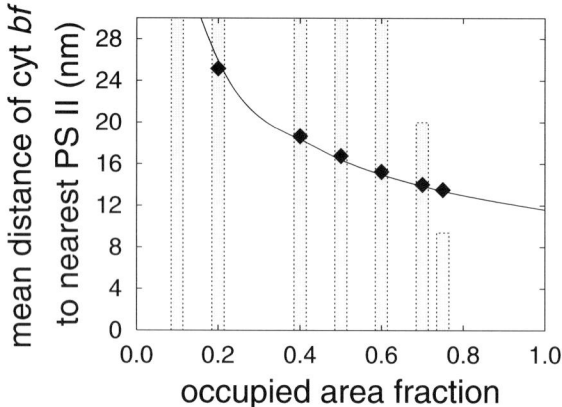

Fig. 23.8 *Line*: Average distance (nm) from a cyt *bf* complex to the nearest PS II in dependence of the area occupation. Data fitted with $y = 11.6 \cdot x^{-0.5}$. *Columns* show the distance of free plastoquinone diffusion. Figure reprinted from reference [87], copyright (2003), with kind permission from Elsevier

was found to be around 0.5 whereas for the geometry of photosynthetic proteins it is between 0.6 and 0.7. However, there are many more factors influencing the percolation threshold, as illustrated in Table 23.3.

Boundary lipids: From spin label studies, it is known that lipid mobility in thylakoids is reduced compared to a protein-free lipid membrane [93]. This suggests that a large number of proteins and protein complexes are surrounded by more or less tightly bound boundary lipids. Little is known about boundary lipids in thylakoids. Depending on the degree of their immobilisation, these lipids could be more or less 'permeable' to PQ.

Possible effects of boundary lipids on PQ diffusion were analysed. In our simulation two extreme cases were considered. In the first case all boundary lipids were assumed to be completely permeable to PQ diffusion, i. e. PQ could rapidly exchange with the boundary lipids. This led to the formation of lipid channels and prevented the formation of diffusion domains. In the second case it was assumed that the boundary lipids are tightly bound and cannot rapidly exchange with plastoquinone and thus impede PQ diffusion. Accordingly this kind of immobile boundary lipids led to a much lower percolation threshold. For both cases only boundary lipids of LHC II were considered since this is the most abundant protein in thylakoids (data

Table 23.3 Factors influencing the percolation threshold c_p and thus domain formation

decreasing c_p	increasing c_p
boundary lipids if tightly bound (cannot exchange with PQ) aggregation in elongated forms attractive protein–protein interaction	boundary lipids if loosely bound (can exchange with PQ) aggregation in compact forms mobility of obstacles

23.4 Crowding Effects in Photosynthetic Membranes

not shown). The two extreme cases have opposite effects and most likely boundary lipids in the real membrane are neither completely permeable nor completely impermeable. Rather, one might expect a mixture of both. Furthermore, different proteins could be surrounded by different lipid boundaries. How boundary lipids influence plastoquinone diffusion will certainly depend on the composition and position of the lipids at the proteins. We know too little about boundary lipids in the thylakoids, but we demonstrate here that they may be an important feature of thylakoid structure[4].

Protein–protein interaction: As described in Sect. 23.2.2.2 molecular crowding may enhance aggregation of proteins into compact complexes. Larger, more compact obstacles, however, are less efficient in impeding diffusion. Thus, aggregation into more compact forms may increase the percolation threshold while aggregation into elongated forms may decrease the percolation threshold. For a more elaborate treatment of protein-protein interaction and its influence on PQ diffusion see reference [88].

Obstacle mobility: The influence of obstacle mobility was also investigated in [87]. Figure 23.9 shows the normalised diffusion coefficient for tracers diffusing between mobile obstacles. It can be seen that the diffusion coefficient approaches a constant (concentration dependent) value as r, the distance travelled, increases. In this case, no percolation threshold exists and long-range diffusion can occur at all concentrations, see Fig. 23.9. This is in accordance with previous simulations [69, 76, 91].

How effective mobile obstacles are in hindering plastoquinone diffusion certainly depends on the diffusion coefficients of the obstacles [91]. However, the mobility

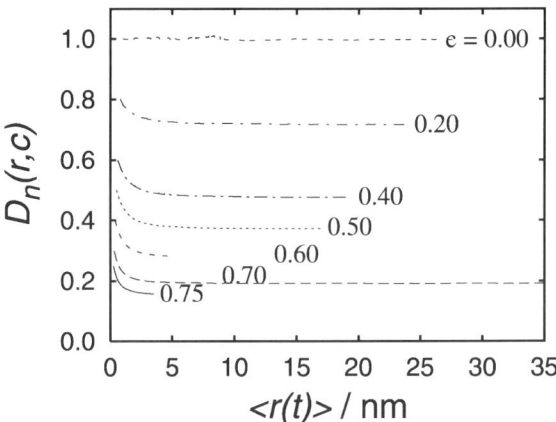

Fig. 23.9 Similar to Fig. 23.5 but this figure shows the normalised diffusion coefficient of the tracer in the presence of *mobile* point obstacles. The diffusion coefficient of the point obstacles was assumed to be the same as that of the tracer. Area occupation as indicated. Figure reprinted from reference [87], copyright (2003), with kind permission from Elsevier

---

[4] For other treatments of lipids surrounding proteins see also the 'dynamic boundary layer model' mentioned in [4, 91] or [54].

of LHC II, the most abundant protein in thylakoids, is probably restricted by lateral interaction and transverse interaction across adjacent membranes in a grana stack. Furthermore, a fraction of LHC II is known to form large supercomplexes with PS II [12, 95]. No information exists about the mobility of such aggregates. It is likely that they are highly immobile because of their size and multiple lateral and cross-membrane interactions. Therefore, the diffusion coefficient of LHC II *in vivo* may be not large enough to prevent domain formation on a time scale relevant to photosynthetic electron transport. Clearly, the influence of protein mobility on PQ diffusion is an important factor in understanding PQ percolation in thylakoids. More information about protein mobility in thylakoids is thus required.

23.4.2 Occupation of Binding Sites

Another interesting effect of molecular crowding is that it may result in a certain fraction of binding sites on PS II or cyt *bf* that are obstructed by other proteins. Assuming a size and position of the binding sites as shown in Fig. 23.10 the percentages of obstructed binding sites are as shown in Table 23.4.

For all occupied area fractions, the percentage of obstructed Q_B binding sites on PS II was very similar to that of Q_r binding sites on cyt *bf*. In contrast to these binding sites the fraction of obstructed Q_o binding sites on cyt *bf* is relatively low as it is less exposed (see Fig. 23.10). Obviously the differential obstruction of different binding sites is depends strongly on their exact position at the protein. In this respect, protein shapes may play a major role and should be taken into consideration.

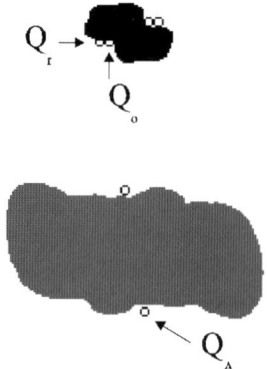

Fig. 23.10 Proteins with binding sites, *grey*: PS II, *black*: cyt *bf*. The *shapes* shown here were used to determine the percentage of occupied binding sites in Table 23.4. Figure reprinted from reference [87], copyright (2003), with kind permission from Elsevier

Table 23.4 Obstructed binding sites in a random protein arrangement as affected by the occupied area fraction. The values represent the average of 500 different random arrangements in a 200 × 200 nm² matrix. The size and position of the binding sites are assumed to be as shown in Fig. 23.10. It should be noted that only binding sites directly blocked by a protein are counted. Table reprinted from reference [87], copyright (2003), with kind permission from Elsevier

Occupied area fraction	Percentage of obstructed bindingsites		
	Q_B	Q_o	Q_r
0.0	0.0	0.0	0.0
0.2	0.8	0.2	0.7
0.4	2.1	0.6	2.2
0.5	3.4	0.8	3.7
0.6	6.7	1.8	7.0
0.7	13.2	4.2	13.2

23.4.3 Reaction Mechanisms in Crowded Thylakoids

Most published models of electron transport assume a common plastoquinone pool shared by all PS II. A shared plastoquinone pool implicitly assumes fast plastoquinone exchange throughout the membrane. However, it was shown above that long range diffusion of plastoquinone in thylakoids may be severely restricted by the integral proteins (for further reference see also references [11, 87]). Furthermore, it has been shown that such restricted diffusion may strongly influence kinetic rates [24,25,78] and the time course of enzymatic reactions [8]. High concentrations of background macromolecules that do not participate directly in a particular considered reaction, like e. g. LHC II in photosynthetic electron transport, have been observed to induce order-of-magnitude or larger changes in the rates and equilibria of numerous investigated reactions [61].

However, the contribution of crowding and diffusion limitation to reaction kinetics is not only determined by the degree of crowding and the apparent diffusion coefficient of mobile educts, but also by the reaction mechanism (see also references [24, 25, 33, 61, 78]).

Despite detailed knowledge of a great deal of rate constants of photosynthetic electron transport in thylakoids is available, the mechanisms of how the electrons are actually transferred are in many cases still unknown. Therefore, two extreme cases for the binding mechanism for the presumably rate limiting electron transfer from PQ to cyt *bf* at the Q_o binding site were investigated [89], tight binding and a collisional mechanism. The tight binding mechanism includes an irreversible binding of PQ to the Q_o site of cyt *bf* before its slow oxidation takes place. The collisional mechanism implies immediate electron transfer after a successful encounter (see also reference [66]). Both mechanisms lead to an exponential decay with the same rate constant k when diffusion processes are not limiting. In thylakoids however, where diffusion may be highly restricted, the two mechanisms were expected to lead to different behaviours. Therefore both mechanisms were simulated, and the results compared.

The experimental approach to the role of the diffusion of plastoquinol was to follow the re-reduction of P700, the reaction centre chlorophyll of PS I, by electrons induced by a light-flash at PS II. This is a convenient method of accurately measuring whole-chain electron transport. The rate constant of the sigmoidal reduction kinetics and the initial lag give information about the rate-determining oxidation of plastoquinol [83] and reactions preceding this step [31], respectively.

The experimental data obtained were then compared with the results of the simulation. The measured data were matched by varying rate constants used in the simulation. Parameters that may be varied in the simulation are the rate constants of the following reactions:

$$QA^- + QB \xrightarrow{k_0} QA + QB^- \qquad \text{transfer of first electron,}$$

$$QA^- + QB^- \xrightarrow{k_1} QA + QB^{2-} \qquad \text{transfer of second electron,}$$

$$PQH_2 + FeS_{ox} \xrightarrow{k_3} PQ^- + FeS_{red} + 2H^+ \qquad \text{oxidation of PQH}_2 \text{ at } Q_o,$$

$$PQ + \text{cyt } b_h^- \xrightarrow{k_4} PQ^- \qquad \text{reduction of PQ at } Q_r,$$

$$FeS_{red} + P700^+ \xleftrightarrow{k_5} FeS_{ox} + P700 \qquad \text{equilibrium,}$$

and k_2 is the dissociation of plastoquinol. The rates of electron transfer at the Q_B binding site are reasonably well known. Therefore, the rate constants for the first (k_0) and the second electron transfer (k_1) at PS II were chosen to be 6670 and 2500 s^{-1}, respectively. These values are within the range of those given by Diner and coworkers [20].

The equilibrium constant k_5 can in principle be calculated from the equilibrium constants for the different involved reactions. However, equilibrium values for these reactions as found in the literature, vary greatly. A collection of references to these equilibrium constants can be found in Berry and Rumberg [9]. Nevertheless, the redox state of P700 only declines to ca. 80% during the first 10 ms considered. Therefore, electrons will be always transferred from the Rieske centre to P700 if the equilibrium constant is larger than 150, which is in accordance with the values found in the literature.

The remaining free parameters (k_2, k_3, k_4) were varied to minimise the least sum of errors squared when compared with the experimental data. k_2 was allowed to vary between 200 and 5000 s^{-1}, k_3 between 100 and 100,000 s^{-1}, and k_4 between 100 and 100,000 s^{-1}. The stoichiometries used were the ones calculated in [87]: PS II:cyt bf:LHC II = 2.56:1:14.12. Additionally two PS I (monomers) per PS II (dimer) were assumed.

23.4.3.1 Tight Binding Mechanism

For the tight binding mechanism an irreversible tight binding of PQ at the Q_o site of cyt bf was assumed followed by the slow oxidation of PQ.

A good fit to the measured data could be obtained (see Fig. 23.11). An occupied area fraction of 0.70 was chosen, corresponding to the value estimated for grana thylakoids in [43, 87]. For the conditions used in the simulation, this area fraction is above the percolation threshold and hence diffusion domains bounded by the integral photosynthetic proteins are formed [87].

The parameters resulting from the best fit agree well with the range of published data measured on thylakoids (see Table 23.5). Three different random protein configurations were investigated and parameters varied to match the measured data.

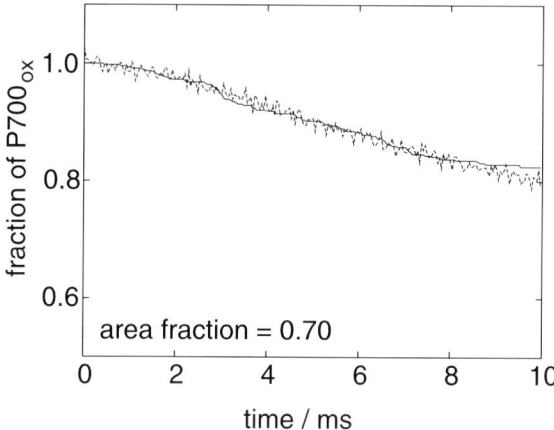

Fig. 23.11 Comparison of the simulation (*solid line*) with the observed P700 re-reduction (*dashed line*) for a tight binding mechanism. An occupied area fraction of 0.70 (i. e. above the percolation threshold) was assumed. Rate constants were $k_2 = 1206.1$ s^{-1}, $k_3 = 294.9$ s^{-1}, and $k_4 = 335.2$ s^{-1} (nomenclature as described in methods). Figure reprinted from [89], copyright (2007), with kind permission from Elsevier

Table 23.5 Rate constants for several electron transfer steps from literature and obtained from simulations. Table reprinted in modified form from [89], copyright (2007), with kind permission from Elsevier

rate constant in s^{-1}	literature data	reference	tight binding ratio of occupied area			collisional ratio of occupied area
			0.70	0.60	0.70 (permeable)	0.70
k_2	495–2310 1000	[66] [38]	1206	1641	243	1352
k_3	305 200–330	[37] [19, 35]	295	100	100	48806
k_4	400	[37]	335	250	369	49562

To investigate the influence of the protein density and hence the retardation of PQ migration, the simulation was repeated for an occupied area fraction of 0.60, which is below the percolation threshold. As with an area fraction of 0.70 good fits were obtained (data not shown). However, the rate constants obtained were not in good agreement with published data (see Table 23.5). To match the experimental data, the reaction at the Q_o site would need to be very slow ($k_3 = 100\,\text{s}^{-1}$, the lowest value allowed in the simulation). This is probably to compensate for the much faster 'finding' of the binding site due to the less restricted diffusion of PQ.

23.4.3.2 Collisional Mechanism

In the collisional mechanism it is assumed that electrons are transferred immediately at each successful encounter. The reaction constant determines the probability of an encounter to be successful. This mechanism reflects a more diffusion limited scenario compared to the tight binding mechanism.

Good fits were only obtained if the rate constant for oxidation of PQH_2 at the Q_o site (k_2) was more than two orders of magnitude larger than measured values in thylakoids (see Fig. 23.12 and Table 23.5). Accordingly, the simulation predicts that for isolated complexes a collisional mechanism would require rate constants that are about 100-fold higher than those measured on thylakoids, which seems unlikely since rate constants for the reactions at the Q_o site on cyt bf measured on *C. reinhardtii in vivo* and *in vitro* are very similar [68]. Equal rates *in vivo* and *in vitro* would fit better with the tight binding mechanism (reaction limited and not diffusion limited).

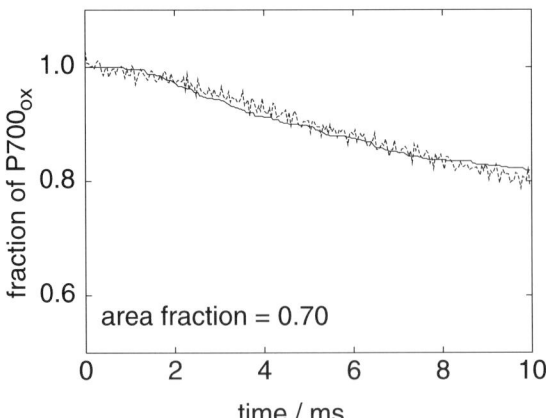

Fig. 23.12 Comparison of the simulation (*solid line*) with the observed P700 re-reduction (*dashed line*) for a collisional mechanism. An occupied area fraction of 0.70 (i.e. above the percolation threshold) was assumed. Rate constants were $k_2 = 1352\,\text{s}^{-1}$, $k_3 = 48{,}806\,\text{s}^{-1}$, and $k_4 = 49{,}562\,\text{s}^{-1}$ (nomenclature as described in methods). Figure reprinted from [89], copyright (2007), with kind permission from Elsevier

23.4.3.3 Discussion of Possible Reaction Mechanisms Realised in Thylakoids

In summary the experimental data were best matched when restriction of plastoquinone diffusion was pronounced (area fraction 0.70) and a tight binding mechanism was assumed. In contrast, it seems that a collisional mechanism was not suitable to describe the measured data on P700 re-reduction kinetics. For both reaction mechanisms lower area occupations or the assumption that proteins do not act as obstacles to PQ diffusion (0.70 (permeable) in Table 23.5) do not agree with published data on reaction constants. Altogether, the simulations clearly point out the importance of structural characteristics of thylakoids and molecular crowding in models of electron transport.

An influence of diffusion processes on reaction rates is also reported for the activation of transducin by rhodopsin [79]. Higher concentrations of reactants are expected, on one hand, to lead to a decreased rate of reaction by lowering the diffusion coefficient of the reactants (here only PQH_2 was considered to be mobile but this does not alter the principal results). On the other hand, they also lead to an increased reaction rate by the law of mass action. The law of mass action, however, may be invalid in a membrane close to the percolation threshold where diffusion of PQ is severely restricted as described in Sect. 23.2.3. Instead, the apparent rate 'constant' becomes fractal and decreases with time. The time dependence of the reaction constant increases with increasing obstacle densities [8].

It should be noted that the tight binding mechanism and the collisional mechanism are both simplifications. More generally, reactions should be described by the rates of binding (k_{on}), unbinding (k_{off}) and the internal electron transfer rate (k_3). However, no firm figures exist for plant cyt *bf* complexes. Therefore the simplifications of a tight binding and a collisional mechanism were introduced, reflecting two extreme cases.

Despite the simplifying assumptions on the reaction mechanism, the simulations clearly point out that the spatial organisation of the proteins within the thylakoid membrane may severely influence photosynthetic rate constants. Hence, more information about the thylakoid organisation is required. Furthermore, our results suggest that PQ oxidation at the Q_o site of cyt *bf* taking place in a heavily crowded membrane is expected to be reaction limited rather than diffusion limited. For a diffusion limited reaction, great differences between apparent (fractal) rate constants and rate constants found in dilute solutions *in vitro* would be predicted. Unfortunately, data for rate constants of photosynthetic electron transport measured on isolated complexes are sparse. More data on isolated complexes would be desirable but isolated complexes may suffer the loss or modification of subunits and show severely artificial kinetics.

23.5 Summary and Outlook

23.5.1 Summary

As illustrated in Sect. 23.2, molecular crowding is ubiquitous in living systems. One example for a physiological process proceeding in a crowded environment is photosynthesis. Although photosynthetic electron transport takes place in a highly crowded membrane and crowding may effect many biochemical reactions, the impact of crowding is hardly considered in investigations of photosynthetic electron transport. For this reason the present work aims at elucidating possible physiological consequences of crowding in thylakoids and at a deeper understanding of how to interpret experimental results in terms of molecular crowding.

One prominent effect of molecular crowding is its influence on diffusion. The diffusion coefficient D of a mobile particle becomes dependent on the concentrations of crowders impeding diffusion and on the distance travelled. Accordingly, any experimental determination of diffusion coefficients in a crowded environment needs to take into account a possible distance dependency of the diffusion coefficient $D(r)$. Differing experimental setups involve measurements of different travelled distances and may lead to different estimates of $D(r, c)$. In measurements by fluorescence photobleaching recovery the distance considered is typically $r \approx 1$ μm, whereas in measurements by eximer formation or fluorescence quenching the analysed distance in much smaller and $r \approx 1 - 10$ nm (see e. g. [22, 77, 93]), which may consequently lead to larger estimates of $D(r)$.

With increasing crowding, diffusion may become further obstructed. Finally, the long range (large r) diffusion coefficient may decline to zero when the area fraction c of obstacles exceeds the percolation threshold c_p. As illustrated in Sect. 23.4 (see in particular Table 23.3), factors that influence the organisation of proteins within the membrane may also shift the percolation threshold and alter the characteristic of diffusion, and hence, reactions in a crowded membrane. Many of these factors, however, are not well known for photosynthetic proteins. Similarly, even the exact area occupation in thylakoids is not known.

Recent estimates determined the area occupation to approximately 0.7. Interestingly, for randomly distributed, non-interacting, immobile proteins this value is very close to the percolation threshold, beyond which fast long range PQ migration is disrupted (Fig. 23.6). However, the arrangement of proteins in the thylakoids is not known in detail, and thus it is still unknown whether the percolation threshold is met. According to the simulations presented here, PQ can travel over large distances on a time scale relevant to photosynthesis if the area occupation in thylakoids is below the percolation threshold. However, already a slight change of the percolation threshold leads to a trapping of PQ in diffusion domains, which it cannot leave on short time scales. Considering these very different scenarios, it becomes obvious that more detailed knowledge of the thylakoid architecture is needed for a thorough understanding of photosynthesis.

23.5 Summary and Outlook

Furthermore, crowding has opposing effect on the biochemical reaction rates. On the one hand, under crowding conditions the thermodynamic activities of the reactants increase. On the other hand, the increased tortuosity of the paths left for molecular motion limits the rate of encounter of two defined species (diffusion control). On the evolutionary scale, every organism is constantly struggling through molecular evolution to reach the right balance between both effects, and/or maintain it (e. g., control of cell volume).

Due to the opposing effects of molecular crowding on chemical reactions, one additional factor determining the impact of molecular crowding on a reaction is the reaction mechanism. For a diffusion limited process the impact of molecular crowding is much more pronounced than for a reaction limited process. However, the reaction mechanism of the rate limiting step of photosynthetic electron transport is not known to date. For this reason, in the simulations two extreme cases for a reaction mechanism were considered; a diffusion limited collisional mechanism and a reaction limited tight binding mechanism. Matching experimental data with the simulation suggests a reaction limited process rather than a diffusion limited reaction. Neither of the two analysed mechanisms coincides with an area occupation clearly below the percolation threshold as determined for randomly distributed proteins.

The finding that the area occupation in thylakoids is close to the percolation threshold on the one hand raises questions like 'Have thylakoids evolved to exploit effects of crowding for fine regulation of photosynthetic electron transport?', or 'Is trapping of PQ in diffusion domains beneficial to avoid harmful over reduction of photosynthetic proteins?'. On the other hand one may ask whether some assumptions commonly made, such as random distribution of proteins within the membrane, might not be correct and additional information is needed. Thus, the simulations point out which factors (see also Table 23.3) need to be investigated experimentally in more detail to evaluate their influence on reactions in a crowded membrane.

Restricted diffusion not only occurs in thylakoids. Diffusion coefficients for both, large molecules (e. g. green fluorescent protein) and small molecules (e. g. carboxyfluorescein) have been measured in various cells, and are found to be reduced from three- to tenfold compared to their values in water [23, 24, 26, 53]. However, in mitochondria the mobility of some enzymes was found to be unexpectedly large, suggesting that these organelles contain relatively uncrowded channels where some proteins move more rapidly than expected for a uniformly congested medium [92].

Biological macromolecules have evolved to function in crowded environments [24], thus raising biologically important questions. Why did cells evolve such a highly packed interior? Are there any advantages in being crowded? How do macromolecules fold, associate and travel through the crowded intracellular medium [13, 28, 73, 97]?

23.5.2 Treatments of Molecular Crowding In Vivo, In Vitro and In Silico

A crowded physico-chemical environment bears several fundamental implications for the structure, function, and evolution of cellular systems, as well as their experimental study.

In vivo the development of several experimental techniques during the last few years enables certain aspects of the behaviour of selected macromolecules – typically labelled and/or overexpressed – to be monitored within living cells (among many others see e.g. [39, 55]). It is hoped that these techniques can be used to investigate the influence of molecular crowding upon the selected species within their native environments. However, one must be very cautious when interpreting the results of such experiments. In particular, it is necessary to design control experiments that can reveal potential artefacts due to labelling and/or overexpression.

In vitro it should be possible, in principle, to measure the effects of crowding on the interactions of extracted macromolecules by adding high concentration of inert background molecules (crowders) to the solution. The ideal crowder should be highly soluble and not be prone to self-association. Further, it should be globular rather than extended to prevent solutions from becoming too viscous to handle. It must not change pH, ionic strength, or redox potential, or other important control factors. Furthermore, the crowder should be easily available in purified form, such that the use of high concentrations does not lead to contamination artefacts. The most important criterion, however, is that the crowder should not interact with the system under consideration, except via steric repulsion. This last requirement is the most difficult to meet, and in practice it is necessary to use a variety of crowding agents to reduce the risk that the observed effect is caused by specific interactions (reviewed in [96], see also [24]).

In silico simple statistical-thermodynamic models for predicting the influence of excluded volume on macromolecular equilibria and reaction rates in crowded solutions were applied. These are based on the representation of a macromolecular solute as an equivalent hard convex particle of similar size and shape. Such models have proved useful in accounting for a variety of experimental observations, in many cases quantitatively [33].

However, the effects of macromolecular crowding may strongly depend on many different factors like e.g. the concentration of background molecules, their shape, interactions other than purely steric repulsion, or the reaction mechanism considered. Furthermore, effects of crowding may be opposing. Although crowding reduces diffusion, it increases thermodynamic activities. Since the net result of all effects due to crowding depends on the precise nature of each system, it is necessary to examine the effects of crowding quantitatively in all studies of macromolecular interactions if these studies are to be regarded as physiologically relevant [59]. To this end, the impact of crowding on photosynthetic electron transport has been simulated and analysed specifically, taking into account realistic concentrations and shapes of photosynthetic proteins.

Acknowledgements The authors wish to thank Thomas Strauß for the extended Powell-routine and countless helpful discussions about the software architecture and for his help with trouble shooting.

References

1. Albertsson P-Å (1995) The structure and function of the chloroplast photosynthetic membrane – a model for the domain organization. Photosynth Res 46:141–149
2. Albertsson P-Å (2000) The domain structure and function of the thylakoid membrane. Recent Res Devel Bioener 1:143–171
3. Albertsson P-Å (2001) A quantitative model of the domain structure of the photosynthetitc membrane. Trends in Plant Science 6:349–354
4. Almeida PFF, Vaz WLC (1995) Handbook of Biological Physics, chapter 6. Lateral Diffusion in Membranes, Elsevier Science BV, Amsterdam, pp 305–357
5. Anderson JM (1982) The significance of grana stacking in chlorophyll b containing chloroplasts. Photobiochem Photophys 3:225–241
6. Anderson JM, Melis A (1983) Localization of different photosystems in seperate regions of chloroplast membranes. Proc Natl Acad Sci USA 80:745–749
7. P-Arvidson O, Sundby C (1999) A model for the topology of the chloroplast thylakoid membrane. Aust J Plant Physiol 26:687–694
8. Berry H (2002) Monte Carlo simulations of enzyme reactions in two dimensions: fractal kinetics and spatial segregation. Biophys J 83(4):1891–901
9. Berry S, Rumberg B (2000) Kinetic modeling of the photosynthetic electron transport chain. Biochemistry 53:35–53
10. Blackwell MF, Gibas C, Gygax S, Roman D, Wagner B (1994) The plastoquinone diffusion coefficient in chloroplasts and its mechanistic implications. Biochem Biophys Acta 1183:533–543
11. Blackwell MF, Whitmarsh J (1989) Examination of plastoquinone diffusion in lipid vesicles. Biophys J 58:1259–1271
12. Boekema EJ, van Breemen JFL, van Roon H, Dekker JP (2000) Arrangement of photosystem II supercomplexes in crystalline macrodomains within the thylakoid membrane of green plant chloroplasts. J Mol Biol 301:1123–1133
13. Bray D (1998) Signalling complexes: biophysical constraints on intercellular communication. Annu Rev Biophys Biomol Struct 27:59–75
14. Breyton C (2000) Conformational changes in the cytochrome b6f complex induced by inhibitor binding. J Biol Chem 275:13195–13201
15. Broadbent SR, Hammersley J-M (1957) Percolation processes I. Crystals and mazes. Proc Cambr Phil Soc 53:629–641
16. Burg MB (2002) Macromolecular crowding as a cell volume sensor. Cell Physiol Biochem 10:251–256
17. Chapman DJ, Barber J (1990) Analysis of plastoquinone-9 levels in appressed and non-appressed thylakoid membrane regions. Biochim Biophys Acta 850:170–172
18. Chebotareva NA, Kurganov BI, Livanova NB (2004) Biochemical effects of molecular crowding. Biochemistry (Mosc) 69(11):1239–51

19. Cramer WA, Soriano GM, Ponomarev M, Huang D, Zhang H, Martinez SE, Smith JL (1996) Some new structural aspects and old controversies concerning the cytochrome *b6f* complex of oxygenic photosynthesis. Annu Rev Plant Physiol Plant Mol Biol 47:477–508
20. Diner AB, Babcock GT (1996) Oxygenic photosynthesis: The light reactions, chapter 12. Structure, dynamics, and energy conversion efficiency in photosystem II. Kluwer Academic Publishers, pp 213–247
21. Eggers DK, Valentine JS (2001) Molecular confinement influences protein structure and enhances thermal protein stability. Prot Sci 10:250–261
22. Eisinger J, Flores J, Petersen WP (1986) A milling crowd model for local and long–range obstructed lateral diffusion. Biophys J 49:987–1001
23. Ellis RJ (2001a) Macromolecular crowding: an important but neglected aspect of the intracellular environment. Curr Opin Struct Biol 11:114–119
24. Ellis RJ (2001b) Macromolecular crowding: obvious but underappreciated. TiBS 26(10):597–604
25. Ellis RJ, Minton AP (2003) Join the crowd. Nature 425:27–28
26. Elowitz MB, Surette MG, Wolf P-E, Stock JB, Leibler S (2003) Protein mobility in the cytoplasm of *Escherichia coli*. J Bacteriol 181:197–302
27. Flory PJ (1941) Molecular size distribution in three dimensional polymers. I, II, III. J Am Chem Soc 63:3083–3100
28. Fulton AB (1982) How crowded is the cytoplasm? Cell 30:345–347
29. Grima R, Schnell S (2006) A systematic investigation of the rate laws valid in intracellular environments. Biophys Chem 124:1–10
30. Guttman HJ, Anderson CF Jr, Record TM (1995) Analyses of thermodynamic data for concentrated hemoglobin solutions using scaled particle theory: implications for a simple two-state model of water in thermodynamic analyses of crowding in vitro and in vivo. Biophys J 68:835–846
31. Haehnel W (1976) The reduction kinetics of chlorophyll a_I as an indicator for proton uptake between the light reactions in chloroplasts. Biochim Biophys Acta 440:506–521
32. Haehnel W (1984) Photosynthetic electron transport in higher plants. Ann Rev Plant Physiol 35:659–693
33. Hall D, Minton AP (2003) Macromolecular crowding: qualitative and semiquantitative successes, quantitative challenges. Biochim Biophys Acta 1649(2):127–39
34. Hankamer B, Barber J, Boekema EJ (1997) Structure and membrane organization of photosystem II in green plants. Annu Rev Plant Physiol Plant Mol Biol 48:641–671
35. Hauska G, Schütz M, Büttner M (1996) Oxygenic Photosynthesis: The Light Reactions, chapter 19. The Cytochrome $b6f$ Complex–Composition, Structure and Function. Kluwer Academic Publishers, pp 377–398
36. Herzfeld J (1996) Entropically-driven order in crowded solutions: from liquid crystals to cell biology. Acc Chem Res 29:31–37
37. Hope AB, Huligol RR, Panizza M, Thompson M, Matthews DB (1992) The flash induced turnover of cytochrome b-563, cytochrome f and plastocyanin in chloroplasts. Models and estimation of kinetic parameters. Biochim Biophys Acta 1100:15–26
38. Hsu B-D (1992) A theoretical study on the fluorescence induction curve of spinach in the absence of DCMU. Biochim Biophys Acta, 1140:30–36
39. Ignatova Z, Gierasch LM (2004) Quantitative protein stability and aggregation *in vivo* by real time fluorescent labeling. Proc Natl Acad Sci USA 101:523–528
40. Joliot P, Joliot A (1992) Electron transfer between photosystem II and the cytochrome bf complex: mechanistic and structural implications. Biochim Biophys Acta 1102:53–61
41. Joliot P, Lavergne J, Béal D (1992) Plastoquinone compartmentation in chloroplasts. I. evidence for domains with different rates of photo-reduction. Biochim Biophys Acta 1101:1–12
42. Kirchhoff H, Horstmann S, Weis E (2000) Control of the photosynthetic electron transport by PQ diffusion microdomains in thylakoids of higher plants. Biochim Biophys Acta, Bioenergetics 1459(1):148–168
43. Kirchhoff H, Mukherjee U, Galla H-J (2002) Molecular architecture of the thylakoid membrane: lipid diffusion space for plastoquinone. Biochemistry 41:4872–4882

44. Kirchhoff H, Tremmel I, Haase W, Kubitscheck U (2004) Supramolecular photosystem II organization in grana thylakoid membranes: evidence for a structured arrangement. Biochemistry 43:9204–13
45. Kopelman R (1986) Rate-processes on fractals: theory, simulations, and experiments. J Stat Phys 42:185–200
46. Kopelman R (1988) Fractal reaction kinetics. Science 241:1620–1626
47. Kühlbrandt W, Wang DN (1991) Three-dimensional structure of plant light-harvesting complex determined by electron crystallography. Nature 350:130–134
48. Lang F, Busch GL, Ritter M, Völkl H, Waldegger S, Gulbins E, Häussinger D (1998) Functional significance of cell volume regulatory mechanisms. Physiol Rev 78:247–306
49. Lavergne J, Bouchaud J-P, Joliot P (1992) Plastoquinone compartmentation in chloroplasts. II. theoretical aspects. Biochim Biophys Acta 1101:13–22
50. Lavergne J, Briantais J-M (1996) Oxygenic Photosynthesis: The Light Reactions, chapter 14. Photosystem II Heterogeneity. Kluwer Academic Publishers, pp 265–287
51. Lavergne J, Joliot P (1991) Restricted diffusion in photosynthetic membranes. TiBS 16:129–134
52. Lebowitz JL, Helfand E, Praestgaard E (1965) Scaled particle theory of fluid mixtures. J Chem Phys 43:774–779
53. Luby-Phelps K (2000) Cytoarchitecture and physical properties of cytoplasm: volume, viscosity, diffusion, intracellular surface area. Int Rev Cytol 192:189–221
54. Marcelja S (1999) Towards a realistic theory of the interaction of membrane inclusions. Biophys J 76:593–594
55. Mc Nulty BC, Young GB, Pielak GJ (2006) Macromolecular crowding in the Escherichia coli periplasm maintains alpha-synuclein disorder. J Mol Biol 355:893–897
56. Michaelis L, Menten M (1913) Die Kinetik der Invertinwirkung. Biochem Z 49:333–369
57. Minton AP (1981) Excluded volume as a determinant of macromolecular structure and reactivity. Biopolymers 20:(2093)–2120
58. Minton AP (1983) The effect of volume occupancy upon the thermodynamic activity of proteins: some biochemical consequences. Mol Cell Biochem 55:119–140
59. Minton AP (1997) Influence of excluded volume upon macromolecular structure and associations in 'crowded' media. Curr Opin Biotechnol 8:65–69
60. Minton AP (2000) Implications of macromolecular crowding for protein assembly. Curr Opin Struct Biol 10:34–39
61. Minton AP (2001) The influence of macromolecular crowding and macromolecular confinement on biochemical reactions in physiological media. J Biol Chem 276(14):10577–10580
62. Minton AP (2006a) How can biochemical reactions within cells differ from those in test tubes. J Cell Sci 119(14):2863–2869
63. Minton AP (2006b) Macromolecular crowding. Curr Biol 16(8):269–271
64. Minton AP, Colclasure GC, Parker JC (1992) Model for the role of macromolecular crowding in regulation of cellular volume. Proc Natl Acad Sci USA 89(21):10504–10506
65. Mitchell P (1976) Possible molecular mechanisms of the protonmotive function of cytochrome systems. J Theor Biol 62:327–367
66. Mitchell R, Spillmann A, Haehnel W (1990) Plastoquinol diffusion in linear photosynthetic electron transport. Biophys J 58:1011–1024
67. Ovchinnikov AA, Zeldovich YB (1978) Role of density fluctuations in bimolecular reaction kinetics. Chem Phys 28:215–218
68. Pierre Y, Breyton C, Tribet C, Kramer D, Olive J, Popot JL (1995) Purification and characterization of the cytochrome $b6f$ complex from *Chlamydomonas reinhardtii*. J Biol Chem 270:29342–29349
69. Pink DA (1985) Protein lateral movement in lipid bilayers. simulation studies of its dependence upon protein concentration. Biochim Biophys Acta 818:200–204
70. Ralston GB (1990) The effect of crowding in protein solutions. J Chem Educ 67:857–860
71. Rivas G, Fernandez JA, Minton AP (1999) Direct observation of the self-association of dilute proteins in the presence of inert macromolecules at high concentration via tracer sedimentation equilibrium: Theory, experiment, and biological significance. Biochemistry, 38(29):9379–9388

72. Rivas G, Fernandez JA, Minton AP (2001) Direct observation of the enhancement of non-cooperative protein assembly by macromolecular crowding: indefinite self-association of the bacterial cell division protein. Proc Natl Acad Sci USA 98:3150–3155
73. Rivas G, Ferrone F, Herzfeld J (2004) Life in a crowded world. EMBO Reports 5(1):23–27
74. Rohwer JM, Postma PW, Kholodenko BN, Westerhoff HV (1998) Implications of macromolecular crowding for signal transduction and metabolite channeling. Biochemistry 95(18):10547–10552
75. Ross PD, Minton AP (1977) Analysis of non-ideal behavior in concentrated hemoglobin solutions. J Mol Biol 112(3):437–452
76. Saxton MJ (1987) Lateral diffusion in an archipelago. The effect of mobile obstacles. Biophys J 52:989–997
77. Saxton MJ (1989) Lateral diffusion in an archipelago. Distance dependence of the diffusion coefficient. Biophys J 56:615–622
78. Saxton MJ (2002) Chemically limited reactions on a percolation cluster. J Chem Phys 116(1):203–208
79. Saxton MJ, Owicki JC (1989) Concentration effects on reactions in membranes: rhodopsin and transducin. Biochim Biophys Acta 979:27–34
80. Segal HL (1959) The Enzymes, 2nd ed., chapter: The development of enzyme kinetics. Academic Press, New York, pp 1–48
81. Somalinga B, Roy R (2002) Volume exclusion effect as a driving force for reverse proteolysis. J Biol Chem 277:43253–43261
82. Staehelin LA, van der Staay GWM (1996) Oxygenic Photosynthesis: The Light Reactions, chapter 2. Structure, composition, functional organization and dynamic properties of thylakoid membranes. Kluwer Academic Publishers, pp 11–30
83. Stiehl HH, Witt HT (1969) Quantitative treatment of the function of plastoquinone in photosynthesis. Z Naturforsch 24:1588–1598
84. Stockmayer WH (1943) Theory of molecular size distribution and gel formation in branched chain polymers. J Chem Phys 11:45
85. Torquato S, Truskett P, Debendetti P (2000) Is random close packing of spheres well defined? Phys Rev Lett 84:2064–2067
86. Toussaint D, Wilczek F (1983) Particle-antiparticle annihilation in diffusive motion. J Chem Phys 78:2642–2647
87. Tremmel IG, Kirchhoff H, Weis E, Farquhar GD (2003) Dependence of plastoquinol diffusion on the shape, size, and density of integral proteins. Biochim Biophys Acta 1607:97–109
88. Tremmel IG, Weis E, Farquhar GD (2005) The influence of protein-protein interactions on the organisation of proteins within thylakoid membranes. Biophys J 88:2650–2660
89. Tremmel IG, Weis E, Farquhar GD (2007) Macromolecular crowding and its influence on possible reaction mechanisms in photosynthetic electron flow. Biochim Biophys Acta 1767:353–361
90. Uversky VN, Cooper EM, Bower KS, Li J, Fink AL (2001) Accelerated α-synuclein fibrillation in crowded milieu. FEBS Letters 515:99–103
91. Vaz WLC, Almeida PFF (1993) Phase tolopogy and percolation in multi–phase lipid bilayers: is the biological membrane a domain mosaic? Curr Opinion Struct Biol 3:482–488
92. Verkman AS (2002) Solute and macromolecular diffusion in cellular aqueous compartments. TiBS 27:27–32
93. Williams WP (1998) Lipids in Photosynthesis: Structure, Function and Genetics, chapter: The physical properties of thylakoid membrane lipids and their relation to photosynthesis. Kluwer Academic Publishers, pp 103–118
94. Wollenberger L, Stefansson H, Yu S-G, Albertsson P-Å (1994) Isolation and characterization of vesicles originating from the grana margins. Biochim Biophys Acta 1184:93–102
95. Yakushevska AE, Jensen PE, Keegstra W, van Roon H, Scheller HV, Boekema EJ, Dekker JP (2001) Supermolecular organization of photosystem II and its associated light-harvesting antenna in *arabidopsis thaliana*. Eur J Biochem 268:6020–6028
96. Zimmerman SB, Minton AP (1991) Estimation of macromolecule concentration and excluded volume effects for the cytoplasm of *Escherichia coli*. J Mol Biol 222:599–620

97. Zimmermann SB, Minton AP (1993) Macromolecular crowding: biophysical, biochemical, and physiological consequences. Annu Rev Biophys Biomol Struct 22:27–65

Chapter 24
Higher Plants as Bioreactors. Gene Technology with C3-Type Plants to Optimize CO_2 Fixation for Production of Biomass and Bio-Energy

Fritz Kreuzaler, Christoph Peterhänsel, and Heinz-Josef Hirsch

RWTH Aachen University, Institute of Biology I, Worringer Weg 1, 52074 Aachen, Germany, Molbio@bio1.rwth-aachen.de

Abstract The application of biomass as a renewable energy source is currently a matter of lively discussions. Higher plants use photosynthesis as a non-exhaustible source for production of organic compounds. Virtually all life on earth depends on the energy provided by this process. Despite its importance, the photosynthetic process is associated with several flaws that limit the productivity of plants under agricultural growth conditions. The first part of this article describes these inefficiencies and strategies that were developed by plants and algae during evolution to overcome the problem. In the second part, an example of a successful bioengineering approach towards higher plant productivity is presented. Prospects for the use of higher plants in energy production are discussed.

24.1 Introduction

Almost all life on earth depends on sunlight and can be described in the words of an engineer as hydrogen technology: photons are produced by nuclear fusion in the sun and minute amounts of them reach the earth. H_2O is cleaved by photosynthesis into hydrogen and oxygen. Hydrogen is transferred to carrier molecules like $NADP^+$, NAD^+ and FAD and is then able to reduce CO_2 to carbohydrates or lipids and thereby stored as a source for energy [1]. In order to do some mechanical- or intellectual work energy is needed. To provide this energy, hydrogen is again mobilized and reacts with oxygen to produce H_2O and ATP a substance containing energy-rich bonds. These can be used for a variety of processes the most obvious of which is muscle movement. All the food we eat and all the fossil fuel we use is a product of photosynthesis. This important process is performed by many different organisms, ranging from bacteria to plants. Photosynthesis is composed of three stages: a) the primary photochemical reaction, splitting H_2O into H_2 and $1/2\ O_2$, b) the electron transport and photophosphorylation, and c) the CO_2 assimilation (Kaplan and Reinhold 1999).

In higher plants, three biochemical pathways are involved in CO_2 reduction and assimilation – the C3-pathway (Calvin-cycle), the C4-pathway (hatch-slack), and the crassulacean acid metabolism (CAM) (Hatch 1987). The C3- and the C4-pathway are more important than the CAM because the first two pathways fix more than 90% of CO_2 in higher plants.

The C3 CO_2 fixation mechanism is presumably the oldest pathway and developed at a time where no oxygen was present in the atmosphere. One of the most important and probably the most abundant enzyme in the world is the ribulose-1,5-bisphosphate carboxylase/oxygenase (Rubisco) which is located in the chloroplasts of plants. Rubisco catalyses the following reactions (Ogren 1984):

$$CO_2 + \text{ribulose-1,5-bisphosphate} \rightarrow 2 \times \text{3-phosphoglycerate(3-PGA)} \qquad (24.1)$$

as a carboxylation reaction and

$$O_2 + \text{ribulose-1,5-bisphosphate} \rightarrow 1 \times \text{3-PGA} + 1 \times \text{2-phosphoglycolate(2-PG)} \qquad (24.2)$$

as a second reaction, an oxygenase reaction.

In the Calvin cycle the two molecules 3-PGA of the carboxylation reaction are converted back to the acceptor molecule ribulose-1,5-bisphosphate and after six cycles one molecule of glucose has been formed in addition which is stored as starch or cellulose. The products of the oxygenase reaction are 3-PGA and 2-PG. 3-PGA can be converted back to ribulose-1,5-bisphosphate. Two molecules of 2-PG are converted to 3-PGA in a complex process by loss of one CO_2 in the mitochondria. This reaction is called photorespiration (Sage 2001). The released CO_2 may be re-captured and transported into the chloroplasts or it is lost from the plants.

The CO_2 concentration in the atmosphere is approximately 0.036% which is a bottleneck for plant growth. C3 plants possess a CO_2 assimilation mechanism, in which a large amount of already fixed CO_2 can be lost by photorespiration. C4 plants instead possess a CO_2 assimilation mechanism which shows no loss of CO_2 by photorespiration. These plants grow much better than C3 species specifically in warm and semi-dry climates and produce more biomass per time (Sharkey 2001).

During the next decades the human population will grow from about 6 to 9 billion. They have to be nourished. Biomass will also be used as renewable energy to produce fuels and biogas for chemical syntheses. Important crop plants show photorespiration. Therefore it would be of great benefit, to modify these plants from a C3-type kind of CO_2 fixation to plants with a C4-like mechanism.

It might also be possible to integrate new biosynthetic pathways into C3 plants which allow an efficient re-capture of CO_2 in the chloroplasts.

24.2 The C3 and C4 CO_2 Fixation Mechanisms

24.2.1 The C3 CO_2 Fixation Pathway, the Calvin Cycle and Photorespiration

More than 95% of the terrestrial plant species including mayor crops such as wheat and rice assimilate CO_2 exclusively by the C3-pathway and thus are named "C3 plants" (Leegood 1995). Figure 24.1 shows the CO_2 fixation mechanism in C3 plants.

CO_2 (0.036% in the atmosphere) reaches the area of fixation by passive diffusion: Into plant leaves through open stomata, into plant cells through plasma membranes, and into chloroplasts through the chloroplast membranes. Rubisco is present in high concentration in the chloroplasts stroma. CO_2 binds to the enzyme and is ligated to ribulose-1,5-bisphosphate to form a non-stable C6 intermediate (β-keto-acid). This branched sugar-acid is cleaved non-enzymatically into two molecules of 3-PGA. Two PGAs are transformed by transaldo/transketolase reactions to produce ribulose-1,5-bisphosphate in the so called Calvin cycle (Furbank and Taylor 1995).

If six CO_2 molecules are cycled through the Calvin pathway, one molecule of glucose is formed and stored as polysaccharides in starch grains. 3-PGA is also used to synthesize the other organic molecules required in the cycle. As mentioned before Rubisco, the key enzyme of photosynthesis catalyses both the carboxylation and the oxygenation of ribulose 1,5-bisphosphate (Edwards and Coruzzi 1989). CO_2 and O_2 compete for the same active site of the enzyme.

CO_2 assimilation occurs when ribulose-1,5-bisphosphate is carboxylated by Rubisco's carboxylase activity and the products, 2* PGA molecules are processed into carbohydrates and also used to regenerate ribulose 1,5-bisphosphate in a reaction sequence requiring ATP and $NADPH^+$.

Photorespiration begins with the oxygenation of ribulose-1,5-bisphosphate by the oxygenase activity of Rubisco to form one 3-PGA and one molecule of 2-PG (Fig. 24.1).

3-PGA directly enters the Calvin cycle (Fig. 24.2) whereas 2-PG enters the photorespiration pathway. 2-PG is toxic for the plant and has to be removed from the biosynthetic pathway. In some algae this is achieved by export of the toxic molecule but in higher plants two molecules of 2-PG are converted to one molecule of 3-PGA

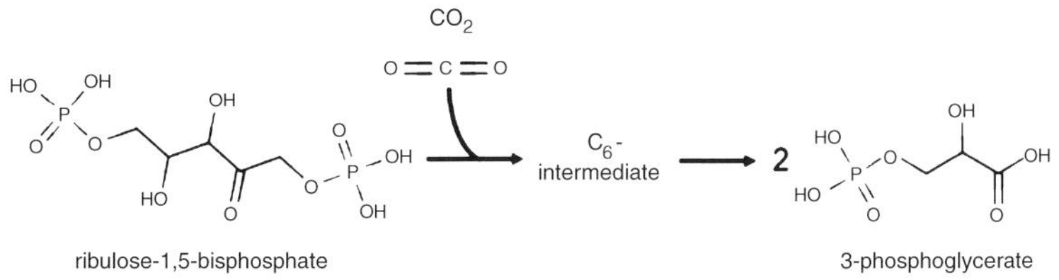

Fig. 24.1 CO_2 fixation in C3 plants. Ribulose-1,5-bisphosphate is carboxylated and an intermediate with 6 C-atoms is formed (2′-carboxy-3-keto-arabinitol-1,5-bisphosphate). This compound spontaneously decades to two molecules of 3-phosphoglycerate

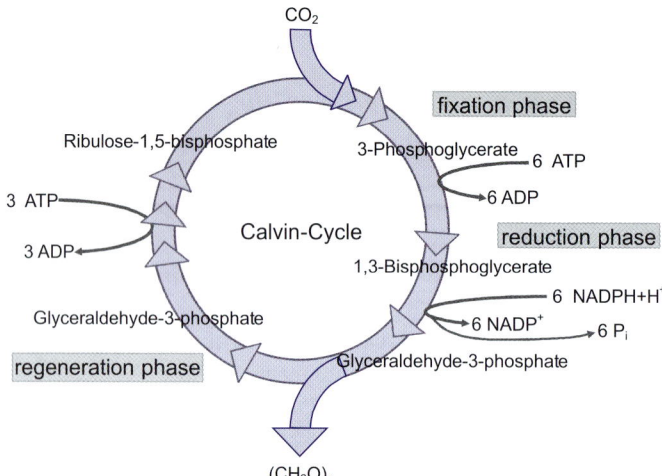

Fig. 24.2 The Calvin cycle. The calvin cycle serves the integration of the fixed carbon into the basal plant metabolism and the regeneration of the acceptor molecule. 3-phosphoglycerate is reduced to glyceraldehyde-3-phosphate. This process consumes reducing equivalents and ATP. The reduced compound can be either used for the regeneration of ribulose-1,5-bisphosphate or as a precursor for the synthesis of carbohydrates such as glucose, sucrose, and starch

and one molecule of CO_2 part of which is released to the atmosphere (loss of CO_2 in C3 plants). The whole pathway is shown in Fig. 24.3.

The oxygenation of ribulose 1,5-bisphosphate is a wasteful process. For every O_2 reacting with the C5 sugar the following tribute has to be paid by the plant:

- One molecule NADPH + H^+ to reduce 3-phosphoglycerate to glceraldehyde-3-phosphate (G3P).
- One quarter of the carbon which passes through the photorespiration pathway is lost to the atmosphere and has to be re-fixed (Edwards and Coruzzi 1989; Douce and Neuburger 1999).
- For two oxygenase reactions one NH_3-molecule is lost and has to be re-fixed at costs of ATP.

But at least three quarters of the carbon is stored in the cell and the starter molecule ribulose 1,5-bisphosphate can be synthesized. If oxygen levels are reduced from 21% to 2% or CO_2 concentration increased from 0.036 to 0.1% C3 plants can increase the net CO_2 assimilation by up to 50% (Evans and Von Caemmerer 1996).

The principal factors influencing the rate of photorespiration are the ratio of CO_2 to O_2 and the temperature (Arp et al. 1998). The ratio of the rate of photorespiration to photosynthesis in air (350 ppm CO_2, 21% O_2) at a temperature of 10 °C is approximately 0.1, rising to about 0.3 at 40 °C. Under water stress, when the stomata close, low intercellular concentrations of CO_2 can result in even higher ratios.

In summary, many problems on earth could be solved if the photosynthetic productivity of C3 plants could be improved. More biomass could be produced on the same space, used as fuel, food and renewable energy.

24.2 The C3 and C4 CO$_2$ Fixation Mechanisms

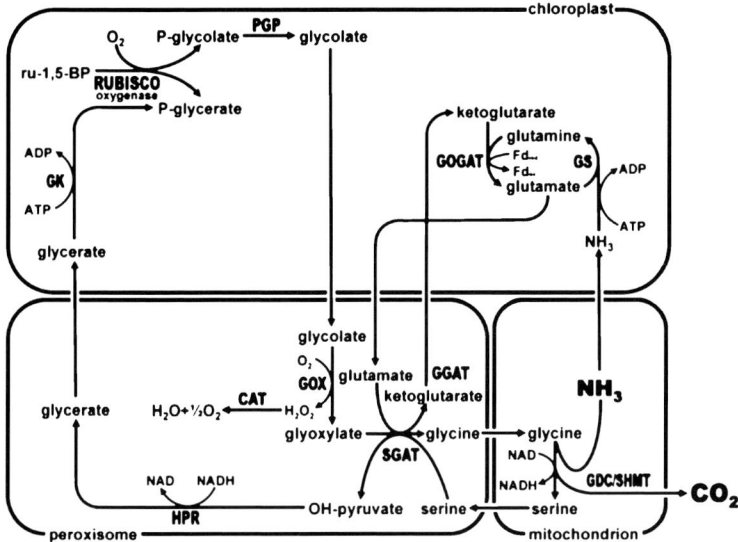

Fig. 24.3 The photorespiratory pathway. Oxygenation of ribulose-1,5-bisphosphate leads to the formation of phosphoglycolate, a toxic compound inhibiting photosynthesis. Phosphoglycolate is recycled in the process of photorespiration. Each on molecule of CO$_2$ and NH$_3$ are lost during the mitochondrial conversion of two molecules of glycine to one molecule of serine. This is a net loss of afore fixed carbon and nitrogen that can limit plant growth. Rubisco, ribulose-1,5-bisphosphate carboxylase/oxygenase; PGP, phosphoglycolate phosphatase; GOX, glycolate oxidase; CAT, catalase; GGAT, glyoxylate/glutamate aminotransferase; GDC/SHMT, glycine decarboxylase/serine hydroxymethyl transferase complex; SGAT, serine/glyoxylate aminotransferase; HPR, hydroxypyruvate reductase; GK, glycerate kinase; GS, glutamine synthetase; GOGAT, glutamate/ oxoglutarate aminotransferase

24.2.2 The C4 Photosynthetic Pathway

On an evolutionary scale, the C4 type CO$_2$ fixation is much younger than the C3 mechanism. It developed during the tertiary when the CO$_2$ concentration decreased from about 1500 to about 300/350 ppm (Sage 1999).

C4 plants have an advantage over C3 species because the CO$_2$ fixation mechanism promotes ecological success in warm (20 – 30 °C), low latitude habitats (Sage 2001, 2004). They have higher maximum efficiency in terms of radiation, water and nitrogen use, and generally they have higher photosynthetic capacity. The enhancement of photosynthetic performance results from the ability of C4 plants to increase the CO$_2$ concentration [CO$_2$] in the vicinity of Rubisco up to saturation. As mentioned before CO$_2$ and O$_2$ compete for the same active centre of the enzyme, but at high [CO$_2$] in the vicinity of Rubisco CO$_2$ bind almost exclusively to the active site. C4 species posses all enzymes for photorespiration but flow of carbon through this pathway – starting with 2-phosphoglycolate – usually is hardly detectable.

C4 plants possess two cell types, mesophyll- and bundle sheath cells, both containing chloroplasts. These two cell types form two layers around the vascular tissue: The bundle sheath cells (BSC) forming the inner and the mesophyll cells (MC) the outer layer. MC and BSC are connected to each other by many plasmodesmata which allow an extensive exchange of small molecules. This arrangement of cells is known as Kranz-anatomy and is schematically presented in Fig. 24.4.

The primary CO_2 (as HCO_3^-) fixation in the C4 pathway is catalysed by phospho*enol*pyruvate carboxylase (PEPC) forming the C4 dicarboxylic acid oxaloacetate (OAA).

OAA can be subsequently converted to malate (NADP-ME type CO_2 fixation as shown in Fig. 24.4) or aspartate (in NAD-ME and PCK CO_2 fixation types) (Edwards and Walker 1983) and directly or indirectly decarboxylated in the vicinity of Rubisco where CO_2 is re-fixed by Rubisco. The primary fixation by PEPC and refixation by Rubisco are separated spatially between the mesophyll and the bundle sheath cells.

Compartmentalisation of the enzymes forming the C4 pathway is regulated mainly at the transcriptional level (Sheen 1999). For incorporation of one CO_2 molecule into carbohydrates C4 plants require more energy compared with C3 species (3.5 ATP in C4 plants and 2 ATP in C3 plants). What are then the benefits for C4 when competing with C3 plants in a certain area and in a specific climate? In moderate or warm climates and under a medium- or high light regime the amount of ATP does not seem to be limited for plant growth. The bottle-neck for plant growth under these conditions seems to be the [CO_2] which is directly dependent to the water supply. C3 grasses have an advantage over C4 grasses, if during growing seasons daytime temperature is about 20 °C on an average and the [CO_2] is about 320 ppm

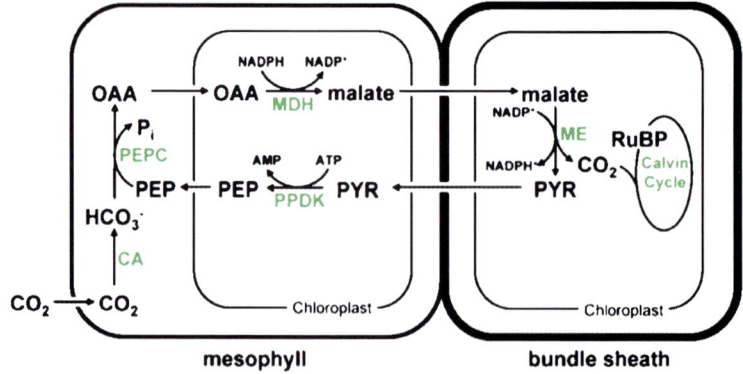

Fig. 24.4 C_4-type CO_2 concentration mechanism. C_4 plants developed means to reduce oxygen fixation by enhancing the CO_2 concentration in the vicinity of Rubisco. Primary carbon fixation takes place in the mesophyll by an oxygen-insensitive enzyme. The resulting C_4 acid diffuses into the bundle sheath where it is decarboxylated and CO_2 is released. Bundle sheath cells have tight cell walls and little contact to the intercellular air space. CO_2 can therefore efficiently be concentrated in this cell type. The C_3 acid resulting from the decarboxylation reaction diffuses back into the mesophyll where the primary acceptor molecule is regenerated. CA, carbonic anhydrase; PEPC, phosphoenolpyruvate carboxylase; MDH, malate dehydrogenase; ME, malic enzyme; PPDK, pyruvate orthophosphate dikinase; PEP, phosphoenolpyruvate; OAA, oxaloacetate; PYR, pyruvate

or at 30 °C and a [CO$_2$] of 560 ppm. C4 grasses have an advantage if [CO$_2$] is about 290 ppm at a temperature of 20 °C or 430 ppm at 30 °C (Leegood and Walker 1999). The picture is different, if plants are under water stress during the growing season. If water irrigation is suboptimal, C4 plants have an advantage over C3 species even at 20 °C and 320 ppm [CO$_2$]. These data are supported by the finding that C4 plants developed about 15 million years ago when the CO$_2$ concentration decreased from between 1000–1500 ppm during the cretaceous time to between 490 to 200 ppm at the end of the Miocene (Cerling 1999).

C4 dicotyledons are not as abundant as C4 monocotyledons. From approximately 10,000 grass species 50% possess C4 photosynthesis and C4 CO$_2$ fixation. The grasses developed at the end of the Cretaceous about 70 million years ago. Less than 5% of the dicots use the C4 pathway. We can only speculate why only a few dicotyledonous plants possess the C4 CO$_2$ fixation mechanism. A few but very important crop plants like maize and sugar cane belong to the C4 species. In tropical- or even in modest climates or under suboptimal irrigation these C4 species can produce high amounts of biomass. The world population but also the climate on earth would benefit, if the CO$_2$ fixation mechanism of important crop plants like wheat, barley, rice, potato or beets could be modified from C3 to C4. More food and animal feed could be harvested and more renewable energy could be produced.

24.3 The Metabolism of Glycolate in *Escherichia coli* and in Some Green Algae

24.3.1 *The Metabolism of Glycolate in E. coli*

Glycolate is a compound which is produced in nature in very large quantities. As we have mentioned earlier (Sect. 24.2.1) the oxygenase reaction of the Rubisco produces one molecule 3-phosphoglycerate (3-PGA) and one molecule 2-phosphoglycolate (2-PG). This organic acid possesses on alcoholic group which still contains endogenous energy for the production of ATP. Many bacteria therefore have evolved biochemical pathways to metabolize 2-PG, to produce energy rich bonds during catabolism or cell material during anabolism (the anaplerotic pathway starting with 2-PG is a follows: 2-phosphoglycolate → glycolate → glyoxylate → glycine → serine → amino acids, pyruvate, sugars, lipids). For *E. c.* the anabolic- and catabolic pathways have been described in detail (Hansen and Hayashi 1962; Kornberg and Sadler 1961). The first enzyme involved in the glycolate pathway is the glycolate dehydrogenase (GDH) which produces glyoxylate described in Eq. (24.3)

$$\text{Glycolate} + \text{NAD}^+ \rightarrow \text{Glyoxylate} + \text{NADH} + \text{H}^+ \,. \tag{24.3}$$

In the photorespiration pathway oxidation of glycolate to glyoxylate is catalysed by glycolate oxidase inside the glyoxysomes. The products of this reaction are glyoxylate plus H_2O_2 (Eq. (24.4)).

$$\text{Glycolate} + 2O_2 \rightarrow \text{Glyoxylate} + 2H_2O_2 \qquad (24.4)$$

H_2O_2 is poisonous for a cell but is detoxified by the enzyme catalase (Eq. (24.5)).

$$H_2O_2 \rightarrow H_2O + 1/2 O_2 \,. \qquad (24.5)$$

The oxidation of glycolate to glyoxylate does not occur in the chloroplast because H_2O_2 is very toxic, but instead in the glyoxysomes.

Glycolate dehydrogenase (GDH) from *E. coli* is a heterotrimeric complex encoded by the three genes *glc*D, *glc*E, *glc*F, which are located in the *glc* operon (Lord 1972; Pellicer et al. 1996). From glyoxylate two biosynthetic pathways can branch off: 1. Glyoxylate reacts with Acetyl-CoA, catalysed by malate-synthase to form malate. This reaction is part of the glyoxylate-cycle, an important anaplerotic pathway. 2. Two molecules glyoxylate ($2 \times C2$) are ligated, catalysed by glyoxylate carboligase (GCL). Tartronic semialdehyde ($1 \times C3$) is formed and CO_2 is released during this reaction (Eq. (24.6)) (Chang et al. 1993).

$$\text{glyoxylate}(C2) + \text{glyoxylate}(C2) \rightarrow \text{tartronic semialdehyde}(C3) + CO_2 \qquad (24.6)$$

Tartronic semialdehyde (C3) is further reduced to glycerate by the enzyme tartronic semialdehyde reductase (TSR) and can subsequently be phosphorylated to 3-phosphoglycerate (3-PGA) by the enzyme glycerate kinase (Gotto and Kornberg 1961; Nelson and Tolbert 1970). 3-PGA is located in the centre of many biosynthetic pathways and can be used for the production of ATP but also for anaplerotic reactions (e. g. formation of glucose). The three enzymes (glycolate dehydrogenase (GDH), glyoxylate carboligase (GCL) and tartronic semialdehyde reductase (TSR)) constitute what is known as the glycerate pathway in E. coli (Gotto and Kornberg 1961).

24.3.2 The Metabolism of Glycolate in Green Algae

As mentioned before glycolate is oxidized to glyoxylate in higher plants in the glyoxysomes by oxygen. H_2O_2 is formed during this process and the toxic compound is detoxified by the enzyme catalase. Some algae possess alternative routes of photorespiratory glycolate metabolism. In many algae glycolate is oxidized to glyoxylate by the enzyme glycolate dehydrogenase on the inner mitochondrial membrane (see Eq. (24.2)). In theory, two molecules of ATP can be formed from NADH and H^+ by phosphorylation of ADP in the respiratory chain. This is energetically less wasteful than the mechanism in terrestrial plants involving glycolate oxidase. Some green algae (chlorophyceae) and some cyanobacteria possess the same mechanism for producing glycerate from glycolate as has been demonstrated in E. coli. This mechanism is shown in the Eqs. (24.5), (24.6) and (24.7) (Eqs. (24.7)–(24.9)) (Anderson and Beardall 1991).

$$\text{glycolate} + NAD^+ \rightarrow \text{glyoxylate} + NADH + H^+ \quad (24.7)$$

$$2 \times \text{glyoxylate} \rightarrow \text{tartronic semialdehyde} + CO_2. \quad (24.8)$$

Tartronic semialdehyde is reduced to glycerate in a second reaction, catalysed by tartronic semialdehyde reductase (Eq. (24.9)).

$$\text{Tartronic semialdehyde} + NADH + H^+ \rightarrow \text{Glycerate} + NAD^+. \quad (24.9)$$

In the photorespiration pathway found in higher plants CO_2 and NH_3 is released in the mitochondria (see Fig. 24.3) during the formation of N^5, N^{10}-methylene-tetrahydrofolate (Eq. (24.10)).

$$\begin{aligned}&\text{Glycine} + NAD^+ + \text{tetrahydrofolate} \rightarrow \\ &N^5, N^{10} - \text{methylene-tetrahydrofolate} + CO_2 + NH_3\end{aligned} \quad (24.10)$$

This is wasted ATP since we need energy to re-assimilate CO_2 (CO_2 is often a bottleneck for plant growth) and for re-fixation of nitrogen. As mentioned earlier in this chapter, glycolate is oxidized in the inner of the mitochondrial membrane. CO_2 is then released too in mitochondria. This situation is not optimal. If the whole glycolate acid pathway would be located in the chloroplasts, CO_2 would be released in the same compartment, where Rubisco is localized. CO_2 could be re-fixed easier and energy losses should be very low. Such a mechanism would even be better than the one in green algae.

24.4 Increased Biomass Production in Transgenic *Arabidopsis* Plants Containing the *E. coli* Glycolate Pathway in the Chloroplasts

24.4.1 The Strategy to Improve CO_2 Fixation and Hence Photosynthesis

As mentioned earlier, most of our important crop plants possess the C3 CO_2 fixation mechanism in which primary CO_2 fixation is catalysed by the enzyme Rubisco. The protein complex is localized inside the chloroplast and catalyzes both, the carboxylation and the oxygenation of ribulose-1,5-bisphosphate (Andrews and Lorimer 1987). The equilibrium between these two activities depends mainly on the CO_2/O_2 ratio in the leaves. We pointed out earlier, that the oxygenase reaction only has disadvantages for the plant. One molecule of CO_2 is released in the mitochondria for every two molecules of 2-phosphoglycolate (2-PG) produced, a net loss of fixed carbon that reduces the production of carbohydrates and biomass. NH_3 is also released in this reaction and needs to be re-fixed through energy-consuming reactions in the chloroplasts (Fig. 24.3). The negative impact of photorespiration on plant growth and biomass yield has been demonstrated by experiments, where the CO_2 concentration has been increased from about 0.036 to 0.1%. This dramatically increases growth of several crops (Long et al. 2006). However, such a strategy cannot be implemented in the field, where losses by photorespiration are exacerbated by high temperature and suboptimal water supply. Under these conditions the stomata close, no CO_2 can enter the cells and the balance between CO_2 and O_2 increases in favor of O_2. In this situation photorespiration increases and the plants loose even more CO_2. We only can overcome this problem by modification of the photorespiratory pathway or of the CO_2-fixating enzyme, Rubisco. Many attempts have been made to change the characteristics, the oxygenase activity of Rubisco. These experiments were not successful. If it would be possible to mutate the important enzyme so that it gains a lower K_m value for CO_2 and a higher half substrate saturation point for O_2, the problems with photorespiration would be solved. Mutation experiments failed in which scientists tried to improve the carboxylation – and at the same time decreased the oxygenation reaction. Up to now, the only possibility to optimize the balance between CO_2 fixation and oxygenation is to increase the CO_2 concentration in the vicinity of the Rubisco. This could be done by the following strategy:

- Integrating a carbonic anhydrase (CA) into the inner chloroplast membrane, which pumps CO_2 or HCO_3^- from the cytoplasm into the chloroplast.
- Expressing a CA in the chloroplasts, with an affinity to Rubisco. This CA produces CO_2 in the vicinity of Rubisco and inhibits photorespiration.
- Directing the following bacterial enzymes to the chloroplasts: glycolate dehydrogenase (GDH), glyoxylate carboligase (GCL) and tartronic semialdehyde reductase (TSR). In this case the photorespiration is placed in the chloroplast and CO_2 is released and increased in the vicinity of Rubisco.

24.4.2 Establishment of the E. coli Glycolate Pathway in Arabidopsis Chloroplasts

24.4.2.1 The General Strategy

The proposed biochemical pathway consists of three enzymatic functions, GDH with three polypeptides (GlcD, GlcE, GlcF), GCL, and TSR activities, with one polypeptide each. Cloning into a plasmid and integration of intact GDH seemed to be the most problematic. Therefore it was decided that *TSR* and *GCL* genes should be transferred first to the nuclear genome of *A. thaliana* wild type plants. To achieve this, all the genes necessary for the establishment of the novel biochemical pathway were cloned into different plant expression vectors and fused with the N-terminus to a chloroplast targeting peptide (cTP). The expression of the foreign genes in all vectors used is under the control of a derivative of the constitutive Cauliflower Mosaic Virus (CaMV) 35S promoter.

24.4.2.2 Generation of GCL, TSR and GCL-TSR Transgenic *A. thaliana* Plants

The *TSR* and *GCL* genes were cloned separately and as a tandem into the plant expression vector pTRA-K-rbcS1-cTP (Fig. 24.5). The expression of GCL in transgenic plants was measured on the protein level by western blot, using an antibody specific for the His-tag, whereas the expression of *TSR* was analyzed using an anti-TSR antibody which was isolated from chicken after injecting the isolated protein from bacteria (data not shown).

A 6xHis-protein marker was used to determine the molecular weight of the proteins. A single band was visible in some samples corresponding in size to the GCL protein fused to the His-tag (65 kDa; data not shown). The TSR protein was detected with a specific antibody prepared from the eggs of immunized hens. The analyzed plants that showed expression for GCL also showed expression for TSR. Two transgenic plants were used to generate homozygous lines by selfing plants expressing *GCL* and *TSR*.

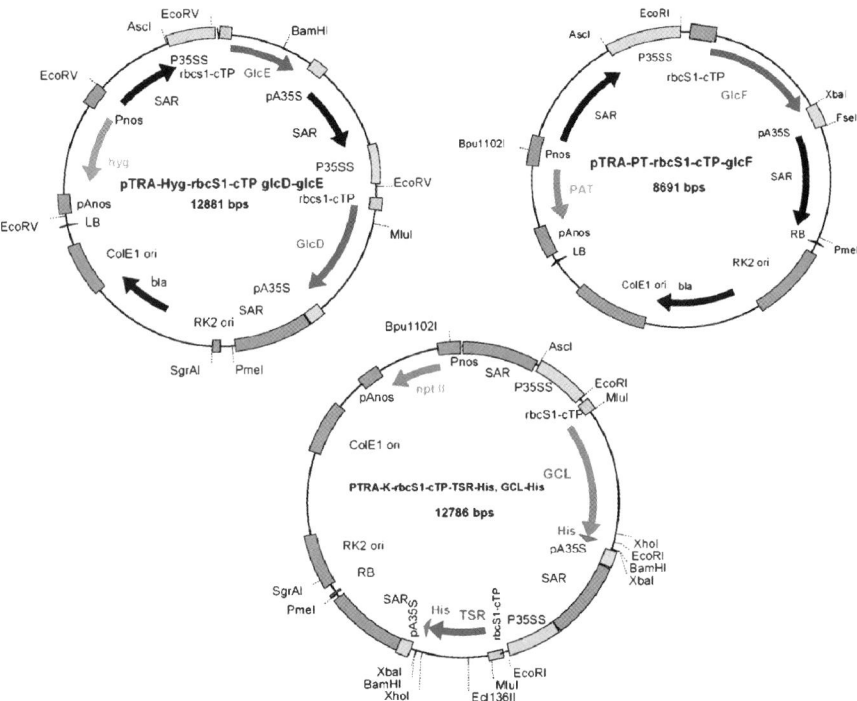

Fig. 24.5 Vector constructs used in this study. The genes are as given in the text. All constructs contain the 35S promoter and the chloroplast targeting sequence of the rbcs1 gene from potato. The genes are flanked by scaffold attachment regions to stabilize gene expression. Each vector construct contains a different resistance marker to allow multiple re-transformations of transgenic plants. Stable transformation of *Arabidopsis thaliana* plants, ecotype Columbia (Col-0), with constructs carrying both genes, together and separately, was carried out by *Agrobacterium* mediated floral dip transformation, details of which are described in Chap. 24.4.2.5

24.4.2.3 Generation of Transgenic *A. thaliana* Plants Expressing the Genes glcD, glcE and glcF

The *E. coli* glycolate dehydrogenase enzyme is formed from three different polypeptides which are encoded by three different open reading frames named *glc*D, *glc*E and *glc*F (Lord 1972; Pellicer et al 1996). To build up a bacterial glycolate dehydrogenase activity inside the chloroplasts of *A. thaliana*, the following strategy was used:

- The coding sequence for GlcF subunit of the *E. coli* GDH operon was cloned into the plant expression vector (pTRA-PT-rbcS1-cTP) that confers resistance to phosphinothricin (BASTA). *A. thaliana* plants were transformed with this plasmid and gene expression was tested and homozygous lines were generated.
- The *glc*D and *glc*E genes were cloned in tandem into the plant expression vector (pSuper-PAM-Sul-rbcS1-cTP) which confers resistance to sulfadiazine antibiotics.
- The homozygous lines expressing *glc*F were retransformed with the double construct containing *glc*D and *glc*E (pSuper-PAM-Sul-rbcS1-cTP-GlcD, GlcE). This final step resulted in the production of transgenic plants that express all *E. coli* GDH genes (i. e. *glc*D, *glc*E and *glc*F genes) called DEF-plants.

16 DEF transgenic plants were analyzed at the RNA level by RT-PCR. In five of the tested plants all three transgenes were expressed. However, only plant 5 and 7 showed comparable expression levels for all three transgenes. For the majority of all plants, the *glc*F expression was clearly lower compared to the expression of *glc*D and *glc*E. The lines DEF 5 and DEF 7 were chosen to be crossed to the offspring of GT1-4-5 homozygous GT-transgenic plants in order to complete the installation of the novel biochemical pathway in *A. thaliana* plants. The *E. coli* glycolate dehydrogenase using an organic compound as an electron acceptor ($NADP^+/NADPH + H^+$) preferentially oxidizes D(−) over L(+) lactate as an alternative substrate and is sensitive to cyanide. In contrast, higher plant peroxisomal glycolate oxidase using oxygen as electron acceptor, oxidizes preferentially L(+) lactate over D(−) lactate and is insensitive to cyanide. In order to find out whether the GlcD, GlcE and GlcF protein subunits are correctly assembled in the chloroplasts of DEF plants and form an active enzyme, the following assays were made: Glycolate dehydrogenase (+/− KCN), D-lactate dehydrogenase (+/− KCN). The enzyme was active (data not shown).

24.4.2.4 Construction of GT-DEF Plants

The generation of transgenic lines is illustrated in Fig. 24.6.

- A plasmid construct containing the genes *GCL*, *TSR* and *NPT*II, the gene for kanamycin resistance, was transferred into the nuclear genome of *A. thaliana* plants by *Agrobacterium* mediated floral dip transformation. Homozygous lines were generated expressing GCL and TSR. The enzymatic activity of both enzymes was measured.

24.4 Increased Plant Biomass

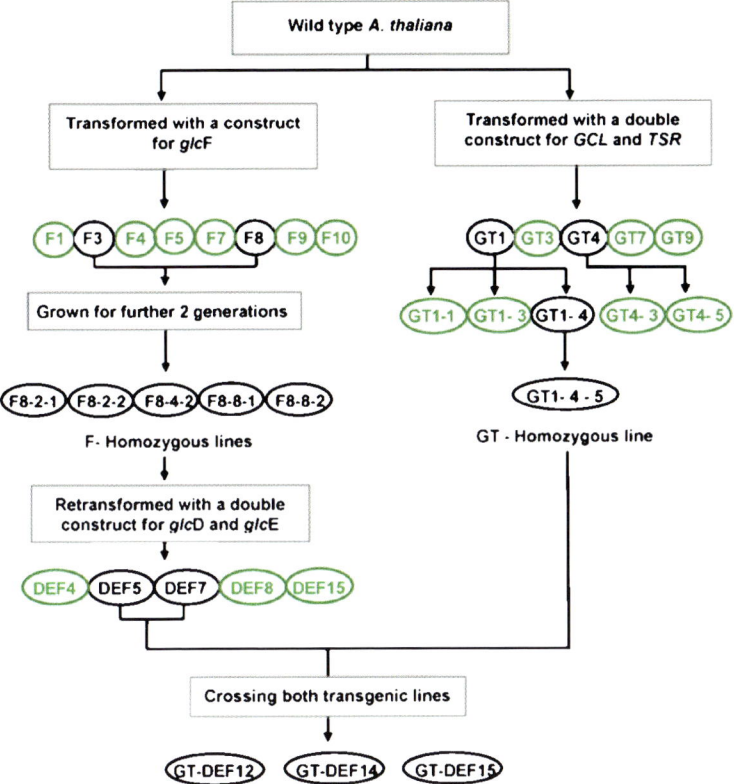

Fig. 24.6 Establishment of transgenic lines containing all vector constructs. *Transgenic lines* overexpressing the complete pathway were generated by a combination of genetic crossing and retransformation as indicated in the figure. *Lines in black circles* were used for pathway establishment, *lines in green circles* serve as controls

- An *Arabidopsis* line was made expressing the three genes for *glc*D, *glc*E and *glc*F subunits of *E. coli* GDH. This line confers resistance to phosphinothricin and to sulfadiazine. The enzymatic activity of the GDH was measured.
- Two off-springs of the DEF (GDH) lines DEF 5 and DEF 7 (in those lines the genes *glc*D, *glc*E and *glc*F were expressed to almost the same level) were crossed to the descendants of GT (GT1-4-5) in order to generate transgenic plants expressing all genes necessary to establish the novel photorespiration pathway in the chloroplasts of *A. thaliana*

24.4.2.5 Plant Transformation

Genetic transformation is a prerequisite for a fast, efficient and stable modification of plant cells and plants. During the last 30 years methods have been improved continuously (Vain 2007) though still today, for many applications there is a need for more efficiency of the transformation process, more precision with regard to the integration sites of the foreign DNA and higher yield of recombinant proteins or secondary metabolites. From a methodical point of view three kinds of transformation are applied: a) Transformation by *Agrobacterium tumefaciens* or viral vectors, biological methods b) transformation by particle bombardment, microinjection or electroporation, physical methods and c) transformation of protoplasts, a chemical method (for reviews see: (Newell 2000; Birch 1997).

The cheapest and most employed method is the transformation with *Agrobacterium tumefaciens*. Originally it was confined to dicotyledons. During the last years modifications of the original methods have overcome a variety of obstacles, so that many important monocotyledons are successfully transformed since the mid-1990s. Plant cells growing in liquid suspension cultures are easily transformed by *A. tumefaciens*. Even algae (Kumar et al. 2004), fungi (de Groot et al. 1998) and mammalian cells (Kunik et al. 2001) are now accessible to this technique.

The *A. tumefaciens* procedure used in our approach for *A. thaliana* transformation is briefly outlined:

The basic cloning steps like fusing plant regulatory functions to coding regions and combining different gene constructs in one plasmid were performed in *Escherichia coli*. The plasmids constructed were derivatives of pPAM (gi13508478) with resistant genes for selection in *E. coli* (ampicillin), *A. tumefaciens* (GV3101) (carbenicillin) and the plant cells (kanamycin, sulfadiazine or phosphinothricin). Plasmids were then transferred to *A. tumefaciens* by electroporation. After three to four days of growth on selection medium in Petri dishes, colonies were checked for the presents of the correct plasmids by PCR. Positive lines are propagated in medium or on fresh plates and suspended in a sucrose solution containing the wetting agent (e. g. Silvet L-77). Floral buds of *A. thaliana* were wetted with the bacterial suspension either by spraying, by dipping the inflorescence into the solution or just by treating the floral buds with tiny droplets of the solution (Clough and Bent 1998; Logemann et al. 2006). Essential for successful transformation seemed to be the developmental stage of the floral buds and the use of a wetting agent. A second treatment of the floral buds may follow after some days. Ripe seeds were collected, surface-sterilized with 70% and 96% ethanol and allowed to germinate on an agar medium. Up to three percent of the seeds may develop into transgenic *Arabidopsis* plants.

24.5 Analysis of the GT-DEF Transgenic Plants. DNA, RNA, Proteins, Physiology, Growth and Production of Biomass

Crossing between GT-homozygous and DEF-plants resulted in the production of GT-DEF transgenic plants expressing all the five polypeptides involved in this pathway. All three enzymes (GDH, GCL and TSR) were active.

DNA from GT-DEF plants was isolated and analyzed by PCR-methods. A multiplex system was developed and optimized in order to test many transgenic plants

24.5 Transgene Analysis

in a short time with low effort. A mixture of plasmids carrying the GDH, TSR and DEF genes was used as a positive control. Some plants showed positive signals for all five genes (data not shown). Other plants showed only bands corresponding to GCL, TSR and glcF. In total 26 plants were analyzed by this method. Three of them contained all genes (GT-DEF) in their genome (GT-DEF 12, 14, 15).

Expression of the genes *GCL*, *TSR*, *glc*D, *glc*E, and *glc*F were tested by RT-PCR (data not shown). All the three GT-DEF plants showed varying expression levels for each single gene. The expression levels of *glc*F, *GCL* and *TSR* are lower compared to *glc*D and *glc*E. The reason for this is not known. During the construction of a new transgenic plant generation methods have to be developed the all genes are expressed approximately with the same rate.

Protein extracts from isolated chloroplasts were isolated and the enzymatic activities for GDH, GCL and TSR were measured. Cell lines showing expression of the genes involved in the new pathway also showed positive enzymatic activities.

In order to check the efficiencies of the novel pathway in GT-DEF plants the ratio of the amino acids glycine to serine was measured. The Gly/Ser ratios in C3 plants in considered to as a marker for the rate of photorespiration. Wild type- and transgenic GT-DEF plants were grown under ambient (350 ppm) and low (100 ppm) CO_2 concentrations. The amounts of glycine and serine from leaf samples were quantified using gas chromatography and mass spectroscopy (GC/MS). When grown under ambient conditions transgenic and wild type plants showed more or less similar levels of Gly/Ser ratios. By growing the same plants in low CO_2 concentration (100 ppm; i.e. enhanced photorespiration) for four hours the Gly/Ser ration increased in WT plants from 0.5 to 4.4 and in GT-DEF plants from 0.6 to 2.0. These results clearly demonstrate that photorespiration is decreased to about 50% in GT-DEF compared to wild type plants.

Ammonia is released in the photorespiratory pathway during the conversion of glycine to serine in the mitochondria. The released ammonia is re-fixed by the GS-GOGAT (GS = Glutamine synthase; GOGAT = Glutamate: glyoxylate aminotransferase) (Tachibana et al. 1986; Lacuesta et al. 1989; Wild and Ziegler 1989). It has been shown that the herbicide phosphinothricin irreversibly inhibits glutamine synthase activity and as a result ammonia accumulates inside the plant leaf tissue. A multi-well ammonium evolution bioassay was introduced by (Deblock et al. 1995) which allows a quantitative assessment of the ammonia accumulated in the plant leaf tissues. The assay works on the basis that leaf tissues incubated in medium supplemented with phosphinothricin are not able to assimilate ammonium. The ammonia accumulating in the leaf tissue diffuses into the surrounding medium where a colorimetric reaction indicates the ammonium concentration. Under phosphinothricin treatment higher ammonia accumulation means higher photorespiratory rates and vice versa. Based on these facts it was decided to measure the amounts of ammonia released from transgenic plant leaf tissues as another marker for the rate of photorespiration. Plants transgenic for the new photorespiration pathway show less ammonia release compared to the wild type plants (ng ammonia/mg leaf material: wild type: 220 ng; transgenic plants: 80 ng). These results also show that the photorespiration rate is reduced in transgenic plants.

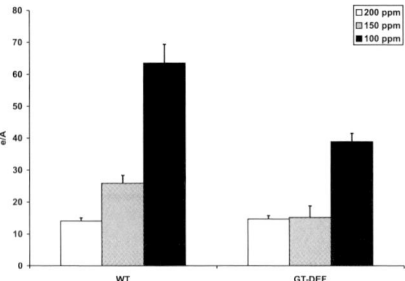

Fig. 24.7 Electron use for CO_2 fixation. Transgenic lines overexpressing the complete glycolate catabolic pathway from *E. coli* (GT-DEF) require less energy compared tot the wild type (WT) for CO_2 fixation under conditions were the CO_2 concentration is limiting photosynthesis

Determination of the electron requirements for CO_2 assimilation (e/A) in plants. In order to study the physiological consequences of the novel pathway we decided to measure chlorophyll a fluorescence parameters. Chlorophyll fluorescence measurements help to determine the electron requirements for CO_2 assimilation (e/A). The e/A is correlated with the carboxylation/oxygenation ratio of Rubisco. The e/A decreases when the rate of Rubisco oxygenation is decreased relative to carboxylation and reaches minimum values in the absence of photorespiration. The e/A therefore is a very helpful tool for measuring the rate of Rubisco carboxylation in plants. At 200 ppm external [CO_2] the e/A values are similar in all the plants tested (see Fig. 24.7). The transgenic plants showed a decrease in the e/A values compared to wild types when plants are grown at 150 and 100 ppm. GT-DEF plants always have the lowest e/A values compared to other plants tested at 150 and 100 ppm CO_2. Taken together the results clearly demonstrate that over-expression of the novel pathway genes in the chloroplasts of *A. thaliana* enhances plant photosynthesis.

24.5.1 Enhanced Biomass Production in DEF and GT-DEF Plants

Representative photographs of wild type, DEF and GT-DEF plants are shown in Fig. 24.8. It is apparent from the photograph that the transgenic lines DEF and GT-DEF both display a significant increase in size. This first view can be confirmed by a more detailed analysis of parameters like rosette diameter, leaf area, total fresh, and the increase of weight during a defined period. The average rosette diameters of 8 weeks old DEF and GT-DEF plants was 11 cm, those of wild type plants of the same age were 7 cm. Leaf area increased from 5.7 cm^2 for wild type to 6.9 and 7.3 cm^2 in DEF and GT-DEF respectively, determined on seven weeks old plants. The most pronounced difference between wild type and the transgenic plants was observed when looking at the increase of weight. After 6 weeks the dry weights of shoots were nearly 40% higher in DEF plants and more than 70% higher in GT-DEF plants than in wild type plants. The effect was even more dramatic for the root system: 150% and 230% more weight in DEF and GT-DEF plants, respectively.

Fig. 24.8 Phenotype of transgenic lines. Transgenic plants show enhanced growth of both leaf and root tissues compared to the wild type (WT). This effect is already partially obtained by installation of glycolate dehydrogenase activity (DEF), but enhanced by establishment of the complete glycolate catabolic pathway (GT-DEF)

It is clear from these experiments that some limits for growth have been overcome in the transgenic DEF plants and to a larger extent in the GT-DEF plants. Yet, besides the glycine/serine ratio only a few metabolite concentrations have been determined. While the level of starch per leaf area was not changed, sucrose levels increased from 330 μmol m^{-2} in wild type to 370 μmol m^{-2} in DEF plants and 400 μmol m^{-2} in GT-DEF plants. How exactly the metabolic fluxes inside the transgenic plants are influenced and shifted to generate such striking growth enhancement remains to be examined in the future.

At least it seems clear that CO_2 fixation capacity is accessible to bioengineering. As mentioned at the beginning of this article, the evolution of improved CO_2 fixation mechanisms is a relatively young invention of nature and occurred independently in different plant families. Increasing the CO_2 concentration at the site of Rubisco is the common feature which is achieved in nature by various forms of primary CO_2 fixation mechanisms. Enhancing the CO_2 concentration at the site of Rubisco seems to be one of the essential steps. This would suggest that steps towards these pathways must be relatively simple and should be accomplished by bioengineering, too.

24.6 Summary

As we have pointed out in the introduction the demand for food, feed and biomass for energy conversion will increase dramatically in the next decades. Today, already the prices for crops but also lumber increase because the energy industry is buying large amounts of biomass. We can face these economic developments in different ways:

The most important possibility to limit the demand for biomass is to save energy. Experts declare – and this finds public acceptance – that a lot of energy can be saved. This will only be done, if the prizes are high enough. Too high prices for energy could strangle the economy. Therefore we are on a small track.

The second possibility to damp the increase of prices for biomass is to produce additional amounts of plant material. Several possibilities seem to exist to reach this goal.

- Improvement of productivity on farmland by better fertilization. This can be done in several parts of the world, but in most agricultural areas fertilization reaches top levels (Europe, United States, Japan, parts of Asia, Australia).
- Improvement of the productivity of cultivated plants by classical breeding. Experts believe that for example the productivity of rice can neither be improved by classical breeding nor by optimized agricultural methods.
- Production of biomass on devastated land which has been re-cultivated. Experts believe that there are about 4 million km^2 of devastated land which might be re-cultivated, but the price for this is enormous.
- Another possibility to produce more biomass is to use gene technology to improve the efficiency of photosynthesis and CO_2-fixation. We believe that there are many possibilities to modify metabolism in plants to produce more biomass and to improve the quality of this biomass to store energy.

References

1. Anderson JW, Beardall J (1991) Molecular Activities of Plant Cells, pp 185–94, Blackwell, Oxford
2. Andrews TJ, Lorimer GH (1987) RUBISCO: structure, mechanism and prospects for improvement. In: Hatch MD, Boardman NK (eds) Photosynthesis, pp 131–218, Academic Press, New York
3. Arp WJ, Van Mierlo JEM, Berendse F, Snijders W (1998) Interactions between elevated CO2 concentration, nitrogen and water: effects on growth and water use of six perennial plant species. Plant Cell Environ 21:1–11
4. Birch RG (1997) Plant Transformation: Problems and Strategies for Practical Application. Annu Rev Plant Phys 48:297–326
5. Cerling TE (1999) Paleorecords of C4 plants and ecosystems. In: Sage RF, Monson RK (eds) C4 Plant Biology, pp 445–69, Academic Press, San Diego
6. Chang YY, Wang AY, Cronan JE Jr (1993) Molecular cloning, DNA sequencing, and biochemical analyses of *Escherichia coli* glyoxylate carboligase. An enzyme of the acetohydroxy acid synthase-pyruvate oxidase family. J Biol Chem 268:3911–19
7. Clough SJ, Bent AF (1998) Floral dip: a simplified method for *Agrobacterium*-mediated transformation of *Arabidopsis thaliana*. Plant J 16:735–43
8. de Groot MJ, Bundock P, Hooykaas PJ, Beijersbergen AG (1998) *Agrobacterium tumefaciens*-mediated transformation of filamentous fungi. Nat Biotechnol 16:839–42
9. Deblock M, Desonville A, Debrouwer D (1995) The selection mechanism of phosphinothricin is influenced by the metabolic status of the tissue. Planta 197:619–26
10. Douce R, Neuburger M (1999) Biochemical dissection of photorespiration. Curr Opin Plant Biol 2:214–22
11. Edwards G, Walker DA (1983) C_3, C_4: Mechanisms, and cellular and environmental regulation, of photosythesis. Blackwell Scientific Publications, London
12. Edwards JW, Coruzzi GM (1989) Photorespiration and light act in concert to regulate the expression of the nuclear gene for chloroplast glutamine synthetase. Plant Cell 1, 241–48
13. Evans JR, Von Caemmerer S (1996) Carbon dioxide diffusion inside leaves. Plant Physiol 110:339–46
14. Furbank RT, Taylor WC (1995) Regulation of photosynthesis in C-3 and C-4 plants: A molecular approach. Plant Cell 7:797–807
15. Gotto AM, Kornberg HL (1961) The metabolism of C2 compounds in micro-organisms. 7. Preparation and properties of crystalline tartronic semialdehyde reductase. Biochem J 81:273–84
16. Hansen RW, Hayashi JA (1962) Glycolate metabolism in *Escherichia coli*. J Bacteriol 83:679–87
17. Hatch MD (1987) C_4 photosynthesis: a unique blend of modified biochemistry, anatomy and ultrastructure. Biochim Biophysica Acta / General Subjects 895:81–106

18. Kaplan A, Reinhold L (1999) CO_2 concentrating mechanisms in photosynthetic microorganisms. In: Jones R L (ed) Annu Rev Plant Phys 50:539–70
19. Kornberg HL, Sadler JR (1961) The metabolism of C2-compounds in micro-organisms. VIII. A dicarboxylic acid cycle as a route for the oxidation of glycolate by *Escherichia coli*. Biochem J 81:503–13
20. Kumar SV, Misquitta RW, Reddy VS, Rao BJ, Rajam MV (2004) Genetic transformation of the green alga *Chlamydomonas reinhardtii* by *Agrobacterium tumefaciens*. Plant Sci 166:731–38
21. Kunik T, Tzfira T, Kapulnik Y, Gafni Y, Dingwall C, Citovsky V (2001) Genetic transformation of HeLa cells by *Agrobacterium*. Proc Natl Acad Sci U S A 98:1871–1876
22. Lacuesta M, Gonzalezmoro B, Gonzalezmurua C, Aparicotejo P, Munozrueda A (1989) Effect of phosphinothricin (glufosinate) on activities of glutamine-synthetase and glutamate-dehydrogenase in *Medicago sativa* L. J Plant Physiol 134:304–07
23. Leegood RC, Walker RP (1999) Regulation of the C_4 pathway. In: RF Sage, RK Monson (eds) C_4 Plant Biology, pp 89–131, Academic Press, San Diego
24. Logemann E, Birkenbihl R, Ulker B, Somssich I (2006) An improved method for preparing *Agrobacterium* cells that simplifies the *Arabidopsis* transformation protocol. Plant Meth 2:16
25. Long SP, Ainsworth EA, Leakey AD, Nosberger J, Ort DR (2006) Food for thought: lower-than-expected crop yield stimulation with rising CO_2 concentrations. Science 312:1918–21
26. Lord JM (1972) Glycolate oxidoreductase in *Escherichia coli*. Biochim Biophys Acta 267:227–237
27. Nelson EB, Tolbert NE (1970) Glycolate dehydrogenase in green algae. Arch Biochem Biophys 141:102–110
28. Newell CA (2000) Plant transformation technology. Developments and applications. Mol Biotechnol 16:53–65
29. Ogren WL (1984) Photrespiration: Pathways, regulation and modification. Annu Rev Plant Phys 35:415–42
30. Pellicer MT, Badia J, Aguilar J, Baldoma L (1996) glc locus of *Escherichia coli*: characterization of genes encoding the subunits of glycolate oxidase and the glc regulator protein. J Bacteriol 178:2051–59
31. Sage RC (1999) Why C4 photosynthesis? In: Sage RF, Monson RK (eds) C4 Plant Biology, pp 3–16, Academic Press, San Diego
32. Sage RF (2001) Environmental and evolutionary preconditions for the origin and diversification of the C4 photosynthetic syndrome. Plant Biol 3:202–13
33. Sage RF (2004) The evolution of C4 photosynthesis. New Phytol 161:341–70
34. Sharkey TD (2001) Photorespiration. Encyclopedia of Life Sciences, 1–5
35. Sheen J (1999) C_4 gene expression. Annu Rev Plant Phys 50:187–217
36. Tachibana K, Watanabe T, Sekizawa Y, Takematsu T (1986) Action mechanism of bialaphos.1. Inhibition of glutamine-synthetase and quantitative changes of free amino-acids in shoots of bialaphos-treated Japanese barnyard millet. J Pesticide Sci 11:27–31
37. Vain P (2007) Thirty years of plant transformation technology development. Plant Biotech J 5:221–29
38. Wild A, Ziegler C (1989) The effect of bialaphos on ammonium-assimilation and photosynthesis. 1. Effect on the enzymes of ammonium-assimilation. Z Naturforsch C 44:97–102

Chapter 25
Controlling Microbial Adhesion: A Surface Engineering Approach

Ilya Digel

Laboratory of Cell Biophysics, Aachen University of Applied Sciences,
Ginsterweg, 1, 52428 Jülich, Germany,
digel@fh-aachen.de

Abstract Vast numbers of microorganisms on the Earth are sessile and therefore cannot be detected and counted by most of traditional microbiological methods. Nevertheless, the ecological, medical and biotechnological impacts of attached microorganisms are enormous. That is why the process of microbial adhesion deserves rapt attention from professionals working on different fields.

Gaining control over all these versatile cell-surface interactions requires considering physicochemical, biochemical and ecological aspects that makes this issue extremely intricate. This paper gives a general overview of physical, chemical and molecular biological aspects of microbial adhesion in connection to their practical applicability. Comprehension of complex nature of adhesion processes allows elaboration of efficient approaches in order to modulate relationships between microorganisms and surfaces.

25.1 The Lost World of Sessile Microorganisms

Like tissue cells growing in in-vitro culture, microorganisms prefer to grow on available surfaces rather than in the surrounding aqueous phase, which was proved by Zobell in 1943 (Zobell 1943). Virtually any surface – animal, mineral, or vegetable – is fair game for microbial colonization and biofilm formation, including contact lenses, ship hulls, dairy and petroleum pipelines, rocks in streams and all varieties of biomedical implants (Dunne Jr 2002; Neu 1996).

The inclination for microorganisms to become surface bound is so ubiquitous in diverse ecosystems that it suggests a strong survival and/or selective advantage for surface dwellers over their free-ranging counterparts (Bhinu 2005; Zobell and Mathews 1936). Indeed, the propensity for microbes to colonize surfaces is advantageous from ecological and biotechnological standpoints because it preferentially targets specialized microorganisms to specific locations, encouraging symbiotic relationships and therefore increasing productivity. From an evolutionary standpoint, the selective advantage of microbial adhesion has been postulated to favor the localization of surface-bound microbial populations in nutritionally favorable, non-hostile environments and at the same time provide some level of protection from external predation (Dunne Jr 2002; Marshall 1986).

Therefore, in natural and artificial aquatic ecosystems, surface-associated microorganisms vastly outnumber organisms in suspension and often organize into complex sessile communities with features that differ dramatically from those of planktonic cells (Zobell and Mathews 1936). The research of microbial adhesion and its significance is a large field covering different aspects of nature and human life, such as soil and plant ecology, the food industry, and importantly, the biomedical field.

It was discovered that the surface colonization by microorganisms consists of two distinct stages, referred as reversible and irreversible attachment, correspondingly. Several terms are usually used to describe microorganism-surface interactions. *Adhesion* is a situation where microorganisms adhere firmly to a surface by complete physicochemical interactions between them, including an initial phase of reversible physical contact and a time-dependent phase of irreversible chemical and cellular adherence. *Immobilization* is rather a biotechnological term, indicating that the cells or molecules are confined, bound with solid component of multi-phase system. *Sorption* is an archaic synonym for adhesion. *Adherence* is a general description of microbial adhesion (the initial process of attachment of microorganisms directly to the surface) and is a less scientific term for microbial adhesion. *Attachment* can be defined as the initial stage of microbial adhesion, referring more to physical contact than complicated chemical and cellular interactions, and is usually reversible.

The expression "irreversible attachment" is used when microbes can no longer move perpendicularly away from the surface (Busscher et al. 1998). It is assumed that most microbes become irreversibly attached only after a period of unstable, reversible adhesion, during which the cells can show motion such as the well-known rotating motion of apically adherent cells of *P. fluorescens* (Korber et al. 1994). The sessile microorganisms can form a *biofilm* – a dense colony of microorganisms affixed to a surface within a protein and sugar-chain matrix (Fig. 25.1). The cells in biofilms are able to divide and produce new swarming cells. Many specialized structures and complex signaling pathways designed specifically for surface recognition and biofilm formation evolved in microbes is another clue supporting the importance of microbial adhesion (Ben-Jacob and Levine 2006; Favre-Bonte et al. 2003; Rasmussen and Givskov 2006).

Most kinds of adhesive bacteria and fungi have been shown to produce adhesion forces of nanonewtons or piconewtons (Tang et al. 2004). However, a bacterial glue has been discovered recently that is stronger than any known adhesive – more than three times as strong as for example super glue. This super-sticky, water-resistant glue is produced by bacteria called *Caulobacter crescentus*. In aquatic environments, *C. crescentus* is one of the first colonizers of submerged surfaces. Using a micromanipulation technique, the adhesion force of single *C. crescentus* cells attached to borosilicate substrates have been measured by Jay Tang with co-workers to be in the range from 0.1 to 2.26 μN (Bodenmiller et al. 2004; Tsang et al. 2006).

25.1 The Lost World of Sessile Microorganisms

Based on the calculation of stress distribution with the finite element analysis method, the adhesion strength between the holdfast and the substrate was estimated to be $>68\,N/mm^2$ in the central region of contact. A few micronewtons may not sound like a great deal of force, but this means that a layer of *C. crescentus* holdfast covering $1\,cm^2$ could hold 680 kg weight which is equivalent to three or four cars balanced on square inch. This strength of adhesion seems to be the strongest ever measured for biological adhesives (Tsang et al. 2006). The combination of great strength and biocompatibility could have applications in joint-replacement, bone and cartilage repair, and eye surgery.

Microbial biofilms play notorious roles in the environment, infectious pathology and industry, where biofilm formation is regarded usually as highly undesirable effect. Sessile microbial communities including pathogenic bacteria growing inside the human body, e. g. in lungs or on implant surfaces (Donlan and Costerton 2002; Dunne Jr 2002) or in drinking-water distribution systems can threaten human health (Szewzyk et al. 2000). In industrial processes adhered microorganisms cause malfunction of equipments, lower the efficiency of heat exchangers, and lower the end-product quality or safety in food industry (Carpentier and Cerf 1993). In some instances, however, the corrosive damage caused by bacterial biofilms provides a never-ending source of economic benefit, as most of us have appreciated after a trip to the dentist.

One should also not forget the protective roles of skin and intestinal bacterial biofilms. This implies that our outer surfaces are extremely populated and highly developed ecological niches. The existing ecological equilibrium is quite sturdy but can be destroyed by systematic misinterpretation of hygiene. For example, regular treatment of skin with antibiotics, triclosan, ethanol, etc. results in natural selection

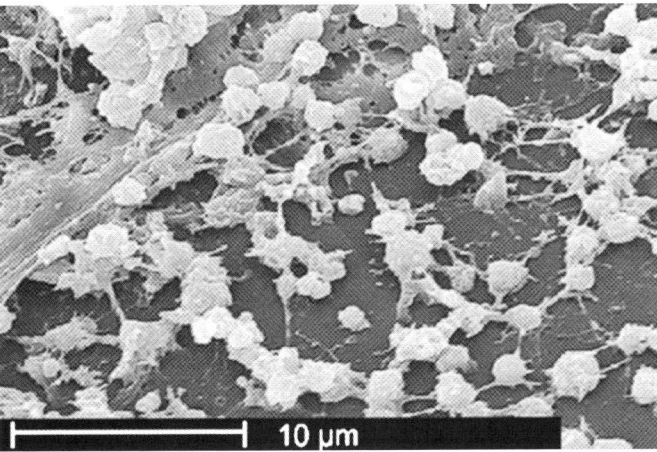

Fig. 25.1 Scanning electron micrograph of a *Staphylococcus* biofilm on the inner surface of a needleless connector. A distinguishing characteristic of biofilms is the presence of extracellular polymeric substances, primarily polysaccharides, surrounding and encasing the cells. Here, these polysaccharides have been visualized by scanning electron microscopy

of respectively resistant microorganisms and therefore has potentially hazardous consequences.

According to our measurements, the surface concentration of microorganisms on the skin of twice washed hands still averages 10^5 cells per cm^2, with species diversity exceeding 50 bacterial and fungal species. Interestingly, the composition of individual's microflora is to great extent defined by his/her genetic peculiarities and this is one of the reasons why close relatives have similar body smell. The organisms belonging to "normal" dermal and intestinal microflora generally do no allow any foreign invaders (often potentially pathogenic, to settle and colonize skin and mucosa). A mature microbial biofilm with its complex architecture typically provides niches with distinct physicochemical conditions, differing e. g. in oxygen availability, in concentration of diffusible substrates and metabolic side products, in pH and in the cell density. Consequently, cells in different regions of a biofilm can exhibit different patterns of gene expression (Costerton 1999). Mixed-species biofilms (like those on our teeth) can contain niches with distinct groups of bacteria having metabolic cooperation (Kuchma and O'Toole 2000). Watnick and Kolter summarized that a mixed species biofilm is a dynamic community harboring bacteria that stay and leave with purpose, compete and cooperate, share their genetic material, and fill distinct niches within the biofilm. They stated, "The natural biofilm is a complex, highly differentiated, multicultural community much like our own city" (Watnick and Kolter 2000).

One interesting phenomenon concerning biofilms is that bacteria in a mature biofilm are far more resistant to antimicrobials (biocides and antibiotics) and to unfriendly physicochemical conditions (pH, temperature, salt concentration) than freely swimming cells. Different mechanisms have been proposed to account for this increased resistance that is most likely multifactorial (Bodenmiller et al. 2004; Marshall 1986; Neu 1996). Extracellular microbial polysaccharides (EPS) may form permeability barriers or make complexes with the antimicrobials thus interfering with the antimicrobial action. As a result of that, reactive oxidants and free radicals may be deactivated in the outer layers of EPS faster than they diffuse. Furthermore, microbial cells of the biofilm phenotype may have reduced susceptibility because of altered cellular permeability, metabolism or growth rate that is still poorly understood (Bodenmiller et al. 2004; Costerton 1999; Watnick and Kolter 2000; Zobell 1943; Dunne Jr 2002).

25.2 Biotechnological Potential of Adhered Microorganisms and Its Limitations

All aforementioned suggests the attractiveness of adhered cells applications for biotechnological applications. In spite of the fact that the first bioreactors using immobilized microbial cells for vinegar production appeared in Pasteur's times (Nabe et al. 1979), the most intent scientific interest and popularity this approach became in 80s. The rise of the attention to immobilized bio-catalysators in particularly this period of time had been provoked by a group of factors related to a new stage in biotechnology development. These factors, on our opinion, were: 1) elaboration of industrial large-scale manufacture of biotechnology products (amino acids, vitamins, antibiotics etc.); 2) conversion of microbial technology towards continuous cultivation, allowing achievement of higher production rates: 3) rigid environmental constraints and 4) development of quality management in microbial biotechnology.

Thus, the evolution of biotechnological science demanded further development and optimization of manufacturing processes, focusing on new technological approaches and new bioreactors' designs. As a matter of fact, immobilization of microbial cells was a sequel of enzyme immobilization approach and all existing methods of cell immobilization originate from engineering enzymology (Ohmiya et al. 1977).

The group of adsorption immobilization methods is based on creation of bonds between a solid surface (referred as a carrier or a substratum) and a biocatalisator (enzymes, immunoglobulins, microbial cells, etc.). These approach features technical and methodological simplicity, low costs and absence of diffusion limitations (Junter and Jouenne 2004). In this case the immobilization process comes to simple contact of cell suspension and the carrier. The desired properties of the carrier include large surface, mechanical stability and non-toxicity. There are plenty of materials fulfilling these requirements and granting additional flexibility to bioreactor design and processing parameters. For example, nitrocellulose membranes, zeolytes, ultra-thin basalt fibers, acetate cellulose threads, hollow polystyrol and acrylontryl tubes, porous polycarbonate disks, clays, talcum, silica gel, polyurethane foam etc. have been successfully used as carrier materials for microbial immobilization (Branyik et al. 2000; Cabuk et al. 2006; Rangsayatorn et al. 2004).

Immobilized cells possess a number of useful properties that are of great practical importance. Most of these advantages are connected with reduction of losses of biomass in the liquid outflow of a bioreactor. This allows achieving higher concentrations of active cells in the working zone of the bioreactor that leads, in turn, to higher production efficacy (Marshall 1986; Rangsayatorn et al. 2004; Watnick and Kolter 2000). It is worthy to mention also that the final product in such a bioreactor is not contaminated with microbial cells and therefore requires less filtering and purification efforts (Branyik et al. 2000, 2004; Ohmiya et al. 1977). Thus, the use of immobilized cells opens new opportunities in bioreactors' design and construction, not applicable otherwise.

The influence of cells adhesion on their growth kinetics (lag-phase shortening, higher biomass production) has been repeatedly proved (Donlan and Costerton 2002; Habash et al. 1997; Sheth et al. 1985). Apparently, the cell attachment dramatically influences the viability and resistance of microorganisms to unfriendly and extreme environments. Higher specific physiological activity and stress-resistivity of immobilized cells compared to suspended ones was reported by many research groups (Bhinu 2005; Branyik et al. 2000; Cabuk et al. 2006; Dunne Jr 2002; Habash et al. 1997; Sheth et al. 1985; Zobell 1943). It has been demonstrated that nitrogenase activity of *Bacillus polymyxa* cells significantly increases after immobilization onto $CaCO_3$ or ceramics, whereas polyacrylamide gel inhibits nitrogenase activity completely (Seyhan and Kirwan 1979). *Acetobacter* cells produce acetic acid even three times faster being immobilized onto titanium-containing gel (Davidenko 1990; Kennedy and Cabral 1990).

Adsorbents also show significant influence on oxygen consumption by immobilized microbial cells. Partial inhibition of *Azotobacter* sp., *E. coli* and *Pseudomonas fluorescens* respiration was noted when ion-exchange resins were used as a solid phase. The exact roles of surface chemical composition and carrier's spatial properties in bacterial growth stimulation are still not well understood. It has been suggested (Bhinu 2005; Ohmiya et al. 1977; Watnick and Kolter 2000; Zobell 1943) that one of the mechanisms stimulating cellular metabolism is the modification of cellular surface structures during adsorption immobilization. This modification influences the permeability of cell envelopes, which leads, in turn, to decrease of the sensitivity threshold of signaling systems responsible for enzyme activity regulation.

Thus, first appeared as just a methodological approach, adsorption cell immobilization has been transformed into a valuable instrument which allows to control and manipulate metabolic activity of the cells. Its practical application and development created a group of principally new microbial technologies (Ben-Jacob and Levine 2006; Branyik et al. 2000, 2004; Li et al. 2005). The advantages and benefits described above significantly reduce labor and time consumption for industrial-scale bioreactors and therefore have made the adsorption immobilization one of the favorite topics for practical biotechnologists. Operating aspects of such large-scale setups in waste water management and drink water purification, were extensively described and repeatedly discussed in a number of publications (Busscher and Van Der Mei 2006; Cabuk et al. 2006; Marshall 1986; Szewzyk et al. 2000).

One distinct problem impeding many biotechnological applications of adhered cells is the inability of many microorganisms to establish strong irreversible contacts with respective surface. As a matter of fact, very important biotechnological microorganisms like brewery yeasts are lacking most of known adhesion factors. This fact limits and hinders dramatically their biotechnological potential because in this case the adsorption immobilization cannot guarantee durable and lasting retention of microbial cells on the surface. Under these conditions, the system "cells-surface" is in dynamic equilibrium that means that in each moment of time some cells attach to the surface and the same amount of cells detach from it.

Together with the lack of adhesiveness, the main reason for poor attachment of yeasts is that the local interactions keeping the cells attached to the surface are effectively counteracted by a number of opposed forces (tearing cells off with the liquid flow and stirring, electrostatic repulsion, Brownian motions and cells' own active motions). Another known problem is limited carrier surface capacity in respect of cells. As a matter of fact, the cells often don't occupy all spatially and sterically available places on the surface but rather attach themselves to some preferable locations which are referred as "centers of adsorption" (Bodenmiller et al. 2004; Marshall 1986). These sites, having high surface energy, are limited in number and therefore cannot provide high surface density of immobilized cells.

All these undesirable conditions indicate the seriousness of the cell detachment problem for biotechnological applications. To be able to develop possible strategies overcoming the aforementioned drawbacks, profound understanding of physical, chemical and biological determinants of cellular adhesion is of great importance. The following two parts will be devoted to short overview of physicochemical and biological aspects of microbial adhesion, respectively.

25.3 Physicochemical Aspects of Microbial Adhesion

There may be an obvious explanation for microbial adhesion, because nutrients in an aqueous environment tend to concentrate near a solid surface. However, the particular physicochemical forces contributing to microbial adhesion are complex.

From physicochemical point of view, a microbial cell can be imagined as a charged (typically negative) particle of 0.5 – 10 μm size, possessing defined surface energy, hydrophobicity, etc. Such abstraction and simplification are quite expedient in many instances, allowing describing adsorption and attachment processes by means of known physical and mathematical models. In many cases, such rough calculations are very helpful and efficient in prediction and description of the system's behavior by evaluation of strength and character of the interactions between involved surfaces (Bodenmiller et al. 2004; Kennedy and Cabral 1990; Marshall 1986). The following characteristics and parameters are mainly applicable in frames of physicochemical approach to describe adhesion: adsorbent capacity r, equilibrium constant K as well as specific surface energy E, characterizing surface wettability.

In general, the speed of adsorption W_a is proportional to adsorbent surface S; the relative free surface $(1 - \Theta)$ and cells concentration in liquid phase C:

$$W_a = k_a \cdot S \cdot C(1 - \Theta), \quad (25.1)$$

where k is velocity constant. Noteworthy, the speed of desorption W_d does not depend on cells concentration in liquid phase:

$$W_d = k_d \cdot S \cdot \Theta. \quad (25.2)$$

These idealized Eqs. (25.1) and (25.2) describe adsorption in frames of classical Longmuir model which is applicable only to a few real processes because it deals with absolutely uniform surface and monolayer adsorption. Taking into account the real surface heterogeneity, this model adequately describes adsorption behavior for extremely small part of the surface only. More realistic approximation requires differentiation with following integration the equation for all small areas having different specific adsorption heat indexes, which is cumbersome.

Adhesion can be described as a balance between attractive Van der Waals forces, electrostatic forces (often repulsive as most microorganisms and immersed surfaces are negatively charged), short range Lewis acid-base interactions, hydrophobic interactions and Brownian motion, etc. These physical interactions might be roughly classified as long-range interactions and short-range interactions respectively (Gottenbos et al. 2002). Microbial cells are transported to the surface by the so-called long-range interactions and upon closer contact, short-range interactions become more important. The long-range interactions (non-specific, distances approx. 150 nm) between cells and material surfaces are defined as a function of the distance and free energy. Short-range interactions become effective when the cell and the surface come into close contact (approx. 3 nm), maintained by chemical bonds (such as hydrogen bonding), ionic and dipole interactions, and hydrophobic interactions. This initial attachment of microorganisms to material surfaces is the primary stage of adhesion, making the following molecular and cellular phases possible.

The interaction between the cell and the surface is dictated by physicochemical variables, which are explained by several different theories (Korber et al. 1997; Marshall 1986). According to Derjaguin, Landau, Verwey and Overbeek (DLVO) theory, unspecific cellular adhesion is caused by combination of dispersion (van der Waals) attraction forces and electrostatic repulsion forces (Busscher et al. 1998; Korber et al. 1997; Marshall 1986; Poortinga et al. 2001). Magnitudes of these forces are affected by the distance of the bacterium from the surface (long-range and short-range forces) and by ionic strength. Thermodynamically, spontaneous cell adsorption onto a surface results in decrease of Gibbs free energy but there is a significant energy barrier because of electrostatic repulsion. The theory predicts that there are two regions where attraction may occur (primary and secondary minimum, correspondingly <1 nm and 5–10 nm from the substratum). Generally is assumed that microbes adhere reversibly to the "secondary minimum" and irreversibly to the "primary minimum" with the aid of cell surface appendages that can pierce the repulsive energy barrier. Microorganisms forced to adhere in the primary minimum by application of a high positive electrode potential are hardly to desorb, indicating strong irreversible adhesion.

Development of a net charge at the particle surface affects the distribution of ions in the surrounding interfacial region, resulting in an increased concentration of counter ions (ions of opposite charge to that of the particle) close to the surface. The surface charge attracts ions of opposite charge in the medium and results in the formation of an electric double layer. This layer surrounding the particle exists as two parts; an inner region (Stern layer) where the ions are strongly bound and an outer (diffuse) region where they are less firmly associated. Within this diffuse layer is a notional boundary known as the slipping plane, within which the particle acts as a single entity. The surface charge is usually characterized by the isoelectric point, the electro-kinetic potential (or zeta potential), or electrophoretic mobility.

The zeta potential is the overall charge a particle acquires in a specific medium. If the particles have low zeta potential values then there is no force to prevent the particles coming together and there is dispersion instability. In general, increase in electrolyte concentration (i.e. ionic strength) decreases electrostatic surface potential because of counter-ions sedimentation and hence, thinning of double electric layer. Hence, one of the important predictions of DLVO theory is promotion of cell adsorption by increase of salt concentration in liquid medium (Marshall 1986; Poortinga et al. 2001). Indeed, at appropriate electrolyte concentration, attraction forces between cells and surface become dominating and result in cellular adhesion. In case of particularly high salt concentration the cells can reach the primary free energy minimum, which causes irreversible binding.

The alternative "wetting" or the "surface free energy/hydrophobicity" theory is also based on surface thermodynamics. This theory relies on determining critical surface tension of the microorganisms and substratum, and is not taking electrostatic interactions into account (Carpentier and Cerf 1993). If the total free energy of the system is reduced by cell contact with a surface, then adsorption will occur. Hydrophobic interactions between cells and surfaces do not themselves result in irreversible adsorption. They just create favorable thermodynamic conditions promoting adhesion. The cells' bounding to surfaces becomes irreversible as a result of covalent, ion and van-der-Waals short-range interactions in the sites of nearest surface contacts (Busscher et al. 1998).

In theory, both physical and chemical adsorption events are stimulated also by decrease of temperature of the system. This follows from le Chatelier principle of dynamic equilibrium, since cellular adsorption is an exothermic (heat-producing) process. Nevertheless, the temperature dependence of cell adsorption has more complicated shape because of many mechanisms involved. Temperature effects of adsorption seem to be connected mainly with surface chemical reactions, when heat effects can reach 400 kJ/mole and more.

25.4 Biological Aspects of Microbial Adhesion

The physicochemical models mentioned above have turned to be very fruitful in many practical applications. Their relevancy has been repeatedly demonstrated in estimation of cell adsorption under variable hydrophobicity, pH and ionic strength values. However, pure physical models of cellular adhesion are often not adequate and accurate enough since they don't take into account the complex self-regulating behavior of a living cell. It is caused by uncertainty of many parameters crucially important for adsorption, for example, variations in surface exposure of adhesion-related molecules (adhesins, surfactants, receptors, lectins). Importantly, microbes are not constant inanimate particles, but living organisms that can adapt to the requirements set by their changing environment and alter their surface composition. Bacterial cell surfaces are generally hydrophobic, but the cells can possess hydrophilic appendages. The net charge of most bacteria is negative, but e. g. *Stenotrophomonas maltophilia* has positive surface charge at physiological pH. Additionally, the cells can also have polarity in their charge (Dunne Jr 2002).

Microbial binding especially to living surfaces is mainly accomplished by biochemical interactions based on specific affinity such as enzyme-substrate, lectin-carbohydrate, and antigen-antibody pair forming. The strength and specificity of these interactions are mainly affected by the protein structure, charge and conformational state of corresponding binding sites. The typical examples of such interactions are adhesion of symbiotic nitrogen-fixing bacteria to roots, and binding of pathogenic microorganisms to hosts surfaces (Busscher and Van Der Mei 2006; Soto and Hultgren 1999). In case of such highly biologically specialized interactions, variations in physical and chemical parameters (temperature, pH, ion strength) as well as changes in culture media composition play relatively unimportant role in adsorption process, having rather ancillary effect.

Molecular biological and biochemical aspects of cell adhesion focus predominantly on identification, isolation and structural analysis of attachment-responsible biological molecules and their genetic determinants. Physiological direction of cellular adsorption concerns mainly the influence of cultivation parameters (temperature, nutrition compounds, oxygen concentration, presence of antibiotics and vitamins) on bacterial adherence-related phenotype, adhesion molecules metabolism and surface structural organization (Habash et al. 1997; Junter and Jouenne 2004).

Once in initial contact with a surface, microbes develop different types of attachment behaviors. Motile attachment behavior of *P. fluorescens* allows the flagellated cells to move along surfaces in a semi-attached condition within the hydrodynamic boundary layer, independent of the flow direction (Korber et al. 1994). Reversible adhesion of *E. coli* cells with residence times of over 2 minutes on a surface has been described as "near-surface swimming" (Vigeant and Ford 1997). The term "irreversible attachment" is used when microbes can no longer move perpendicularly away from the surface (Busscher et al. 1998).

It is assumed that most microbes become irreversibly attached only after a period of unstable, reversible adhesion, during which the cells can show motion such as the rotating motion of apically adherent cells of *P. fluorescens* (Korber et al. 1994). *Vibrio cholerae* and *E. coli* first utilize the flagella to spread across the surface, and then anchor onto the surface with pili and possibly outer membrane proteins (Davey and O'Toole 2000; Davey et al. 2003; Kuchma and O'Toole 2000). Microbes can also attach irreversibly, while retaining active motility by mechanisms known as gliding, swarming, twitching, swimming, darting and sliding. Uropathogenic *E. coli* cells were shown to attach irreversibly and yet actively migrate along solid surfaces. *P. aeruginosa* requires type IV pili for twitching motion on a surface and for the subsequent build-up of stagnant microcolonies (O'Toole et al. 2000).

Very intriguing results reported by a multidisciplinary teams from bioengineering and microbiology suggest that fluid forces in the human body, as occur with salivating, swallowing, sneezing, urinating, and weeping, may in some cases even strengthen bacterial adhesion instead of detaching and flushing away the infectious agents (Thomas et al. 2002; Christersson et al. 1988; Brooks and Trust 1983). The researchers described a mechanism by which a certain bacterial adhesion protein, FimH, responds to high shear conditions by buckling down the bacteria onto the invaded surface. The protein reacts to low shear conditions by releasing the bind to allow bacterial spread across the surface. FimH is the second adhesion protein, after fibronectin, for which these scientists have described a structural mechanism whereby mechanical forces regulate protein function.

25.4 Biological Aspects of Microbial Adhesion

The recent progress in development of more sensitive and specific molecular biological methods (two-dimensional SDS PAGE, DNA- and protein microarrays, polymerase chain reaction) has significantly stimulated interest to physiological and biochemical aspects of cellular adsorption. Whiteley et al. (Whiteley et al. 2001) demonstrated by DNA microarray technology that in mature biofilms of *P. aeruginosa* genes e. g. for synthesis of flagella and the sigma factor RpoS were repressed, whereas genes encoding proteins for e. g. temperate bacteriophage, urea metabolism, membrane transport, translation and gene regulation were up-expressed. Another study on *Pseudomonas* attachment indicated expression of 15 unique proteins, supposedly required in the initial stages of biofilm development. The planktonic cells had high levels of 11 proteins that were absent in 2- and 18-h-old biofilms. Only 10% of the proteins were present in all growth phases and at the same level (no up- or down-regulation). Cells from the 18-h-old biofilm contained 7 unique proteins, whose identity suggested for e. g. fermentative metabolism in the biofilm (Oosthuizen et al. 2002).

Similarly, by 2D electrophoresis, *P. aeruginosa* has been shown to display at least three phenotypes with different protein profiles: a planktonic, a mature biofilm, and a dispersing biofilm phenotype (Stoodley et al. 2002). Davies et al. (1998) published the first study that showed a role for quorum sensing in the formation of biofilms, and launched a period of active research of cell-to-cell signaling in biofilms. This researcher group showed also that *lasI*-mutant cells of *P. aeruginosa* that were unable to synthesize 3OC12-HSL (3-oxo-dodecanoyl-homoserine lactone) were able to attach and initiate the biofilm formation similar to the wild type cells, but the mature biofilms were continuous sheets lacking the differentiated architecture with microcolonies and water channels (Davies et al. 1998).

As mentioned above, microbial adhesion to a material surface can be described as a two-phase process including an initial, instantaneous, and reversible physical phase (phase one) and a time-large pendent and irreversible molecular and cellular phase (phase two), which was first proposed by Marshall and colleagues (Marshall 1986) and has been accepted by the majority of researchers. The exact mechanism of two-stage nature of microbial attachment is still under discussion. Many authors believe that the second stage is undoubtedly mediated by new-synthesized adhesion molecules, creating strong chemical bounds between contacting surfaces. This point of view has found support in several studies that revealed cooperativeness in cellular attachment, as well as increased expression of specific gene groups prior to the second stage (Costerton 1999; Davey and O'Toole 2000; Korber et al. 1997; Rasmussen and Givskov 2006). Adhesin can be defined as a substance (a surface macromolecule, commonly lectins or lectin-like proteins or carbohydrates) produced by microorganisms and is thought to be a specific material for specific adhesion. Bacteria may have multiple adhesins for different surfaces (different receptors). Generally, any structures responsible for adhesive activities can be called adhesins. The molecules responsible for attachment are of different chemical nature and include glycoproteins, lectins, teichoic acids etc. (Bhinu 2005; Korber et al. 1997; Soto and Hultgren 1999).

Biologically defined expression level of adhesins can dramatically vary even inside of the same microbial population (Flint et al. 2001; Peng et al. 2002). This results in significant population heterogeneity in respect of cell surface hydrophobicity, charge, adhesiveness etc. It is known also that environmental changes induce reorganization and restructuring of cell surface components. Understanding of the mechanisms determining biological response is a key element for maximally adequate description, reliable prediction and control of cellular adsorption.

Other researchers are the opinion that biosynthesis of adhesins, though important, doesn't play critical role in irreversible binding to the surface. These authors explain the presence of the second phase as a result of multiple sites of polymer binding to the surface, so the adsorption is irreversible because of kinetic, not biological reasons. This hypothesis also has strong arguments, for example the fact that dead cells and even large synthetic polymers demonstrate similar two-stage adsorption kinetics (Davey and O'Toole 2000; Marshall 1986; Poortinga et al. 2001). Moreover, in many cases the irreversible stage begins almost immediately (several minutes) after the first stage and therefore the cell simply does nor enough time to synthesize any considerable amount of adhesion molecules (Marshall 1986; Poortinga et al. 2001). The most probable scenario is that early stages of microbial adsorption are guided and driven by non-biological mechanisms, whereas later formation of bacterial biofilm is definitely biologically controlled and regulated.

25.5 Surface Conditioning as a Tool Facilitating Microbial Adhesion

All abovementioned considerations give important clues in developing appropriate strategies for forced adhesion of intrinsically suspended cells. Stimulation of unspecific cells' adsorption is of great interest as a complex of measures concerning on the one hand strain-specific biological differences and on the other hand physicochemical parameters of adsorption system (pH, presence of bi- and trivalent metal ions, temperature, shear stress).

Direct bonding of microbial cells to many inorganic surfaces is difficult due to relative inertness of the surface in its ground state. However, better attachment can be achieved after surface activation/modification. The term "modification" relates to all the processes that lead to change in chemical composition of the surface. Modification via functional group immobilization provides unique opportunity to engineer the interfacial properties of solid substrates while retaining their basic geometry and mechanical strength. There are extensive reports on immobilization of modifiers like chelate-forming organic reagents, polymers, metal salts, natural compounds and some microorganisms on solid matrices like ion-exchange resins, cellulose, fibers, activated carbon, sand, clay, zeolytes, polymers, metal oxides and highly dispersed silica (Herrera-Alonso et al. 2006; Qhobosheane et al. 2001).

Immobilization of molecules on organic/inorganic support has been studied with much attention given to establishment of new covalent bond on the desired sur-

25.5 Surface Conditioning as a Tool Facilitating Microbial Adhesion

face (Li and Takahashi 2004). Such investigations of immobilization of groups or compounds depend on substitution reaction between the modifiers and the surface of the carrier material. Many attempts have been made to stabilize cellular attachment by means of multifunctional reagents (glutaraldehyde, cyanuric choride (2,4,6-trichloro-1,3,5-triazine chlorotriazine trichlorocyanidine) and others) but they failed for the most part because of high toxicity of the compounds used.

An activated surface should have along with low toxicity, good sorption capacity, chemical stability under experimental condition and, optionally, high selectivity (Preinerstorfer et al. 2006). With this aim in view, we have developed and put to the test two different approaches aimed to increase the efficiency of adsorption immobilization based of surface modification/conditioning. The first strategy is based on modifying the surface by transition metals' ions; the second uses positively charged water-soluble polymers to eliminate electrostatic repulsion effects and to create more centers of adsorption (Fig. 25.2).

Prior to controlled surface modification experiments, several types of common adsorption materials were checked in order to select chemically inert and mechanically stable carriers having developed surface and satisfactory sorption activity. The parameters have being checked during this phase included adsorption rate, desorption rate, cell viability and material availability for later large-scale applications. Biotechnologically valuable yeast strains *Torulopsis kefyr var. kumis* (capable to lactose fermentation), *Rhodotorula glutinis* (important carotene producer) and *Saccharomyces cerevisiae* (used in beer production) were selected as test microorganisms. The yeasts' adsorption rate curves obtained for the most promising materials are shown in Fig. 25.3.

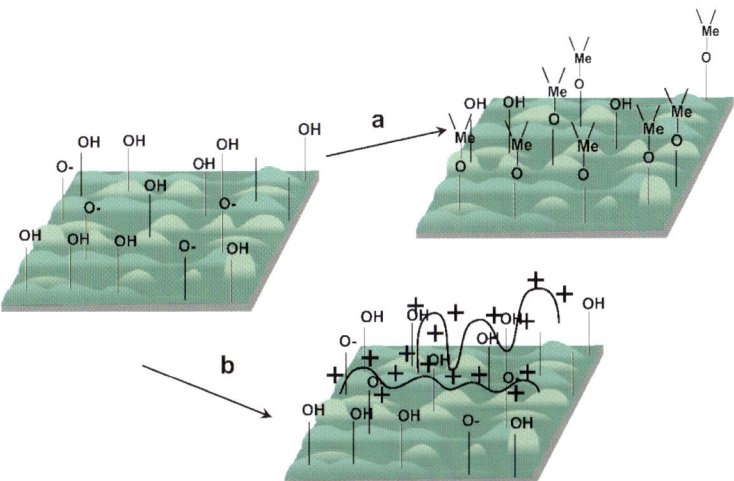

Fig. 25.2 Two main surface engineering approaches have been suggested to control and manipulate the microbial adsorption process. **a** Engraftment of transition metal ions; **b** preconditioning of surface by charged polymers

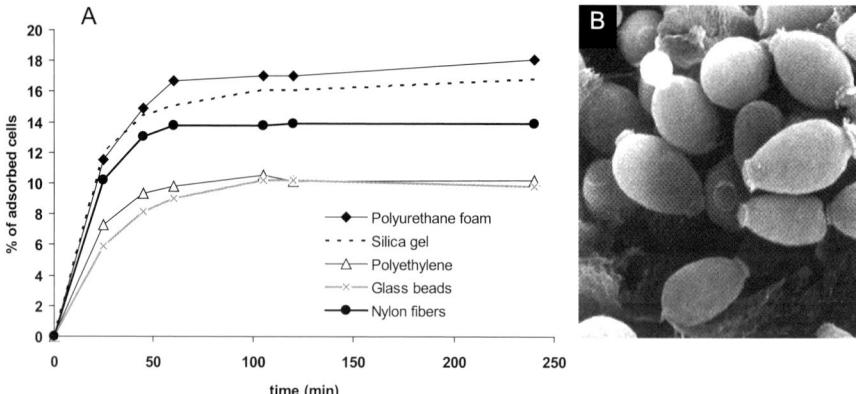

Fig. 25.3 a The Torulopsis kefyr var. kumis adsorption rates on different surfaces. Standard error of mean did not exceed 5% and is not shown. **b** Yeast cells attached to the surface

As follows from the data obtained, several chemically different materials showed similar adsorption/desorption kinetics, which is mainly connected with their large surface (large number of pores and surface peculiarities, providing surface-active sites). This group of five carrier materials was taken into the further modification stages.

25.5.1 Adsorption Facilitation by Transition Metal Ions

IUPAC defines of a transition metal as an element whose atom has an incomplete d sub-shell, or which can give rise to cations with an incomplete d sub-shell. The group of transition metals includes approximately 40 elements such as Fe, Mn, Co, Ni, Cu, Mo, Cr, Cd. The interesting property of transition metals is that their valence electrons, or the electrons they use to combine with other elements, are present in more than one shell. This is the reason why they often exhibit several common oxidation states, reaching sometimes very high oxidation states like 7+ and even 8+. Transition metals in high oxidation states are usually bonded covalently to more electronegative elements like oxygen, sulfur, carbon or fluorine, forming polyatomic groups such as chromates, carbonyls or permanganates (Fig. 25.4). This ability to form very complex compounds, especially with organic molecules, defines their role as catalysts, chromogens and electron-transferring molecules in many biological and chemical processes, making them especially interesting objects for bioorganic chemistry and biochemistry.

Surface chemical bonding of transition metals ions offers a unique advantage since the grafted ion detachment is prevented due to strong covalent bonding of the ion to the substrate. This approach, based on engraftment of multivalent metal ions into the substratum, involves aimed chemical modification of the surface and therefore puts special requirements on substratum stability and chemical compo-

25.5 Surface Conditioning as a Tool Facilitating Microbial Adhesion

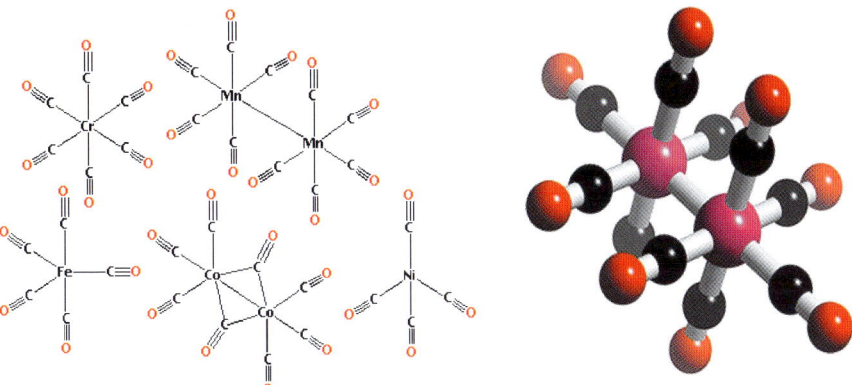

Fig. 25.4 The transition metals can form various carbonyl structures because of large number of valence electrons

sition. Among the initially selected materials (Fig. 25.5) only silica gel and glass beads have shown appropriate characteristics after ion engraftment procedure, combining good metal ions engraftment rates with constant mechanical properties. In spite of good adsorption properties, many other potentially valuable adsorbents like polyurethane foam and calcium alginate were unfortunately partially destroyed by transition metal salts due to metal-induced chemical reactions in polymer matrix. Thus, mainly silica gel and glass beads were later used in further experiments on microbial cells immobilization using surface conditioning by means of transition metals.

Among the different adsorbents for microbial cells, silica gel especially modified with various compounds has received great attention. Silica is a polymer of silicic

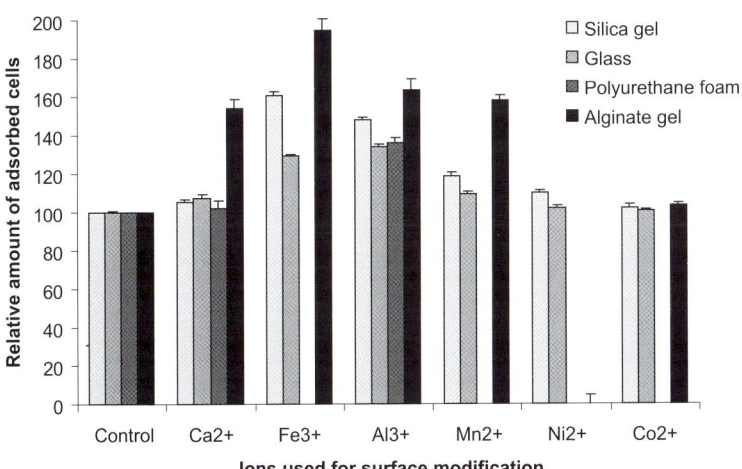

Fig. 25.5 Increase in T. kefyr adsorption after surface activation by transition metals. *Missing bars* indicate immediate material destruction followed transition-metal engraftment. The *error bars* show SEM

acid, consisting of inter-linked SiO_4 in tetrahedral fashion, which has the stoichiometry SiO_2. Silica gel is a porous, granular form of silica, synthetically manufactured from sodium silicate or silicon tetrachloride or substituted chlorosilane/orthosilicate solution. The active silica surface with large specific surface area is of great importance in adsorption and ion exchange. These properties are well studied, even though shape of silica surface is basically unknown. At the surface, the structure terminates in either siloxane group ($\equiv$Si–O–Si$\equiv$) with the oxygen atom on the surface, or one of the several forms of silanol groups ($\equiv$Si–OH). The silanol groups could be isolated (free silanol groups), where the surface silicon atom has three bonds into the bulk structure and the fourth to OH group and the vicinal or bridged silanols, where two isolated silanol groups attached to two different silicon atoms are bridged by H-bond. A third type of silanols called geminal silanols consists of two hydroxyl groups attached to one silicon atom (Preinerstorfer et al. 2006; Qhobosheane et al. 2001).

Silica surface can be modified either by *physical treatment* (thermal or hydrothermal) that leads to change in ratio of silanol and siloxane concentration of the silica surface or by *chemical treatment* that leads to change in chemical characteristics of silica surface. By the modification the adsorption properties are significantly affected. Chemisorption of specific functional groups or ions on silica surface provides immobility, mechanical stability and water insolubility, thereby increasing the efficiency, sensitivity and selectivity of the analytical and biotechnological applications (Preinerstorfer et al. 2006; Qhobosheane et al. 2001).

As shown in the Fig. 25.6, metal-grafted silica has given a set of properties to the surface, which differs considerably from the original one. Attachment of the metal ion to silica surface via sequences of reaction among with much better adsorption properties, dramatically changed pH- and ionic strength responses of the attached

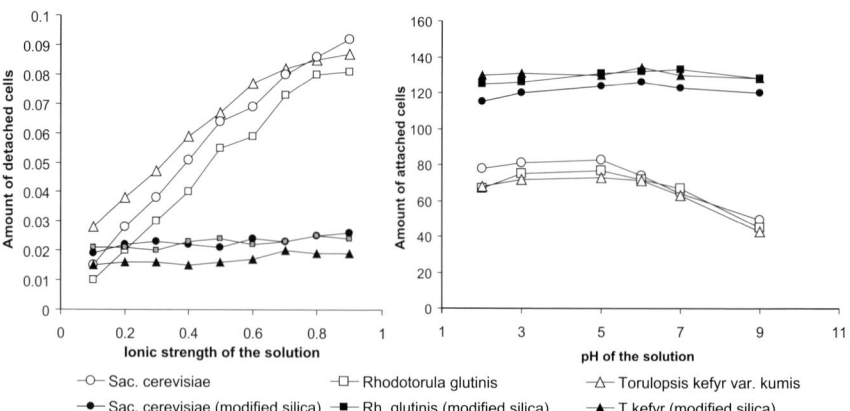

Fig. 25.6 Surface modification by transition metal salts caused dramatic changes in pH-dependence and ionic strength-dependence of microbial adsorption/desorption. SEM did not exceed 5% and is not shown

cells. This implies that the nature of cell-surface contact has changed completely, now involving metal-organic chemical bounds rather than charged groups interactions. Analysis of adsorption/desorption kinetics curves supports this hypothesis, strongly suggesting the appearance of new, much stronger type of chemical bonds after surface modification.

One important question to address is known toxicity of transition (heavy) metals caused by almost irreversible inhibition of many vitally important cellular enzymes. Our experiments on ethanol production, cell morphology, viability and proliferation rates showed no toxic effects on the yeasts immobilized with help of transition metals. To exhibit toxicity, heavy metals must enter the cell and directly contact with cellular enzymes. The surface bound cations obviously lack this property, realizing their high binding potential solely superficially.

In our laboratory, the surface modification using transition metals salts was successfully applied for immobilization of yeast cells under conditions of flow-through bioreactor. For the laboratory-scale bioreactor, very encouraging economic parameters (maintenance costs, economic coefficient, etc.) were obtained for the case of metal-aided adhesion.

25.5.2 Surface Preconditioning with Water-soluble Charged Polymers

An important natural phenomenon in the initial adhesion of microorganisms to non-living surfaces, e. g. to stainless steel, worthy of consideration is *surface conditioning* (Carpentier and Cerf 1993; Korber et al. 1994). It means that when a pristine surface is immersed, interactions between the surface and the liquid phase begin immediately. The surface will be modified by the adsorption of inorganic salts, proteins, glycoproteins, and humic compounds etc., depending on the environment. The conditioning film is formed rapidly, as significant organic deposits have been detected after only 15 min. Film thicknesses ranging from 10 to 80 nm have been measured (Carpentier and Cerf 1993; Korber et al. 1994). In practice, most microbial cells moving from the bulk aqueous phase towards a surface have their primary contact with such a preconditioned surface.

Most particles acquire an electric charge in aqueous suspension due to the ionization of their surface groups. Long-range electrostatic forces influence significantly the initial phase of bacterial adhesion onto solid surfaces. Eventually, the surface charge properties of any novel "colonization-resistant" surface will be gradually altered by adsorption of salts, proteins, glycoproteins and other molecules from the environment. Once a surface has been conditioned, its properties such as hydrophobicity are often permanently altered, so that the affinity of an organism for a native and a conditioned surface can be quite different (Dunne Jr 2002). Bacterial charged exopolysaccharides are the main component of the biofilm glycocalyx, which has also been coined the slime layer. It is known that many biotechnologically important yeast strains lacking such positively charged extracellular glycopolysaccharides and therefore adhere reversibly and to less extent.

The immense ability of *C. crescentus* to stick so firmly to surfaces mentioned above is the result of special excreted polysaccharides. These molecules are composed of long chains of *N*-actetylglucosamine (GlcNac), which are essential for the bacterial holdfast's remarkable stickiness (Li et al. 2005). Although the exact chemical makeup of the adhesive gel is not yet known, GlcNac polymers have been shown to be critical components of *C. crescentus* holdfast. On the other hand, Fletcher et al. found that adsorbed gelatine or pepsin impaired the attachment of marine pseudomonades to polystyrene Petri dishes. The basic proteins such as histone and poly-L-lysin facilitated *S. mutans* adherence to a hydroxyapatite disk and the acidic proteins such as phosvitin, *b*-lactoglobulin, or poly-L-glutamate inhibited adhesion of *S. epidermidis* adhesion to Teflon (Fletcher 1976).

Being spontaneous, surface preconditioning is highly unpredictable in terms of controlling microbial adhesion, so artificial manipulation with macromolecules seems to be inviting to gain more control over adsorption system. Charged polymers are of particular interest as charged groups on the surface of cells may be important physical factor for microbial adhesion. Polyethyleneimine (PEI) and other synthetic positively charged water-soluble polymers present a cheap alternative to natural macromolecules, successfully interfering with charge repulsion and even totally changing double electric layer (Fig. 25.7). Binding of PEI with negatively charged surfaces of yeasts and carriers is almost irreversible (measured constant of dissociation $K_d \approx 10^{-8}$ M) because of multiplicity of charge contact points.

Providing yeasts with such molecular "implants" might significantly enhance microbial adhesion, creating new thermodynamic conditions favorable for development of irreversible binding. The capability of this approach has been demonstrated on adsorption rate experiments, where modified surface bound much more yeast cells than the control one (Fig. 25.8). Correspondingly, the stability as well as the operational performance of the immobilized cells was much higher when PEI-aided surface conditioning had been used. These features are reflected in the Fig. 25.8 as

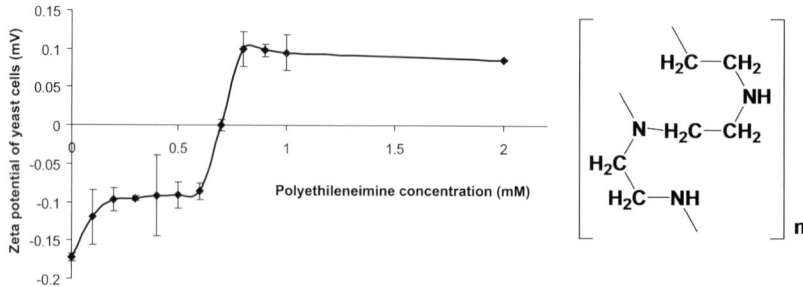

Fig. 25.7 Synthetic water-soluble polymer polyethyleneimine (PEI) is highly positively charged at wide pH range and therefore can cause inversion of double electric layer of colloid particles. *Right*: structural formula of PEI; *left*: recharging of Torulopsis kefyr surface by increased PEI concentration

25.5 Surface Preconditioning with Water-soluble Charged Polymers

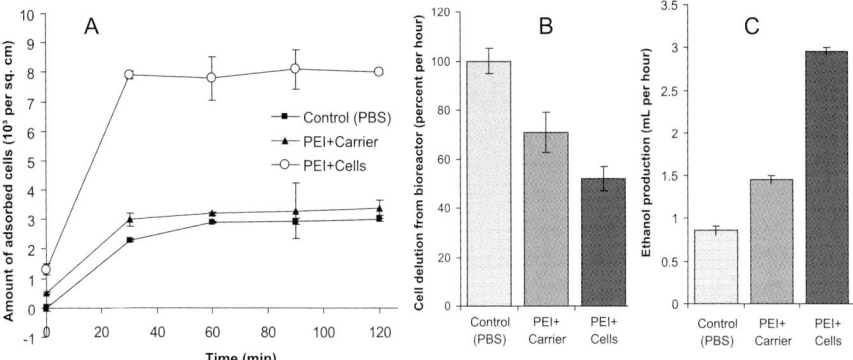

Fig. 25.8 Effects of polyethyleneimine on adsorption of *T. kefyr var. kumis* onto polyurethane foam. **A**: Kinetics of yeasts adsorption in dependence of whether cells or carrier was treated by PEI first. As a control, phosphate buffered saline (PBS) was used. **B**: Influence of PEI-induced preconditioning on detachment and dilution rate of immobilized cells from the bioreactor. **C**: Increase in ethanol production caused by PEI treatment as compared to the control group

reduced detachment of the cells from the substratum as well as the increased overall fermentation activity of the immobilized microbial biomass. This set of properties makes industrial and semi-industrial applications of such engineered surfaces very tempting especially for waste treatment and water purification brunches.

Intriguingly, during these experiments, we have noted a significant difference between the group where PEI was first mixed with yeast cells and the group where PEI was added to carrier beads. Adsorption of PEI on the cellular surface, followed by cell immobilization onto carrier resulted in appreciable increase in cell adhesion, adsorption strength and total biotechnological productivity. At the same time, if PEI was adsorbed onto the carrier surface first, the stimulating effect, though still visible, decreased. The reason for the observed asymmetry seems to be the difference in chemical surface groups of cells and carrier (polyurethane) respectively. Additionally, cellular component has much higher specific surface, providing more opportunities for polymer binding.

Being biologically neutral, PEI does not influence metabolic processes in yeasts and consequently its effects on adsorption system are rather related to effective surface charge neutralization together with creation of additional adhesion-active sites. Indeed, the analysis of adsorption kinetics implies that the surface preconditioning by PEI leads to apparent increase of adsorption-active surface, simultaneously shifting the thermodynamic equilibrium towards irreversible cell adhesion.

Resuming, microbial adhesion is a very complicated process that is affected by many factors, including some characteristics of the bacteria itself, the target material surface, and the environmental factors, such as the presence of macromolecules, ions or bactericidal substances. A better understanding of the unique behaviour of certain microorganisms, the surface characteristics of the carrier material, and the relevant environment conditions would make it possible for one to control the adhesion process by adjusting these factors. The efficacy of surface engineering approach in manipulating microbial adhesion was demonstrated for two different issues. The controllable stimulation or inhibition of adhesion can be achieved either by physical adsorbing or by chemical grafting of functional groups onto a suitable matrix. The feasibility and applicability of these methods has been proved in ethanol production using laboratory-scale bioreactors.

Acknowledgements The author would like to express his admiration and special gratitude to Prof. Dr. A. A. Zhubanova for her ceaseless help.

References

1. Ben-Jacob E, Levine H (2006) Self-engineering capabilities of bacteria. J R Soc Interface 3:197–214
2. Bhinu VS (2005) Insight into biofilm-associated microbial life. J Mol Microbiol Biotechnol 10:15–21
3. Bodenmiller D, Toh E, Brun YV (2004) Development of surface adhesion in Caulobacter crescentus. J Bacteriol 186:1438–1447
4. Branyik T, Kuncova G, Paca J (2000) The use of silica gel prepared by sol-gel method and polyurethane foam as microbial carriers in the continuous degradation of phenol. Appl Microbiol Biotechnol 54:168–172
5. Branyik T, Vicente AA, Kuncova G, Podrazky O, Dostalek P, Teixeira JA (2004) Growth model and metabolic activity of brewing yeast biofilm on the surface of spent grains: a biocatalyst for continuous beer fermentation. Biotechnol Prog 20:1733–1740
6. Brooks DE, Trust TJ (1983) Interactions of erythrocytes with bacteria under shear. Ann N Y Acad Sci 416:319–331
7. Busscher HJ, Poortinga AT, Bos R (1998) Lateral and perpendicular interaction forces involved in mobile and immobile adhesion of microorganisms on model solid surfaces. Curr Microbiol 37:319–323
8. Busscher HJ, Van Der Mei HC (2006) Microbial adhesion in flow displacement systems. Clin Microbiol Rev 19:127–141
9. Cabuk A, Akar T, Tunali S, Tabak O (2006) Biosorption characteristics of Bacillus sp. ATS-2 immobilized in silica gel for removal of Pb(II). J Hazard Mater 136:317–323
10. Carpentier B, Cerf O (1993) Biofilms and their consequences, with particular reference to hygiene in the food industry. J Appl Bacteriol 75:499–511
11. Christersson CE, Glantz PO, Baier RE (1988) Role of temperature and shear forces on microbial detachment. Scand J Dent Res 96:91–98
12. Costerton JW (1999) Introduction to biofilm. Int J Antimicrob Agents 11:217–221
13. Davey ME, Caiazza NC, O'Toole GA (2003) Rhamnolipid surfactant production affects biofilm architecture in Pseudomonas aeruginosa PAO1. J Bacteriol 185:1027–1036
14. Davey ME, O'Toole GA (2000) Microbial biofilms: from ecology to molecular genetics. Microbiol Mol Biol Rev 64:847–867
15. Davidenko TI, Bondarenko GI (1990) The use of microorganism cells immobilised in carrageenan for the synthesis of organic substances. Russian Chemical Reviews 59(3):299–306. Ref Type: Abstract
16. Davies DG, Parsek MR, Pearson JP, Iglewski BH, Costerton JW, Greenberg EP (1998) The involvement of cell-to-cell signals in the development of a bacterial biofilm. Science 280:295–298
17. Donlan RM, Costerton JW (2002) Biofilms: survival mechanisms of clinically relevant microorganisms. Clin Microbiol Rev 15:167–193

18. Dunne WM Jr (2002) Bacterial adhesion: seen any good biofilms lately? Clin Microbiol Rev 15:155–166
19. Favre-Bonte S, Kohler T, Van DC (2003) Biofilm formation by Pseudomonas aeruginosa: role of the C4-HSL cell-to-cell signal and inhibition by azithromycin. J Antimicrob Chemother 52:598–604
20. Fletcher M (1976) The effects of proteins on bacterial attachment to polystyrene. J Gen Microbiol 94:400–404
21. Flint S, Palmer J, Bloemen K, Brooks J, Crawford R (2001) The growth of Bacillus stearothermophilus on stainless steel. J Appl Microbiol 90:151–157
22. Gottenbos B, Busscher HJ, Van Der Mei HC, Nieuwenhuis P (2002) Pathogenesis and prevention of biomaterial centered infections. J Mater Sci Mater Med 13:717–722
23. Habash MB, Van Der Mei HC, Reid G, Busscher HJ (1997) Adhesion of Pseudomonas aeruginosa to silicone rubber in a parallel plate flow chamber in the absence and presence of nutrient broth. Microbiology 143(8):2569–2574
24. Herrera-Alonso M, McCarthy TJ, Jia X (2006) Nylon surface modification: 2. Nylon-supported composite films. Langmuir 22:1646–1651
25. Junter GA, Jouenne T (2004) Immobilized viable microbial cells: from the process to the proteome em leader or the cart before the horse. Biotechnol Adv 22:633–658
26. Kennedy JK, Cabral MS (1990) Use of titanium species for the immobilization of cells. Transition Metal Chemistry 15(3):197–207. Ref Type: Abstract
27. Korber DR, Choi A, Wolfaardt GM, Ingham SC, Caldwell DE (1997) Substratum topography influences susceptibility of Salmonella enteritidis biofilms to trisodium phosphate. Appl Environ Microbiol 63:3352–3358
28. Korber DR, James GA, Costerton JW (1994) Evaluation of Fleroxacin Activity against Established Pseudomonas fluorescens Biofilms. Appl Environ Microbiol 60:1663–1669
29. Kuchma SL, O'Toole GA (2000) Surface-induced and biofilm-induced changes in gene expression. Curr Opin Biotechnol 11:429–433
30. Li B, Takahashi H (2004) New immobilization method for enzyme stabilization involving a mesoporous material and an organic/inorganic hybrid gel. Biotechnology Letters 22(24):1953–1958. Ref Type: Abstract
31. Li G, Smith CS, Brun YV, Tang JX (2005) The elastic properties of the caulobacter crescentus adhesive holdfast are dependent on oligomers of N-acetylglucosamine. J Bacteriol 187:257–265
32. Marshall KC (1986) Adsorption and adhesion processes in microbial growth at interfaces. Adv Colloid Interface Sci 25:59–86
33. Nabe K, Izuo N, Yamada S, Chibata I (1979) Conversion of Glycerol to Dihydroxyacetone by Immobilized Whole Cells of Acetobacter xylinum. Appl Environ Microbiol 38:1056–1060
34. Neu TR (1996) Significance of bacterial surface-active compounds in interaction of bacteria with interfaces. Microbiol Rev 60:151–166
35. O'Toole GA, Gibbs KA, Hager PW, Phibbs PV Jr, Kolter R (2000) The global carbon metabolism regulator Crc is a component of a signal transduction pathway required for biofilm development by Pseudomonas aeruginosa. J Bacteriol 182:425–431
36. Ohmiya K, Ohashi H, Kobayashi T, Shimizu S (1977) Hydrolysis of lactose by immobilized microorganisms. Appl Environ Microbiol 33:137–146
37. Oosthuizen MC, Steyn B, Theron J, Cosette P, Lindsay D, Von HA, Brozel VS (2002) Proteomic analysis reveals differential protein expression by Bacillus cereus during biofilm formation. Appl Environ Microbiol 68:2770–2780
38. Peng JS, Tsai WC, Chou CC (2002) Inactivation and removal of Bacillus cereus by sanitizer and detergent. Int J Food Microbiol 77:11–18
39. Poortinga AT, Bos R, Busscher HJ (2001) Electrostatic interactions in the adhesion of an ion-penetrable and ion-impenetrable bacterial strain to glass. Colloids Surf B Biointerfaces 20:105–117
40. Preinerstorfer B, Lubda D, Lindner W, Lammerhofer M (2006) Monolithic silica-based capillary column with strong chiral cation-exchange type surface modification for enantioselective non-aqueous capillary electrochromatography. J Chromatogr A 1106:94–105

41. Qhobosheane M, Santra S, Zhang P, Tan W (2001) Biochemically functionalized silica nanoparticles. Analyst 126:1274–1278
42. Rangsayatorn N, Pokethitiyook P, Upatham ES, Lanza GR (2004) Cadmium biosorption by cells of Spirulina platensis TISTR 8217 immobilized in alginate and silica gel. Environ Int 30:57–63
43. Rasmussen TB, Givskov M (2006) Quorum-sensing inhibitors as anti-pathogenic drugs. Int J Med Microbiol 296:149–161
44. Seyhan E, Kirwan DJ (1979) Nitrogenase activity of immobilized Azotobacter vinelandii. Biotechnol Bioeng 21:271–281
45. Sheth NK, Franson TR, Sohnle PG (1985) Influence of bacterial adherence to intravascular catheters on in-vitro antibiotic susceptibility. Lancet 2:1266–1268
46. Soto GE, Hultgren SJ (1999) Bacterial adhesins: common themes and variations in architecture and assembly. J Bacteriol 181:1059–1071
47. Stoodley P, Sauer K, Davies DG, Costerton JW (2002) Biofilms as complex differentiated communities. Annu Rev Microbiol 56:187–209
48. Szewzyk U, Szewzyk R, Manz W, Schleifer KH (2000) Microbiological safety of drinking water. Annu Rev Microbiol 54:81–127
49. Tang G, Yip HK, Samaranayake LP, Chan KY, Luo G, Fang HH (2004) Direct detection of cell surface interactive forces of sessile, fimbriated and non-fimbriated Actinomyces spp. using atomic force microscopy. Arch Oral Biol 49:727–738
50. Thomas WE, Trintchina E, Forero M, Vogel V, Sokurenko EV (2002) Bacterial adhesion to target cells enhanced by shear force. Cell 109:913–923
51. Tsang PH, Li G, Brun YV, Freund LB, Tang JX (2006) Adhesion of single bacterial cells in the micronewton range. Proc Natl Acad Sci USA 103:5764–5768
52. Vigeant MA, Ford RM (1997) Interactions between motile Escherichia coli and glass in media with various ionic strengths, as observed with a three-dimensional-tracking microscope. Appl Environ Microbiol 63:3474–3479
53. Watnick P, Kolter R (2000) Biofilm, city of microbes. J Bacteriol 182:2675–2679
54. Whiteley M, Bangera MG, Bumgarner RE, Parsek MR, Teitzel GM, Lory S, Greenberg EP (2001) Gene expression in Pseudomonas aeruginosa biofilms. Nature 413:860–864
55. Zobell CE (1943) The Effect of Solid Surfaces upon Bacterial Activity. J Bacteriol 46:39–56
56. Zobell CE, Mathews HM (1936) A Qualitative Study of the Bacterial Flora of Sea and Land Breezes. Proc Natl Acad Sci USA 22:567–572

Chapter 26
Air Purification Technology by Means of Cluster Ions Generated by Plasma Discharge at Atmospheric Pressure

Kazuo Nishikawa and Matthew Cook

Appliance Systems Product Development Center, Sharp Corporation, Osaka, Japan,
nishikawa.kazuo@sharp.co.jp

Abstract The increased density of our living environment coupled with pollution of the atmosphere has led to a growing need for the removal of harmful molecules in the air (1). As a result research into applying a plasma discharge into the atmosphere and creating ozone and radicals of strong chemical reactivity to purify the air environment has gathered momentum. The removal of airborne particles, such as bacteria, allow for an improvement in indoor air quality so that our environment is healthy and pleasant. Within the medical field, illnesses caused by viruses such as influenza and SARS (2), hospital infections caused by airborne bacteria, fungi and allergic bronchial tube asthma (3), Japanese cedar hay fever caused by inhaling cedar pollen (4) are becoming large social concerns.

In this research article, we discuss how we have applied our novel plasma discharge technology to produce positive and negative "cluster" ions. This ion-generating device operates at a normal atmospheric pressure. Subsequent investigations have permitted characterization of the resultant cluster ions. We have performed a series of experiments to prove the air purification effects of cluster ions, paying close attention to airborne harmful microbes and cedar pollen allergens.

26.1 Ion Generating Device

An electrode is formed on the surface of the flat ceramic dielectric body. This ceramic body is attached to high voltage and applied electrodes. The high AC voltage is applied permitting a plasma discharge state on the surface. If molecular energy from the air is applied by plasma discharge, ionization and dissociation of the airborne molecules occurs. These events all occur at atmospheric pressure (5).

The ion generating device has been designed to allow the discharged electron energy to become a monochrome beam. When the optimal 5 eV voltage is applied each molecule in the atmosphere there is distribution from the electrode.

Fig. 26.1 Photo of an ion-generating device

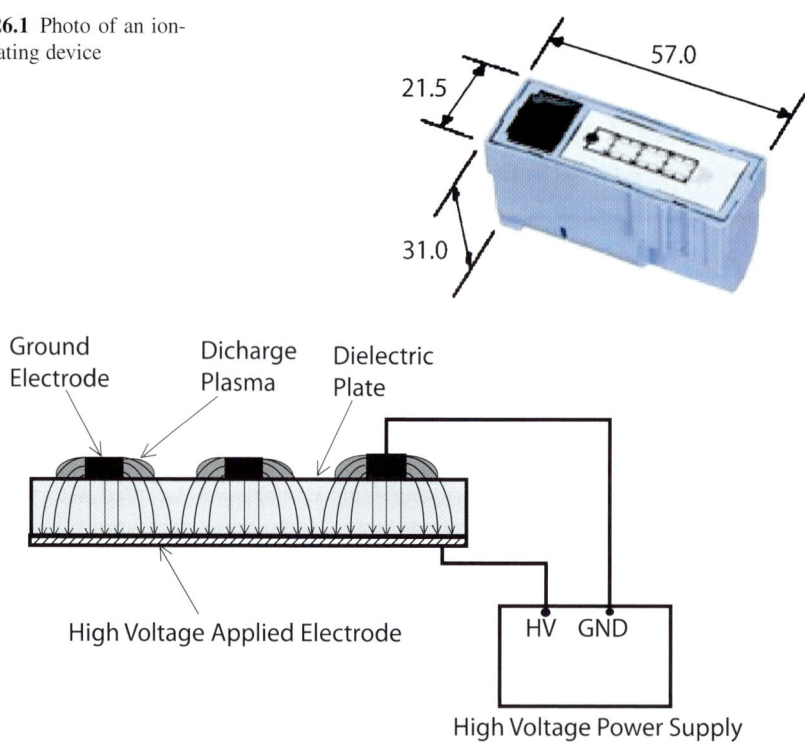

Fig. 26.2 Section diagram of an ion-generating electrode

The concentration of ozone has been confirmed to be less that 0.005 ppm in the vicinity of the device. This value has been deemed safe by a number of health bodies over the world including the Association of Home Appliance Manufacturers, AHAM. We always measure this since there have been numerous reports of the differences in concentration of Ozone from ion-generating devices.

This device is displayed in Fig. 26.1 and shows the actual component used in many SHARP products. In Fig. 26.2, the ion-generating device is shown in cross section and highlights some of the points stated above.

26.2 Characteristics of Positive and Negative Ions

The ion density is measured using an air ion counter (Dan Science 83-1011B) (7) by means of the double concentric circle tube method (6).

For identification of the type of ion we measured the mass spectrum of the positive and negative ions (Fig. 26.3a and b) using a flight time decomposition analysis apparatus. For the positive ions oxonium ions H_3O^+ were generated and around them, cluster ions of $H_3O^+(H_2O)_m$ with a structure of water molecules aligned

26.3 Effect of Removing Airborne Bacteria

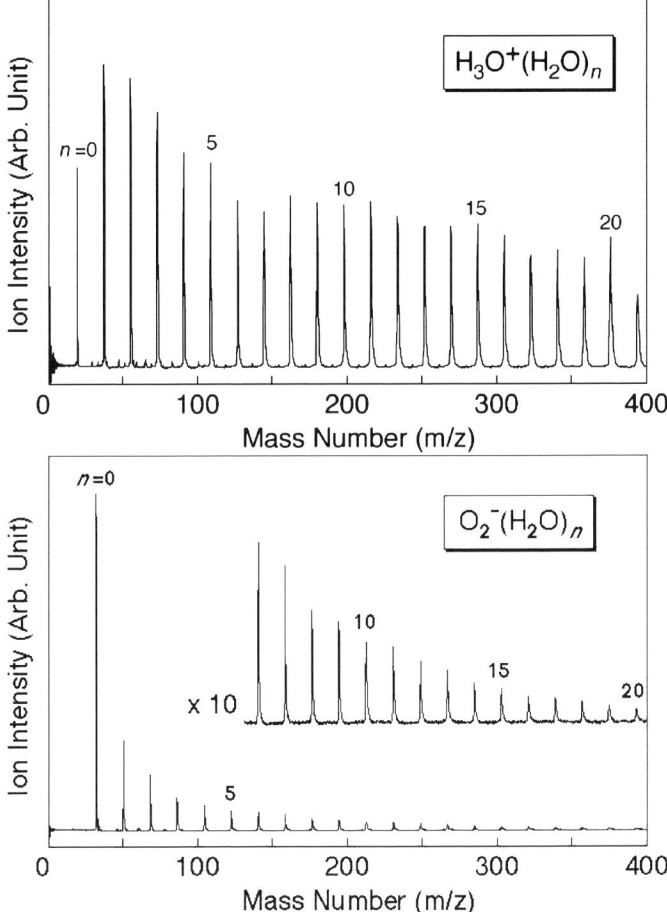

Fig. 26.3 **a** Positive ion mass spectrum. **b** Negative ion mass spectrum

($_m$ is constant), for the negative ions, oxygen molecule ions O_2 were generated and around them were cluster ions of $O_2^-(H_2O)_n$ with a structure of water molecules aligned around them ($_n$ is constant). Other types of ion generation were not found during our studies.

26.3 Effect of Removing Airborne Bacteria

Bacillus coli cultures were sprayed as a mist into a room with a volume of 30 m. Measurements was made to calculate the density of bacillus coli using an air sampler. The results are shown in Fig. 26.4 and display the changes with the passage of time of the amount of bacillus coli in the air. The effect of removing 90% of *bacillus*

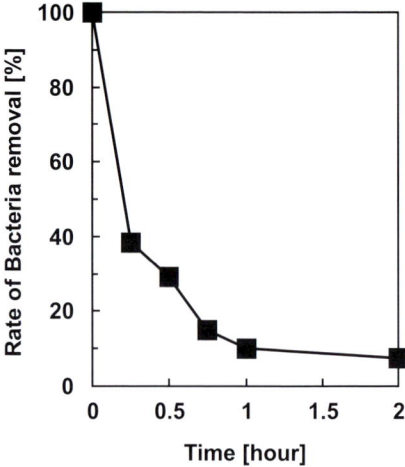

Fig. 26.4 Effect of removing bacillus coli in air by positive and negative cluster ions; Y axis = Rate of Bacteria removal; X axis = Time

coli in the air required one hour when discharging cluster ions at a mean ion density of $3000/cm^3$.

Figure 26.5 shows a photo of a Petri dish of the *bacillus coli* collected cultivated during 24 hours in LB agar medium. Without generating cluster ions, the growth of *bacillus coli* colonies was confirmed. When positive and negative ions were applied, the generation of colonies were not observed. This showed that the growth of *bacillus coli* is prevented resulting from their deactivated by positive and negative cluster ions.

Figure 26.6 shows changes with the passage of time in the density of the *bacillus coli* due to positive and negative ions. In the case of negative ions only (mean density $6000/cm^3$), the ratio of remaining bacteria after one hour was 85% (removal rate 15%), and the effect of removing bacteria was extremely small and even after more time passed, no more were removed. In contrast, in the case of positive and negative ions with added positive ions (mean density $3000/cm^3$), the ratio of re-

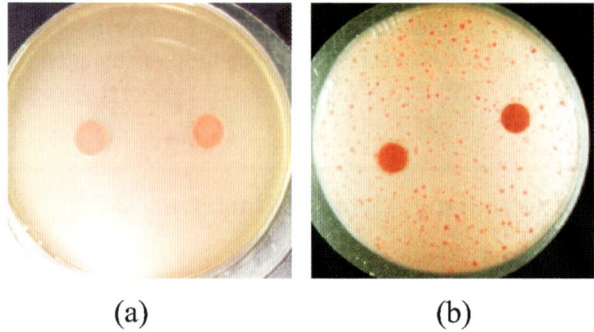

Fig. 26.5 Photo of *bacillus coli* incubation in a Petri dish **a** with ions and **b** without ions

26.3 Effect of Removing Airborne Bacteria

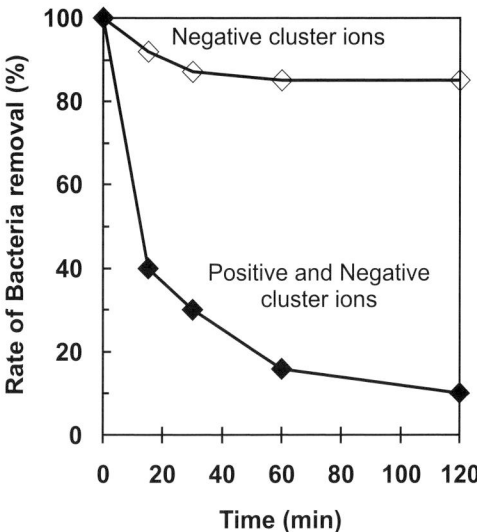

Fig. 26.6 Characteristic of suspended *E. coli* bacteria removal with the positively and negatively charged ions, and the negatively charged ions

maining bacteria after one hour was 16% (removal rate 84%), and a large removal effect was obtained. From this we can see that a removal effect of bacteria can be achieved by positive and negative cluster ions.

MRSA (Methicillin-resistant Staphylococcus aureus) that is a common hospital acquired infection was sprayed into a box with a volume $1\,\text{m}^3$. The density of MRSA in the air measured using an air sampler. Positive and negative cluster ions were discharged into the air (mean density $10{,}000\,\text{cm}^3$) and it was found that after 30 minutes, 90% of the original MRSA concentration in the air had been removed and after 60 minutes no MRSA could be found in the air. Figure 26.7 shows the changes with time of the density of floating MRSA in the air.

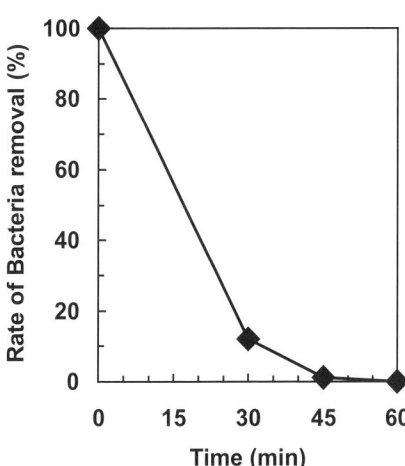

Fig. 26.7 Effect of removing MRSA (*methicillin-resistance staphylococcus aureus*) in air by positive and negative cluster ions

26.4 Effect of Removing Floating Fungi (Mould)

Cladosporium spores were sprayed into a box of 1 m³ and the density of floating fungi in the air measured. Figure 26.8 shows the changes with time in the density of the floating fungi in the air.

By discharging positive and negative cluster ions into the air (mean density 10,000/cm³), it was confirmed that 90% of the fungi were removed after 45 minutes and after 60 minutes more than 99%. Figure 26.9 shows the state of propagation of fungi when positive and negative ions and negative ions only were used. When left for 10 days, propagation of the fungi could be seen when there were only negative ions, but when there were positive and negative ions, the fungi were not found to propagate.

From this we observed that positive and negative cluster ions have the profound effect of restricting propagation of fungi.

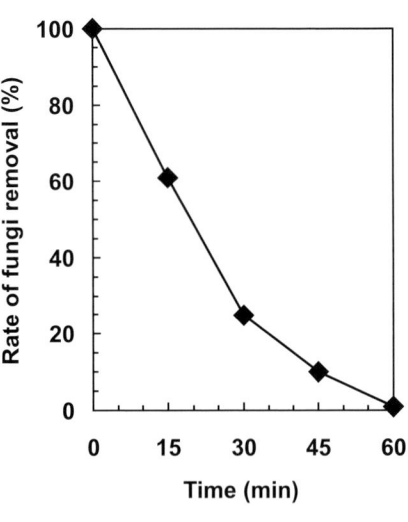

Fig. 26.8 Effect of removing fungi count (*cladosporium*) in air by positive and negative cluster ions

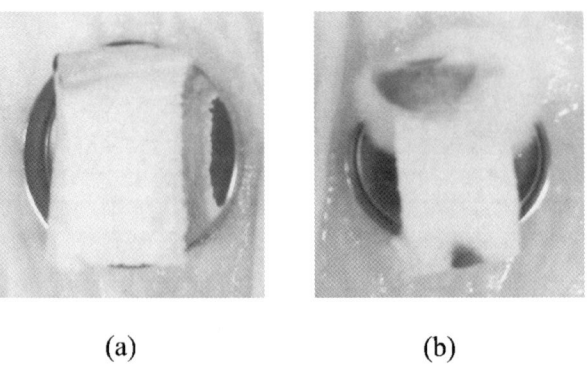

Fig. 26.9 Photographs of the mold growth with the operation of positively and negatively charged cluster ions (**a**), and negatively charged cluster ions (**b**)

26.5 Effect of Deactivating Floating Viruses

The apparatus used to test viruses required a slightly different set up and Fig. 26.10 shows a schematic example of this. The ion-generating device is installed in an acrylic tube of length 200 mm and an external diameter (only if the ion density is $2000/cm^3$) of 170 mm. At one end of the tube an atomizer is attached to spray the virus, while the other side required an impinger to collect the virus. In our extensive experiments, the influenza virus A(H1N1) A/PR8/34 was used. The atomizer contained 10 mL of virus solution and was attached to one end of the cylindrical test apparatus.

The impinger contained 10 mL of phosphate buffer solution (PBS). From an air compressor air supply, air at a of speed 4 m/s was passed into the cylinder (if the ion density if $2000/cm^3$ only the air speed is 0.4 m/s), the virus was sprayed and passed over the ion generating device within the cylinder. The volume of spray was 3.0 mL and the spray speed set at 0.1 mL/min. Taking the case of when the ion generating device is not being operated as the control, we compared the amount of virus when the ion generating device was operated. Tests were carried out with the ion density of positive and negative ions of $200,000\,cm^3$, $100,000\,cm^3$, $50,000\,cm^3$ and $5000\,cm^3$ and $2000\,cm^3$. The air that passed through the tube was collected by the impinger for 30 minutes intervals at a sampling speed of 10 L/min.

The ion density was measured at a distance of 10 cm from the jet part of the cylinder test apparatus. The atmosphere was kept at a temperature of $18 \pm 1\,°C$ and relative humidity $43 \pm 2\%$.

Figure 26.11 shows the ratio of the number of formations of plaque of influenza depending on the ion density. The measurements used the plaque method which uses Madin-Darby canine kidney (MDCK) cells. Taking the number of plaques when the ion-generating device is not in action during the control as 100%, when the ion density of $200,000\,cm^3$, $100,000\,cm^3$, $50,000\,cm^3$ and $5000\,cm^3$ and $2000\,cm^3$ is

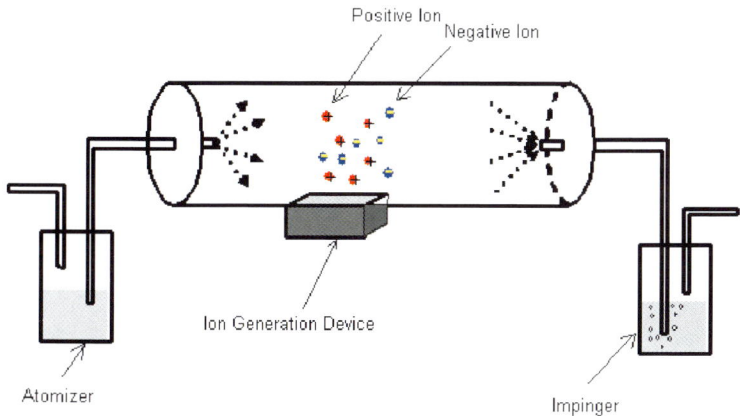

Fig. 26.10 Schematic diagram of the test apparatus

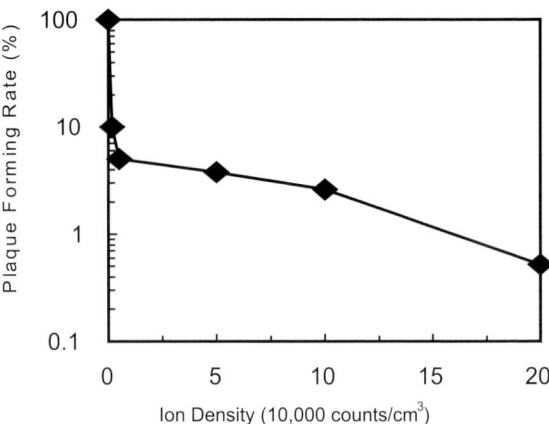

Fig. 26.11 Influenza virus plaque forming rate by the plaque method using MDCK cells

acted on by the influenza virus, a reduction of the number of plaques of 99.5%, 97%, 96%, 95% and 90% was found.

When this was the case the ozone density was less than 0.005 ppm. Also, the when the ion density was less than $100/cm^3$ and the ozone density 0.005 ppm, a reduction of the number of plaques was not found.

Figure 26.12 shows a photo of the MDCK cells that have been vaccinated with the influenza virus acted on by ions and not acted on by ions. When the influenza virus acted on by ions was vaccinated, the virus has been deactivated by the ions so the transmission to the cells could not be seen and the cells maintain their normal shape.

Figure 26.13 shows a photo of the red cell aggregation reaction of the influenza virus acted on by ions and not acted upon. It shows the properties that gather in the centre when red blood cells are placed in a receptacle that has a hollow in the centre (top photo). The red cells are aggregated by the protein on the surface of the virus when the influenza virus that has not been acted on by ions is vaccinated into the

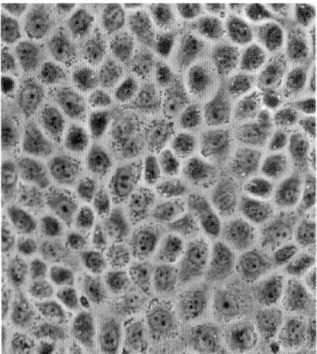
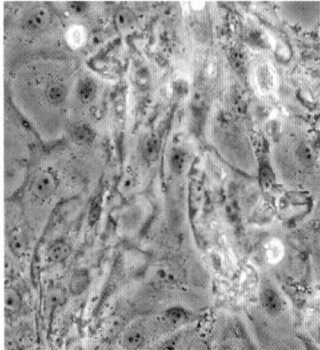

Fig. 26.12 Efficacy evaluation against influenza virus test of viral infection in madin-darby canine kidney (MDCK) cells

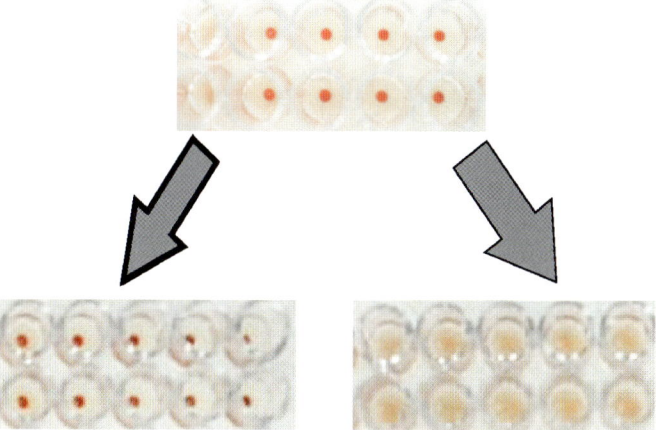

Fig. 26.13 Observation the mode of red blood corpuscle vaccinated with influenza virus

red blood cells, and the red blood cells no longer collect in the centre (bottom right photo). Even if the influenza virus that has been acted on by ions is vaccinated into the red cells, aggregation of the red blood cells due to the virus does not occur and the same properties as normal blood cells were shown (photo bottom left). When the ions are made to act on the influenza virus, it was found that the function of the protein that causes the aggregation reaction (hemagglutinin) and the red cells on the virus surface is reduced.

26.6 Virus Deactivation Model Using Cluster Ions

The effect of deactivating the influenza virus in the air with positive and negative cluster ions generated has been confirmed. Figure 26.14 shows the influenza de-activation model in the air due to positive and negative cluster ions. Further work has been carried out to define this method, however the results we not available for this manuscript. It is thought that the cluster ions collide with viruses in the air and surround the virus.

Positive cluster ions H_3O^+ $(H_2O)_m$ (m is constant) and negative cluster ions $O_2^-(H_2O)_n$ (n is constant) react on the surface of the virus and active species of very high reactivity are generated. Our future work will determine which surface molecules are indeed effected by these cluster ions. However we postulate that these cluster ions alter the protein of the virus surface. In the influenza virus, the protein hemagglutinin that protruded from the surface is altered by cluster ions (8). Hemagglutinin performs a key role in the assay permitting red blood cell (RBC) aggregation. As the hemagglutinin is altered by the ions, it is thought that the influenza virus is deactivated and little or no RBC aggregation was observed.

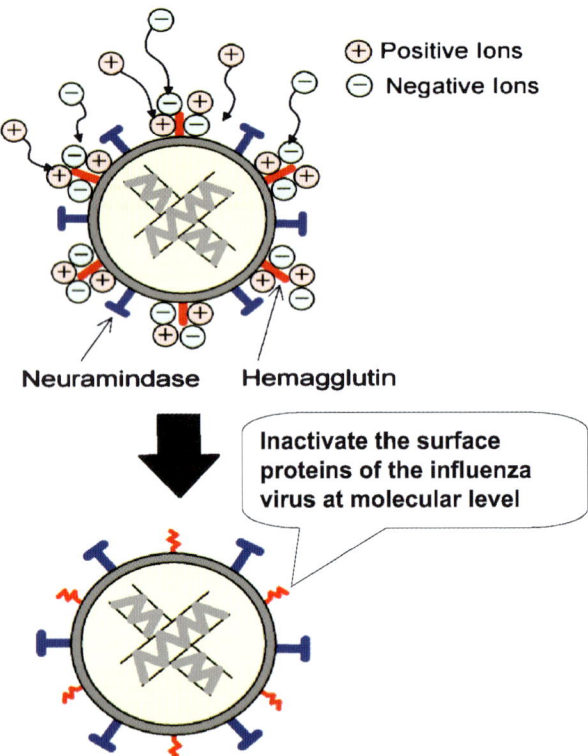

Fig. 26.14 Model for inactivation of infection capacity from viruses

26.7 Allergen Deactivation Effect

Crude antigens (protein density 200 ng/mL) extracted from Japanese Cedar pollen were sprayed in a mist using a nebulizer into the cylindrical receptacle of 0.9 L capacity. By operating an ion generating device placed in the receptacle, positive and negative cluster ions were generated in the receptacle space. The cedar pollen crude antigens exposed in cluster ions were collected for about 90 seconds. Following this 5 sets of analyses were performed: an allergen evaluation reaction, ELISA method, ELISA inhibition method, intradermal reaction and a conjunctival reaction tests.

26.7.1 Allergen Evaluation Reaction

The main antigens found in the cedar pollen are Cry j1, Cry j2. The presence of these proteins in the air determines the extent of reactivity with, for example, humans. The results of the evaluation by the ELISA method (Fig. 26.15) show the reactivity of

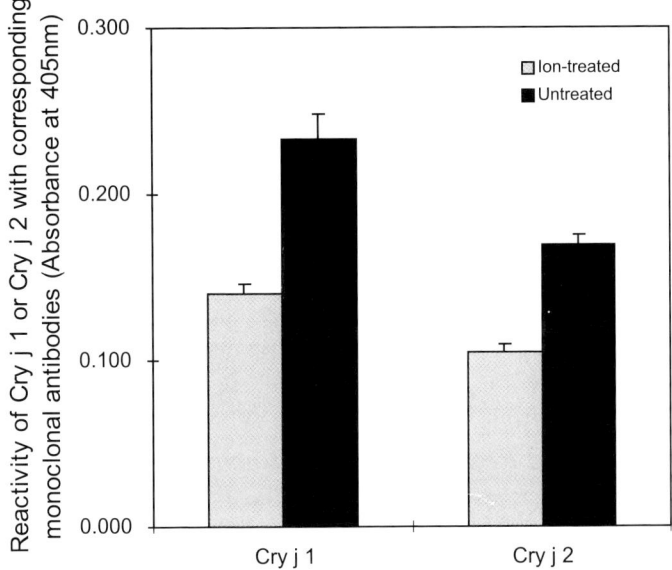

Fig. 26.15 The charged ions reduced the antigenicities of Cry j1 and Cry j2 with cedar pollen allergen

Cry j1, Cry j2. The assay concerns the reactivity of their respective monoclonal antibodies and as shown, their reactivity is reduced by the cluster ions.

26.7.2 ELISA Method

In both Cry j1 and Cryj2, the reactivity with the monoclonal antibodies reduced significantly. In particular the reduction of the reactivity of Cry j1 and its monoclonal antibodies was notable. The ELISA method evaluated whether the reaction between the cedar pollen crude antigens and the Immunoglobulin E (IgE) blood serum of the cedar pollen allergy sufferers changes due to the cluster ions. The results of evaluating the blood serum from 42 different sufferers showed that in 33 patients, due to the cluster ions, regarding the crude antigens, the reactivity with the blood serum IgE of patients reduced significantly. In 2 patients the reactivity with IgE reduced more than 80%, in 3 patients 70–80%, in 5 patients 60–70%, in 8 patients 50–60%, in 6 patients 40–50% and in the final 4 patients it reduced by 30–40%.

26.7.3 ELISA Inhibition Method

To evaluate quantitatively the allergy reaction deactivation rate of the cedar pollen crude antigens and the cedar pollen allergy sufferers' blood serum IgE by cluster ions, the ELISA inhibition method was carried out. The results of this assay can be seen in Fig. 26.16. When the antigenecity at inhibition rate 50% was evaluated, in cluster ion processing crude antigens 13.5 ng is required while in the case of crude antigens without processing 2.83 ng were required. The cluster ion processing antigens showed the same inhibition rate in an amount of about 4.8 so it was confirmed that the reactivity between the cedar pollen crude antigens and the patients' blood serum IgE was reduced by about 79% by the cluster ions.

26.7.4 Intradermal Reaction and Conjunctival Reaction Tests

We evaluated intradermally and by a conjunctival reaction test whether the allergenity of the cedar pollen crude antigens changes *in vivo* due to the effect of cluster ions.

For the intradermal reaction 0.02 mL of a diluted mixture of cedar pollen crude antigens and 0.9% solution of NaCl with a protein density of $0.5\,\mu g/mL$ was injected into the skin of the forearm flexor side of the cedar allergy sufferer using a tuberculin hypodermic needle. The diameter and radius of the wheals and red spots that arose after about 15 minutes were measured and from these mean radii the reactivity was evaluated. We made red spots of less than 10 mm −, red spots of 10 – 20 mm ±, red spots of 20 – 30 mm and wheals of less than 10 mm +, red spots of 30 – 40 mm, wheals of 10 – 14 mm ++, red spots of more than 40 mm, wheals of more than 15 mm and things giving pseudopodia to the wheals +++. The results of

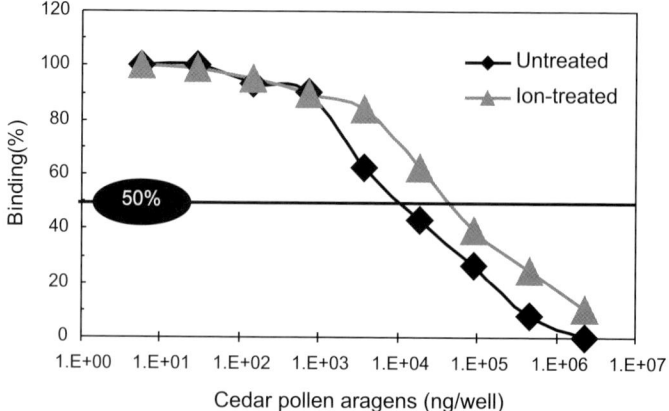

Fig. 26.16 Inhibition of IgE-binding to untreated Japanese cedar pollen allergen with various concentrations of ion-treated or untreated Japanese cedar pollen allergens

carrying out intradermal reaction exams on 6 cedar allergy sufferers showed that in all patients the intradermal reactivity due to cedar pollen crude antigens reduced from +++ to + due to cluster ions.

With respect to the conjunctival reaction, 5 μL of a mixture of cedar pollen crude antigens was diluted in 0.9% NaCl at a protein concentration of 0.5 μg/mL was dropped into the eyes of 6 cedar pollen sufferers using a pipette. In addition, the same test was performed with; however 5 μL of a mixture of cedar pollen crude antigens had been treated with cluster ions before mixing into solution with PBS.

Following incubation for 15 minutes, the extent of any conjunctival reaction was observed. This qualitative test covers a number of different areas which include: meniscus skin wall, the congestion of the eyelid skin and spherical conjunctiva, itchiness and tears. When no congestion was observed, we assessed it as 1, when there was slight congestion and itchiness ±, when congestion was seen on either the lower or upper conjunctiva of the eyeball +, when congestion was seen on both the upper and lower eyeball conjunctiva ++, and when congestion was found on all the eyeball conjunctiva +++ and when edema of the eyelid was seen ++++.

Tests for the conjunctiva reaction were carried out on 6 cedar pollen sufferers and the results showed that in 5 patients the conjunctival reactivity due to cedar pollen antigens was reduced from ++ to − due to cluster ions and we understood that the potency of the conjunctiva reaction was lost.

26.8 Conclusion

We have developed an ion generating device for generating positive ions H_3O^+ and negative ions O_2^-. Also we have demonstrated the effect of removing pollen allergens and airborne harmful microbes (fungi, bacteria and viruses) as well as pollen allergens.

The air purification technology developed that uses positive and negative cluster ions shows many excellent characteristics for purification of the environment and it is expected that it will develop into a wide range of applications in industry types other than household products.

References

1. Tetsuji Oda et al. (2000) The future of Technologies on Environmental Measures by Discharge. Japan Society of Applied Physics 69:263–289
2. Bunshichi Shimizu (1996) Understanding the Virus, Koudansha, pp 129–187
3. Yoichi Minamishima, Yasuo Mizukuchi, Hiroaki Nakayama (1987) Current Microbiology, Nanzando
4. Yozo Saito et al. (1994) The Science of Hay fever
5. Journal of The Japan Research Group of Electrical Discharges (1998) Discharge Handbook, Oum, pp 400–417
6. Shinichiro Kitagawa, Zenichiro Kawasaki, Kazuhiko Miura, Koichiro Domoto (1996) Study of Atmosphere Electricity, Published by Tokai University, pp 45–61
7. Tadashi Akiyama (1997) Clean Technology 8:54–58
8. Artmann et al. (2005) Bactericidal effects of plasma-generated cluster ions. Med Biol Eng Comput 43(6):800–807

Chapter 27
Astrobiology

Gerda Horneck

German Aerospace Center, Institute of Aerospace Medicine, 51170 Koeln, Germany,
gerda.horneck@dlr.de

Abstract Astrobiology addresses questions that have intrigued humans for a long time: "How did life originate?", "Are we alone in the Universe?", and "What is the future of life on Earth and in the Universe?" It is a multidisciplinary approach to understand the processes that lead to life's appearance, evolution and distribution as a planetary phenomenon, embedded in the evolution of the Universe. The ultimate goal is to reach a more universal definition of life. Knowledge of the history of life on Earth and of strategies of life to adapt to extreme environments is used as base for assessing the occurrence of potential habitats beyond the Earth, e. g. within our Solar System on Mars or Jupiter's moon Europa. Planetary protection guidelines have been developed to avoid unintentional introduction of terrestrial life forms to those planets or moons of interest, which may jeopardize future search for life studies. Measures include bioload monitoring and/or reduction, depending on target body and type of mission. Finally, radioastronomy has just started to detect extrasolar planetary systems. More sensitive space telescopes are being developed with the aim to detect and characterize Earth-like planets including possible biosignatures on them.

27.1 Introduction

Astrobiology is a relatively new research area that attempts to reveal the origin, evolution and distribution of life on Earth and throughout the Universe in the context of cosmic evolution. In this attempt, we understand life as a system which is capable of demonstrating evolution by means of natural selection. The final goal of astrobiology is building the foundations for the construction and testing of meaningful axioms to support a theory of life.

To reach this goal, a multidisciplinary approach is required involving disciplines as astronomy, planetary research, geology, paleontology, chemistry, biology and others. Astrobiology extends the boundaries of biological investigations beyond the Earth, to other planets, comets, meteorites, and space at large. Focal points are the different steps of the evolutionary pathways through cosmic history that may be related to the origin, evolution and distribution of life. An excellent overview over this emerging field of astrobiology has been provided by Brack (1998), Horneck and Baumstark-Khan (2002), Gargaud et al. (2005), and Horneck and Rettberg (2007). In the interstellar medium, as well as in comets and meteorites, complex organics in huge reservoirs are detected that eventually may provide the chemical ingredients for life. More and more data on the existence of planetary systems in our Galaxy are being acquired which support the assumption that habitable zones are frequent and are not restricted to our own Solar System. From the extraordinary capabilities of life to adapt to environmental extremes, the boundary conditions for the habitability of other bodies within our Solar System and beyond can be assessed. Hence, astrobiology has the potential to give new impulses to biology much as the development of astronomy has broadened our understanding of the physical world and the spectral analysis of the stars has proven the universality of the concepts of chemistry (Lederberg 1960).

27.2 Origin and History of Life on Earth

On Earth, most of the early geological record has been erased by later events so that we remain ignorant of the true historical facts concerning the origin of life on this planet. However, already the oldest sedimentary rocks show signatures of fossil microorganisms (Westall et al. 2001), indicating that the history of life on Earth goes back over at least 3.5 billion years. Therefore, events leading to the origin of life must have predated this time. Figure 27.1 gives an overview of the evolutionary events from the beginnings towards our present biosphere.

With regard to the chemical prerequisites for the origin of life, the availability of the so-called "biogenic" elements carbon, hydrogen, oxygen, nitrogen, sulfur and phosphorus (CHONSP) as well as relevant "biogenic" organic compounds are con-

Fig. 27.1 Iconic presentation of the origin and history of life on Earth (Credit: NASA, artist's design)

sidered to be indispensable, as well as the presence of liquid H_2O. The biogenic elements, which make up the bulk of terrestrial biomass, are among the most abundant elements in the Universe. Whether the organic starting material relevant to the origin of life came from *in-situ* production on our planet or from delivery by extraterrestrial sources is still an open question.

In laboratory experiments, simulating the conditions of the primitive Earth, it was possible to form amino acids, the building blocks of the proteins, from methane, ammonia and water (Miller 1953). However, these pioneering experiments succeeded only when a reducing gas mixture containing significant amounts of hydrogen was used. Although the true composition of the early terrestrial atmosphere remains unknown, geochemists now favor a non-reducing primitive atmosphere, dominated by CO_2; these conditions would allow only limited *in-situ* production of the essential precursors of life.

With regard to potential delivery of organic material to the early Earth, comets are of special interest, because they contain a large amount of organic molecules and are considered the most pristine celestial bodies bearing witness to the existence of a dynamic organic chemistry from the earliest stages of our Solar System. Hence, comets have been suggested to be the major source of the hydrosphere, the atmosphere, and probably also of the organic compounds of the early Earth (Thomas et al. 2006).

Support for a scenario of an extraterrestrial import of organics can be found in the meteorites collected on the Earth's surface, micrometeorites found embedded deep within ice and cosmic dust sampled at stratospheric and Earth-orbit level. Particularly carbonaceous chondrites contain up to 5% by weight of organic matter. Carbonaceous micrometeorites show a high percentage (30%) of unmelted chondrites from 0.1 to 1 mm in diameter. This observation indicates that many particles cross the terrestrial atmosphere without drastic modification by thermal treatment. It has been estimated that during the early terrestrial bombardment, about 10^{17} tons of carbon have been brought to the early Earth by micrometeorites. This number is 5 orders of magnitude larger than that of the present surface biomass's carbon (Maurette 2006).

There are currently two alternative assumptions about the environmental conditions when life started on Earth. The detection of hydrothermal vents in the late seventies of the last century surrounded by a rich ecosystem well adapted to this submarine hot environment has supported the hypothesis of a hot origin of life (Pace 1991), which has however been critically commented by Gribaldo and Forterre (2005). The ocean floor at a submarine alkaline hot spring has been suggested to provide all prerequisites for the emergence of life on the early Earth about 4 billion years ago. Deep-sea hydrothermal systems are producing sites of hydrocarbons, even today. As energy source for the reduction of CO_2, the oxidative formation of pyrite from FeS and H_2S has been postulated. Pyrite provides an active surface binding for the organic molecules formed and has been proposed as the site of a chemolithoautothrophic origin of life (Wächtershäuser 1998). Support to a hot origin of life comes also from the universal phylogenetic tree of life, based on molecular biology analysis where microorganisms that are adapted to extremely high temperatures (hyperthermophiles) cluster around the "root" of this tree (Stetter 2002).

Alternatively, a cold scenario has been proposed for the origin of life. According to Trinks et al. (2005) a sea ice reactor would consist of a dynamic three phase system of ice crystals, brine channels and gas bubbles with dynamic temperature gradients and energy transport. In laboratory experiments simulating the dynamic variability of real sea ice, the abiotic synthesis of long chain biomolecules (polynucleotides) was achieved. Hence sea ice occurring abundantly at the polar ice caps could provide optimal conditions for the early replication of nucleic acids and the RNA world, a suggested precursor of the first cellular system.

There exist two means to trace back the history of life on Earth, the fossil record and the molecular record. Paleobiology searches for the history of life to its very early stages. The search for fossils spans over a time-period of 3.8 billion years, up to the first evidence of sedimentary rocks. However, as one goes back in time more than 2.5 billion years to the Archean, only few sedimentary rocks have survived without alteration. Therefore, it is sometimes difficult, to establish the authenticity of Archean microfossils. The oldest *bona fide* evidence for life on the early Earth comes from the 3.3 to 3.5 billion years old sediments found in South Africa and Australia (Westall et al. 2001). In these formations well preserved micron-sized fossils of microorganisms and biofilms were discovered which once flourished in hydrothermal shallow ponds.

Because all common biological pathways of autotrophic carbon fixation discriminate against the heavy isotope of carbon ^{13}C, the measurement of the ^{13}C/^{12}C fractionation has been used as a means to differentiate between biogenic (organic) carbon and sedimentary carbonate in the deposits. Figure 27.2 shows that this depletion of ^{13}C is quite conservatively transcribed from the extant biomass through

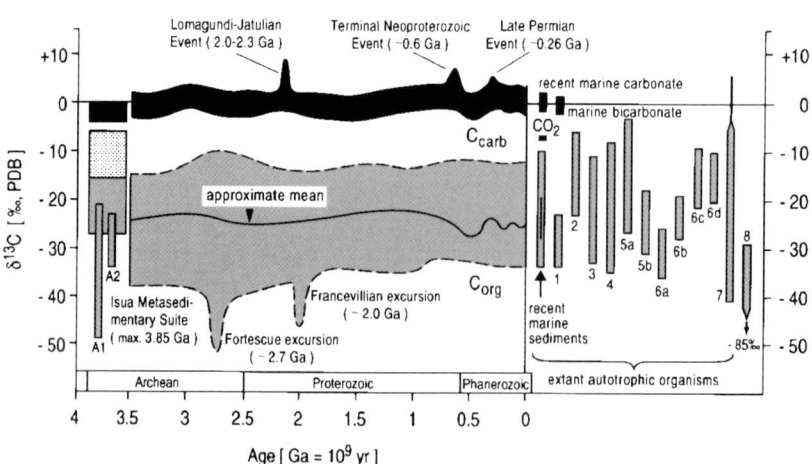

Fig. 27.2 Isotopic fractionation of carbon in contemporary organisms and fossils, given as δ^{13}C values, indicating either an increase (+) or a decrease (−) in the ^{13}C/^{12}C ratio of the respective substance compared to a carbonate standard (from Schidlowki 2002)

recent marine sediments over billions of years into the Archean period, and with a slight modification even back to the 3.8 billion years old Isua formation (Schidlowski 2002).

From the currently available paleobiological and geochemical data, there is evidence that life has emerged very early on the juvenile Earth – with a degree of certainty earlier than 3.5 billion years ago and probably earlier than 3.8 billion years ago. Autotrophic carbon fixation has been extant since at least 3.8 billion years and therefore must have evolved in much older times than are covered by terrestrial rock record.

Complementary to the fossil record, and in the endeavor to trace the history of life back from extant forms to the universal ancestor, molecular phylogeny makes use of the fact that at the level of the genotype changes constantly occur which are fixed randomly in time. By comparing the genotypic information stored in the sequence of e. g., nucleic acids, a universal phylogenetic tree has been constructed

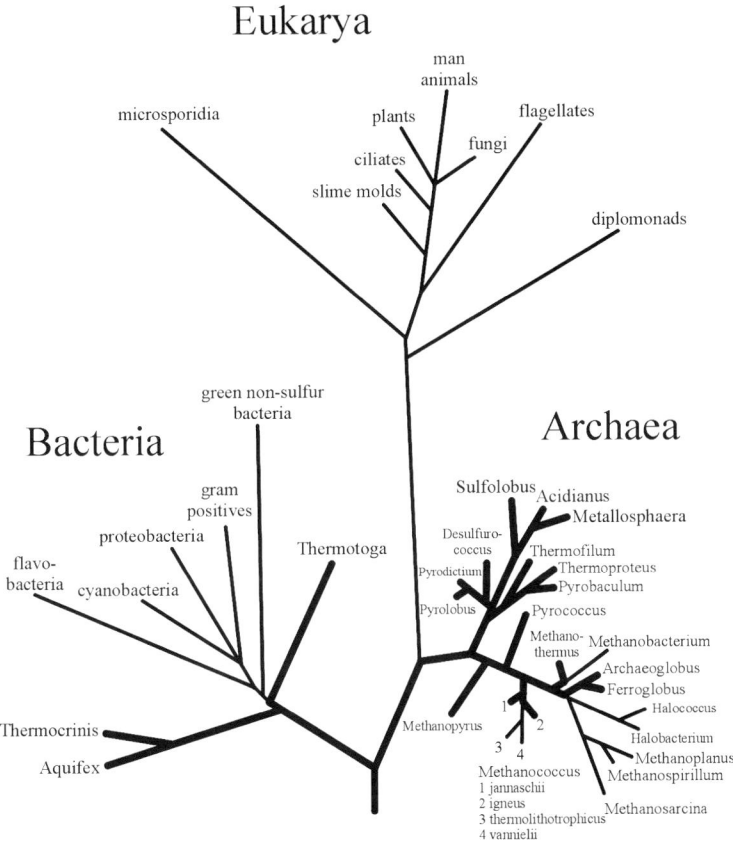

Fig. 27.3 Phylogenetic tree of life, based on the sequence analysis of ribonucleic acid (16S rRNA) of a variety of organisms. Bold lines represent hyperthermophiles (from Stetter 2002)

that groups all known organisms in three domains: Bacteria, Archaea and Eukarya (Woese et al. 1990) (Fig. 27.3).

This molecular record allows also for inferences on the metabolic characteristics of the common ancestor. In Fig. 27.3, hyperthermophilic microorganisms of both, Archaea and Bacteria, are among members forming the deepest branching. These observations support the assumption of a chemolithoautotrophic hyperthermophilic nature of the common ancestor (Stetter 2002).

Recently, the evolution of life through time has been correlated with the rise in atmospheric oxygen generated by oxygenic photosynthesis (Catling et al. 2005). This hypothesis is based on the fact that aerobic metabolism, i. e. respiration, provide about an order of magnitude more energy for a given intake of food, e. g. 1 mol of glucose, than anaerobic metabolism, i. e. fermentation. Consequently, the accessibility of oxygen to the organisms, either by diffusion or by blood circulation, determines their growth and complexity.

The terrestrial atmosphere started with virtually no oxygen and remained anoxic until about 2.4 billion years ago. For this period, fossils give evidence of micron-sized, single cellular organisms, only. Two stepwise increases in atmospheric oxygen occurred, one around 2.4 to 2.3 billion years ago and the next 1.0 to 0.5 billion years ago. Both oxygen increases appear to have been followed by substantial changes in the climate, e. g., built up of an ozone shield, and biota: First larger fossils, visible to the naked eye, are dated for 1.89 billion years ago, animal fossils date back to 600 million years ago.

27.3 Impact Scenario and Interplanetary Transport of Life

The fossil record reveals that microbial autotrophic ecosystems existed on the early Earth already by 3.5 billion years or even 3.8 billion years ago. Before this date, during the Hadean period (before 3.8 billion years ago) the Earth was struck several times by gigantic impacts sufficient to vaporize the entire terrestrial ocean, as has been extrapolated from lunar crater records (Oberbeck and Fogleman 1990). As a result of a runaway greenhouse effect, surface temperatures up to 2000 K have been suggested which would have certainly sterilized the Earth. This impact catastrophe scenario implies that, if life did already exist in the Hadean, it may have been extinguished several times, until the end of the heavy bombardment.

Impactors of sizes larger than 1 km lead to the ejection of a considerable amount of soil and rocks that are thrown up at high velocities, some fraction reaching escape velocity. Meteorites of lunar and some of Martian origin detected within in the last decades are witnesses of these processes. The question arises, whether such rock or soil ejecta could also be the vehicle for life to leave its planet of origin, or, in other words, whether spreading of life in the Solar System via natural transfer of viable microbes is a feasible process. Simulation experiments have shown that some rock populating microorganisms would survive all different steps of such an interplanetary transfer of life, including the escape from the planet, the long journey in space and the entry process on another planet (Mileikowsky et al. 2000; Stöffler et al. 2007).

27.4 Strategies of Life to Adapt to Extreme Environments

Microorganisms have invented several strategies to cope with and adapt to environments of a wide range of physical and chemical parameters. Examples are microbial ecosystems in deep crystalline rock aquifers several hundreds of meters below the surface, in the interior of ice-cores from drillings in the Antarctic ice down to a depth of several km and in cores from drillings in permafrost regions in Siberia at similar depths. It was found that the interior of rocks in cold and hot deserts provides ecological niches for endolithic microbial communities just as crystalline salts from evaporite deposits. Microorganisms have been isolated from extremely cold environments, such as the Antarctic soils as well as from hot environments at temperatures in the range of 80 °C to 113 °C which are usually associated with active volcanism as hot springs, solfataric fields, shallow submarine hydrothermal vents, abyssal hot vent systems ("black smokers") as well as oil-bearing deep geothermally heated soils. Microbial communities are also found buried in groundwater sediments, in marine sediments several 100 m below the sea floor, as well as in the atmosphere where viable microorganisms were collected from altitudes up to 77 km. These examples demonstrate that nearly all sites on Earth are inhabited by microbial communities, where an energy source is available and which are compatible with the chemistry of carbon–carbon bonds (reviewed by Stan-Lotter 2007). Table 27.1 gives the environmental range allowing growth of at least one microbial representative.

Metabolic diversity is one of the approaches microorganisms use for adapting to extreme environments. Whereas photosynthesis is the most common autotrophic

Table 27.1 Environmental range allowing growth or survival of microorganisms

Parameter	Growth	Survival
Temperature (°C)	$-20-+113$	$-262-+150$
Pressure (Pa)	10^5-10^8	$10^{-7}-\geq 10^8$
Ionizing radiation (Gy)	≈ 50	≤ 5000
UV radiation (nm)	terrestrial (≥ 290)	$\approx$terrestrial (≥ 290)
Water stress (a_w)	≥ 0.7	$0-1.0$
Salinity	$\leq 30\%$	salt crystals
pH	1–11	0–12.5
Nutrients	high metabolic versatility, high starvation tolerance	not required, better without
Gas composition	different requirements (oxic or anoxic)	better without oxygen
Time (a)	$\leq 0.5^1$	$\leq (25-40) \times 10^6$

[1] generation time

pathway (and the only one used by eukaryotes), prokaryotes have invented a variety of autotrophic pathways, which either use an energy source different from sunlight (e. g., H_2, Fe^{2+}, Mn^{2+}, reduced or oxidized sulfur or nitrogen compounds) or use an electron donor different from water (e. g., H_2, Fe^{2+}, or H_2S and $S°$). This metabolic versatility enables prokaryotic microorganisms to colonize even deep subsurface sites which can not been reached by sunlight.

Special challenges to microorganisms are environments with fluctuations among extreme conditions, as they are experienced in deserts or on alpine rock surfaces (e. g., rapid changes in temperature and water activity). Other oscillations may concern salinity, pH, redox potential or radiation stress. Microbial mats are especially adapted to cope with these changing environments. They dwell on or inside rocks, in air or under water where energy sources, nutrients or water are only occasionally available. These mats, which are sometimes covered by protective layers of slime, sugars and pigments, are composed of so-called poikilotrophic microbial communities, a mixture of microorganisms capable of outlasting long periods of unfavorable conditions at a reduced metabolic rate.

Several prokaryotes as well as a few eukaryotes possess strategies of surviving unfavorable conditions in a kind of dormant state and are capable of regaining full metabolic activity if conditions change to less hostile ones again. Hence, the limits for microbial survival extend much further than those for growth (Table 27.1). Temporary transition of microbial cells to the dormant, so-called anabiotic state, which involves biochemical, physiological and ultrastructural changes, is a widespread mechanism developed by organisms to promote survival of interim hostile conditions.

Under certain conditions, bacterial cells produce a dormant spore. In these spores, the DNA is extremely well protected against environmental stressors, such

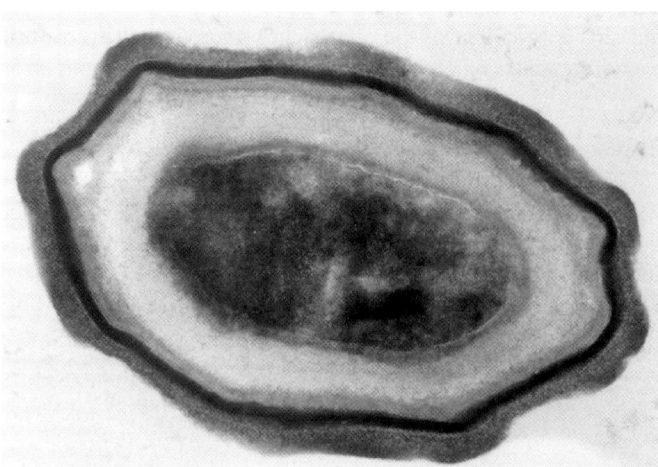

Fig. 27.4 Electonmicrograph of a spore of *Bacillus subtilis* with the inner core containing the DNA surrounded by protective layers, the long axis of the spore is 1.2 μm, the core area 0.25 μm^2 (courtesy of S. Pankratz)

as desiccation, oxidizing agents, UV and ionizing radiation, low and high pH as well as temperature extremes (reviewed in Nicholson et al. 2000). The high resistance of bacterial spores is mainly due to a dehydrated core enclosed in a thick protective envelop, the cortex and the spore coat layers (Fig. 27.4), and the saturation of their DNA with small, acid-soluble proteins whose binding greatly alters the chemical and enzymatic reactivity of the DNA. Bacterial spores have survived for extended periods in space (so far maximum duration tested was 6 years) which is governed by a high vacuum, temperature extremes and an intense radiation of solar and galactic origin. Isolates from Dominican amber suggest that *Bacillus* spores remain viable for several millions of years.

27.5 Signatures of Life

Within the on-going and planned ventures to explore the planets, moons and other bodies of our Solar System by orbiters and robotic landing missions, the search for signatures of life beyond the Earth is one of the major drivers. The techniques planned for the search for life on other planets are mostly inferred from terrestrial applications.

On Earth, the most commonly used direct methods to detect and analyze microbial communities in extreme environments include the following approaches:

- direct observations of structural characteristics in the micro- and macroscale;
- culture techniques for isolating microorganisms in pure culture and then analyzing these cultures for their biochemical properties;
- activity measurements in microcosms which focus on the net effects of microbial processes in the community;
- chemical marker techniques to record characteristic biochemical primary substances or chemical secondary products (Table 27.2).

Based on the fact that during its more than 3.5 billion years old history life on Earth has substantially modified the terrestrial lithosphere, hydrosphere and atmosphere, indirect proofs of life are also widely used (Table 27.2). Examples are the fossil deposits of petroleum and coal, the sediments of shell limestone, the coral reefs, and the deposits of banded iron formation, biomineralization and bioweathering, which all bear witness of biological activity in the geological past. Stromatolites and related biosedimentary built-ups are examples of microbial biomineralization, which are widespread in the geological record (Horneck 2000; Clancy et al. 2005).

Composition and dynamic cycles of the terrestrial hydrosphere and atmosphere are decisively influenced by the terrestrial biosphere. Examples are the water cycle, the CO_2 cycle and the nitrogen cycle. The albedo of our planet is also modified by the surface vegetation. Concerning the water cycle, evapotranspiration, especially of the tropic rain forest, is an important biogenic effect contributing to the release of water. The atmospheric O_2 is largely a product of photosynthetic activity which

Table 27.2 Methods to detect and analyze microbial communities in extreme environments

Method	Approach	Information
Direct observation of structural indications of life	Optical microscopy Phase-contrast microscopy	Dividing cells
	Epifluorescence microscopy	Differentiation between cellular structures and similar structures of abiotic origin
	Confocal laser scanning microscopy	Three-dimensional visualization of microbial communities
	Scanning electron microscopy	Subcellular structures
Detection of microbial activity	Cultivation	Biochemical and phylogenetic analysis of isolates
	DNA probes and biomarker chips	Not yet culturable microorganisms
	In situ biochemical analysis	Metabolically active communities
	Microelectrodes for *in situ* measurements of natural gradients of gas, pH, minerals etc.	Metabolic activity, e. g. photosynthesis, respiration
	Combination of epifluorescence microscopy and television image analysis	Metabolic activity
Detection of chemical signatures and biomarkers	Determination of total biomass, biomolecule contents	Major contribution species; physiological stage, e. g. active, dormant, extinct
	Determination of isotopic ratios	Biogenic origin
	Determination of optical handedness	Biogenic origin
	Fourier transform Raman spectroscopy	Distribution of organic and inorganic components
Indirect fingerprints of life	Macroscopical deposits	Biomineralization, bioweathering, indication of extinct life
	Mineral analyis	Biogenic minerals
	Detection of dynamic cycles of hydrosphere or atmosphere	Indication of extant life

started already in the early history of life with cyanobacteria as the main primary producers. Photodissociation of O_2 in the upper layers of our atmosphere has led to the built up of the UV-protective ozone layer in the stratosphere. Concerning the CO_2 cycle, the marine phytoplankton constitutes a large CO_2 sink, which is essential for maintaining the steady state.

There exist several biogenic minerals, which have distinctive crystallographies, morphologies and isotopic ratios that make them distinguishable from their abiotically produced counterparts of the same chemical composition. Especially those minerals that result from genetically controlled mineralization processes and that are formed within a preformed organic framework have non-interchangeable characteristics, such as orientation of the crystallographic axes and the microarchitecture. Examples are the skeletons of the unicellular marine *Acantharia* that are composed of strontium sulfate, the shells of amorphous silicate of diatoms, and the biogenic magnetite formed by bacteria (Schwartz et al. 1992). It is important to note, that minerals produced under biologically controlled processes are not necessarily in equilibrium with the extracellular environment. In contrast, those minerals that are formed extracellularly by biologically induced mineralization processes can be less easily distinguished from their abiotically formed counterparts. They are formed in an open environment and are in equilibrium with this environment.

In the selection of candidate targets for the search for life in the Solar System, the overarching argument is the putative habitability of the planet or moon under consideration.

27.6 Criteria for Habitability

The criteria for habitability come from the notification of at least three basic prerequisites for a planet or moon to be habitable (see also Schulze-Makuch and Irwin 2004 for a critical interpretation):

- a carbon based chemistry,
- an energy source, and
- water in liquid phase.

Carbon based molecules are the universal building blocks of life as we know it. The ability of carbon to form complex, stable molecules with itself and with other elements, e. g., hydrogen, oxygen, or nitrogen, is unique and is attributed to at least three factors:

- stability of carbon molecules due to the high carbon-carbon bond energy,
- capability to form double and triple bonds in addition to covalent bonds; and
- high activation energy for substitutions and bond cleavage reactions, which support the stability of the molecules to water and oxygen.

Although a wealth of complex organic molecules has been detected at many extraterrestrial places, such as the interstellar medium, comets, meteorites and planetary atmospheres (Ehrenfreund and Menten 2002), they have not been found at the surface of Mars, so far (Tokano 2005). The most plausible explanation for the absence of organics on the surface of Mars is an active surface photochemistry from the energetic solar UV radiation, where peroxides are produced by UV-irradiation of hematite in the presence of traces of water.

Life requires a flow of energy to organize its material, perform the work, and maintain a low state of entropy. The energy sources, used by life on Earth, are narrow bands of visible light for photoautotroph organisms, redox reactions for chemoautotroph organisms or organic molecules for heterotrophic forms of life. Subsurface organisms use the oxidation of inorganic electron donors, such as hydrogen, sulfur, sulfide, ammonia, nitrite, or iron. However, there seems to be no limitation on which redox reactions are used, even relatively rare elements, such as arsenic, selenium, copper, lead, and uranium serve the purpose.

Water has been called "the spring of life" (Brack 2002). In the liquid phase water exhibits several peculiarities that make it indispensable for life. Water serves as

- diffusion milieu,
- selective solvent,
- heat dissipater,
- stabilizer of the biopolymer structure, and
- reaction partner in essential biological processes, e. g., in photosynthesis.

Based on the observation that liquid water is the most essential prerequisite for life, the common definition of a "habitable planet" has been one that can sustain substantial liquid water on its surface. Assuming a tolerable temperature range between about 0 °C and 100 °C at the surface of a planet, our Solar System would provide a habitable zone in an orbit between 0.7 and 2.0 AU, or in more conservative estimate the width of the habitable zone is restricted to the range of 0.95 to 1.37 AU (Fig. 27.5). Venus, Earth and Mars are situated in this habitable zone or in close neighborhood. However, in view of the adaptability of microorganisms to extreme environments and the detection of rich submarine and subsurface microbial ecosystems on Earth, liquid water does not necessarily need to exist at the surface to sustain a biosphere. Therefore, the definition of a "habitable planet" may require revision.

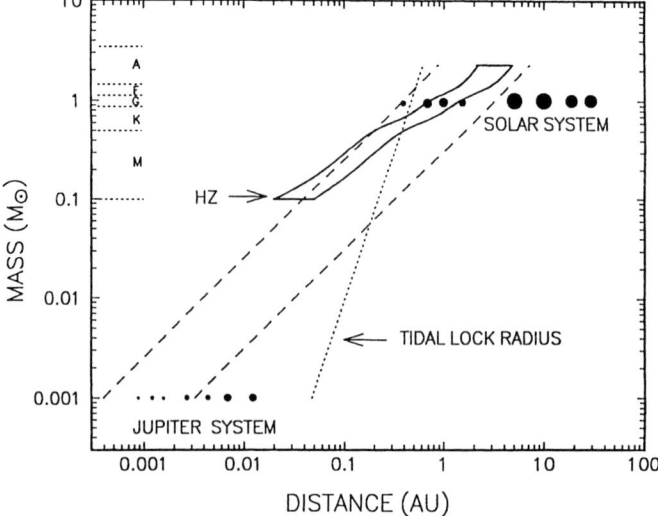

Fig. 27.5 Habitable zone (HZ) as a function of the distance from the star and its mass (from Franck et al. 2002)

27.7 Planets and Moons in Our Solar System That are of Interest to Astrobiology

Based on the availability of liquid water, our neighbour planet Mars as well as Jupiter's moon Europa have been considered as prime candidates for habitability.

Mars (Table 27.3) with a mean distance to the Sun of 1.52 AU is located at the outer border of the habitable zone around the Sun which is estimated under the premise of the presence of liquid water on the planet's surface at some time during its 4.5 billion years lasting history. With the exception of the Earth, Mars is by far the most intensively studied of the planets of our Solar System. In 1972, for the first time a spacecraft, Mariner 9, passed over the younger parts of Mars revealing a wide variety of geological processes, indicated by volcanoes, canyons, and channels that resemble dry river beds (Tokano 2005). These extensive fluvial features confirmed during several follow-on missions to Mars were difficult to reconcile with any origin other than liquid water. They attest to a stable flow of water on Mars at some time in the past, and sporadically even in more recent times.

Understanding the history of water on Mars appears to be one of the clues to the puzzle on the probability of life on Mars. The estimates of the total amount of water that may have existed at the surface of Mars range over two orders of

Table 27.3 Properties of the terrestrial planets and the Moon (Credit: NASA planetary fact sheet, http://nssdc.gsfc.nasa.gov/planetary/factsheet/index.html)

Property	Planet				
	Mercury	Venus	Earth	Moon	Mars
Mass (10^{24} kg)	0.330	4.87	5.97	0.073	0.642
Diameter (km)	4879	12,104	12,756	3475	6794
Density (kg/m^3)	5427	5243	5515	3340	3933
Gravity (m/s^2)	3.7	8.9	9.8	1.6	3.7
Escape velocity (km/s)	4.3	10.4	11.2	2.4	5.0
Rotation period (h)	1407.6	−5832.5[1]	23.9	655.7	24.6
Length of day (h)	4222.6	2802.0	24.0	708.7	24.7
Distance from Sun (10^6 km)	57.9	108.2	149.6	0.384[2]	227.9
Perihelion (10^6 km)	46.0	107.5	147.1	0.363[2]	206.6
Aphelion (10^6 km)	69.8	108.9	152.1	0.406[2]	249.2
Orbital period (days)	88.0	224.7	365.2	27.3	687.0
Orbital velocity (km/s)	47.9	35.0	29.8	1.0	24.1
Orbital inclination (degrees)	7.0	3.4	0.0	5.1	1.9
Orbital eccentricity	0.205	0.007	0.017	0.055	0.094
Axial tilt (degrees)	0.01	177.4	23.5	6.7	25.2
Mean temperature (celsius)	167	464	15	−20	−65
Surface pressure (10^5 Pa)	0	92	1	0	0.01
Number of moons	0	0	1	0	2
Global magnetic field	Yes	No	Yes	No	No (?)

[1] Negative numbers indicate retrograde (backwards relative to the Earth) rotation
[2] Values represent mean apogee and perigee for the lunar orbit. The orbit changes over the course of the year, so the distance from the Moon to Earth roughly ranges from 357,000 km to 407,000 km

magnitude. A low amount of water ranging from 3.6 to 133 m is suggested from the composition of the contemporary atmosphere, e. g. the D/H ratio. On the other hand, the geological flow features, provide evidence of abundant water at the surface of Mars, at least at some time in the past, assuming a global inventory of water of at least 440 m. From the global neutron mapping of the Mars Odyssey mission, the present distribution of water in the shallow subsurface was divided in 4 types of regions (reviewed in Tokano 2005):

- regions with dry soil with a water content of about 2 wt%,
- northern permafrost regions with a high content of water ice (up to 53 wt% of water),
- southern permafrost regions with high content of water ice (>60 wt% of water) covered by a dry layer of regolith,
- regions with water-rich soil at moderate latitudes (about 10 wt% of water) covered by a dry layer of soil.

These latter water-rich regions are well separated from the Martian atmosphere by the rather thick layer of desiccated regolith. Therefore, it was supposed that they were formed long time ago, when the climate allowed liquid water at the surface.

The history of water on Mars suggests a dramatic change in the climate about 3.8 billion years ago. Based on a model by McKay and Davis (1991) 4 different epochs can be distinguished for the history of water on Mars, starting with a wet water-rich planet before 3.8 billion years ago, and becoming gradually more and more dry. In the following, for each of these epochs the probability for indigenous life is discussed taking into consideration the requirements for the emergence of life and current knowledge of terrestrial ecosystems as model systems for a putative Martian biota.

In the first epoch, reaching up to the end of the heavy bombardment about 3.8 billion years ago with a relatively warm surface and liquid water in flat oceans and rivers, the main prerequisites for the origin of life, as we currently assume, did exist. Therefore, by analogy to the early Archean biosphere on Earth, an early Martian biosphere could be postulated with habitats and microenvironments similar to those on the early Earth. The major uncertainty seems to be whether liquid water was available long and abundantly enough for life to arise.

Finally, atmospheric CO_2 was irreversibly lost due to carbonate formation, and the pressure and temperature then declined. However, ice-covered lakes could have persisted over a period of 700 ± 300 million years which provided liquid water habitats on early Mars, analogous to ice-covered lakes in Antarctica or cryoconite holes on glaciers. These terrestrial analogues contain plankton organisms as well as an abundant benthic community forming microbial mats. Such ice-covered lakes might have served as niches for putative life on Mars to retreat by providing both thermal stability against a cooling external environment and enhanced concentrations of CO_2 and N_2 against a thinning atmosphere.

With gradually decreasing pressure and temperature the ice-covered lakes would eventually have dried out due to lack of melt water supply thereby initiating the next epoch where liquid water would be restricted to porous rocks and to the subsurface. In order to cope with these dramatic environmental changes, adaptive steps of a putative Martian biota could have been to withdraw into protected niches, e. g. inside rocks or in the subsurface. Such endolithic habitats exist on Earth in areas of extreme aridity and frigidity, e. g. dry valleys of Antarctica (Friedmann and Ocampo-Friedmann 1985). The cryptoendolithic microbial communities form lichen-dominated ecosystems with cold-adapted nearly exclusively eukaryotic algae and less commonly cyanobacteria as primary producers and fungi as consumers. Other potential biotic oases, to which the putative life on Mars might have withdrawn, are the polar ice caps and permafrost regions, or subsurface hydrothermal areas in connection with volcanic activities.

The present atmosphere is too cold to support liquid water on the surface for long and too thin to support ice – any ice that does form will quickly sublimate into water vapour. The life-threatening surface conditions of Mars were clearly shown by the 2 Viking landing missions, which searched for indications of microbial activity on Mars. Based on the assumption that Martian life, if it exists, will be carbonaceous, that its chemical composition is similar to that of terrestrial life, and that it most likely metabolizes simple organic compounds, a life detection instrument package was installed to detect metabolic activity of potential microbial soil communities (Fig. 27.6). All three Viking biology experiments gave positive results indicative

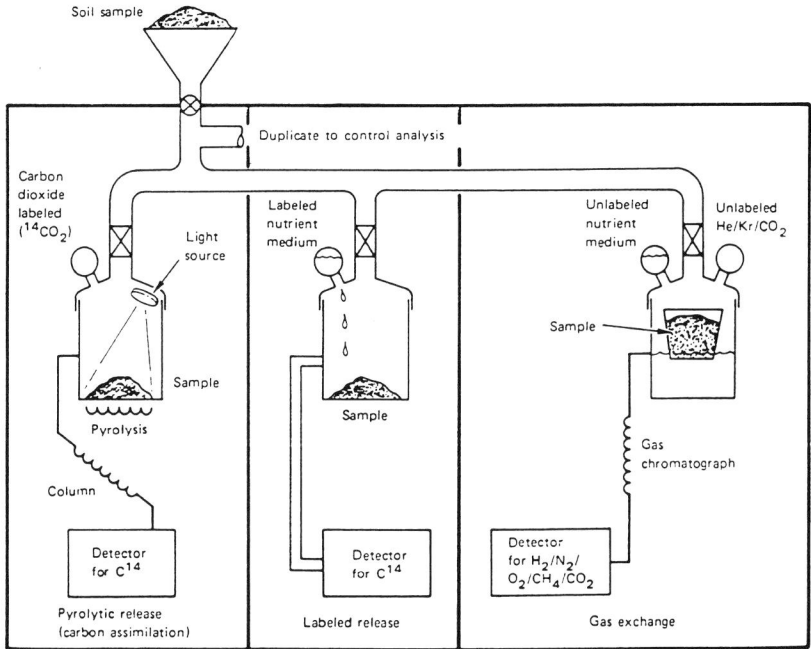

Fig. 27.6 Sketch of the three experiments of the Viking landers that searched for metabolic signatures of putative Martian microorganisms in surface samples of Martian regolith (from *Viking 1, Early Results*, NASA SP-408, 1976)

of active chemical processes when samples of Martian soil were subjected to incubation under the conditions that were imposed to them. However, no organic carbon was found in the Martian soil by the gas chromatography mass spectrometry (GCMS) experiment. So far, the mechanisms underlying the results of the Viking biology experiments are not known. A number of hypotheses have been forwarded in order to explain the enigma of an active chemistry in the absence of organics. The most plausible one is based on photochemical surface processes where the energetic solar UV-irradiation forms peroxides in the regolith when impinging on the hematite in the presence of traces of H_2O.

Considering the open questions with regard to the habitability of Mars, the *in situ* search for signatures indicative for putative extant or extinct life on Mars can only be one of the final steps in the quest for extraterrestrial life. Much information can be obtained from remotely sensed global measurements, such as the seasonal atmospheric and surface water distribution, the mineralogical inventory and distribution, geomorphologic features obtained with high spatial resolution, thermal mapping of potential volcanic regions to determine possible geothermally active sites, and trace gases like H_2, H_2S, CH_4, SO_x, and NO_x. The on-board measurements on current and planned missions to Mars, such as Mars Global Surveyor, Mars Express, Mars Odyssey and Mars Reconnaissance Orbiter efficiently serve these needs. The recent detection of traces of methane in the Martian atmosphere, hints of the presence of deep underground water-ice, and indications of relatively young volcanic activities in the north polar regions are prominent results of the current Mars Express mission of the European Space Agency. The search for possible biological oases will be connected with the detection of areas where liquid water still exists under the current conditions on that planet.

With the Mars Exploration Rover mission NASA intends to unravel the story of water on the red planet. Especially in the Meridiani Planum, the landing site of the rover Opportunity, they detected distinct layering in some rocks which showed that water once flowed there on the surface of Mars, leaving ripple-like curves in the outcrop rocks. Bead-like objects, the so-called "blueberries", turned out to be rich in hematite, a mineral that requires water to form. The detection of sodium chloride which only forms when water has been present is another indication of liquid surface water in the past of Mars.

ESA's exploration program foresees as the next step the ExoMars mission that uses a rover with high autonomy and equipped with the analytical capacity to select suitable drilling site or exposed vertical stratigraphy to find signs of extinct or extant life (Fig. 27.7). To do this requires the development of an efficient Mars drilling system and the correspondent sample analysis suite to be used in the underground exploration of selected sites. In addition, the habitability of these regions will be explored by *in-situ* measurements of the climate, radiation environment and surface and subsurface chemistry in dry and wet state. It is important to understand the mechanisms of the strong oxidative processes present on the surface of Mars which have been identified by the Viking experiments.

Jupiter's moon Europa, is another focus for astrobiology (Table 27.4). More than 95% of the spectroscopically detectable material on its surface is H_2O (Greenberg

27.7 Planets and Moons in Our Solar System that are of Interest to Astrobiology

Fig. 27.7 Artist representation of the ExoMars rover of ESA to be launched in 2013 (Credit: ESA)

Table 27.4 Characteristics of Jupiter's moon Europa

Parameter	Quantity and dimension
Radius	1560 ± 10 km
Density	3018 ± 35 kg/m^3
Gravity acceleration at the surface	1.31 m/s^2
Period of rotation	3.55 days
Distance to Jupiter	0.6709×10^9 m
Mass of Jupiter	1.905×10^{27} kg
Eccentricity	0.96%

2005). It has been established with high probability that this moon of Jupiter harbors an ocean of liquid water, beneath a thick ice crust. In addition to liquid water, carbon and energy sources are requested to support life as we know on Earth. If carbon might have be delivered by impacts of various bodies (although crust resurfacing does not show so many impact craters), the question of energy sources is still open. The existence of liquid water beneath the ice crust might be the result of a deep hydrothermal activity, radioactive decay and/or tidal heating. In this case, conditions allowing prokaryotic-like life as we know on Earth would have been gathered.

27.8 Planetary Protection

The introduction and possible proliferation of terrestrial life forms on other planets by means of orbiters, entry probes or landers could entirely destroy the opportunity to examine the planets in pristine condition. From this concern by the scientific community, the concept of Planetary Protection has evolved. Its intent is twofold:

- to protect the planet being explored and to prevent jeopardizing search for life studies, including precursors and remnants, and
- to protect the Earth from the potential hazards posed by extraterrestrial matter carried on a spacecraft returning from another celestial body (Rummel 1989).

Planetary Protection issues are bound by an international treaty (UN Doc. A/6621, Dec. 17, 1966) and agreement (UN Gen. Ass. Resol. A/34/68, Dec. 5, 1979). Since 1959, the Committee of Space Research (COSPAR) has developed Planetary Protection guidelines that originally were based on relevant information about the probability of survival and release of organisms either contained in or on exposed surfaces of spacecraft, about the surface and atmosphere characteristics of the planet under consideration, and about the probable distribution and growth of types of organisms involved. This concept of probability of contamination of a planet of biological interest was replaced by a concept of contamination control to be elaborated specifically for certain space-mission/target-planet combinations, such as orbiters, landers, or sample return missions (Rummel 1989, recently reviewed by Horneck et al. 2007).

Considering the scientific interest and mission goals of the different targets in our Solar System, COSPAR Planetary Protection guidelines group the missions in five categories (Table 27.5).

Category I missions include any mission (orbiters, fly-by, or landers) to a target body, which is not of direct interest for understanding the process of chemical evolution or origin of life, or other questions of astrobiology. The target bodies include the planets Mercury and Venus, the Moon, as well as undifferentiated, metamorphosed asteroids and the Sun. In this category, no protection of the planets or bodies is warranted and therefore no Planetary Protection requirements are imposed on the spacecraft and mission.

27.8 Planetary Protection

Table 27.5 Current COSPAR Planetary Protection categories with regard to missions to Solar System bodies and types of missions

Category	Category I	Category II	Category III	Category IV	Category V
Type of Mission	Any but Earth Return	Any but Earth Return	No direct contact (fly-by, some orbiters)	Direct contact (lander, probe, some orbiters)	Earth return
Target Body	See text	See text	See text	See text	See text
Degree of Concern	None	Record of planned impact probability and contamination control measures	Limit on impact probability; passive bioload control	Limit on probability of non-nominal impact; limit on bioload (active control)	If restricted Earth return: No impact on Earth or Moon; returned hardware sterile; containment of any sample.
Range of Requirements	None	Documentation only: PP plan; pre-launch report; post-launch report; post-encounter report; end-of-mission report	Documentation (Category II plus): Contamination control; organics inventory; implementing procedures (Trajectory biasing, cleanroom, bioload reduction (as necessary))	Documentation (Category II plus): Pc analysis plan; microbial reduction plan; microbial assay plan; organics inventory; Implementing procedures (Trajectory biasing, cleanroom, bioload reduction, partial sterilization of contacting hardware, bioshield, monitoring of bioload via bioassay)	Outbound: Same category as target body/outbound mission. Inbound: If restricted Earth return: Documentation (Category II plus): Pc analysis plan; microbial reduction plan; microbial assay plan; trajectory biasing; sterile or contained returned hardware; continual monitoring of project activities; project advanced studies/research. If unrestricted Earth return: None

PP = Planetary Protection
Pc = Probability of contamination

Category II refers to all types of missions (orbiters, fly-by, or landers) to those target bodies where there is significant interest relative to the process of chemical evolution, but where there is only a remote chance that contamination carried by a spacecraft could jeopardize future exploration. Target planets and other bodies include the giant planets Jupiter, Saturn, Uranus, and Neptune, the remote dwarf planet Pluto with its moon Charon, the comets, carboneous chondrite asteroids – these asteroids contain a substantial amount of organic material – and the objects of the Kuiper Belt. In this category, the Planetary Protection recommendations require relatively simple documentation, such as on probability of impact or landing on the celestial body and measures of microbiological contamination record.

Category III comprises fly-by or orbiter missions to a target planet or body, which – according to the current understanding – is of interest with regard to chemical evolution and/or origin of life, or for which scientific opinion considers that a potential contamination with terrestrial microorganisms could jeopardize future biological experiments, for instance, those that search for signatures of indigenous life. Our neighbor planet Mars and the Jovian moon Europa are classified in this category. In this category, besides a detailed documentation, (more extensive than in category II) it is required to assemble and test the space craft in a clean room, but also to monitor the bioburden (number of terrestrial microorganisms present) and reduce it if necessary.

Category IV missions are those missions of utmost astrobiological interest. They include mostly entry probe or lander missions to a target planet of interest with regard to chemical evolution and/or origin of life or for which scientific opinion provides a significant chance that contamination by terrestrial microorganisms could jeopardize future biological experiments. As in category III, the celestial targets are Mars and Jupiter's moon Europa, but in this case, missions to the surface or subsurface are considered. For category IV missions, more detailed Planetary Protection documentation is required than for category III missions. The Planetary Protection requirements include bioassays to enumerate the bioload, a probability of contamination analysis, an inventory of the bulk constituent organics and an increased number of implementation procedures. The implementation procedures required include trajectory biasing, clean room assembly, bioburden reduction and monitoring, and possibly partial sterilization of the directly contacting hardware and a bioshield for that hardware. The current bioburden requirements for Mars landers are less than 300 bacterial spores/m^2 and a total surface spore bioburden of less than 3×10^5 spores. This level can mostly be reached without special sterilization using clean room techniques. The standardized, validated, spore assay used by Planetary Protection practitioners is the culture method after heat-shock at 80 °C for 15 min before plating and incubation. Spore-forming microbes are of particular concern in the context of Planetary Protection, because in many species their endospores can be highly resistant to a variety of environmental extremes, including certain sterilization procedures and the harsh environment of outer space or planetary surfaces.

So far, most work on bioload control of spacecraft is restricted to cultivable microorganisms. However, it is well known, that many microorganisms found in environmental samples as not readily cultivable in the laboratory. Using new technologies, such as fluorescent DNA staining of bacterial cells or epifluorescence microscopy, for instance, has shown that the number of microorganisms in aquatic environments is 100 to 1000 times greater than that determined by cultivation techniques. These so-called non-cultivable microorganisms are assumed to represent the majority of terrestrial microorganisms. New techniques of genomics and metagenomics are required to assess their occurrence on spacecraft or landing modules.

If the landing site on the planet or moon is located in a region where liquid water is likely to exist, then sterilization procedures are required to reduce the total spore load by 4 orders of magnitude, i. e. to 30 spores or less. Currently, the following methods of bioburden reduction and sterilization of the spacecraft are used: biocleaning by wiping with alcohol or sporicidals or by use of UV radiation; sterilization by gamma rays, dry heat or H_2O_2 gas plasma sterilization. In view of the current and planned landing activities on Mars, with robotic and finally human visits, the Planetary Protection guidelines are currently under review within ESA and NASA. The import of internal and external microorganisms inevitably accompanying any human mission to Mars, or brought purposely to Mars as part of a bioregenerative life support system needs careful consideration with regard to Planetary Protection issues.

Category V includes all missions within our Solar System which return to the Earth. Missions that bring back to the Earth samples from Mars or Europa are classified as "*restricted for Earth return*". The highest degree of concern is expressed by the absolute prohibition of impact upon return, the need that any returned hardware, which directly contacted the target body is sterile (contact chain breaking if sterilization is not possible) and the need for containment of any non-sterilized sample collected and returned to Earth. In general, this concern is reflected in a range of requirements that encompass those of Category IV plus a continuing monitoring of project activities. Special research is required before the mission, in order to develop suitable and efficient remote sterilization procedures and reliable containment techniques.

27.9 Search for Life Beyond the Solar System

Estimates of the occurrence and frequency of habitable zones outside our Solar System were first mainly based on astronomical concepts of the structure and dynamics of our Galaxy, on planetary atmosphere models and on biological interpretations of the requirements for the emergence and evolution of life. For solar type stars, a habitable belt, ranging from about 0.95 to about 1.5 AU has been suggested (Fig. 27.5). In addition to single main-sequence stars, variable stars, giant stars and binary systems have been examined for supporting a habitable zone. Within our Galaxy, the orbit of our Sun was suggested as especially favorable for supporting life forms mainly due to the availability of heavier elements.

An important step in the search for habitable zones outside of our Solar System was achieved in 1995 with the discovery of the first extrasolar planet orbiting a star similar to our Sun, 51 Pegasi (Udry and Mayor 2002). Its presence was inferred from the induced modulation of the observed stellar radial velocity. In the following years, this field has rapidly evolved and more than 200 extrasolar planets have been detected so far, using different approaches. All planets so far detected are massive, most are Jupiter-class planets, considered unlikely to harbour life as we know it. Many have short orbital periods. If planets like Earth exist, with smaller masses and longer orbital periods, their discovery will require more sensitive instruments and years of precise, sustained observations. The ESA project Darwin is intended to look for spectral signatures of atmosphere constituents, such as CH_4 and O_3 in order to identify Earth-like planets capable of sustaining life (Foing 2002). Another ambitious mission searching for Earth-like planets is NASA's Terrestrial Planet Finder (TPF) as a suite of two complementary observatories, a visible-light coronagraph and a mid-infrared formation-flying interferometer. They will detect and characterize Earth-like planets around as many as 150 stars up to 45 light-years away.

For estimations of the expected number of habitable planets in our Galaxy, a formula, known as the Drake equation has been developed which takes into consideration (1) the rate of star formation, (2) the fraction of stars which have planetary systems, (3) the average number of planets per planetary system which fall in a habitable zone, and (4) the fraction of habitable planets on which life arises (Drake 1974). Because interstellar distances are so vast, radio-communication has been deemed the only way of detecting life beyond our Solar System. This requires that additional terms be included in the Drake formula, namely (5) the fraction of planets with life on which intelligence arises, (6) the fraction of intelligent species that evolve a technological state that enables interplanetary communication, and (7) the lifetime of such technological civilisation. Estimates made on the number of habitable planets in our galaxy range between 2×10^6 and 1×10^{11} (reviewed in Ulmschneider 2006). However, so far, radio-astronomical search for extraterrestrial intelligence has not given any positive indications.

27.10 Outlook

It is owing to astrobiology, that the question on the origins and distribution of life in the Universe is now tackled in a multidisciplinary scientific approach. Major research activities pertain to

- comparing the overall pattern of chemical evolution of potential precursors of life, in the interstellar medium, and on the planets and small bodies of our Solar System;
- tracing the history of life on Earth back to its roots;
- deciphering the environments of the planets in our Solar System and of their satellites, throughout their history, with regard to their habitability;
- testing the impact of space environment and simulated planetary environments on survivability and adaptability of resistant life forms;
- searching for other planetary systems in our Galaxy and for their habitability.

27.10 Outlook

It is important to note the multidisciplinary character of astrobiological research which involves scientists from a wide variety of disciplines, such as astronomy, planetary research, organic chemistry, paleontology and the various subdisciplines of biology including microbial ecology and molecular biology. Pieces of information provided by each discipline have contributed to the conception of the phenomenon of life within the process of cosmic evolution. New techniques that have been developed within the various disciplines are now accessible. Besides space technology and various remote sensing techniques, they include among others radioastronomical molecular spectroscopy, isotope fractionation analysis, nucleic acid and protein sequencing technology, immunofluorescence approach for the detection of hitherto uncultured microorganisms, and sensitive assays in organic chemistry and radiation biochemistry. Their application has already led to several conceptional breakthroughs, especially in the field of early biological evolution and in the detection of extrasolar planetary systems.

As mentioned above, the final goal of astrobiology is building the foundations for the construction and testing of meaningful axioms to support a theory of life. The discovery of a second genesis of life, either directly from planetary missions within our Solar System, e. g., to our neighbour planet Mars or Jupiter's moon Europa, or indirectly by radioastromony, would provide clues necessary to reach a universal definition of life.

References

1. Brack A (2002) Water, the Spring of Life. In: Horneck G, Baumstark-Khan C (eds) Astrobiology, the Quest for the Conditions of Life. Springer, Berlin Heidelberg New York, pp 79–88
2. Brack A (ed) (1998) The Molecular Origins of Life, Assembling the Pieces of the Puzzle. Cambridge University Press, Cambridge, UK
3. Brack A, Fitton B, Raulin F (1999) Exobiology in the Solar System and the Search for Life on Mars, ESA SP-1231, European Space Agency. Noordwijk, The Netherlands
4. Catling DC, Glein CR, Zahnle KJ, McKay CP (2005) Why O_2 is required by complex life on habitable planets and the concept of planetary oxygenation time. Astrobiology 5:415–437
5. Clancy P, Brack A, Horneck G (2005) Looking for Life, Searching the Solar System. Cambridge University Press, Cambridge, UK
6. Drake FD (1974) In: Ponnamperuma C, Cameron AGW (eds) Interstellar Communications, Houghton Mifflin, Boston, p 118
7. Ehrenfreund P, Menten KM (2002) From Molecular Clouds to the Origin of Life. In: Horneck G, Baumstark-Khan C (eds) Astrobiology, the Quest for the Conditions of Life. Springer, Berlin Heidelberg New York, pp 7–23
8. Foing BH (2002) Space Activities in Exo-Astrobiology. In: Horneck G, Baumstark-Khan C (eds) Astrobiology, the Quest for the Conditions of Life. Springer, Berlin Heidelberg New York, pp 389–398
9. Franck S, von Bloh W, Bounama C, Steffen M, Schönberger D, Schellnhuber HJ (2002) Habitable Zones in Extrasolar Planetary Systems. In: Horneck G, Baumstark-Khan C (eds) Astrobiology, the Quest for the Conditions of Life. Springer, Berlin Heidelberg New York, pp 47–56
10. Friedmann EI, Ocampo-Friedmann R (1985) Blue-green algae in arid cryptoendolithic habitats. Archiv für Hydrobiologie 38/39:349–350
11. Gargaud M, Barbier B, Matin H, Reisse J (eds) (2005) Lectures in Astrobiology. Volume 1, Springer, Berlin Heidelberg New York
12. Greenberg R (2005) Europa – The Ocean Moon. Springer and Praxis, Chichester UK
13. Gribaldo S, Forterre P (2005) Looking for the most 'Primitive' Life Forms: Pitfalls and Progresses. In: Gargaud M, Barbier B, Matin H, Reisse J (eds) Lectures in Astrobiology. Volume 1, Springer, Berlin Heidelberg New York, pp 595–615
14. Horneck G (2000) The microbial world and the case for Mars. Planet Space Sci 48:1053–1063
15. Horneck G, Baumstark-Khan C (eds) (2002) Astrobiology, the Quest for the Conditions of Life. Springer, Berlin Heidelberg New York
16. Horneck G, Rettberg P (eds) (2007) Complete Course in Astrobiology, Wiley-VCH, Berlin, New York
17. Horneck G, Debus A, Mani P, Spry JA (2007) Astrobiology Exploratory Missions and Planetary Protection Requirements. In: Horneck G, Rettberg P (eds) (2007) Complete Course in Astrobiology, Wiley-VCH, Berlin, New York, pp 353–398
18. Lederberg J (1960) Exobiology: approaches to life beyond the Earth. Science 132:393–400

19. Maurette M (2006) Micrometeorites and the Mysteries of Our Origins. Springer, Berlin Heidelberg New York
20. McKay CP, Davis WL (1991) The duration of liquid water on early Mars. Icarus 90:214–221
21. Mileikowsky C, Cucinotta F, Wilson JW, Gladman B, Horneck G, Lindegren L, Melosh J, Rickman H, Valtonen M, Zheng JQ (2000) Natural transfer of viable microbes in space, Part 1: From Mars to Earth and Earth to Mars. Icarus 145:391–427
22. Miller SL, (1953) The production of amino acids under possible primitive Earth conditions. Science 117:528
23. Nicholson WL, Munakata N, Horneck G, Melosh HJ, Setlow P (2000) Resistance of Bacillus endospores to extreme terrestrial and extraterrestrial environments. Microb Mol Biol Rev 64:548–572
24. Oberbeck VR, Fogleman G (1990) Impact constrains on the environment for chemical evolution and the origin of life. Origins of Life 20:181–195
25. Pace N (1991) Origin of life – Facing up the physical setting. Cell 65:531–533
26. Rummel JD (1989) Planetary protection policy overview and application to future missions. Adv Space Res 9:(6)181–184
27. Schidlowski M (2002) Search for Morphological and Biogeochemical Vestiges of fossil Life in Extraterrestrial Settings: Utility of Terrestrial Evidence. In: Horneck G, Baumstark-Khan C (eds) Astrobiology, the Quest for the Conditions of Life. Springer, Berlin Heidelberg New York, pp 373–386
28. Schulze-Makuch D, Irwin LN (2004) Life in the Universe, Expectations and Constraints, Springer, Berlin Heidelberg
29. Schwartz DE, Mancinelli RL, Kaneshiro ES (1992) The use of mineral crystals as bio-markers in the search for life on Mars. Adv Space Res 12:(4)117–119
30. Stan-Lotter H (2007) Extremophiles, the Physico-Chemical Limits of Life (Growth and Survival). In: Horneck G, Rettberg P (eds) Complete Course in Astrobiology, Wiley-VCH, Berlin, New York, pp 125–154
31. Stetter KO (2002) Hyperthermophilic microorganisms. In: Horneck G, Baumstark-Khan C (eds) Astrobiology, the Quest for the Conditions of Life. Springer, Berlin Heidelberg New York, pp 169–184
32. Stöffler D, Horneck G, Ott S, Hornemann U, Cockell CS, Moeller R, Meyer C, de Vera J-P, Fritz J, Artemieva NA (2007) Experimental evidence for the potential impact ejection of viable microorganisms from Mars and Mars-like planets. Icarus 186:585–588
33. Thomas PJ, Hicks RD, Chyba CF, McKay CP (eds) (2006) Comets and the Origin and Evolution of Life, 2nd Edition. Springer, Berlin Heidelberg New York
34. Tokano T (ed) (2005) Water on Mars and Life. Springer, Berlin Heidelberg
35. Trinks H, Schröder W, Biebricher CK (2005) Ice and the origin of life. Origins of Life and Evolution of Biospheres 35:429–445
36. Udry S, Mayor M (2002) The Diversity of Extrasolar Planets Around Solar-Type Stars. In: Horneck G, Baumstark-Khan C (eds) Astrobiology, the Quest for the Conditions of Life. Springer, Berlin Heidelberg New York, pp 25–46
37. Ulmschneider P (2006) Intelligent Life in the Universe. 2nd Edition, Springer, Berlin Heidelberg
38. Wächtershäuser G (1998) Origin of Life in an Iron-Sulfur World. In: Brack A (ed) The Molecular Origins of Life, Assembling the Pieces of the Puzzle. Cambridge University Press, Cambridge UK, pp 206–218
39. Westall F, de Witt MJ, van der Graaf S, de Ronde C, Gerneke D (2001) Early Archean fossil bacteria and biofilms in hydrothermally influenced shallow water sediments, Barberton greenstone belt, South Africa. Precambrian Res 1006:93–116
40. Woese CR, Kandler O, Wheelis ML (1990) Towards a natural system of organisms: proposal for the domains Archaea, Bacteria and Eucarya. Proc Natl Acad Sci USA 87:4576–4579

Chapter 28
Authors

Aizawa, Masuo
He earned his Dr. Eng. from Tokyo Institute of Technology (Tokyo Tech.) in 1971. He was Assistant Prof. at Tokyo Tech. in 1971–1980, Associate Prof. at Tsukuba Univ. in 1980–1986, Prof. at Tokyo Tech. in 1986–2001, and Vice President in 2000–2001. Since 2001, he has been President of Tokyo Tech. He is also an Executive Member of the Council for Science and Technology Policy, Cabinet Office.
Tokyo Institute of Technology, O-okayama, Meguro-ku, Tokyo 152-8550, Japan.

Artmann, Gerhard M.
He studied physics at the Technical University of Dresden (1970–1974). He obliged to leave East Germany in 1985 and became a Ph.D. student at RWTH Aachen (1986–1988). Since 1989 he is a Professor at Aachen University of Applied Sciences where he was a dean of the Faculty Applied Physics (2002–2004). He habilitated in 1999 at TU Ilmenau and is chair of the Center of Competence in Bioengineering since 2003.
Aachen University of Applied Sciences, Cellular Engineering, Ginsterweg 1, 52428 Jülich, Germany

Baroud, Gamal
He is a junior Canada Research Chair in Skeletal Reconstruction and Biomedical Engineering. He received his Dr.-Ing. (PhD) in 1997 (Germany). Among others he has been Director of the Biomechanics Laboratory and an Associate Professor at Université de Sherbrooke since September 2003. His research focus is on the biomechanics and micro-instrumentation of minimally-invasive use of medical cement to treat osteoporotic bone.
Laboratoire de biomécanique de l'Université de Sherbrooke, Canada

Chien, Shu
Shu Chien, M.D. Ph.D., University Professor of Bioengineering & Medicine at UCSD, is a leader in integrative engineering & physiology, from molecules to organs. He is Member of U.S. Institute of Medicine, National Academy of Engineering, National Academy of Sciences, American Academy of Arts & Sciences, Academia Sinica, and Chinese Academy of Sciences, as well as President of many societies.
Department of Bioengineering and Whitaker Institute of Biomedical Engineering, University of California, San Diego, La Jolla, CA 92093-0412 and Department of Medicine, University of California, San Diego, La Jolla, CA 92093-0427

Chiu, Jeng-Jiann
Jeng-Jiann Chiu received Ph.D. in fluid mechanics from National Cheng-Kung University and is an associate investigator at National Health Research Institutes in Taiwan. He has won the Distinguished Research Award by National Science Council, Young Investigator Academic Achievement Award by National Health Research Institutes, Tiente Lee Award, and Academia Sinica Young Investigator Research Award.
Division of Medical Engineering Research, National Health Research Institutes, Miaoli 350, Taiwan, ROC

Cook, Matthew
Matthew Cook has a Ph.D. (Biochemistry, 1999). Employment followed at The Technology Partnership (TTP) and then TTP LabTech (2003), MC performed a group product marketing role. In 2004, MC left TTP LabTech and spent 2 years in a technical product marketing role at SHARP Electronics. In 2006, MC returned to head business development at TTP LabTech's renowned consultancy business.
Appliance Systems Product Development Center, Sharp Corporation, Osaka, Japan

Denecke, Bernd
Interdisciplinary Center for Clinical Research on Biomaterials and Material-Tissue Interaction of Implants IZKF BioMAT, RWTH Aachen University. Study in Biology, Ruhr University Bochum. Ph.D. Cell and Tumor Biology, University of Essen. Post Doc, Medical Microbiology, MHH Hannover University Research Associate, RWTH Aachen University.
Interdisciplinary Center for Clinical Research on Biomaterials and Material-Tissue Interaction of Implants "IZKF BioMAT.", RWTH Aachen University, Pauwelsstrasse 30, 52074 Aachen, Germany

Digel, Ilya

He was born in 1973 in Zhambyl (Kazakhstan) and finished secondary school there with a gold medal. In 1995 he finished Kazakh National State University (Biology and Chemistry) in Almaty with a Diploma of Honors. After post-graduate scholarship (1995–1998) at the same University he became PhD in microbiology. Currently works in the Laboratory of Cellular Biophysics at Aachen University of Applied Sciences.
Laboratory of Cell Biophysics, Aachen University of Applied Sciences, Ginsterweg, 1, 52428 Jülich, Germany

Ermert, Helmut

He received the Dr.-Ing. degree from the RWTH Aachen Univ. in 1970 and his Dr.-Ing. habil. degree 1975 from the Engineering Faculty at the Univ. of Erlangen-Nuremberg. Since 1987 he has been Professor of Electrical Engineering at the Ruhr-Univ. in Bochum, Germany. He is continuing research on measurement techniques, diagnostic imaging, sensors in the rf and microwave area as well as in the ultrasonic area for applications in medicine, non-destructive testing, and industry.
Ruhr Center of Excellence for Medical Engineering, KMR eV Bochum, Germany.

Golz, Stefan

He was born 1968 in Gelsenkirchen, Germany. Study of Biology 1989-1995 at University of Cologne. Ph. D. student at the Institute for Genetics; Cologne; Dr. rer. nat. (1999) about protein-protein interactions of T4 endonuclease VII; 1999 Senior Research Scientist at Bayer Healthcare AG. Skills: reportergene technologies, ultra High-Throughput Screening, expression analysis, cell biology, genomics.
Bayer Healthcare AG, Institute for Target Research, 42096 Wuppertal

Hescheler, Jürgen

He studied medicine (1984), followed by his doctoral thesis (1985) and his habilitation (1988, all Universität des Saarlandes). He has been working in several research projects and is the winner of several awards such as Copp-Prize of the German Society for Osteology, Award of the BMJFFG and the Dorothy Hegarty Award. Since 2004 he is Coordinator of the European Integrated Project FunGenES.
University of Cologne, Center of Physiology and Pathophysiology, Institute of Neurophysiology, Robert Koch Str. 39, 50931 Cologne, Germany

Hirsch, Heinz-Josef

Heinz-Josef Hirsch was born in Cologne in 1946. He studied Biology and received his Ph.D. in Genetics in 1973 from the University of Cologne. He worked for six years at the German Research Centre for Biotechnology (GBF) in Braunschweig in the Microbiology Department. In 1982 he became a scientific staff member at the RWTH Aachen University in the Department of Botany and Molecular Genetics.
RWTH Aachen University, Institute of Biology I, Worringer Weg 1, 52074 Aachen, Germany

Horneck, Gerda

Dr. Gerda Horneck is former Head of the Radiation Biology Section and Deputy Director of the Institute of Aerospace Medicine of the DLR. She has been involved in radiobiological and astrobiological space experiments on Spacelab, LDEF, EURECA, FOTON, MIR, ISS, and ExoMars. For her research Dr. Horneck was awarded several honors of ESA, NASA, DLR and the International Academy of Astronautics.
German Aerospace Center, Institute of Aerospace Medicine, D 51170 Koeln, Germany

Hüttmann, Gereon

He was born in Herne, Germany in 1962. He received, in 1988, the Master of Science degree in Physics and, in 1992, his Ph.D. in Physical Chemistry from the University of Göttingen. Since 1992, he is research member of the Medical Laser Center in Lübeck and the Institute of Biomedical Optics of the University Lübeck, Germany. He is leading a group working on nanoparticle cell surgery and optical diagnosis.
Institute of Biomedical Optics University Lübeck, Germany

Jahnen-Dechent, Willi

Study in Biology, University of Mainz, PhD Biochemistry, Max-Planck-Institute and University of Cologne, Post-Doc, Melbourne University, Ludwig Institute for Cancer Research, Melbourne Australia and University of Massachussetts, Amherst, Assistant Professor, Physiological Chemistry, University of Mainz, Full Professor, RWTH Aachen University
Institute for Biomedical Engineering, RWTH Aachen University, Pauwelsstrasse 30, 52057 Aachen, Germany

Kassab, Ghassan S.
He is currently the Thomas J. Linnemeier Guidant Foundation Chair and Professor in Biomedical Engineering at Indiana-Purdue University, the recipient of the NIH Young Investigator Award and the AHA Established Investigator Award. He is also a Fellow of the AIMBE and the APS (CV section) and has published over one hundred peer-reviewed full articles. Skills: the coronary circulation in health and disease.
Department of Biomedical Engineering, Surgery, Cellular and Integrative Physiology; Indiana University Purdue University Indianapolis, IN 46202

Kaunas, Roland
The author received his Ph.D. in Bioengineering from the University of California, San Diego. Dr. Kaunas is the recipient of the American Heart Scientist Development Grant and he is a member of the Biomedical Engineering Society. Dr. Kaunas is currently an Assistant Professor of Biomedical Engineering at Texas A&M University and his scientific interests are focused on vascular mechanobiology.
Department of Biomedical Engineering and the Cardiovascular Research Institute, 337 Zachry Engineering Center, Texas A&M University, College Station, U.S.A.

Kayser, Peter
Peter Kayser studied chemistry at Aachen University of Applied Sciences (AcUAS). His diploma thesis was done at the department of cell biophysics (2003–2004). He took part in different projects, mainly in the field of LPS induced in vitro sepsis models. Since 2006 he is also the Coordinator of the Centre of Competence in Bioengineering, but working as a scientist as well.
Centre of Competence in Bioengineering, Aachen University of Applied Sciences, Ginsterweg 1, 52428 Jülich, Germany

Kelm, Malte
1986–1995 Internship, Residency and Postdoctoral Fellow in Cologne and Düsseldorf, Germany and London, UK, 1995–2005 Cardiology, Heinrich-Heine University Düsseldorf, Germany, since 12/2005 Head of the Medical Clinic, Department of Cardiology, Pulmonary Diseases, Vascular Medicine, and Intensive Care University Hospital Aachen
Department of Medicine, Medical Clinic I, University Hospital RWTH Aachen, Germany

Keymel, Stefanie
09/2005–03/2006 Institute for cardiovascular molecular biology, University Hospital Aachen, since 04/2006 Internship, Medical Clinic, Department of Cardiology, Pulmonary Diseases, Vascular Medicine, and Intensive Care University Hospital Aachen
Department of Medicine, Medical Clinic I, University Hospital RWTH Aachen, Germany

Khaled, Walaa
Walaa Khaled was born in Cairo, Egypt, in 1972. He received the B.E. from the department of Biomedical Engineering Cairo University in 1993. He received his M.E. from the University, Erlangen, Germany 2002 and his PhD from the Ruhr University, Bochum, Germany in 2007. His research interests include biomedical application of ultrasound, Elastography, tissue characterization, inverse problems and signal processing.
Institute of High Frequency Engineering, Ruhr-University Bochum, D-44780 Bochum, Germany.

Kistler, Erik B.
He received his BA in music and BS and MS in electrical engineering from UC Santa Barbara, his PhD in bioengineering from UC San Diego and did his postdoctoral fellowship at The Scripps Research Institute. His medical school education was at The George Washington School of Medicine. He completed his residency in anesthesiology at UC San Diego and is continuing his training in Critical Care at Mass General Hospital in Boston. His outside interests include surfing and music.
Department of Anesthesiology, University of California San Diego, La Jolla, CA

Kleinbongard, Petra
05/1999 Master of Science (Biology), Ruhr-University Bochum, Germany, 10/2003 Ph.D.; Division of Cardiology, Heinrich-Heine-University Düsseldorf, Germany in cooperation with Department of Biology, Ruhr-University Bochum, Germany, Since 11/2003 Postdoctoral Fellow, Division of Cardiology, Heinrich-Heine-University Düsseldorf, Germany and Postdoctoral Fellow Division of Cardiology, University Hospital Aachen
Department of Medicine, Medical Clinic I, University Hospital RWTH Aachen, Germany

Koyama, Sumihiro

He received his Ph.D. in engineering from Tokyo Institute of Technology in 1997. He was researcher at Japan Agency for Marine-Earth Science and Technology (JAMSTEC) in 1997–2007. Since 2007, he has been senior research scientist at Jamstec. His research interests are cellular engineering.
Extremobiosphere Research Center, Japan Agency for Marine-Earth Science and Technology, 2-15 Natsushima-cho, Yokosuka 237-0061, Japan.

Kreuzaler, Fritz

Fritz Kreuzaler, born in Gumbinnen in 1943, studied Biology and Biochemistry in Freiburg i. Br. and received his PhD in Biochemistry in 1974. For three years he worked at Boyler College of Medicine with B.W. O'Malley and from 1980 to 1986 at the MPI of Plant Breeding in Cologne with J. Schell. In 1986 he became head of the Department of Botany and Molecular Genetics at the University Aachen.
RWTH Aachen University, Institute of Biology I, Worringer Weg 1, 52074 Aachen, Germany

Lam, Hayley

Hayley Lam is currently a PhD student in Bioengineering at the University of California, Berkeley. The focus of her research is on human embryonic stem cell differentiation into vascular cells using micropatterning and microfluidic techniques. She also enjoys teaching science and engineering to middle school and high school students.
Department of Bioengineering, University of California, Berkeley

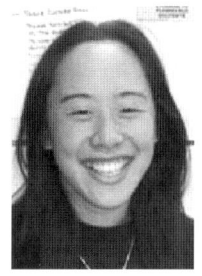

Li, Song

Song Li is an Associate Professor of Bioengineering in University of California, Berkeley. He had B.S. and M.S. from Peking University in China, and he did his Ph.D. research work under the guidance of Dr. Shu Chien at UC San Diego. He joined the bioengineering faculty at UC Berkeley in 2001. His research focuses on mechanobiology, stem cell engineering and micro/nanomaterials.
Department of Bioengineering, University of California, Berkeley

Linder, Peter
Dipl. Ing. Peter Linder M.Sc., born 3 May 1979 in Jülich, Germany, 2000–2004 University of Applied Sciences Aachen (AcUAS), Diploma Biomedical Engineering; 2004–2006 UAS Aachen, M.Sc. Biomedical Engineering; 2006–now UAS Aachen & University College London, PhD studies; 2001–now UAS Aachen, Laboratory for Cell Biophysics, Research and Development
Aachen University of Applied Sciences, Laboratory for Cell Biophysics, Ginsterweg 1, 52428 Jülich, Germany

Meiselman, Herbert J.
Herbert J. Meiselman has worked in the area of blood rheology for 45 years, and is currently professor of Physiology and Biophysics, USC, Los Angeles. Areas of research include the rheological behavior of blood and its formed elements in health and disease, cell-cell interactions, and the in vivo effects of altered blood rheology; areas of interest include restoring his 1965 Porsche 356SC.
Department of Physiology and Biophysics, Keck School of Medicine, University of Southern California, Los Angeles, USA

Neu, Björn
Björn Neu received his Doctorate in Biophysics in 1999 from the Humboldt University in Berlin, then joined Prof. Meiselman, USC, Los Angeles to carry out research in the field of haemorheology. Currently he is a Professor of Bioengineering at Nanyang Technological University, Singapore where his research interests include cell-cell interactions, polymers at biointerfaces and the rheological behavior of blood.
Division of Bioengineering, School of Chemical and Biomedical Engineering, Nanyang Technological University, 70 Nanyang Drive, Singapore 63745

Neuss, Sabine
Sabine Neuss studied Biology at RWTH Aachen University followed by her PhD in Biology at RWTH Aachen University where she is also a Research Associate.
Institute of Pathology, RWTH Aachen University, Pauwelsstrasse 30, 52074 Aachen, Germany

Nishikawa, Kazuo
Senior technical specialist, Appliance Systems Product, Development Center, Appliance Systems Group, Sharp Corporation; Specialization: Air purification engineering, biophysics and chemical engineering; Awards: Prize of Progress by The Institute of Electrical Engineers of Japan (2001).
Appliance Systems Product Development Center, Sharp Corporation, Osaka, Japan

Patel, Shyam
Shyam Patel is currently a PhD candidate in Bioengineering at the University of California, Berkeley. His research focuses on micro/nanopatterning and chemical modification of polymers for cell and tissue engineering. He is also actively translating his laboratory research into clinical therapies.
Department of Bioengineering, University of California, Berkeley

Penn, Alexander H.
Alexander H. Penn completed his Bachelor of Science in Chemical Engineering degree from Rice University in Houston, TX in 1998. He then went on to receive a Master of Science and Doctor of Philosophy in Bioengineering from the University of California, San Diego where he currently continues the work he started there on the role of cytotoxic mediators in shock.
Department of Bioengineering, The Whitaker Institute of Biomedical Engineering

Peterhänsel, Christoph
Christoph Peterhänsel was born in Aachen in 1967. He graduated and passed his Ph.D. in phytopathology at RWTH Aachen. He worked as a PostDoc at the German Cancer Research Centre in Heidelberg before coming back to Aachen as a research group leader working on photorespiration and gene regulation.
RWTH Aachen University, Institute of Biology I, Worringer Weg 1, 52074 Aachen, Germany

Pfannkuche, Kurt

Kurt Pfannkuche studied biology in Cologne and did his diploma thesis in the workgroup of Klaus Rajewsky (2001/2002). Since 2002 he is employed as a PhD student in the laboratory of Jürgen Hescheler at the institute of neurophysiology, Cologne.
University of Cologne, Center of Physiology and Pathophysiology, Institute of Neurophysiology, Robert Koch Str. 39, 50931 Cologne, Germany

Robitzki, Andrea A.

Chair for Molecular biological-biochemical Processing Technology, Chair of Management Board of the Center for Biotechnology and Biomedicine, Director of the Institute of Biochemistry, Member of the Medical Engineering Board of the German Ministry for Education and Research, Skills: Molecular tissue engineering, nanostructured cell/tissue based microarrays, microimplants, bioelectronic drug screening, microlaser dissection/catapulting, neuronal de-/regeneration
Centre for Biotechnology and Biomedicine, University of Leipzig, Germany

Rothermel, Andrée

Andrée Rothermel is a postdoctoral research fellow at the Center for Biotechnology and Biomedicine in Leipzig, Germany. His work focuses on the development of cell and tissue based sensors. He received his M.A. (1996) and Ph.D (2000) from the University of Technology, Darmstadt. Since 2003 he is working at the Center for Biotechnology and Biomedicine.
Centre for Biotechnology and Biomedicine, University of Leipzig, Germany

Ruau, David

David Ruau received his master degree in Bioinformatics from the University of Toulouse, France in 2003 and is currently a Ph.D. student at the University of RWTH Aachen, Germany. Originally trained in database management, he broaden his knowledge during his Ph.D. study. Now his research interest concentrates on developing new approach for DNA microarray data mining by focusing on third party data integration.
Helmholtz Institute for Biomedical Engineering, RWTH Aachen University, Pauwelsstrasse 20, 52074 Aachen, Germany

Sachinidis, Agapios

He studied Chemistry and Biochemistry (Diploma: 1984) and got his PhD (1987) at the University of Münster. He was employed at the University of Zürich and Bonn and became 1998 in Bonn Prof. of Pathobiochemistry. He became 2002 Prof. for Physiology and Pathobiochemistry at the University of Cologne. Current scientific activities: embryonic stem cell genomics in cardiovascular cell types developed from embryonic stem cells.
University of Cologne, Center of Physiology and Pathophysiology, Institute of Neurophysiology, Robert Koch Str. 39, 50931 Cologne, Germany

Schmid-Schönbein, Geert W.

He is Professor of Bioengineering at the University of California San Diego and has a long track record in microvascular research, cell and molecular mechanics. He is President of several societies and member of the US National Academy of Engineering. His recent work is focused on biochemical and biomechanical trigger mechanisms for inflammation and development of a new Auto-Digestion theory for several human diseases.
Department of Bioengineering, The Whitaker Institute of Biomedical Engineering, University of California San Diego, La Jolla California, 92093-0412

Shyy, John

Dr. John Shyy's research is to elucidate the mechanotransduction mechanisms by which fluid shear stress imposing on the vascular endothelial cells. Both *in vitro* experiments involving flow channels and *in vivo* approaches with the use of animal models are used to study the effect of atheroprotective versus atheroprone flow patterns on the vascular biology in health and disease.
Division of Biomedical Sciences, University of California, Riverside, CA 92521

Sponagel, Stefan

He is head of the Biomaterials Laboratory at the Aachen Univ. of Applied Sciences, Germany. He received his Dr. rer. nat. from the Technical Univ. Darmstadt (1982) and his Dr.-Ing. from the Univ. Kaiserslautern (1987, both Germany). He worked in the elastomer industry and became Professor of mathematics and material sciences in 1994. His main research interests include continuum mechanics and material modeling. He is member of the Institute of Bioengineering, Germany.
Biomechanics Laboratory, Aachen University of Applied Sciences, Campus Jülich, Ginsterweg 1, 52428 Jülich, Germany

Staat, Manfred
He is head of the Biomechanics Laboratory at the Aachen Univ. of Applied Sciences, Germany. He received his Dr.-Ing. from the RWTH Aachen Univ. (1988). He worked at the Research Center Jülich before he became Professor of biomechanics (1997). Main research interests: continuum mechanics and optimization methods in biomechanics and plasticity. He is member of the centre of competence in Bioengineering and of the Institute of Bioengineering.
Biomechanics Laboratory, Aachen University of Applied Sciences, Campus Jülich, Ginsterweg 1, 52428 Jülich, Germany

Suh, Hwal (Matthew)
He received D.D.S. and M.S.D. degrees from Yonsei Univ., Seoul, Korea (1978 & 84) and his Ph.D. from Osaka Univ. (1992). At present, he is chair professor of both Department of Medical Engineering in College of Medicine and Program of Nano Science and Technology in the Graduate School at Yonsei Univ., and also leader of National Brain Korea 21 Project Team of Nanobiomaterials for the Cell-based Implants.
Department of Medical Engineering, College of Medicine, Yonsei University, Seoul 120-752, Korea

Temiz Artmann, Ayşegül
Dr. A (Temiz) Artmann was born in Ankara in 1966, married and has a daughter. She graduated from Ankara Univ. School of Medicine in 1990. Following university education did her specialty in Physiology and PhD in Biochemistry. She worked at Akdeniz and Dokuz Eylül Univ. Medical School in Turkey and USC in Los Angeles-USA. Currently working as a Prof. at Aachen, Univ. of Applied Sciences.
Medical & Molecular Medicine, Aachen University of Applied Sciences, Ginsterweg 1, D 52428 Jülich, Germany

Topcu, Murat
He received his MSc. in Biomedical Engineering as well as his Dipl.-Ing. in Aeronautics at Aachen University of Applied Sciences (AcUAS), Germany. Currently he works at the Biomechanics Laboratory of the AcUAS, Campus Jülich.
Biomechanics Laboratory, Aachen University of Applied Sciences, Campus Jülich, Ginsterweg 1, 52428 Jülich, Germany

Tremmel, Ira G.
She completed her Diploma in Physics (1994) as well as her Diploma in Biology (1998) from Philips University Marburg Germany before she received her PhD from the Australian National University. Currently she works as Scientific Officer at the Max-Planck-Institute for biophysical chemistry, Göttingen, Germany. Winner of several scientific awards.
Max-Planck-Institut für biophysikalische Chemie, Computational Biomolecular Dynamics Group, Am Fassberg 11, D-37077 Göttingen, Germany

Trzewik, Jürgen
He finished his studies of Biomedical Engineering – including a DAAD scholarship at the University of California, Dept. of Bioengineering, San Diego, USA – at Aachen University of Applied Sciences, Germany as Dipl.-Ing. and received his PhD from the University of Technology Ilmenau, Germany. Currently he is working as a scientist at Ethicon, a Johnson & Johnson Company.
Ethicon GmbH, Robert-Koch-Straße 1, D-22851 Norderstedt

Usami, Shunichi
Shunichi Usami received M.D., Ph.D. in Physiology from Kyoto Prefectury University. He was a research faculty in Physiology at Columbia University, Institute of Biomedical Sciences at Academia Sinica, and Bioengineering at UCSD. He has made major contributions in blood rheology, microcirculation and mechanotransduction, and has designed many innovative research tools for biomedical research.
Department of Bioengineering and Whitaker Institute of Biomedical Engineering, University of California, San Diego, La Jolla, CA 92093-0412

Vant, Christianne
Christianne Vant (M.A.Sc. Carleton University, B.Sc.(Eng.) University of Guelph) worked as a Research Assistant at the Orthopaedic Research Laboratory at McGill University (Montreal, Canada) and the Biomechanics Laboratory at l'Université de Sherbrooke (Sherbrooke, Canada). Her research interests include biomechanics and orthopaedics with specific interests in spinal biomechanics and osteoporosis.
Laboratoire de biomécanique, Université de Sherbrooke, Sherbrooke, Canada

Wang, Yingxiao
Dr. Yingxiao Wang obtained his bachelor and master degrees in Mechanics from Peking University, P.R. China. He received his Ph.D. degree in Bioengineering and continued his postdoc work together with Drs. Shu Chien and Roger Tsien at UC San Diego. He is currently an assistant professor in the Department of Bioengineering and Beckman Institute for Advanced Science and Technology at UIUC.
Department of Bioengineering and Beckman Institute for Advanced Science and Technology, University of Illinois, Urbana-Champaign, Urbana, IL 61801

Wöltje, Michael
He studied Biology at the University of Bremen and received his PhD in Cell and Molecular Biology at the University Magdeburg and University of Bremen. Currently he works as Research Associate at RWTH Aachen University
Interdisciplinary Center for Clinical Research on Biomaterials and Material-Tissue Interaction of Implants IZKF BioMAT, RWTH Aachen University, Pauwelsstrasse 30, 52074 Aachen, Germany

Zenke, Martin
He received his PhD from Ruprecht-Karls-University Heidelberg (1982). After different employments, e. g. as EMBL fellow (European Molecular Biology Laboratory, Heidelberg, Germany), as Junior Scientist and Group Leader at Institute for Molecular Pathology, Vienna, Austria and as Research Group Leader at the Max-Delbrück-Center for Molecular Medicine, Berlin he became Professor of Cell Biology and Chairman at RWTH Aachen University, Germany.
Institute for Biomedical Engineering, Department of Cell Biology, RWTH Aachen University Medical School, Pauwelsstrasse 30, 52074 Aachen and Helmholtz Institute for Biomedical Engineering, Pauwelsstrasse 20, 52074 Aachen, Germany

Zerlin, Kay F.
Dr. Kay Zerlin, born in 1977, studied chemistry at the University of Wuppertal and received a PhD in chemical microbiology in 2004. Afterwards he had a postdoctoral research fellow in the laboratory of biophysics at the Aachen University of Applied Sciences. Since the beginning of 2007 Dr. Zerlin is working on a scientific officer position at the research centre juelich.
PTJ Jülich, Research Centre Juelich GmbH, 52425 Jülich, Germany

Index

Symbols

1,1,1,3,3,3-hexafluoro-2-propanol (HFP) 168
1-ethyl-3(3-dimethylaminopropyl) carbodi-imide (EDC) 166

A

absorption, optical
  blood 94
  melanin 94
  proteins 94
actin
  dynamics 319
  networks 319
  persistence length of 325
  stretch 319–322
activated leukocyte 511
activated protein C (APC) 501
activator protein-1 (AP-1) 39, 43, 50, 52, 481
activity
  coefficient 547, 548, 551
  thermodynamic 550, 571, 572
adeno-associate virus 161
adenovirus 161
adenylate cyclase 15
adherence 602
adhesins 609, 611, 612
adhesion 163, 183, 184, 187, 192, 601
  biological aspects 610
  influence on growth kinetics 605
  influence on physiological activity 605, 607
  physicochemical aspects 601, 607
  strength 603, 610
adsorbent 607, 615

adsorption 605
  equation 607
adult respiratory distress syndrome (ARDS) 498
adult stem cells 193, 198, 199, 214
Aequorea victoria 6
affine strain approximation 330
AFM   *see* Atomic Force Microscopy (AFM)
aggregates 138
aggregation 544, 549, 562, 563
Agrobacterium tumefaciens 591, 594
ahesion
  biological aspects 611
albumin 501
aldehydes 165, 166
aldol condensation 165
alkaline phosphatase 11
allogeneic tissue 163
α-dispersion 238
alpha-fetoprotein (AFP) 185, 188
amine to amine (NH-NH) bonding 166
amino acid 587, 643
aminoglycoside phosphotransferase 138
ammonia 595
amnion 299, 302
anabolic pathway 587
angiotensin II 236
anhydrate succinyl acid 168
anisotropy in cell behaviour 324, 325
anti-inflammatory cytokines 499
antibody 591
anticoagulant 501
apatite 175
APC 501
Arabidopsis thaliana 591
ARDS 498

arginine–glycine–aspartic acid (RGD)
    sequence   163
aromatic residues   166
artery
    abdominal   472, 480, 483
    aortic arch   472, 480, 484
    carotid   472, 478, 479, 482–484
    coronary   472
    femoral   472, 479
    iliac   478, 479
    infrarenal   472
    thoracic   472, 476, 480, 484
artificial coronary artery   97
artificial vascular structures   98
assay development   16
association   548, 550, 551, 572
astroglial cell   41, 50
atherosclerosis   378, 383, 384, 453, 455,
    469–471, 473, 479, 480, 482, 483,
    485–487
atherosclerotic plaques   317
Atomic Force Microscopy (AFM)   70, 323
ATP   552, 554, 583, 586
attachment
    irreversible   602, 610
    reversible   602, 610

B

Bacillus polymyxa   605
balance of mass   275
balance of momentum   275
basal epithelial layer   170
beating
    cells   143, 147
    clusters   138
    frequency   138, 147, 148
    tissue   137
Beroe abyssicola   7
Berovin   7
β-dispersion   238
β-Galactosidase   11
biaxial testing of adherent cells   335
bio-inert   163
bioactive agents   163
bioartificial heart tissue   147
biocompatibility   162, 163, 193
biodegradable materials   162, 163
bioengineering   453
biofilm   602–604, 611
biogenic elements   642
    CHONSP   642
bioimpedance   232
bioinformatics   23, 25, 27, 32, 34

biological agents   161
bioluminescence   5
biomass   581, 598
biomaterials   159–163, 167, 176, 193, 194,
    204, 206, 209, 212
bioMEMS   231, 233
biomolecules   163
biophysical force   355
bioreactors   581
biosensor   231, 233, 239
biotechnology   453
2′-carboxy-3-keto-arabinitol-1,5-bisphosphate
    583
block matching   110, 113, 114, 125, 127
blockade of digestive enzymes   514
blood
    cell   452, 459, 460, 463
        deformability   453
        function   453, 459, 460
    flow   452, 454, 459, 461, 463, 464
    pressure   452
    vessels   451, 452
blood components   420
blood plasma   415, 426, 429, 430, 432, 433
blood viscosity   395
body temperature   415, 418, 419, 421, 434,
    438
Bolinopsin   8
bone mineral density   527
boundary lipids   558, 562
bridging
    flocculation   396
    model   397
bulging test   292, 296
bundle sheath cells   586

C

c-Fos   43, 52
c-fos   481
c-fos   41, 50–54
c-Jun   43, 52
c-jun   481
c-jun   41, 50–52
c-reactive protein   510
C3 plants   582, 595
C3-pathway   582, 583
C4 plants   582, 585
C4-pathway   582, 586
calcium
    mobilization   16
    readout   17
calcium phosphate (CaP)   531, 534
calcium sulfate (CaS)   531

Index

cAMP response element (CRE)   14
CaMV   590
cantilevers for measuring cellular forces   325
carbonic anhydrase   586
carboxyl groups   303
carboxylation   582, 596
Cardano's formula   290
cardiac infarction   136, 143
cardiomyocyte   135–139, 235, 242, 353, 363
cardiovascular system   371
cartilage   97
catabolic pathway   587, 596
Cauliflower Mosaic Virus   590
Caulobacter crescentus   602, 618
CD14   499
cell   136, *see* endothelial cell
　adhesion   96
　based therapies   135
　migration   96
　permeable proteins   136, 154
　proliferation   96
cell blebs   517
cell cycle   136, 141, 152
cell encapsulation   162, 188
cell hybridization   165
cell migration   219
cell therapy   160, 162
cell tracker   362
cell-based implant   159, 161, 165
Cell-tak   359
CellDrum   353, 356
cement
　acrylic cement   534
　calcium phosphate cement   534
　cement cohesiveness   531, 533
　cement delivery device   530
　cement leakage   528, 529
　cement polymerization   531
　cement viscosity   531
characteristic polynomial   279
charge separation   553
charged soluble polymers   613, 617
chemiosmotic theory   554
chitin   163
chitosan   163
chitosan scaffolds   96
chlorophyceae   588
chlorophyll fluorescence   596
chloroplast   552–554
chloroplast targeting peptide   590, 591
chloroplasts   582, 590
chondrocytes   136, 138
chondroitin-6-sulphate   163
chromosome

chromosome elimination cassette (CEC)   141, 142
　elimination   141
　single chromosomes and pair chromosomes   141
circular dichroism   424, 425, 437, 439
circulation   451, 454, 456, 463
Clytia gregaria   6
Clytin   7
CO_2-fixation   582, 587
CO_2 fixation   552
coagulation   498, 499, 501
coelenterazine dependent luciferase   9
collagen   163–189, 294, 297, 300–302
　gel   148
　matrix   150
　ring   148
　sponge   149
　type I   149
collagen matrix   366
colloid   498, 501, 504, 506
colloid osmotic pressure (COP)   415, 432, 498, 505, 506
comets   643
cone-and-plate viscometer   473, 481
confocal microscopy   86, 359, 361
conger eel cell   49
connexin (Cx)   477, 478, 480, 482, 483, 485
constitutive relation
　for adherent cells   332
　hyper-viscoelastic   327
constraint   277
continuum model of adherent cells
　based on the mixture theory   *see* mixture theory
　bio-chemo-mechanical   329–331
contractility
　actin   *see* actin contractility
　and Rho small GTPase   *see* stress fibers and rho activity
　cell   335
　Hill model   330
contrast agents   95
COP   498, 505, 506
core   499, 505
cosine distance   27
COSPAR   658
crassulacean acid metabolism   582
CRE   18
Cre recombinase   141, 142, 152–154
creep   260, 261, 271
　creep compliance   267
　creep retardation   267
　creep test   266, 271

cross-correlation   103, 110–113, 122, 125, 127
crosslink   167
Cryozootech   140
crystalloid   501
current configuration   272, 273, 275
cyanobacteria   588
cyclin-dependent kinase (CDK)   478
cytochrome oxidase   98
cytochrome bf   553–556, 560, 564–569
cytokines   161, 163, 185, 189
cytoplasm viscosity   319
cytoskeletal
   actin   *see* actin
   intermediate filaments   *see* intermediate filament dynamics
   microtubules   *see* microtubules
cytoskeleton   476, 479, 480, 485
   actin   477, 479, 480
   microtubule   477
   stress fibers   477
cytosol   *see* cytoplasm
cyt bf   *see* cytochrome bf

D

data mining   23–25
dedifferentiation   136
deep-sea eel cell   48, 49
deformation   255–257, 267, 271, 276, 277, 281–286, 291, 295, 302, 307, 309, 311
   deformation gradient   273, 276, 306
   deformation history   272–274, 276, 277, 282
   quasi-static   103, 106, 107, 110, 127
   relative deformation gradient   273
   static   105, 108
deformation gradient tensor   327, 328
deformation tensor   273
   left Cauchy–Green deformation tensor   273
   right Cauchy–Green deformation tensor   273, 274
dendritic cells (DC)   23, 29–32
depletion
   aggregation   396
   effect   396
   electro-osmotic flow   398
   flocculation   397
   layer   396, 397
      thickness   402, 406
   model   397
determinism   276
development   135, 143, 145
   embryonic   136, 151
   fetal   136, 138
   terratomas   139
differentiation   137–140, 143, 148, 151, 152
diffusion   543, 544, 550 ff, 556–563, 565–569, 569 ff
   domains   556, 560, 562, 567, 570, 571
   limited   549, 550, 565, 568, 569, 571
digestive enzymes   509
displacement estimation   110, 113
disseminated intravascular coagulation   519
DLVO theory   608
DMAC(dimethyl-acrylamide) solution   179
DNA microarrays   484
   one-color microarrays   25
   tiling array   24
   two-color microarrays   25, 26
Drake equation   662
drug development   231
drug discovery   5, 18
drugability   4
dual energy X-ray absorptiometry (DEXA)   527
dynamic
   dynamic stiffening   268
   dynamic stiffness   268
   dynamic viscosity   268

E

E. coli   497, 499, 502, 506
echidna   434, 435
ECM   354
ecosystems   647
   microbial   647
eight-chain-network model   302
ektacytometry   460, 461
elasticity
   elastic-visco-elastic solid   271
elastin   174, 294, 301
elastography   103, 105–107, 109, 111, 114, 115, 117, 121, 123, 124, 127
   reconstructive ultrasound   107, 114
   transient   107
   ultrasound   103, 105–107, 114, 115, 121–124, 127
electric impedance spectroscopy (EIS)   237
electrical potential   37, 39, 42, 45–47, 53, 54
electrical stimulation device   147
electron acceptor   553, 554, 592
electron requirements   596
electrophysiological
   capabilities   146
   parameters   145

properties 145
studies 143
electrophysiological recording 245
electrostatic interaction 304
ellipticity 425, 426, 436, 437
embryoid bodies (EBs) 138, 143
embryonic development 151, 152
embryonic stem cell (ES)
  human stem (hES) cells 135
  murine (mES) cells 135, 137
endocytosis 162
endosomal escape 162
endothelial
  dysfunction 453, 455
  function 454, 455, 458
endothelial cell (EC)
  cell cycle 478
  DNA synthesis 478, 485, 486
  dysfunction 470, 471, 485, 486
  fluid shear stress on 317
  mechanical stress on 317
  migration 478, 479
  morphology 317, 472, 476, 483, 485
  permeability 470, 471, 479, 480, 485, 486
  proliferation 478, 486
endothelium 453, 454, 458, 460, 461
endotoxin 497–501
energy
  deformation energy 298, 300, 301
  free energy 287, 305–307, 548, 549
  internal energy 286, 289, 300, 302
  renewable 581, 587
energy elastic term 291, 295
engineered skin 97
enhanced green fluorescent protein (eGFP) 138, 154
  green fluorescent protein (GFP) 154
  green fluorescent proteine (GFP) 141, 149
entropic term 288
entropy 548, 549
entropy elastic 291, 295
epidermal growth factor (EGF) 186
epidermal growth factor receptor (EGFR)
  tyrosine kinase 40, 43
epithel 500
equibiaxial loading 291
equilibria 545, 547, 548, 550, 551, 565, 572
equilibrium constant 547, 548, 556, 566
ERK 40, 44, 45
erythrocyte 500, 502, *see also* RBC
ESA 656, 657, 662
Escherichia coli 497, 499, 502, 506, 587, 592
Euclidean distance 26
Europa 656, 660

Jupiter's moon 656
extracellular matrix (ECM) 44, 163, 216, 470, 479
extracellular recording 234, 237
extracellular-signal regulated kinase (ERK) 478
extrasolar planet 661
extravasation 533
extreme environments 647, 650

F

FAD 581
fever (pyrexia) 419, 434, 440
fibrin 163
fibrinolysis 498
fibroblasts 96, 136–138, 140–142, 151, 153
fibronectin 183, 184, 186, 187, 360
field potential 234
filtration method 460
fingerprints patterns 276
finite element 116, 118–120
flow
  complex 469, 470, 472, 473, 477–480
  fully developed 474–480
  laminar 469, 470, 473–486
  Poiseuille 475
  reattachment 472–481
  reciprocating 480
  recirculation 469, 472–477, 481, 483–485
  separation 472–475, 480, 481, 483–485
  stagnant 474–477
  step flow 470, 473–482, 484
  turbulence 474
  uniform 469, 471, 472, 475
flow injection analysis 457
flow measurements 95
flow of energy 651
flow of matter 256
flow-mediated dilation (FMD) 454, 455
flow-overload 379
fluid therapy 501
fluorescence 12
  confocal fluorescence microscopy 69, 70
  epi-fluorescence microscopy 69, 70
  fluorescence lifetime imaging microscopy (FLIM) 65, 74, 77, 78
  fluorescence proteins (FPs) 65–71
  fluorescence recovery after photobleaching (FRAP) 65, 74, 76, 77
  fluorescence resonance energy transfer (FRET) 65, 68, 74–78
  green fluorescence protein (GFP) 66–68, 71–75, 79

total internal reflection fluorescence microscopy (TIRF)   70
fluoroscopy   529
focal adehesions, stress upon   324
folding   549
force   253–255, 261, 274, 277, 288–290
  body force   275, 276, 286, 287
  cohesive force   295
  contracting force   304
  dimensionless force   306, 307
  entropic force   305
  expanding force   304
  external force   286
  muscle force   301
  tensile force   288, 289
  tension force   309
  tensional force   300
  thermodynamic force   304
  traction force   309
force measurements at cell matrix interface   324, 325
force-stretch relation   297
1536-well format   17
four-chain-network model   302
$^{13}C/^{12}C$ fractionation   644
FRET   13
functional imaging   94

G

Gal4 System   15
γ-dispersion   238
gas chromatography   595
gas-phase chemiluminescence   457
gel   303, 304
  polyelectrolytic gel   304
gene expression
  chemokine receptor 4 (CXCR4)   483
  connective tissue growth factor (CTGF)   483
  E-selectin   482
  early growth response factor-1 (Egr-1)   481
  endothelial nitric oxide (eNOS)   483–486
  guanylate cyclase 1 α_3 (GUCY1A3)   483
  intercellular adhesion molecule-1 (ICAM-1)   481, 482, 486
  interleukin-8 (IL-8)   483
  krüppel-like factor 2 (KLF2)   483, 484
  matrix metalloproteinase (MMP)   483
  monocyte chemotactic protein-1 (MCP-1)   470, 481, 486
  platelet-derived growth factor (PDGF)   481
  sterol regulatory element binding protein (SREBP)   482, 485
  thrombospondin (THBS)   483
  toll-like receptor (TLR)   482
  tumor necrosis factor receptor superfamily (TNFRSF)   483
  vascular adhesion molecule-1 (VCAM-1)   482, 486
gene expression and stretch   339
Gene Ontology (GO)
  over-representation analysis   25, 32, 33
gene therapy   160
GFP   96
glutamate/oxoglutarate aminotransferase   585
glutamine synthase   595
glutamine synthetase   585
glutaraldehyde   165, 167
glycerate kinase   585, 588
glycine   587, 595
glycine decarboxylase/serine hydroxymethyl transferase complex   585
glycolate   587, 590
glycolate dehydrogenase   588, 592
glycolate oxidase   585, 588
glycoside   505
glyoxylate   587, 588
glyoxylate carboligase   588, 590
glyoxylate/glutamate aminotransferase   585
glyoxysomes   587, 588
GPCR   14, 18
gram negative   497–499
grana   554–556, 567
green algae   588
green fluorescent protein (GFP)   7, 12

H

habitability   651
  habitable planet   652
  habitable zone   652, 661
haemodynamic   500
hatch-slack-pathway   582
heart   135, 143, 145, 147, 150–152, 154
  blood vessels   136, 148
  engineered heart tissue   148, 151
  slice   136, 145–147
heat shock   37, 39, 40, 55
heavy water (D_2O)   429
HEK293   154
HeLa cell   41, 45–47, 54
hemodynamic force   471
  cyclic stretch   471
  hydrostatic pressure   471

shear stress
  gradient   475–481, 485, 486
  high   469–471, 475–480, 482–484, 486
  laminar   478, 479, 481, 482, 485, 486
  low   469, 470, 472, 476, 478, 480, 483, 484
  non-uniform   469, 471, 473
  oscillatory   469, 470, 472, 473, 480, 484, 486, 487
hemoglobin (Hb)   303, 309, 456–461, 464
  aggregation   428, 431, 436, 439, 440
  amino acid sequence   440, 441
  concentration   420, 423, 425, 433
  denaturation   425, 432, 434, 440
  gel   415, 417, 426
  hydration shell   429
  radius   431
  temperature transition   418, 420, 423, 427, 429, 433, 438
  thermal unfolding   428, 439
  viscosity   423
hepatocyte growth factor (HGF)   184
hereditary integral   327
hierarchical clustering
  agglomerative   24, 28, 31
  divisive   28
high performance liquid chromatography (HPLC)   456, 457
high throughput real time screening   232
high throughput screening   3, 19
  HTS   16
  uHTS   14, 16, 19
hill force-velocity relationship   330
His$_6$-TAT-NLS-Cre (HTNC)   153, 154
histocompatibility   139
history of life   642, 644, 649
  fossil record   644
  molecular record   644
homeostasis   318, 334, 344, 371
  mechanical   372
  strain   385
homeothermal animals   418, 419
Hooke's law   107, 116, 117, 255, 262, 263, 299
hsp70   41, 45, 54, 55, 57
human leukocyte antigen (HLA)   160
hyaline cartilage   303, 309
hyaluronic acid   163, 171, 173, 182, 184, 186
hydrodynamic radius (Rh)   431, 434, 436
hydrogen   581
hydrophilic   259, 299
hydrophilicity   168
hydrophobic   258, 259, 299
hydrophobicity   168

hydrostatic pressure   37, 40, 47–49
hydrothermal vents   643
hydroxypyruvate reductase   585
hyper-viscoelastic relation   327
hypoalbuminemia   501
hypotension   497, 498, 501
hypovolemic shock   500
hysteresis   261

I

imaging depth   93
immobilization   605, 606
impact catastrophe scenario   646
impedance recording   238
impedance spectroscopy   238, 241, 245
implantation   160, 162, 165, 183
in vitro   195
  cell integration   147
  culture   136, 148
  tissue   139, 147, 148, 151, 152, 155
  transplantation   143, 145
in vivo
  tissue   150, 151
  transplantation   143
  tumorigenic potential   138
  vascularisation   151
incompressible   277
inflammation   498, 499, 501, 509
inflammatory cascade   509
integrability condition   285, 288
integrin   40, 44
interleukin   499, 500
interleukin-1α (IL-1α)   40, 57
interleukin-1β (IL-1β)   40, 57
interleukin-12   57
interleukin-6 (IL-6)   40, 42, 44, 56, 57
interleukin-8 (IL-8)   40, 42, 56, 57
intermediate filaments dynamics   319
internal ribosome entry site (IRES)   138
interplanetary transfer of life   647
intervertebral disk   305, 309, 310
intestinal homogenates   519
intestinal ischemia   509
intestine   513
intima/media thickness ratio   472
inverse problem   103, 114–119, 127
ionomers   304
irreversible processes   257, 286
ischemia   242, 501
isochoric   273
  isochoric processes   305
isothermal process   287, 305

J

JNK 40, 44, 45
JNK activation by stretch 339–344

K

κB 499
k-means 24
Kdo 505
Kendall's τ rank correlation 27
keratinocytes 170
Kranz-anatomy 586
Kyoto Encyclopedia of Genes and Genomes (KEGG) 32

L

lactate 592
lamellipodia 479
laminin 171, 173
laser doppler perfusion imaging 462, 463
laser-based manipulation 242
law of mass action 547, 549, 550, 569
leukaemia inhibitory factor (LIF) 137
leukocyte 501, 502
leukocytosis 497
light harvesting complex (LHC) 544, 553, 559, 565
light reaction 552
light scattering 438
linkage methods 28, 31
lipid A 499, 505
lipids 587
lipopolysaccharide 497–506
lipoprotein 471, 479, 480, 485
　low density 471
liquid-to-powder ratio 529, 531
local action 276
loss
　loss modulus 268–270
　loss tangent 268
loxP sites 141, 142, 152–154
LPS 497–506
luciferase
　Firefly 9, 10
　Gaussia 8, 9
　Metridia 8, 9, 18
　Renilla 8, 9
　Vargula 10
luciferin 10
lymphocytic choriomeningitis virus (LCMV) 139

M

magnetic cell sorting (MACS) 138
magnetic twisting cytometry (MTC) 323
maize 587
malate dehydrogenase 586
malic enzyme 586
Manhattan distance 26
Mars 653, 660
　ExoMars mission 656
　Exploration Rover 656
　habitability of 656
　history of water 653
　Viking biology experiments 655
mass spectroscopy 595
material 267–271
　Kelvin–Voigt material 271
　material dependent 255
　material points 272
　Poynting–Thomson material 262, 265, 269–271
matrix metalloproteinase (MMP) 45
meander model 301
mechanical load 354, 364
　biaxial load 362, 364
　uniaxial load 356, 362
mechanical properties
　actin 319, 325
　microtubules 325
　nucleus 318, 319
　plasma membrane 319
mechanically regulated cellular processes 355
mechanics 253–255, 257, 301, 309
　biomechanics 253, 254, 276, 286, 298, 299, 311
　continuum mechanics 253–255, 263, 271, 272
　rational mechanics 257
mechanobiology 65, 66
mechanotransduction 317, 318, 322, 331, 332, 339, 343, 384, 385
mesenchymal stem cell (MSC) 184
mesenchymal stromal cells (MSC) 205
mesodermal 136, 151
mesophyll 586
meteorites 643
metrics, mechanical and biochemical 332
Michaelis–Menten 549
micro systems technology (MST) 231, 233
micro-cavity 240
micro-domain 556, 558
micro-fluidic systems 240
microbial communities 648, 650

Index 689

poikilotrophic 648
microcirculation 451, 460–464, 500
microelectrode array-based impedance spectroscopy (MAIS) 239
microelectrode arrays 234
microelectrode cavity array (MCA) 241
microfabrication 216
microfabrication techniques 233
micron-resolution particle image velocimetry (μPIV) 475
micropipette aspiration 415, 417, 420–422, 425, 460, 461, 498, 504, 506
microsporon septicum 499
microtubule(s)
 compression 325
 dynamics 325
 mechanical properties of 325
 persistence length of 325
migration 162
miniaturization 4, 16, 18
minimum energy hypothesis 372, 373
minimum essential medium (MEM) 173
Minkowski metrics 26
Miocene 587
mitogen-activated protein kinases (MAPKs) and stretch 339
mixture theory 326–329, 333, 345
MKN45 cell 41, 47, 54
modulus
 modulus of elasticity 262
 modulus of viscosity 262
molecules
 carbon based 651
monocyte chemotactic protein-1 (MCP-1) 40, 43, 56, 57
monolayer 363
Mooney–Rivlin law 291
motion 255, 256, 272, 273, 277, 282, 311
 body motion 300
 oscillating motion 267
motions 262, 300, 302
mouse 3T3-L1 cell 48, 49, 55
MRT 85
mucosal barrier 509
multi-organ failure 510
multielectrode arrays (MEAs) 145
multiplexing 18
multipotent 193, 201, 202, 213, 214
murine
 cells 144
 embryoid (embryonic bodies) 144, 151
 embryos 137
 heart 138, 146
 model system 143

Murray's law 373
myocardial infarction 242
myosin 301
 heavy chain (α-MHC and β-MHC) 138, 143
 light chain 143

N

N^5, N^{10}-methylene-tetrahydrofolate 589
NAD^+ 581, 587, 589
$NADP^+$ 553, 554, 581, 592
NADPH 552, 554, 583, 592
nanofibers 225
nanotechnology (NT) 216, 233
NASA 656, 662
nerve growth factor (NGF) 39, 41, 50–53
network 253, 294
 Gauß network 294
 Kilian network 253, 294
 weak network 261
NFAT 15
NH_2
 to COOH covalent bonding 166
nitric oxide
 nitrate 452, 453, 456, 458
 nitrite 456–459
 nitrosated/nitrosylated species 456, 458
nitric oxide (NO) 481, 483, 485, 486
nitric oxide synthase (NOS) 40, 43
 endothelial (eNOS) 452, 454, 455, 457–459
 red blood cell NOS (RBC NOS) 459–464
nitrogen 585, 589
nitroglycerine 452–454, 463
NO 40
non-biodegradable materials 162
non-mammalian substances 163
non-viral vectors 161, 162
nuclear factor (NF)κ B 40, 43
nuclear factor-κ B (NF-κ B) 481, 484, 485
nuclear magnetic resonance (NMR) 415, 429, 430

O

O-chain 505
Obelia longissima 6
objectivity 277
obstacles 550, 556–559, 563, 569, 570
oncogenesis 160, 173
optical coherence tomography 85
 Doppler 95
 optical frequency-domain imaging 89

polarization sensitive   94
  spectral-domain   88
  spectroscopic   94
  swept-source   89
  time-domain   88
organ replacement   194, 202
origin of life   642, 644
orthogonal polarized spectral imaging (OPS)   463
osmolarity   506
osteogenesis   176
osteoporosis   527, 535
oxaloacetate   586
oxygen   646
oxygenase   8
oxygenation   584, 596
ozone oxidation   178

P

p38   40
palpation   103–105, 123
pancreatic digestive enzymes   513
parallel-plate flow chamber   473, 481
PC12 cell   41, 42, 53, 54
PCT   500
Pearson correlation   26, 27, 31
percolation   556, 560, 564
  threshold   556, 558, 560, 562, 567, 569–571
phantom   103, 106, 107, 113, 115, 116, 119, 121
phase separation   531
phase transition   253, 298, 304, 308, 309, 422, 423, 431, 433
phenotype   136, 149
phosphinothricin   592, 595
phosphoenolpyruvate   586
phosphoenolpyruvate carboxylase   586
3-phosphoglycerate   582, 584
phosphoglycolate   582, 585
2-phosphoglycolate   585, 587
phosphoglycolate phosphatase   585
photolithography   217
photoprotein   7
photorespiration   590, 593
photosynthesis   543, 552, 570, 581, 585, 650
photosynthetic capacity   585
photosynthetic electron transport   543, 545, 552, 554, 560, 564, 565, 569–572
photosystem (PS)   553–556, 559, 560, 562 ff
phylogenetic tree   645
physiological shock   510
planetary protection   658

plant transformation   593
plasticity   194, 198, 199, 213
plastocyanin   553–556
plastoquinol (PQH$_2$)   553, 554, 556, 560, 566
plastoquinone (PQ)   553, 554, 556, 558–560, 562, 565–567, 569–571
platelet   470–472, 486
platypus   434, 435, 438, 439
pluripotency   137, 143, 155
pluripotent   196, 197, 201, 202, 210–213
poikilothermal animals   418, 419
Poiseuille's law   471
poloxamer   408
Poly Vinyl Alcohol (PVA)   115, 116, 121, 122
  (PVA-c)   115, 122
  cryogel   115
poly(dimethylsiloxane)   217
poly(ethylene glycol)   407
poly(l-lactic acid) (PLLA) matrix   96
polyconvex energy function   286
polyethyleneimine (PEI)   618, 619
polymer   257–259, 262, 301, 304–306
  biopolymer   253, 258, 259, 277, 291, 293, 294, 297, 303
  technical polymer   294, 296, 297
polymer penetration   401, 403
polymerization   178, 181
polyurethane (PU)   176, 178
polyvinyl chloride (PVC)   303
porous collagen-hyaluronic acid (HA-Col) membrane   182
porous collagen-laminin membrane   171
PQ   *see* plastoquinone (PQ)
preeclampsia   237
pressure   529–531
  delivery pressure   529
  extravertebral pressure   531
  infiltration pressure   529
  injection pressure   529–531
pressure-overload   380
principal invariants   279
principle component analysis (PCA)   24, 29–32
principle of determinism   276
pro-inflammatory cytokines   499
procalcitonin (PCT)   500
procurement   160
proliferation   136, 138, 147, 151, 161, 162, 170, 173, 178, 184, 188
35S promoter   590, 591
prostate   104, 123, 124
  cancer   104, 122, 127
  prostatectomy   123

tumor   123
protease inhibitor   513
protein activation
  conformational changes in   337
  correlating the magnitude of stretch with   340
protein C   501
protein kinase C (PKC)   38–42, 49, 52–54, 57
proteins   259
PS   *see* photosystem (PS)
puromycin acetyl transferase (PAC)   138, 139, 141
pyruvate   586, 587
pyruvate orthophosphate dikinase   586

Q

QT interval   236

R

radiopacifier   534
radiopacity   533
random walk simulation   558
Ras   40, 44
rate   544, 545, 550, 565–569, 571
  constant   551, 565–569
  law   549, 550
  limiting   565, 571
RBC   497, 498, 502, 504, 506
  RBC aggregability   503
  RBC aggregation   497, 502
  RBC deformability   497, 502, 503
  RBC shape   503
  RBC tongue length   498, 504, 506
re-capture of CO_2 582
reaction
  constant   568, 569
  equilibria   543, 544, 547, 551
  limited   568, 569, 571
  rate   543, 544, 548, 550, 568, 569, 571, 572
reaction mechanism   565, 571, 572
  collisional   565, 568, 569, 571
  diffusion limited   571
  tight binding   565, 567–569, 571
reactive oxygen species (ROS)   481
real-time monitoring   231, 233
recombinant   149, 153
red blood cells (RBC)   305, 309, 497, 498, 502, 504, 506
  passage through micropipettes   420, 421
  shape   416
  transition from blockage to passage   418
  volume   415, 426–428
Reference configuration   272
reference configuration   273
regeneration   136, 151
regenerative medicine   159
relaxation
  relaxation curve   265
  relaxation function   265, 266, 270, 296, 302
  relaxation kernel   265, 284
  relaxation test   264
  relaxation time   262, 264, 269
relaxation time   415, 420, 430, 434
remodeling
  of cells   318, 322
  of tissue   318, 322
reporter gene   5
  flash-light   6, 19
  glow-light   6, 8
retardation time   262, 264, 267
retrovirus   161
Reynolds number   475–477
RGD   204
rheology   256–258, 263
rho family small GTPase   479
Rho GTPase   *see* stress fibers and rho activity
ribulose-1,5-bisphosphate   582, 589
ribulose-1,5-bisphosphate carboxylase/oxygenase   582, 585
right Cauchy–Green tensor   327, 328
Robot-System   19
rouleaux (of RBC)   497

S

scaffold   148–150, 193, 202, 203, 207, 211, 213
search for life   651
self assembly   549, 551
self organisation   543, 549, 550
self organizing maps (SOMs)   32
sepsis   497–502
septic shock   497–502
serine/glyoxylate aminotransferase   585
severe sepsis   498–501
shear stress   37, 40, 66, 71–76, 78
shock factor   516
signatures of life   649
silica gel   605, 615, 616
silk fibroin   163
Silvet L-77   594
single nucleotide polymorphisms (SNPs)   24
SIRS   497, 498
skin   276, 291, 297–299, 302, 303

smooth muscle cell (SMC) 470, 471, 485, 486
sorption 602
space 279
　Euclidean space 272
　Euclidean vector space 279, 282
　Hilbert space 280–282
Spearman rank correlation 27
spheroids 240, 242
spore 648
stenosis 483, 484
　post-stenotic 480, 483, 484
steric repulsion 544, 545, 547, 572
steric stabilization 396
stiffness 301
　dynamic stiffness 268
　equilibrium stiffness 265
stomata 583, 584, 590
storage modulus 268
strain 375–378
strain energy
　fiber 320, 321, 327, 328, 342
　for isotropic linear elastic Hooke's law 330
　function 327
strain history 263, 264, 281, 302
strain imaging 103, 106, 107, 109, 113, 122, 124, 127
strain rate 330
strain retardation 260, 271
strain, fiber 338, 341–343
stress 255, 260, 263, 265, 268, 276, 281, 293, 371, 375–378
　Cauchy 327–331
　Cauchy stress 275
　fluid shear 317
　mechanical normal 317
　on actin filaments 329, 330
　on focal adhesions 324
　on intermediate filaments 330
　Piola–Kirchhoff stress 275, 297
　residual stress 276, 293
　stress history 266
　stress relaxation 260, 261
　stress state 274, 276, 282, 296
　stress tensor 275
　stress-strain curves 299
　stress-strain relation 259, 260
stress fiber(s)
　and Rho activity 321
　and stretch 319–322
　and substrate stiffness 334
　assembly 334
　mechanics 337

model 320, 321
organization 320, 321, 337
pre-stretch 334
stability of 321
strain energy of 320
tension 325
stress tensor 277, 278, 281
stretch 71, 73, 76
　and mechanotransduction 331, 332, 338–344
　and stress fiber orientation 319–322
　and traction forces 335–337
stretch retardation 267
stroma 552–556
succinylated collagen 168
sugar cane 587
sugars 587
sulfadiazine 592, 593
super-coiled triple helical peptide chains 163
surface conditioning 615, 617, 618
surface oxidization 176, 178
surface topography 215
survival function 329, 334
systemic inflammatory reaction response (SIRS) 497
systemic inflammatory response syndrome (SIRS) 498

T

tachycardia 497
tachypnea 497
tartronic semialdehyde 588, 589
tartronic semialdehyde reductase 588, 589
technical polymers 259, 302
telopeptides 165
tendons 300–302
tensegrity
　and prestress 326
　mathematical models based on 325
tensor 273, 274, 277–280, 282, 283, 285, 286
　rotation tensor 273
tetrahydrofolate 589
theorem
　Cayley–Hamilton theorem 279
　polar decomposition theorem 273
therapeutic cells 160–162
thermodynamics 255, 257, 286, 287, 305, 309–311
　phenomenological thermodynamics 286
　thermodynamics of irreversible processes 257
thermoregulation 418, 419
thrombin-thrombomodulin (TM) 501

Index

thrombosis 498
thylakoid 543–545, 552–558, 560, 562–571
3D tissue construct 363
tissue engineered implants 161
tissue engineering 193–196, 201, 202, 206, 207
tissue repair 147, 152
TM 501
Toll-like receptor (TLR) 40, 45
toxicity 531
traction forces 222
traction microscopy 324–326, 335, 336
transcription
   brachyury (T) 151
   Isl1 (islet 1) 151
   natriuretic factor (ANF) 143
   NKX2.5 143
transcription factor 499
transcriptome 135
transformation 591, 592
transforming growth factor (TGF-β1) 23, 29–34
transition metals 613, 614, 617
transition temperature 304
translocase 418
Transmissible Spongiform Encephalopathy (TSE) 163
transplantation 159, 160
tumor necrosis factor (TNF) 501
two-photon microscopy 86

U

ultrasound 85, 104–106, 108, 121
unfolding 549
uniform shear hypothesis 374
UV irradiation 167, 168, 170

V

vascular tissue 585
vascularisation 194, 202, 203
vasodilation/vasorelaxation 452–454, 459

VE-cadherin 480, 483, 485
vertebral compression fractures 527
vertebrogram 529
vertebroplasty 528
   percutaneous vertebroplasty 528
video microscopy 463, 464
vinculin 338, 339
viral vectors 161
visco-hyperelastic relation 327
viscoelasticity 260, 271, 284
   viscoelastic liquid 261
   viscoelastic solid 267
viscometer 535
viscosity 531–535
volume
   available 546–549
   excluded 545–548, 551, 572
   occupied 544

W

wall shear stress 372, 374, 378, 379, 382
water
   liquid 652
water stress 587
white blood cell (WBC) 470, 472, 486, 500–502
   foam cell 471
   lymphocyte 472
   macrophage 472
   monocyte 470, 471
Wormlike-Chain-Model (WCM) 301
wound dressing 170

X

X-ray 85
xenogeneic 163, 165, 188

Z

zeta potential 608
ZKM model 373

Printing: Krips bv, Meppel, The Netherlands
Binding: Stürtz, Würzburg, Germany